Cytogenetic Abnormalities

Cytogenetic Abnormalities

Chromosomal, FISH and Microarray-Based Clinical Reporting

Susan Mahler Zneimer, Ph.D., FACMGG
CEO and Scientific Director, MOSYS Consulting
Adjunct Professor, Moorpark College, Moorpark, California

WILEY Blackwell

Registered Office
John Wiley & Sons, Ltd, The Atrium, Southern Gate, Chichester, West Sussex, PO19 8SQ, UK

Editorial Offices
9600 Garsington Road, Oxford, OX4 2DQ, UK
The Atrium, Southern Gate, Chichester, West Sussex, PO19 8SQ, UK
111 River Street, Hoboken, NJ 07030–5774, USA

For details of our global editorial offices, for customer services and for information about how to apply for permission to reuse the copyright material in this book please see our website at www.wiley.com/wiley-blackwell.

Library of Congress Cataloging-in-Publication Data

Zneimer, Susan Mahler, author.
Cytogenetic abnormalities : chromosomal, FISH, and microarray-based clinical reporting / Susan Mahler Zneimer.
p. ; cm.
Includes bibliographical references and index.
ISBN 978-1-118-91249-2 (paperback)
I. Title.
[DNLM: 1. Chromosome Aberrations. 2. Cytogenetic Analysis. 3. In Situ Hybridization, Fluorescence. 4. Microarray Analysis. 5. Neoplasms–pathology. 6. Terminology as Topic. QS 677]
QH600
572.8′7–dc23

2014018423

A catalogue record for this book is available from the British Library.

Wiley also publishes its books in a variety of electronic formats. Some content that appears in print may not be available in electronic books.

Set in 10/12pt Adobe Garamond by SPi Publisher Services, Pondicherry, India

1 2014

Dedication

To my husband, Martin Chetlen, for his guidance, love and support, without which this book neither could nor would ever have been written.

And to my parents, Nadine and Joel, who raised me to think, and to imagine all the possibilities. May their memory be forever a blessing.

To all three, I raise my glass and say… thank you!

Contents

Preface

Numerous books and an untold number of publications in the scientific literature describe various chromosome abnormalities, both common and rare. One can search for any chromosome abnormality in a book index or an online scientific database and find every possible chromosome change that has been identified. Most of these abnormalities are reported, rightly so, in order to make a correlation with a clinical phenotype, diagnosis or prognosis of a disease. In other words, the emphasis has always been on directly correlating chromosome abnormalities with disease.

However, to date, little emphasis has been placed on standardizing how to designate chromosome abnormalities with their nomenclature and the related interpretation of these results with patient diagnoses. This book attempts to address the lack of standardization for writing the nomenclature of chromosome abnormalities and interpretive comments regarding those abnormalities. With over 250 cytogenetic laboratories in the United States alone, and maybe 500 laboratories worldwide, it is indeed time to develop more established guidelines for writing and interpreting cytogenetic, fluorescence *in situ* hybridization (FISH) and microarray test results.

The International System for Human Cytogenetic Nomenclature (ISCN) is the only resource (albeit a great one) to use in order to designate clear nomenclature in writing test results. Even so, if one is to compare a cytogenetic result from 10 different laboratories in the designation of a marker chromosome, for example, there could easily be five different ways to either write the ISCN nomenclature or interpret the result.

This book will be useful to cytogeneticists who write test results, to other geneticists, physicians, allied professionals and scientists who need to read these reports or use them in their clinical and/or research endeavors, and to those medical and human genetic students who are required to understand cytogenetic test results.

In this book, for each normal and abnormal cytogenetic result, the ISCN nomenclature and an interpretive comment with related recommendations are given. These abnormalities are divided among constitutional disorders and acquired malignancies, and are described by their genetic composition, genetic nomenclature and associated clinical features. Each chapter also contains a bibliography from which the information was obtained as well as pertinent research articles, databases and/or online web addresses for the reader's reference.

This book attempts to discuss as many genetic and malignant diseases as possible that have a known and prominent underlying genetic defect; however, it is not possible to list all the cytogenetic abnormalities in one manuscript. Readers who wish to see other abnormalities not included in the manuscript are encouraged to contact the author for reference in a future edition of the book.

Acknowledgments

I wish to thank those who have graciously read my manuscript for content and coherence. Special thanks go to Nancy Wold and Martin Chetlen for all their assistance and guidance throughout the duration of this product. Nancy Wold, in particular, has been a great help in making this manuscript readable and understandable. She has played a large role in seeing this project come to fruition through her encouragement and support. My thanks are inadequate for all her endeavors.

I would also like to thank Robert Winning for all his help with formatting the images.

I would also like to thank the many cytogenetic directors and laboratory staff who have helped me with their wisdom, experience and advice. All of these people and their laboratories have provided me with invaluable information, including many of the images in the book.

My deep appreciation goes to the directors and staff of the following laboratories.

- ARUP Laboratories: Dr Sarah South, Lynda Kerby, Tamara Davis and Bonnie Issa.
- Ascend Genomics (formerly PathCentral): Dr Mansoor Mohamed and Dr Shelly Gunn.
- City of Hope: Dr Joyce Murata-Collins, Popsie Gaytan, David Eve, Gina Alvarez, Maria Cruz.
- CombiMatrix: Dr Karine Hovanes, Dr Richard Hockett Jr.
- Kaiser Permanente Laboratories: Dr Lauren Jenkins, Dr Mehdi Jahmedhor, Dr Xu Li, Lloyd Maxwell, Michael Tiffert, Virginia Nottoli, Grace Santiago.
- LabCorp of America: Dr Peter Papenhausen, Dr James Tepperberg, Dr William Kearns, Dr Bing Huang, Martin Sasaki, Ati Girgin, Monika Skapino, Sharlene Anderson, Kristen Trujillo, AnnMarie Bell, Kenny Xi, Rosa Thompson and Jose Navarro.
- Natera: Dr Zackary Demko, Dr Matt Hill, Dr Megan Hall and Sallie McAdoo.

It is such a pleasure to have great colleagues with whom to work. I am forever indebted to all of you for your assistance in the completion of this manuscript.

I would also like to thank my colleagues, Dr Lauren Jenkins and Dr Paula Berry, for their long-term friendship, treasured advice and encouragement in completing this manuscript.

Lastly, I would like to thank my editor, Justin Jeffryes and all those at Wiley Publishing, for their encouragement and help in initiating and completing this enormous project.

I have spent most of the day putting in a comma and the rest of the day taking it out.

Oscar Wilde

About the companion website

This book is accompanied by a companion website:

www.wiley.com/go/zneimer/cytogenetic

The website includes:

- PDFs of all Example Report Boxes from the book for downloading. The Example Reports are the most important downloads that are needed for the book
- Powerpoints of all figures from the book for downloading
- PDFs of all tables from the book for downloading.

Introduction

Genetic testing is complex, and it is often difficult to understand the meaning of its results. One part of genetic testing is cytogenetic testing, often called chromosome analysis, which uses whole cells that are grown in culture in the laboratory to isolate DNA and identify differences of the chromosomes that would yield a genetic abnormality and lead to a genetic disorder or cancer. Cytogenetic analysis requires extensive manipulation of cells taken from an individual's body to isolate and analyze the chromosomes microscopically for the identification of chromosome aberrations. The complexity of this testing continues with writing a comprehensive and cohesive laboratory report that contains the correct information and correct nomenclature, and is understandable by the professional community that is conveying these results to patients. All too often, without assistance from genetic counselors, the professional receiving cytogenetic results from the genetics laboratory does not understand the nomenclature or the interpretive comments that explain a cytogenetics abnormality. These results are then not appropriately conveyed to the patient, widening the divide between medical science and the general population.

This book begins with an overview of genetics in general, cytogenetics in particular, and how its significance pertains to the general population. The discussion continues in subsequent chapters, in which specific cytogenetic abnormalities are discussed, and includes how cytogenetic reports are written for each abnormality, and how professionals and individuals with these disorders should interpret cytogenetic results.

Overview of cytogenetic testing in the laboratory

Cytogenetic testing is the study of chromosomes and their genetic composition, which is studied at different genetic levels. The "Gestalt" view of cytogenetics is the largest overview of chromosomes in which the banding level is important in identifying gross versus subtle, but visible, genetic changes. This level is generally referred to as standard or conventional cytogenetics. Conventional cytogenetics allows for the identification of large DNA changes that are visible on a chromosome, at least 1 million base pairs, whether the genetic change is a balanced or unbalanced abnormality. It is still the method of choice for many types of indications for genetic testing, such as cancer diagnosis and prognosis,

Cytogenetic Abnormalities: Chromosomal, FISH and Microarray-Based Clinical Reporting, First Edition. Susan Mahler Zneimer.

history of spontaneous abortions, newborn dysmorphology, prenatal diagnosis and endocrinology disorders. There are limitations to conventional cytogenetics, one of which is the inability to visualize small abnormalities under a microscope, <1–5 million base pairs, eliminating the possibility of identifying submicroscopic genetic abnormalities. This limitation led to the development of new methodologies, including fluorescence *in situ* hybridization and microarray techniques, which enable the identification of smaller genetic changes and do not depend on the level of chromosome banding and morphology.

The next level of chromosome analysis is the level at which DNA probes can be used to identify regions of a chromosome using *in situ* hybridization (ISH). Although the first ISH analysis was done using radioactive tritium as a DNA probe, now the conventional approach uses fluorescent dyes attached to a small segment of DNA, hundreds to thousands of DNA base pairs long, that is specific to chromosomal regions of interest. This fluorescence *in situ* hybridization (FISH) is very useful in identifying DNA change on the chromosome that is specific to a disease region, a locus-specific region or a part of a chromosome that is used for chromosomal identification, such as a centromere, subtelomere or whole chromosome paint probe. FISH is very useful in identifying deletions, duplications and rearrangements of small disease regions where DNA probes can be made, or for chromosomal identification, when the presence of a chromosome is unidentifiable by standard chromosome analysis. FISH analysis also has its limitations, due to the small range of DNA size that supports a probe for hybridization. This range is generally from DNA segments from 1000 to 200,000 base pairs long. Another limitation of FISH is the need to know which region of a chromosome with which to probe. This is a targeted DNA test of the genome and only segments of interest are utilized in this methodology. Only with whole chromosome paint probes will the total genome be visible with the FISH methodology.

The next level of chromosome analysis is microarray technology. This method allows for a whole genome analysis at the molecular level in which small segments of DNA probes (oligo DNA probes) down to single base pair analyses (single nucleotide polymorphisms, SNPs) are used to identify all the segments of a genome for DNA imbalances in individuals. This methodology has become quite prevalent in all aspects of cytogenetic analysis, since large and small imbalances, including unbalanced rearrangements, duplications and deletions of any size, can be identified in an individual. Therefore, this type of analysis is replacing many aspects of the standard chromosome and FISH analyses, especially when specific indications for cytogenetic testing are known to be submicroscopic, and no clear disease state is in the differential diagnosis.

Table 1 Levels of DNA resolution from standard chromosome analysis by specimen type

Specimen type	Average band level	Average number of genes per band	Average number of base pairs per band	Band width in Mb
Bone marrow	350	100	9×10^6	9
Prenatal (AF and CVS)	450	75	7×10^6	7
Routine blood	550	60	6×10^6	6
High-resolution blood	800	45	4×10^6	4

AF, amniotic fluid; CVS, chorionic villus sampling.

Table 2 Abnormality detection by methodology

Abnormality	G-band	FISH (targeted sites)	Chromosome microarray analysis (CMA)
Whole chromosome imbalance (aneuploidy)	+	+	+
Balanced rearrangement (reciprocal translocations, inversions)	+	+	-
Deletions, duplications >10 Mb	+	+	+
Deletions, duplications <5~10 Mb (submicroscopic)	-	+	+
Mosaicism, if present at these levels	>20%	>5%	>20%
Marker (unidentifiable) chromosomes (non-mosaic)	+	+	+
Triploidy	+	+	+

FISH, fluorescence *in situ* hybridization.

The most significant use of microarray technology is for non-specific disorders, such as indications of autism spectrum disorders, mental impairment, developmental delay and brain or other organ dysfunction. It has also become prevalent for prenatal diagnosis and cancer disorders when many non-random, recurrent genetic changes are possible or, for example, in leukemias where many cytogenetic changes are known to be the underlying genetic change causing disease. A single microarray analysis is a good method to identify any of the possible abnormalities of gain or loss of genetic material known to be involved in specific diseases. This is in contrast to the many FISH probes that may be needed to test for a single disease or chromosome analysis, which is difficult to perform on neoplastic cells.

Microarray analysis, however, also has some limitations, including that it is most effective for identifying unbalanced rearrangements, whereas balanced rearrangements are not detectable (though this will probably be developed for clinical use in the near future). Microarray analysis is also too new to be the standard of care for most indications for genetic testing. However, ongoing development of this methodology will most likely make this type of testing more prevalent in the future.

See Tables 1 and 2 for a summary of the detection of chromosome abnormalities at each level and methodology employed.

Laboratory procedures for each type of methodology can be found in other sources, which are listed at the end of this chapter.

Genetic testing in most countries is generally governed by at least one agency. Some information regarding governmental and other regulatory agency requirements is provided but regulations vary depending on the state and country or residence of the laboratory or patient. Each government and agency has specific guidelines or laws that guide the laboratory for ethical, quality and monetary aspects. Other regulations in the United States include statements regarding whether a test is FDA approved or is an assay specific reagent (ASR) or for research use only (RUO). For the needed statement on reports, see Part 1, Section 2, where this is applicable.

Other rules and regulations will be discussed throughout the book when applicable.

Bibliography

Gardner RJM, Sutherland GR. *Chromosome Abnormalities and Genetic Counseling.* Oxford Monographs on Medical Genetics. Oxford University Press, Oxford, 2003.

Gersen S, Keagle M (eds). *Principles of Clinical Cytogenetics.* Humana Press, Totowa, New Jersey, 1999.

Grewal SI, Jia S. Heterochromatin revisited. Nat Rev Genet 2007; 8: 35–46.

McKusick VA. Mendelian Inheritance in Man and its online version, OMIM. Am J Hum Genet 2007; 80: 588–604.

Rooney DE, Czepulkowski BH. *Human Cytogenetics. A Practical Approach.* Oxford University Press, New York, 1992.

Shaffer LG, McGowan-Jordan J, Schmid M (eds). *ISCN 2013: An International System for Human Cytogenetic Nomenclature.* Karger Publishers, Unionville, CT, 2013.

Tobias E, Connor M, Ferguson Smith M. *Essential Medical Genetics*, 6th edn. Wiley-Blackwell, Oxford, 2011.

Tolmie JL, MacFadyen U. Down syndrome and other autosomal trisomies. In: Rimoin D, O'Connor J, Pyeritz R, Korf B (eds) *Emery and Rimoin's Principles and Practice of Medical Genetics*, 5th edn. Churchill Livingstone, Edinburgh, 2006, pp.1015–1037.

Part 1 Constitutional Analyses

Section 1
Chromosome Analysis

CHAPTER 1

Components of a standard cytogenetics report, normal results and culture failures

1.1 Components of a standard cytogenetics report

All cytogenetic reports should have specific information which helps to standardize that each laboratory is performing a minimum standard of competency and accuracy of results. Clinical laboratory improvement amendments (CLIA), College of American Pathologists (CAP) and various US states have placed requirements on each report. The information below is required for CLIA, CAP, NY State and CA State for regulatory compliance.

- Specimen type
- Indication for testing
- Number of cells counted
- Number of cells analyzed
- Number of cells karyotyped
- Banding technique
- ISCN nomenclature
- Interpretation

1.1.1 Specimen type

Specimen type refers to the source of tissue that is being analyzed for cytogenetic testing. The most common specimen types are:

- amniotic fluid and chorionic villus sampling (CVS) for prenatal studies
- peripheral blood for studies of liveborn individuals
- fetal tissue for products of conception (fetal demise) studies
- bone marrow, bone core or peripheral blood for leukemias
- bone marrow or lymph nodes for lymphomas
- muscle or skin biopsies for possible mosaic studies
- tumor biopsies for acquired or inherited malignancies.

Cytogenetic Abnormalities: Chromosomal, FISH and Microarray-Based Clinical Reporting, First Edition. Susan Mahler Zneimer.

1.1.2 Indication for testing

Obtaining relevant clinical information about the patient is important in order to correlate cytogenetic results with the diagnosis. It sometimes becomes necessary for the laboratory to determine the appropriate set-up conditions of the specimen and the types of testing to perform, due to the various possibilities that exist. Therefore, in order for the laboratory to know what specific testing to perform, it needs all relevant patient information. Without the necessary patient and family clinical information, it may become a guessing game for the laboratory on the correct processing step to take. This is especially significant when it applies to cancer cytogenetics. Since certain cancer cells, including acute leukemias and myeloid disorders, divide continuously and do not require a B-cell or T-cell mitogen stimulant for cells to go through mitosis, the cultures that are initiated should be unstimulated 24-hour and 48-hour cultures. This is in contrast to chronic leukemias and other lymphoproliferative disorders, which do better with a B- or T-cell mitogen (e.g. IL4, TPA) to stimulate the cells to divide to have enough metaphases for analysis and which contain the abnormal cell type rather than normal lymphocytes. Also, knowing if acute lymphoblastic leukemia (ALL) is an indication for a patient will require only direct, overnight or 24-hour unstimulated cultures for analysis. Otherwise, there will be an overgrowth of normal cells dividing by the second day, and the abnormal lymphoblasts that are indicative of ALL will die off and not be present for analysis.

Culture initiation or set-up is also specific for the tumor type in question. No one culture medium is sufficient for all tumor types and so the culture medium should be specifically tailored for the proper growth of the abnormal tumor cells. For a guide on cancer cell culture media and growth factors for neoplastic cell growth, see the bibliography for detailed information.

1.1.3 Number of cells counted and analyzed

Counted cells refer to identifying a single cell and counting the number of chromosomes present plus identifying the sex chromosomes of that cell. Analyzed cells refer to identifying each chromosome homolog, band for band, to determine if any abnormalities exist within any of the chromosomes present.

Colonies refer to amniotic fluid cells that are cultured *in situ* on a small culture vessel, such as a coverslip. Colonies originate from single amniotic fluid cells that will grow and divide near each other in a colony, visibly separated from other originating amniotic fluid colonies. This type of culture allows for a greater distinction of progenitor cells in analysis versus allowing cells to congregate, grow and divide without spatial distinction, in which there is no knowledge of which cells are progenitor cells and which are the result of cell division and clones of progenitor cells. Without colonies, the cells in culture may be growing and dividing from only a very few hardy cells, and could possibly result in only a small number of original cells being analyzed, excluding possible mosaicism at a lower level.

The standard number of cells to be counted and analyzed depends on the specimen type. See Table 1.1 for a guide to the most common guidelines for cells counted and analyzed.

Table 1.1 Standard number of cells counted and analyzed per specimen type

	Postnatal peripheral blood	Prenatal amniotic fluid	Prenatal chorionic villus sampling	Neoplastic bone marrow and blood	Fetal demise and liveborn tissues	Neoplastic tumors	Mosaic studies
Cells counted	20	15 colonies or 20 cells	20	20	20	20–30	30–50
Cells analyzed	5	5 colonies or 5 cells	5	20	5	20–30	5

1.1.4 Number of cells karyotyped

The number of cells to be karyotyped is generally two per cell line. Exceptions to this rule include karyotyping only one cell of sideline clones in a neoplastic study, which will be discussed in greater detail in the cancer section of the book. More than two cells may be karyotyped if an abnormality is subtle and requires more than two cells to clarify the abnormality present.

1.1.5 Banding techniques

The standard banding techniques include those that clearly distinguish the significant bands identified by the International System for Human Cytogenetic Nomenclature (ISCN). The most common banding techniques which show the best banding patterns include G-banding, R-banding and Q-banding. Each technique uses different staining procedures to visualize the differential staining of cytosine/guanine (CG)-rich and adenosine/thymine (AT)-rich DNA. In each staining procedure, the bands observed are the same, but are visualized by AT with dark bands and CG with light bands or vice versa.

Other banding techniques are used to enhance specific regions of the chromosome, such as the centromere with C-banding, satellite regions of acrocentric chromosomes with nuclear organizer region (NOR) staining or telomeric regions with T-banding.

For a comprehensive discussion of banding techniques, refer to the bibliography at the end of the chapter.

1.1.6 Band levels

The banding level refers to an estimated total number of black, gray and white bands throughout the genome as it would appear in an ideogram of each chromosome. In the ISCN 2013 edition, on pages 16–31, ideograms of the chromosomes are described by band levels. There are a few reports in the literature of standardizing approaches to count the total number of bands in a karyotype. One approach is to count bands including the telomere, centromere and all the dark and light bands on chromosome 10. Table 1.2 details the correlation between the number of bands with the band level, using chromosome 10 as a reference.

Another approach for estimating band level is to count segments of specific chromosomes. For two different approaches, see a summary of these band estimations in Tables 1.3 and 1.4. Examples of cells with their corresponding karyotypes of each band level are depicted in Figure 1.1.

In a cytogenetics report, recording band level is generally a requirement. There is some debate on whether the highest band level observed in the best cell should be recorded in the report, or whether the band level of the best karyotype should be reported, or an average of the cells or karyotypes. Many laboratories record the best band level seen in a karyotype, which is easily documented for regulatory purposes and which may be corroborated if that karyotypic image is placed in the report itself.

Typically, the band level of a normal prenatal specimen of amniotic fluid and chorionic villus sampling is approximately 450 bands. For peripheral blood on liveborns, the typical band level is 500–550 bands. Hematological malignancies and solid tumors typically have fewer bands, generally in the range of 300–400 bands, reflecting the difficulty in analyzing dividing cells from abnormal cell types in malignancies.

When performing a high-resolution study, in which the minimum band level is 550–650 bands, a comment in the interpretation may be useful in order for the reader to know at what level chromosome analysis was achieved. It is also useful to report when the banding level did not reach the minimum requirement established by the laboratory or regulatory agency.

Table 1.2 Band level by counting the bands on chromosome 10 (adapted from Welborn and Welborn 1993)

Number of bands on chromosome 10	Estimated band level
12	375
13–14	400
15–16	425
17–18	450
19–21	475
22–23	500
24–25	525
26–28	550
29–30	600
31–32	650
33–34	700
35–36	750
37–38	800
39–40	850

Table 1.3a Tabulated band resolution of chromosomal segments (adapted from Josifek et al. 1991)

Chromosome region	Total bands counted for each band level		
	Band level 350–400	Band level 550	Band level 850
Chromosome region 1 from p31-p32	1	3	3
Whole chromosome 10	5	12	19
Short arm of chromosome 11	2	5	6
Long arm of chromosome 12	4–5	8	14
Whole chromosome X	6–8	12	18
Total bands counted	**18–21**	**40**	**60**

Table 1.3b Correlation of total bands with band level

Total bands	Band level
18	350
21	400
28	450
34	500
40	550
47	650
54	750
60	850

Table 1.4 Counting gray G-positive bands on chromosomes 10, 18q and 19

Band level	Chromosomes with the number of G-positive gray bands		
	10	**18q**	**19**
150–200	6	3	3
200–425	7–9	4	4
425–700	10–12	4	5–6
>700	13	5	7

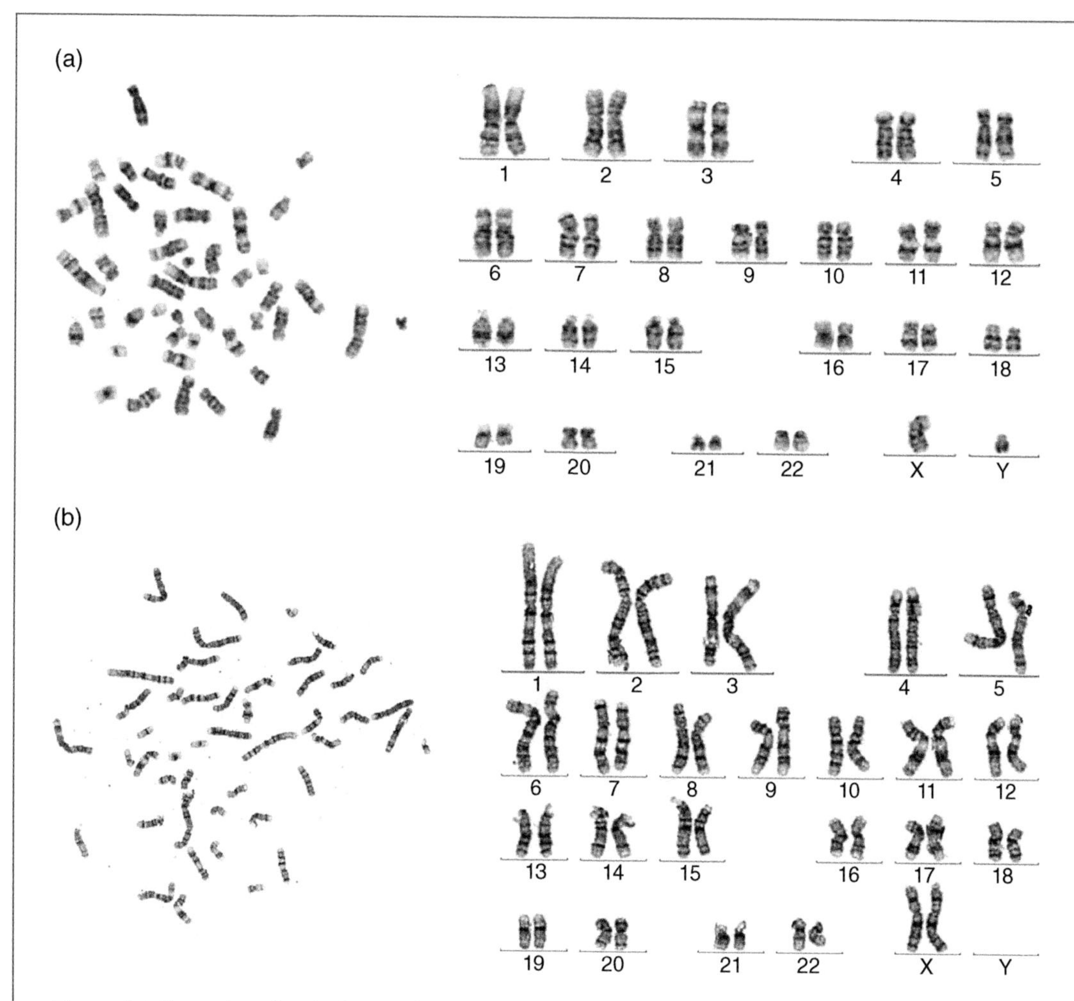

Figure 1.1 Examples of metaphase cells with their corresponding karyotypes of each band level. (a) 46,XY estimated at a 350 band level. (b) 46,XX estimated at a 400 band level.

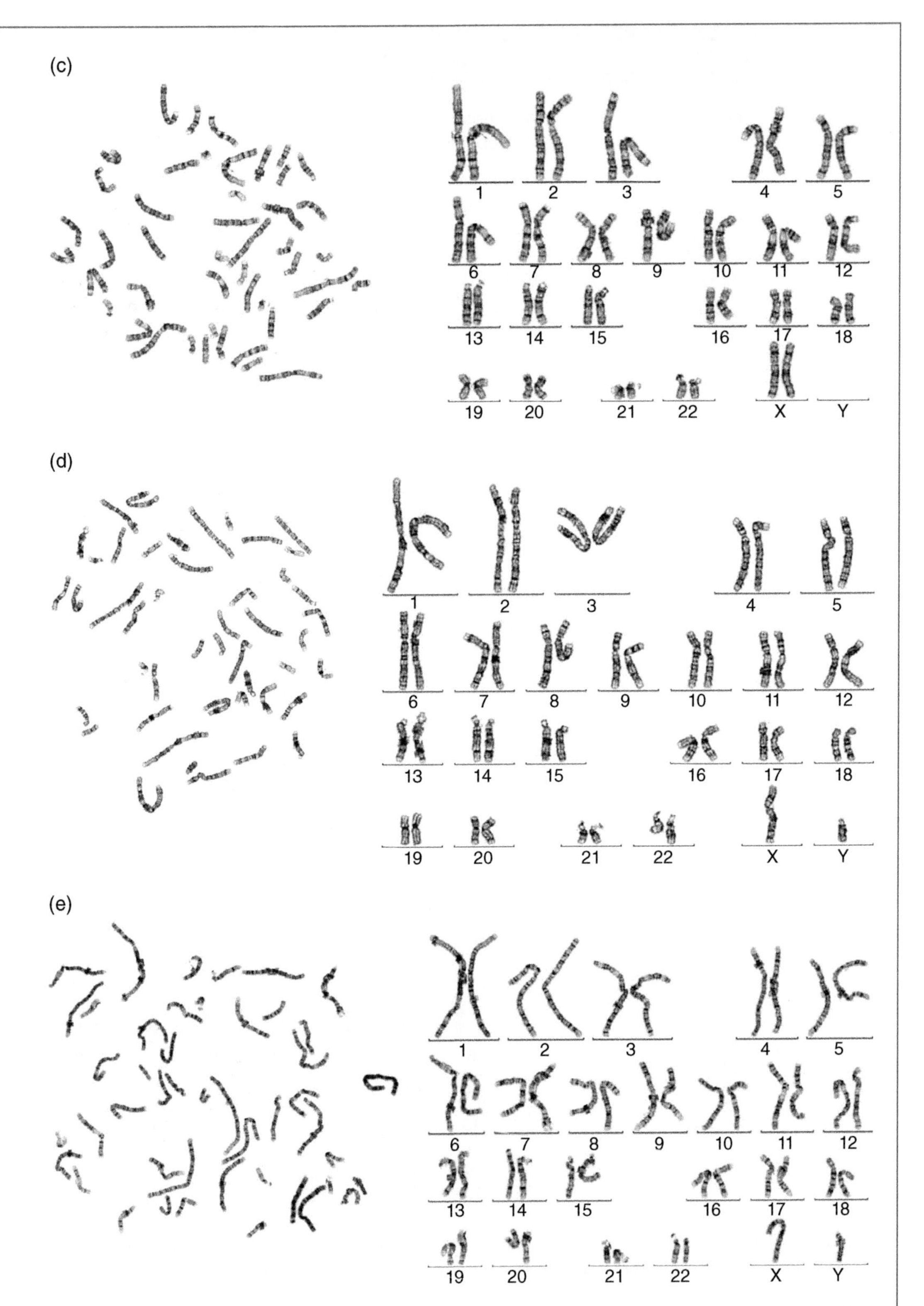

Figure 1.1 (*Continued*) (c) 46,XX estimated at a 450 band level. (d) 46,XY estimated at a 550 band level. (e) 46,XY estimated at a 750 band level. Courtesy of Sarah South PhD, ARUP Laboratories.

TYPICAL COMMENTS FOR BAND LEVEL IN A REPORT MAY INCLUDE

- Chromosome analysis at a band level of 550 was achieved in this study, which is within the optimum range for a high-resolution study.
- Chromosome analysis at a band level of 500 was achieved in this study, which is below the optimum range of 550 bands for a high-resolution study.

1.1.7 Summary of ISCN nomenclature

The International System for Human Cytogenetic Nomenclature (Shaffer et al. 2013) is the best resource for understanding how to write cytogenetic nomenclature. There should be little to no variation in the way laboratories write the ISCN nomenclature of an abnormality; however, it is not always easy to figure out how to write or understand ISCN nomenclature without some tutoring. This book will explain how to write each type of abnormality that exists and give examples of reports on how it is interpreted.

The following are the main concepts in writing any cytogenetic result. First is the number of chromosomes present in any person's cells or genome. The usual number of chromosomes (modal chromosome number) for the human species is 46, so all normal karyotypes (another word for cytogenetic designation) have a graphic organization of chromosomes by number, and so the nomenclature begins with 46 in humans.

The second part of karyotypic designation is the sex chromosome complement. Normal human females contain two X chromosomes and normal human males contain one X and one Y chromosome.

ISCN RULES FOR NORMAL RESULTS

- First write the modal chromosome number followed by a comma.
- Then write the sex chromosome complement.

Therefore, normal results would be either 46,XX or 46,XY for a female or male, respectively.

Notice that punctuation is very important in cytogenetic nomenclature. All chromosome numbers are followed by a comma and no spaces are used to separate chromosome number from sex chromosome content.

For all cytogenetic results that require a description of a variant (polymorphism) or an abnormality, chromosomes are broken down into regions for unique identification. The ISCN gives a very nice depiction of chromosome regions (also called bands) as seen when chromosomes are chemically manipulated in the laboratory. The most common form of manipulation or "banding" of chromosomes is Giemsa, or G-banding. This manipulation subjugates chromosomes with an enzyme such as trypsin or pancreatin, followed by Giemsa, Wright's or Leischman's stain which will degrade the AT-rich DNA content more than the GC-rich DNA content, giving rise to a black/gray/white appearance of chromosomes, in which DNA is systematically destroyed. Chromosomes appear like a bar code, such as one sees on products in a store, in which the black/gray/white regions are consistent from cell to cell and person to person. Therefore, all apparently normal cytogenetically individuals will have a karyotype that looks identical, comparing the black/gray/white banding pattern of each chromosome band.

The ISCN has specific terms describing different types of abnormalities. These terms are generally abbreviated in the karyotypic designation. Pages 36–38 of the ISCN book describe all the symbols and terms for the karyotypic designation. The ISCN also describes the difference in result designations for constitutional, neoplastic disorders, metaphase and interphase fluorescence *in situ* hybridization (FISH) analyses and array comparative genomic hybridization analyses. As each abnormality is described in this book, the correct ISCN designation will be discussed, so as to help with writing the correct nomenclature.

1.1.8 Report formatting

Various regulatory agencies require certain information on each report, including the following.

- Patient name
- Ordering physician
- Collection date
- Report date
- Unique patient identifier (usually the unique laboratory case number or specimen identification number), and the date of birth is suggested, not required
- Indication for study
- Specimen type
- Number of cells counted
- Number of cells analyzed
- Number of cells karyotyped
- Chromosome banding level
- Type of banding performed
- Types of cultures used for analysis
- ISCN result
- Interpretation
- Name and signature of reporting personnel
- Name of the medical director
- Laboratory identification, which may include the name of the laboratory and regulatory agency number (such as a CLIA ID number)
- Location of the laboratory and locations of each part of the analysis if there is more than one location involved

One example reporting format includes three sections: one for the demographic information of the patient, one for the clinical information associated with testing, and one for the results and interpretation. A fourth area of the report may include an image or images of the karyotype if desired.

EXAMPLE REPORTING OF A NORMAL RESULT

Laboratory Name

Cytogenetics Report

Patient Name: Jane Doe	Collection Date: 1/1/2014
Ordering Physician: Dr Smith	Report Date: 1/10/2014
Date of Birth: 1/1/2000	Laboratory Number: A14-000021

Specimen Type: Peripheral blood
Indication for Study: Down syndrome

Number of cells counted: 20	Banding: G-banding
Number of cells analyzed: 5	Banding resolution: 550
Number of cells karyotyped: 2	Cultures performed: 72-hour PHA stimulated

ISCN Result: 46,XX Normal Female Karyotype

Interpretation

Chromosome analysis revealed a normal female chromosome complement in all cells examined.

There was no evidence of a chromosome abnormality within the limits of the current technology.

Date: January 10, 2014

Susan Zneimer, PhD, FACMG
Cytogenetics Laboratory Director

(Any disclaimer here you may want to provide.)

This test was performed at: Address of Laboratory

Dr Jones, Medical Director
Laboratory License Number: 00001

(Cytogenetic Image)

1.2 Prenatal normal results

Prenatal studies differ from postnatal studies in the type of cells that are analyzed. Prenatal specimens are composed of fibroblast cells, epithelial cells and other cells sloughed off from the growing fetus in amniotic fluid. These cells, when put in a culture vessel, will adhere to the bottom of the vessel and grow until confluent and there is no more room for growth expansion. The two specimen types currently used for prenatal diagnosis in the first or second trimester are CVS and amniotic fluid. These two specimen types differ in the timing of collection and analysis, the clinical procedure in obtaining a specimen, and how the cells are grown and analyzed in the laboratory.

Prenatal samples may be obtained from percutaneous umbilical cord sampling (PUBS), generally obtained in the third trimester, if a fetal karyotype is needed, usually due to abnormal ultrasound findings. Testing with PUBS is the same as for postnatal peripheral blood chromosome analysis. Details for this testing will be described subsequently in Section 1.3.

1.2.1 Prenatal cell analysis

For CVS, a minimum of 20 cells are counted and a minimum of five cells are analyzed for each specimen. This analysis is usually divided among at least two primary cultures to ensure a minimum of two independent sources of cells for evaluation. The number of 20 cells was derived due to the possibility of mosaicism – that is, more than one cell line present in culture. Twenty cells will rule out the possibility of 14% mosaicism at a 95% confidence interval or 21% mosaicism with 99% confidence. This is now a standard procedure in most laboratories worldwide.

For amniotic fluid (AF), a minimum of 15 colonies are counted with a minimum of five cells analyzed for each patient. Also, a minimum of two primary cultures are analyzed when possible. Since clinicians generally divide an amniotic fluid specimen into two 15 mL conical tubes, the convention is to grow each tube of cells independently and analyze at least some cells from each of the tubes received in the laboratory. A total of a 15 colony analysis has been designated as a standard procedure, since each of the 15 colonies will account for known independent cells, which is a better method for ruling out mosaicism than by culturing cells indiscriminately.

When AF specimens need to be subcultured, a total of 20 cells are counted and the process is analogous to CVS analysis.

Therefore, results are reported somewhat differently between CVS and AF; namely, by the cell count and by the possible presence of maternal cell contamination in CVS. Example reporting of results usually requires some variation between the specimen types.

1.2.2 Amniotic fluid

Amniotic fluid specimens are generally obtained at 14–18 weeks' gestation, with some clinicians performing early amniocentesis at 12–14 weeks' gestation. AF is generally easier to obtain than CVS and does not have the risk of maternal cell contamination unless it is a very bloody sample, implying that either placental decidua or maternal tissue is present.

Most amniotic fluid analyses are performed with *in situ* cultures, which allow for a discrete number of original progenitor cells to be cultured. This will ensure that a known number of independent cells will be evaluated in the cytogenetic analysis. This process is accomplished ideally by seeding the amniotic cells on coverslips in a small culture vessel, usually petri dishes, that allow for cell growth separately from neighboring cells. The original progenitor cells will then grow and divide through mitotic division approximately once each day and form a colony of identical cells. Each progenitor cell will form its own colony of cells independently from the other colonies, thus ensuring the growth of different cells to be analyzed. Cells, once confluent, are then processed *in situ* for cytogenetic analysis, keeping each colony of cells intact for the chromosomal evaluation.

In some instances, there are an inadequate number of colonies available for analysis. When this happens, the cells need to be subcultured whereby all the colonies are collected and distributed evenly in the culture vessel to promote growth. In this case, the cells are then treated as a normal cell culture and not a colony culture.

EXAMPLE REPORTING OF NORMAL FEMALE AF RESULTS

ISCN Result: 46,XX Normal Female Karyotype

Interpretation

Chromosome analysis revealed a normal female chromosome complement in all 15 colonies examined from multiple cultures of amniocytes.

There was no evidence of a chromosome abnormality within the limits of the current technology.

EXAMPLE REPORTING OF NORMAL MALE AF RESULTS

ISCN Result: 46,XY Normal Male Karyotype

Interpretation

Chromosome analysis revealed a normal male chromosome complement in all 15 colonies examined from multiple cultures of amniocytes.

There was no evidence of a chromosome abnormality within the limits of the current technology.

When only one tube of amniotic fluid is received, some laboratories may add the following comment to the interpretation.

EXAMPLE REPORTING OF RESULTS FROM ONLY ONE TUBE OF AF

ISCN Result: 46,XY Normal Male Karyotype

Interpretation

Chromosome analysis revealed a normal male chromosome complement in all 15 colonies examined from multiple cultures of amniocytes. However, all cells analyzed were obtained from a single tube of amniotic fluid, which is below our laboratory standard of analyzing cells from at least two tubes received.

There was no evidence of a chromosome abnormality within the limits of the current technology.

1.2.3 Chorionic villus sampling

Chorionic villus sampling specimens are usually obtained clinically between weeks 10 and 14 of gestation. This was developed in order to do prenatal diagnosis in the first trimester of pregnancy as an earlier diagnostic genetic test rather than second trimester amniotic fluid testing. Though CVS has a slightly higher risk of miscarriage and fetal defects compared to amniotic fluid testing, it is a preferred test for some individuals who want results sooner.

The cells obtained from CVS are villi from the trophoblast of the growing fetus and may at times include decidua, which may be maternal or placental in origin, and must be carefully processed in the laboratory to ensure that the fetal cells and not maternal cells are analyzed. The villi need to be carefully disseminated into small pieces and placed in a culturing vessel. The cells obtained are fibroblasts, which will adhere to the surface of the vessel in which the cells are placed. These cells will then grow and divide and, when cells are confluent, will be lifted off the bottom of the vessel and collected for the process of cytogenetic analysis. Since there is a possibility of maternal cell contamination, depending on the quality of the specimen received, some reports may want to include information regarding the quality of that specimen.

EXAMPLE REPORTING OF NORMAL FEMALE CVS RESULTS – GOOD QUALITY OF FETAL VILLI RECEIVED

ISCN Result: 46,XX Normal Female Karyotype

Interpretation

Chromosome analysis revealed a normal female chromosome complement in all 20 cells examined from multiple cultures of chorionic villus sampling.

There was no evidence of a chromosome abnormality within the limits of the current technology. These results most likely reflect the fetal rather than the maternal karyotype. However, due to the nature of the tissue submitted, the possibility of maternal cell contamination cannot be excluded.

EXAMPLE REPORTING OF NORMAL FEMALE CVS RESULTS – PREDOMINANTLY MATERNAL DECIDUA RECEIVED

ISCN Result: 46,XX Normal Female Karyotype

Interpretation

Chromosome analysis revealed a normal female chromosome complement in all 20 cells examined from multiple cultures of chorionic villus sampling.

There was no evidence of a chromosome abnormality within the limits of the current technology. The sample appeared to contain only maternal decidua. Therefore, this result may represent maternal cell contamination rather than the fetal karyotype.

EXAMPLE REPORTING OF NORMAL FEMALE CVS RESULTS – POOR QUALITY OF FETAL VILLI RECEIVED

ISCN Result: 46,XX Normal Female Karyotype

Interpretation

Chromosome analysis revealed a normal female chromosome complement in all 20 cells examined from multiple cultures of chorionic villus sampling.

The specimen consisted entirely of membranous tissue. While this tissue is most likely fetal in origin, the results could reflect maternal cell contamination rather than the fetal karyotype.

Some laboratories like to report the amount of villi received. This information could be added in the interpretation, including the following possible comments.

FOR ADEQUATE MATERIAL RECEIVED (MINIMUM 10 mg)

This amount of chorionic villi received was approximately _ mg of cleaned sample. This is considered adequate sampling for cytogenetic analysis.

FOR INADEQUATE MATERIAL RECEIVED (LESS THAN 10 mg)

This amount of chorionic villi received was approximately _ mg of cleaned sample. This is considered less than adequate sampling for cytogenetic analysis, i.e. less than our general minimum requirement of 10 mg.

FOR INADEQUATE MATERIAL RECEIVED (LESS THAN 10 mg) AND A NORMAL FEMALE RESULT

This amount of chorionic villi received was approximately _ mg of cleaned sample. This is considered less than adequate sampling for cytogenetic analysis, i.e. less than our general minimum requirement of 10 mg. Due to this result showing a normal female karyotype, maternal cell contamination cannot be excluded.

1.2.4 Maternal cell contamination

Maternal cell contamination (MCC) is defined as a co-mixture of maternal cells with those of the fetus. This only becomes apparent when the fetus is either male or has a cytogenetic abnormality. Even with a cytogenetic abnormality of a female, one cannot exclude maternal cells from true fetal mosaicism. Therefore, minimizing MCC is critical in prenatal specimens, or there is a risk of analyzing the mother rather than the fetus. In CVS, diligently separating out the chorionic villi from decidua, which may be maternally derived, is critical. Approximately 1–2% of CVS specimens show MCC. One can estimate this amount to be 2–4%, considering the equal likelihood of having MCC in female fetuses that are not distinguishable.

In CVS, the incidence of MCC is higher in cultured specimens than in direct preparations, making direct preparations an advantage for this purpose alone. However, for complete chromosome analysis, cultured specimens are preferred due to the better quality of chromosome morphology. The literature also reports a higher incidence of MCC in transcervical CVS versus the transabdominal method.

For amniotic fluid specimens, there is a very low rate of MCC (approximately 0.2–0.4%), which may be present in bloody samples or those with a posterior placenta. Posterior placentas may interfere with the amniocentesis procedure, increasing the likelihood of puncturing and obtaining placental material with the fluid.

For fetal demises when only placental tissue with no known fetal parts is obtained for cytogenetic analysis, the risk of MCC is substantial. Products of conception (POC) tissue in which no fetal parts are distinguishable may also yield a high degree of MCC. To best diagnose a POC for MCC, FISH analysis may be performed, which has a better outcome with interphase analysis to visualize even a small degree of XX versus XY cells (to be discussed further in Chapter 14, Interphase analysis).

For peripheral blood chromosome studies, MCC is very rare and not usually considered as a high probability when mosaicism is seen.

EXAMPLE REPORTING OF MATERNAL CELL CONTAMINATION WITH NORMAL MALE RESULTS IN AF

ISCN Result: 46,XY[12]/46,XX[3]

Interpretation

Chromosome analysis showed two cell lines present in multiple cultures of amniotic fluid. One cell line showed a normal male chromosome complement in 12 cells. The remaining three cells showed a normal female chromosome complement. The normal female cells most likely represent maternal cell contamination.

Maternal cell contamination occurs in less than 1% of amniotic fluid specimens. Other explanations may include a reabsorbed twin or true chimerism.

Recommendations:

1. Genetic counseling.
2. Ultrasound may be useful in visualizing male genitalia as well as detecting a viable or reabsorbed twin.
3. A repeat amniocentesis could be considered if clinically indicated.

EXAMPLE REPORTING OF MATERNAL CELL CONTAMINATION WITH NORMAL MALE RESULTS IN CVS

ISCN Result: 46,XY[15]/46,XX[5]

Interpretation

Chromosome analysis showed two cell lines present in multiple cultures of chorionic villus sampling. One cell line showed a normal male chromosome complement in 15 cells. The remaining five cells showed a normal female chromosome complement. The normal female cells most likely represent maternal cell contamination; however, a twin pregnancy or chimerism cannot be excluded.

EXAMPLE REPORTING OF RESULTS OF A TWIN PREGNANCY WHEN BOTH TWINS HAVE THE SAME KARYOTYPE

ISCN Result: 46,XX or 46,XY (for each reported twin) Normal Female/Male Karyotype

Interpretation

Chromosome analysis revealed a normal female/male chromosome complement in all 20 cells examined from multiple cultures of amniocytes/chorionic villus sampling.

There was no evidence of a chromosome abnormality within the limits of the current technology.

Note: Accurate cytogenetic results from twin pregnancies depend on the successful sampling of both fetal sacs. Although accurate in most cases, it is possible that the same results for twin pregnancies represent the sampling of only one of the two fetuses.

1.3 Neonatal normal results

Cytogenetic studies of newborns and liveborns in general are performed on peripheral blood specimens and at times, if a very fast result is needed on a newborn, bone marrow may be obtained. Bone marrow studies are being phased out more and more, since the only recommended need for a fast turnaround time on newborns is for cases of possible cri-du-chat syndrome or Down syndrome, in which a heart defect is present and a physician needs to decide whether or not to perform surgery within hours to save the baby. If it is known that the baby has a genetic defect where surgery will not ameliorate the problem, surgery is usually not performed. However, if the baby has an isolated heart defect unrelated to a chromosomal abnormality, then surgery may save the baby's life. Consequently, a quick cytogenetic study may be useful, if not critical, in these cases. However, FISH analysis for the disorder, if available, may be a better approach than cytogenetic analysis, since cells would not need to be cultured, but rather analysis may be performed on a direct preparation with interphase cells.

Therefore, for the most part, cytogenetic analysis on peripheral blood is the preferred sample type for postnatal chromosome analysis and yields better chromosome morphology and banding level than other specimen types. Peripheral blood, like fibroblasts, goes through cell division approximately once every 24 hours, and cells can be synchronized to divide simultaneously to yield a large number of metaphase cells for analysis. Peripheral blood samples have the advantage over fibroblasts as it is easier to obtain cells in metaphase (harvested cells) and samples may be subjected to high-resolution techniques more easily than fibroblasts, thus yielding not only a large number of cells, but also cells with longer chromosomes. One other advantage of peripheral blood is the ability to collect (harvest) cells in metaphase, days sooner than fibroblasts, thus achieving a far better culture turnaround time of 2–3 days versus 5–10 days for fibroblast cells.

Whole blood samples and even harvested cells may be saved easily for further studies such as microarray or FISH analyses when a clinical indication for further testing is useful or even preferred. For example, with a clinical indication of mental impairment, developmental delay and autistic features, genomic microarrays are now considered a first-tier test for cytogenetic analysis rather than metaphase chromosome studies by the American College of Medical Genetics (ACMG). In such cases, when only chromosome studies are requested, a comment may be added in the cytogenetic report of a normal chromosome study to suggest microarray analysis, which may give further information at a higher level of chromosome resolution.

Below is an example of a recommended comment to add microarray analysis after a normal chromosome study when enough material is left for further analysis.

Chromosomal microarray analysis (CMA) may prove informative in this case. The current specimen, if available, may be used for this analysis.

1.4 Normal variants in the population

The definition of a normal genetic variation is one that occurs at 1% or greater in the general population. This is true for chromosomal variants as well as molecular genetic variation changes. These variants are generally inherited but can occur *de novo.*

1.4.1 Large heterochromatic regions (Figure 1.2)

One of the most common chromosome genetic variants is a variable amount of heterochromatin below the centromere in the proximal long arm of specific chromosomes: namely, chromosomes 1, 9 and 16. These variants have no clinical consequence and, therefore, laboratories do not even include them in clinical reports, unless desired.

ISCN RULES FOR REPORTING A LARGE HETEROCHROMATIC VARIANT

- First write the modal chromosome number of 46, followed by the sex designation.
- Then write the chromosome involved followed by the abbreviation "qh+."

Note: there are no parentheses around the chromosome or the qh+ designation.

For example: 46,XY,9qh+ represents a large heterochromatic region on the proximal long arm of chromosome 9.

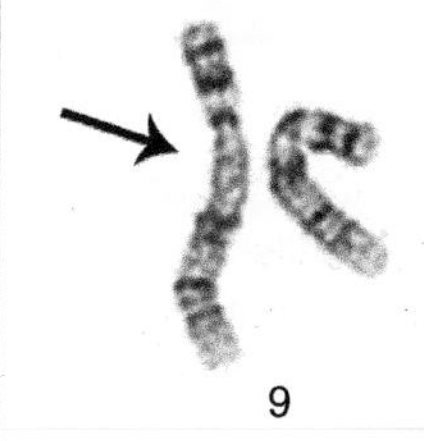

Figure 1.2 Large heterochromatic region of chromosome 9 – 9qh+. Courtesy of Sarah South PhD, ARUP Laboratories.

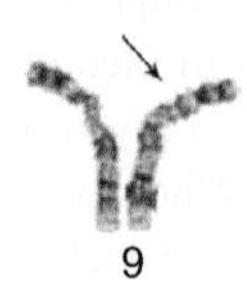

Figure 1.3 Inversion of the heterochromatic region of chromosome 9 - inv(9)(p12q13). Courtesy of Sarah South PhD, ARUP Laboratories.

1.4.2 Variant pericentric inversions around the centromere (Figure 1.3)

One of the most common chromosome genetic variants is the inversion around the centromere of chromosome 9: inv(9)(p12q13). An inversion of this region is a change in the heterochromatin to the proximal short arm. Similarly, an inversion of the centromeric region of chromosomes 1, 2 and 16 is also seen. These variants generally have no clinical consequence and, therefore, some laboratories do not even include them in clinical reports, although some do add a comment in the interpretation of the report.

ISCN RULES FOR REPORTING A NORMAL VARIANT PERICENTRIC INVERSION

- First write the modal chromosome number of 46, followed by the sex designation.
- Then write the abbreviation "inv" followed by the chromosome involved, enclosed by parentheses.
- Then write the breakpoints of the inversion, starting with breakpoints of the short arm (p arm) followed by breakpoints of the long arm (q arm) enclosed by parentheses.

For example: 46,XY,inv(9)(p12q13) for a pericentric inversion around the centromere of chromosome 9.

When both an inversion and a large heterochromatic region are observed on a single chromosome, for example on chromosome 9, then the ISCN nomenclature is written as: 46,XY,inv(9)(p12q13)9qh++ (Figure 1.4).

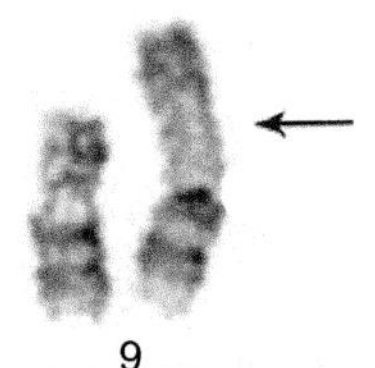

Figure 1.4 Inversion around the centromere of chromosome 9 with a large heterochromatic region: inv(9)(p12q13)9qh++. Courtesy of Sarah South PhD, ARUP Laboratories.

EXAMPLE REPORTING OF A VARIANT PERICENTRIC INVERSION

ISCN Result: 46,XX,inv(9)(p12q13) Normal Female Karyotype

Interpretation

Chromosome analysis revealed a normal female chromosome complement in all 20 cells examined. However, a pericentric inversion of chromosome 9 was present in all cells. This rearrangement is considered a normal variant in the population with no known clinical significance.

1.4.3 Variant satellite regions

Acrocentric chromosomes are chromosomes that contain a long arm and a short arm containing no euchromatic region and that have stalks and satellite regions on the short arm (Figure 1.5). The stalk region contains DNA that codes for ribosomal RNA (rRNA) and is described as the nuclear organizer region (NOR). When these regions are expressing the genes for rRNA, the NOR may be visualized with specialized staining, called silver staining (Figure 1.6). These will be seen as large dark regions in the short arms. When seen by R- or G-banding, the stalk regions appear as an unstained area. The region above, or distal to, the stalk is the satellite region, seen as a black band at the top. Variant stalk and satellite regions may look much larger or possibly smaller than normal, or the satellites may appear as double satellites (Figure 1.7), but are still considered as a normal variant if no euchromatic DNA material is present. A shortened stalk region, seen by the distance from the centromere to the satellite, is generally not mentioned in a report, because lack of a stalk would refer to rRNA that is not currently being expressed in the cell and would not have any noticeable distance from the centromere to the satellite.

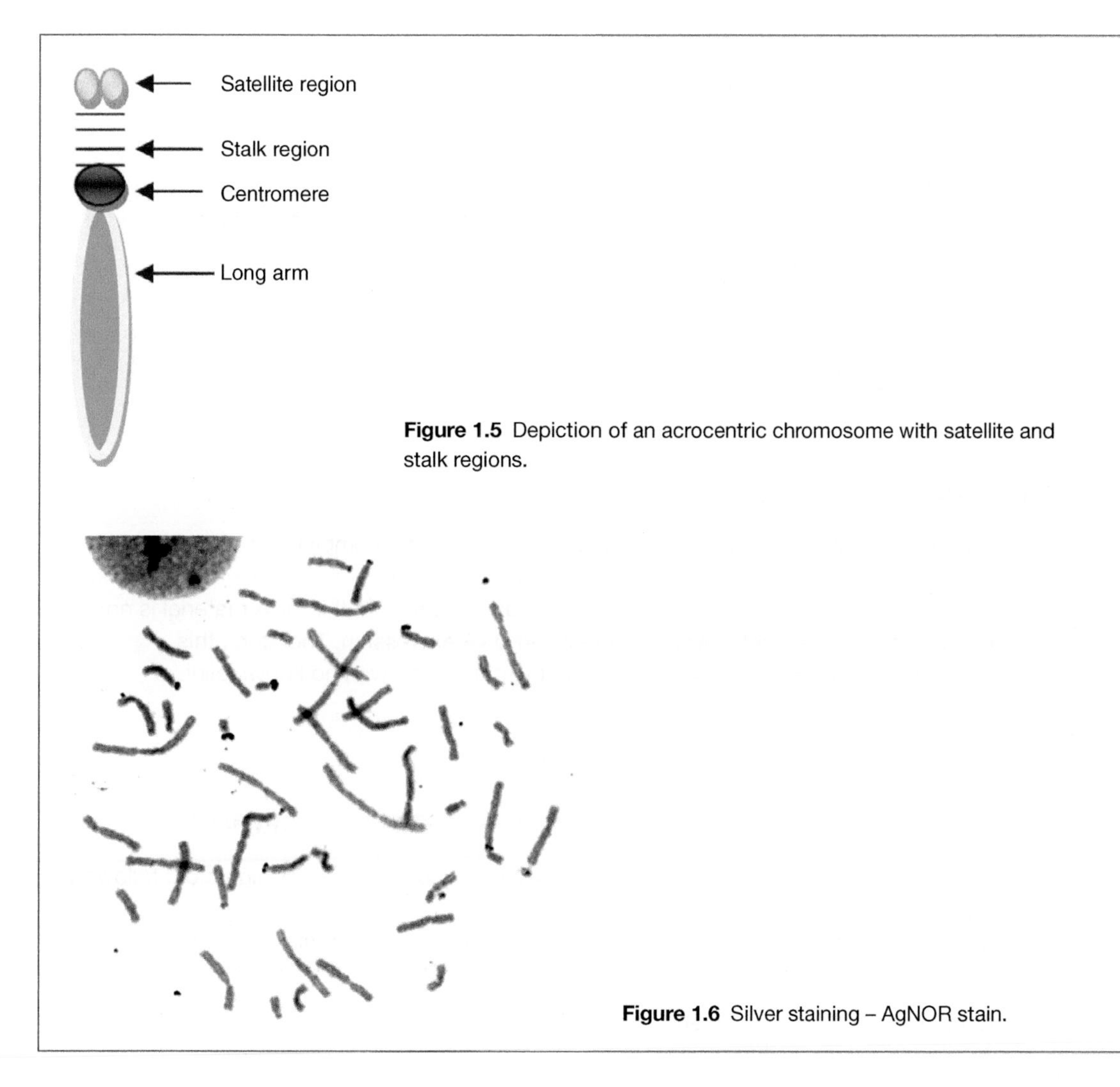

Figure 1.5 Depiction of an acrocentric chromosome with satellite and stalk regions.

Figure 1.6 Silver staining – AgNOR stain.

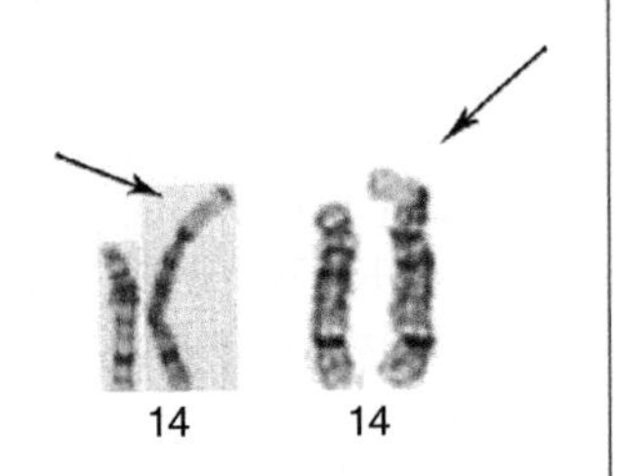

Figure 1.7 Large stalk (stk+) region on chromosome 14 in the short arm. Courtesy of Mehdi Jamehdor, Kaiser Permanente Genetics Laboratory.

However, these variants may be added to a report to distinguish these regions if they appear too large or different to ignore, but are known by various banding techniques not to have any euchromatic material that may be of clinical significance.

ISCN RULES FOR THE VARIATION OF THE STALK REGION

- First write the modal chromosome number of 46, followed by the sex designation, followed by a comma.
- Then write the chromosome number involved followed by the short arm designation "p" followed by "stk."
- Then write the designation "+" for an enlarged stalk or "−" for a shortened stalk.

For example: 46,XY,14pstk+ for a large stalk region on chromosome 14.

EXAMPLE REPORTING OF AN ENLARGED STALK REGION

ISCN Result: 46,XY,14pstk+ Normal Male Karyotype

Interpretation

Chromosome analysis revealed a normal male chromosome complement in all 20 cells examined. However, a large stalk region on the short arm of chromosome 14 was present in all cells. Specific banding techniques were performed to verify that this DNA material is not euchromatin and, consequently, would not result in DNA expression. Therefore, this rearrangement is considered a normal variant in the population with no known clinical significance.

ISCN RULES FOR THE VARIATION OF THE SATELLITE REGION

- First write the modal chromosome number of 46, followed by the sex designation, followed by a comma.
- Then write the chromosome number involved followed by the short arm designation "p" followed by the designation "s+" for an enlarged stalk or "s−" for a shortened stalk.

For example: 46,XY,22ps+ for an enlarged satellite region on chromosome 22 (Figure 1.8).

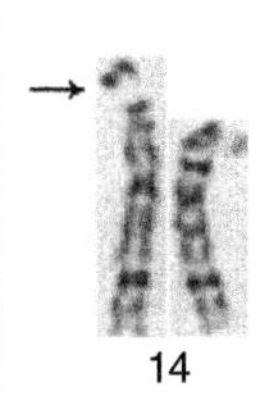

Figure 1.8 Large satellite (ps+) region on chromosome 14 in the short arm. Courtesy of Mehdi Jamehdor, Kaiser Permanente Genetics Laboratory.

EXAMPLE REPORTING OF AN ENLARGED SATELLITE REGION

ISCN Result: 46,XY,22 ps+   Normal Male Karyotype

Interpretation

Chromosome analysis revealed a normal male chromosome complement in all 20 cells examined. However, a large satellite region on the short arm of chromosome 22 was present in all cells examined. Specific banding techniques were performed to verify that this DNA material is not euchromatin and, consequently, would not result in DNA expression. Therefore, this rearrangement is considered a normal variant in the population with no known clinical significance.

1.4.4 Variant Yqh

The Y chromosome is significant due to the unique genes on its short arm and proximal long arm, but much of the DNA material in this small chromosome has only heterochromatin in the distal long arm (Yqh region). This Yqh region is quite variable in size, but as long as there is some noticeable heterochromatin, the chromosome is considered normal, regardless of the amount of the Yqh region. C-banding is a good method for detecting the presence or absence of heterochromatin on the long arm of the Y chromosome (Figure 1.9). However, there may be a need or desire to comment in the report regarding this region.

For example, when the Yqh region is very small or unnoticeable but known to be present via extra banding techniques, such as C-banding, the Yqh- designation may be used. Alternatively, when the Yqh region is exceptionally large, beyond the size of chromosome 18, a comment may be given in the report describing a Yqh + region (Figure 1.10).

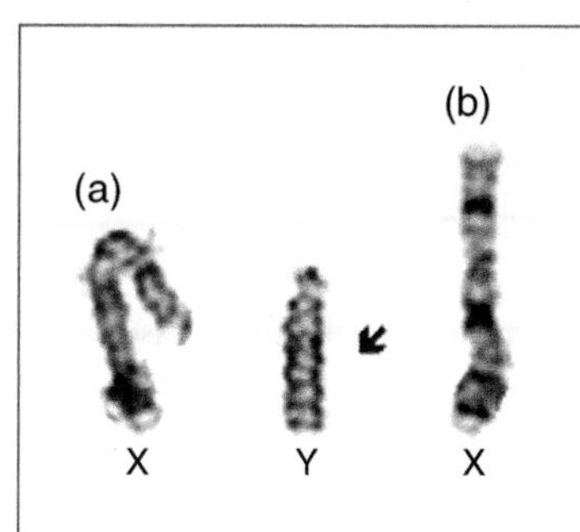

Figure 1.9 Variation in the heterochromatic region of the Y chromosome. (a) Large heterochromatic region – Yqh+. (b) Small heterochromatic region – Yqh-.

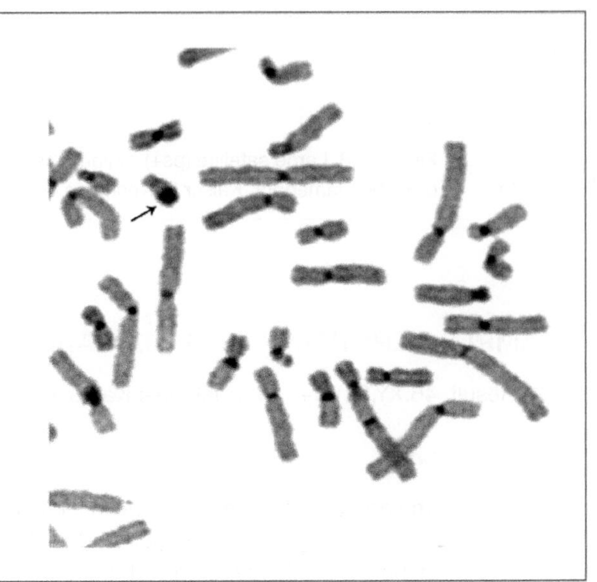

Figure 1.10 C-banding showing the heterochromatic regions of the chromosomes in a cell. The arrow depicts the Y chromosome with a significant heterochromatic region.

ISCN RULES FOR A VARIANT PERICENTRIC INVERSION OF THE Y CHROMOSOME

- First write the modal chromosome number of 46, followed by XY.
- Then write qh + for enlarged heterochromatin or qh– for diminished heterochromatin region.

Note: There is no comma separating the X and Y or between Y and qh.

EXAMPLE REPORTING OF A DIMINISHED SIZE OF THE HETEROCHROMATIN REGION ON THE Y CHROMOSOME

ISCN Result: 46,XYqh– Normal Male Karyotype

Interpretation

Chromosome analysis revealed a normal male chromosome complement in all 20 cells examined. However, a very small heterochromatic region in the long arm of the Y chromosome was observed in all cells examined. A specific banding technique (C-banding) was performed to verify that this DNA material does not result in a deletion of euchromatin, which might result in loss of DNA coding genes. Therefore, this rearrangement is considered a normal variant in the population with no known clinical significance.

It may be possible to see satellites on a chromosome region, which is not normal. In these cases, the abnormality will be discussed in the corresponding chapters. For example, satellites on the Y chromosome are discussed in Chapter 10 on sex chromosome abnormalities. For satellites on unidentifiable chromosomes, see Chapter 7 on marker chromosomes.

1.5 Disclaimers and recommendations

Some laboratories include disclaimers and recommendations in their reports. Disclaimers are used to indicate potential issues which may affect the accuracy of a result, such as issues associated with the limits of the current technology. Possible disclaimers may contain the following information.

EXAMPLE REPORTING OF DISCLAIMERS

- Based on the chromosome morphology and banding resolution of this study, standard cytogenetic methodology does not routinely detect subtle or submicroscopic rearrangements, low-level mosaicism, and some artifacts including maternal cell contamination.
- Testing performed in the absence of an accurate clinical indication may not give conclusive results.
- Chromosome analysis will not detect genetic conditions with Mendelian, multifactorial or environmental etiologies.
- Any image that accompanies this report is a representative image only and should not be used for the diagnosis of a patient.

Recommendations are helpful in a report to give pertinent information to clinicians regarding further action or testing that is based on either the result itself or the result in combination with the indication.

EXAMPLE REPORTING OF COMMON RECOMMENDATIONS

- Genetic counseling.
- Examine parental chromosomes to determine a potential origin of this abnormality.
- If clinically indicated, a follow-up study may be considered.
- If an increased incidence of fetal loss has been observed in this couple, peripheral blood studies on the parents may give further information.
- Monitor subsequent pregnancies with prenatal diagnosis.
- Consider chromosome studies on at-risk family members.
- A confirmatory cytogenetic study should be performed at the time of delivery or termination.
- Chromosome analysis of the patient's first-degree biological relatives is suggested to identify other carriers of this rearrangement who are also at risk.
- Due to the location of the breakpoints in this rearrangement and the nature of the unbalanced derivative chromosomes that could be inherited in future pregnancies, this patient may be at an increased risk for liveborn offspring with congenital anomalies and miscarriages.
- Familial chromosomal rearrangements are generally not associated with an increased risk of phenotypic and/or developmental abnormalities; however, submicroscopic deletions, duplications, or disruption of a gene or regulatory element at the breakpoints that may cause abnormalities cannot be excluded.
- Genetic counseling is recommended and prenatal diagnosis should be offered for future pregnancies.

- Consultation with a clinical geneticist or cytogeneticist is suggested to determine if these findings address the clinical indication of the patient.
- As regulatory agencies require that follow-up information on all prenatal studies be correlated with laboratory findings, please provide the outcome information at the time of delivery or termination.
- Submit a specimen at the outcome of this pregnancy for cytogenetic confirmation as a quality control laboratory monitor (at no additional charge).
- Prenatal diagnosis is suggested in future pregnancies.
- Note that the patient does not wish to know the sex of the fetus.
- Genomic microarray analysis may prove informative for further characterization of this rearrangement. As recent literature suggests, imbalances at the breakpoints or elsewhere in the genome may be found in approximately 30–40% of individuals with a cytogenetically balanced rearrangement and an abnormal phenotype.
- Additional testing with microarray analysis is suggested to further characterize this abnormal chromosome finding.
- If clinically indicated, consider uniparental disomy (UPD) studies.
- Note that maternal cell contamination is common in products of conception specimens. Therefore, a normal female chromosome complement should be interpreted with caution, as it may not be representative of the fetal karyotype.
- These results are based on the analysis of cells from a single primary culture.
- No clinical information was provided with this specimen. Although no chromosome abnormalities were detected, the accuracy or appropriateness of the testing performed is compromised in the absence of clinical information. Consultation with a clinical geneticist or cytogeneticist is suggested to determine if the testing reported here addresses the patient's clinical circumstances.
- High-resolution chromosome analysis at a band level of ___ was achieved in this study.
- Chromosome analysis at a band level of ___ was achieved in this study, below the optimum of 650 bands for a high-resolution study.

1.6 Culture failures

Cultures are initiated differently for each specimen type, depending on the source of the cell type. For prenatal specimens and solid tissues, including products of conception, the cell type is generally fibroblastoid, thereby requiring long-term tissue culture in flasks or another sterile vessel for days to weeks in an incubator. These cells may be grown on coverslips, flaskets (T-12 size) or flasks (T-25 size), depending on laboratory preference. In all cases, fibroblast cells adhere to the surface of the growth vessel within 24 hours and after this time require daily culture medium to be added or changed for cell growth and division. Cells are harvested in the growing stage when they are subconfluent.

Optimal specimen acquisition by the clinician is critical to cell growth in culture in the laboratory. Bloody prenatal specimens, inadequate material received, non-viable tissues obtained, non-sterile techniques used and contamination before the specimen is received in the laboratory all contribute to poor cell growth and possible culture failure in the laboratory.

Transport of the specimen is also critical for cell viability in the laboratory. Transport for longer than 24 hours, in too hot or cold a temperature, or cells transported without proper containment in a sterile vessel all contribute to poor growth of cells in culture and possible culture failure.

Table 1.5 Typical acceptable failure rates by specimen type

Specimen type	Accepted culture failure rate
Amniotic fluid and chorionic villus sampling	5–10%
Solid tissues (biopsies)	5–10%
Products of conception	Not estimated due to viability of specimen received
Peripheral blood	1–5%

Once a specimen is in the laboratory, other factors contribute to culture failures. Since fibroblast cells require constant changes in culture media and sterile culture techniques to maintain growth and prepare the cells for collection or harvest for chromosome analysis, culture failures may occur due to problems in laboratory technique. Many of these culture failures can be ameliorated by good training, sterile technique and proper standard operating procedures.

The other specimen type generally obtained for constitutional chromosome analysis is peripheral blood. Peripheral blood is usually grown in conical or round-bottom tubes in which the cells float in culture medium for 48–72 hours. The original culture medium placed in the tube with the cells is usually sufficient for cell growth and division for this time period before the cells are ready to be harvested for chromosome analysis. Culture failures are rare and predominantly due to the specimen itself, rather than being related to laboratory techniques. Culture failures may occur due to patient medication or radiation that inhibits cell growth, an indication of severe anemia causing a paucity of cells obtained or a low white blood cell count related to a hematological disease.

Culture failures caused by laboratory techniques may be due to poor quality control (QC) measures for monitoring culture media and other factors relating to the equipment in the laboratory, such as incubators, refrigerators, freezers and biological safety hoods for maintaining sterility. Once proper QC measures are taken in the laboratory, culture failures will be minimized. Table 1.5 shows approximate acceptable culture failure rates for each specimen type.

EXAMPLE REPORTING OF CULTURE FAILURES IN POCs

The sample received failed to grow in culture. It is not unusual for this type of specimen from a fetal demise to fail to grow in culture.

Recommendation

If an increased incidence of fetal loss has been observed in this couple, peripheral blood studies on the parents may prove informative.

EXAMPLE REPORTING OF CULTURE FAILURES IN PRENATAL SPECIMENS

The sample received failed to grow in culture. It is not unusual for bloody samples or specimens with suboptimal fetal material received to fail to grow in culture.

Recommendation

A repeat CVS specimen or amniotic fluid specimen for chromosome analysis may give further results.

EXAMPLE REPORTING OF CULTURE FAILURES IN PERIPHERAL BLOOD

The sample received failed to grow in culture. It is not unusual for peripheral blood specimens to fail to grow in culture with patients on medication that inhibit cell growth.

Recommendation

A repeat specimen for chromosome analysis may give further results.

EXAMPLE REPORTING OF CULTURE FAILURES IN SOLID TISSUE (BIOPSY)

The sample received failed to grow in culture. The tissue biopsy received was below the optimum size for cytogenetic analysis.

Recommendation

A repeat specimen for chromosome analysis may give further results.

1.7 Contamination

Contamination of cultures may be due to the presence of bacteria or fungi that occurs within the specimen itself, at the time of collection, during transportation or in the laboratory. If a specimen is contaminated within the individual or before it is received in the laboratory, standard laboratory procedures may eradicate or lessen the degree of contamination enough to perform cytogenetic analysis. However, this possibility decreases the longer the specimen remains in culture.

Even if the specimen appears free of contamination at the time it is received in the laboratory, it may subsequently acquire bacteria or fungi. Again, treatment protocols may reduce or eradicate the problem to the extent that chromosome analysis may be successful.

Standard laboratory techniques usually call for adding an antibacterial and antifungal reagent, such as gentamicin, to culture media to decrease the possibility of contamination in cell cultures. Usually a 1% reagent will forestall any contamination if the specimen is clean at the time it is received in the laboratory.

By far the worst type of contamination is fungal. Fungus tends to be hearty and grows very quickly, such that typical antifungal reagents will not be sufficient to eradicate the fungus from the specimen. Bacterial contamination may be slower to grow and may not kill the specimen as rapidly as fungus, but is still a danger to cell viability. The best approach for saving the specimen and the laboratory from widespread contamination is to remove the contaminated specimen from the others and isolate it in a holding incubator. Treating the specimen with double the amount of antibiotic/antimycotic reagent and treating daily are suggested. Also, harvesting the specimen as soon as possible rather than waiting for proper confluency will actually increase the probability of a successful number of viable cells for chromosome analysis.

The worst-case scenario is to leave the contaminated specimen in the same incubator as the other specimens, thereby compromising all the samples in the laboratory. Once contamination is widespread in one or more incubators, it becomes increasingly difficult to eradicate the source of contamination without a laborious, time-consuming process.

To proactively prevent contamination, use a biological safety hood with an ultraviolet (UV) light source. The hood should be used at all times when working with long-term cultures. Additionally, each day (or night), the UV light should remain on for a minimum of 15 minutes to destroy any possible contaminants. After each use in the hood, wiping down the surface with isopropyl alcohol will kill any contaminant from the area. Bleach (10%) may be used daily to kill any remaining source of contaminant for hoods and tabletop counters in the laboratory to protect against contaminants.

EXAMPLE REPORTING OF CONTAMINATION WITHIN 24 HOURS OF RECEIPT IN THE LABORATORY

The sample received showed signs of bacterial/fungal contamination within 24 hours of receipt in the laboratory. This indicates that the sample had been contaminated prior to its arrival in the laboratory.

Repeated attempts to induce the cultures to grow were not successful; therefore, cytogenetic analysis was not possible. Send a repeat sample to attempt further cytogenetic analysis, if possible.

EXAMPLE REPORTING OF CONTAMINATION OF UNKNOWN SOURCE

The sample received failed to grow in culture due to the presence of bacterial/fungal contamination.

Repeated attempts to induce the cultures to grow were not successful; therefore, cytogenetic analysis was not possible. Send a repeat sample to attempt further cytogenetic analysis, if possible.

References

Shaffer LG, McGowan-Jordan J, Schmid M (eds). *ISCN 2013: An International System for Human Cytogenetic Nomenclature*. Karger Publishers, Unionville, CT, 2013.

Welborn JL, Welborn R. Banding resolution of human chromosomes: a method of accuracy and simplicity. Am J Med Genet 1993; 47: 1180–1183.

Bibliography

Blumberg BD, et al. Minor chromosomal variants and major chromosomal anomalies in couples with recurrent abortion. Am J Hum Genet 1982; 34(6): 948–960.

Centers for Disease Control and Prevention (CDC). Improved national prevalence estimates for 18 selected major birth defects – United States, 1999–2001. Morb Mortal Wkly Rep 2006; 54: 1301.

Fiorentino F, Spizzichino L, Bono S, et al. PGD for reciprocal and Robertsonian translocations using array comparative genomic hybridization. Hum Reprod 2011; 26(7): 1925–1935.

Gardner RJM, Sutherland GR. *Chromosome Abnormalities and Genetic Counselling*. Oxford Monographs on Medical Genetics. Oxford University Press, Oxford, 2003.

Gersen S, Keagle M (eds). *Principles of Clinical Cytogenetics*. Humana Press, Totowa, New Jersey, 1999.

Greisman HA, Hoffman NG, Yi HS. Rapid high-resolution mapping of balanced chromosomal rearrangements on tiling CGH arrays. J Mol Diagn 2011; 13(6): 621–633.

Grinberg KN, Terekhov SM: Chromosome imbalance and the cell proliferative potential in vitro. Bull Exp Biol Med 1985; 99: 191–193.

Gunn SR, Robetorye RS, Mohammed MS. Comparative genomic hybridization arrays in clinical pathology: progress and challenges. Mol Diagn Ther 2007; 11(2): 73–77.

Hsu LYF. Prenatal diagnosis of chromosomal abnormalities through amniocentesis. In: Milunsky A (ed) *Genetic Disorders and The Fetus*, 4th edn. Johns Hopkins University Press, Baltimore, MD, 1998, p.179.

Jacobs PA, Browne C, Gregson N, et al. Estimates of the frequency of chromosome abnormalities detectable in unselected newborns using moderate levels of banding. J Med Genet 1992; 29: 103.

Josifek K, Haessig C, Pantzar T. Evaluation of chromosome banding resolution: a simple guide for laboratory quality assurance. Applied Cytogenet 1991; 17: 101–105.

Lam YH, Tang MHY, Sin SY, Ghosh A. Clinical significance of amniotic fluid cell culture failure. Prenat Diagn 1998; 18: 343–347.

Lebedev IN, et al. Features of chromosomal abnormalities in spontaneous abortion cell culture failures detected by interphase FISH analysis. Eur J Hum Genet 2004; 12: 513–520.

Lindstrand A, Schoumans J, Gustavsson P, Hanemaaijer N, Malmgren H, Blennow E. Improved structural characterization of chromosomal breakpoints using high resolution custom array-CGH. Clin Genet 2010; 77(6): 552–562.

Reid R, Sepulveda W, Kyle PM, Davies G. Amniotic fluid culture failure: clinical significance and association with aneuploidy. Obstet Gynecol 1996; 87: 588–592.

Rooney DE, Czepulkowski BH. *Human Cytogenetics. A Practical Approach*. Oxford University Press, New York, 1992.

Schluth-Bolard C, Delobel B, Sanlaville D, et al. Cryptic genomic imbalances in de novo and inherited apparently balanced chromosomal rearrangements: array CGH study of 47 unrelated cases. Eur J Med Genet 2009; 52(5): 291–296.

Sismani C, Kitsiou-Tzeli S, Ioannides M, et al. Cryptic genomic imbalances in patients with de novo or familial apparently balanced translocations and abnormal phenotype. Mol Cytogenet 2008; 1: 15.

Staebler M, Donner C, van Regemorter N, et al. Should determination of the karyotype be systematic for all malformations detected by obstetrical ultrasound? Prenat Diagn 2005; 25: 567.

CHAPTER 2

Mosaicism

True mosaicism is defined as two or more different genetic cell lines observed in a specimen or an individual. When mosaicism occurs within a single zygote, the two separate cell lines are present due to a postfertilization or mitotic error early in embryogenesis. All the subsequent cell divisions will result in daughter cells derived from each of the separate progenitor cells.

A non-disjunction event giving rise to mosaicism may be the result of a meiotic II error (Figure 2.1). For example, trisomy 21 may be present in all cells of an individual, but subsequently the extra chromosome 21 is lost in a mitotic division early in embryogenesis due to anaphase lag, in which a third chromosome 21 was lost when daughter cells divide. The opposite may also occur, in which a normal zygote was formed in meiosis, but a subsequent non-disjunction mitotic event occurred giving rise to an abnormal cell line, such as trisomy 21. Two daughter cells are derived with each cell division; therefore, one daughter cell would be trisomy 21 and the other daughter cell would be monosomy 21. Monosomy 21 is lethal and is consequently lost. This second scenario is a less frequent event, but still occurs.

When discussing true mosaicism, it is usually in the context of an abnormal cell line with the presence of a normal cell line. However, two or more abnormal cell lines may be present. In these cases, the principle of mosaicism remains the same; however, the mechanism of how the cell lines arise may differ.

Another true mosaicism arises when a single individual forms from two separate zygotes. This results in true chimerism. A true chimera is only observed by cytogenetics when the two cell lines are of different sexes, resulting in 46,XX and 46,XY cell lines. Two separate 46,XX cell lines or two separate 46,XY cell lines resulting in a true chimera would not be noticeable during cytogenetic analysis. A true chimera is either a dizygotic twin pregnancy that did not divide into two individuals or dizygotic twins that exchange hematopoietic stem cells during pregnancy. A true chimera with XX and XY cells may result in abnormalities of sexual differentiation and will have specific sexual clinical manifestations. The incidence of true chimeras is very low, approximately 1 in 100,000.

XX/XY mosaicism may also need to be evaluated for maternal cell contamination. Maternal cell contamination and confined gonadal mosaicism are discussed in Chapter 10.

Cytogenetic Abnormalities: Chromosomal, FISH and Microarray-Based Clinical Reporting, First Edition. Susan Mahler Zneimer.

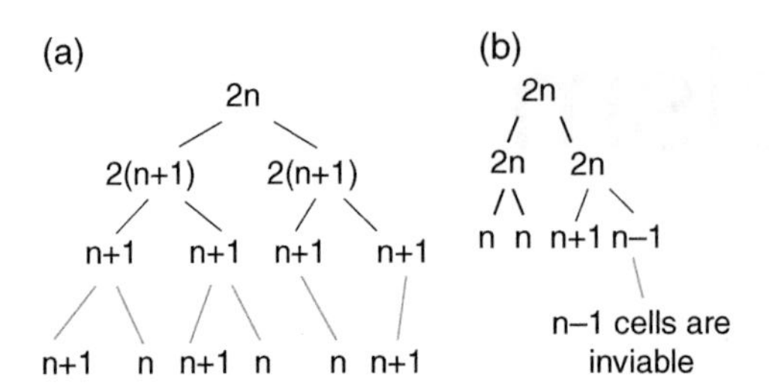

Figure 2.1 Non-disjunction in meiosis in a diploid cell (2n) leading to mosaic trisomy. (a) Trisomy may be present in all cells of an individual at meiosis II due to a non-disjunction in meiosis I, but subsequently the extra chromosome is lost in a mitotic division early in embryogenesis due to anaphase lag, in which a third chromosome 21 was lost when daughter cells divide. (b) Cells may start out normal from meiosis I, but some of the gametes are affected by non-disjunction in meiosis II, giving rise to a trisomy. This diagram depicts one gamete with a single extra chromosome; one gamete is missing a single chromosome and two gametes are normal.

More than one cell line observed but confined to the gonads (gonadal mosaicism) is another form of mosaicism. This is usually detectable when a child, or more than one child in a family, is studied cytogenetically and found to have a chromosome defect but the parents appear cytogenetically normal in peripheral blood. The parent may in fact be masking gonadal mosaicism, where the cytogenetic abnormality is only within the gonad cells. In this event, the parent may be phenotypically normal but because the gonad is abnormal, may pass on the cytogenetic abnormality to their offspring in their cells.

Mosaicism may be pseudomosaicism rather than true mosaicism. Pseudomosaicism occurs when the definition of a true mosaic is not met. Pseudomosaicism is defined at different levels in a cytogenetic laboratory based on the cells examined.

Level I mosaicism (pseudomosaicism) is defined as a single cell abnormality that cannot be confirmed in any other cells observed, regardless of the cell count, or from any number of primary cultures established from the specimen source. This pseudomosaicism is usually not reported in constitutional cases unless it is clinically significant and can be studied by other means in order to attempt to confirm the abnormality as an abnormal cell line. For example, one cell of 50 cells examined is found to be trisomy 21 in a child with mild phenotypic features of Down syndrome. This might not be observed in dividing cells in peripheral blood, or it may be confined to specific tissues of the body; therefore, peripheral blood chromosome analysis may not yield a true picture of the prevalence of an abnormal cell line in an individual. Fluorescent *in situ* hybridization (FISH) for the abnormality, if probes are available, may be a better means to confirm a possible cell line in either peripheral blood or by studying a different cell type, such as muscle or buccal smear (FISH will be discussed in section 2.1).

However, the vast majority of cases with a single abnormal cell are regarded as cultural artifacts, especially when a large number of cells have been analyzed, and cultured artifacts are generally not reported by the laboratory.

Level II mosaicism (pseudomosaicism) is defined as two or more cells observed with the same abnormality, but is confined to a single primary culture of the specimen. Due to this possibility, it is essential to initiate more than one culture from any specimen type when performing chromosome analysis in order to distinguish true mosaicism from Level II pseudomosaicism. In a prenatal specimen of amniotic fluid set up as *in situ* colonies, even two or more colonies within a single primary vessel are considered Level II pseudomosaicism. Level II mosaicism is usually not reported by the laboratory, but clinically relevant abnormalities may warrant additional studies to try and confirm a possible abnormal cell line.

Level III mosaicism is defined as the presence of two or more cells with the same chromosome abnormality present in two or more primary cultures of a specimen. Level III mosaicism is considered true mosaicism.

There is a difficulty in following these levels of mosaicism as definitive, since it is always possible that Level I or II mosaics may be true mosaics, but masked by present laboratory techniques. Therefore, discretion is advised when interpreting low-level mosaicism, regardless of the specimen type. Extended studies may be warranted to try and confirm a possible abnormal cell line, or at least minimize the probability of true mosaicism with a large set of cells examined, or a separate means of analysis may be advisable, such as FISH analysis.

2.1 Normal results with 30–50 cells examined

EXAMPLE REPORTING OF NORMAL RESULTS WITH 30 CELLS ANALYZED

ISCN Result: 46,XX   Normal Female Karyotype

Interpretation

Chromosome analysis revealed a normal female chromosome complement in all 30 cells examined. There was no evidence of a chromosome abnormality within the limits of the current technology.

With a total of 30 metaphase cells examined, this should exclude mosaicism of greater than 15% (99% confidence interval) in this tissue. If mosaicism is still of concern, consider testing an alternative tissue sample as levels of mosaicism may vary across different tissues.

EXAMPLE REPORTING OF NORMAL RESULTS WITH 50 CELLS ANALYZED

ISCN Result: 46,XX   Normal Female Karyotype

Interpretation

Chromosome analysis revealed a normal female chromosome complement in all 50 cells examined. There was no evidence of a chromosome abnormality within the limits of the current technology.

With a total of 50 metaphase cells examined, this should exclude mosaicism of greater than 9% (99% confidence interval) in this tissue. If mosaicism is still a concern, consider testing an alternative tissue sample as levels of mosaicism may vary across different tissues.

2.2 Normal and abnormal cell lines

ISCN RULES FOR REPORTING NORMAL AND ABNORMAL CELL LINES TOGETHER

- Always write the abnormal cell line first, followed by the normal cell line separated by a slash.
- In each cell line, the number of cells per cell line is enclosed by a bracket.

For example, a female with mosaic Turner syndrome would be written as: 45,X[12]/46,XX[8].

2.2.1 Mosaic trisomies

ISCN RULES FOR REPORTING OF MOSAIC TRISOMIES

- First write the modal chromosome number of the abnormal cell line, followed by the sex designation.
- Then write "+" with the chromosome number involved, followed by the total number of abnormal chromosomes in brackets.
- A slash separates the abnormal from the normal cell line.
- Then write the normal cell line nomenclature, followed by the number of cells with the normal cell line in brackets.

EXAMPLE REPORTING OF MOSAIC TRISOMY 21 – MOSAIC DOWN SYNDROME

ISCN Result: 47,XX,+21[#]/46,XX[#]

Interpretation

Chromosome analysis revealed two cell lines present. # number of cells of 20 cells examined showed an abnormal female chromosome complement with an extra chromosome 21. In # cells, a normal female chromosome complement was observed. These findings are consistent with the clinical diagnosis of mosaic Down syndrome.

The clinical features of mosaic Down syndrome are variable depending on the total percentage of cells with an extra chromosome 21 and the frequency of trisomy 21 in specific cell types and tissues. The severity of phenotypic features may therefore be mild to resembling a full Down syndrome phenotype.

Genetic counseling is recommended.

2.2.2 Mosaicism for XX and XY

EXAMPLE REPORTING OF XX/XY IN A PRENATAL SPECIMEN

ISCN Result: 46,XX[#]/46,XY[#]

Interpretation

Chromosome analysis revealed two cell lines present. # number of cells of 20 cells examined showed a normal female chromosome complement. In # cells, a normal male chromosome complement was observed. These findings are consistent with maternal cell contamination of a normal male fetus, a reabsorbed twin or a true chimera.

The clinical features of XX/XY mosaicism may range from a normal female to a normal male phenotype, or include manifestations of abnormal sexual differentiation.

Further studies are recommended to help clarify the origin of this mosaicism. Fluorescence *in situ* hybridization, molecular studies and a repeat prenatal sample are suggested to evaluate further cells for a final cytogenetic description of these possible cell lines.

Genetic counseling is recommended.

EXAMPLE REPORTING OF A TRUE CHIMERA IN A CHILD

ISCN Result: 46,XX[#]/46,XY[#]

Interpretation

Chromosome analysis revealed two cell lines present. # number of cells of 20 cells examined showed a normal female chromosome complement. In # cells, a normal male chromosome complement was observed. These findings are consistent with maternal cell contamination, the clinical diagnosis of mixed sexual differentiation or true chimerism.

The clinical features of XX/XY mosaicism may range from a normal female to a normal male phenotype, or include manifestations of abnormal sexual differentiation.

To confirm the possibility of a true chimera, molecular studies of this child and both parents are required to evaluate if a double contribution of maternal and paternal alleles is present in the two cell lines.

Genetic counseling is recommended.

2.3 Two or more abnormal cell lines

ISCN RULES FOR REPORTING TWO OR MORE ABNORMAL CELL LINES

- First, always write the abnormal cell line that contains the most number of cells in the cell line.
- Then write the second abnormal cell line with the next highest number of cells in the cell line.
- Each cell line follows with decreasing number of cells in the cell line.
- If there is a normal cell line, this is always the last cell line written.
- For each cell line, write the number of cells per cell line enclosed by brackets after the nomenclature.

For example, a female with mosaic variant Turner syndrome would be written as: 45,X[18]/47,XXX[10]/46,i(X)(q10)[2].

EXAMPLE REPORTING OF THREE RELATED ABNORMAL CELL LINES DETECTED

ISCN Result: 45,X[18]/47,XXX[10]/46,X,i(X)(q10)[2]

Interpretation

Chromosome analysis revealed three related abnormal cell lines present. The first cell line, seen in 18 of 30 cells examined, showed the loss of one X chromosome. The second cell line, seen in 10 cells, showed the gain of one X chromosome. The third cell line, seen in two cells, showed an isochromosome of the long arm of one X chromosome. These findings are consistent with the clinical diagnosis of variant Turner syndrome.

Females with this mosaic karyotype may be expected to have increased risk of infertility and taller mean adult height. Other Turner syndrome manifestations include short stature, primary amenorrhea, broad chest, wide-spaced nipples, low hairline and webbed neck.

Note: Due to the risk of gonadoblastoma in cases of Turner syndrome with cryptic Y chromosome mosaicism, FISH analysis with X and Y probes is suggested.

Genetic counseling is recommended.

Bibliography

Chial H. Somatic mosaicism and chromosomal disorders. Nature Educ 2008; 1: 1.

De Marchi M, Carbonara AO, Carozzi F, et al.True hermaphroditism with XX/XY sex chromosome mosaicism: report of a case. Clin Genet 2008; 10(5): 265–272.

Fitzgerald PH, Donald RA, Kirk RL. A true hermaphrodite dispermic chimera with 46,XX and 46,XY karyotypes. Clin Genet 1979; 15(1): 89–96.

Gardner RJM, Sutherland GR. *Chromosome Abnormalities and Genetic Counselling.* Oxford Monographs on Medical Genetics. Oxford University Press, Oxford, 2003.

Gersen S, Keagle M (eds). *Principles of Clinical Cytogenetics.* Humana Press, Totowa, New Jersey, 1999.

Hook EB. Exclusion of chromosomal mosaicism: tables of 90%, 95% and 99% confidence limits and comments on use. Am J Hum Genet 1977; 29: 94–97.

Jones KL. *Smith's Recognizable Patterns of Human Malformations*, 5th edn. Saunders, Philadelphia, 1997.

Kalousek DK, Dill FJ. Chromosomal mosaicism confined to the placenta in human conceptions. Science 1983; 221(4611): 665–7.

Rieger R, Michaelis A, Green MM. *Glossary of Genetics and Cytogenetics: Classical and Molecular*, 4th edn. Springer-Verlag, Berlin, 1976.

Robinson WP, et al. Molecular studies of chromosomal mosaicism: relative frequency of chromosome gain or loss and possible role of cell selection. Am J Hum Genet 1995; 56: 444–451.

Rooney DE, Czepulkowski BH. *Human Cytogenetics. A Practical Approach.* Oxford University Press, New York, 1992.

Russel LM, Strike P, Browne CE, Jacobs PA. X chromosome loss and aging. Cytogenet Genome Res 2007; 116(3): 181–185.

Schinzel A. *Catalogue of Unbalanced Chromosome Aberrations in Man*, 2nd edn. Walter de Gruyter, Berlin, 2001.

Shaffer LG, McGowan-Jordan J, Schmid M (eds). *ISCN 2013: An International System for Human Cytogenetic Nomenclature.* Karger Publishers, Unionville, CT, 2013.

Stetten G, Escallon CS, South ST, McMichael JL, Saul DO, Blakemore KJ. Reevaluating confined placental mosaicism. Am J Med Genet 2004; 131: 232–239.

Strachan T, Read AP. Chromosome abnormalities. In: *Human Molecular Genetics*, 2nd edn. Wiley-Liss, New York, 1999.

Tobias ES, Connor M, Ferguson Smith M. *Essential Medical Genetics*, 6th edn. Wiley-Blackwell, Oxford, 2011.

Wilson GN. *Clinical Genetics, A Short Course.* Wiley-Liss, New York, 2000.

Wolff DJ, van Dyke DL, Powell CM, Working Group of the ACMG Laboratory Quality Assurance Committee. Laboratory guideline for Turner syndrome. Genet Med 2010; 12: 52–55.

Youssoufian H, Pyeritz RE. Mechanisms and consequences of somatic mosaicism in humans. Nature Reviews Genetics 2002; 3: 748–758.

CHAPTER 3

Autosomal trisomies – prenatal and livebirths

3.1 Introduction

Autosomal trisomies refer to the gain of a single chromosome that does not involve the sex chromosomes. The only non-mosaic viable autosomal trisomies in humans are the gains of chromosomes 13, 18 and 21. These and other autosomal trisomies may be viable in a mosaic form, including gains of chromosomes 8, 9, 20 and 22. Other autosomal trisomies exist but these are generally not viable and will be discussed in Chapter 11.

One common aspect of autosomal trisomies is that they carry an increased risk with advanced maternal age, due to the aging of the female eggs, giving a preponderance of non-disjunction (malsegregation of chromosomes) events during gametogenesis. That is, as women age, the risk of having an egg that contains a cell division error resulting in an extra chromosome increases. Most meiotic errors result in lethality of those eggs but a few chromosomes survive, resulting in a chromosomal imbalance once fertilized with a normal sperm gamete. Even most trisomies for chromosomes 13, 18 and 21 will end in a fetal demise; however, a small proportion of zygotes result in a viable fetus and in a chromosomal genetic defect. Trisomy 21 is by far the most common of the viable trisomies but approximately 90% of trisomy 21 zygotes result in fetal death.

Another common feature among autosomal trisomies is that the trisomy may be the result of either a gain of an intact chromosome 21 or a translocation event that is either *de novo* (a new mutation) or inherited in a balanced form from a parent. For chromosomes 13 and 21, due to these chromosomes being acrocentric and not carrying any coding, euchromatic DNA on the short arm, these are referred to as Robertsonian translocations in which the long arm of an acrocentric chromosome is joined with the long arm of another acrocentric chromosome. Descriptions of both trisomy by gain of an intact chromosome and translocation trisomies will be discussed below.

Another common factor among autosomal trisomies is the generic manifestations seen in most, if not all, patients, including growth deficiency, mental impairment and multiple congenital anomalies. However, each trisomy has unique features that make a clinical diagnosis fairly distinctive. Each of the predominant clinical features of these trisomies will be presented in this chapter.

Another factor among autosomal trisomies in general is the presence of mosaicism. That is, a trisomic cell line is seen in conjunction with a normal cell line. In these instances, the phenotypic manifestations

Cytogenetic Abnormalities: Chromosomal, FISH and Microarray-Based Clinical Reporting, First Edition. Susan Mahler Zneimer.

tend to be milder than in the "full-blown" trisomy in which all cells have the abnormality. Mosaic trisomies will also be discussed here.

ISCN RULES FOR REPORTING TRISOMIES

- First write the modal chromosome number, followed by the sex designation.
- Then write the plus sign, "+", followed by the chromosome number involved.

For example, a male with trisomy 21 is written as 47,XY,+21.

3.2 Trisomy 21 – Down syndrome

SPECIFIC FEATURES OF DOWN SYNDROME

Down syndrome is the most common genetic cause of mental impairment. The overall incidence of Down syndrome is approximately 1 in 660 livebirths, but the incidence is much higher at prenatal diagnosis and with advanced maternal age.

The most prominent clinical features of Down syndrome at birth include hypotonia, short stature, brachycephaly, flat occiput, short neck with some webbing, low-set ears, epicanthal folds, upslanting palpebral fissures and protruding tongue. Diagnosis in childhood also includes mental impairment and developmental delay. Other manifestations that have a high risk in these patients are congenital heart disease, duodenal atresia and tracheo-esophageal fistula. Patients also carry a high risk of developing acute leukemia.

Trisomy 21 accounts for approximately half of all the chromosomal abnormalities detected prenatally. It is estimated that up to 80–90% of conceptuses die *in utero*. The fetuses that survive are most likely either mosaic for a normal cell line or those without severe heart defects.

The recurrence risk for Down syndrome, once Down syndrome has been diagnosed in a child, is approximately 1% overall. However, for younger mothers (30 years and younger), the risk is about 1.4% and for mothers older than 35 years, the risk is age related. The risks of second-degree relatives with a child with Down syndrome and paternal age do not alter the recurrence risk.

Translocation Down syndrome carries a different set of recurrence risks and will be discussed separately.

EXAMPLE REPORTING OF TRISOMY 21 – STANDARD GAIN OF CHROMOSOME 21 (FIGURE 3.1)

ISCN Result: 47,XX,+21 or 47,XY,+21

Interpretation

Chromosome analysis revealed an abnormal female/male chromosome complement in all cells examined with an extra chromosome 21 (trisomy 21), consistent with the clinical diagnosis of Down syndrome.

The main clinical features of Down syndrome include hypotonia, flat facies, down-slanting eyes, small ears and mental deficiency. The incidence is 1 in 660 livebirths.

The recurrence risk of future pregnancies with a chromosomal abnormality is approximately 1%, but increases with maternal age.

Genetic counseling is recommended.

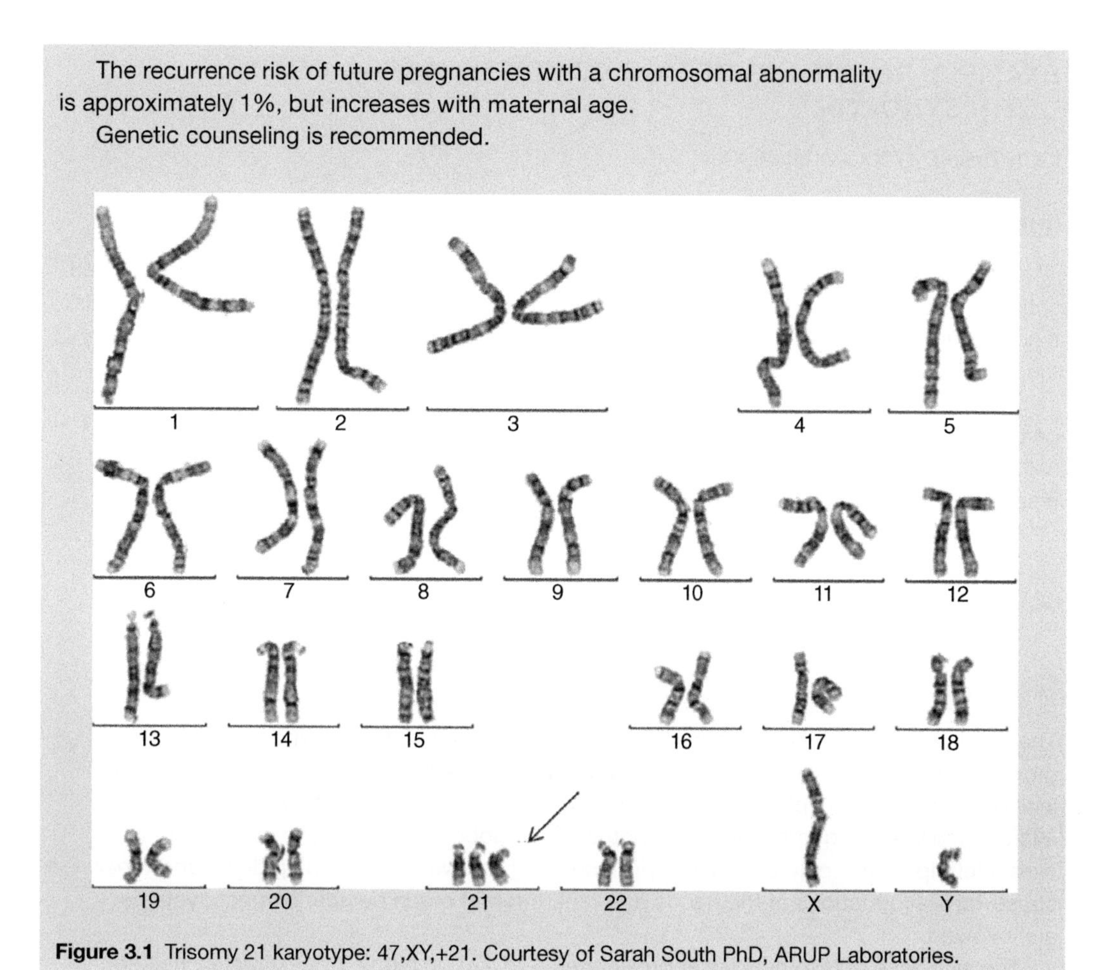

Figure 3.1 Trisomy 21 karyotype: 47,XY,+21. Courtesy of Sarah South PhD, ARUP Laboratories.

For translocation Down syndrome, refer to Chapter 4.

3.3 Mosaic trisomy 21 – mosaic Down syndrome

Only a small percentage (about 2%) of Down syndrome patients are mosaic for trisomy 21. The phenotype is usually milder than when seen in all cells, and manifestations may vary depending on the proportion of abnormal cells in different tissues formed during embryogenesis. Some of these patients have a carrier parent, usually the mother, who also carries some eggs with an extra chromosome 21, which is normal in somatic cells. Another mechanism leading to mosaic Down syndrome is the loss of "full-blown" trisomy 21 in which trisomy rescue takes place in some cells, reverting cells to disomy 21.

Another mechanism of mosaicism is somatic non-disjunction in which a normal embryo is formed but during subsequent cell divisions early on in embryogenesis, a malsegregation of chromosome 21 occurs, giving rise to some cells with an extra chromosome 21 and other cells with nullisomy 21 (which is lethal). Other somatic cells that do not undergo this error in cell division remain as normal; thus, only some cells result in the gain of chromosome 21.

EXAMPLE REPORTING OF MOSAIC TRISOMY 21 – MOSAIC DOWN SYNDROME

ISCN Result: 47,XX,+21[#]/46,XX[#]

Interpretation

Chromosome analysis revealed two cell lines present. # number of cells of 20 cells examined showed an abnormal female chromosome complement with an extra chromosome 21. In # cells, a normal female chromosome complement was observed. These findings are consistent with the clinical diagnosis of mosaic Down syndrome.

The clinical features of mosaic Down syndrome are variable depending on the total percentage of cells with an extra chromosome 21 and the frequency of trisomy 21 in specific cell types and tissues. The severity of phenotypic features may, therefore, be mild to resembling a full Down syndrome phenotype.

Genetic counseling is recommended.

3.4 Trisomy 13 – Patau syndrome

SPECIFIC FEATURES OF TRISOMY 13

The most prominent phenotypic manifestations of trisomy 13 Patau syndrome include growth retardation, severe mental impairment and severe nervous system malformations, including holoprosencephaly and heart defects. Other clinical features include microcephaly, sloping forehead, microphthalmia, coloboma of the iris or absence of the eyes, cleft lip and/or palate, polydactyly, clenching fists with overlapping digits and other structural malformations of the hands and feet. Internal organ defects of most systems are involved.

The incidence of trisomy 13 is approximately 1 in 12,000 births. Approximately half of the individuals who survive to birth will die within the first month of life, due to the severity of malformations present. Trisomy 13 follows the same increased risk due to maternal age as do other trisomies, i.e. equally at approximately 1%. The risk of trisomy 13 due to a translocation event is about 20% of cases reported. Mosaic trisomy 13 has been reported in a subset of individuals with a milder phenotype and longer survival.

EXAMPLE REPORTING OF TRISOMY 13 – PATAU SYNDROME

ISCN Result: 47,XX,+13

Interpretation

Chromosome analysis revealed an abnormal female chromosome complement in all cells examined with an extra chromosome 13 (trisomy 13), consistent with the clinical diagnosis of Patau syndrome.

The main clinical features of trisomy 13 include holoprosencephaly, narrow hyperconvex fingernails, defects of the eye, nose and lip, and severe mental deficiency. The incidence is 1 in 12,000 livebirths.

The recurrence risk of future pregnancies with a chromosomal abnormality is approximately 1%, but increases with maternal age.

Genetic counseling is recommended.

EXAMPLE REPORTING OF MOSAIC TRISOMY 13 – MOSAIC PATAU SYNDROME

ISCN Result: 47,XX,+13[#]/46,XX[#]

Interpretation

Chromosome analysis revealed two cell lines present. # number of cells of 20 cells examined showed an abnormal female chromosome complement with an extra chromosome 13. In # cells, a normal female chromosome complement was observed. These findings are consistent with the clinical diagnosis of mosaic Patau syndrome.

The clinical features of mosaic trisomy 13 are variable depending on the total percentage of cells with an extra chromosome 13 and the frequency of trisomy 13 in specific cell types and tissues. The severity of phenotypic features may, therefore, be mild to resembling a full trisomy 13 syndrome phenotype.

Genetic counseling is recommended.

3.5 Trisomy 18 – Edwards syndrome

SPECIFIC FEATURES OF TRISOMY 18 – EDWARDS SYNDROME

The most common features of trisomy 18 are growth retardation and severe mental impairment (as in trisomy 13 syndrome), but also hypertonia, small mouth, narrow palate, prominent occiput, micrognathia, clenched fists with overlapping digits, and prominent rocker-bottom feet and other structural malformations of the hands and feet. Internal organ defects of most systems are involved, including severe cardiac defects.

The incidence of trisomy 18 is approximately 1 in 6000 births, with a higher frequency among females (male:female ratio equals 1:3–4). Approximately 10% of the individuals who survive to birth will die within the first year of life; up to 40% will die within 1 week of birth. Trisomy 18 follows the same increased risk due to maternal age as do other trisomies, and the recurrence risk is low, approximately 1–2% or less. Mosaic trisomy 18 has been reported in a subset of individuals with a milder phenotype and longer survival.

EXAMPLE REPORTING OF TRISOMY 18 – EDWARDS SYNDROME

ISCN Result: 47,XX,+18

Interpretation

Chromosome analysis revealed an abnormal female chromosome complement in all cells examined with an extra chromosome 18 (trisomy 18), consistent with the clinical diagnosis of Edwards syndrome.

The main clinical features of trisomy 18 include clenched hand, short sternum, feeble fetal activity, weak cry, polyhydramnios, growth deficiency and mental deficiency. The incidence is 1 in 6000 livebirths, with a 3:1 preponderance of females to male.

The recurrence risk of future pregnancies with a chromosomal abnormality is approximately 1%, but increases with maternal age.

Genetic counseling is recommended.

EXAMPLE REPORTING OF MOSAIC TRISOMY 18 – MOSAIC EDWARDS SYNDROME

ISCN Result: 47,XX,+18[#]/46,XX[#]

Interpretation

Chromosome analysis revealed two cell lines present. # number of cells of 20 cells examined showed an abnormal female chromosome complement with an extra chromosome 18. In # cells, a normal female chromosome complement was observed. These findings are consistent with the clinical diagnosis of mosaic trisomy 18 – Edwards syndrome.

The clinical features of mosaic trisomy 18 are variable depending on the total percentage of cells with an extra chromosome 18 and the frequency of trisomy 18 in specific cell types and tissues. The severity of phenotypic features may, therefore, be mild to resembling a full trisomy 18 syndrome phenotype.

Genetic counseling is recommended.

3.6 Trisomy 8 – mosaic

SPECIFIC FEATURES OF TRISOMY 8

Trisomy 8 is generally seen in the mosaic form, most likely due to the severity of malformations when seen in all cells. The primary manifestations include variable growth retardation and mental impairment, specific craniofacial abnormalities including prominent forehead, deep-set eyes, strabismus, broad nasal bridge, upturned nares, cleft palate, micrognathia and dysplastic ears. Skeletal abnormalities are also common, including spina bifida, hip dysplasia, scoliosis, joint contractures and other malformations. Internal organ defects of most systems are involved, including severe cardiac defects.

There is no direct correlation between the number of cells with trisomy 8 and the severity of phenotypic features. Mosaicism also varies between tissues studied, with a larger proportion of abnormal cells seen in fibroblasts than in peripheral blood.

The incidence and recurrence risks are not known due to the rarity of these cases in the scientific literature.

EXAMPLE REPORTING OF MOSAIC TRISOMY 8

ISCN Result: 47,XX,+8[#]/46,XX[#]

Interpretation

Chromosome analysis revealed two cell lines present. # number of cells of 20 cells examined showed an abnormal female chromosome complement with an extra chromosome 8. In # cells, a normal female chromosome complement was observed. These findings are consistent with the clinical diagnosis of mosaic trisomy 8.

Since reported cases of livebirth trisomy 8 are mosaic, the clinical features may be variable. The main clinical manifestations include thick lips, deep-set eyes, prominent ears, camptodactyly (permanent flexion of one or more fingers), poor coordination and mental deficiency.

Genetic counseling is recommended.

3.7 Trisomy 9 – mosaic

SPECIFIC FEATURES OF TRISOMY 9

This trisomy is also seen in the mosaic form, also likely due to the severity of malformations when seen in all cells. The primary manifestations include growth retardation and severe mental impairment, specific craniofacial abnormalities including high narrow forehead, deep-set eyes, upturned palpebral fissures, prominent nose, micrognathia and prominent upper lip. Skeletal abnormalities are also common, including bone dysplasia, narrow chest and other malformations. Internal organ defects of most systems are involved, including severe cardiac defects. Most individuals die within the postnatal period.

There is no known correlation between the number of cells with trisomy 9 and the severity of phenotypic features.

The incidence and recurrence risks are not known due to the rarity of these cases in the scientific literature.

EXAMPLE REPORTING OF MOSAIC TRISOMY 9

ISCN Result: 47,XX,+9[#]/46,XX[#]

Interpretation

Chromosome analysis revealed two cell lines present. # number of cells of 20 cells examined showed an abnormal female chromosome complement with an extra chromosome 9. In # cells, a normal female chromosome complement was observed. These findings are consistent with the clinical diagnosis of mosaic trisomy 9.

Since reported cases of livebirth trisomy 9 are mosaic, the clinical features may be variable. The main clinical manifestations include joint contractures, congenital heart defects, low-set malformed ears, micrognathia and severe mental deficiency.

Genetic counseling is recommended.

3.8 Trisomy 20 – mosaic, prenatal

SPECIFIC FEATURES OF TRISOMY 20

This trisomy is always seen in the mosaic form and is primarily detected prenatally. When detected prenatally, most cases are not confirmed at birth and have normal chromosomal results. The abnormal cells most likely originate from, and are confined to, extraembryonic tissue that is not present in liveborn individuals. Fetuses with mosaic trisomy 20 have many phenotypic abnormalities, including microcephaly, craniofacial dysmorphism, cardiac defects and kidney malformations. Cytogenetic studies on fetal tissues show abnormal cells in most of the major organs; postnatal studies right after birth have shown abnormal cells in cultured fibroblasts, but not in peripheral blood.

EXAMPLE REPORTING OF MOSAIC TRISOMY 20

ISCN Result: 47,XX,+20[#]/46,XX[#]

Interpretation

Chromosome analysis revealed two cell lines present. # number of cells of 20 cells examined showed an abnormal female chromosome complement with an extra chromosome 20. In # cells, a normal female chromosome complement was observed. The trisomy 20 cells were found in at least two primary cultures, representing true mosaicism; however, these cells may also represent extraembryonic tissue and not be representative of the fetus.

Trisomy 20 is generally found only in prenatal specimens and is rarely found in peripheral blood. Few cases have been shown to have an adverse clinical outcome. The abnormal trisomic cells may exist harmlessly in amnion cells, giving rise to a phenotypically and chromosomally normal fetus, even with true mosaicism. However, there are reports in the literature of true mosaic trisomy 20 with phenotypic defects.

Genetic counseling and level II ultrasound examination to identify possible fetal abnormalities are recommended. Fetal blood sampling is not recommended, as the trisomic cells do not appear in blood.

3.9 Trisomy 22 – mosaic, prenatal

SPECIFIC FEATURES OF TRISOMY 22

This trisomy is often seen in the mosaic form and is primarily detected prenatally. When detected prenatally, many cases are not confirmed at birth and have normal chromosomal results. Liveborns have been reported whose phenotypic abnormalities include intrauterine growth retardation, facial dysmorphism, low-set ears, microcephaly, micrognathia, cleft palate, epicanthal folds, webbed neck and cardiac defects. Most liveborns die in the first few months after birth. Postnatal studies right after birth have shown abnormal cells in cultured fibroblasts and in a smaller proportion of abnormal cells in peripheral blood.

EXAMPLE REPORTING OF MOSAIC TRISOMY 22

ISCN Result: 47,XX,+22[#]/46,XX[#]

Interpretation

Chromosome analysis revealed two cell lines present. # number of cells of 20 cells examined showed an abnormal female chromosome complement with an extra chromosome 22. In # cells, a normal female chromosome complement was observed. The trisomy 22 cells were found in at least two primary cultures, representing true mosaicism; however, these cells may also represent extraembryonic tissue and not be representative of the fetus.

Trisomy 22 is generally found only in prenatal specimens and is rarely found in peripheral blood. Few cases have been shown to have an adverse clinical outcome. The abnormal trisomic cells may exist harmlessly in amnion cells, giving rise to a phenotypically and chromosomally normal fetus even with true mosaicism. However, there are reports in the literature of true mosaic trisomy 22 with phenotypic defects.

Genetic counseling and level II ultrasound examination to identify possible fetal abnormalities are recommended. Fetal blood sampling is not recommended, as the trisomic cells rarely appear in blood.

Bibliography

Antonarakis SE, Adelsberger PA, Petersen MB, et al. Analysis of DNA polymorphisms suggests that most de novo dup(21q) chromosomes in patients with Down Syndrome are isochromosomes and not translocations. Am J Hum Genet 1990; 47: 968.

Arnold GL, Kirby RS, Stern TP, Sawyer JR. Trisomy 9: review and report of two new cases. Am J Med Genet 1995; 56: 252.

Bacino CA, Schreck R, Fischel-Ghodsian N, et al. Clinical and molecular studies in full trisomy 22: further delineation of the phenotype and review of the literature. Am J Med Genet 1995; 56: 359.

Centers for Disease Control and Prevention (CDC). Improved national prevalence estimates for 18 selected major birth defects – United States, 1999–2001. Morb Mortal Wkly Rep 2006; 54: 1301.

Cimini D, Degrassi F. Aneuploidy: a matter of bad connections. Trends Cell Biol 2005; 15: 442–551.

Collins VR, Muggli EE, Riley M, et al. Is Down Syndrome a disappearing birth defect? J Pediatr 2008; 152: 20.

Fujimoto A, Allanson J, Crowe CA, et al. Natural history of mosaic trisomy 14 Syndrome. Am J Med Genet 1992; 44: 189.

Gardner RJM, Sutherland GR. *Chromosome Abnormalities and Genetic Counselling*. Oxford Monographs on Medical Genetics. Oxford University Press, Oxford, 2003.

Gersen S, Keagle M (eds). *Principles of Clinical Cytogenetics*. Humana Press, Totowa, New Jersey, 1999.

Hassold T, Chiu D. Maternal age-specific rates of numerical chromosome abnormalities with special reference to trisomy. Hum Genet 1985; 70: 11.

Hassold T, Hall H, Hunt P. The origin of human aneuploidy: where we have been, where we are going. Hum Mol Genet 2007; 16(Spec No. 2): R23–R208.

Hsu LY. Prenatal diagnosis of chromosomal abnormalities through amniocentesis. In: Milunsky A (ed) *Genetic Disorders and the Fetus*, 4th edn. Johns Hopkins University Press, Baltimore, 1998, p.179.

Hsu LY, Shapiro LR, Gertner M, et al. Trisomy 22: a clinical entity. J Pediatr 1971; 79: 12.

Hsu LY, Yu MT, Neu RL, et al. Rare trisomy mosaicism diagnosed in amniocytes, involving an autosome other than chromosomes 13, 18, 20, and 21: karyotype/phenotype correlations. Prenat Diagn 1997; 17: 201.

Hulten MA, Patel SD, Tankimanova M, et al. On the origin of trisomy 21 Down syndrome. Molec Cytogenet 2008; 121.

Jacobs PA, Browne C, Gregson N, et al. Estimates of the frequency of chromosome abnormalities detectable in unselected newborns using moderate levels of banding. J Med Genet 1992; 29: 103.

Jones KL. *Smith's Recognizable Patterns of Human Malformation*, 6th edn. Saunders, London, 2006.

Kratzer PG, Golbus MS, Schonberg SA, et al. Cytogenetic evidence for enhanced selective miscarriage of trisomy 21 pregnancies with advancing maternal age. Am J Med Genet 1992; 44: 657.

Oliver-Bonet M, Navarro J, Carrera M, Egozcue J, Benet J. Aneuploid and unbalanced sperm in two translocation carriers: evaluation of the genetic risk. Molec Hum Reprod 2002; 8(10): 958–963.

Mokate T, Leask K, Mehta S, et al. Non-mosaic trisomy 22: a report of 2 cases. Prenat Diagn 2006; 26: 962.

Nussbaum R, McInnes R, Huntington F (eds). *Thompson and Thompson Genetics in Medicine*, 6th edn. Saunders, Philadelphia, 2001, pp. 157–162.

Pellestor F, Andreo B, Anahory T, Hamamah S. The occurrence of aneuploidy in human: lessons from the cytogenetic studies of human oocytes. Eur J Med Genet 2006, 49: 103–116.

Riccardi VM. Trisomy 8: an international study of 70 patients. Birth Defects Orig Artic Ser 1977; 13: 171.

Rooney DE, Czepulkowski BH. *Human Cytogenetics. A Practical Approach*. Oxford University Press, New York, 1992.

Schinzel A. Trisomy 8 and trisomy 9 are distinctly different clinical entities. Am J Med Genet 1993; 46: 603.

Shaffer LG, McGowan-Jordan J, Schmid M (eds). *ISCN 2013: An International System for Human Cytogenetic Nomenclature*. Karger Publishers, Unionville, CT, 2013.

Staebler M, Donner C, van Regemorter N, et al. Should determination of the karyotype be systematic for all malformations detected by obstetrical ultrasound? Prenat Diagn 2005; 25: 567.

Tinkle BT, Walker ME, Blough-Pfau RI, et al. Unexpected survival in a case of prenatally diagnosed non-mosaic trisomy 22: clinical report and review of the natural history. Am J Med Genet A 2003; 118A: 90.

Tobias ES, Connor M, Ferguson Smith M. *Essential Medical Genetics*, 6th edn. Wiley-Blackwell, Oxford, 2011.

Tolmie JL. Down Syndrome and other autosomal trisomies. In: Rimoin DL, Connor JM, Pyeritz RE (eds) *Emery and Rimoin's Principles and Practice of Medical Genetics*, 3rd edn. Churchill Livingstone, Edinburgh, 1996, p.925.

Warburton D, Dallaire L, Thangavelu M, et al. Trisomy recurrence: a reconsideration based on North American data. Am J Hum Genet 2004; 75: 376.

Weijerman ME, van Furth AM, Vonk Noordegraaf A, et al. Prevalence, neonatal characteristics, and first-year mortality of Down Syndrome: a national study. J Pediatr 2008; 152: 15.

Wooldridge J, Zunich J. Trisomy 9 Syndrome: report of a case with Crohn disease and review of the literature. Am J Med Genet 1995; 56: 258.

CHAPTER 4
Translocations

The definition of a translocation is the interchange of chromosomal material between two or more chromosomes. This interchange of genetic material requires the breakage and reunion at the chromosome regions involved at gametogenesis of either the sperm or egg. A translocation may be either inherited from a parent or a new mutation (*de novo*).

When a translocation is balanced, there is no loss or gain of genetic material when breakage and reunion occur. This results in one normal homolog of the chromosomes involved and one homolog of the rearranged chromosomes.

A translocation may be unbalanced in which a parental carrier of a balanced translocation passes on one of the chromosomes involved in the rearrangement but without the other chromosome involved. This will result in an imbalance with both a deletion and duplication.

It is also possible in *de novo* translocations that an apparently balanced rearrangement results in a disruption of genes at the breakpoints of the chromosomes involved or non-equal crossing over during meiosis, both of which can lead to deletions, duplications or mutations of genes at the region of breakage and reunion.

4.1 Reciprocal (balanced) translocations

SPECIFIC FEATURES OF RECIPROCAL TRANSLOCATIONS

Individuals who carry a balanced rearrangement often have no knowledge of their genetic change and only find out about this condition when a problem arises with infertility, the inability to carry a child to term, by giving birth to a child with an unbalanced chromosomal abnormality or with prenatal diagnosis. It may also be possible that their genetic change is carried by other family members, and was ascertained through a recommendation to have cytogenetic analysis.

Carriers of balanced translocations are at risk of producing unbalanced gametes in their offspring during gametogenesis. There are various possibilities of transmitting this genetic material to offspring. Due to meiotic segregation of chromosomes in meiosis during gametogenesis, there is a possibility of transmitting a combination of the two rearranged chromosomes, which results in an imbalance of genetic material.

Cytogenetic Abnormalities: Chromosomal, FISH and Microarray-Based Clinical Reporting, First Edition. Susan Mahler Zneimer.

Table 4.1 Theoretic translocation t(1;2)(p32;q21) describing the different possible scenarios of transmitting chromosomes 1 and 2 in a female carrier

Cytogenetic scenarios	Outcomes of offspring (male example)
Both normal chromosomes 1 and 2	46,XY Normal chromosome 1 and 2 complement
Both rearranged chromosomes 1 and 2	46,XY,t(1;2)(p32;q21)mat Balanced translocation t(1;2) inherited from the mother
One rearranged chromosome 1 and one normal chromosome 2	46,XY,der(1)t(1;2)(p32;q21)mat Unbalanced translocation with monosomy for chromosome 1 from band 1p32 to the short arm terminus, and trisomy for chromosome 2 from band 2q21 to the long arm terminus inherited from the mother
One rearranged chromosome 2 and one normal chromosome 1	46,XY,der(2)t(1;2)(p32;q21)mat Unbalanced translocation with monosomy for chromosome 2 from band 2q21 to the long arm terminus, and trisomy for chromosome 1 from band 1p32 to the short arm terminus inherited from the mother

The theoretic probability of each imbalanced translocation scenario is 25%. However, empirically, such is not the case. Most carriers of balanced translocations end in a miscarriage early in gestational age, usually in the first trimester, depending on the amount of the imbalanced genetic material (for details of the mechanism of how the segregation of gametes are formed, please see the bibliography).

Table 4.1 shows an example of a theoretic translocation t(1;2)(p32;q21), describing the different possible scenarios of transmitting chromosomes 1 and 2, with the karyotype designation 46,XX, t(1;2)(p32;q21), where the female is the carrier and the offspring is a male.

The risk of such an imbalance varies depending on the size of the chromosomal regions involved in the imbalance and which parent is the carrier. When a translocation region is very large, the number of genes would be great, decreasing the risk of viable unbalanced offspring. Instead, it increases the risk of early pregnancy loss or spontaneous abortions usually in the first trimester. The smaller the region of the rearrangement, the likelier it is that a fetus would be able to tolerate the small number of genes either lost or gained, thus increasing the risk of abnormal liveborn offspring.

Assigning risk factors to abnormal offspring for each specific translocation event is difficult to assess. Gardner and Sutherland (2003) attempt to assign risk values to specific abnormalities, but their sample sizes are extremely small. Most abnormalities fall within a 2–20% risk of having abnormal offspring rather than a fetal demise. Since translocations result from any possible exchange of genetic material between two chromosomes in gametogenesis, most possible translocations have not been studied to any degree and only a few reports of each rearrangement have been documented. Based on one study (Warburton 1991), the overall risk of a carrier of balanced translocation is approximately 0.1–13.8%. A female carrier has a greater risk for transmitting an unbalanced rearrangement over a male carrier, though the risks are not well established.

General risk factors for pregnancy loss in the general population are approximately 15%, and being a carrier of a balanced rearrangement increases that risk to a range of 20–30% for pregnancy loss. The risk may be as high as 50% if the genetic material involved is large.

One example of a documented, known reciprocal translocation in families is t(11;22)(q23;q11.2) (Figure 4.1). This is a scenario which involves a translocation with a small chromosome (chromosome 22); hence, only a small amount of genetic material from this one chromosome is rearranged. This gives

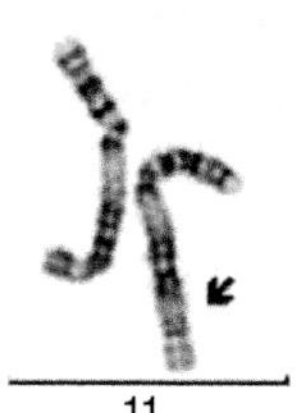

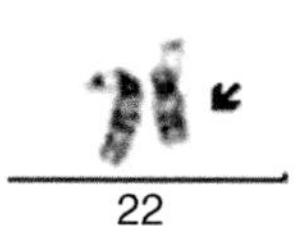

Figure 4.1 Partial karyotype showing a balanced translocation: t(11;22)(q23.3;q11.2). Courtesy of Sarah South PhD, ARUP Laboratories.

rise to a possible 3:1 malsegregation during gametogenesis that results in an extra small chromosome. In this case, the resultant chromosomal complement is 47,XY,+der(22)t(11;22)(q23;q11.2). This finding results in partial trisomy 22q and is consistent with the clinical diagnosis of Emanuel syndrome. More unbalanced karyotypic anomalies will be discussed in the following chapters.

However, the most common ramification for a carrier of a balanced translocation is the risk for miscarriages. Since most of the translocation events will carry enough genetic material on one or both of the chromosomes involved, there is a very high risk of this genetic material being lethal to the fetus. Infertility or, more appropriately termed, a history of multiple miscarriages is one of the most common indications for a constitutional chromosome analysis. Infertility would imply having the incapacity to conceive. Therefore, a history of miscarriages is a more appropriate term for carriers of balanced rearrangements.

ISCN RULES FOR REPORTING TRANSLOCATIONS

- First write the modal chromosome number, followed by the sex designation.
- Then write "t" for translocation, followed by, in parentheses, the two chromosomes involved, in which the first chromosome is the smallest chromosome number separated by a semicolon.
- Then write, in parentheses, the breakpoints involved for each chromosome, separated by a semicolon.

For example, a male with a translocation between chromosome 14 at a breakpoint in the long arm at band q13 and chromosome 22 at a breakpoint in the long arm at band q11.2 is written: 46,XY,t(14;22)(q13;q11.2).

For unbalanced translocations, see derivative chromosomes.

4.1.1 Balanced translocation – prenatal

EXAMPLE REPORTING OF A BALANCED TRANSLOCATION – AF/CVS

ISCN Result: 46,XX/XY,t(#;#)(p/q#;p/q#)

Interpretation

Chromosome analysis derived from # of culture vessels examined revealed an abnormal male/female fetal chromosome complement in all cells examined with an apparently balanced reciprocal translocation between the short/long arm of chromosome # and the short/long arm of chromosome #, with breakpoints at bands _ and _, respectively.

Chromosome analysis of both parents is indicated to determine if the rearrangement is familial or *de novo* in origin. A familial rearrangement is unlikely to have a phenotypic effect. *De novo* rearrangements, however, are associated with a significant risk for clinical abnormalities.

Genomic microarray analysis may prove informative for further characterization of this rearrangement. As recent literature suggests, imbalances at the breakpoints or elsewhere in the genome may be found in approximately 30–40% of individuals with a cytogenetically balanced rearrangement and an abnormal phenotype.

Genetic counseling is recommended.

4.1.2 Balanced translocation – postnatal

EXAMPLE REPORTING OF A BALANCED TRANSLOCATION IN A CHILD

ISCN Result: 46,XX/XY,t(#;#)(p/q#;p/q#)

Interpretation

Chromosome analysis revealed an abnormal male/female fetal chromosome complement in all cells examined with an apparently balanced reciprocal translocation between the short/long arm of chromosome # and the short/long arm of chromosome #, with breakpoints at bands _ and _, respectively.

Chromosome analysis of both parents is indicated to determine if the rearrangement is familial or *de novo* in origin. A familial rearrangement is unlikely to have a phenotypic effect. *De novo* rearrangements, however, are associated with a significant risk for clinical abnormalities.

Genomic microarray analysis may prove informative for further characterization of this rearrangement. As recent literature suggests imbalances at the breakpoints or elsewhere in the genome may be found in approximately 30–40% of individuals with a cytogenetically balanced rearrangement and an abnormal phenotype.

Genetic counseling is recommended.

4.1.3 Balanced translocation carrier

EXAMPLE REPORTING OF PARENTAL CARRIERS OF A BALANCED TRANSLOCATION (FIGURE 4.2)

ISCN Result: 46,XY,t(14;22)(q13;q11.2)

Interpretation

Chromosome analysis revealed an abnormal male chromosome complement in all cells examined with an apparently balanced reciprocal translocation between the long arm of chromosome 14 and the long arm of chromosome 22. This finding is consistent with a history of spontaneous abortions.

Due to the location of the breakpoints in this rearrangement and the nature of the unbalanced derivative chromosomes that could be inherited in future pregnancies, this patient may be at an increased risk for liveborn offspring with congenital anomalies and miscarriages.

Genetic counseling is recommended and prenatal diagnosis should be offered for future pregnancies.

Chromosome analysis of the patient's first-degree biological relatives is suggested to identify other carriers of this rearrangement who are also at risk.

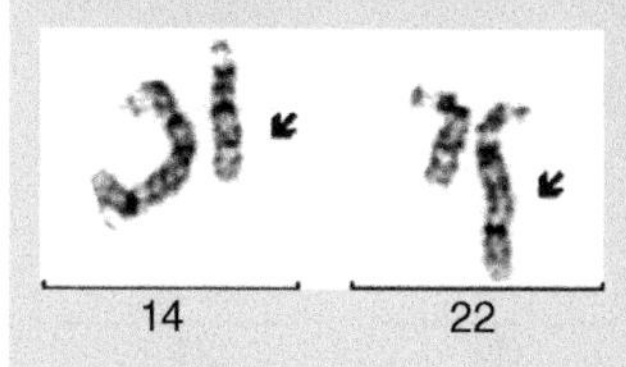

Figure 4.2 Partial karyotype showing a balanced translocation: t(14;22)(q13;q11.2). Courtesy of Sarah South PhD, ARUP Laboratories.

4.1.4 Balanced translocations – inherited

Inherited translocations are those that are seen in parents who carry a balanced rearrangement which is passed on to offspring in the same form. This reciprocal translocation should have no clinical consequences to the offspring (assuming the carrier parent is phenotypically normal).

ISCN RULES FOR REPORTING INHERITED TRANSLOCATIONS

- When parents of a proband with a translocation are studied cytogenetically, and one of the parents is found to carry the same translocation, then "pat" or "mat" is added after the breakpoints for paternal or maternal inheritance, respectively.

For example, a maternally inherited translocation t(11;21)(p13;q11.2) from a daughter would be written as: 46,XX,t(11;21)(p13;q11.2)mat.

- When parents of a proband with a translocation are studied cytogenetically, and both parents are cytogenetically normal, then "dn", for *de novo*, is added after the breakpoints.

For example, a new mutation of translocation t(1;3)(p13;q24) in a female would be written as: 46,XX,t(11;21)(p13;q11.2)dn.

Note: There is no space between the breakpoints and the parental or *de novo* designation.

EXAMPLE REPORTING OF A PROBAND WITH A TRANSLOCATION THAT IS INHERITED FROM A PARENT (FIGURE 4.3)

ISCN Result: 46,XX,t(11;21)(p13;q11.2)mat

Interpretation

Chromosome analysis revealed an abnormal female chromosome complement in all cells examined with an apparently balanced reciprocal translocation between the short arm of chromosome 11 and the long arm of chromosome 21.

This rearrangement was also seen in the cytogenetic study of the mother; therefore, this translocation is maternally inherited.

Genomic microarray analysis may prove informative for further characterization of this rearrangement. As recent literature suggests, imbalances at the breakpoints or elsewhere in the genome may be found in approximately 30–40% of individuals with a cytogenetically balanced rearrangement and an abnormal phenotype.

Chromosome analysis of the patient's first-degree biological relatives is suggested to identify other carriers of this rearrangement who are also at risk.

Genetic counseling is recommended.

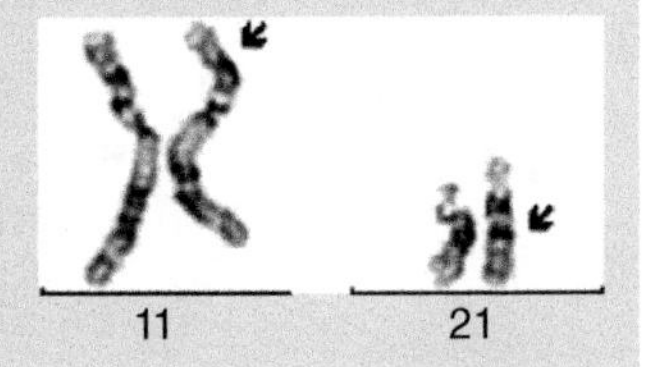

Figure 4.3 Partial karyotype showing a balanced translocation which is maternally inherited: t(11;21)(p13;q11.2)mat. Courtesy of Sarah South PhD, ARUP Laboratories.

4.1.5 Balanced translocations – *de novo*

SPECIFIC FEATURES OF *DE NOVO* TRANSLOCATIONS

De novo translocations vary from inherited translocations in their risk of having a clinical consequence due to a possible submicroscopic imbalance of DNA at the breakpoints of the rearrangement with the chromosomes involved. Among individuals with mental impairment who have been studied cytogenetically, there is a seven-fold increase of these individuals having an apparently balanced translocation compared to the general population. The detection of balanced translocations in amniotic fluid specimens shows a 2–3-fold increase of congenital anomalies compared to the general population.

EXAMPLE REPORTING OF A PROBAND WITH A *DE NOVO* TRANSLOCATION (FIGURE 4.4)

ISCN Result: 46,XX,t(1;14)(q31;q24.3)dn

Interpretation

Chromosome analysis revealed an abnormal female chromosome complement in all cells examined with an apparently balanced reciprocal translocation between the long arm of chromosome 1 and the long arm of chromosome 14.

This rearrangement was not seen in the cytogenetic studies of the parents; therefore, this translocation is presumably a new mutation (*de novo* in origin).

Genomic microarray analysis may prove informative for further characterization of this rearrangement. As recent literature suggests, imbalances at the breakpoints or elsewhere in the genome may be found in approximately 30–40% of individuals with a cytogenetically balanced rearrangement and an abnormal phenotype.

Chromosome analysis of the patient's first-degree biological relatives is suggested to identify other carriers of this rearrangement who are also at risk.

Genetic counseling is recommended.

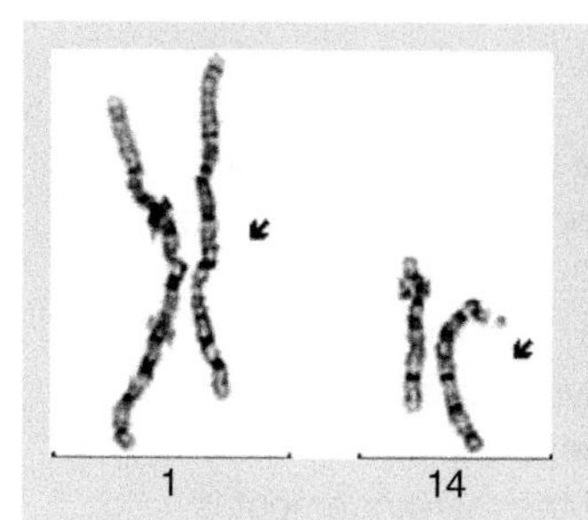

Figure 4.4 Partial karyotype showing a balanced translocation: t(1;14)(q31;q24.3). Courtesy of Sarah South PhD, ARUP Laboratories.

4.1.6 Balanced translocations with aneuploidy

Reciprocal translocations may also be present with aneuploidy, such as trisomy 21. In such cases, reports need to include the possibility of an inherited translocation in addition to the trisomy 21.

EXAMPLE REPORTING OF A BALANCED TRANSLOCATION WITH ANEUPLOIDY (FIGURE 4.5)

ISCN Result: 47,XY,t(14;21)(q24.1;q22.1),+21

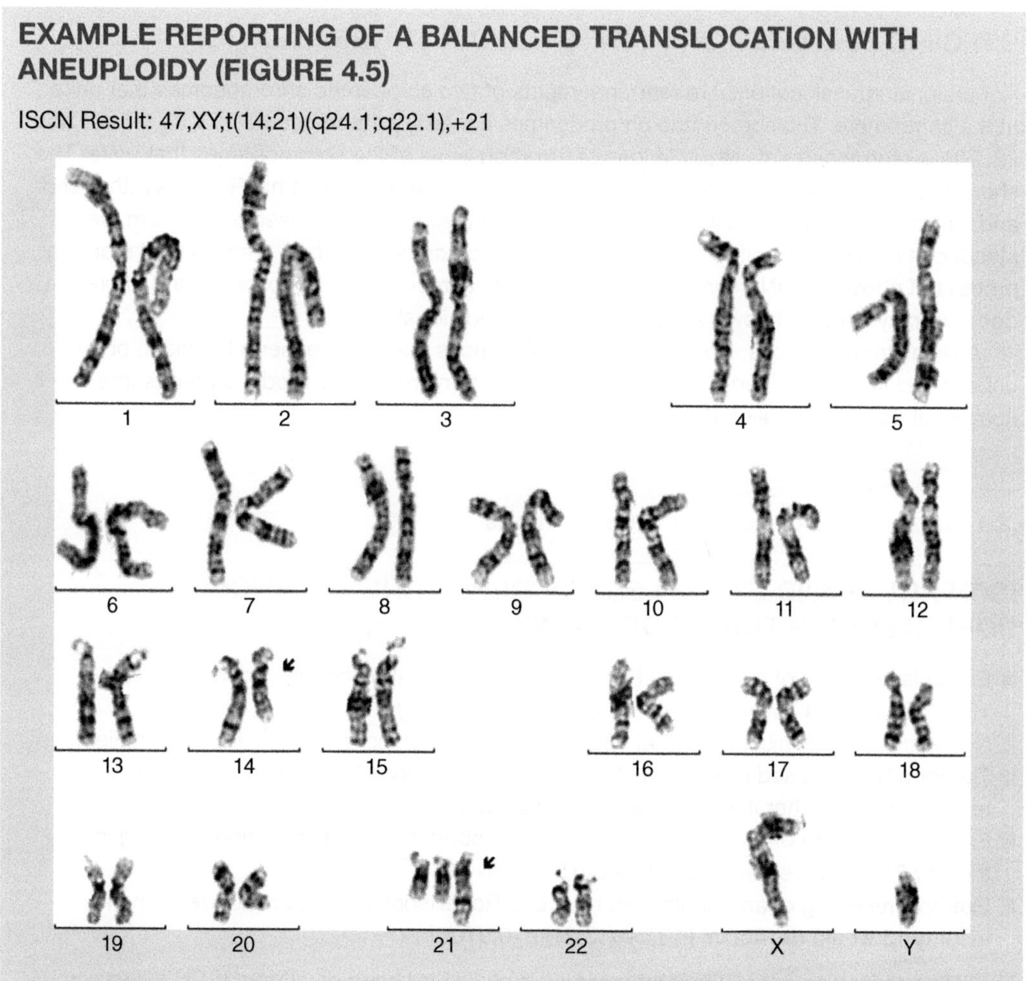

Figure 4.5 Reciprocal translocation with Down syndrome: 46,XY,der(14;21)(q24.1;q13.2),+21. Courtesy of Sarah South PhD, ARUP Laboratories.

Interpretation

Chromosome analysis revealed an abnormal male chromosome complement in all cells examined with an apparently balanced reciprocal translocation between the long arms of chromosomes 14 and 21, with breakpoints at bands q24.1 and q22.1, respectively.

In addition, all cells examined show the gain of chromosome 21, consistent with the diagnosis of Down syndrome. The main clinical features of Down syndrome include hypotonia, flat facies, down-slanting eyes, small ears and mental deficiency.

Chromosome analysis of both parents is indicated to determine if the rearrangement is familial or *de novo* in origin. A familial rearrangement is unlikely to have a phenotypic effect. *De novo* rearrangements, however, are associated with a significant risk for clinical abnormalities.

Genetic counseling is recommended.

4.2 Robertsonian translocations

SPECIFIC FEATURES OF ROBERTSONIAN TRANSLOCATIONS

Robertsonian translocations are rearrangements of two acrocentric chromosomes that unite at the centromere. The acrocentric chromosomes consist of chromosomes 13, 14, 15, 21 and 22. These rearrangements result in loss of the short arms of the chromosomes that unite. The short arms of acrocentric chromosomes are the regions where ribosomal RNA is synthesized and do not play a role in coding regions of DNA, so their loss is inconsequential from the standpoint of clinical manifestations. The loss of two acrocentric short arms will not stop the process of ribosomal RNA synthesis from the other acrocentric chromosomes and, therefore, does not play a role in known abnormalities in an individual.

As in all translocations, Robertsonian translocations may be in either a balanced or an unbalanced form. Balanced Robertsonian translocations result in a modal chromosome number of 45 and are called derivative chromosomes.

4.2.1 Balanced Robertsonian translocations

ISCN RULES AND EXAMPLES FOR REPORTING BALANCED ROBERTSONIAN TRANSLOCATIONS

- Example reporting of a balanced Robertsonian translocation between chromosomes 13 and 21 in a male would be: 45,XY,der(13;21)(q10;q10).
- No mention of the missing normal 13 and 21 chromosomes is needed, but it is implied.
- The term “der” is used rather than “t” for translocation for all Robertsonian translocations, regardless of whether it is balanced or unbalanced.
- Example reporting of an unbalanced form of a Robertsonian translocation resulting in trisomy 21 would be: 46,XY,der(13;21)(q10;q10),+21.
- Example reporting of an unbalanced form of a Robertsonian translocation resulting in trisomy 13 would be: 46,XY,+13,der(13;21)(q10;q10).

The rationale for where the extra chromosome is placed is based on current ISCN guidelines (2013), which state that abnormalities are described in the nomenclature by chromosome order, smaller to larger, and that chromosome aneuploidy precedes structural abnormalities.

Therefore, with trisomy 13, the aneuploid gain of chromosome 13 precedes the translocation involving chromosomes 13 and 21 whereas the aneuploid gain of chromosome 21 follows a translocation involving chromosome 13, since 13 is a smaller number than 21.

EXAMPLE REPORTING OF ROBERTSONIAN TRANSLOCATIONS – BALANCED, GENERAL

ISCN Result: 45,XY,der(#;#)(q10;q10)

Interpretation

Chromosome analysis revealed an abnormal male chromosome complement in all cells examined with an apparently balanced Robertsonian translocation between the long arms of chromosomes # and #.

Carriers of Robertsonian translocations generally do not exhibit phenotypic or developmental abnormalities; however, there is an increased risk of conceptuses with a genetic imbalance or uniparental disomy. Errors during meiosis may lead to decreased fertility, repetitive pregnancy loss or chromosomally abnormal offspring.

Genetic counseling is recommended.

Given the familial origin of most translocations, chromosome analysis of this patient's first-degree biological relatives should be considered.

Balanced Robertsonian translocations in a child

EXAMPLE REPORTING OF ROBERTSONIAN TRANSLOCATIONS – BALANCED, GENERAL IN A CHILD

ISCN Result: 45,XY,der(#;#)(q10;q10)

Interpretation

Chromosome analysis revealed an abnormal male chromosome complement in all cells examined with an apparently balanced Robertsonian translocation between the long arms of chromosomes # and #.

Carriers of Robertsonian translocations generally do not exhibit phenotypic or developmental abnormalities; however, there is an increased risk of conceptuses with a genetic imbalance or uniparental disomy. Errors during meiosis may lead to decreased fertility, repetitive pregnancy loss or chromosomally abnormal offspring.

Parental chromosome analysis is indicated to determine if this rearrangement is inherited or a *de novo* event. A familial rearrangement is unlikely to have a phenotypic effect; however, *de novo* rearrangements are associated with a significant risk of clinical abnormalities.

Genomic microarray analysis may prove informative for further characterization of this rearrangement. As recent literature suggests, imbalances at the breakpoints or elsewhere in the genome may be found in approximately 30–40% of individuals with a cytogenetically balanced rearrangement and an abnormal ph)enotype.

Genetic counseling is recommended.

Balanced Robertsonian translocations in an adult

EXAMPLE REPORTING OF ROBERTSONIAN TRANSLOCATIONS – t(14;21) BALANCED IN AN ADULT

ISCN Result: 45,XY,der(14;21)(q10;q10)

Interpretation

Chromosome analysis revealed an abnormal male chromosome complement in all cells examined with an apparently balanced Robertsonian translocation between the long arms of chromosomes 14 and 21.

Because the rearrangement appears to be balanced, it is most likely of no clinical significance to the patient; however, unbalanced gamete formation during meiosis may lead to pregnancy loss or the production of abnormal offspring.

Carriers of a balanced t(14;21) have a risk for trisomy 21 in their liveborn children of approximately 10% for female carriers and less than 1% for male carriers, and they have an increased risk for miscarriages.

Genetic counseling and prenatal diagnosis in future pregnancies are recommended.

Given the familial origin of most translocations, chromosome analysis of this patient's first-degree biological relatives should be considered.

EXAMPLE REPORTING OF ROBERTSONIAN TRANSLOCATIONS – t(13;14) BALANCED IN AN ADULT (FIGURE 4.6)

ISCN Result: 45,XY,der(13;14)(q10;q10)

Interpretation

Chromosome analysis revealed an abnormal male chromosome complement in all cells examined with an apparently balanced Robertsonian translocation between the long arms of chromosomes 13 and 14.

Because the rearrangement appears to be balanced, it is most likely of no clinical significance to the patient; however, unbalanced gamete formation during meiosis may lead to pregnancy loss or the production of abnormal offspring.

Carriers of a balanced t(13;14) have a risk for trisomy 13 in their liveborn children of approximately 10% for female carriers and less than 1% for male carriers, and they have an increased risk for miscarriages.

Genetic counseling and prenatal diagnosis in future pregnancies are recommended.

Given the familial origin of most translocations, chromosome analysis of this patient's first-degree biological relatives should be considered.

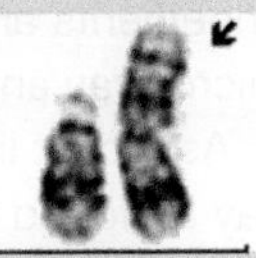

Figure 4.6 Partial karyotype showing a balanced Robertsonian translocation: der(13;14)(q10;q10).

Balanced Robertsonian translocations – prenatal

EXAMPLE REPORTING OF ROBERTSONIAN TRANSLOCATION WITH CHROMOSOMES 14 OR 15 INVOLVED – AMNIOTIC FLUID ANALYSIS

ISCN Result: 45,XY,der(14;21)(q10;q10)

Interpretation

Chromosome analysis revealed an abnormal male chromosome complement in all cells examined with an apparently balanced Robertsonian translocation between chromosomes 14 and 21.

Given the familial origin of most translocations, chromosome analysis of this patient's first-degree biological relatives should be considered.

Study of the amniocytes to determine whether the chromosome 14s were inherited from a single or both parents (uniparental disomy 14 study) is also strongly recommended.

Uniparental disomy for chromosome 14 may be associated with congenital anomalies and mental retardation. Once parental cytogenetic analyses and the uniparental disomy study are complete, it will be possible to predict the effect of the Robertsonian translocation on the phenotype of the fetus with greater accuracy.

Genetic counseling is recommended.

4.2.2 Unbalanced Robertsonian translocations

EXAMPLE REPORTING OF UNBALANCED ROBERTSONIAN TRANSLOCATIONS – GENERAL

ISCN Result: 46,XX,der(#;#)(q10;q10),+#

Interpretation

Chromosome analysis revealed an abnormal female chromosome complement in all cells examined with two normal chromosome #s plus a Robertsonian translocation between chromosome # and #, resulting in trisomy #.

Parental cytogenetic analyses are indicated to determine if this is an inherited or *de novo* rearrangement.

Genetic counseling is recommended.

Robertsonian translocation Down syndrome

SPECIFIC FEATURES OF ROBERTSONIAN TRANSLOCATION DOWN SYNDROME

Translocation Down syndrome is seen in about 4% of Down syndrome patients and has the same chromosomal result as standard trisomy 21, that is, the gain of one chromosome 21; however, the origin of the chromosomal gain differs. With translocation Down syndrome, the long arms of two acrocentric chromosomes join (Robertsonian translocations). These rearrangements result in 46 chromosomes, with two normal copies of chromosome 21 plus the gain of chromosome 21 joined with another acrocentric chromosome, including either chromosomes 13, 14, 15, 21 or 22. The most common translocation partner is chromosome 14.

There is no correlation between translocation Down syndrome and maternal age; however, if the translocation is inherited from a parent, especially the mother, the recurrence risk is much higher than in standard Down syndrome. The theoretical risk of carriers of Robertsonian translocations having a Down syndrome gamete is 33% with the chances equal among normal gametes, balanced translocation gametes and unbalanced translocation gametes. However, empirical data have shown only 10–15% of liveborn progeny of carrier mothers and 5% or less with carrier fathers.

Two Robertsonian translocations will be discussed: translocation (21;21) and (14;21), as these are two distinct types of recurrence risks. The other autosome partners are more rare and follow the pattern of t(14;21).

Translocation (21;21) is unique in that carrier parents of this rearrangement will by necessity pass this abnormal chromosome to *all* offspring, making the recurrence risk of translocation Down syndrome 100%. Although the carrier rate of t(21;21) is very low, once the identification of t(21;21) in a child is known, the parents should be chromosomally studied to determine if a parent carries the rearrangement in the balanced form.

For t(14;21) and the other acrocentric chromosomal partners, the recurrence risk is 10–15% of maternal carriers and ≤5% for paternal carriers; however, the risk of being a balanced carrier is higher than that of t(21;21).

EXAMPLE REPORTING OF UNBALANCED ROBERTSONIAN TRANSLOCATION TRISOMY 21 – DOWN SYNDROME, t(21;21) (FIGURE 4.7a)

ISCN Result: 46,XY,+21,der(21;21)(q10;q10)

Interpretation

Chromosome analysis revealed an abnormal male chromosome complement in all cells examined with one normal chromosome 21 plus a Robertsonian translocation between two chromosome 21s, resulting in trisomy 21. These abnormalities are consistent with the diagnosis of translocation Down syndrome.

The main clinical features of Down syndrome include hypotonia, flat facies, down-slanting eyes, small ears and mental deficiency.

Parental chromosome analyses are required in order to clarify if this translocation is inherited or *de novo* in origin. If *de novo* in origin, the recurrence risk is generally considered to be 1% or less. The risk of a parent carrying the (21;21) chromosome is very small, and carriers of (21;21) translocations are most likely isochromosomes and not translocations. However, the outcome of genetic material is still the same. It is of importance that carriers of (21;21) chromosomes will always have gametes containing the (21;21) chromosome. Therefore, 100% of offspring will have three copies of chromosome 21 (Down syndrome).

Genetic counseling is critical to convey to t(21;21) carriers that they will not be able to have genetically normal children.

Genetic counseling is recommended.

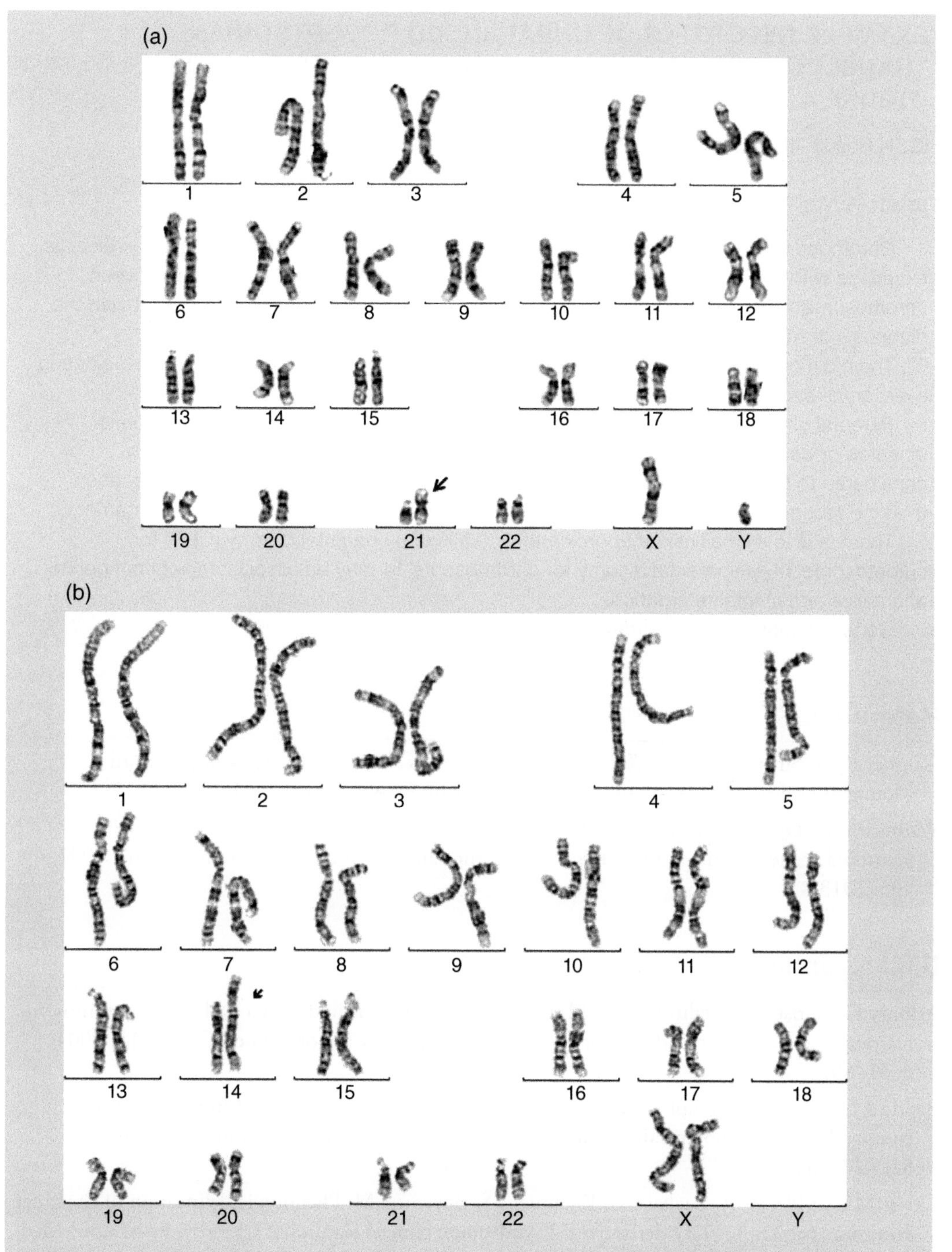

Figure 4.7 Unbalanced Robertsonian translocations with Down syndrome. (a) 46,XY,+21,der(21;21)(q10;q10). (b) 46,XX,der(14;21)(q10;q10),+21.

EXAMPLE REPORTING OF UNBALANCED ROBERTSONIAN TRANSLOCATION TRISOMY 21 – DOWN SYNDROME, t(14;21) (FIGURE 4.7b)

ISCN Result: 46,XX,der(14;21)(q10;q10),+21

Interpretation

Chromosome analysis revealed an abnormal female chromosome complement in all cells examined with two normal chromosome 21s plus a Robertsonian translocation between chromosome 14 and 21, resulting in trisomy 21. These abnormalities are consistent with the diagnosis of translocation Down syndrome.

The main clinical features of Down syndrome include hypotonia, flat facies, down-slanting eyes, small ears and mental deficiency.

Parental chromosome analyses are required in order to clarify if this translocation is inherited or *de novo* in origin. If it is *de novo* in origin, the recurrence risk is generally considered to be 1% or less. However, if a parent carries a balanced (14;21) translocation, then the recurrence risk is approximately 10% for female carriers and 1% for male carriers.

There is also a small risk of approximately 0.5% of uniparental disomy (UPD) for chromosome 14. Uniparental disomy for chromosome 14 may be associated with congenital anomalies and mental retardation.

Genetic counseling is recommended.

References

Gardner RJM, Sutherland GR. *Chromosome Abnormalities and Genetic Counselling*. Oxford Monographs on Medical Genetics. Oxford University Press, Oxford, 2003.

Warburton D. De novo balanced chromosome rearrangements and extra marker chromosomes identified at prenatal diagnosis: clinical significance and distribution. Am J Hum Genet 1991; 49: 995–1013.

Bibliography

Astbury C, Christ LA, Aughton DJ, et al. Detection of deletions in de novo "balanced" chromosome rearrangements: further evidence for their role in phenotypic abnormalities. Genet Med 2004; 6: 81–89.

Baptista J, Mercer C, Prigmore E, et al. Breakpoint mapping and array CGH in translocations: comparison of a phenotypically normal and an abnormal cohort. Am J Hum Genet 2008; 82: 927–936.

Carter MT, St Pierre SA, Zackai EH, Emanuel BS, Boycott KM. Phenotypic delineation of Emanuel syndrome (supernumerary derivative 22 syndrome): clinical features of 63 individuals. Am J Med Genet 2009; 149A: 1712–1721.

De Gregori M, Ciccone R, Magini P, et al. Cryptic deletions are a common finding in "balanced" reciprocal and complex chromosome rearrangements: a study of 59 patients. J Med Genet 2007; 44: 750–762.

Fraccaro M, Lindsten J, Ford CE, et al. The 11q;22q translocation: a European collaborative analysis of 43 cases. Hum Genet 1980; 56: 21–51.

Gersen S, Keagle M (eds). *Principles of Clinical Cytogenetics.* Humana Press, Totowa, New Jersey, 1999.

Gribble SM, Prigmore E, Burford DC, et al. The complex nature of constitutional de novo apparently balanced translocations in patients presenting with abnormal phenotypes. J Med Genet 2005; 42: 8–16.

Jones KL. *Smith's Recognizable Patterns of Human Malformation*, 6th edn. Saunders, London, 2006.

Mackie Ogilvie C, Scriven PN. Meiotic outcomes in reciprocal translocation carriers ascertained in 3-day human embryos. Eur J Hum Genet 2002; 10(12): 801–806.

Midro AT, Stengel-Rutkowski S, Stene J. Experiences with risk estimates for carriers of chromosomal reciprocal translocations. Clin Genet 1992; 41(3): 113–122.

Oliver-Bonet M, Navarro J, Carrera M, Egozcue J, Benet J. Aneuploid and unbalanced sperm in two translocation carriers: evaluation of the genetic risk. Molec Hum Reprod 2002; 8(10): 958–963.

Robertson WRB. Chromosome studies. I. Taxonomic relationships shown in the chromosomes of Tettigidae and Acrididae. V-shaped chromosomes and their significance in Acrididae, Locustidae and Gryllidae: chromosome and variation. J Morph 1916; 27: 179–331.

Rooney DE, Czepulkowski BH. *Human Cytogenetics. A Practical Approach.* Oxford University Press, New York, 1992.

Shaffer LG, McGowan-Jordan J, Schmid M (eds). *ISCN 2013: An International System for Human Cytogenetic Nomenclature.* Karger Publishers, Unionville, CT, 2013.

South S, Rector L, Aston E, Rowe L, Yang SP. Large clinically consequential imbalances detected at the breakpoints of apparently balanced and inherited chromosome rearrangements. J Mol Diagn 2010; 12(5): 725–729.

Therman E, Susman B, Denniston C. The nonrandom participation of human acrocentric chromosomes in Robertsonian translocations. Ann Hum Genet 1989; 53: 49–65.

Zackai EH, Emanuel BS. Site-specific reciprocal translocation, t(11;22)(q23;q11), in several unrelated families with 3:1 meiotic disjunction. Am J Med Genet 1980; 7: 507–521.

CHAPTER 5

Inversions and recombinant chromosomes

5.1 Risks of spontaneous abortions and liveborn abnormal offspring

Inversions occur when a single chromosome breaks at two locations, inverts 180° and then reaneals together. Presumably, this occurs without loss or gain of genetic material if it occurs without a crossing-over event during meiosis. However, the breaks themselves may interrupt a gene or a sequence of DNA that plays a functional role, thereby potentially leading to loss of function of that DNA sequence, which may have phenotypic consequences.

An inversion is considered pericentric when the inverted region contains the centromere, whereas a paracentric inversion does not contain the centromere, but is within a single chromosomal arm.

Inversions will usually be inherited in offspring without clinical consequences, once it is seen in a family member, and that individual is phenotypically normal. However, there is always a risk that the inversion will undergo a meiotic crossing-over event during gametogenesis, causing a recombinant chromosome. The clinical consequences of recombinant chromosomes, without regard to the DNA sequence involved, differ between pericentric and paracentric inversions.

5.2 Pericentric inversions and their recombinants

Pericentric inversions that contain a crossing-over event in gametogenesis will lead to duplications and deletions of the regions lying *outside* the inverted segment of DNA (recombinant chromosomes). This may result in an abnormal liveborn or increased spontaneous abortions. The risk of liveborn abnormal offspring will be greater with large inversions since the regions outside the inverted segment will be small and not as lethal as small inversions with large deletions and duplications, which lead to a fetal demise.

Recombinant chromosomes, which arise with a meiotic crossing-over event, are distinct cytogenetically because there will be a deletion of a portion of either the short or long arm and a duplication of the opposite short or long arm at the ends of the chromosomes, appearing as a mirror image of that region. For example, an inversion of chromosome 8, 46,XX,inv(8)(p21q22), with a recombination event during meiosis, may result in either a deletion of the short arm from bands p21 to the telomere

Cytogenetic Abnormalities: Chromosomal, FISH and Microarray-Based Clinical Reporting, First Edition. Susan Mahler Zneimer.

(p21−>pter) and a duplication of the long arm from bands q22 to the telomere (q22 −>qter) or vice versa. Consequently, when looking at the karyotype, one will see either the p21−>pter or q22 −>qter on both ends of the chromosomes and a deletion of the opposite bands. If the inversion is small, the deleted and duplicated regions are large, giving a clear visual depiction of the duplicated and deleted regions. If the inversion is large, the duplicated and deleted regions will be small and harder to detect.

Therefore, if a proband with phenotypic abnormalities is being evaluated for cytogenetic abnormalities and a duplication of a region on both ends of a single chromosome is observed, then parental chromosome studies are recommended to discern if this rearrangement is a recombinant form of an inversion inherited from a parent. The original study of the proband with this duplicated material on both ends of a chromosome would first be written as a derivative chromosome (see Chapter 7 for details), and only be written as a recombinant chromosome once parental studies show this rearrangement as being inherited by a parental inversion.

ISCN RULES FOR REPORTING PERICENTRIC INVERSIONS

- First write the modal chromosome number, followed by the sex designation.
- Then write “inv”, followed, in parentheses, by the chromosome number involved.
- Then, in parentheses, write the breakpoints involved.

The breakpoints always start at the most short arm proximal band. This is the same nomenclature whether it is a pericentric or paracentric inversion.

Note that there is no semicolon in between the breakpoints, since only one chromosome is involved in the rearrangement, and a semicolon is only used to separate breakpoints of different chromosomes.

For example, a male with a pericentric inversion of chromosome 12 at breakpoints p11.2 and q13.3 would be written as 46,XY,inv(12)(p11.2q13.3) (Figure 5.1).

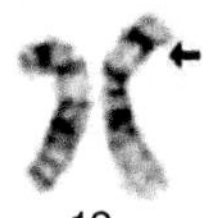

Figure 5.1 Partial karyotype showing a pericentric inversion of chromosome 12: inv(12)(p11.2q13.3). Courtesy of Karine Hovanes PhD, CombiMatrix, Inc.

5.2.1 Pericentric inversion – adult

EXAMPLE REPORTING OF A PERICENTRIC INVERSION IN AN ADULT

ISCN Result: 46,XY,inv(12)(p11.2q13.3)

Interpretation

Chromosome analysis revealed an abnormal male chromosome complement in all cells examined with the presence of a pericentric inversion of chromosome 12.

Due to the location of the breakpoints in this rearrangement and the nature of the unbalanced recombinant chromosomes that could be inherited in future pregnancies, this patient may be at an increased risk for liveborn offspring with congenital anomalies and miscarriages.

Chromosome analysis of the patient's first-degree biological relatives is suggested to identify other carriers of this rearrangement who are also at risk.

Genetic counseling is recommended and prenatal diagnosis should be offered for future pregnancies.

5.2.2 Pericentric inversion – child

EXAMPLE REPORTING OF A PERICENTRIC INVERSION IN A CHILD

ISCN Result: 46,XY,inv(12)(p11.2q13.3)

Interpretation

Chromosome analysis revealed an abnormal male chromosome complement in all cells examined with the presence of a pericentric inversion of chromosome 12.

Parental cytogenetic analyses are recommended to determine if this inversion is familial or *de novo* in origin. Familial chromosomal rearrangements are generally not associated with an increased risk of phenotypic and/or developmental abnormalities; however, submicroscopic deletions, duplications or disruption of a gene or regulatory element at the breakpoints that may cause abnormalities cannot be excluded.

Genomic microarray analysis may prove informative for further characterization of this rearrangement. As recent literature suggests, imbalances at the breakpoints or elsewhere in the genome may be found in approximately 30–40% of individuals with a cytogenetically balanced rearrangement and an abnormal phenotype.

Chromosome analysis of the patient's first-degree biological relatives is suggested to identify other carriers of this rearrangement who are also at risk.

Genetic counseling is recommended.

EXAMPLE REPORTING OF A PERICENTRIC INVERSION IN A PARENT OF A CHILD WITH THE SAME INVERSION

ISCN Result: 46,XX,inv(12)(p11.2q13.3)

Interpretation

Chromosome analysis revealed an abnormal female chromosome complement in all cells examined with the presence of a pericentric inversion of chromosome 12.

This is the same pericentric inversion seen in the child of this individual; therefore, the child's inversion is maternally inherited.

Genetic counseling is recommended.

5.2.3 Pericentric inversion – inherited

ISCN RULES WITH PARENTAL STUDIES PERFORMED

The ISCN result in the proband's report should be modified to include either that it is *de novo* or inherited from a parent. The ISCN describes *de novo* mutations as "dn", and "pat" or "mat" for paternal or maternal inheritance, respectively, following the inversion breakpoints.

EXAMPLE REPORTING OF PERICENTRIC INVERSIONS KNOWN TO BE INHERITED FROM A PARENT

ISCN Result: 46,XY,inv(12)(p11.2q13.3)mat

Interpretation

Chromosome analysis revealed an abnormal male chromosome complement in all cells examined with the presence of a pericentric inversion of chromosome 12 that is inherited from the mother.

Familial chromosomal rearrangements are generally not associated with an increased risk of phenotypic and/or developmental abnormalities; however, submicroscopic deletions, duplications or disruption of a gene or regulatory element at the breakpoints that may cause abnormalities cannot be excluded.

Chromosomal microarray analysis may prove informative for further characterization of this rearrangement. As recent literature suggests, imbalances at the breakpoints or elsewhere in the genome may be found in approximately 30–40% of individuals with a cytogenetically balanced rearrangement and an abnormal phenotype.

Genetic counseling is recommended.

This is a modified report to add information regarding the maternal chromosome results.

5.2.4 Pericentric inversion – *de novo*

EXAMPLE REPORTING OF PERICENTRIC INVERSIONS KNOWN *NOT* TO BE INHERITED FROM A PARENT

ISCN Result: 46,XY,inv(12)(p11.2q13.3)dn

Interpretation

Chromosome analysis revealed an abnormal male chromosome complement in all cells examined with the presence of a pericentric inversion of chromosome 12 that is *de novo* in origin. Both maternal and paternal chromosome analysis showed no inversion present (see case #s ___ and ___).

Chromosomal microarray analysis may prove informative for further characterization of this rearrangement. As recent literature suggests, imbalances at the breakpoints or elsewhere in the genome may be found in approximately 30–40% of individuals with a cytogenetically balanced rearrangement and an abnormal phenotype.

Genetic counseling is recommended.

This is a modified report to add information regarding the parental chromosome results.

5.2.5 Recombinant chromosomes from a pericentric inversion

Reporting of recombinant chromosomes requires parental studies to be performed, and the known inheritance pattern is incorporated into the result of the proband. Otherwise, without parental studies, the abnormal chromosome is written as a derivative chromosome (see Chapter 7 for details).

ISCN RULES FOR REPORTING RECOMBINANT CHROMOSOMES

- First write the modal chromosome number as 46, followed by the sex designation.
- Then write "rec" followed by, in parentheses, the chromosome number involved.
- Then write "dup" followed by, in parentheses, the chromosome number and arm that is duplicated, followed by the original inversion event.
- "mat" or "pat" is added at the end.

One need not write in the ISCN that a deletion is also present; it is implied. However, it is prudent to describe the duplication and deletion regions in the interpretation for a better understanding of the imbalance present.

For example, a male with a recombinant chromosome 8 inherited from a mother who carries a pericentric inversion of chromosome 8 would be written as rec(8)dup(8p)inv(8)(p21q24)mat.

EXAMPLE REPORTING OF A RECOMBINANT CHROMOSOME FROM A PERICENTRIC INVERSION

ISCN Result: 46,XY,rec(8)dup(8p)inv(8)(p21q24)mat

Interpretation

Chromosome analysis revealed an abnormal male chromosome complement in all cells examined with the presence of a recombinant chromosome resulting from a pericentric inversion of chromosome 8 inherited from the mother. This recombinant chromosome results in a duplication of part of the short arm of chromosome 8 from bands p21 to pter and a deletion of part of the long arm from bands q24 to qter.

Chromosomal microarray analysis may prove informative for further characterization of this rearrangement and more specific gene regions involved.

Genetic counseling is recommended.

5.3 Paracentric inversions and their recombinants

Paracentric inversions, with a crossing-over event, will lead to acentric and dicentric marker chromosomes that ordinarily decrease the viability of the individual's offspring and result in increased spontaneous abortions and fetal demises.

Inverted chromosomes with a crossing-over event that leads to viable but abnormal offspring are recombinant chromosomes, if it is known to be inherited from a parent who carries the balanced form of the inversion. These recombinant chromosomes will vary in phenotypic expression based on the region of imbalance. Knowing the chromosomal region of inversions is critical to extrapolate the DNA segments of imbalance to help predict the clinical consequences to the individual.

Therefore, if a proband with phenotypic manifestations is being evaluated for cytogenetic abnormalities, and a duplication of a region on both ends of a single chromosome is observed, then parental chromosome studies are recommended to discern if this rearrangement is a recombinant form of an inversion inherited from a parent. The original study of the proband with the dicentric chromosome would first be written as a derivative chromosome (see Chapter 7 for details), and only be written as a recombinant chromosome once parental studies show this rearrangement as inherited by a parental inversion.

ISCN RULES FOR REPORTING PARACENTRIC INVERSIONS

- First write the modal chromosome number, followed by the sex designation.
- Then write "inv", followed, in parentheses, by the chromosome number involved.
- Then, in parentheses, write the breakpoints involved.

The breakpoints always start at the most short arm proximal band. This is the same nomenclature whether it is a pericentric or paracentric inversion.

Note that there is no semicolon between the breakpoints, since only one chromosome is involved in the rearrangement, and a semicolon is only used to separate breakpoints of different chromosomes.

For example, a male with a paracentric inversion within the short arm of chromosome 1 would be written as 46,XY,inv(1)(p32.1p34.3) (Figure 5.2).

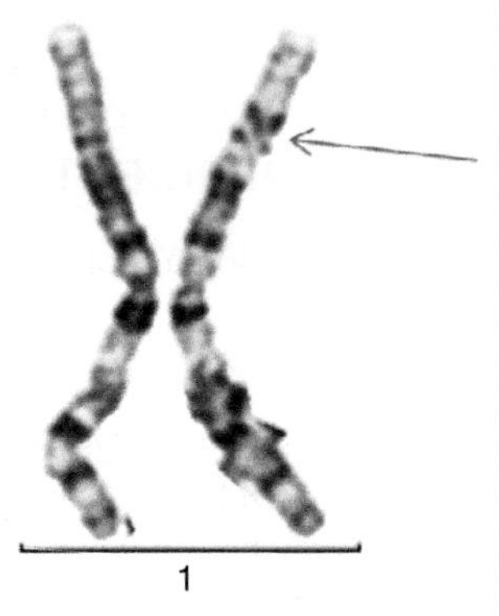

Figure 5.2 Partial karyotype showing a paracentric inversion of chromosome 1: inv(1)(p32.1p34.3).

5.3.1 Paracentric inversions – adult

EXAMPLE REPORTING OF PARACENTRIC INVERSIONS IN AN ADULT

ISCN Result: 46,XY,inv(1)(p32.1p34.3)

Interpretation

Chromosome analysis revealed an abnormal male chromosome complement in all cells examined with the presence of a paracentric inversion of the short arm of chromosome 1.

Due to the location of the breakpoints in this rearrangement and the nature of the unbalanced recombinant chromosomes that could be inherited in future pregnancies, this patient may be at an increased risk for liveborn offspring with congenital anomalies and miscarriages.

Chromosome analysis of the patient's first-degree biological relatives is suggested to identify other carriers of this rearrangement who are also at risk.

Genetic counseling is recommended and prenatal diagnosis should be offered for future pregnancies.

5.3.2 Paracentric inversions – child

EXAMPLE REPORTING OF PARACENTRIC INVERSIONS – CHILD

ISCN Result: 46,XY,inv(1)(p32.1p34.3)

Interpretation

Chromosome analysis revealed an abnormal male chromosome complement in all cells examined with the presence of a paracentric inversion of the short arm of chromosome 1.

Parental cytogenetic analyses are recommended to determine if this inversion is familial or *de novo* in origin. Familial chromosomal rearrangements are generally not associated with an increased risk of phenotypic and/or developmental abnormalities; however, submicroscopic deletions, duplications or disruption of a gene or regulatory element at the breakpoints that may cause abnormalities cannot be excluded.

Chromosomal microarray analysis may prove informative for further characterization of this rearrangement. As recent literature suggests, imbalances at the breakpoints or elsewhere in the genome may be found in approximately 30–40% of individuals with a cytogenetically balanced rearrangement and an abnormal phenotype.

Chromosome analysis of the patient's first-degree biological relatives is suggested to identify other carriers of this rearrangement who are also at risk.

Genetic counseling is recommended.

5.3.3 Recombinant chromosomes of a paracentric inversion

Recombinant forms of a paracentric inversion result in dicentric chromosomes, acentric fragments and single centromere chromosomes with a duplication and deletion. All scenarios are generally not viable and, therefore, are not detected in offspring when they occur. However, if a dicentric chromosome is present in a proband, then parental chromosome studies are recommended. Once parental studies are performed and known to show an inherited paracentric inversion, then the recombinant chromosome may be written. Paracentric inversions are written the same as pericentric inversions, only with the breakpoint in a single arm.

Below is a modified report of a known parental paracentric inversion and the resulting recombinant chromosome in the child.

EXAMPLE REPORTING OF A RECOMBINANT CHROMOSOME FROM A PARACENTRIC INVERSION

ISCN Result: 46,XY,rec(1)dic(1;1)inv(1)(p32.1p34.3)mat

Interpretation

Chromosome analysis revealed an abnormal male chromosome complement in all cells examined with the presence of a recombinant chromosome resulting from a paracentric inversion of chromosome 1 inherited from the mother. This recombinant chromosome results in a dicentric chromosome with a duplication of the long arm and the short arm from the centromere to band p32.1 and a deletion of the short arm from p34.2 to pter.

Genomic microarray analysis may prove informative for further characterization of this rearrangement and more specific gene regions involved.

Genetic counseling is recommended.

Bibliography

Allderdice PW, Browne N, Murphy DP. Chromosome 3 duplication q21 leads to qter deletion p25 leads to pter Syndrome in children of carriers of a pericentric inversion inv(3) (p25q21). Am J Hum Genet 1975; 27(6): 699–718.

Boué J, Taillemite JL, Hazael-Massieux P, Léonard C, Boué A. Association of pericentric inversion of chromosome 9 and reproductive failure in ten unrelated families. Humangenetik 1975 ; 30(3): 217–224.

Bowen P, Ying KL, Chung GS. Trisomy 9 mosaicism in a newborn infant with multiple malformations. J Pediatr 1974; 85(1): 95–97.

De la Chapelle A, Schröder J, Stenstrand K, et al. Pericentric inversions of human chromosomes 9 and 10. Am J Hum Genet 1974; 26(6): 746–766.

Gardner RJM, Sutherland GR. *Chromosome Abnormalities and Genetic Counselling*. Oxford Monographs on Medical Genetics. Oxford University Press, Oxford, 2003.

Gersen S, Keagle M (eds). *Principles of Clinical Cytogenetics*. Humana Press, Totowa, New Jersey, 1999.

Herva R, de la Chapelle A. A large pericentric inversion of human chromosome 8. Am J Hum Genet 1976; 28(3): 208–212.

Jacobs PA, Frackiewicz A, Law P, Hilditch CJ, Morton NE. The effect of structural aberrations of the chromosomes on reproductive fitness in man .II. Results. Clin Genet 1975; 8(3): 169–178.

Madan K, Bobrow M. Structural variation in chromosome No 9. Ann Genet 1974; 17(2): 81–86.

Rooney DE, Czepulkowski BH. *Human Cytogenetics. A Practical Approach*. Oxford University Press, New York, 1992.

Schinzel A, Hayashi K, Schmid W. Mosaic-trisomy and pericentric inversion of chromosome 9 in a malformed boy. Humangenetik 1974; 25(3): 171–177.

Shaffer LG, McGowan-Jordan J, Schmid M (eds). *ISCN 2013: An International System for Human Cytogenetic Nomenclature*. Karger Publishers, Unionville, CT, 2013.

Tobias ES, Connor M, Ferguson Smith M. *Essential Medical Genetics*, 6th edn. Wiley-Blackwell, Oxford, 2011.

Van der Linden AG, Pearson PL, van de Kamp JJ. Cytological assessment of meiotic exchange in a human male with a pericentric inversion of chromosome No. 4. Cytogenet Cell Genet 1975; 14(2): 126–139.

Vine DT, Yarkoni S, Cohen MM. Inversion homozygosity of chromosome no. 9 in a higly inbred kindred. Am J Hum Genet 1976; 28(3): 203–207.

CHAPTER 6

Visible deletions, duplications and insertions

6.1 Definitions

The definition of a deletion is the loss of genetic material on a chromosome, while a duplication is added genetic material on a chromosome. Visible deletions and duplications usually require at least 1–4 Mb of missing or added DNA to be visible under a microscope. Deletions and duplications can either be interstitial or terminal. An interstitial deletion/duplication is one in which a segment of DNA is lost/gained in the middle of a chromosome. A terminal deletion/duplication refers to a region that is from the telomere of either the short arm or the long arm of a chromosome to a region lost/gained within the chromosome.

Deletions/duplications could result from unequal meiotic recombination, in which regions of a pair of homologous chromosomes that are undergoing recombination during meiosis I do not proceed normally and leave out or duplicate a portion of DNA as the crossing-over event occurs.

Another possibility for a deletion or duplication is with the presence of a parentally inherited translocation that results in an unequal allocation of DNA from the chromosomes inherited in the offspring. Each time a translocation carrier goes through embryogenesis, the possibility exists of generating an unbalanced form of the translocation, resulting in a deletion/duplication. The possibility of a pericentric inversion going through a crossing-over event during meiosis I will also lead to a deletion/duplication event.

It may appear obvious that a deletion/duplication is terminal or interstitial. However, by molecular cytogenetic analysis, these regions may not coincide with what is seen visibly by standard chromosome analysis. In order to clearly define whether a deletion/duplication is interstitial or terminal, microarray analysis could be performed, to clarify the exact size and placement of the abnormality.

For information regarding submicroscopic deletions, see Sections 2 and 3.

Cytogenetic Abnormalities: Chromosomal, FISH and Microarray-Based Clinical Reporting, First Edition. Susan Mahler Zneimer.

6.1.1 Visible deletions

SPECIFIC FEATURES OF VISIBLE DELETIONS

Deletions observed in liveborn individuals often do not have distinct abnormalities, but include growth and developmental delay, mental impairment and multiple congenital anomalies. These abnormalities are usually present in individuals due to the large size of a deletion that is cytogenetically visible. However, since deletions may occur anywhere in the genome as well as vary in the size of the deletion, the possible phenotypic manifestations can vary greatly. Once the region of the deletion is defined, correlation with medical resources may be used to help match the region involved with phenotypic abnormalities described in other patients to infer the possible clinical consequences for an individual.

Certain deletions may not have clinical consequences, such as the loss of short arm material of acrocentric chromosomes, since these regions do not have DNA that is unique and expresses a protein product other than ribosomal RNA (rRNA). Since rRNA is expressed on all the acrocentric chromosomes, loss at one of these chromosomes has not been shown to have clinical manifestations.

Certain visible deletions in the genome are recurring, including del(4p), del(5p), del(18p) and del(18q), which give rise to specific syndromes, such as Wolf–Hirschhorn and cri du chat syndromes for chromosomes 4 and 5, respectively. Since these regions are now generally studied with FISH techniques, a further discussion of these syndromes appear in Section 2.

It is important to identify the possible inheritance causing a deletion, since it is possible for a translocation carrier to produce unbalanced gametes that could lead to an unbalanced form of the translocation resulting in a deletion, especially when a terminal deletion is seen.

ISCN RULES FOR REPORTING DELETIONS

- First write the modal chromosome number, followed by the sex designation.
- Then write "del", followed, in parentheses, by the chromosome involved.
- Then, in parentheses, write the breakpoints from the proximal to distal bands.

For example, an interstitial deletion would be written as: 46,XX del(22)(q11.21q11.23). Note that there is no semicolon separating the breakpoints when an abnormality occurs within the same chromosome.

For a terminal deletion, the term "pter" or "qter" is not written as a breakpoint for short and long arms, respectively.

For example, a long arm terminal deletion would be written as: 46,XX,del(3)(q13.2). A terminal deletion is implied if there is no second breakpoint in the karyotypic designation.

6.1.2 Interstitial deletions (Figure 6.1)

Interstitial deletion – 17p11.2, Smith–Magenis syndrome

EXAMPLE REPORTING OF INTERSTITIAL DELETION – 17p11.2p11.2 (FIGURE 6.1a)

ISCN Result: 46,XX,del(17)(p11.2p11.2)

Interpretation

Chromosome analysis revealed an abnormal female chromosome complement in all cells examined with an interstitial deletion within the short arm of chromosome 17 of band p11.2.

This deletion is consistent with the diagnosis of Smith–Magenis syndrome. FISH analysis for the SMS critical region probe is suggested to confirm this finding.

It is possible that this deletion is an unbalanced form of a rearrangement inherited from a parent. Therefore, parental chromosomal analysis is recommended to help clarify the origin of this abnormality.

Genetic counseling is recommended.

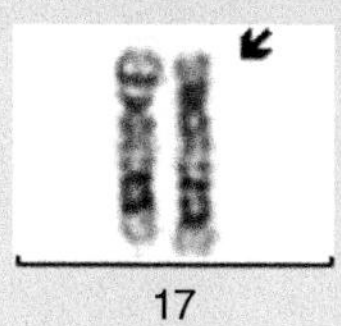

Figure 6.1a Partial karyotype showing an interstitial deletion of chromosome 17: del(17)(p11.2p11.2).

Interstitial deletion – 22q11.2, DiGeorge/velocardiofacial syndrome

EXAMPLE REPORTING OF INTERSTITIAL DELETION – 22q11.21q11.23 (FIGURE 6.1b)

ISCN Result: 46,XY,del(22)(q11.21q11.23)

Interpretation

Chromosome analysis revealed an abnormal male chromosome complement in all cells examined with an interstitial deletion within the long arm of chromosome 22 from band q11.21 to q11.23.

This deletion is consistent with the diagnosis of DiGeorge/velocardiofacial syndrome. FISH analysis for the DiGeorge critical region probe is suggested to confirm this finding.

Genetic counseling is recommended.

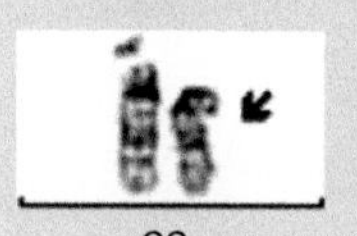

Figure 6.1b Partial karyotype showing an interstitial deletion of chromosome 22: del(22)(q11.21q11q23). Courtesy of Sarah South PhD, ARUP Laboratories.

6.1.3 Terminal deletions (Figure 6.2)

Terminal deletion – 1p36

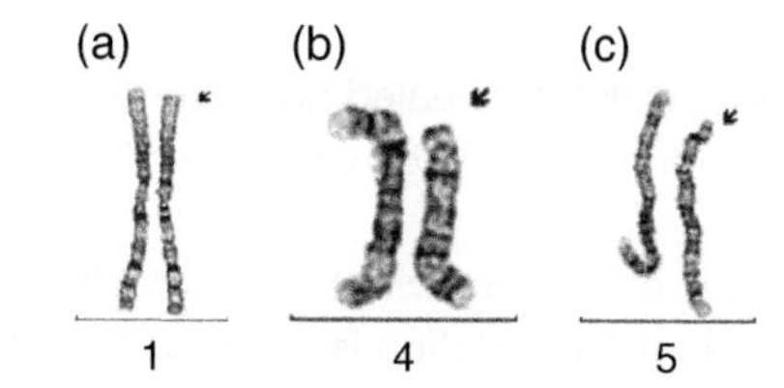

Figure 6.2 Partial karyotypes showing terminal deletions. (a) del(1)(p36.13). (b) del(4)(p15.1). (c) del(5)(p15.1). Courtesy of Sarah South PhD, ARUP Laboratories.

EXAMPLE REPORTING OF TERMINAL DELETION – 1p36

ISCN Result: 46,XX,del(1)(p36.13)

Interpretation

Chromosome analysis revealed an abnormal female chromosome complement in all cells examined with a terminal deletion of the distal short arm of chromosome 1 from band p36.13 to the terminus.

This deletion is consistent with the diagnosis of a chromosome 1p36 deletion.

It is possible that this deletion is an unbalanced form of a rearrangement inherited from a parent. Therefore, parental chromosomal analysis is recommended to help clarify the origin of this abnormality.

Genetic counseling is recommended.

Terminal deletion – 4p15, Wolf–Hirschhorn syndrome

EXAMPLE REPORTING OF TERMINAL DELETION – 4p15

ISCN Result: 46,XX,del(4)(p15.1)

Interpretation

Chromosome analysis revealed an abnormal female chromosome complement in all cells examined with a terminal deletion of the distal short arm of chromosome 4 from band p15.1 to the terminus.

This deletion is consistent with the diagnosis of Wolf–Hirschhorn syndrome.

It is possible that this deletion is an unbalanced form of a rearrangement inherited from a parent. Therefore, parental chromosomal analysis is recommended to help clarify the origin of this abnormality.

Genetic counseling is recommended.

Terminal deletion – 5p14, cri du chat syndrome

EXAMPLE REPORTING OF TERMINAL DELETION – 5p14

ISCN Result: 46,XX,del(5)(p14)

Interpretation

Chromosome analysis revealed an abnormal female chromosome complement in all cells examined with a terminal deletion of the distal short arm of chromosome 5 from band p14 to the terminus.

This deletion is consistent with the diagnosis of cri du chat syndrome.

It is possible that this deletion is an unbalanced form of a rearrangement inherited from a parent. Therefore, parental chromosomal analysis is recommended to help clarify the origin of this abnormality.

Genetic counseling is recommended.

6.2 Visible duplications

SPECIFIC FEATURES OF VISIBLE DUPLICATIONS

Since duplications imply partial trisomy for a segment of DNA, the phenotype will depend on the DNA region that is trisomic. Very few duplications have a distinctive phenotype that is syndromic. The only known recurrent duplications include partial 3q trisomy, corresponding to features seen in Cornelia de Lange syndrome, and partial 11p trisomy, which is seen in patients with Beckwith–Wiedemann syndrome. Other regions of duplications correspond to the same chromosomal region as deletions, which exemplify the paradigm of unequal crossing over during meiosis I. Since these regions are now generally studied with FISH techniques, a further discussion of these syndromes appears in Section 2.

Duplications, like deletions, may be either interstitial or terminal (Figure 6.3). Unlike deletions, duplications may be seen in tandem, with more than one extra copy present. Although more typically found as an acquired change in malignancies, tandem duplications in constitutional disorders have been reported.

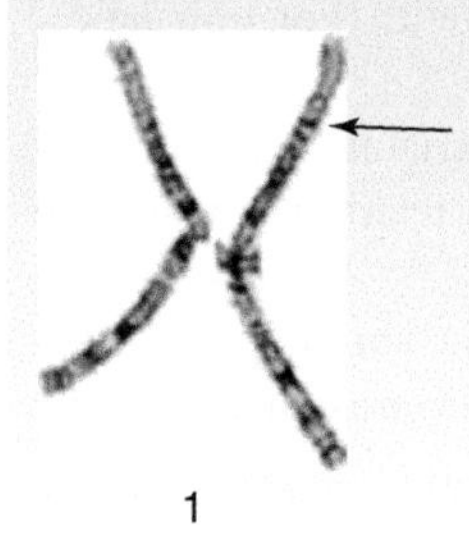

Figure 6.3 Partial karyotype showing an interstitial duplication of chromosome 1: dup(1)(p32.2p34.1). Courtesy of Sarah South PhD, ARUP Laboratories.

ISCN RULES FOR REPORTING DUPLICATIONS

- First write the modal chromosome number, followed by the sex designation.
- Then write "dup", followed, in parentheses, by the chromosomes involved.
- Then, in parentheses, write the breakpoints from the proximal to distal bands.

For example, an interstitial deletion would be written as: 46,XX,dup(3)(q13.2q26.2). Note that there is no semicolon separating the breakpoints when an abnormality occurs within the same chromosome.

For a terminal duplication, the term "pter" or "qter" is not written as a breakpoint for short and long arms, respectively.

For example, a long arm terminal duplication would be written as: 46,XX,dup(3)(q13.2). A terminal duplication is implied if there is no second breakpoint in the karyotypic designation.

EXAMPLE REPORTING OF INTERSTITIAL DUPLICATIONS

ISCN Resort: 46,XY,dup(3)(q13.2q26.2)

Interpretation

Chromosome analysis revealed an abnormal male chromosome complement in all cells examined with an interstitial duplication within the long arm of chromosome 3 from band q13.2 to q26.2.

Duplications of this size usually result in developmental delay, mental impairment and multiple congenital anomalies.

Genetic counseling is recommended.

6.3 Balanced Insertions

Insertions refer to a segment of chromosomal material moved and then inserted into a chromosome, interrupting the normal chromosome. The inserted material may be either interchanged from its original place within a single chromosome or interchanged from one chromosome into another chromosome. In either case, the inserted region is considered a balanced rearrangement. (For unbalanced insertions, refer to Chapter 7.)

Insertions may be either directly inserted, in which the inserted genetic material is "right side up," or inversely inserted, in which the inserted genetic material is "upside down" in relation to where the genetic material originated.

While it is optional to use the direct (dir) or inverted (inv) abbreviations in the karyotypic designation, it may be helpful for the reader to better understand the change in the chromosome. However, the ISCN designation of the breakpoints clarifies in which direction the insertion takes place.

6.3.1 Insertions within the same chromosome

ISCN RULES FOR REPORTING INSERTIONS WITHIN THE SAME CHROMOSOME (FIGURE 6.4)

- First write the modal chromosome number, 46, followed by the sex designation.
- Then write "ins" followed, in parentheses, by the chromosome involved. When the inserted material is within the same chromosome, then there is only one chromosome designation.
- Next write the breakpoints, enclosed in parentheses. There are three breakpoints to list. The first breakpoint refers to the band at which the inserted material is now located. This is followed by the breakpoints of the inserted material. When only one chromosome is involved, the breakpoints are not separated by semicolons.

The breakpoints of the insertion are listed proximal to distal in a direct insertion, and distal to proximal if the inserted region is inverted.

For example, a direct insertion within one chromosome would be written as: 46,XY,ins(4)(p15q21q31). This refers to long arm material of chromosome 4 from bands q21 → q31 directly inserted into the short arm of chromosome 4 at band p15.

An inverted insertion within one chromosome would be written as: 46,XY,ins(4)(p15q31q21). This refers to long arm material of chromosome 4 from bands q21 → q31 inversely inserted into the short arm of chromosome 4 at band p15.

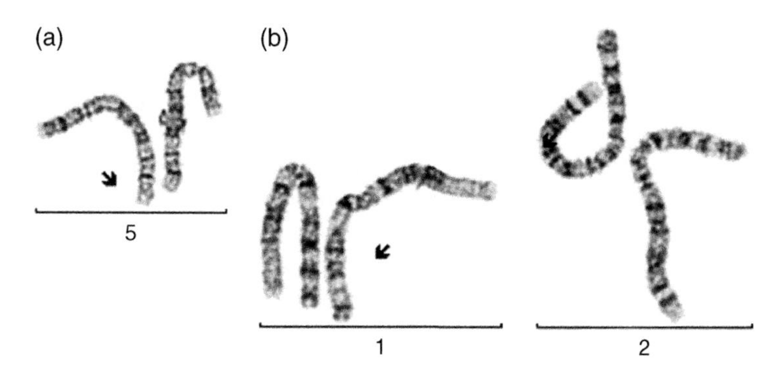

Figure 6.4 Partial karyotypes showing insertions. (a) ins(5)(q35;q31.1q31.3). (b) der(1)del(1)(q23.1q43) ins(1;2)(q23.1;q21.1q32.1). Courtesy of Sarah South PhD, ARUP Laboratories.

EXAMPLE REPORTING OF INSERTION OF MATERIAL WITHIN ONE CHROMOSOME

ISCN Result: 46,XX,ins(5)(q35q31.1q31.3)

Interpretation

Chromosome analysis revealed an abnormal female chromosome complement in all cells examined with an insertion of genetic material from the long arm of chromosome 5, from bands q31.1 → q31.3, into the long arm of chromosome 5 at band q35. This appears to be a balanced rearrangement. However, it is possible that genes involved at the breakpoints of this rearrangement were disrupted, causing phenotypic abnormalities.

Further testing with chromosomal microarray analysis is suggested to identify possible gene disruptions or small unbalanced rearrangements.

Genetic counseling is recommended.

6.3.2 Insertions from one chromosome into a different chromosome

ISCN RULES FOR REPORTING INSERTIONS FROM ONE CHROMOSOME INTO A DIFFERENT CHROMOSOME

- First write the modal chromosome number, 46, followed by the sex designation.
- Then write "ins" followed, in parentheses, by the chromosomes involved.
- When the inserted material is added to a different chromosome, then both chromosomes are within the parentheses, with the recipient chromosome listed before the chromosome with the inserted material.
- Next write the breakpoints within parentheses. There are three breakpoints to list. The first breakpoint refers to the band at which the inserted material is now located. This is followed by the breakpoints of the inserted material.

With two chromosomes involved, the breakpoint of the inserted chromosome is separated by a semicolon before the insertion breakpoints are written.

The breakpoints of the insertion are listed proximal to distal in a direct insertion, and distal to proximal if the inserted region is inverted.

For example, a direct insertion with two autosomes involved would be written as: 46,XY,ins(4;7)(p15;q11.2q36). This refers to long arm material of chromosome 7 from bands q11.2 → q36 directly inserted into the short arm of chromosome 4 at band p15.

An inverted insertion of an autosome into the X chromosome would be written as: 46,Y,ins(X;7)(q22;q36q11.2). This refers to a male with long arm material of chromosome 7 from bands q11.2 → q36 inversely inserted into the short arm of the long arm of the X chromosome at band q22. Note that the designation of the insertion of an X chromosome is written after the normal Y chromosome.

EXAMPLE REPORTING OF AN INSERTION IN AN INDIVIDUAL WITH PHENOTYPIC ABNORMALITIES

ISCN Result: 46,Y,ins(X;7)(q22;q36q11.2)

Interpretation

Chromosome analysis revealed an abnormal male chromosome complement in all cells examined with an inversely directed insertion of genetic material from the long arm of chromosome 7, from bands q36 → q11.2, into the long arm of chromosome X at band q22. This appears to be a balanced rearrangement. However, it is possible that genes involved at the breakpoints of this rearrangement were disrupted, causing phenotypic abnormalities.

Further testing with microarray analysis is suggested to identify possible gene disruptions or small unbalanced rearrangements.

Genetic counseling is recommended.

EXAMPLE REPORTING OF AN INSERTION IN AN INDIVIDUAL WITH MULTIPLE MISCARRIAGES

ISCN Result: 46,Y,ins(X;7)(q22;q36q11.2)

Interpretation

Chromosome analysis revealed an abnormal male chromosome complement in all cells examined with an inversely directed insertion of genetic material from the long arm of chromosome 7, from bands q36 → q11.2, into the long arm of chromosome X at band q22. This appears to be a balanced rearrangement. However, carriers of insertions have an increased risk of infertility, miscarriages and offspring with an unbalanced form of this rearrangement.

Genetic counseling is recommended.

Bibliography

Alfi O, Donnell GN, Crandall BF, et al. Deletion of the short arm of chromosome no.9 (46,9p-): a new deletion Syndrome. Ann Genet 1973; 16: 17.

Anderlid BM, Schoumans J, Annerén G, et al. FISH-mapping of a 100-kb terminal 22q13 deletion. Hum Genet 2002; 110: 439.

Brewer C, Holloway S, Zawalnyski P, et al. A chromosomal deletion map of human malformations. Am J Hum Genet 1998; 63: 1153.

Brown S, Russo J, Chitayat D, Warburton D. The 13q- Syndrome: the molecular definition of a critical deletion region in band 13q32. Am J Hum Genet 1995; 57: 859.

Christ LA, Crowe CA, Micale MA, et al. Chromosome breakage hotspots and delineation of the critical region for the 9p-deletion Syndrome. Am J Hum Genet 1999; 65: 1387.

Church DM, Bengtsson U, Nielsen KV, et al. Molecular definition of deletions of different segments of distal 5p that result in distinct phenotypic features. Am J Hum Genet 1995; 56: 1162.

Drumheller T, McGillivray BC, Behrner D, et al. Precise localisation of 3p25 breakpoints in four patients with the 3p-Syndrome. J Med Genet 1996; 33: 842.

Estabrooks LL, Rao KW, Driscoll DA, et al. Preliminary phenotypic map of chromosome 4p16 based on 4p deletions. Am J Med Genet 1995; 57: 581.

Francke U. Williams-Beuren Syndrome: genes and mechanisms. Hum Mol Genet 1999; 8: 1947.

Gardner RJM, Sutherland GR. *Chromosome Abnormalities and Genetic Counselling*. Oxford Monographs on Medical Genetics. Oxford University Press, Oxford, 2003.

Gersen S, Keagle M (eds). *Principles of Clinical Cytogenetics*. Humana Press, Totowa, New Jersey, 1999.

Gersh M, Goodart SA, Pasztor LM, et al. Evidence for a distinct region causing a cat-like cry in patients with 5p deletions. Am J Hum Genet 1995; 56: 1404.

Girirajan S, Elsas LJ 2nd, Devriendt K, Elsea SH. RAI1 variations in Smith-Magenis Syndrome patients without 17p11.2 deletions. J Med Genet 2005; 42: 820.

Greenberg F. Contiguous gene Syndrome. Growth: Genetics and Hormones 1993; 9: 5.

Heilstedt HA, Ballif BC, Howard LA, et al. Physical map of 1p36, placement of breakpoints in monosomy 1p36, and clinical characterization of the Syndrome. Am J Hum Genet 2003; 72: 1200.

Jones KL. *Smith's Recognizable Patterns of Human Malformation*, 6th edn. Saunders, London, 2006.

Lejeune J, et al. [3 Cases of partial deletion of the short arm of a 5 chromosome.] C R Hebd Seances Acad Sci 1963; 257: 3098.

Lin AE, Garver KL, Diggans G, et al. Interstitial and terminal deletions of the long arm of chromosome 4: further delineation of phenotypes. Am J Med Genet 1988; 31: 533.

Lowery MC, Morris CA, Ewart A, et al. Strong correlation of elastin deletions, detected by FISH, with Williams Syndrome: evaluation of 235 patients. Am J Hum Genet 1995; 57: 49.

Lu X, Meng X, Morris CA, Keating MT. A novel human gene, WSTF, is deleted in Williams Syndrome. Genomics 1998; 54: 241.

Lupski JR. Charcot-Marie-Tooth polyneuropathy: duplication, gene dosage, and genetic heterogeneity. Pediatr Res 1999; 45: 159.

Mainardi PC, Perfumo C, Calì A, et al. Clinical and molecular characterisation of 80 patients with 5p deletion: genotype-phenotype correlation. J Med Genet 2001; 38: 151.

Maranda B, Lemieux N, Lemyre E. Familial deletion 18p Syndrome: case report. BMC Med Genet 2006; 7: 60.

Marinescu RC, Mainardi PC, Collins MR, et al. Growth charts for cri-du-chat Syndrome: an international collaborative study. Am J Med Genet 2000; 94: 153.

Niebuhr E. The Cri du Chat Syndrome: epidemiology, cytogenetics, and clinical features. Hum Genet 1978; 44: 227.

Nienhaus H, Mau U, Zang KD. Infant with del(3) (p25-pter): karyotype-phenotype correlation and review of previously reported cases. Am J Med Genet 1992; 44: 573.

Overhauser J, Huang X, Gersh M, et al. Molecular and phenotypic mapping of the short arm of chromosome 5: sublocalization of the critical region for the cri-du-chat Syndrome. Hum Mol Genet 1994; 3: 247.

Phelan MC. Deletion 22q13.3 syndrome. Orphanet J Rare Dis 2008; 3: 14.

Pilz DT, Macha ME, Precht KS, et al. Fluorescence in situ hybridization analysis with LIS1 specific probes reveals a high deletion mutation rate in isolated lissencephaly sequence. Genet Med 1998; 1: 29.

Pober BR. Williams-Beuren Syndrome. N Engl J Med 2010; 362: 239.

Roelfsema JH, White SJ, Ariyürek Y, et al. Genetic heterogeneity in Rubinstein-Taybi Syndrome: mutations in both the CBP and EP300 genes cause disease. Am J Hum Genet 2005; 76: 572.

Rooney DE, Czepulkowski BH. *Human Cytogenetics. A Practical Approach*. Oxford University Press, New York, 1992.

Sarda P, Lefort G, Fryns JP, et al. Interstitial deletion of the distal long arm of chromosome 4. J Med Genet 1992; 29: 259.

Shaffer LG, Kennedy GM, Spikes AS, Lupski JR. Diagnosis of CMT1A duplications and HNPP deletions by interphase FISH: implications for testing in the cytogenetics laboratory. Am J Med Genet 1997; 69: 325.

Shaffer LG, McGowan-Jordan J, Schmid M (eds). *ISCN 2013: An International System for Human Cytogenetic Nomenclature*. Karger Publishers, Unionville, CT, 2013.

Shprintzen RJ. Velo-cardio-facial Syndrome: 30 years of study. Dev Disabil Res Rev 2008; 14: 3–10.

Silverman GA, Schneider SS, Massa HF, et al. The 18q- Syndrome: analysis of chromosomes by bivariate flow karyotyping and the PCR reveals a successive set of deletion breakpoints within 18q21.2-q22.2. Am J Hum Genet 1995; 56: 926.

Strathdee G, Sutherland R, Jonsson JJ, et al. Molecular characterization of patients with 18q23 deletions. Am J Hum Genet 1997; 60: 860.

Tam E, Young EJ, Morris CA, et al. The common inversion of the Williams-Beuren Syndrome region at 7q11.23 does not cause clinical symptoms. Am J Med Genet A 2008; 146A: 1797.

Tazelaar J, Roberson J, van Dyke DL, et al. Mother and son with deletion of 3p25-pter. Am J Med Genet 1991; 39: 130.

Tobias ES, Connor M, Ferguson Smith M. *Essential Medical Genetics*, 6th edn. Wiley-Blackwell, Oxford, 2011.

Turleau C. Monosomy 18p. Orphanet J Rare Dis 2008; 3: 4.

Valente KD, Koiffmann CP, Fridman C, et al. Epilepsy in patients with angelman Syndrome caused by deletion of the chromosome 15q11-13. Arch Neurol 2006; 63: 122.

Van Buggenhout GJ, Pijkels E, Holvoet M, et al. Cri du chat Syndrome: changing phenotype in older patients. Am J Med Genet 2000; 90: 203.

Williams CA, Beaudet AL, Clayton-Smith J, et al. Angelman Syndrome 2005: updated consensus for diagnostic criteria. Am J Med Genet A 2006; 140: 413.

Wong AC, Ning Y, Flint J, et al. Molecular characterization of a 130-kb terminal microdeletion at 22q in a child with mild mental retardation. Am J Hum Genet 1997; 60: 113.

Wright TJ, Ricke DO, Denison K, et al. A transcript map of the newly defined 165 kb Wolf-Hirschhorn Syndrome critical region. Hum Mol Genet 1997; 6: 317.

CHAPTER 7

Unidentifiable marker chromosomes, derivative chromosomes, chromosomes with additional material and rings

7.1 Marker chromosomes

A marker is defined as a chromosome whose banding pattern cannot clearly identify its chromosomal origin. If subsequent banding or testing by FISH or microarray analyses is performed, the origin of the marker may then become known, in which case a modified report may be written to describe this chromosome by its chromosomal origin. Markers are generally supernumerary chromosomes. If 46 chromosomes are present with one chromosome as unidentifiable, then this chromosome may be written as a derivative chromosome. However, this is usually seen as an acquired change and not constitutional in origin. (See Part 2.)

ISCN RULES FOR REPORTING A MARKER CHROMOSOME (FIGURE 7.1)

- First write the modal chromosome number, followed by the sex designation.
- Then add "+mar".
- If more than one marker chromosome is present, add "+" and then the number of markers, then "mar".

For example, with two markers present, the ISCN would be: 48,XY,+2mar.

Cytogenetic Abnormalities: Chromosomal, FISH and Microarray-Based Clinical Reporting, First Edition. Susan Mahler Zneimer.

Markers may be distinguished from one another, when seen in different cell lines, by designating them as mar1, mar2, etc. For example: 47,XY+mar1[14]/47,XY,+mar2[6]. This may be used when it is clear that the two marker chromosomes are different morphologically.

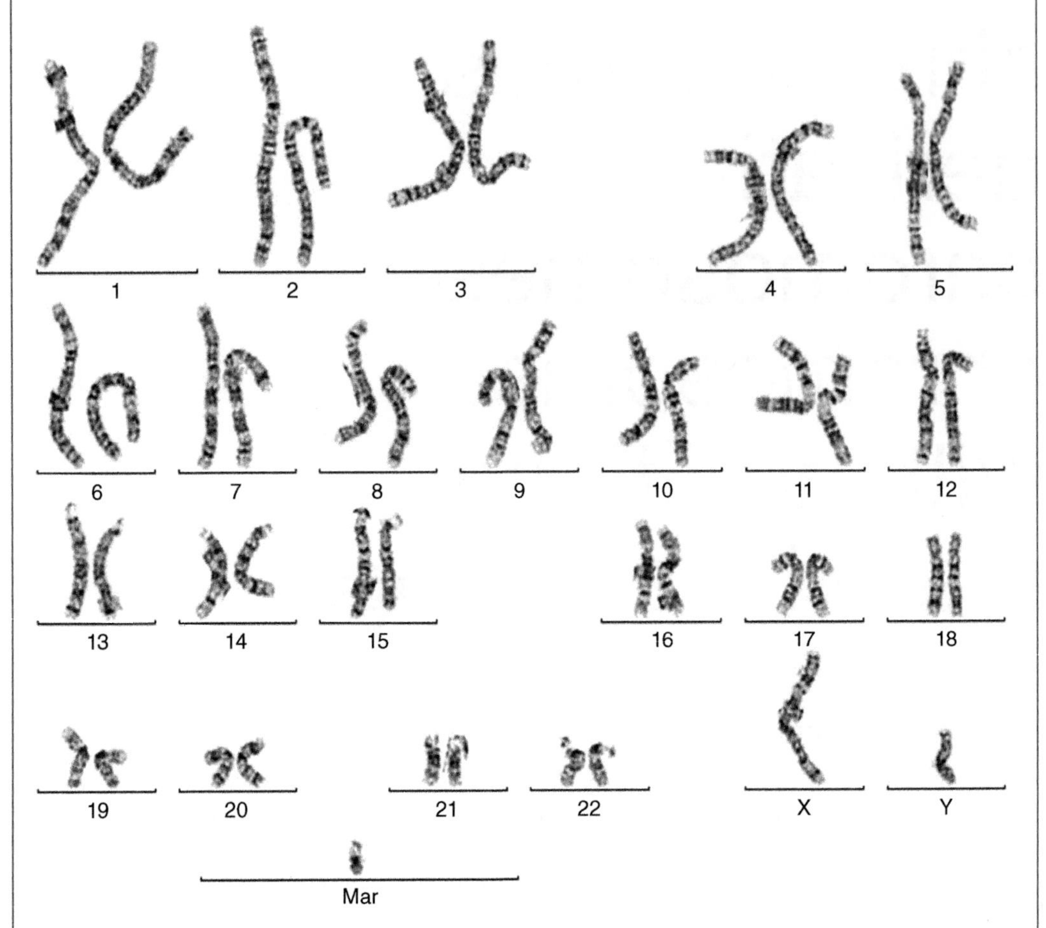

Figure 7.1 Karyotype showing a small marker chromosome: 46,XY,+mar. Courtesy of Karine Hovanes, Ph.D., CombiMatrix, Inc.

EXAMPLE REPORTING OF A SUPERNUMERARY MARKER CHROMOSOME OF UNKNOWN ORIGIN

ISCN Result: 47,XY,+mar

Interpretation

Chromosome analysis revealed an abnormal male chromosome complement in all cells examined with a small extra supernumerary marker chromosome of unknown origin. This extra chromosome results in trisomy for a small amount of genetic material, but of unknown etiology.

Further testing with FISH and chromosomal microarray analysis is suggested to further characterize this marker chromosome.

Genetic counseling is recommended.

7.2 Derivative chromosomes

Derivative chromosomes are defined in three categories. The first category refers to an unbalanced rearrangement involving two or more chromosomes; the second refers to two or more structural abnormalities within a single chromosome. The third category refers to a whole arm translocation, including Robertsonian translocations. The ISCN describes both balanced and unbalanced Robertsonian translocations as "der." (For more details, see Chapter 4.)

Derivative chromosomes generally result from unbalanced rearrangements. One of the most common forms of a derivative chromosome is from an unbalanced translocation. However, with derivative chromosomes involving more than one rearrangement that involves two or more chromosomes, then it may lead to both balanced and unbalanced rearrangements.

7.2.1 Derivative chromosomes – one rearrangement with two chromosomes

ISCN RULES FOR DESCRIBING DERIVATIVE CHROMOSOMES – ONE REARRANGEMENT WITH TWO CHROMOSOMES INVOLVED

- First write the modal chromosome number, followed by the sex designation.
- Then add "der" followed, in parentheses, by the chromosome involved.
- Then write the abbreviation for the rearrangement involved.
- Next write the chromosomes involved, enclosed in parentheses, followed by the breakpoints involved, separated by a colon and enclosed in parentheses.

For example, for an unbalanced translocation between chromosomes 9 and 17, in which there are two normal 9s, one normal 17 and one abnormal 17, then the ISCN would be written as: 46,XY,der(17)t(9;17)(p13;q11.2).

7.2.2 Derivative chromosomes – more than one rearrangement with two chromosomes

ISCN RULES FOR DESCRIBING DERIVATIVE CHROMOSOMES – MORE THAN ONE REARRANGEMENT WITH TWO CHROMOSOMES INVOLVED

- First write the modal chromosome number, followed by the sex designation.
- Then add "der" followed, in parentheses, by the first chromosome involved, beginning with the smallest chromosome number.
- Next write the abnormalities present in that chromosome, listed in order from the p arm terminus to the q arm terminus; follow the steps in the above description, and end with a comma.
- Then write "der" followed by the next chromosome involved, repeating the procedure used for the first chromosome.

For example, for a balanced translocation between chromosomes 9 and 17 that also includes a deletion within the short arm of chromosome 17, the ISCN is written as: 46,XY,der(9)t(9;17)(p13;q11.2),der(17)del(17)(p11.2p13)t(9;17).

Note that even though this is a balanced translocation, both chromosomes 9 and 17 are written as derivative chromosomes.

Also note that the breakpoints of the translocation are listed only the first time it is written, in this case for the der(9), and the same breakpoints are implied thereafter, for the der(17).

Also note that each chromosome involved is separated by a comma. However, within each chromosome, the aberrations are listed continuously without a comma and from p arm to q arm, rather than alphabetically.

Furthermore, in this scenario, if a third abnormality exists that is also involved with one of these abnormalities, then the third abnormality would follow the string of that abnormality.

For example, with the 46,XY,der(9)t(9;17)(p13;q11.2),der(17)del(17)(p11.2p13)t(9;17) karyotype, if the derivative chromosome 17 also contains another abnormality, such as a duplication of chromosome 9 material on the translocated chromosome 17, then the duplication nomenclature should follow the t(9;17), since it required the translocation to occur mechanistically before the duplication of the chromosome 9p material could occur. The ISCN nomenclature would then be: 46,XY,der(9)t(9;17)(p13;q11.2),der(17)del(17)(p11.2p13)t(9;17)dup(9)(p13p22).

EXAMPLE REPORTING OF A DERIVATIVE CHROMOSOME – WITH MORE THAN ONE REARRANGEMENT WITH TWO CHROMOSOMES INVOLVED

ISCN Result: 46,XY,der(9)t(9;17)(p13;q11.2),der(17)del(17)(p11.2p13)t(9;17).

Interpretation

Chromosome analysis revealed an abnormal male chromosome complement in all cells examined with a translocation between chromosomes 9 and 17, which also contains a deletion within the short arm of chromosome 17. Although the translocation appears balanced, it is possible that genes involved at the breakpoints of this rearrangement were disrupted, causing phenotypic abnormalities. Also, this finding results in a deletion within the short arm of chromosome 17.

Further testing with chromosomal microarray analysis is suggested to further characterize this abnormality.

Genetic counseling is recommended.

One example of a derivative chromosome involving an unbalanced translocation between two chromosomes is: 46,XY,+der(14)t(14;20)(q11.2;p13). This rearrangement originally was designated as an extra small supernumerary marker chromosome of unknown origin (see Figure 7.1). After further testing with FISH and microarray analyses, this rearrangement was then characterized as the nomenclature above. This extra chromosome results in trisomy or duplication of the proximal short arm of chromosome 14 and the distal short arm of chromosome 20. The indication for cytogenetic analysis for this individual was developmental delay in a child.

Figure 7.2 Partial karyotype showing a derivative chromosome 5: 46,XY,der(5)dup(5)(p14.2q35.3)r(5)(p14.2q35.3). Courtesy of Sarah South PhD, ARUP Laboratories.

7.2.3 Supernumerary derivative chromosome

EXAMPLE REPORTING OF A SUPERNUMERARY DERIVATIVE CHROMOSOME

ISCN: 47,XY,+der(14)t(14;20)(q11.2;p13)

Interpretation

Chromosome analysis revealed an abnormal male chromosome complement in all cells examined with a small extra supernumerary derivative chromosome resulting from an unbalanced translocation between the long arm of chromosome 14 and the short arm of chromosome 20. This extra chromosome results in trisomy or duplication of the proximal short arm of chromosome 14 and the distal short arm of chromosome 20.

This is a revised report to add information from the FISH and microarray analyses that further characterized this extra chromosome which was first reported as a marker chromosome (see corresponding FISH and microarray reports).

Genetic counseling is recommended.

7.2.4 Derivative chromosome with more than one rearrangement in a single chromosome

One example of a derivative chromosome resulting from more than one structural abnormality in a single chromosome is: 46,XY,der(5)dup(5)(p14.2q35.3)r(5)(p14.2q35.3) (Figure 7.2). This abnormal chromosome 5 is composed of a ring plus a duplication of the region involved in the ring: 5p14.2q35.3. This is one of two cell lines observed in a child to rule out cri-du-chat syndrome. The other cell line showed the ring without the duplication. Since there are two structural abnormalities in this chromosome, it is written as a derivative chromosome. (For an example report, see under Ring chromosomes.)

7.2.5 Derivative chromosome with a whole arm translocation

One example of a derivative chromosome resulting from a whole arm translocation is 45,XY,der(12;18)(q10;q10). This abnormality consists of the whole long arm of chromosome 12 translocated to the whole long arm of chromosome 18. This is an unbalanced rearrangement in which the short arms of chromosomes 12 and 18 are lost. In constitutional studies, seeing an unbalanced whole arm translocation that is not a Robertsonian translocation is rare, as only chromosomes with loss of a small amount of material will be viable. When occurring in submetacentric chromosomes containing euchromatic material on both the short and long arms, this type of rearrangement most likely results in fetal demise and is usually not seen in liveborns.

EXAMPLE REPORTING OF A DERIVATIVE CHROMOSOME WITH A WHOLE ARM TRANSLOCATION

ISCN Result: 45,XY,der(12;18)(q10;q10)

Interpretation

Chromosome analysis revealed an abnormal male chromosome complement in all cells examined with a whole arm translocation between the long arms of chromosomes 12 and 18. This is an unbalanced rearrangement in which the short arms of chromosomes 12 and 18 are lost, resulting in monosomy for chromosomes 12p and 18p.

Genetic counseling is recommended.

7.3 Chromosomes with additional material

Added material on chromosomes is material that is unidentifiable as to its chromosomal origin and which is translocated to a known chromosome. This added material is located at the end of one of the arms of the known chromosome, thereby possibly replacing the telomeric end of the known chromosome. These abnormalities will result in trisomy for the added material and possibly monosomy for the distal portion of the known chromosome if it is proximal to the telomere. As with marker chromosomes, if subsequent banding or testing by FISH or microarray analyses is performed, the origin of the added material may become known, in which case a modified report may be written to describe this chromosome by its chromosomal origin. When added material is on a known chromosome, the modal chromosome number does not change. It is also implied that there is no second centromere present which contains the added material, in which case the designation would be written as a dicentric chromosome.

ISCN RULES FOR REPORTING ADDED MATERIAL ON A CHROMOSOME

- First write the modal chromosome number, followed by the sex designation.
- Then write "add" followed, in parentheses, by the known chromosome.
- Then, in parentheses, write the breakpoint of the known chromosome where the added material is present.

For example, with added material on the long arm of chromosome 14, the ISCN would be: 46,XY,add(14)(q32).

EXAMPLE REPORTING OF A CHROMOSOME WITH ADDED MATERIAL (FIGURE 7.3)

ISCN Result: 46,XY,add(14)(q32)

Interpretation

Chromosome analysis revealed an abnormal male chromosome complement in all cells examined with added material of unknown origin on the long arm of chromosome 14. This abnormality results in the gain of genetic material of unknown etiology, and the loss of chromosome 14 from band q32 to the telomere.

Further testing with FISH and microarray analysis is suggested to further characterize this abnormal chromosome.

Genetic counseling is recommended.

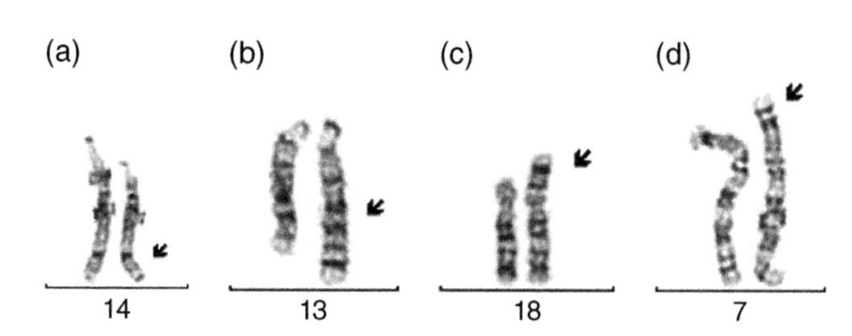

Figure 7.3 Partial karyotypes showing chromosomes with additional material of unknown origin. (a) add(14)(q32). (b) 46,XX,add(13)(q34). (c) 46,XX,add(18)(p11.2). (d) 46,XX,add(7)(p21). Courtesy of Sarah South PhD, ARUP Laboratories.

7.4 Ring chromosomes

Ring chromosomes are defined as the break of a chromosome at two points and the rejoining of that chromosome during meiosis or mitotic cell division. This generally involves loss of genetic material at the ends of the chromosomes. Since the ring chromosome would essentially need to break open to replicate itself at each successive cell cycle, the possibility of breakage and reunion of the ring may not succeed appropriately. Consequently, ring chromosomes become unstable and possibly modify the genetic material in each cell division. Thus, an individual with a ring chromosome may actually have more than one ring configuration, resulting in mosaic forms, or some cells with loss of the ring altogether.

Autosomal rings seen as a constitutional abnormality are rare and often not viable due to their instability and possible loss of too much genetic material. However, ring composition results in various phenotypes, which is determined in part by whether the ring is deleted for telomeric and other chromosomal regions, and due to the consequence of having a circular configuration. Rings going through sister chromatid exchange (SCE) have a detrimental consequence. A single SCE can yield a double ring after mitosis, whereas two or more SCE events can lead to complex ring patterns which most likely result in breakage of the ring.

X chromosome rings are more common and are discussed in Chapter 10.

7.4.1 Rings that replace a normal homolog

Ring chromosomes may either replace a normal homolog of a chromosome if the ring has not lost much genetic material or may be an extra numerary chromosome, which is often small and may be mosaic.

Since ring chromosomes lose genetic material on the distal ends, certain rings are associated with constitutional genetic disorders, such as Wolf–Hirschhorn and cri-du-chat syndromes, due to a distal short arm deletion of these chromosomes. Other rings may involve chromosomes not associated with a deletion syndrome and present with non-definitive phenotypic findings.

ISCN RULES FOR REPORTING A RING CHROMOSOME (FIGURE 7.4)

- First write the modal chromosome number, followed by the sex designation.
- Then write “r” followed, in parentheses, by the chromosome number involved.
- Then, in parentheses, write the breakpoints involved.

Establishing breakpoints may be difficult and more than one size ring may be present, making the nomenclature arduous to write. Often only one to two forms of a ring are stable, allowing for an easier description in the report.

For example, in the case with cri-du-chat described previously as a derivative chromosome, the ISCN would be: 46,XY,r(5)(p14.2q35.3)[18]/46,XY,der(5)dup(5)(p14.2q35.3) r(5)[2]. Note that in the second cell line, it is considered a derivative chromosome since more than one abnormality is present in a single chromosome, and that the r(5) does not need the breakpoints repeated, as this is implied in the original cell line.

Further note that all abnormalities of a single chromosome are not separated by commas, but are strung out in a single phrase. (More details of derivative chromosomes are given in the next section). For a depiction of the der(5)dup(5)r(5) abnormalities see Figure 7.2.

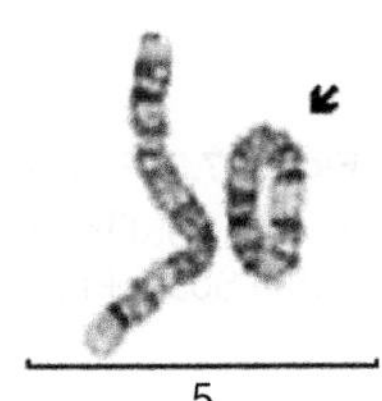

Figure 7.4 Partial karyotype showing a ring chromosome 5: 46,XY,r(5)(p14.2q35.3). Courtesy of Sarah South PhD, ARUP Laboratories.

EXAMPLE REPORTING OF A RING WITH 46 CHROMOSOMES

ISCN Result: 46,XY,r(5)(p14.2q35.3)[18]/46,XY,der(5)dup(5)(p14.2q35.3)r(5)[2]

Interpretation

Cytogenetic analysis revealed two related abnormal male cell lines present. The first cell line, seen in 18 of 20 cells, showed an abnormal chromosome complement with a ring chromosome 5, from bands p14.2 to q35.3, replacing a normal chromosome 5 homolog.

The second cell line, seen in 2 of 20 cells, showed a similar ring chromosome 5, which also includes a duplication of the bands seen in the first cell line. Both ring chromosomes result in a deletion of the distal short and long arms of chromosome 5; the second cell line also results in a duplication of most of chromosome 5, not including the distal short and long arms.

Loss of chromosome 5 from bands p14.2 to pter is consistent with the diagnosis of cri-du-chat syndrome.

Additional testing with chromosomal microarray analysis is suggested to further characterize these ring chromosomes.

Genetic counseling is recommended.

7.4.2 Rings as a supernumerary chromosome

Ring chromosomes identified in an individual may present as an extra chromosome and are most often small, but will result in the gain of a sufficient quantity of genetic material to be trisomic for that region. Consequently, they will be associated with phenotypic manifestations or, if carried by a normal parent, have the risk of passing the ring to offspring in a different form with clinical manifestations.

ISCN RULES FOR REPORTING AN EXTRA RING CHROMOSOME

- First write the modal chromosome number, followed by the sex designation.
- Then add "+r" followed, in parentheses, by the chromosome involved.
- Finally, write, in parentheses, the breakpoints involved, if identifiable.

EXAMPLE REPORTING OF A SUPERNUMERARY RING

ISCN Result: 47,XY,+r(18)(p11.2q22)

Interpretation

Chromosome analysis revealed an abnormal male chromosome complement in all cells examined with an extra ring chromosome derived from chromosome 18 from bands p11.2 to q22. This finding results in a duplication of the proximal short arm and most of the long arm of chromosome 18.

Additional testing with microarray analysis is suggested to further characterize this ring chromosome.

Genetic counseling is recommended.

7.4.3 Rings with unknown chromosomal origin or breakpoints

Some rings may be distinctly of a certain origin, such as chromosome 7 or 21, when replacing a normal homolog, but whose breakpoints are not identifiable. In these cases, the ISCN is written as r(7) or r(21) only. Other rings that are not clearly identifiable as any chromosome, especially when they are small, or as a supernumerary chromosome, would be written as "r" alone.

EXAMPLE REPORTING OF A RING OF KNOWN CHROMOSOMAL ORIGIN WITHOUT IDENTIFIABLE BREAKPOINTS

ISCN Result: 46,XY,r(7)

Interpretation

Chromosome analysis revealed an abnormal male chromosome complement in all cells examined with a ring chromosome 7 replacing a normal chromosome 7 homolog; however, the breakpoints are not discernible. This finding results in a deletion of the distal short and long arms of chromosome 7.

Additional testing with chromosomal microarray analysis is suggested to further characterize this ring chromosome.

Genetic counseling is recommended.

EXAMPLE REPORTING OF A RING OF UNKNOWN CHROMOSOMAL ORIGIN

ISCN Result: 47,XY,+r

Interpretation

Chromosome analysis revealed an abnormal male chromosome complement in all cells examined with a small extra supernumerary ring chromosome of unknown origin. This extra chromosome results in trisomy for a small amount of genetic material, but of unknown etiology.

Additonal testing with chromosomal microarray analysis is suggested to further characterize this ring chromosome.

Genetic counseling is recommended.

More complex ring chromosomes are present in neoplastic syndromes, and are described in Part 2.

7.5 Homogenously staining regions

Homogenously staining regions (HSRs) are regions of a chromosome that do not have a distinct banding pattern. This region may be small or extend a chromosome band to a great length.

ISCN RULES FOR DESCRIBING HOMOGENOUSLY STAINING REGIONS

- First write the modal chromosome number, followed by the sex designation.
- Then add "hsr" followed, in parentheses, by the chromosome involved.
- Finally, in parentheses, write the breakpoints involved.
- With more than one hsr region on the same chromosome, each region is designated as a derivative chromosome and both hsr regions are described separately.
- If an hsr region is seen on two chromosomes as part of a chromosomal rearrangement, then the hsr is described on both chromosomes.

For example, for a single hsr, the nomenclature is written as: 46,XY,hsr(14)(q11.2).

EXAMPLE REPORTING OF A HOMOGENOUSLY STAINING REGION (FIGURE 7.5)

ISCN Result: 46,XY,hsr(14)(q11.2)

Interpretation

Chromosome analysis revealed an abnormal male chromosome complement in all cells examined with a homogenously staining region of unknown origin within the long arm of chromosome 14. This finding results in amplification of a portion of the genome, but of unknown etiology.

Additional testing with chromosomal microarray analysis is suggested to further characterize this abnormal chromosome.

Genetic counseling is recommended.

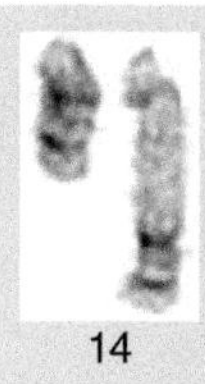

Figure 7.5 Partial karyotype showing a homogenously staining region on the long arm of chromosome 14: hsr(14)(q11.2).

Bibliography

Biedler JL, Albrechta M, Spenglerb A. Non-banding homogeneous chromosome regions in cells with very high dihydrofolate reductase levels. Genetics 1974; 77(Suppl): 4–5.

Blennow E, Bui TH, Kristoffersson U, et al. Swedish survey on extra structurally abnormal chromosomes in 39 105 consecutive prenatal diagnoses: prevalence and characterization by fluorescence in situ hybridization. Prenat Diagn 1994; 14: 1019.

Conlin LK, Kramer W, Hutchinson AL, et al. Molecular analysis of ring chromosome 20 syndrome reveals two distinct groups of patients. J Med Genet 2011; 48: 1e9.

Fickelscher I, Starke H, Schulze E, et al. A further case with a small supernumerary marker chromosome (sSMC) derived from chromosome 1 – evidence for high variability in mosaicism in different tissues of sSMC carriers. Prenat Diagn 2007, 27: 783–785.

Gardner RJM, Sutherland GR. *Chromosome Abnormalities and Genetic Counselling*. Oxford Monographs on Medical Genetics. Oxford University Press, Oxford, 2003.

Gersen S, Keagle M (eds). *Principles of Clinical Cytogenetics*. Humana Press, Totowa, New Jersey, 1999.

Kosztolanyi G. The genetics and clinical characteristics of constitutional ring chromosomes. J Assoc Genet Technol 2009; 35: 44e8.

Kosztolanyi G. Does "ring syndrome" exist? An analysis of 207 case reports on patients with a ring autosome. Hum Genet 1987; 75: 174e9.

Kosztolanyi G, Mehes K, Hook EB. Inherited ring chromosomes: an analysis of published cases. Hum Genet 1991; 87: 320e4.

McDermid HE, Duncan AM, Brasch KR, et al. Characterization of the supernumerary chromosome in cat eye Syndrome. Science 1986; 232: 646.

Mears AJ, el-Shanti H, Murray JC, et al. Minute supernumerary ring chromosome 22 associated with cat eye Syndrome: further delineation of the critical region. Am J Hum Genet 1995; 57: 667.

Murthy SK, Malhotra AK, Jacob PS, et al. Analphoid supernumerary marker chromosome characterized by aCGH and FISH as inv dup(3)(q25.33qter) de novo in a child with dysmorphic features and streaky pigmentation: case report. Mol Cytogenet 2008; 1: 19.

Nunberg JH, Kaufman RJ, Schimke RT, Urlaub G, Chasin LA. Amplified dihydrofolate reductase genes are localized to a homogeneously staining region of a single chromosome in a methotrexate resistant Chinese hamster ovary cell line. Proc Natl Acad Sci USA 1978; 75: 5553–5556.

Pezzolo A, Gimelli G, Cohen A, et al. Presence of telomeric and subtelomeric sequences at the fusion points of ring chromosomes indicates that the ring syndrome is caused by ring instability. Hum Genet 1993; 92: 23e7.

Rodriguez L, Liehr T, Martinez-Fernandez ML, Lara A, Torres A, Martinez-Frias ML. A new small supernumerary marker chromosome, generating mosaic pure trisomy 16q11.1–q12.1 in a healthy man. Mol Cytogenet 2008; 1: 4.

Rooney DE, Czepulkowski BH. *Human Cytogenetics. A Practical Approach*. Oxford University Press, New York, 1992.

Rossi E, Messa J, Zuffardi O. Ring syndrome: still true? J Med Genet 2008; 45: 766e8.

Sodre CP, Guilherme RS, Meloni VF, et al. Ring chromosome instability evaluation in six patients with autosomal rings. Genet Mol Res 2010; 9: 134e43.

Shaffer LG, McGowan-Jordan J, Schmid M (eds). *ISCN 2013: An International System for Human Cytogenetic Nomenclature*. Karger Publishers, Unionville, CT, 2013.

Shimizu N, Shingaki K, Kaneko-Sasaguri Y, Hashizume T, Kanda T. Where and how the bridge breaks: anaphase bridge breakage plays a crucial role in gene amplification and HSR generation. Exp Cell Res 2005; 302(2): 233–243.

Tobias ES, Connor M, Ferguson Smith M. *Essential Medical Genetics*, 6th edn. Wiley-Blackwell, Oxford, 2011.

Trifonov V, Fluri S, Binkert F, et al. Complex rearranged small supernumerary marker chromosomes (sSMC), three new cases; evidence for an underestimated entity? Mol Cytogenet 2008; 1: 6.

Windle B, Draper B, Yin B, O'Gorman YX, Wahl S. A central role for chromosome breakage in gene amplification, deletion formation, and amplicon integration. Genes Dev 1991; 5(2): 160–174.

CHAPTER 8

Isochromosomes, dicentric chromosomes and pseudodicentric chromosomes

8.1 Isochromosomes/dicentric chromosomes

Isochromosomes refer to an identical chromosome arm on both sides of a centromere, with subsequent loss of the other chromosome arm. The isochromosome may replace a normal homolog or be an supernumerary chromosome. Isochromosomes may contain a single centromere or two centromeres. With two centromeres present, it is referred to as an isodicentric chromosome.

If an isochromosome forms by replacing a normal homolog, then the resulting genetic material would be trisomic for the isochromosome material and monosomic for the arm that is lost. If the isochromosome is an supernumerary chromosome, then there would be tetrasomy for the isochromosome arm, while the other arm remains diploid. For the isochromosome to be present in liveborns, it would need to be small; otherwise, the amount of genetic material gained or lost would be too large to be viable. Also, with a postzygotic non-disjunction, mosaicism for a normal cell line is possible, ameliorating the severity of the imbalance of genetic material.

Mechanistically, there is more than one proposed method by which isochromosomes may form. Briefly, one mechanism is a duplication of one arm of a chromosome during meiosis that undergoes a lack of proper centromere division, whereby division of the chromosome occurs laterally rather than longitudinally along a chromosome. When the centromere is present on one half of the split chromosome, then the other half of the chromosome may be lost. It is also possible that the centromere duplicates along with the chromosome arm, giving rise to a dicentric isochromosome.

Another mechanism proposed is that the isochromosome is really the combining of two sister chromatids rather than a duplication of a single chromosome arm. This is referred to as a U-shaped exchange. Since the genetic material of sister chromatids is identical, these two scenarios would be indistinguishable on the cytogenetic level.

With lateral centromere misdivision, it would be possible for complementary isochromosomes to be present, in which both arms of a chromosome form separate isochromosomes and both remain in the cell rather than being lost.

Cytogenetic Abnormalities: Chromosomal, FISH and Microarray-Based Clinical Reporting, First Edition. Susan Mahler Zneimer.

The common isochromosomes seen in liveborns include supernumerary chromosomes with i(9p), i(12p), i(18p), i(20p) and i(22q); isochromosomes replacing a homolog are commonly seen with i(21q) and i(Xp/Xq). With the exception of the X chromosome, all the other isochromosomes contain very small duplicated arms. Some of these isochromosomes will be discussed in more detail due to their association with syndromes.

Another related abnormality is the inverted duplicated isochromosome 15, which will also be discussed. The i(Xp/Xq) will be discussed in Chapter 10.

A dicentric chromosome, as the term implies, contains two distinct centromeres of an abnormal chromosome. The abnormal chromosome may be the result of a translocation, most commonly seen in Robertsonian translocations, or of a paracentric inversion with a meiosis crossing-over event. Chromosomes only require one active centromere, and the second centromere becomes inactivated; otherwise, two active centromeres would split chromosomes apart during cell division. Therefore, most dicentric chromosomes are truly pseudodicentric with one inactive centromere. However, distinguishing between dicentric and pseudodicentric is less relevant in the nomenclature, unless cytogenetically obvious. An inactive centromere will appear more as a band rather than a dark constriction, which is typical of an active centromere.

8.1.1 Isochromosome 12p – Pallister–Killian syndrome

The i(12p) is associated with Pallister–Killian syndrome and is generally seen as a supernumerary chromosome, thereby resulting in tetrasomy 12p, and is predominantly seen in the mosaic form.

SPECIFIC FEATURES OF PALLISTER–KILLIAN SYNDROME

The clinical features of i(12p) include hypotonia, intellectual disability, developmental delay, sparse hair, areas of unusual pigmentation, speech anomalies and distinctive facial features, including high, rounded forehead, broad nasal bridge, short nose, widely spaced eyes, low-set ears, wide mouth, large tongue and cleft or high-arched palate. Other birth defects may be present.

Pallister–Killian syndrome is a rare condition, although the incidence in the general population is not known. This disorder may be underdiagnosed because it can be difficult to detect in people with mild signs and symptoms.

ISCN RULES FOR REPORTING A SINGLE CENTROMERE ISOCHROMOSOME AS A SUPERNUMERARY CHROMOSOME

- First write the modal chromosome number of 47, followed by the sex designation.
- Then write a plus sign "+" followed by "i".
- Finally write, in parentheses, the chromosome involved, followed, in parentheses, by the breakpoint p10 for the short arm or q10 for the long arm duplicated.

For example, mosaic isochromosome 12p, seen in 12 of 20 cells, would be: 47,XX,+i(12)(p10)[12]/46,XX[8].

Note: If mosaic, separate the normal cell line from the i(12p) with a slash, and write the abnormal cell line first.

EXAMPLE REPORTING OF ISOCHROMOSOME 12p (FIGURE 8.1)

ISCN Result: 47,XX,+i(12)(p10)[12]/46,XX[8]

Interpretation

Cytogenetic analysis revealed two cell lines present. Twelve of the 20 cells examined showed an abnormal female chromosome complement with an extra isochromosome consisting of the short arm of chromosome 12, resulting in tetrasomy (4 copies) of chromosome 12p. The remaining 8 cells showed a normal female chromosome complement.

Mosaic isochromosome (12p) is consistent with the diagnosis of Pallister–Killian syndrome. Clinical features of Pallister–Killian syndrome include hypotonia, intellectual disability, developmental delay, sparse hair, areas of unusual pigmentation, speech anomalies and distinctive facial features.

Genetic counseling is recommended.

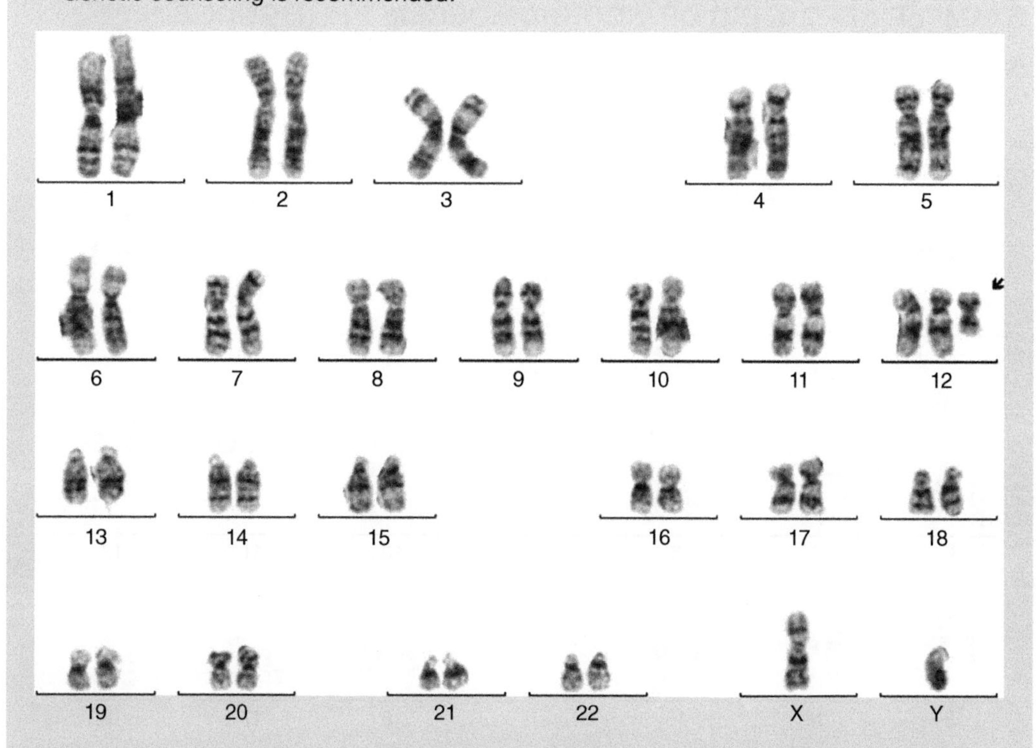

Figure 8.1 Karyotype showing 47,XY,i(12)(p10), consistent with Pallister–Killian syndrome. Courtesy of Sarah South Ph.D., ARUP Laboratories.

8.1.2 Isochromosome 21q – translocation Down syndrome

Down syndrome has been described previously (see Chapter 3). The isochromosome 21q, although it results in trisomy 21, occurs by a different method than the non-disjunction resulting in aneuploidy, as discussed previously. The clinical implications of Down syndrome due to i(21q) are critical for the risk assessment for future pregnancies of a couple with a child who carries this abnormality. If a parent carries a balanced version of this abnormality, namely, with the loss of a normal chromosome 21 and the gain of i(21q), then all future pregnancies of this individual will pass on this abnormal chromosome, reuslting in 100% of offspring with trisomy 21. For t(21q;21q) translocation Down syndrome see Chapter 4 (Robertsonian translocations).

ISCN RULES FOR REPORTING A SINGLE CENTROMERE ISOCHROMOSOME IN ALL CELLS EXAMINED WHEN REPLACING A NORMAL HOMOLOG

- First write the modal chromosome number of 46, followed by the sex designation.
- Then write "i" followed, in parentheses, by the chromosome involved.
- Then, in parentheses, write the breakpoint p10 for the short arm or q10 for the long arm duplicated.

For example, isochromosome 21q would be: 46,XX,i(21)(q10).

EXAMPLE REPORTING OF ISOCHROMOSOME 21q (FIGURE 8.2)

ISCN Result: 46,XX,i(21)(q10)

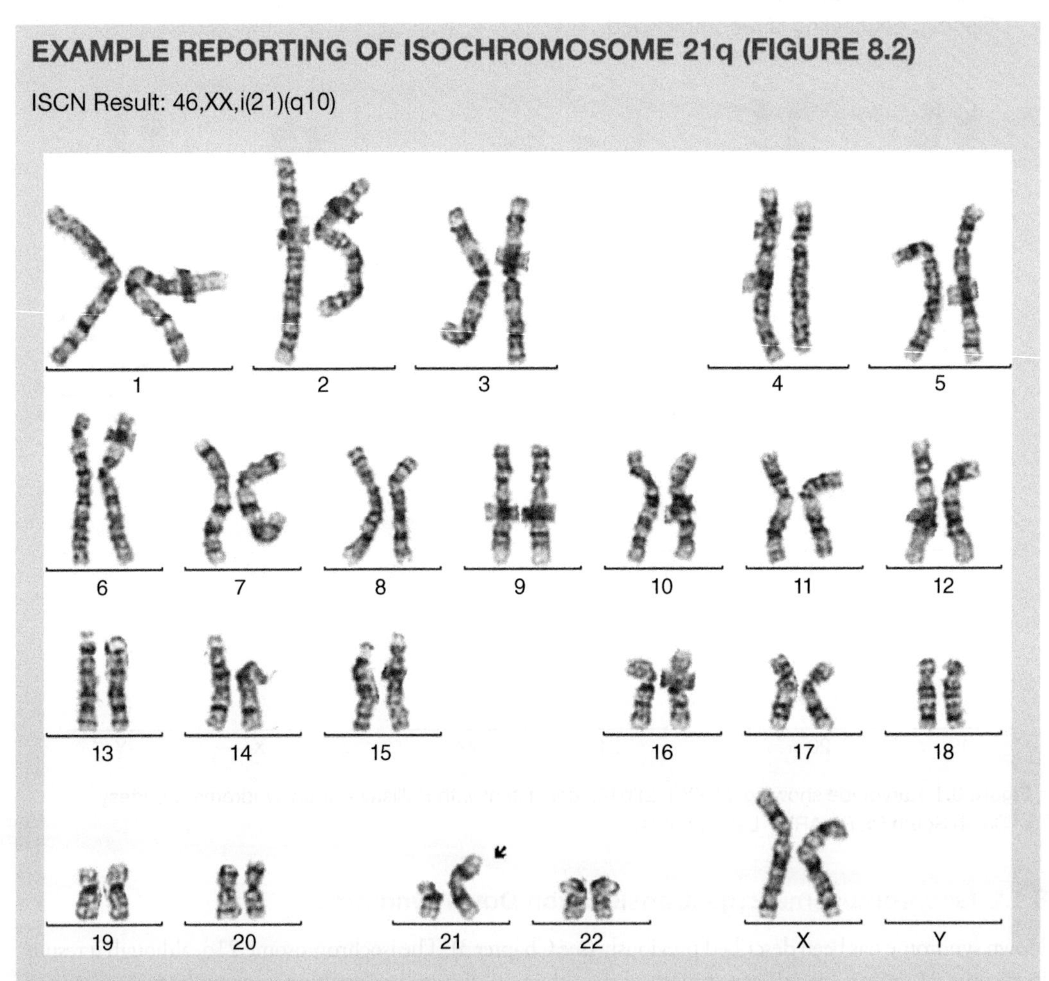

Figure 8.2 Karyotype showing an isochromosome of the long arm of chromosome 21, replacing one normal chromosome 21, resulting in trisomy 21 (Down syndrome): 46,XX,i(21)(q10).

Interpretation

Chromosome analysis revealed an abnormal female chromosome complement in all cells examined with an isochromosome of the long arm of chromosome 21, replacing a normal chromosome 21, resulting in trisomy 21 (Down syndrome).

The main clinical features of Down syndrome include hypotonia, flat facies, down-slanting eyes, small ears and mental deficiency.

Parental chromosome analyses are recommended in order to clarify if this translocation is inherited or *de novo* in origin. If *de novo* in origin, the recurrence risk is generally considered to be 1% or less. The risk of a parent carrying the i(21q) chromosome is very small. However, carriers of i(21q) chromosomes will always have gametes containing the i(21q) chromosome; therefore, 100% of offspring will have three copies of chromosome 21 (Down syndrome).

Genetic counseling is critical to convey to i(21q) carriers that they will not be able to have genetically normal children.

Genetic counseling is recommended.

8.1.3 Isodicentric chromosome 15q – inverted duplication 15q

The isodicentric 15q is also described as an inverted duplication (inv dup) and may be written either way. However, recent studies have shown a possible mechanism of this chromosome as the consequence of recombination of homologous chromosome 15s; therefore, idic is probably a better choice for the nomenclature.

The idic(15q) is always seen as a supernumerary chromosome, which is also called an extra structurally abnormal chromosome (ESAC). This is a small bisatellited extra chromosome, and results in tetrasomy for a small segment of the long arm of chromosome 15. It is important to clarify if this ESAC is of chromosomal 15 origin, since the region on the long arm just distal to the centromere is the locus for SNRPN and other genes associated with Prader–Willi and Angelman syndromes. Once a bisatellited marker is identified by cytogenetic analysis, FISH analysis is needed to determine the chromosome origin and what, if any, genes are present. This will greatly clarify the genotypic identification and phenotypic consequences for the patient.

More details, both cytogenetically and clinically, of the idic 15q are discussed in Section 2. However, often one needs to write the ISCN and report this possible abnormality cytogenetically before FISH analysis is completed. Therefore, when writing the ISCN of an ESAC of unknown origin, it is appropriate to use the "marker" designation (discussed in Chapter 7). However, if one observes a bisatellited ESAC, then one may write the chromosomes as a possible idic.

ISCN RULES FOR REPORTING A BISATELLITED ESAC

- First write the modal chromosome number, 47, followed by the sex designation.
- Then write "+idic" followed, in parentheses, by the chromosome involved.
- Then, in parentheses, write the breakpoints observed.

For example: 47,XX,+idic(15)(q12). This nomenclature implies an extra chromosome 15 with the bands from the short arm terminus to band q12 in the long arm, with *no* material from 15q12 to the long arm terminus. This abnormality would result in tetrasomy for the centromere of chromosome 15 to 15q12.

EXAMPLE REPORTING OF A BISATELLITED QUESTIONABLE ISODICENTRIC 15q

ISCN Result: 47,XX,+?idic(15)(q12)

Interpretation

Chromosome analysis revealed an abnormal female chromosome complement in all cells examined with an extra small bisatellited chromosome that may correspond to an isodicentric chromosome 15, containing the short arm to the long arm at band q12, resulting in tetrasomy (4 copies) of chromosome 15pter to 15q12.

It is possible that if this chromosome is of chromosome 15 origin and involves euchromatic material on the long arm, it may result in phenotypic manifestations that include developmental delay, autism, epilepsy and mild dysmorphic features.

To clarify if this extra small chromosome is of chromosomal 15 origin, FISH analysis with chromosome 15 probes is recommended.

Genetic counseling is recommended.

8.1.4 Isodicentric 21q

Another isodicentric chromosome that is common is the idic 21q. This chromosome is usually not an extra chromosome but replaces a normal chromosome 21 homolog, resulting in trisomy for chromosome 21. This chromosome is distinct because it is the joining of two chromosome 21s at the terminal long arm, not at the centromere. Therefore, it appears as upside down 21s with the satellites at either end. This bisatellited chromosome is larger than the bisatellited chromosome 15q. The bands of chromosome 21 are clearly visible and look duplicated from end to end, not centromere to centromere.

The centromere to centromere chromosome 21 would be an isochromosome 21q or der(21;21)(q10;q10), as discussed in Chapter 4. This chromosome also looks very similar to the i(12p) abnormality, and one must be careful to have distinct banding or do further testing with FISH or microarrays to distinguish between the two.

ISCN RULES FOR REPORTING AN ISODICENTRIC CHROMOSOME FROM END TO END (NOT CENTROMERE TO CENTROMERE)

- First write the modal count of 46, followed by the sex designation.
- Then write "idic" followed, in parentheses, by the chromosome involved.
- Then write the terminal long arm band, enclosed in parentheses.

For example, with chromosome 21: 46,XX,idic(21)(q22.3). This implies a duplication of chromosome 21 from the short arm to the telomere of the long arm.

EXAMPLE REPORTING OF ISODICENTRIC 21q FROM END TO END (NOT CENTROMERE TO CENTROMERE)

ISCN Result: 46,XX,idic(21)(q22.3)

Interpretation

Chromosome analysis revealed an abnormal female chromosome complement in all cells examined with an isodicentric chromosome 21 replacing a normal chromosome 21, which is an end-to-end fusion of two chromosome 21s, resulting in trisomy 21 (Down syndrome).

The main clinical features of Down syndrome include hypotonia, flat facies, down-slanting eyes, small ears and mental deficiency.

Parental chromosome analyses are recommended in order to clarify if this translocation is inherited or *de novo* in origin. If *de novo* in origin, the recurrence risk is generally considered to be 1% or less. The risk of a parent carrying the idic(21q) chromosome is very small, and carriers of idic(21q) chromosomes are most likely isochromosomes and not translocations. However, the outcome of genetic material remains the same.

Of importance is that carriers of (21;21) chromosomes will always have gametes containing the (21;21) chromosome; therefore, 100% of offspring will have three copies of chromosome 21 (Down syndrome).

Genetic counseling is critical to convey to (21;21) carriers that they will not be able to have genetically normal children.

Genetic counseling is recommended.

8.1.5 Isodicentric 22q – cat-eye syndrome

Isodicentric 22q is associated with cat-eye syndrome. This disorder is also generally observed in the mosaic form, which may be seen either as a supernumerary chromosome or replacing a normal chromosome 22, thereby resulting in chromosome 22q trisomy or tetrasomy. The mechanism of this abnormality may be unequal crossing over seen as a dicentric inverted duplication, since many reports in the literature show a variable duplicated genetic region and a wide variation in phenotypic manifestations. Generally the phenotype corresponds to the amount of chromosome 22 that is duplicated, encompassing the few genes in the cat-eye critical regions (CECR1 and CECR2).

SPECIFIC FEATURES OF CAT-EYE SYNDROME

This disorder is named for the distinct ocular anomalies, including coloboma of the iris, or partial absence of ocular tissue, and other associated symptoms and findings that may vary greatly in range and severity to the point of mild symptoms that may not be recognized as a syndrome. However, other manifestations may include mild intrauterine growth retardation, mild mental deficiency, craniofacial malformations, congenital heart defects, malformations of the heart, the kidneys and/or the anal region.

EXAMPLE REPORTING OF ISODICENTRIC 22q

ISCN Result: 47,XX,+idic(22)(q11.1)

Interpretation

Chromosome analysis revealed an abnormal female chromosome complement in all cells examined with an extra isodicentric chromosome of the long arm of chromosome 22, resulting in tetrasomy (4 copies) of chromosome 22q.

Isodicentric (22q) is consistent with the diagnosis of cat-eye syndrome. Clinical features of cat-eye syndrome include coloboma of the iris, and other associated symptoms and findings that may vary greatly in range and severity. Other manifestations may include mild mental deficiency, craniofacial malformations, congenital heart defects, malformations of the heart, the kidneys and/or the anal region.

Genetic counseling is recommended.

8.2 Pseudodicentric chromosomes

ISCN RULES FOR REPORTING PSEUDODICENTRIC CHROMOSOMES

- These are written similarly to isodicentrics; however, they usually involve two chromosomes as a rearrangement, where one normal homolog of each chromosome is replaced by the fusion of the two chromosomes involved.
- First write the modal count of 45, followed by the sex designation.
- Then write "psu dic" followed, in parentheses, by the chromosomes involved, starting with the chromosome with the active centromere, followed by the chromosome with the inactive centromere.
- Then write the breakpoints of each chromosome, enclosed in parentheses.

Note: When a pseudodicentric chromosome replaces two normal chromosomes, there are a total of 45 chromosomes, and there is no designation to write a missing chromosome.

For example, a translocation involving chromosomes 13 and 22, where the chromosome 22 centromere appears as the primary constriction and therefore active, would be written as: 45,XY,psu dic(22;13)(p12;p12) (Figure 8.3). This corresponds to chromosome 22 from p12 to qter, joined with chromosome 13 from p12 to qter, with one normal homolog of both chromosomes 13 and 22.

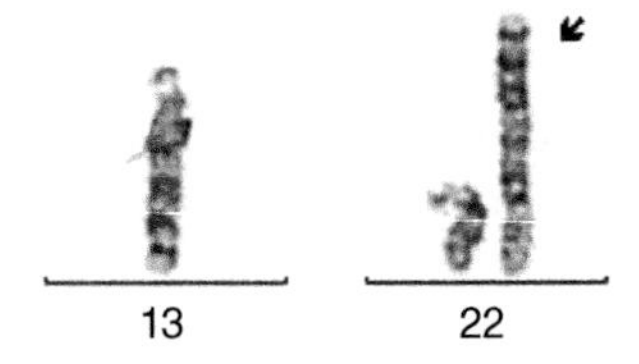

Figure 8.3 A partial karyotype showing a pseudodicentric chromosome involving chromosomes 13 and 22. This chromosome is the fusion of chromosomes 13 and 22 at the short arms, leaving no apparent imbalance for long arm material of either chromosome: 45,XY,psu dic(22;13)(p12;p12). Chromosome 22 appears to contain the active centromere, and so is written first in the nomenclature.

EXAMPLE REPORTING OF A PSEUDODICENTRIC CHROMOSOME

ISCN Result: 46,XY,psu dic(22;13)(p12;p12)

Interpretation

Chromosome analysis revealed an abnormal male chromosome complement in all cells examined with a pseudodicentric chromosome involving chromosomes 13 and 22. This chromosome fused together the complete chromosomes of 13 and 22 at the short arms, leaving no apparent imbalance for long arm material.

Chromosome analysis of the parental chromosomes is indicated to determine a potential origin of this translocation.

Chromosomal microarray analysis may prove informative for further characterization of this rearrangement. As recent literature suggests, imbalances at the breakpoints or elsewhere in the genome may be found in approximately 30–40% of individuals with a cytogenetically balanced rearrangement and an abnormal phenotype.

Genetic counseling is recommended.

Bibliography

Darlington CD. The origin of iso-chromosomes. J Genet 1939; 39: 351–361.

De la Chapelle A, Hortling H, Wenstrom J, Ockey CH. Isochromosome-X in Man. Part 1. Hereditas 1966; 54: 260–276.

Driscoll CJ, Darvey NL. Chromosome pairing: effect of colchicine on an isochromosome. Science 1970; 169: 290–291.

Gardner RJM, Sutherland GR. *Chromosome Abnormalities and Genetic Counselling.* Oxford Monographs on Medical Genetics. Oxford University Press, Oxford, 2003.

Gersen S, Keagle M (eds). *Principles of Clinical Cytogenetics.* Humana Press, Totowa, New Jersey, 1999.

Jones KL. *Smith's Recognizable Patterns of Human Malformation*, 6th edn. Saunders, London, 2006.

Liehr T, Pfeiffer RA, Trautmann U. Typical and partial cat eye syndrome: identification of the marker chromosome by FISH. Clin Genet 1992; 42: 91–96.

McDermid HE, Duncan AM, Brasch KR, et al. Characterization of the supernumerary chromosome in cat eye Syndrome. Science 1986; 232: 646.

McTaggart KE, Budarf ML, Driscoll DA, et al. Cat eye syndrome chromosome breakpoint clustering: identification of two intervals also associated with 22q11 deletion syndrome breakpoints. Cytogenet Cell Genet 1998; 81: 222–228.

Mears AJ, el-Shanti H, Murray JC, et al. Minute supernumerary ring chromosome 22 associated with cat eye Syndrome: further delineation of the critical region. Am J Hum Genet 1995; 57: 667.

Rooney DE, Czepulkowski BH. *Human Cytogenetics. A Practical Approach.* Oxford University Press, New York, 1992.

Rosias P, Sijstermans J, Theunissen P, et al. Phenotypic variability of the cat eye syndrome, case report and review of the literature. Genet Counsel 2001; 12: 273–282.

Schinzel A. *Human Cytogenetics Database.* Oxford Medical Databases Series. Oxford University Press, Oxford, 1994.

Shaffer LG, McGowan-Jordan J, Schmid M (eds). *ISCN* 2013: *An International System for Human Cytogenetic Nomenclature.* Karger Publishers, Unionville, CT, 2013.

Tobias ES, Connor M, Ferguson Smith M. *Essential Medical Genetics*, 6th edn. Wiley-Blackwell, Oxford, 2011.

CHAPTER 9

Composite karyotypes and other complex rearrangements

9.1 Composite karyotypes

Composite karyotypes are used when there is some variation from cell to cell in the abnormalities present; therefore, a composite of all the abnormalities is listed in the nomenclature. This genetic heterogeneity is usually seen in neoplastic syndromes and not constitutional disorders, unless a patient has undergone some radiation or chemotherapeutic treatment when a constitutional chromosome analysis is performed. Many examples of composite karyotypes will be given in Part 2 on hematological malignancies. The discussion here will apply to only constitutional composite examples.

Only clonal abnormalities are reported in composites, and it may be possible to put the abnormalities in subclones that consist of related cell lines. For example, all the abnormalities in common in all the cells would be considered the stemline. The additional abnormalities that vary in the other cells are sidelines. However, if no such stemline or sidelines can be derived from the heterogeneity observed, a composite karyotype would be appropriate to use in reporting these abnormalities.

ISCN RULES FOR WRITING COMPOSITE KARYOTYPES

- First write the modal chromosome number or range of numbers, if the variation of the cells includes gains and/or losses. Since composites list all the abnormalities present, it may be possible to have loss and gain of the same chromosome listed in the nomenclature.
- When writing a range for the modal number, use the ~ symbol, e.g. 45~49 chromosomes. However, do not include random loss in the modal chromosome number.
- Write the sex designation next, including any losses, gains or structural rearrangements of the X and then the Y chromosomes.
- Next write the autosomal abnormalities in numerical order, starting with chromosome 1 and ending with chromosome 22.
- Write the gains and losses of a particular chromosome, followed by the structural abnormalities of that chromosome in alphabetical order.

Cytogenetic Abnormalities: Chromosomal, FISH and Microarray-Based Clinical Reporting, First Edition. Susan Mahler Zneimer.

- There is a choice of adding the number of cells present with each abnormality by adding brackets with the cell number after each abnormality. However, the bracket with the cell number is not required.
- At the end of the string of abnormalities, write a bracket with "cp" and the total number of cells examined.

For example, with 30 cells examined containing the following karyotypes:

8 cells showing 48,XX,+8,der(13;21)(q10;q10),+21
7 cells showing 49,XX,+8,+16,+21
6 cells showing 48,XX,+8,der(13;21)(q10;q10),+16
5 cells showing 47,XX,+8,der(13;21)(q10;q10)
2 cells showing 46,XX,der(13;21)(q10;q10),+16,−20
2 cells showing 47,XX,−15,+16,+21

The composite designation would be written as:

46~49,XX,+8,der(13;21)(q10;q10),+16,+21[cp30]

or

46~49,XX,+8[26],der(13;21)(q10;q10)[21],+16[17],+21[17][cp30]

Note that the −15 and −20 abnormalities are not written in the nomenclature, since loss of two cells is not equivalent to a clonal abnormality and is considered random loss. Loss of a single chromosome must be seen in three or more cells to be considered clonal. Gain of a single chromosome needs to occur in only two or more cells to be clonal, as do structural rearrangements.

EXAMPLE REPORTING OF A COMPOSITE KARYOTYPE SEEN IN A FETAL DEMISE

ISCN Result: 46~49,XX,+8[26],der(13;21)(q10;q10)[21],+16[17],+21[17][cp30]

Interpretation

Chromosome analysis revealed an abnormal female chromosome complement in all cells examined with both structural and numerical chromosomal abnormalities. There was some variation from cell to cell in the abnormalities; therefore, a composite of all the abnormalities is listed in the nomenclature. The number of cells seen for each abnormality is seen in brackets.

The clonal abnormalities observed include:

- gain of chromosomes 8, 16 and 21
- a Robertsonian translocation between the long arms of chromosomes 13 and 21.

Autosomal trisomies are the most frequent type of cytogenetic abnormality observed in pregnancy loss and are usually sporadic. The risk of recurrence approximates 1% or the maternal age-associated risk, whichever is higher. Therefore, prenatal diagnosis is recommended for any subsequent pregnancies.

Given the familial origin of most Robertsonian translocations, chromosome analysis of this patient's first-degree biological relatives should be considered.

Genetic counseling is recommended.

9.2 Complex rearrangements

Complex abnormalities in a single cell line are seen in both constitutional and neoplastic disorders. Only constitutional complex rearrangements will be discussed here. For more information regarding complex rearrangements observed in neoplastic disorders, see Part 2.

Complex chromosomal rearrangements (CCRs) are defined as any structural rearrangement involving three or more chromosomal breaks and exchange of genetic material between at least two chromosomes. Different classifications of CCRs have been described.

- Two breaks in a single chromosome and a single break in one or more chromosomes. This would yield, for example, an interchromosomal insertion of genetic material or two translocations involving the same chromosome.
- A single break in three or more chromosomes. This would yield, for example, three or more chromosomes involved in a related translocation event, or two or more unrelated translocation events involving at least three chromosomes.
- At least three breaks in a single chromosome with at least one break in an additional chromosome. This would yield, for example, one chromosome with an inversion and a translocation that involves at least a second chromosome.

There are relatively few reports of CCRs in the literature, which implies that these events are quite rare. One report shows that in a study of 12,538 consecutive postnatal chromosome analyses, nine CCR cases were observed (<0.1%), accounting for 1.0% of the total number of abnormal cytogenetic results. Multiple, but simple, two-way translocations account for the majority of CCRs reported, accounting for 20% of the total number of abnormal cytogenetic cases in this study.

Most CCRs in constitutional studies appear as balanced rearrangements or with individuals who have mild to severe phenotypic abnormalities. Individuals may be phenotypically normal with reproductive problems, but may have affected offspring with the same or an unbalanced form of the CCR. It appears that the genomic material itself and not the number of breaks and chromosomes involved may influence phenotype. Although CCRs have the potential for giving rise to unbalanced forms, a high rate of gamete loss may explain their scarcity in the literature.

Types of CCRs vary, as in the following:

- an inversion of one chromosome that also contains a translocation with another chromosome, for example, 46,XX,der(2)inv(2)(q14.1q23)t(2;3)(p15;q27) (Figure 9.1)
- a three-way translocation, for example, 46,XY,t(1;7;14)(q32.3;p21.2;q21.2) (Figure 9.2)
- an insertion of genetic material from one chromosome into another
- two translocation events, for example, 46,XY,der(7)t(7;13)(p11.2;q14.3)t(7;21)(q11.2;q22.3) (Figure 9.3).

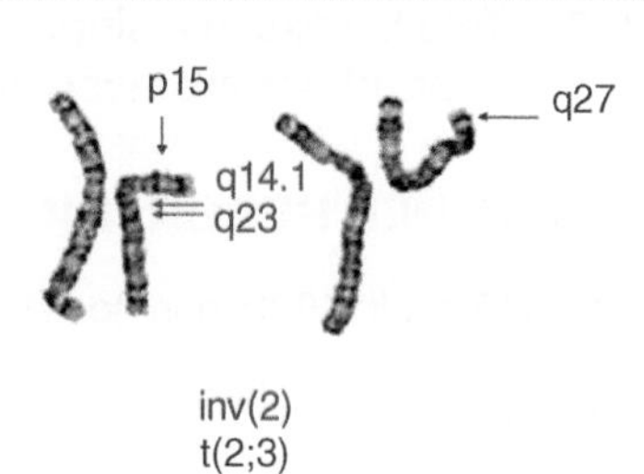

Figure 9.1 Partial karyotype showing a complex chromosome rearrangement with an inversion of chromosome 2 that also contains a translocation event with chromosome 3: 46,XX,der(2)inv(2)(q14.1q23)t(2;3)(p15;q27).

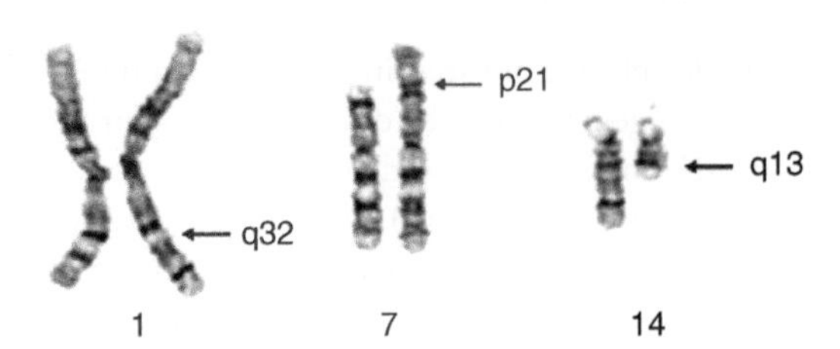

Figure 9.2 Partial karyotype showing a complex chromosome rearrangement involving a three-way translocation with one break per chromosome: 46,XY,t(1;7;14)(q32;p21;q13).

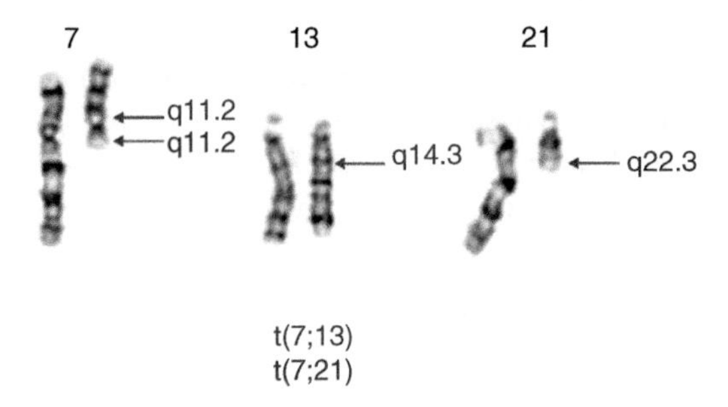

Figure 9.3 Partial karyotype showing a complex chromosome rearrangement with two translocation events involving chromosomes 7, 13 and 21: 46,XY,der(7)t(7;13)(p11.2;q14.3)t(7;21)(q11.2;q22.3).

ISCN RULES FOR REPORTING COMPLEX REARRANGEMENTS

- When writing the abnormalities present, start with sex chromosome abnormalities, followed by the autosomes in order from the lowest to largest chromosome number involved, i.e. chromosome 1→ chromosome 22.
- When numerical and structural abnormalities are present in the same chromosome homologs, then the numerical aberration goes before the structural aberration.
- Structural abnormalities of a single chromosome homolog are written in alphabetical order; for example, derivatives go before duplications since "de" is alphabetically before "du".

For example, for two abnormalities within a single chromosome, consisting of a deletion and an insertion, the ISCN would be: 46,XY,der(1)del(1)(q23.1q43)ins(1;2)(q23.1;q21.1q32.1).

Another example is an inversion of one chromosome that also contains a translocation event with another chromosome: 46,XX,der(2)inv(2)(q14.1q23)t(2;3)(p15;q27). See Figure 9.1.

9.2.1 Two breaks in a single chromosome and a single break in another chromosome

EXAMPLE REPORTING OF COMPLEX REARRANGEMENTS

ISCN Result: 46,XY,der(1)del(1)(q23.1q43)ins(1;2)(q23.1;q21.1q32.1)

Interpretation

Chromosome analysis revealed an abnormal male chromosome complement in all cells examined with an abnormal chromosome 1, consisting of a deletion within the long arm from bands q23.1 to q43, and an insertion of chromosome 2 long arm material from bands q21.1 to q32.1 into the long arm of chromosome 1 at band q23.1.

This abnormality results in an imbalance of genetic material of both chromosome 1 and chromosome 2.

Further testing with chromosomal microarray analysis is suggested to identify possible gene disruptions and to clarify the chromosome regions involved.

Chromosome analysis of the parental chromosomes is indicated to determine a potential origin of this translocation.

Genetic counseling is recommended.

EXAMPLE REPORTING OF AN INVERSION OF ONE CHROMOSOME THAT ALSO CONTAINS A TRANSLOCATION EVENT WITH ANOTHER CHROMOSOME

ISCN Result: 46,XX,der(2)inv(2)(q14.1q23)t(2;3)(p15;q27)

Interpretation

Chromosome analysis revealed an abnormal female chromosome complement in all cells examined with an abnormal chromosome 2, which contains a translocation in the short arm with the long arm of chromosome 3, and a paracentric inversion of chromosome 2 in the long arm.

Both these abnormalities appear balanced; however, genomic microarray analysis may prove informative for further characterization of this rearrangement. Recent literature suggests that imbalances at the breakpoints or elsewhere in the genome may be found in approximately 30–40% of individuals with a cytogenetically balanced rearrangement and an abnormal phenotype.

Chromosome analysis of the parental chromosomes is indicated to determine a potential origin of this translocation.

Genetic counseling is recommended.

9.2.2 A single break in three chromosomes

ISCN RULES FOR WRITING A THREE-WAY TRANSLOCATION

- First write the modal chromosome number, followed by the sex designation. However, if an X or Y chromosome is involved in the translocation, then the intact sex chromosome is written, followed by a comma. The sex chromosome in the translocation follows the comma, enclosed in parentheses. For example, 46,X,t(X;1;5).
- In three-way translocations the first chromosome described is a sex chromosome, if applicable. Otherwise, it is the autosome with the lowest number.
- The second chromosome described in the translocation is the chromosome that receives the translocated material of the first chromosome.
- Then the third chromosome described is the chromosome that receives the translocated material from the second chromosome, etc.

- It then becomes apparent that the first chromosome described has received the translocated material from the third chromosome.

For example: 46,XY,t(1;7;14)(q32;p21;q13). In this case, chromosome 1 is the lowest chromosome number and is listed first. At chromosome 7p21 is the long arm material from chromosome 1 (1q32 → qter). At chromosome 14q13 is the short arm material of chromosome 7 (p21 → pter). At chromosome 1q32 is the long arm material from chromosome 14 (q13 → qter). See Figure 9.2.

EXAMPLE REPORTING OF A THREE-WAY TRANSLOCATION

ISCN Result: 46,XY,t(1;7;14)(q32;p21;q13)

Interpretation

Chromosome analysis revealed an abnormal male chromosome complement in all cells examined with a three-way translocation between the long arm of chromosome 1, the short arm of chromosome 7 and the long arm of chromosome 14.

This abnormality appears balanced. However, genomic microarray analysis may prove informative for further characterization of this rearrangement. Recent literature suggests imbalances at the breakpoints or elsewhere in the genome may be found in approximately 30–40% of individuals with a cytogenetically balanced rearrangement and an abnormal phenotype.

Chromosome analysis of the parental chromosomes is indicated to determine a potential origin of this translocation.

Genetic counseling is recommended.

9.2.3 Two translocation events in a single cell line

ISCN RULES FOR WRITING TWO TRANSLOCATION EVENTS IN A SINGLE CELL LINE

- First write the modal chromosome number, followed by the sex designation. However, if an X or Y chromosome is involved in the translocation, then the intact sex chromosome is written, followed by a comma. The sex chromosome in the translocation follows the comma. For example, 46,X,t(X;5).
- With two translocation events, the first translocation described is a sex chromosome, if applicable. Otherwise, it is the autosome with the lowest number and its translocation partner.
- The second translocation described is the translocation that has the higher chromosome number, followed by its translocation partner.

For example: 46,XY,t(1;14)(q32;q13),t(7;8)(p21;q11.2). In this case, chromosome 1 is the lowest chromosome number and is listed first, followed by the translocation partner, chromosome 14. The second translocation is written next, using the same nomenclature rules for a typical two-way translocation, with the lower chromosome number (chromosome 7) written before the larger chromosome number (chromosome 8).

- When two translocations involve a single chromosome in common, then it becomes a derivative chromosome and each translocation follows the rules described above.

For example: 46,XY,der(7)t(7;13)(p11.2;q14.3)t(7;21)(q11.2;q22.3). This CCR contains two translocation events. One involves the short arm of one chromosome 7 with the long arm of chromosome 13, while the second translocation involves the same chromosome 7 with the long arm of chromosome 21. See Figure 9.3.

EXAMPLE REPORTING OF TWO TRANSLOCATION EVENTS IN A SINGLE CELL LINE

ISCN Result: 46,XY,der(7)t(7;13)(p11.2;q14.3)t(7;21)(q11.2;q22.3)

Interpretation

Chromosome analysis revealed an abnormal male chromosome complement in all cells examined with two translocation events, both involving a single chromosome 7 with two other chromosomes involved. The first translocation is between the short arm of chromosome 7 with the long arm of chromosome 13, and the second translocation is between the long arm of the same chromosome 7 with the long arm of chromosome 22.

These rearrangements appear balanced. However, genomic microarray analysis may prove informative for further characterization of this rearrangement. Recent literature suggests imbalances at the breakpoints or elsewhere in the genome may be found in approximately 30–40% of individuals with a cytogenetically balanced rearrangement and an abnormal phenotype.

Chromosome analysis of the parental chromosomes is indicated to determine a potential origin of this translocation.

Genetic counseling is recommended.

Bibliography

Chen CP, Chern SR, Lee CC, et al. Prenatal diagnosis of de novo t(2;18;14)(q33.1;q12.2;q31.2), dup(5)(q34q34), del(7)(p21.1p21.1), and del(10)(q25.3q25.3) and a review of the prenatally ascertained de novo apparently balanced complex and multiple chromosomal rearrangements. Prenat Diagn 2006; 26(2): 138–146.

Gardner RJM, Sutherland GR. *Chromosome Abnormalities and Genetic Counselling*. Oxford Monographs on Medical Genetics. Oxford University Press, Oxford, 2003.

Gersen S, Keagle M (eds). *Principles of Clinical Cytogenetics*. Humana Press, Totowa, New Jersey, 1999.

Giardino D, Corti C, Ballarati L, et al Prenatal diagnosis of a de novo complex chromosome rearrangement (CCR) mediated by six breakpoints, and a review of 20 prenatally ascertained CCRs. Prenat Diagn 2006; 26(6): 565–570.

Gorski JL, Kistenmacher ML, Punnett HH, Zackai EH, Emanuel ES. Reproductive risks for carriers of complex chromosome rearrangements: analysis of 25 families. Am J Med Genet 1988; 29: 247–261.

Kausch K, Haaf T, Kohler J, Schmid M. Complex chromosomal rearrangement in a woman with multiple miscarriages. Am J Med Genet 1988; 31: 415–420.

Kleczkowska A, Fryns JP, van den Berghe H. Complex chromosomal rearrangements (CCR) and their genetic consequences. J Genet Hum 1982; 30(3): 199–214.

Kousseff BG, Papenhausen P, Essig Y-P, Torres M. Complex chromosome rearrangement with ankyloblepharon filiforme adnatum. J Med Genet 1993; 30: 167–170.

Lurie IW, Wulfsberg EA, Prabhakar G, Rosenblum-Vos LS, Supovitz KR, Cohen MM. Complex chromosomal rearrangements: some breakpoints may have cellular adaptive significance. Clin Genet 1994; 46: 244–247.

Patsalis PC. Complex chromosomal rearrangements. Genet Counsel 2007; 18(1): 57–69.

Pellestor F, Anahory T, Lefort G, et al. Complex chromosomal rearrangements: origin and meiotic behavior. Hum Reprod Update 2011; 17(4): 476–494.

Rooney DE, Czepulkowski BH. *Human Cytogenetics. A Practical Approach.* Oxford University Press, New York, 1992.

Shaffer LG, McGowan-Jordan J, Schmid M (eds). *ISCN 2013: An International System for Human Cytogenetic Nomenclature.* Karger Publishers, Unionville, CT, 2013.

Tobias ES, Connor M, Ferguson Smith M. *Essential Medical Genetics*, 6th edn. Wiley-Blackwell, Oxford, 2011.

CHAPTER 10

Sex chromosome abnormalities

Sex chromosome abnormalities as a whole are the most common chromosomal abnormalities seen in liveborns, with a frequency of approximately 1 in 1000 male and female individuals. Both aneuploidy and structural rearrangements are common in both the X and Y chromosomes. Both loss and gain of the X and Y are observed at a high frequency when compared to autosomal aneuploidy. For the X chromosome in females, this is mainly due to X chromosome compensation with X inactivation. Meiotic Y chromosome loss has the same result as X chromosome loss, whose product is a 45,X conceptus.

Structural rearrangements of both the X and Y chromosomes are also common, giving rise to various phenotypic consequences based on the regions that are rearranged, lost or gained when unbalanced. The following are the most common abnormalities observed, divided into sections of aneuploidy and structural rearrangements and also separated by female and male phenotypes for each abnormality.

10.1 X chromosome aneuploidies – female phenotypes

Aneuploidy of the X chromosome in females results in monosomy X, trisomy X or gains of more than one X chromosome. In this section, common disease associations with X chromosome loss and gain are discussed.

10.1.1 45,X – Turner syndrome and its variants

SPECIFIC FEATURES OF TURNER SYNDROME AND ITS VARIANTS

The clinical manifestations of Turner syndrome and its variants include newborns with large amounts of neck skin, resulting in webbing of the neck, and lymphedema. Diagnoses made in later childhood years include short stature, primary amenorrhea, broad chest, wide-spaced nipples, low hairline and webbed neck. Hypopigmented nevi and hypoplasia of the nails are seen in a proportion of girls. One primary feature of Turner syndrome is sterility, as the ovaries begin to degenerate *in utero* and by birth they are streaked and result in failure to develop secondary sexual characteristics. Another major feature is cardiac defects, the most common of which is a bicuspid aortic valve, but can also include coarctation of the aorta, atrial septal defect and aortic stenosis.

Cytogenetic Abnormalities: Chromosomal, FISH and Microarray-Based Clinical Reporting, First Edition. Susan Mahler Zneimer.

Mosaicism for a normal cell line with a 45,X cell line may decrease the severity of defects. Current treatments for some of these defects include sex hormone replacement for the development of secondary sexual characteristics and growth hormone for increased height development. Fertilization and childbirth have been achieved with *in vitro* fertilization using donor oocytes.

The incidence of Turner syndrome is approximately 1 in 5000 female births. The rate of Turner syndrome conceptuses is much higher; however, it is estimated that over 99% of fetuses are spontaneously aborted. The recurrence risk of Turner syndrome is no greater than the general population.

Due to the risk of gonadoblastoma in cases of Turner syndrome with cryptic Y chromosome mosaicism, the American College of Medical Genetics (ACMG) recommends fluorescence *in situ* hybridization (FISH) analysis with X and Y probes if a 30-cell analysis reveals an apparently non-mosaic 45,X karyotype.

ISCN RULES FOR REPORTING MONOSOMY X

- First write the modal chromosome number of 45.
- Then write the sex chromosome complement, which is "X" only, without any further description, such as 0 or O: 45,X.

For any loss or gain of constitutional sex chromosomes, no minus or plus signs are used.

EXAMPLE REPORTING OF 45,X – TURNER SYNDROME WITH 20–30 CELLS ANALYZED

ISCN Result: 45,X

Interpretation

Chromosome analysis revealed an abnormal female chromosome complement in all cells examined with a single X chromosome, resulting in monosomy X. This result is consistent with the clinical diagnosis of Turner syndrome.

Note: Due to the risk of gonadoblastoma in cases of Turner syndrome with cryptic Y chromosome mosaicism, the ACMG recommends fluorescence *in situ* hybridization (FISH) analysis with X and Y probes if a 30-cell analysis reveals an apparently non-mosaic 45,X karyotype.

Genetic counseling is recommended.

ISCN RULES FOR REPORTING MOSAIC MONOSOMY X WITH A NORMAL FEMALE CELL LINE

- First write the modal chromosome number of 45.
- Then write the sex chromosome complement, which is "X" only, without any further description, such as 0 or O: 45,X.
- Then write the number of cells in brackets, followed by a slash.
- Then write 46,XX, followed by the number of cells in brackets.

With one abnormal cell line and one normal cell line, always write the abnormal cell line first.

EXAMPLE REPORTING OF MOSAIC 45,X/46,XX – WITH MORE THAN 10% 45,X CELLS

ISCN Result: 45,X[# cells]/46,XX[# cells]

Interpretation

Cytogenetic analysis revealed two cell lines present. The first cell line, seen in # cells examined, showed an abnormal female chromosome complement with loss of one X chromosome, resulting in monosomy X. The remaining # cells showed a 46,XX, normal female karyotype.

This mosaic chromosome complement is consistent with the diagnosis of variant Turner syndrome.

Genetic counseling is indicated.

EXAMPLE REPORTING OF MOSAIC 45,X/46,XX – WITH LESS THAN 10% 45,X CELLS IN AN ADULT

ISCN Result: 45,X[# cells]/46,XX[# cells]

Interpretation

Cytogenetic analysis revealed two cell lines present. The first cell line, seen in # cells examined, showed a female chromosome complement with loss of one X chromosome, resulting in monosomy X. The remaining # cells showed a 46,XX, normal female karyotype.

X chromosome aneuploidy of 10% or less in peripheral blood is not unusual in stimulated lymphocyte cultures from females of reproductive age, and is not associated with an increased risk for recurrent miscarriages or chromosomally abnormal liveborn offspring. It is generally considered a normal characteristic of aging with no known reproductive significance in phenotypically normal females.

ISCN RULES FOR REPORTING MOSAIC MONOSOMY X WITH A SECOND ABNORMAL CELL LINE

- First write the modal chromosome number of the cell line with the most number of cells (mainline).
- Then write the sex chromosome complement.
- Next write the number of cells in brackets, followed by a slash.
- Finally write the modal chromosome number of the cell line with the smaller number of cells involved, followed by the number of cells in brackets.

With two abnormal cell lines, always write the abnormal cell line with the most number of cells first, followed by the cell line with the smaller number of cells.

If there are two abnormal cell lines and a normal cell line, always write the normal cell line last, regardless of the number of cells involved.

For example: 45,X[39]/47,XXX[11]
47,XXX[19]/45,X[11]/46,XX[20]

EXAMPLE REPORTING OF MOSAIC TURNER SYNDROME – 45,X/47,XXX

ISCN Result: 45,X[# cells]/47,XXX[# cells]

Interpretation

Cytogenetic analysis revealed two abnormal cell lines present. The first cell line, seen in # cells examined, showed an abnormal female chromosome complement with loss of one X chromosome, resulting in monosomy X. The second cell line, seen in # cells examined, showed an abnormal female chromosome complement with gain of one copy of the X chromosome.

These findings are consistent with the diagnosis of variant Turner syndrome. Females with this mosaic karyotype may be expected to have increased risk of infertility and taller mean adult height.

Genetic counseling is recommended.

EXAMPLE REPORTING OF MOSAIC TURNER SYNDROME – 45,X/46,X,+mar

ISCN Result: 45,X[# cells]/46,X,+mar[# cells]

Interpretation

Cytogenetic analysis revealed two abnormal cell lines present. The first cell line, seen in # cells examined, showed an abnormal female chromosome complement with loss of one X chromosome, resulting in monosomy X. The second cell line, seen in # cells examined, showed an abnormal female chromosome complement with one copy of the X chromosome and one small unidentified marker chromosome, which is presumably either X or Y chromosome material.

These findings are consistent with the diagnosis of variant Turner syndrome. Females with this mosaic karyotype may be expected to have increased risk of infertility and taller mean adult height.

Note: Due to the risk of gonadoblastoma in cases of Turner syndrome with cryptic Y chromosome mosaicism, fluorescence *in situ* hybridization (FISH) analysis with X and Y probes is suggested.

Genetic counseling is recommended.

10.1.2 47,XXX syndrome

SPECIFIC FEATURES OF 47,XXX SYNDROME

The incidence of 47,XXX syndrome is approximately 1 in 1000 female births. Advanced maternal age increases the frequency of this syndrome. The recurrence risk of 47,XXX is no greater than the general population. Most females appear phenotypically normal, and have normal fertility, with the only observed manifestation being mild learning defects.

In a small proportion of individuals, more than one extra X chromosome is observed. For example, 48,XXXX, 49,XXXXX and 50,XXXXXX karyotypes have been reported. Trisomy, tetrasomy and quintasomy X have been shown to have more severe phenotypic abnormalities, including significant learning disabilities.

ISCN RULES FOR REPORTING GAIN OF X

- First write the modal chromosome number of 47.
- Then write the sex chromosome complement, which is "XXX", without any further description.

For any loss or gain of constitutional sex chromosomes, no minus or plus signs are used.

EXAMPLE REPORTING OF 47,XXX

ISCN Result: 47,XXX

Interpretation

Chromosome analysis revealed an abnormal female chromosome complement in all cells examined with the presence of an extra X chromosome.

The phenotype associated with this karyotype is variable. While most 47,XXX females are clinically normal, their IQ may be lower than that of their siblings with a small risk of mild-to-moderate developmental problems in motor, speech or language development. Some reports have noted premature ovarian failure; however, pubertal development is normal and most have normal fertility. The majority of individuals are physically normal as adults and have moderately tall stature.

Genetic counseling is recommended.

EXAMPLE REPORTING OF MOSAIC 47,XXX/46,XX – WITH LESS THAN 10% 47,XXX CELLS IN AN ADULT

ISCN Result: 47,XXX[# cells]/46,XX[# cells]

Interpretation

Cytogenetic analysis revealed two cell lines present. The first cell line, seen in # cells examined, showed a female chromosome complement with gain of one X chromosome. The remaining # cells showed a 46,XX, normal female karyotype.

X chromosome aneuploidy of 10% or less in peripheral blood is not unusual in stimulated lymphocyte cultures from females of reproductive age, and is not associated with an increased risk for recurrent miscarriages or chromosomally abnormal liveborn offspring. It is generally considered a normal characteristic of aging with no known reproductive significance in phenotypically normal females.

10.1.3 Gain of more than one copy of the X chromosome

Gains of the X chromosome are also referred to as X chromosome polysomy. Each additional X chromosome will show an increase in some of the phenotypic features or severity of the features manifested. The most common clinical manifestations include mental impairment (from mild to moderate), speech delay, hypertelorism, epicanthal folds, micrognathia and hypoplasia of the mid-face. In addition, stature may be above average and incomplete secondary sexual development may occur, with small breasts, scant pubic and axillary hair, and gonadal dysgenesis. Behavioral problems may also occur. Infertility or reduced fertility may be present, but there are reports of normal fertility and normal offspring with mothers carrying additional X chromosomes. With the presence of five or more X chromosomes, other features including microcephaly, congenital heart and renal anomalies may be present.

10.2 X and Y chromosome aneuploidies – male phenotypes

This section discusses diseases associated with the male phenoytpes and mosaic cell lines with loss of the X chromosome.

10.2.1 47,XXY – Klinefelter syndrome

SPECIFIC FEATURES OF KLINEFELTER SYNDROME

The incidence of Klinefelter syndrome is 1 in 1000 male births. Most males are only diagnosed in later adolescent age with the lack of normal secondary sexual development or at adult age due to infertility. Klinefelter syndrome is the most common cause of hypogonadism and male infertility. The most common phenotypic features include low levels of testosterone leading to small testes, gynecomastia and lack of other secondary sexual characteristics. However, males are usually tall, at the 75th percentile or higher. Current treatments to mitigate the phenotypic problems include testosterone replacement therapy in early adolescence to improve secondary sexual characteristics. Some fertility has been achieved with testicular sperm aspiration and intracytoplasmic sperm injection. Although learning difficulties and behavioral problems may be present, they are usually mild.

The recurrence risk of Klinefelter increases with advanced maternal age and with azoospermic males and, to a lesser degree, with males with significant learning difficulties.

ISCN RULES FOR REPORTING GAIN OF X OR Y IN A MALE

- First write the modal chromosome number of 47.
- Then write the sex chromosome complement, which is "XXY" or "XYY", without any further description.

For any loss or gain of constitutional sex chromosomes, no minus or plus signs are used.

EXAMPLE REPORTING OF 47,XXY – KLINEFELTER SYNDROME

ISCN Result: 47,XXY

Interpretation

Chromosome analysis revealed an abnormal male chromosome complement in all cells examined with the presence of an extra X chromosome. This result is consistent with the clinical diagnosis of Klinefelter syndrome.

Genetic counseling is recommended.

10.2.2 47,XYY syndrome

SPECIFIC FEATURES OF 47,XYY SYNDROME

The incidence of 47,XYY syndrome is 1 in 1000 male births. There is a large degree of variability of phenotypic manifestations, ranging from asymptomatic males to males with a high level of learning disabilities and criminal behavior. IQs have been reported to be 10–15 points lower than those of other siblings. Males tend to be tall but are otherwise physically normal. The recurrence risk is no greater than the general population.

EXAMPLE REPORTING OF 47,XYY

ISCN Result: 47,XYY

Interpretation

Chromosome analysis revealed an abnormal male chromosome complement in all cells examined with the presence of an extra Y chromosome.

The phenotype associated with this karyotype is variable. While most 47,XYY males are clinically normal, their IQ may be lower than that of their siblings. Patients with a 47,XYY complement have an increased risk for mild-to-moderate developmental problems in motor, speech or language development. The majority are physically normal as adults and have moderately tall stature. Pubertal development is normal, and most have normal fertility.

Genetic counseling is recommended.

10.2.3 45,X/46,XY

SPECIFIC FEATURES OF 45,X/46,XY

This mosaicism shows a range of clinical features from a female with classic Turner syndrome to a newborn with ambiguous genitalia to a normal male or a male with phenotypic and/or fertility problems.

With the presence of the Y chromosome, there is a risk of gonadoblastoma in females of approximately 15–20% and surgical removal of the gonads is recommended. Clinical features of a female with this mosaicism may include normal or ambiguous or abnormal external genitalia. These females may have features of Turner syndrome, ranging from mild phenotypic manifestations to a normal female phenotype.

Clinical features of a male with this mosaicism may include normal genitalia, which is the most common phenotype with these individuals after birth, or may involve abnormal genitalia with hypospadias, micropenis and abnormal scrotum.

ISCN RULES FOR REPORTING MOSAIC MONOSOMY X WITH A NORMAL MALE CELL LINE

- First write the modal chromosome number of 45.
- Then write the sex chromosome complement, which is "X" only, without any further description, such as 0 or O: 45,X.

- Next write the number of cells in brackets, followed by a slash.
- Finally write 46,XY, followed by the number of cells in brackets.

With one abnormal cell line and one normal cell line, always write the abnormal cell line first.

EXAMPLE REPORTING OF MOSAIC 45,X/46,XY

ISCN Result: 45,X[# cells]/46,XY[# cells]

Interpretation

Cytogenetic analysis revealed two cell lines present. The first cell line, seen in # cells examined, showed one X chromosome and the loss of one sex chromosome. The remaining # cells showed a 46,XY, normal male karyotype.

The phenotype associated with this mosaicism ranges from infants with ambiguous genitalia to females with Turner syndrome stigmata, hypospadias and gonadal dysgenesis, as well as sterile male phenotypes. However, phenotypic features cannot be accurately predicted by this peripheral blood study alone, as mosaicism varies in different somatic tissues.

In a phenotypic female with a Y chromosome-bearing cell line, there is a significant risk for gonadoblastoma. Appropriate monitoring and management are indicated.

Genetic counseling is recommended.

10.2.4 Gain of additional X and/or Y chromosomes

The gain of one X and one Y chromosome (48,XXYY) is a common variant of Klinefelter syndrome, though not a common occurrence (1 in 50,000 males). As in Klinefelter syndrome, tall stature is common; hypogonadism with small testes and gynecomastia are also observed. There may also be mental impairment and psychosocial or behavioral problems.

With additional X or Y chromosomes, the severity of clinical features increases, with increased mental and behavioral problems in addition to gonadal manifestations. It is more common to add X chromosomes than Y chromosomes with sex chromosome polysomy.

10.3 X chromosome structural abnormalities

SPECIFIC FEATURES OF X CHROMOSOME ABNORMALITIES

X chromosome abnormalities may be the result of numerous abnormalities involving the X chromosome. Structural changes to the X chromosome will lead to deletions and duplications of regions of the X chromosome, which can result in a range of phenotypic manifestations. The most common abnormalities involving structural X chromosome changes include:

- X-X translocations
- X-Y translocations
- isochromosomes of the short or long arms
- deletions
- duplications
- rings.

X-X translocations will result in deletions and duplications of X chromosome material, as will isochromosomes of the X and ring X. Deletions of the X chromosome are significant depending on the chromosomal region lost. Since X inactivation occurs in all females with two or more X chromosomes, and many "housekeeping" genes are on the X chromosome, deletions of part of the X chromosome may have severe clinical consequences. The gene that controls X inactivation is the XIST (X inactive specific transcript) gene at the X inactivation center (XIC) at Xq13. Only the inactive X chromosome will express the XIST gene, in order to control the inactivation of the genes on that chromosome. However, recent studies have shown that there are genes on the inactive X that escape the inactivation process and may instead be expressed on both X chromosomes in a normal female.

The X inactivation process is important in structurally abnormal X chromosomes, because if the abnormal X does not contain the XIST gene, then inactivation of that chromosome is not possible and the abnormal X remains active, thereby inactivating the normal X, and increasing phenotypic manifestations in the individual.

10.3.1 Isochromosomes of the X chromosome

Isochromosomes of the X chromosome may be either of the short arm or the long arm: i(X)(p10) or i(X)(q10), respectively. These chromosomes result in duplications and deletions of whole arms of the X chromosome. For an isochromosome of the short arm, i(Xp), a karyotype of 46,X,i(X)(p10) in the non-mosaic form results in a duplication of the short arm (3 copies) and a deletion of the long arm (1 copy). The opposite is true for an isochromosome of the long arm: 46,X,i(X)(q10), in which there is a duplication of the long arm and a deletion of the short arm in all cells. Isochromosomes of the X are variants of Turner syndrome and exhibit many of the clinical features of Turner syndrome.

ISCN RULES FOR REPORTING A SINGLE CENTROMERE ISOCHROMOSOME X WHEN REPLACING A NORMAL X CHROMOSOME

- First write the modal chromosome number of 46, followed by X and then a comma.
- Then write "i" followed, in parentheses, by X.
- Next, in parentheses, write the breakpoint p10 for the short arm or q10 for the long arm duplicated.

For example, isochromosome X of the long arm would be: 46,X,i(X)(q10). Isochromosome X of the short arm would be: 46,X,i(X)(p10).

i(X)(q10)

EXAMPLE REPORTING OF i(X)(q10) (FIGURE 10.1)

ISCN Result: 46,X,i(X)(q10)

Interpretation

Chromosome analysis revealed an abnormal female chromosome complement in all cells examined with one normal X chromosome and an isochromosome of the long arm of the second X chromosome. This finding results in three copies of the long arm and one copy of the short arm of the X chromosome.

This chromosome complement is consistent with the diagnosis of variant Turner syndrome. Genetic counseling is recommended.

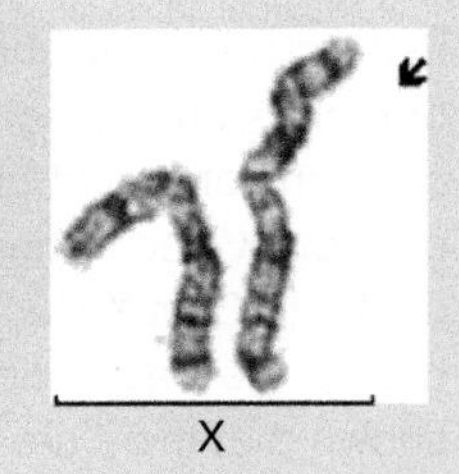

Figure 10.1 Partial karyotype showing an isochromosome of the long arm of the X chromosome: 46,X,i(X)(q10).

ISCN RULES FOR REPORTING MOSAIC 45,X WITH i(Xq)

- With two abnormal cell lines, always write the abnormal cell line with the most cells first, followed by the cell line with the smaller number of cells.
- If 45,X is seen in more cells, then first write the modal chromosome number of 45, followed by "X", followed by the number of cells in brackets, followed by a slash.
- Next write the modal chromosome number of 46, followed by X and then a comma.
- Then write "i" followed, in parentheses, by X.
- Finally write, in parentheses, the breakpoint p10 for the short arm or q10 for the long arm duplicated.

For example: 45,X[19]/46,X,i(X)(q10)[11] or 46,X,i(X)(q10)[19]/45,X[11].

EXAMPLE REPORTING OF MOSAIC 45,X/i(Xq)

ISCN Result: 45,X[# cells]/46,X,i(X)(q10)[# cells]

Interpretation

Cytogenetic analysis revealed two cell lines present. The first cell line, seen in # cells, showed loss of one X chromosome, resulting in monosomy X. The remaining # cells showed an isochromosome of the long arm of the X chromosome.

This mosaic chromosome complement is consistent with the diagnosis of variant Turner syndrome.

Genetic counseling is recommended.

i(X)(p10)

EXAMPLE REPORTING OF i(X)(p10)

ISCN Result: 46,X,i(X)(p10)

Interpretation

Chromosome analysis revealed an abnormal female chromosome complement in all cells examined with one normal X chromosome and an isochromosome of the short arm of the second X chromosome. This finding results in three copies of the short arm and one copy of the long arm of the X chromosome.

This chromosome complement is consistent with the diagnosis of variant Turner syndrome.

Genetic counseling is recommended.

EXAMPLE REPORTING OF MOSAIC i(X)(p10)

ISCN Result: 45,X[# cells]/46,X,i(X)(p10)[# cells]

Interpretation

Cytogenetic analysis revealed two cell lines present. The first cell line, seen in # cells, showed loss of one X chromosome, resulting in monosomy X. The remaining # cells showed an isochromosome of the short arm of the X chromosome.

This mosaic chromosome complement is consistent with the diagnosis of variant Turner syndrome.

Genetic counseling is recommended.

Three related abnormal cell lines

EXAMPLE REPORTING OF THREE RELATED ABNORMAL CELL LINES DETECTED

ISCN Result: 45,X[14]/47,XXX[10]/46,X,i(X)(p10)[2]/46,XX[2]

Interpretation

Chromosome analysis revealed three related abnormal cell lines and a normal female chromosome complement present. The first cell line, seen in 14 cells examined, showed loss of one X chromosome, resulting in monosomy X. The second cell line, seen in 10 cells, showed gain of one X chromosome. The third cell line, seen in two cells, showed an isochromosome of the short arm of one X chromosome. The remaining two cells showed a normal female chromosome complement. These findings are consistent with the clinical diagnosis of variant Turner syndrome.

Females with this mosaic karyotype may be expected to have increased risk of infertility and taller mean adult height. Other Turner syndrome manifestations include short stature, primary amenorrhea, broad chest, wide-spaced nipples, low hairline and webbed neck.

Note: Due to the risk of gonadoblastoma in cases of Turner syndrome with cryptic Y chromosome mosaicism, fluorescence *in situ* hybridization (FISH) analysis with X and Y probes is suggested.

Genetic counseling is recommended.

10.3.2 Deletions of the X chromosome

SPECIFIC FEATURES OF del(Xp) AND del(Xq)

X chromosomes may have deletions in either the short or long arms and they may be either interstitial or terminal. Some deletions may be inherited but often the size of the deletion will determine the viability of the individual and phenotypic consequences. Reports of gonadal mosaicism exist in which a child shows a deletion in either the mosaic form or as the only cell line, while the mother shows an apparently normal female chromosome complement. However, the mother may actually have a low level of mosaicism for the abnormality in the gonads alone, in which the deletion is not present or present in only a small proportion of cells in other tissue types.

Deletions of the short arm may also show other X chromosome abnormalities, such as duplications, mosaicism for a 45,X cell line or other abnormal cell lines, which will complicate the clinical phenotype and make it difficult to assess the overall outcome of features when detected in a child.

Gene defects resulting from specific deletions may give rise to specific clinical features, such as mutations or loss of the SHOX gene on Xp22.3, which are associated with short stature; X-linked ichthyosis; Kallman syndrome, which is associated with anosmia and hypogonadism; and X-linked mental retardation syndrome and chondrodysplasia punctata, which are associated with skeletal dysplasias. A variety of other X-linked disorders also exist, as the X chromosome contains thousands of essential functional genes in humans which, when deleted, may have clinical consequences.

Generally, when a deletion of the X chromosome is seen in females, the deleted X will be preferentially inactivated, lessening the severity of the clinical manifestations. Also, this X inactivation will usually not confer the phenotype of an X-linked recessive disorder, and allow the intact second X chromosome to be active. Xp deletions usually have short stature and features of Turner syndrome, especially when the pseudoautosomal region is deleted, in part due to the loss of the SHOX gene. When the deletion consumes the whole short arm, then complete ovarian failure is usually observed whereas, with partial short arm deletions, such as Xp21 to the terminus, only premature ovarian failure is observed.

Long arm X chromosome deletions have variable phenotypic manifestations in females. They usually do not exhibit Turner syndrome features as strongly as Xp deletions, but many females do show short stature, gonadal dysgenesis and premature ovarian failure. The severity of these manifestations increases with larger deletions.

Males with X chromosome deletions are usually not viable, unless the deletion region is small or in the mosaic form with a normal male cell line. When viable, a risk of an X-linked recessive disorder is present.

Deletions of Xp

EXAMPLE REPORTING OF del(Xp)

ISCN Result: 46,X,del(X)(p11.2)

Interpretation

Chromosome analysis revealed an abnormal female chromosome complement in all cells examined with one normal X chromosome and a deletion of the short arm of the second X chromosome. This chromosome complement is consistent with the diagnosis of variant Turner syndrome.

Females with deletions of the short arm of the X chromosome are often less affected than those with monosomy X. There is a higher percentage of patients with greater height and decreased chance of menstruation and fertility. Deletions involving Xp21-pter are generally associated with short stature.

Genetic counseling is recommended.

Deletions of Xq

EXAMPLE REPORTING OF del(Xq)

ISCN Result: 46,X,del(X)(q11.2)

Interpretation

Chromosome analysis revealed an abnormal female chromosome complement in all cells examined with one normal X chromosome and a deletion of the long arm of the second X chromosome. This chromosome complement is consistent with the diagnosis of variant Turner syndrome.

Long arm X chromosome deletions have variable phenotypic manifestations in females. They usually do not exhibit Turner syndrome features as strongly as Xp deletions, but many females do show short stature, gonadal dysgenesis and premature ovarian failure.

Genetic counseling is recommended.

10.3.3 Ring X

SPECIFIC FEATURES OF RING X

Generally, rings of the X chromosome are mosaic with a normal cell line and are seen predominantly in females and rarely in males. Rings of the X chromosome will differ in the clinical phenotype depending on the size of the X chromosome material that is deleted and whether the XIST gene is present or absent. When the ring X in a female is large with only a small amount of material deleted from the telomeric ends of the short and long arms, then the phenotypic consequences are less severe. Usually the phenotype only includes variant Turner syndrome manifestations or may appear normal. In a female with a very small ring X that has lost the XIST gene, then the phenotype usually involves severe mental retardation and other clinical features.

When rings of the X chromosome are seen in males, they will necessarily need to be very large rings with a small amount of X chromosome material deleted or it will be unviable, due to the nature of males inheriting only a single X chromosome. However, the exception is when the ring X in a male is due to a supernumerary chromosome and thus has a 47,XY,+ r(X) karyotype, which may be either a small or large ring X with variable phenotypic consequences.

EXAMPLE REPORTING OF r(X)

ISCN Result: 46,X,r(X)

Interpretation

Chromosome analysis revealed an abnormal female chromosome complement in all cells examined with one normal X chromosome plus a small ring X chromosome, r(X), resulting in a deletion of part of the X chromosome.

The presence of a small ring X chromosome is associated with mental retardation and congenital malformations. This is due to the functional disomy that results from the failure of these chromosomes to undergo X inactivation.

Note: Due to the risk of gonadoblastoma in cases of Turner syndrome with cryptic Y chromosome mosaicism, fluorescence *in situ* hybridization (FISH) analysis with X and Y probes is suggested.

Genetic counseling is recommended.

10.4 Y chromosome structural abnormalities

SPECIFIC FEATURES OF Y CHROMOSOME ABNORMALITIES

Abnormalities of the Y chromosome may include deletions, isochromosomes, rings and dicentric chromosomes. Mosaicism for a 45,X cell line is observed often with an abnormal Y chromosome, which increases the variability of the phenotypic consequences. Also, a different proportion of abnormal cells in various tissues may exist, complicating the clinical picture of these individuals. One main distinction of Y chromosome abnormalities is whether the sex-determining (SRY) gene is present or absent. With the presence of the SRY gene, the phenotype will be male. When the SRY gene is missing, the phenotype will revert to a female. Loss of much of the short arm of the Y chromosome will lead to a variant of Turner syndrome phenotype.

Loss of long arm material of the Y chromosome in males, including the AZF genes, will result in infertility. If mosaicism is also present for a 45,X cell line, genitalia may be ambiguous or possibly female. Gonadal dysgenesis is common in Y chromosome abnormalities, as is a risk of gonadoblastoma, and should be a consideration in these cases.

10.4.1 Isochromosome of the Y short arm

If a small metacentric chromosome is present with a 45,X karyotype, it is important to identify whether this small chromosome is of X or Y origin. If it appears to be an isochromosome of the short arm of the Y chromosome, then further banding and FISH analysis are recommended. If C-banding shows the Y chromosome to have a single centromere region and no long arm heterochromatic material, then this is consistent with an isochromosome of the Y short arm. Patients with this chromosome complement show a broad range of phenotypic expression, ranging from individuals with ambiguous genitalia to normal-appearing males, as well as females with Turner syndrome features. In phenotypic males, this finding can be associated with small or abnormal testes, incomplete masculinization and short stature.

FISH results using an SRY probe should confirm the presence of the SRY gene for this karyotypic designation to be correct.

EXAMPLE REPORTING OF i(Yp) WITH THE SRY GENE CONFIRMED BY C-BANDING AND FISH

ISCN Result: 46,X,i(Y)(p10)

Interpretation

Chromosome analysis revealed an abnormal male chromosome complement in all cells examined with the presence of a small metacentric Y chromosome.

C-banding of the Y chromosome showed a single centromere region and no long arm heterochromatic material.

Fluorescent *in situ* hybridization (FISH) analysis was performed with the SRY gene probe, which is specific for the Y short arm region at Yp11.3 (see separate FISH report). The FISH results are consistent with the interpretation of the presence of an isochromosome of the Y short arm.

Patients with this chromosome complement show a broad range of phenotypic expression. Manifestations range from individuals with ambiguous genitalia to normal-appearing males, as well as females with Turner syndrome features. In phenotypic males, this finding can be associated with small or abnormal testes, incomplete masculinization and short stature.

Most cases of i(Yp) are *de novo* in origin. However, paternal blood chromosome studies are indicated to clarify whether this chromosome is inherited.

With the presence of the Y chromosome, there is a risk of gonadoblastoma in females, and surgical removal of the gonads is recommended. The risk of gonadoblastoma is approximately 15–20% in females with the presence of the Y chromosome.

Genetic counseling is recommended.

EXAMPLE REPORTING OF i(Yp) WITH THE SRY GENE WHEN NOT CONFIRMED BY C-BANDING OR FISH

ISCN Result: 46,X,i(Y)(p10)

Interpretation

Chromosome analysis revealed an abnormal male chromosome complement in all cells examined with the presence of a small metacentric Y chromosome.

C-banding and FISH studies are recommended to determine the extent of the Y euchromatic (coding DNA) region and if the SRY gene is present.

Patients with this chromosome complement show a broad range of phenotypic expression. Manifestations range from individuals with ambiguous genitalia to normal-appearing males, as well as females with Turner syndrome features. In phenotypic males, this finding can be associated with small or abnormal testes, incomplete masculinization and short stature.

Most cases of i(Yp) are *de novo* in origin. However, paternal blood chromosome studies are indicated to clarify whether this chromosome is inherited.

With the presence of the Y chromosome, there is a risk of gonadoblastoma in females, and surgical removal of the gonads is recommended. The risk of gonadoblastoma is approximately 15–20% in females with the presence of the Y chromosome.

Genetic counseling is recommended.

10.4.2 Y chromosome with satellites on the long arm

Y chromosomes with satellites on the long arm (Yqs) are not uncommon in the general male population and are generally considered a normal variant, without clinical consequences, especially when inherited from the father. Satellites on the short arm are also possible, but have rarely been reported. The satellites are most likely small arm material from an acrocentric chromosome, which was the result of a translocation event. This can be tested by doing silver staining to identify if a nuclear organizer region (NOR) is present. When satellites are seen on the Y chromosome, it is suggested that a chromosome study be performed on the father to determine whether it is inherited or *de novo* in origin. An inherited abnormality would not generally carry a risk of abnormalities, but there may be a risk if it is *de novo* in origin, especially if Y chromosome genes are interrupted. However, this possibility is remote.

EXAMPLE REPORTING OF Yqs

ISCN Result: 46,X,Yqs

Interpretation

Chromosome analysis revealed a male chromosome complement in all cells examined in which satellites are present on the terminus of the long arm of the Y chromosome. This finding is generally considered a heritable normal variant in the population with no clinical significance. Paternal blood chromosome analysis is suggested to confirm if this rearrangement is familial or *de novo* in origin.

Genetic counseling is recommended.

10.4.3 Inversion of the Y chromosome

Inversions of the Y chromosome may be either pericentric or paracentric in origin. Pericentric inversions are not uncommon in the general population and does not generally have any adverse clinical consequences. However, its prevalence in the population has not been determined so as to identify if this is a normal variant in the population or more rarely observed. These males usually do not have reproductive issues, conferring no risk of having an abnormal child. Very few paracentric inversions of the Y chromosome have been reported. When inherited from the father, they should confer no clinical consequence. Otherwise, clinical manifestations may be coincidental.

The exception to this apparently innocuous cytogenetic abnormality is when a clinically significant gene, such as the SRY or AZF, is disrupted in a *de novo* inversion. In such cases, ambiguous genitalia and other fertility and gonadal manifestations may be present. When clinical manifestations exist in an individual and show an inversion of the Y chromosome, paternal chromosome studies, and possibly molecular and SRY testing, may be required to help identify the origin of the abnormality present.

EXAMPLE REPORTING OF inv(Y) (FIGURE 10.2)

ISCN Result: 46,X,inv(Y)(p11.2q11.23)

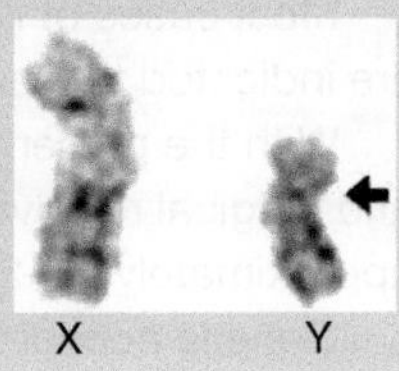

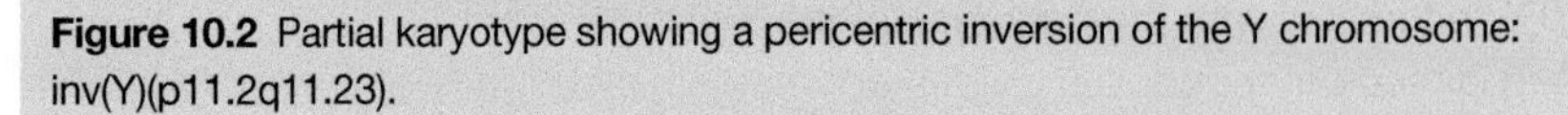

Figure 10.2 Partial karyotype showing a pericentric inversion of the Y chromosome: inv(Y)(p11.2q11.23).

Interpretation

Chromosome analysis revealed a male chromosome complement in all cells examined with a pericentric inversion of the Y chromosome. This finding is generally considered a heritable normal variant in the population with no clinical significance, occurring in approximately 1 in 1000 males. Paternal blood chromosome analysis is suggested to confirm if this rearrangement is familial or *de novo* in origin.

Genetic counseling is recommended.

10.4.4 Isochromosomes of the Y chromosome with 45,X

SPECIFIC FEATURES OF THE Y ISOCHROMOSOME

Isochromosomes and isodicentric chromosomes of the Y vary clinically in whether the SRY gene is present or absent, and the vast majority of cases are mosaic with the presence of a 45,X cell line. The resulting deletions and duplications of isochromosomes will consequently confer various phenotypic manifestations. With the loss of the SRY gene, female development will occur and may result in Turner syndrome phenotypic abnormalities. Gonadoblastoma may also be a risk factor and needs to be considered in writing cytogenetic reports. Another factor is whether the AZF genes on Yq12 are present or absent. Loss of AZF loci is associated with infertility. Therefore, phenotypic features may include gonadal dysgenesis, short stature and infertility. Further studies for the presence or absence of SRY, AZF and other Y sex-determining genes are suggested to help clarify the significance of the cytogenetic abnormality detected.

EXAMPLE REPORTING OF 45,X/46,X,idicY (FIGURE 10.3)

ISCN Result: 45,X[# cells]/46,X,idic(Y)(p11.2q12)

Interpretation

Cytogenetic analysis revealed two cell lines present. The first cell line, seen in # cells, showed loss of one X chromosome, resulting in monosomy X. The remaining # cells showed one normal X chromosome, plus an isodicentric chromosome of the short arm and a portion of the long arm of the Y chromosome.

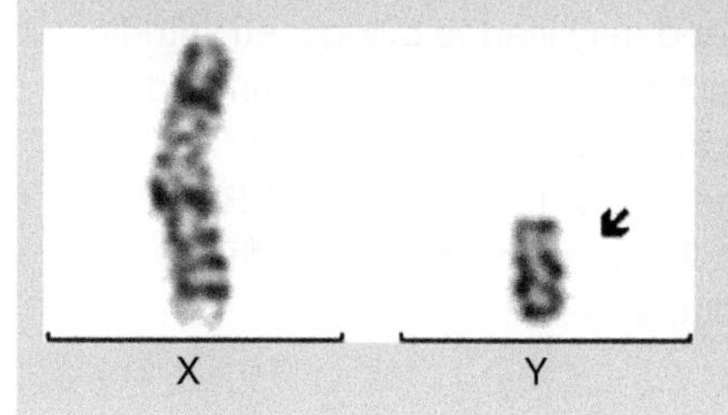

Figure 10.3 Partial karyotype showing an isodicentric Y chromosome: 46,X,idic(Y)(q11.23).

This mosaic chromosome complement is associated with a variable phenotypic spectrum that includes males with normal genitalia and short stature, and may include infertility due to azoospermia. The probability of fertility and short stature may be refined by molecular genetic analysis of the abnormal Y chromosome.

Increased surveillance, including periodic testicular ultrasound and biopsy, is indicated in males with abnormal genitalia.

Genetic counseling is recommended.

10.5 46,XX males and 46,XY females

10.5.1 46,XX male

SPECIFIC FEATURES OF 46,XX MALES

An apparently normal XX, female karyotype in males is due to various causes. One is an exchange of Xp/Yp material during meiosis crossing over, resulting in the presence of the SRY gene on the X chromosome, which is cryptic with cytogenetic analysis. This mechanism is seen in the majority of XX males. These males typically have the SRY gene from Yp11.2 inserted into the pseudoautosomal region of the short arm of the X chromosome which is homologous to the Yp chromosome at Xp22.3. If this region of chromosomal exchange takes place and is visible cytogenetically, then the ISCN would not be XX, but rather der(X)t(X;Y)(p22.3;p11.2). This translocation is discussed later in the chapter.

A second cause is due to cryptic mosaicism for an XY or XXY cell line with the majority of cells seen as XX, also resulting in an SRY-positive male. If mosaicism is detected cytogenetically, then the ISCN would be designated as XX/XY or XX/XXY and distinguished from a complete XX male karyotype.

A third cause is described as SRY-negative XX males, in which male development is present in various degrees with either true hermaphroditism or the involvement of other male-determining factors, such as SOX9 on chromosome 17, or other sex-determining genes.

In the cases of SRY-positive XX males, the clinical manifestations are similar to Klinefelter syndrome, since both Klinefelter syndrome and XX males with SRY contain the SRY gene and two X chromosomes. However, the phenotype of these XX males may not involve the same tallness or diminished intelligence as Klinefelter males, but do contain testicular dysfunction and azoospermia.

XX males who do not have the SRY gene have variable clinical manifestations, depending on the gene(s) involved in developing male characteristics. Certain features have been described with the absence of SRY, including intellectual impairment, gynecomastia, hypogenitalia and testicular atrophy, with varying degrees of masculinity.

Many XX males are not diagnosed until there is a lack of pubertal development. However, some phenotypic manifestations at birth or in early childhood may lead to cytogenetic studies in prepubertal males to identify sex chromosome abnormalities. It is certainly recommended to have SRY studies performed on these males to identify the origin of this genotype/phenotype discrepancy. It may also require paternal cytogenetic studies when the SRY gene is present in an XX male to see if it is an inherited or sporadic event. In XX males with the absence of SRY, other genetics studies are recommended to ascertain if other sex-determining genes are present.

For true hermaphroditism the karyotype is usually 46,XX, though there is the presence of both ovarian and testicular gonads. Other chromosome complements exist, however, including true 46,XX/46/XY cell lines, or 46,XY in a few cases. The lack of mosaicism may reflect the inability to see a second cell line in the tissue sample tested or a low level of mosaicism undetected in 20–50 cells. Other tissue types may be necessary to identify a second cell line, or FISH analysis with X and Y probes may be prudent to identify mosaicism and the presence of SRY. Uniparental disomy may also be the cause of XX male true hermaphrodites, as is discussed in Chapter 12. True hermaphrodites with a 46,XX karyotype may also be caused by the presence of male-determining genes that do not include SRY, as discussed above.

ISCN RULES FOR REPORTING AN XX MALE

If a normal chromosome complement is seen in XX males, then the ISCN nomenclature is written as a normal result. However, the report should contain a phrase that implies the genotype does not match the phenotype, to alert the reader that there is an abnormality present.

For example: 46,XX in a phenotypic male.

EXAMPLE REPORTING OF 46,XX MALES WITH POSITIVE FISH ANALYSIS FOR THE SRY GENE

ISCN Result: 46,XX in a phenotypic male

Interpretation

Chromosome analysis revealed a normal female chromosome complement in all cells examined. This finding suggests a diagnosis of an XX male, which usually results in a phenotype similar to that of Klinefelter syndrome. Stature and intelligence are generally closer to that of normal males.

Fluorescent *in situ* hybridization (FISH) testing (reported separately) reveals the presence of the SRY gene on one of the X chromosomes, confirming the male phenotype in this patient.

Genetic counseling is recommended.

EXAMPLE REPORTING OF 46,XX MALES WITH NEGATIVE FISH ANALYSIS FOR THE SRY GENE

ISCN Result: 46,XX in a phenotypic male

Interpretation

Chromosome analysis revealed a normal female chromosome complement in all cells examined. This finding suggests a diagnosis of an XX male, which usually results in a phenotype similar to that of Klinefelter syndrome. Fluorescent *in situ* hybridization (FISH) testing (reported separately) reveals no evidence of a SRY gene signal. SRY-negative 46,XX males are at increased risk for ambiguous genitalia and true hermaphroditism.

These defects presumably arise through inappropriate activation of the testis-determining cascade. Recurrence risk is dependent upon the genes and etiology involved. Molecular genetic analysis to identify this gene(s) is recommended. Ultrasound examination and examination of this patient by a clinical geneticist or endocrinologist are also suggested.

Genetic counseling is recommended.

EXAMPLE REPORTING OF 46,XX MALES WITH NO FISH ANALYSIS FOR THE SRY GENE PERFORMED

ISCN Result: 46,XX in a phenotypic male

Interpretation

All cells examined show a normal female chromosome complement. This finding suggests a diagnosis of an XX male, which usually results in a phenotype similar to that of Klinefelter syndrome. Stature and intelligence are generally closer to that of normal males.

Molecular or FISH studies for the SRY gene are suggested.

Genetic counseling is recommended.

10.5.2 46,XY female

SPECIFIC FEATURES OF 46,XY FEMALES

Complete sex reversal of a 46,XY female may be due to single gene defects or an inherited chromosome with a deletion of the SRY gene. Gene defects can vary including, for example, the SOX9 gene on chromosome 17q, resulting in campomelic dysplasia with skeletal dysplasia and female genital tract development. Another gene defect is androgen insensitivity syndrome with the loss of the androgen receptor gene on chromosome Xq12, resulting in a female with primary amenorrhea, infertility and primary gonadal dysgenesis. Other true gonadal dysgenesis may arise due to a mutation in the SRY gene itself, resulting in variable clinical manifestations ranging from a normal fertile male to a female with streaked gonads and failure to reach puberty.

Since the mechanism may not be apparent at the time the patient has a diagnosis of a chromosomal abnormality, gonadoblastoma is a risk and needs to be addressed in the cytogenetics report. Further molecular testing for specific genetic mutations may be helpful in determining the etiology of the sex reversal.

ISCN RULES FOR REPORTING AN XY FEMALE

If a normal chromosome complement is seen in XY females, then the ISCN nomenclature is written as a normal result. However, the report should contain a phrase that implies the genotype does not match the phenotype to alert the reader that there is an abnormality present.

For example: 46,XY in a phenotypic female.

EXAMPLE REPORTING OF XY FEMALE

ISCN Result: 46,XY in a phenotypic female

Interpretation

Chromosome analysis revealed a normal male chromosome complement in all cells examined. In a phenotypic female, this finding is consistent with a sex reversal disorder such as androgen insensitivity or gonadal dysgenesis.

As the risk of gonadoblastoma is variable, further testing to determine the etiology of this sex reversal is indicated.

Genetic counseling is recommended.

10.6 X chromosome translocations

One other special case for carriers of reciprocal translocations is the involvement of the X chromosome. The X chromosome may be involved in a translocation with an autosome, another X chromosome or the Y chromosome. In each case, there would be a difference in the outcome for female versus male individuals.

Since the X chromosome undergoes random X inactivation early in embryogenesis in all female cells, X-autosome translocations will result in non-random X inactivation with the inactivation of the normal X, because the X chromosome that is attached to the autosome must remain active along with the autosomal chromosome material. This being the case, all females with an X-autosome translocation will have only the part of the X chromosome active that is translocated to the autosome, resulting in partial monosomy X with active genes, even when the rearrangement is balanced.

10.6.1 X-autosome translocations

Translocations of the X chromosome with an autosome often result in a female with primary or secondary ovarian failure and some clinical manifestations of Turner syndrome, depending on the region of the X chromosome affected. Translocations of the X chromosome involving the long arm from bands Xq13 to Xq26 include the critical region for Turner syndrome that will show Turner syndrome features to a greater degree than other regions of the X chromosome. Since most females with an X-autosome translocation will have the normal X chromosome inactivated in order to keep the genes on the autosome active, this will subsequently leave only the translocated part of the X chromosome active, resulting in monosomy for the region of the X chromosome that is not involved in the translocation.

When an X-autosome balanced translocation occurs that involves a region proximal to the Xq13 region, the female may be phenotypically normal if no genes are interrupted. This is also assuming X inactivation does not interfere with normal gene expression. However, if non-random X inactivation is present, then phenotypic consequences may result. It has been shown that even a phenotypically normal carrier mother with an X-autosome translocation may have a daughter with the same translocation with varying phenotypic abnormalities, since X inactivation occurs differently for each female even from mother to daughter.

Another consequence of an X-autosome translocation is when the translocation event disrupts a gene on the X chromosome, leading to loss of function of that gene, resulting in a female manifesting an X-linked recessive disorder. Since there are many significant X-linked genes, this is not a remote possibility. One published study reported a girl with a normal karyotype, but with a mutation in the Duchenne muscular dystrophy (DMD) gene, giving rise to the disease, while her monozygotic twin sister was phenotypically normal. X inactivation studies showed skewed X inactivation with 80% of the affected sister with the abnormal X chromosome active, and with 80% of the normal X chromosome active in the normal sister. Studies of the mother showed a 50–50% random X inactivation pattern.

One consequence of a carrier female with an X-autosome translocation is the possibility of an unbalanced rearrangement in offspring, or multiple miscarriages, due to the loss of viability of the unbalanced rearrangement. Due to non-random X inactivation, there is also a risk of the balanced translocation showing phenotypic manifestations, as described above. It is estimated that X-autosome translocation carriers have a risk of 20–40% of having offspring with either a structural or aneuploid abnormality. The phenotypic manifestations of having an X-autosome translocation may vary from normal or mild effects, to mental impairment and significant birth defects.

A male may also carry a balanced X-autosome translocation and would typically be normal, except for fertility issues and possibly other genital abnormalities. There is also a risk of the disruption of a gene on the X chromosome, giving rise to an X-linked disorder.

For those with an unbalanced X-autosome translocation, either 45 chromosomes with a derivative chromosome and loss of the normal X chromosome may result, or 46 chromosomes with a derivative chromosome and the presence of a normal X chromosome may result. In the case of 45 chromosomes, if there is no imbalance of the autosome, then the consequence would be loss of the X chromosome of the region not involved in the translocation event and trisomy for the autosome involved in the translocation.

Examples of X-autosome translocations include the following.

- A female with a balanced translocation between the X chromosome (outside the critical region) and chromosome 2. The karyotype would be: 46,X,t(X;2)(p11.2;q33). This female would most likely be phenotypically normal, unless there is a gene interrupted, and also have an increased risk of multiple miscarriages.
- The karyotype of a male with the same translocation as above would be: 46,Y,t(X;2)(p11.2;q33). This male would most likely be phenotypically normal, unless there is a gene interrupted, and also have an increased risk of multiple miscarriages in the partner.
- A female with a balanced translocation between the X chromosome (inside the critical region) and chromosome 2. The karyotype would be: 46,X,t(X;2)(q22;q33). This female may have clinical manifestations of Turner syndrome, abnormalities if a gene is interrupted, and also an increased risk of multiple miscarriages.
- The karyotype of a male with the same translocation as above would be: 46,Y,t(X;2)(q22;q33). This male may have mild-to-severe genital abnormalities, as well as abnormalities if a gene is interrupted, and also have an increased risk of multiple miscarriages in the partner.
- A female with an unbalanced translocation between the X chromosome and chromosome 2, in which there is *loss* of the normal X chromosome. Then the two karyotype possibilities would be: 45,der(X)t(X;2)(p11.2;q33), and opposite derivative, 45,der(2)t(X;2)(p11.2;q33). This female would most likely be unviable due to either having the der(X) with two normal copies of chromosome 2, or no normal copies of the X chromosome, and be trisomic for chromosome 2 from bands q33 to qter. Since there is no normal X, this female would be nullisomy for the short arm from p11.2 to pter, and monosomy for the proximal short arm from band p11.2 to the centromere and for the entire long arm of the X chromosome. The opposite derivative with the der(2) would most likely be unviable with monosomy for chromosome 2 from bands q33 to qter, and be nullisomy for the long arm and proximal short arm of the X chromosome.

- A female with an unbalanced translocation between the X chromosome and chromosome 2, in which there is *one normal* X chromosome. Then the two karyotype possibilities would be: 46,X,der(X)t(X;2)(p11.2;q33) and 46,X,der(2)t(X;2)(p11.2;q33). This female would have two normal copies of chromosome 2, one normal copy of the X chromosome, and with the der(X) would be trisomic for chromosome 2 from bands q33 to qter. Since there is one normal X, this female would be monosomy for the proximal short arm from band p11.2 to the telomere. This female would only be viable if the region of the autosome is small enough to sustain life. The phenotypic consequences may include mild Turner syndrome manifestations and trisomy for 2q33-2qter. The opposite derivative, 46,X,der(2)t(X;2)(p11.2;q33), would most likely be unviable with monosomy for chromosome 2q33 to qter and for the proximal short arm and the entire long arm of the X chromosome.
- A female with an unbalanced translocation between the X chromosome and chromosome 2, in which there are *two normal* X chromosomes. Then the two karyotype possibilities would be: 47,XX,+der(X)t(X;2)(p11.2;q33) and 47,XX,+der(2)t(X;2)(p11.2;q33). This female would have two normal copies of chromosomes X and 2, and with the + der(X) would be trisomic for chromosome 2 from bands q33 to qter. Since there are two normal Xs, this female would be trisomy for the proximal short arm and the entire long arm of the X chromosome. The phenotypic consequences may include manifestations of partial trisomy X and further manifestations of trisomy for 2q33-2qter. The opposite derivative, 47,XX,+der(2)t(X;2)(p11.2;q33), would be unviable with trisomy for most of chromosome 2 (disomy of q33 to qter) and partial trisomy for Xp11.2 to pter.

ISCN RULES FOR X-AUTOSOME BALANCED TRANSLOCATIONS

- First write the modal chromosome number of 46, followed by "X".
- Then write "t" followed, in parentheses, by X and the autosome involved.
- Next, in parentheses, write the breakpoints involved in the X and the autosome.

With balanced X-autosome chromosome translocations, the normal X chromosome is written alone enclosed by commas, i.e. 46,X, followed by the translocation event.

For example: 46,X,t(X;2)(q22;q33).

EXAMPLE REPORTING OF BALANCED X-AUTOSOME TRANSLOCATIONS

ISCN Result: 46,X,t(X;2)(q22;q33)

Interpretation

Chromosome analysis revealed an abnormal female chromosome complement in all cells examined with one normal X chromosome and a translocation between the long arms of one X chromosome and chromosome 2. This translocation involves the critical region of Turner syndrome, and it may result in clinical manifestations of Turner syndrome if there is non-random X inactivation.

Other abnormalities are possible if a gene is interrupted. There is also a risk of multiple miscarriages due to the possibility of unbalanced gametes inherited from a balanced translocation carrier.

Genetic counseling is recommended.

10.6.2 X-X translocations

Females with a balanced translocation between two X chromosomes would presumably be phenotypically normal, except when the translocation event disrupts a gene on the X chromosome, leading to loss of function of that gene, resulting in a female manifesting an X-linked recessive disorder. There is also a risk for multiple miscarriages due to the possibility of unbalanced gametes inherited from a balanced translocation carrier.

Unbalanced X-X translocations will give rise to nullisomy, monosomy or trisomy for the X chromosome, depending on the number of normal X chromosomes present. The portion of the X chromosome that is in an imbalance (monosomy) needs to be ascertained in order to fully identify if it is in the critical region of Turner syndrome (Xq13-Xq26), which will imply Turner syndrome clinical manifestations versus milder effects if only the short arm is affected.

ISCN RULES FOR X-X TRANSLOCATIONS

- First write the modal chromosome number of 46.
- Then write "t" followed by "X;X", enclosed by parentheses.
- Then write the breakpoints involved in the two X chromosomes, enclosed in parentheses.

With X-X chromosome translocations, there is no normal X chromosome. Therefore, after the modal chromosome number, the translocation event is written.

For example: 46,t(X;X)(p22;q33).

EXAMPLE REPORTING OF X-X TRANSLOCATIONS IN A FEMALE ADULT WITH INFERTILITY

ISCN Result: 46,t(X;X)(p22;q33)

Interpretation

Chromosome analysis revealed an abnormal female chromosome complement in all cells examined with an apparently balanced translocation between the long arm of one X chromosome and the short arm of the homologous X chromosome. There is a risk for multiple miscarriages due to the possibility of unbalanced gametes inherited from a balanced translocation carrier.

Genetic counseling is recommended.

EXAMPLE REPORTING OF X-X TRANSLOCATIONS IN A FEMALE CHILD WITH PHENOTYPIC ABNORMALITIES

ISCN Result: 46,t(X;X)(p22;q33)

Interpretation

Chromosome analysis revealed an abnormal female chromosome complement in all cells examined with a translocation between the long arm of one X chromosome and the short arm of the homologous X chromosome. This translocation appears to be balanced. However, abnormalities are possible if a gene is interrupted, leading to loss of function of that gene, thereby resulting in a female manifesting an X-linked recessive disorder.

Chromosome analysis of both parents is indicated to determine if the rearrangement is familial or *de novo* in origin. A familial rearrangement is unlikely to have a phenotypic effect. *De novo* rearrangements, however, are associated with a significant risk for clinical abnormalities.

Chromosomal microarray analysis may prove informative for further characterization of this rearrangement. Recent literature suggests imbalances at the breakpoints or elsewhere in the genome may be found in approximately 30–40% of individuals with a cytogenetically balanced rearrangement and an abnormal phenotype.

Genetic counseling is recommended.

10.6.3 X-Y translocations

Since X-Y translocations will need to have a normal X or Y in addition to the translocation chromosome, the most common form of the translocation will involve the region of the X in the short arm distal to Xp22 due to the need to have the region of the X with the rest of the chromosome present for viability. It also most commonly involves the region of the Y chromosome at the long arm distal to the euchromatic region or at Yq12.

With an X-Y translocation that contains a normal X chromosome, the abnormality is a derivative X chromosome, 46,X,der(X)t(X:Y), and results in a female with loss of the region outside the X-Y translocation region. Phenotypic consequences will result in Turner syndrome features, including short stature. If the translocation is very distal on the short arm, then few clinical features may be present. With loss of the short arm region of Xp21, including the steroid sulfatase, Kallman and Duchenne muscular dystrophy genes, a deletion of this region may result in one of these X-linked disorders with severe phenotypic consequences.

With an X-Y translocation that contains a normal Y chromosome, the abnormality is a derivative X chromosome, 46,Y,der(X)t(X:Y), and results in a male with short stature and infertility, if the region deleted includes Yq12. Loss of more of the Y chromosome may result in a more severe phenotype or unviability.

ISCN RULES FOR X-Y TRANSLOCATIONS IN A FEMALE

- First write the modal chromosome number of 46, followed by “X”.
- Next write “der(X)” followed by “t(X;Y)”.
- Then write the breakpoints involved in the X and then the Y chromosomes, enclosed by brackets.

With X-Y chromosome translocations in a female with a normal X chromosome, the normal X chromosome is written alone, enclosed by commas, i.e. 46,X, followed by der(X) and then the translocation event.

For example: 46,X,der(X)t(X;Y)(p22.3;q11.2).

EXAMPLE REPORTING OF der(X)t(X;Y) IN A FEMALE

ISCN Result: 46,X,der(X)t(X;Y)(p22.3;q11.2)

Interpretation

Chromosome analysis revealed an abnormal female chromosome complement in all cells examined with one normal X chromosome and a translocation between the short arm of one X chromosome and the long arm of the Y chromosome.

Fluorescent *in situ* hybridization (FISH) testing (reported separately) reveals the absence of the SRY gene and centromere of the Y chromosome, confirming the presence of a translocation between the distal region of the long arm of the Y chromosome to the short arm of one X chromosome. This abnormality results in a female phenotype with loss of the X chromosome from Xp22.3 to the short arm telomere, and gain of the long arm of the Y chromosome from Yq12 to the telomere.

Females with this type of translocation usually present with clinical manifestations of Turner syndrome including short stature, infertility and possibly other clinical manifestations.

Genetic counseling is recommended.

ISCN RULES FOR X-Y TRANSLOCATIONS IN A MALE

- First write the modal chromosome number of 46, followed by “Y”.
- Next write “der(X)” followed by “t(X;Y)”.
- Then write the breakpoints involved in the X and then the Y chromosomes, enclosed by brackets.

With X-Y chromosome translocations in a male with a normal Y chromosome, the normal Y chromosome is written alone, enclosed by commas, i.e. 46,Y, followed by der(X) and then the translocation event.

For example: 46,Y,der(X)t(X;Y)(p22.3;q11.2).

EXAMPLE REPORTING OF der(X)t(X;Y) IN A MALE

ISCN Result: 46,Y,der(X)t(X;Y)(p22.3;q11.2)

Interpretation

Chromosome analysis revealed an abnormal male chromosome complement in all cells examined with one normal Y chromosome and a translocation between the short arm of one X chromosome and the long arm of the Y chromosome.

Fluorescent *in situ* hybridization (FISH) testing (reported separately) reveals the presence of the SRY gene and centromere of the Y chromosome, confirming a normal Y chromosome with a translocation between the distal region of the long arm of the Y chromosome to the short arm of one X chromosome. This abnormality results in a male phenotype with loss of the X chromosome from Xp22.3 to the short arm telomere, and gain of the long arm of the Y chromosome from Yq12 to the telomere.

Males with this type of translocation usually present with short stature and infertility and possibly other clinical manifestations.

Genetic counseling is recommended.

Bibliography

Barr ML, Sergovich FR, Carr DH, Saver EL. The triplo-X female: an appraisal based on a study of 12 cases and a review of the literature. Can Med Assoc J 1969; 101: 247.

Blaschke RJ, Rappold G. The pseudoautosomal regions, SHOX and disease. Curr Opin Genet Dev 2006; 16: 233–239.

Donaldson MDC, Gault EJ, Tan KW, Dunger DB. Optimising management in Turner Syndrome: from infancy to adult transfer. Arch Dis Child 2006; 91: 513–520.

Ferlin A, Moro E, Garolla A, Foresta C. Human male infertility and Y chromosome deletions: role of the AZF-candidate genes DAZ, RBM and DFFRY. Hum Reprod 1999; 14: 1710.

Gardner RJM, Sutherland GR. *Chromosome Abnormalities and Genetic Counselling*. Oxford Monographs on Medical Genetics. Oxford University Press, Oxford, 2003.

Gersen S, Keagle M (eds). *Principles of Clinical Cytogenetics*. Humana Press, Totowa, New Jersey, 1999.

Hook EB, Warburton D. The distribution of chromosomal genotypes associated with Turner's Syndrome: livebirth prevalence rates and evidence for diminished fetal mortality and severity in genotypes associated with structural X abnormalities or mosaicism. Hum Genet 1983; 64: 24.

Hsu LY. Phenotype/karyotype correlations of Y chromosome aneuploidy with emphasis on structural aberrations in postnatally diagnosed cases. Am J Med Genet 1994; 53: 108.

Iourov IY, Vorsanova SG, Liehr T, Monakhov VV, Soloviev IV, Yurov YB. Dynamic mosaicism manifesting as loss, gain and rearrangement of an isodicentric Y chromosome in a male child with growth retardation and abnormal external genitalia. Cytogenet Genome Res 2008; 121 302–306.

Jacobs PA, Brunton M, Melville MM, et al. Aggressive behavior, mental sub-normality and the XYY male. Nature 1965; 208: 1351.

Jones KL. *Smith's Recognizable Patterns of Human Malformations*, 6th edn. Saunders, London, 2006.

Kassai R, Hamada I, Furuta H, et al. Penta X Syndrome: a case report with review of the literature. Am J Med Genet 1991; 40: 51.

Linden MG, Bender BG, Robinson A. Sex chromosome tetrasomy and pentasomy. Pediatrics 1995; 96: 672.

Palermo GD, Schlegel PN, Sills ES, et al. Births after intracytoplasmic injection of sperm obtained by testicular extraction from men with nonmosaic Klinefelter's Syndrome. N Engl J Med 1998; 338: 588.

Peet J, Weaver DD, Vance GH. 49,XXXXY: a distinct phenotype. Three new cases and review. J Med Genet 1998; 35: 420.

Rao E, Weiss B, Fukami M, et al. Pseudoautosomal deletions encompassing a novel homeobox gene cause growth failure in idiopathic short stature and Turner Syndrome. Nat Genet 1997; 16: 54.

Robinson A, de la Chapelle A. Sex chromosome abnormalities. In: Rimoin DL, Connor JM, Pyeritz RE (eds) *Emery and Rimoin's Principles and Practice of Medical Genetics*, Churchill Livingstone, Edinburgh, 1996.

Robinson A, Bender BG, Linden MG, Salbenblatt JA. Sex chromosome aneuploidy: the Denver Prospective Study. Birth Defects Orig Artic Ser 1990; 26: 59.

Robinson A, Linden MG, Bender BG, et al. Prenatal diagnosis of sex chromosome abnormalities. In: Milunsky A (ed) *Genetic Disorders and the Fetus*, 4th edn. Johns Hopkins University Press, Baltimore, 1998.

Rooney DE, Czepulkowski BH. *Human Cytogenetics. A Practical Approach.* Oxford University Press, New York, 1992.

Ross JL, Stefanatos GA, Kushner H, et al. Persistent cognitive deficits in adult women with Turner Syndrome. Neurology 2002; 58: 218.

Saenger P. Clinical review 48: the current status of diagnosis and therapeutic intervention in Turner's Syndrome. J Clin Endocrinol Metab 1993; 77: 297.

Senner CE, Brockdorff N. *Xist* gene regulation at the onset of X inactivation. Curr Opin Genet Dev 2009; 19: 122–126.

Shaffer LG, McGowan-Jordan J, Schmid M (eds). *ISCN* 2013: *An International System for Human Cytogenetic Nomenclature.* Karger Publishers, Unionville, CT, 2013.

Shanske A, Sachmechi I, Patel DK, et al. An adult with 49,XYYYY karyotype: case report and endocrine studies. Am J Med Genet 1998; 80: 103.

Shchelochkov OA, Cooper ML, Ou Z, et al. Mosaicism for r(X) and der(X)del(X)(p11.23)dup(X)(p11.21p11.22) provides insight into the possible mechanism of rearrangement. Mol Cytogenet 2008, 1: 16.

Tobias ES, Connor M, Ferguson Smith M. *Essential Medical Genetics*, 6th edn. Wiley-Blackwell, Oxford, 2011.

Tournaye H, Staessen C, Liebaers I, et al. Testicular sperm recovery in nine 47,XXY Klinefelter patients. Hum Reprod 1996; 11: 1644.

Wallis M, Waters P, Graves JAM. Sex determination in mammals – before and after the evolution of SRY. Cell Mol Life Sci 2008; 65: 3182–3195.

Warburton D, Kline J, Stein Z, Susser M. Monosomy X: a chromosomal anomaly associated with young maternal age. Lancet 1980; 1: 167.

Wilhelm D, Koopman P. The makings of maleness: towards an integrated view of male sexual development. Nat Rev Genet 2006; 7: 620–631.

Witkin HA, Mednick SA, Schulsinger F, et al. Criminality in XYY and XXY men. Science 1976; 193: 547.

Wolff DJ, van Dyke DL, Powell CM, A Working Group of the ACMG Laboratory Quality Assurance Committee. Laboratory guideline for Turner Syndrome. Genet Med 2010; 1: 52–55.

Zneimer SM, Schneider NR, Richards CS. In situ hybridization shows direct evidence of skewed X inactivation in one of monozygotic twin females manifesting Duchenne muscular dystrophy. Am J Med Genet 1993; 45: 601–605.

CHAPTER 11

Fetal demises/ spontaneous abortions

The overall incidence of chromosome abnormalities in conceptuses is approximately 20%, with the vast majority ending in a spontaneous abortion (SAB). Only 0.6% survive to term. Chromosomal abnormalities are seen more often in the first trimester, in which 60% of first-trimester SABs show a chromosomal aberration. By the 16th week of gestation, most fetuses with a chromosome aberration will have demised, with only 5% of fetuses showing a chromosome abnormality in late SABs or stillbirths. Table 11.1 shows the common various cytogenetic results seen in fetal demises.

Blood chromosome analyses of couples may be useful after three SABs to establish if an individual carries a chromosome abnormality causing infertility. In males with azoospermia or oligospermia, approximately 12% are karyotypically abnormal. Sex chromosome abnormalities are the most common in cases of male infertility, including XXY, XXY/XY mosaicism and 45,X from either parent. Less frequent are autosomal abnormalities, including aneuploidies and structural rearrangements. Small rings and markers, which are unstable, can cause problems in cell division in the sperm, resulting in azoospermia. Other rearrangements, such as translocations and inversions, may also result in loss of gametes causing infertility. One subset of azoospermia due to a chromosomal abnormality is deletions in the Y chromosome, including the Yq11.23 region, with loss of the YRRM or DAZ genes.

11.1 Aneuploid rate

The overall spontaneous abortion rate is approximately 15%. From a paper by Hsu et al. (1972), spontaneous abortions that contain chromosome abnormalities can be summarized with these proportions:

- autosomal trisomies = 52%
- autosomal monosomies = <1%
- 45,X = 19%
- triploidy = 16%
- tetraploidy = 6%
- other = 7%.

Cytogenetic Abnormalities: Chromosomal, FISH and Microarray-Based Clinical Reporting, First Edition. Susan Mahler Zneimer.

Table 11.1 Common cytogenetic results seen in fetal demises (adapted from Tobias et al. 2011)

Cytogenetic results	Cytogenetic frequency
Normal chromosomes	40%
Autosomal trisomies	30%
Monosomy X	10%
Triploidy (69,XXX, 69,XXY, 69,XYY)	10%
Tetraploidy (92,XXXX, 92,XXYY, 92,XXXY)	5%
Other (unbalanced findings)	5%
Total	**100%**

The overall rate of chromosome abnormalities in spontaneous abortions versus those seen as livebirths can be summarized as:

- triploid/tetraploid = 100% SABs
- 45,X = 99%
- +16 = 100%
- +18 = 95%
- +21 = 78%
- other trisomies = 99.5%
- 47,XXY/47,XXX/47,XYY = 21%
- unbalanced rearrangements = 85%
- balanced rearrangements = 16%.

11.2 Confined placental mosaicism

Confined placental mosaicism (CPM) is defined as mosaicism within the placenta for a chromosome abnormality that is not present in the fetus. The majority of these abnormalities are trisomies. For example, 47,XX,+15/46,XX may be seen in the placenta, but the fetal tissue is only 46,XX.

Even with a 46,XX karyotype in fetal tissue, the fetus may have phenotypic abnormalities due to poor placental growth or uniparental disomy (UPD) in the fetus. UPD is discussed in Chapter 12.

EXAMPLE REPORTING OF MATERNAL CELL CONTAMINATION IN A FETAL DEMISE

ISCN Result: 46,XX

Interpretation

Chromosome analysis revealed a normal female chromosome complement in all cells examined.

Note that maternal cell contamination is common in products of conception cultures. Therefore, a normal female karyotype should be interpreted cautiously, as it may represent maternal cells and not the chromosome complement of the fetus.

COMMENTING ON MATERNAL CELL CONTAMINATION IN A REPORT WITH A FETAL DEMISE WITH AN ABNORMAL RESULT

Add to the interpretation:

The abnormal cells may be fetal in origin or may represent confined placental mosaicism.

11.3 Hydatidiform moles

Hydatid moles form in an abnormal pregnancy when the placenta is converted into a mass of tissue due to abnormal growth of the chorionic villi, which undergo a degenerative hydropic (fluid-filled) change, with the appearance of a bunch of grapes.

A complete mole is seen when the genetic origin of the mole is completely paternal. Most complete moles are 46,XX, normal female karyotype. The chromosomes are all paternal in origin and homozygous. This originates from a 23,X sperm with an ovum that lacks a nucleus. The chromosome complement then doubles. The result is that the fetus is usually absent and with hyperplasia of the trophoblast.

A partial mole is seen when there is a contribution of an extra haploid set of chromosomes. Therefore, partial moles are triploid, 69,XXX or 69,XXY. In two-thirds of the cases, the extra set of chromosomes is paternal in origin. An extra paternal set results in a large trophoblast and poor embryonic development whereas an extra set of maternal chromosomes results in severe intrauterine embryonic growth retardation and a small, fibrotic placenta.

One consequence of hydatidiform moles is the risk of choriocarcinomas, in which a malignancy forms from the molar tissue of paternal origin. The reciprocal form is an ovarian teratoma, which is a benign tumor that arises from 46,XX cells that contain only maternal chromosomes with no paternal contribution.

EXAMPLE REPORTING OF MOLAR PREGNANCIES IN A REPORT WITH NORMAL RESULTS

ISCN Result: 46,XX

Interpretation

Chromosome analysis revealed a normal female chromosome complement in all cells examined.

Cytogenetic analysis can detect an abnormality in cases of molar pregnancies only when due to triploidy.

Alternative testing by molecular or immunohistochemical methods is suggested for those cases in which a normal diploid chromosome complement is detected.

Molecular testing for a molar pregnancy requires both maternal and fetal tissue and can be performed on formalin-fixed, paraffin-embedded tissue specimens. Testing of cultured villous tissue in combination with a maternal blood sample can be performed if the laboratory is notified within one week of the report date.

11.4 Monosomy X in a fetus

SPECIFIC FEATURES OF MONOSOMY X IN A FETUS

Although 45,X results in Turner syndrome in liveborns and has an incidence of 1 in 5000 females in the population, the frequency at conception is much higher and results in spontaneous abortions in over 99% of fetuses. Generally, aneuploidies are the result of maternal non-disjunction in meiosis. However, 45,X can result from non-disjunction from either parent, resulting in the loss of either one X or Y chromosome. Approximately 80% of 45,X fetuses show the X chromosome arising from paternal non-disjunction during spermatogenesis or as a postfertilization error. Thus, this is not a maternal age effect as in autosomal trisomies. The presence of two cell lines, such as 45,X/46,XX, would imply a postfertilization error, as in other mosaic karyotypes. There is no known recurrence risk for 45,X fetuses, and parental studies usually result in normal karyotypes. Other variants of 45,X are common in fetuses, just as they are in liveborns. See Table 11.2 for a summary of other variants.

Table 11.2 Karyotypic abnormalities associated with Turner syndrome or its variants

Karyotype	Description	% of total abnormal variant Turner syndrome karyotypes
45,X	Loss of X or Y – monosomy X	50%
Mosaic 45,X	Partial monosomy X	24%
46,X,i(X)(q10)	Trisomy Xq and monosomy Xp	17%
46,X,r(X)	Partial monosomy X	7%
46,X,i(X)(p10)	Trisomy Xp and monosomy Xq	2%

EXAMPLE REPORTING OF 45,X IN A FETUS

ISCN Result: 45,X

Interpretation

Chromosome analysis revealed an abnormal chromosome complement in all cells examined with a single X chromosome, resulting in monosomy X.

Monosomy X is the single most common cytogenetic abnormality in spontaneous abortions. The vast majority of conceptuses with monosomy X miscarry and do not survive to term. There is no increased risk of recurrence of this abnormality.

Genetic counseling is recommended.

11.5 Trisomies in a fetus

SPECIFIC FEATURES OF TRISOMIES IN A FETUS

Only autosomal trisomies and not monosomies are discussed here since all autosomal monosomies end as a spontaneous abortion and most often are too early in gestational age to be recognized as a true pregnancy. Autosomal trisomies most often result in a spontaneous abortion, accounting for approximately 97% in pregnancy loss, and are the most common autosomal cytogenetic abnormality seen in fetal loss by cytogenetics.

Aneuploidies in general can be either maternal or paternal in origin, but are increasingly common in maternal oogenesis with advanced maternal age, although smaller chromosomes may not have the same maternal effect as larger chromosomes. Non-disjunction may occur in either the first or second stage of meiosis, or postfertilization mitosis, which may include mosaicism. Most trisomies are spontaneously aborted, usually in the first trimester of pregnancy, leaving only a small proportion that survive to term although frequencies of particular trisomies vary with gestational age.

The most common autosomal trisomy seen in fetal demises by cytogenetics is trisomy 16, resulting in 100% fetal loss, where none will survive to term. It is estimated that 31% of all trisomies detected by cytogenetics are trisomy 16, and trisomy 16 accounts for over 7% of all fetal loss.

Trisomy 21 and trisomy 22 each account for approximately 11% of autosomal trisomies in fetal loss. Each of the other chromosomes shows less than 5% fetal loss.

EXAMPLE REPORTING OF GENERAL AUTOSOMAL TRISOMIES IN A FETUS

ISCN Result: 47,XY,+__

Interpretation

Chromosome analysis revealed an abnormal male chromosome complement in all cells examined with an additional chromosome __ (trisomy __).

Autosomal trisomies are the most frequent type of cytogenetic abnormality observed in pregnancy loss and are usually sporadic.

Genetic counseling is recommended.

11.5.1 Trisomy 21 in a fetus

EXAMPLE REPORTING OF TRISOMY 21 (DOWN SYNDROME) IN A FETUS

ISCN Result: 47,XY,+21

Interpretation

Chromosome analysis revealed an abnormal male chromosome complement in all cells examined with an additional chromosome 21 (trisomy 21). This finding is consistent with a clinical diagnosis of Down syndrome in liveborn individuals.

Autosomal trisomies are the most frequent type of cytogenetic abnormality observed in pregnancy loss and are usually sporadic. The risk of recurrence is approximately 1% or the maternal age-associated risk, whichever is higher. Therefore, prenatal diagnosis is recommended for any subsequent pregnancies.

Genetic counseling is recommended.

11.5.2 Trisomy 13 in a fetus

EXAMPLE REPORTING OF TRISOMY 13 (PATAU SYNDROME) IN A FETUS

ISCN Result: 47,XX,+13

Interpretation

Chromosome analysis revealed an abnormal female chromosome complement in all cells examined with an extra chromosome 13 (trisomy 13), consistent with the clinical diagnosis of Patau syndrome in liveborn individuals.

Autosomal trisomies are the most frequent type of cytogenetic abnormality observed in pregnancy loss and are usually sporadic. The risk of recurrence is approximately 1% or the maternal age-associated risk, whichever is higher. Therefore, prenatal diagnosis is recommended for any subsequent pregnancies.

Genetic counseling is recommended.

11.5.3 Trisomy 18 in a fetus

EXAMPLE REPORTING OF TRISOMY 18 (EDWARD SYNDROME) IN A FETUS

ISCN Result: 47,XX,+18

Interpretation

Chromosome analysis revealed an abnormal female chromosome complement in all cells examined with an extra chromosome 18 (trisomy 18), consistent with the clinical diagnosis of Edwards syndrome in liveborn individuals.

Autosomal trisomies are the most frequent type of cytogenetic abnormality observed in pregnancy loss and are usually sporadic. The risk of recurrence is approximately 1% or the maternal age-associated risk, whichever is higher. Therefore, prenatal diagnosis is recommended for any subsequent pregnancies.

Genetic counseling is recommended.

11.5.4 Trisomy 16 in a fetus

EXAMPLE REPORTING OF TRISOMY 16 IN A FETUS

ISCN Result: 47,XX,+16

Interpretation

Chromosome analysis revealed an abnormal female chromosome complement in all cells examined with an extra chromosome 16 (trisomy 16). Trisomy 16 is the most common autosomal aneuploidy among abortuses. Recurrence risks for future pregnancies, given a trisomy 16 spontaneous abortion, are not well established.

Genetic counseling is recommended.

11.6 Double trisomy

SPECIFIC FEATURES OF DOUBLE TRISOMIES

Double trisomies occur when there are two meiotic errors of either the maternal, paternal or both gametes during gametogenesis. Therefore, the incidence of double trisomies is the product of two separate events, which make any double trisomy very rare. All three instances of gamete non-disjunctions have been reported. Double trisomies may occur with either two autosomes, two (or more) sex chromosomes, or one autosome and one sex chromosome. Gain of an X or Y chromosome, plus gain of chromosomes 18 and 21 are the most commonly reported. One report described trisomy 14 and 21 in the same fetus, in which each extra chromosome was inherited from a different parent in gamete formation. Another implication of double trisomies is uniparental disomy, which will be discussed in Chapter 12.

The risk of fetal demises with a double trisomy is greater than with a single trisomy, and very few cases have liveborn offspring. Also, double trisomies show a strong association with advanced maternal age, even more so than single trisomies.

ISCN RULES FOR REPORTING DOUBLE TRISOMIES

- First write the modal chromosome number of 48, followed by the sex designation.
- Next write "+" followed by the aneuploid chromosome numerically smallest.
- Then write "+" followed by the larger aneuploid chromosome.
- If one of the aneuploid chromosomes is a sex chromosome, the extra sex chromosome is placed with the sex chromosome complement.

Examples:

- Gain of an X chromosome and chromosomes 18 would be: 48,XXX,+18.
- Gain of chromosomes Y and 21 would be: 48,XYY,+21.
- A male with the gain of chromosomes 8 and 21 would be: 48,XY,+8,+21.

EXAMPLE REPORTING OF DOUBLE TRISOMIES IN A FETUS (FIGURE 11.1)

ISCN Result: 48,XXX,+18

Interpretation

Chromosome analysis revealed an abnormal female chromosome complement in all cells examined with an additional X chromosome and chromosome 18. Trisomies are the most common abnormal findings in products of conception.

Genetic counseling is recommended.

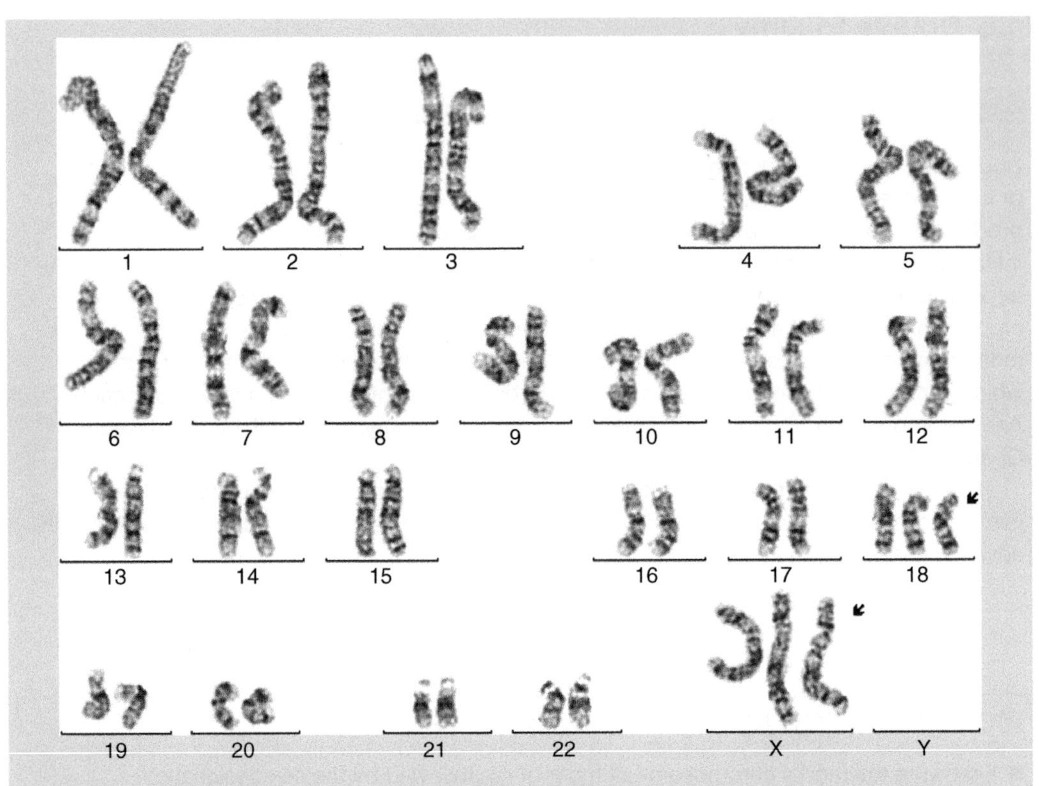

Figure 11.1 Karyotype showing a double trisomy, with the gain of chromosomes X and 18: 48,XXX,+18.

11.7 Triploidy

SPECIFIC FEATURES OF TRIPLOIDY

Triploidy results in the presence of an extra haploid set of chromosomes, which can be seen as a non-disjunction from either parent although most are of paternal origin. Triploidy most often arises due to double fertilization. A minority arises from fertilization of a diploid sperm or a diploid egg. Triploidy can either be 69,XXX or 69,XXY, with very few being 69,XYY and none being 69,YYY.

Triploidy is seen in approximately 1–2% of all pregnancies and 18% of all cytogenetically abnormal fetuses. The vast majority results in a spontaneous abortion. The few fetuses that survive to term generally have intrauterine growth retardation (IUGR), disproportionately small trunk-to-head size, syndactyly and multiple congenital malformations.

Most triploid fetuses have a large placenta with manifestations of a partial hydatidiform mole, which arises with double paternal chromosomes. Those fetuses with triploidy that are due to maternal non-disjunction show intrauterine growth retardation, macrocephaly and a small, non-cystic placenta. Therefore, maternal triploidy results in a small fetus with a more normal placenta whereas paternally derived triploid fetuses are more normal-sized with an abnormal placenta. These findings are the result of uniparental disomy or genomic imprinting, where there is a differential expression of genes associated with the parental origin of the extra set of chromosomes. UPD and imprinting will be discussed in greater detail in Chapter 12.

Cases of mosaicism with triploidy and a normal cell line have been reported, and individuals have been known to survive until 10 years of age. Clinical findings generally include IUGR, psychomotor impairment, asymmetric growth, syndactyly, genital anomalies and irregular skin pigmentation. The recurrence risk of triploidy is not known, but it is thought to be no greater than the general population risk.

ISCN RULES FOR REPORTING TRIPLOIDY

- First write the modal chromosome number of 69, followed by the sex chromosome complement, which is either XXX, XXY or XYY.

Therefore, a female triploid fetus is written as 69,XXX, and a male triploid fetus is written as 69,XXY or 69,XYY.

EXAMPLE REPORTING OF TRIPLOIDY IN A FETUS

ISCN Result: 69,XXY

Interpretation

Chromosome analysis revealed an abnormal male chromosome complement in all cells examined with 69 chromosomes due to gain of an entire, extra haploid (23) set of chromosomes, resulting in triploidy.

Triploidy is a common abnormality in products of conception and has not been associated with a higher risk for subsequent abnormal liveborn offspring.

If paternally derived (seen in 70% of cases), gross phenotypic features of a partial hydatidiform mole may result. Transformation of a partial mole, although rare, has been reported. Therefore, follow-up monitoring of beta-hcg levels should be considered.

Genetic counseling is recommended.

11.8 Tetraploidy

SPECIFIC FEATURES OF TETRAPLOIDY

Tetraploidy results in the presence of four haploid sets of chromosomes, rather than the normal two haploid sets, with a chromosome number of 92. This usually arises from failure to complete the first zygotic division, resulting in duplication of a diploid set of chromosomes. The sex chromosome complement is either XXXX or XXYY, and not XYYY or XXXY, supporting the mechanism of malsegregation at the first meiotic division.

The incidence of spontaneous abortions with tetraploidy is approximately 6–7%. All abort spontaneously with very few known cases of liveborns reported, although mosaic tetraploidy with a normal cell line has been reported in which survival depends on the proportion and distribution of abnormal cells in the body. Specific phenotypic findings among the abortuses include intrauterine growth retardation, cranial malformations, limb abnormalities and hypotonia.

ISCN RULES FOR REPORTING TETRAPLOIDY

- First write the modal chromosome number of 92, followed by the sex designation, which is either XXXX or XXYY.

Therefore, a female tetraploid fetus is written as 92,XXXX, and a male tetraploid fetus is written as 92,XXYY.

EXAMPLE REPORTING OF TETRAPLOIDY IN A FETUS

ISCN Result: 92,XXYY

Interpretation

Chromosome analysis revealed an abnormal male chromosome complement in all cells examined with 92 chromosomes due to gain of an entire, extra diploid (46) set of chromosomes, resulting in tetraploidy.

Tetraploidy is a non-random finding in spontaneous abortions and is generally considered to be postconceptual and sporadic in origin. Therefore, the recurrence risk is negligible.

Genetic counseling is recommended.

Reference

Tobias ES, Connor M, Ferguson Smith M. *Essential Medical Genetics*, 6th edn. Wiley-Blackwell, Oxford, 2011.

Bibliography

Alberman ED, Creasy MR. Frequency of chromosomal abnormalities in miscarriages and perinatal deaths. J Med Genet 1977; 14: 313.

Cowles TA, Zneimer SM, Elder FB. A prospective cytogenetic study in third trimester placentas of fetuses with unexplained intrauterine growth restriction. J Maternal-Fetal Invest 1996; 6: 207–209.

Gardner RJM, Sutherland GR. *Chromosome Abnormalities and Genetic Counselling*. Oxford Monographs on Medical Genetics. Oxford University Press, Oxford, 2003.

Gersen S, Keagle M (eds).*Principles of Clinical Cytogenetics*. Humana Press, Totowa, New Jersey, 1999.

Hook EB. Spontaneous deaths of fetuses with chromosomal abnormalities diagnosed prenatally. N Engl J Med 1978; 299: 1036.

Hook EB. Chromosome abnormalities and spontaneous fetal death following amniocentesis: further data and associations with maternal age. Am J Hum Genet 1983; 35: 110.

Hsu LYF. Prenatal diagnosis of chromosomal abnormalities through amniocentesis. In: Milunsky A (ed) *Genetic Disorders and the Fetus*, 4th edn. Johns Hopkins University Press, Baltimore, 1998.

Hsu LYF, Garcia F, Grossman D, Kutinsky E, Hirschhorn K. Fetal wastage and maternal mosaicism. Obstet Gynecol 1972; 40: 98–103.

Jones KL. *Smith's Recognizable Patterns of Human Malformations*, 6th edn. Saunders, London, 2006.

McFadden DE, Kalousek DK. Two different phenotypes of fetuses with chromosomal triploidy: correlation with parental origin of the extra haploid set. Am J Med Genet 1991; 38: 535.

McWeeney DT, Munné S, Miller RC, et al. Pregnancy complicated by triploidy: a comparison of the three karyotypes. Am J Perinatol 2009; 26: 641.

Miny P, Koppers B, Dworniczak B, et al. Parental origin of the extra haploid chromosome set in triploidies diagnosed prenatally. Am J Med Genet 1995; 57: 102.

Reijo R, Lee T, Salo P, et al. Diverse spermatogenic defects in humans caused by Y chromosome deletions encompassing a novel RNA-binding protein gene. Nat Genet 1995; 10(4): 383–393.

Robinson W, McFadden D, Stephenson M. The origin of abnormalities in recurrent aneuploidy/polyploidy. Am J Hum Genet 2001; 69(6): 1245–1254.

Rooney DE, Czepulkowski BH. *Human Cytogenetics. A Practical Approach*. Oxford University Press, New York, 1992.

Shaffer LG, McGowan-Jordan J, Schmid M (eds). *ISCN 2013: An International System for Human Cytogenetic Nomenclature*. Karger Publishers, Unionville, CT, 2013.

Warburton D, Kline J, Stein Z, et al. Cytogenetic abnormalities in spontaneous abortions of recognized conception. In: Porter IH, Willey A (eds) *Perinatal Genetics Diagnosis and Treatment*. Academic Press, New York, 1986.

Yilmaz Z, Sahin FI, Tarim E, Kuscu E. Triploidies in first and second trimesters of pregnancies in Turkey. Br J Med Genet 2007; 10: 71.

Zaragoza MV, Surti U, Redline RW, et al, Parental origin and phenotype of triploidy in spontaneous abortions: predominance of diandry and association with the partial hydatidiform mole. Am J Hum Genet 2000; 66: 1807–1820.

CHAPTER 12

Uniparental disomy

Uniparental disomy (UPD) is defined as the inheritance of chromosome homologs or segments of both homologs from a single parent and loss of the corresponding homolog from the other parent. Normally, a parent passes a single homolog of each of the pairs of chromosomes to their offspring, thus ensuring one-half of the genome is maternally inherited and one-half is paternally inherited.

There are two main variations of UPD that are known to exist: uniparental isodisomy (iUPD) and uniparental heterodisomy (hUPD). Isodisomy refers to a single homolog, which is the exact same chromosome inherited in duplicate by a single parent. Heterodisomy refers to both homologs of a chromosome inherited by a single parent. In both cases, there is no contribution of this homologous pair from the other parent. However, it is important to distinguish the two types of UPD. Isodisomy refers to the inheritance of a single duplicated chromosome whereas, in heterodisomy, two different chromosomes are inherited. In both scenarios, these chromosomes are derived from a single parent.

Uniparental disomy may also be present as a complete chromosomal complement in a fetus, which is either completely maternal or paternal in origin. This is reported as UPDmat and UPDpat, respectively. UPDmat in a fetus results in an ovarian cyst. UPDpat in a fetus results in a complete hydatidiform mole.

Uniparental disomy of a single chromosome may be derived from three known mechanisms. The first is a non-disjunction event resulting in a trisomy fetus, which then undergoes "trisomy rescue" in order to eliminate the third chromosome and revert back to a disomy for that homologous pair. However, the one chromosome lost leaves the two remaining from a single parent. When a non-disjunction occurs during meiosis I division, the resulting UPD will be heterodisomy whereas a meiotic II division error will lead to homologs that are identical, resulting in isodisomy.

The second mechanism of UPD is a zygote that has undergone a meiotic non-disjunction, leading to a monosomy fetus that subsequently undergoes "monosomy rescue," which is a duplication of the single chromosome to yield a disomy complement. Structurally abnormal chromosomes, such as Robertsonian translocations and other reciprocal translocations, increase the chance of UPD due to an increased risk of malsegregation of homologs during meiosis.

The third mechanism of UPD is due to gamete complementation in which the union of two gametes results in two copies of a chromosome from a single gamete with no copies from the other gamete. This leads to disomy of a chromosome from a single parent.

Cytogenetic Abnormalities: Chromosomal, FISH and Microarray-Based Clinical Reporting, First Edition. Susan Mahler Zneimer.

Chromosome analysis cannot detect UPD, as the homologous chromosomes will appear normal. UPD is distinguishable only by DNA markers. However, the topic is discussed here due to the importance of pursuing UPD studies if the cytogenetic results appear either normal or abnormal, in which the implication of UPD is a possibility, which will have clinical consequences. The clinical consequences of UPD are seen through genomic imprinting of specific chromosomal regions in which parent-specific phenotypes are observed. Most of the chromosomes now show some clinical effects of UPD. These chromosomes have at least some regions associated with specific disorders that show specific phenotypic consequence, when present from a single parent, generally by the loss of the other parental genomic DNA.

Approximately 30–35% of clinical UPD cases also have visible chromosomal abnormalities. Abnormalities include balanced and unbalanced rearrangements and additional marker chromosomes. The remaining 65–70% of UPD cases have either reportedly normal chromosome analyses or no known chromosome analysis performed.

Table 12.1 shows the most common cytogenetic regions of concern for UPD. Table 12.2 shows the frequency of common UPD cases. Although the disorders listed in the tables are the most common syndromes known with UPD, there are over 200 imprinted genes in the human genome. Most of these genes are not yet known to be associated with specific syndromes or phenotypic consequences, but may play a role in the increased incidence of pregnancy complications observed in assisted reproduction technologies (ART).

Table 12.1 Uniparental disomy (UPD) of the most common cytogenetic regions and their corresponding clinical disorders

Chromosome region	Clinical disorder with paternal UPD	Clinical disorder with maternal UPD
Upd(2)mat		CPM for trisomy 2
Upd(6)pat	Transient neonatal diabetes with low birth rate and other reported anomalies	
Upd(7)mat		Cystic fibrosis, Silver–Russell syndrome
Upd(9)mat		CPM for trisomy 9
Upd(11)pat	CPM for trisomy 11, Beckwith–Wiedemann syndrome	
Upd(14)mat		Specific findings include MR, DD, short stature, precocious puberty
Upd(14)pat	MR, polyhydramnios, low birth weight, and other specific features	
Upd(15)mat		Prader–Willi syndrome
Upd(15)pat	Angelman syndrome	
Upd(16)mat		Trisomy 16 mosaicism, IUGR, heart defects and other anomalies
Upd(21)pat	Robertsonian translocation	
Upd(22)mat		Robertsonian translocation
Upd(X)mat		DMD

CPM, confined placental mosaicism; DD, developmental delay; DMD, Duchenne muscular dystrophy; IUGR, intrauterine growth retardation; MR, mental retardation.

Table 12.2 Frequency of common uniparental disomy (UPD) cases

UPD clinical association	Chromosome region	Incidence	Frequency of UPD
UPD detected in a newborn overall		1 in 3500–5000	0.029%
Russell Silver syndrome	matUPD(7)	1 in 100,000	5%
AS	patUPD(15)	1 in 16,000	7%
PWS	matUPD(15)	1 in 17,500	25%
BWS	UPD(11p)	1 in 13,700	20%
Transient neonatal diabetes	patUPD(6)	1 in 500,000	40%
Maternal UPD syndrome	matUPD(14)	Not estimated	>95%

Adapted from Amor and Halliday (2008).
AS, Angelman syndrome; BWS, Beckwith–Wiedemann syndrome; PWS, Prader–Willi syndrome.

Since chromosomes 14 and 15, especially, show a high frequency of UPD, it is important to include a comment in a cytogenetics report to imply this possibility and suggest testing for UPD, if applicable. This is especially true when an individual is known to be a Robertsonian translocation carrier, and there is a risk of UPD in the offspring. When UPD studies are performed, then a comment for the UPD studies may be added to the report.

12.1 Uniparental disomy of chromosome 14

EXAMPLE REPORTING OF POSSIBLE UPD FOR CHROMOSOME 14 DUE TO A ROBERTSONIAN TRANSLOCATION

ISCN Result: 45,XY,der(14;21)(q10;q10)

Interpretation

Chromosome analysis revealed an abnormal male chromosome complement in all cells examined with an apparently balanced Robertsonian translocation between the long arms of chromosomes 14 and 21.

Because the rearrangement appears to be balanced, it is most likely of no clinical significance to the patient. However, unbalanced gamete formation during meiosis may lead to pregnancy loss or to the production of abnormal offspring. Given the familial origin of most translocations, chromosome analysis of this patient's first-degree biological relatives should be considered.

Carriers of a balanced t(14;21) have a risk for trisomy 21 in their liveborn children (approximately 10% for female carriers and less than 1% for male carriers), an increased risk for miscarriages, and approximately a 0.5% risk of uniparental disomy (UPD) for chromosome 14. If clinically indicated, consider UPD studies for chromosome 14.

The abnormal phenotype associated with maternal UPD chromosome 14 includes mild developmental delay, short stature and precocious puberty. The abnormal phenotype associated with paternal UPD chromosome 14 includes mental retardation, dwarfism, skeletal dysplasia and characteristic facial features.

Genetic counseling is recommended, and prenatal diagnosis is suggested in future pregnancies.

ISCN RULES FOR REPORTING UNIPARENTAL DISOMY FROM A KNOWN PARENT

- First write the modal chromosome number of 46, followed by the sex designation.
- Then write “upd” followed by the abnormality involved, followed, in parentheses, by the chromosome involved and then, in parentheses, write the breakpoints involved.
- Then write either “mat” or “pat” for the maternally or paternally derived UPD, respectively.

EXAMPLE REPORTING OF KNOWN UPD FOR CHROMOSOME 14 DUE TO A ROBERTSONIAN TRANSLOCATION

ISCN Result: 45,XY,upd der(14;21)(q10;q10)mat

Interpretation

Chromosome analysis revealed an abnormal male chromosome complement in all cells examined with an apparently balanced Robertsonian translocation between the long arms of chromosomes 14 and 21. Genetic analysis showed that this rearrangement results in uniparental disomy inherited from the mother.

The abnormal phenotype associated with maternal UPD chromosome 14 includes mild developmental delay, short stature and precocious puberty. The abnormal phenotype associated with paternal UPD chromosome 14 includes mental retardation, dwarfism, skeletal dysplasia and characteristic facial features.

Genetic counseling is recommended, and prenatal diagnosis is suggested in future pregnancies.

12.2 Uniparental disomy of chromosome 15

EXAMPLE REPORTING OF POSSIBLE UPD FOR CHROMOSOME 15 WITH NORMAL CHROMOSOMES

ISCN Result: 46,XX

Interpretation

Chromosome analysis revealed a normal female chromosome complement in all cells examined. However, due to the indication of Angelman syndrome, the possibility of uniparental disomy exists for chromosome 15 that may account for the phenotype of this patient.

If clinically indicated, consider UPD studies for chromosome 15.

12.3 Uniparental disomy of chromosome 11p15

EXAMPLE REPORTING OF POSSIBLE UPD FOR CHROMOSOME 11 WITH A dup(11)(p15.1p15.3)

ISCN Result: 46,XX,dup(11)(p15.1p15.3)

Interpretation

Chromosome analysis revealed an abnormal female chromosome complement in all cells examined with a duplication of the distal part of the short arm of chromosome 11. This abnormality is consistent with the diagnosis of Beckwith–Wiedemann syndrome.

Uniparental disomy for chromosome 11 is also a possibility. Consider UPD studies for chromosome 11.

Genetic counseling is recommended.

Reference

Amor DJ, Halliday J. A review of known imprinting syndromes and their association with assisted reproduction technologies. Hum Reprod 2008; 23: 2826–2834.

Bibliography

Betz A, Turleau C, de Grouchy J. Heterozygosity and homozygosity for a pericentric inversion of human chromosome 3. Ann Genet 1974; 17: 79–80.

Carpenter NJ, Say B, Barber ND. A homozygote for pericentric inversion of chromosome 4. J Med Genet 1982; 19: 469–471.

Eggermann T, Meyer E, Ranke MB, et al. Diagnostic proceeding in Silver-Russell Syndrome. Mol Diagn 2005; 9: 205–209.

Engel E. A new genetic concept: uniparental disomy and its potential effect, isodisomy. Am J Med Genet 1980; 6: 137–143.

Gardner RJM, Sutherland GR. *Chromosome Abnormalities and Genetic Counselling*. Oxford Monographs on Medical Genetics. Oxford University Press, Oxford, 2003.

Gersen S, Keagle M (eds). *Principles of Clinical Cytogenetics*. Humana Press, Totowa, New Jersey, 1999.

Lalande M, Calciano MA. Molecular epigenetics of Angelman Syndrome. Cell Mol Life Sci 2007; 64: 947–960.

Liehr T. Cases with uniparental disomy (UPD). http://ssmc-tl.com/Start.html, accessed 23 January 2014.

Liehr T. Cytogenetic contribution to uniparental disomy. Mol Cytogenet 2010; 3: 3–8.

Nicholls RD, Knoll JH, Butler, Karam S, Lalande M. Genetic imprinting suggested by maternal heterodisomy in nondeletion Prader-Willi Syndrome. Nature 1989; 342: 281–285.

Online Mendelian Inheritance of Man (OMIM). www.ncbi.nlm.nih.gov/omim

Robinson WP. Mechanisms leading to uniparental disomy and their clinical consequences. Bioessays 2000; 22: 452–459.

Rooney DE, Czepulkowski BH: *Human Cytogenetics. A practical approach*. New York: Oxford University Press; 1992.

Ruggeri A, Dulcetti F, Miozzo M, et al. Prenatal search for UPD 14 and UPD 15 in 83 cases of familial and de novo heterologous Robertsonian translocations. Prenat Diagn 2004; 24: 997–1000.

Shaffer LG, McGowan-Jordan J, Schmid M (eds). *ISCN 2013: An International System for Human Cytogenetic Nomenclature*. Karger Publishers, Unionville, CT, 2013.

Spence JE, Perciaccante RG, Greig GM, et al. Uniparental disomy as a mechanism for human genetic disease. Am J Hum Genet 1988; 42: 217–226.

Tobias ES, Connor M, Ferguson Smith M. *Essential Medical Genetics*, 6th edn. Wiley-Blackwell, Oxford, 2011.

Warburton D. Uniparental disomy: a rare consequence of the high rate of aneuploidy in human gametes. Am J Hum Genet 1988; 42: 215–216.

Weksberg R, Squire JA. Molecular biology of Beckwith-Wiedemann Syndrome. Med Pediatr Oncol 1996; 27: 462–469.

Wood AJ, Oakey RJ. Genomic imprinting in mammals: emerging treatments and established theories. PLoS Genet 2006; 2: e147.

Woodage T, Prasad M, Dixon JW, et al. Bloom Syndrome and maternal uniparental disomy for chromosome 15. Am J Hum Genet 1994; 55: 74–80.

Section 2 Fluorescence *In Situ* Hybridization (FISH) Analysis

CHAPTER 13

Metaphase analysis

13.1 Introduction

This chapter describes fluorescence *in situ* hybridization (FISH) for metaphase analysis. This includes microdeletions and microduplications that are too small to detect by standard chromosome analysis, 24-color FISH paint analysis to detect markers and rearrangements of unidentifiable chromosome origin, and subtelomere analysis. All these analyses have different ISCN nomenclatures and interpretations in a cytogenetics report. Often these analyses are combined with standard cytogenetics in a single report, but some laboratories prefer to do each test as a separate report. Some laboratories will integrate the separate reports when all testing is completed. Example reports of both types will be described. Table 13.1 gives a listing of the common microdeletions and microduplications that can be diagnosed by metaphase FISH analysis to date. Most regulatory agencies require a comment in the report that genetic counseling is recommended for all abnormal results.

Fluorescence *in situ* hybridization analyses are, for the most part, not FDA-approved tests and require a special disclaimer on reports to warn physicians of this fact. Further, each laboratory is responsible for validating the hybridization process and specific criteria for each of the FISH probes used for testing. FISH probes that are not FDA approved will either be analyte specific (ASRs) or research only (RUOs). Both may be obtained from a manufacturer or "homebrewed," i.e. made by a laboratory. In either case, the following disclaimer, written word for word, must be placed in each report using non-FDA probes.

ASR/RUO disclaimer

This test was developed and its performance characteristics determined by the "name of laboratory here" as required by the CLIA'88 regulations. It has not been cleared or approved by the U.S. Food and Drug Administration (FDA). The FDA has determined that such clearance or approval is not necessary. These results are provided for informational purposes only, and should be interpreted only in the context of established procedures and/or diagnostic criteria.

Fluorescence *in situ* hybridization probes may have different DNA composition depending on the manufacturer or laboratory that developed the probe. Therefore, the examples given below for each probe are designed to be generic and not specific to a particular company or laboratory. However, there may be descriptions of probes that appear to be specifically designed, especially

Cytogenetic Abnormalities: Chromosomal, FISH and Microarray-Based Clinical Reporting, First Edition. Susan Mahler Zneimer.

Table 13.1 List of microdeletions and microduplications detectable by metaphase FISH analysis

Chromosomal region	Syndrome	Genes associated with disorder	Deletion/ duplication
1p36.3	1p36 deletion syndrome	Multiple	Deletion
4p16.3	Wolf–Hirschhorn	WHSC1	Deletion
5p15.2	Cri-du-chat		Deletion
5q35	Sotos	NSD1	Deletion
7q11.23	Williams	ELN	Deletion
8q24.1	Langer–Gideon	EXT1, TRPS1	Deletion
15q11.2	Prader–Willi	SNRPN1	Deletion
15q11.2	Angelman	UBE3A	Deletion
15q11.2	15q11.2 duplication syndrome	SNRPN1, D15S10	Duplication
17p11.2	Smith–Magenis	SMS	Deletion
17p11.2	Potocki–Lupski	SMS	Duplication
17p13.3	Miller–Dieker	LIS1	Deletion
20p12.2	Alagille	JAG1, NOTCH2	Deletion
22q11.2	DiGeorge/velocardiofacial	HIRA	Deletion
22q11.2	22q11.2 duplication syndrome and cat-eye syndrome	HIRA	Duplication
22q13.3	22q13.3 deletion syndrome	ARSA	Deletion
Xp22.3	Kallmann	KAL1	Deletion
Xp22.3	Steroid sulfatase	STS	Deletion
Xq13.2	Turner and other sex chromosome disorders	XIST	Deletion
Yp11.3	Turner and other sex chromosome disorders	SRY	Deletion

FDA-approved probes, but this is for the purpose of clarity and not to be misconstrued as the only possible configuration for the particular probe described.

13.2 Reporting normal results

ISCN RULES FOR REPORTING A NORMAL FISH METAPHASE TEST

- Metaphase FISH nomenclature always begins with “ish” followed by a space.
- Next write the chromosome involved, followed by the chromosome band of the region tested.
- Then write the probe name, which can be referred to by the clone name or, if unavailable, use the locus designation used by the Genome Database (GDB). If that probe is

unavailable, then use the Human Genome Organization (HUGO)-approved nomenclature. It is also acceptable to use the GDB nucleotide number in a specific genome build, such as NCBI build 35 (B35).

- The probe name is usually an acronym, so it is written as all capitalized and not italicized.
- With normal FISH results, the probe name is followed by the multiplication sign "x" followed by the number of signals observed (typically two signals, one per chromosome with normal results).

Note that for normal results, the chromosome and chromosome band region are not surrounded by parentheses, but the probe and signal number are surrounded by parentheses.

Also note that even though a control probe may be present in the analysis, only the disease gene is written in the nomenclature, although the control probe may be mentioned in the interpretation for clarity.

For example, for the DiGeorge/velocardiofacial syndromes, a normal FISH result with the gene name would be written as: ish 22q11.2(HIRAx2).

EXAMPLE REPORTING OF NORMAL RESULTS

ISCN Result: ish __ chromosome band(probex2)

Interpretation

Fluorescence *in situ* hybridization (FISH) analysis was performed with the __ chromosome region (probe).

This analysis showed no evidence for a deletion at _ in _ metaphase cells scored, which is commonly observed in ___ syndrome.

Since FISH showed only normal results, consider microarray analysis to detect smaller deletions or regions not covered by this FISH probe.

EXAMPLE REPORT FORMAT FOR NORMAL METAPHASE FISH ANALYSIS

Laboratory Name
Cytogenetics Report

Patient Name: Jane Doe
Ordering Physician: Dr. Smith
Date of Birth: 1/1/2000

Collection date: 1/1/2014
Report date: 1/10/2014
Laboratory Number: A14-000021

Specimen Type: Peripheral Blood
Indication for Study: Rule out a 1p36 microdeletion

Number of cells analyzed by FISH: 10 metaphase cells

ISCN Result: ish 1p36.3(CDC2L2x2)

Interpretation

Fluorescence *in situ* hybridization (FISH) analysis was performed with the 1p36.3 probe (CDC2L2) and a 1p subtelomeric control probe (CEB108/T7).

This analysis showed no evidence for a deletion or rearrangement in 10 metaphase cells scored, which is commonly observed in 1p36.3 deletion syndrome.

Since FISH showed only normal results, consider microarray analysis to detect smaller deletions or regions not covered by these FISH probes.

Date: January 10, 2014

Susan Zneimer, PhD, FACMG
Cytogenetics Laboratory Director

(Any disclaimer here you may want to provide.)

This test was performed at: Address of Laboratory

Dr Jones, Medical Director
Laboratory License Number: 00001

(Cytogenetic Image)

13.3 Common disclaimers

Laboratories often add their own disclaimer to FISH reports to address the limitations of this type of testing. That is, FISH testing does not detect any abnormality that is not covered by the DNA probe being performed, and does not replace standard chromosome analysis. Therefore, to some, it is important that those reading these reports understand what is *not* being assessed by FISH testing so as to not be misled by the information given.

EXAMPLE GENERAL DISCLAIMER

FISH testing only analyzes the presence or absence of the DNA sequences of the FISH probe analyzed. A normal result does not preclude the presence of cytogenetic or molecular changes elsewhere in the genome. Therefore, FISH testing should be considered as an adjunct test to other genome testing, including standard chromosome analyses and other molecular genetic assays.

The following will not be detected by this analysis: uniparental disomy, mutations affecting imprinting, deletions that are contiguous to or smaller than this probe, or point mutations.

EXAMPLE DISCLAIMER FOR NORMAL CYTOGENETICS REFLEXING TO METAPHASE FISH

Since chromosome analysis on metaphase cells did not reveal a cytogenetically visible deletion at _, fluorescence *in situ* hybridization (FISH) is recommended to rule out a microdeletion/microduplication at the _ locus on _ associated with _. The preparations from the current specimen can be used for this testing.

EXAMPLE GENERAL DISCLAIMER WITH AN ADDED COMMENT FOR ABNORMAL FISH RESULTS

In addition, identification of an abnormality by FISH analysis may require standard chromosome analysis or other molecular genetic analyses to determine the mutational mechanism accounting for this result.

13.4 Microdeletions

Microdeletions discussed in this section refer to submicroscopic deletions that are detected by FISH analysis and result in a recognizable clinical disorder. Many of these deletions are now diagnosed by microarrays, but FISH analysis is still used by many laboratories as a simpler approach to identify these disorders on a singular basis. Many of the microdeletion disorders are contiguous gene syndromes, which involve more than one gene deletion, resulting in variable phenotypic consequences depending on the genes involved in the deletion region. The mechanisms for many of these gene syndromes are due to non-allelic homologous recombination in which the homologous pair of chromosomes, when they undergo the recombination process in meiosis, form an unequal crossing-over event resulting in a deletion or duplication of a small chromosomal region. These regions are considered "hot spots" in which these crossing-over events are non-random, giving rise to specific and distinctive disorders.

As genomic technologies continue to improve the resolution of genomic regions of chromosomes, more microdeletion syndromes will be identified, but they may also be diagnosed at the cytogenetic molecular level of microarrays. Other small deletions and regions not detectable by FISH analysis will be discussed in Section 3.

13.4.1 ISCN rules for microdeletions

ISCN RULES FOR REPORTING AN ABNORMAL (DELETED) MICRODELETION FISH TEST

- Metaphase FISH nomenclature always begins with "ish" followed by a space.
- Next write "del" followed by the chromosome involved, surrounded by parentheses.
- Then write the chromosome band of the region tested, surrounded by parentheses.
- The band region that is deleted is usually a single interstitial band, so the nomenclature needs to have the band written twice (without a semicolon) to reflect that it is not a terminal deletion.
- Then write the probe name, which can be referred to by the clone name or, if unavailable, use the locus designation used by the GDB. If that probe is unavailable, use the HUGO-approved nomenclature. It is also acceptable to use the GDB nucleotide number in a specific genome build, such as NCBI build 35 (B35).
- The probe name is usually an acronym, so it is written as all capitalized and not italicized.
- With abnormal FISH results, the probe name is followed by a minus sign "-".
- The probe name and minus sign are both surrounded by parentheses.

Note that for abnormal results, the chromosome, chromosome band region, probe and signal number are all surrounded by separate parentheses.

Also note that even though a control probe may be present in the analysis, only the disease gene is written in the nomenclature, although the control probe may be mentioned in the interpretation of the report for clarity.

For example, for the DiGeorge/velocardiofacial syndromes, an abnormal FISH result using the GDB D-number would be written as: ish del(22)(q11.2q11.2)(D22S75-).

When both homologs of chromosome 22 show a deletion with the probe tested, then the nomenclature would be written as: ish del(22)(q11.2q11.2)(D22S75-),del(22)(q11.2q11.2)(D22S75-).

13.4.2 Example reporting of microdeletions

EXAMPLE REPORTING OF ABNORMAL RESULTS

ISCN Result: ish del(chromosome)(bands)(probe-)

Interpretation

Fluorescence *in situ* hybridization (FISH) analysis was performed with the __chromosome region probe (probe) and a control __ probe (probe).

This analysis showed a deletion of chromosome __ region in 10 metaphase cells scored. This result is consistent with the diagnosis of __ syndrome.

Parental chromosome and FISH studies are recommended to rule out either the presence of a balanced rearrangement or whether this abnormality is *de novo* in origin.

Genetic counseling is recommended.

EXAMPLE REPORT FORMAT FOR ABNORMAL METAPHASE FISH ANALYSIS

Laboratory Name
Cytogenetics Report

Patient Name: Jane Doe
Ordering Physician: Dr. Smith
Date of Birth: 1/1/2000

Collection Date: 1/1/2014
Report Date: 1/10/2014
Laboratory Number: A14-000021

Specimen Type: Peripheral Blood
Indication for Study: Rule out a 1p36 microdeletion

Number of cells analyzed by FISH: 10 metaphase cells

ISCN Result: ish del(1)(p36.3p36.3)(CDC2L2-)

Interpretation

Fluorescence *in situ* hybridization (FISH) analysis was performed with the 1p36.3 probe (CDC2L2) and a 1p subtelomeric control probe (CEB108/T7).

This analysis showed a deletion of chromosome 1p36.3 region in 10 metaphase cells scored. This result is consistent with the diagnosis of chromosome 1p36.3 deletion syndrome.

Parental chromosome and FISH studies are recommended to rule out either the presence of a balanced rearrangement or whether this abnormality is *de novo* in origin.
Genetic counseling is recommended.

Date: January 10, 2014

Susan Zneimer, PhD, FACMG
Cytogenetics Laboratory Director

(Any disclaimer here you may want to provide.)

This test was performed at: Address of Laboratory

Dr Jones, Medical Director
Laboratory License Number: 00001

(Cytogenetic Image)

13.4.3 Microdeletion disorders by chromosome order

Chromosome 1p36.3 deletion

SPECIFIC FEATURES OF 1P36.3 DELETION

Chromosome 1p36.3 deletions contain a deletion region close to the short arm telomere of chromosome 1. The incidence is approximately 1 in 5000–10,000 newborns, making it one of the most common microdeletions currently known. The clinical features of this syndrome include severe mental impairment, absence of speech or speech delay, vision or hearing problems, behavior problems and numerous malformations throughout the body. The most common structural malformations include brain anomalies and seizures, hypotonia, dysphagia, microcephaly, distinct facial features including deep-set eyes and mid-face hypoplasia, a broad, flat nose, broad philtrum, low-set and abnormally shaped ears. Other organ abnormalities may include those of the skeleton, heart, gastrointestinal system, kidneys or genitalia.

EXAMPLE REPORTING OF CHROMOSOME 1p36.3 DELETION – NORMAL

ISCN Result: ish 1p36.3(CDC2L2x2)

Interpretation

Fluorescence *in situ* hybridization (FISH) analysis was performed with the 1p36.3 probe (CDC2L2) and a 1p subtelomeric control probe (CEB108/T7).

This analysis showed no evidence for a deletion or rearrangement in 10 metaphase cells scored, which is commonly observed in 1p36.3 deletion syndrome.

Since FISH showed only normal results, consider microarray analysis to detect smaller deletions or regions not covered by these FISH probes.

EXAMPLE REPORTING OF CHROMOSOME 1p36.3 DELETION – ABNORMAL

ISCN Result: ish del(1)(p36.3p36.3)(CDC2L2-)

Interpretation

Fluorescence *in situ* hybridization (FISH) analysis was performed with the 1p36.3 probe (CDC2L2) and a 1p subtelomeric control probe (CEB108/T7).

This analysis showed a deletion of chromosome 1p36.3 region in 10 metaphase cells scored. This result is consistent with the diagnosis of 1p36.3 deletion syndrome.

Parental chromosome and FISH studies are recommended to rule out either the presence of a balanced rearrangement or whether this abnormality is *de novo* in origin.

Genetic counseling is recommended.

Chromosome 4p16.3 deletions – Wolf–Hirschhorn syndrome

SPECIFIC FEATURES OF WOLF–HIRSCHHORN SYNDROME (WHS)

Wolf–Hirschhorn syndrome is a microdeletion syndrome involving the chromosome 4p16.3 region. This deletion syndrome may be large enough to be visible by standard cytogenetic analysis. However, it may be worthwhile to confirm a deletion of the critical region by FISH analysis. Other deletions may be too small to be clearly defined by chromosome studies alone, as many deletions are less than 2 Mb in size. Small terminal and interstitial deletions have been reported. The critical region is defined as a 165 Kb region between the GDB probes D4S166 and D4S3247. The incidence of WHS is approximately 1 in 50,000 newborns and occurs in approximately twice as many females as males.

The clinical features of this syndrome are distinctive, including a characteristic facial appearance, delayed growth and development, intellectual disability and seizures. Phenotypic manifestations include a broad, flat nasal bridge and a high forehead, often described as a "Greek warrior helmet." Other characteristic facial features include a short philtrum, down-turned mouth, micrognathia, poorly formed ears with pits or tags, and microcephaly. Other clinical manifestations include failure to thrive, delayed growth and development, hypotonia, short stature, seizures and moderate-to-severe mental impairment. Additional organ involvement includes skeletal abnormalities, dental problems including missing teeth, cleft palate and/or cleft lip, and abnormalities of the eyes, heart, genitourinary tract and brain.

Pitt–Rogers–Danks syndrome has features which overlap the features of WHS. It is now recognized that these two conditions are actually part of a single syndrome, and that Pitt–Rogers–Danks syndrome has a less severe phenotype than WHS.

The severity of disease is related to the loss of multiple genes in the WHS critical region, including the WHSC1, LETM1 and MSX1 genes. This is described as a contiguous gene deletion syndrome. The WHSC1 gene is associated with many of the characteristic features of WHS, including the distinctive facial appearance and developmental delay. Deletion of the LETM1 gene is associated with seizures or other abnormal electrical activity in the brain. Deletion of the MSX1 gene may be responsible for dental abnormalities and cleft lip and/or palate.

EXAMPLE REPORTING OF CHROMOSOME 4p16.3 DELETIONS – WOLF–HIRSCHHORN SYNDROME – NORMAL

ISCN Result: ish 4p16.3(WHSC1x2)

Interpretation

Fluorescence *in situ* hybridization (FISH) analysis was performed with the 4p16.3 probe (WHSC1) and a control CEP4 probe.

This analysis showed no evidence for a deletion or rearrangement in 10 metaphase cells scored, which is commonly observed in Wolf–Hirschhorn syndrome.

Since FISH showed only normal results, consider microarray analysis to detect smaller deletions or regions not covered by this FISH probe.

EXAMPLE REPORTING OF CHROMOSOME 4p16.3 DELETIONS – WOLF–HIRSCHHORN SYNDROME – ABNORMAL

ISCN Result: ish del(4)(p16.3p16.3)(WHSC1-)

Interpretation

Fluorescence *in situ* hybridization (FISH) analysis was performed with the 4p16.3 probe (WHSC1) and a control CEP4 probe.

This analysis showed a deletion of chromosome 4p16.3 region in 10 metaphase cells scored. This result is consistent with the diagnosis of Wolf–Hirschhorn syndrome.

Parental chromosome and FISH studies are recommended to rule out either the presence of a balanced rearrangement or whether this abnormality is *de novo* in origin.

Genetic counseling is recommended.

Chromosome 5p15.2 deletions – cri-du-chat syndrome (Figure 13.1)

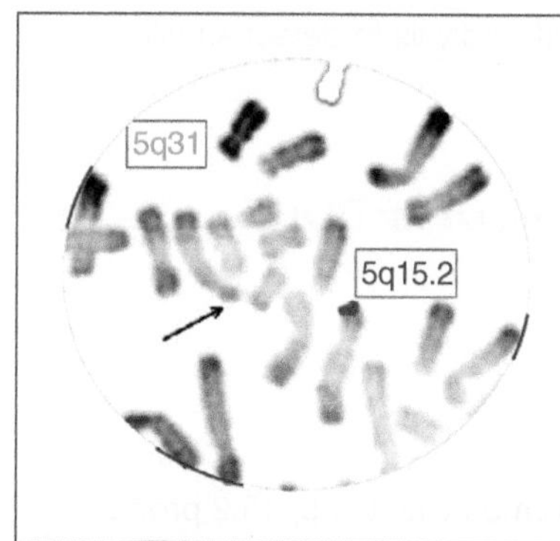

Figure 13.1 Metaphase FISH analysis resulting in cri-du-chat syndrome: ish del(5)(p15.2p15.2)(D5S23-). Courtesy of Lauren Jenkins PhD, Kaiser Permanente Regional Laboratory, Northern California.

SPECIFIC FEATURES OF CRI-DU-CHAT SYNDROME

Cri-du-chat syndrome is a microdeletion syndrome involving the chromosome 5p15.2 region. This deletion syndrome may be large enough to be visible by standard cytogenetic analysis, but the breakpoints are variable and may be worthwhile to confirm a deletion of the critical region by FISH analysis. Other deletions may be too small to be clearly defined by chromosome studies alone. The critical region is considered to be proximal to p15.3 and as a contiguous gene deletion syndrome, certain genes are correlated with specific phenotypic manifestations.

The incidence of this syndrome is approximately 1 in 20,000–50,000 newborns. One major clinical feature of these newborns is a high-pitched cry that sounds like that of a cat. Therefore, this syndrome is appropriately named for the French meaning of "cat's cry." Other clinical features of this syndrome include severe mental impairment, developmental delay, microcephaly, hypotonia, hypertelorism, low-set ears, a small jaw and a rounded face. One of the most severe anomalies is a heart defect, which makes this syndrome critical to diagnose within hours of birth in order to determine whether surgery is a viable option or not. Usually, with a life-threatening heart defect that is not associated with cri-du-chat syndrome, surgery may save the baby's life whereaswith a chromosome 5 deletion, repairing the heart will not ameliorate the conditions of the syndrome.

The correlation of clinical features is related to the region of the deletion. For example, a deletion of the interstitial bands of 5p15.31 to 5p15.33 shows manifestations including mild mental impairment, speech delay and typical cry, but with no other findings. Terminal deletions of the region of 5p15.1–5p15.2 to the telomere are associated with typical cry and facial anomalies, with moderate mental impairment. Deletions of 5p14.3–p14.1 interstitial region are associated with typical cry and facial anomalies, with moderate-to-severe mental impairment.

EXAMPLE REPORTING OF CHROMOSOME 5p15.2 DELETIONS – CRI-DU-CHAT SYNDROME – NORMAL

ISCN Result: ish 5p15.2(D5S23x2)

Interpretation

Fluorescence *in situ* hybridization (FISH) analysis was performed with the 5p15.2 probe (D5S23) and a control D5S721 probe.

This analysis showed no evidence for a deletion or rearrangement in 10 metaphase cells scored, which is commonly observed in cri-du-chat syndrome.

Since FISH showed only normal results, consider microarray analysis to detect smaller deletions or regions not covered by this FISH probe.

EXAMPLE REPORTING OF CHROMOSOME 5p15.2 DELETIONS – CRI-DU-CHAT SYNDROME – ABNORMAL

ISCN Result: ish del(5)(p15.2p15.2)(D5S23-)

Interpretation

Fluorescence *in situ* hybridization (FISH) analysis was performed with the 5p15.2 probe (D5S23) and a control D5S721 probe.

This analysis showed a deletion of chromosome 5p15.2 region in 10 metaphase cells scored. This result is consistent with the diagnosis of cri-du-chat syndrome.

Parental chromosome and FISH studies are recommended to rule out either the presence of a balanced rearrangement or whether this abnormality is *de novo* in origin.

Genetic counseling is recommended.

Chromosome 5q35 deletions – Sotos syndrome

SPECIFIC FEATURES OF SOTOS SYNDROME

Sotos syndrome is a microdeletion syndrome involving the chromosome 5p15.2 region. It is the result of haploinsufficiency of the NSD1 gene (nuclear receptor binding SET domain protein 1). Mutations, partial NSD1 gene deletions and microdeletions of the 5q35 chromosomal region have been reported in this syndrome. The majority of microdeletions in Sotos syndrome are seen in patients of Japanese heritage (50% of patients) whereas only a minority of non-Japanese patients have this microdeletion.

Sotos syndrome, also known as cerebral gigantism, is characterized by distinctive facial features, including prominent forehead, downward-slanting palpebral fissures, pointed chin and macrocephaly. Other manifestations include learning disabilities, development delay, social and motor delays, and behavior problems including anxiety, withdrawal, depression, phobias, sleep apnea, tantrums, inappropriate speech and irritability.

The incidence is approximately 1 in 14,000 livebirths, with most abnormalities seen as a *de novo* event.

EXAMPLE REPORTING OF CHROMOSOME 5q35 DELETIONS – SOTOS SYNDROME – NORMAL

ISCN Result: ish 5q35(D5S721)

Interpretation

Fluorescence *in situ* hybridization (FISH) analysis was performed with the 5q35 probe (D5S721) and a 5p15.2 control (D5S23) probe.

This analysis showed no evidence for a deletion or rearrangement in 10 metaphase cells scored, which is commonly observed in Sotos syndrome.

Since FISH showed only normal results, consider microarray analysis to detect smaller duplications or regions not covered by this FISH probe.

EXAMPLE REPORTING OF CHROMOSOME 5q35 DELETIONS – SOTOS SYNDROME – ABNORMAL

ISCN Result: ish del(5)(q35q35)(D5S721-)

Interpretation

Fluorescence *in situ* hybridization (FISH) analysis was performed with the 5q35 probe (D5S721) and a 5p15.2 control (D5S23) probe.

This analysis showed a deletion of chromosome 5q35 region in 10 metaphase cells scored. This result is consistent with the diagnosis of Sotos syndrome.

Parental chromosome and FISH studies are recommended to rule out either the presence of a balanced rearrangement or whether this abnormality is *de novo* in origin.

Genetic counseling is recommended.

Chromosome 7q11.23 deletions – Williams syndrome (Figure 13.2)

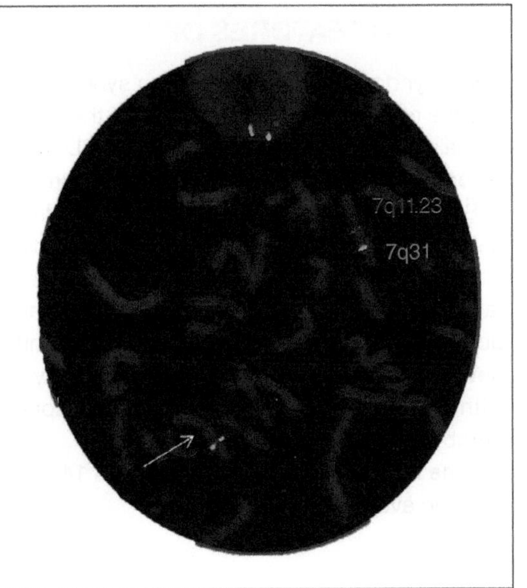

Figure 13.2 Metaphase FISH analysis resulting in Williams syndrome: ish del(7)(q11.23q11.23)(ELN-). Courtesy of Mehdi Jamehdor MD, Kaiser Permanente Regional Laboratory, Southern California.

SPECIFIC FEATURES OF WILLIAMS SYNDROME

Williams Syndrome (WS) is a microdeletion syndrome involving the chromosome 7q11.23 region. Deletions of the WS critical region include the elastin gene, but there are other genes within this 1.6 Mb deletion region. It is now known that non-homologous recombination between the GTF2IP1/NCF1 locus and the GTF2IP1/NCF1P1 locus occurs causing this abnormality. There are other rare intrachromosomal exchanges between loci which can also account for deletions. The incidence is approximately 1 in 7500–20,000 individuals in the population.

Clinical findings include mild-to-moderate intellectual disability, learning problems, unique personality characteristics and distinctive facial features including a broad forehead, short nose with a broad tip, full cheeks, wide mouth and dental problems, such as small, widely spaced teeth and teeth that are crooked or missing.

Individuals with WS typically have difficulty with visual-spatial tasks, but do well on tasks that involve spoken language, music and learning by repetition (rote memorization). Affected individuals have outgoing, engaging personalities, although attention deficit disorder (ADD), problems with anxiety and phobias are common. Cardiovascular disease is common, including supravalvular aortic stenosis (SVAS), hypertension, joint problems and soft, loose skin, hypercalcemia and short stature. Medical problems involving the eyes and vision, the digestive tract and the urinary system have also been reported.

The genes involved in the deletion region include CLIP2, ELN, GTF2I, GTF2IRD1 and LIMK1. Loss of the ELN gene is associated with connective tissue abnormalities and cardiovascular disease, specifically supravalvular aortic stenosis. Deletions of CLIP2, GTF2I, GTF2IRD1, LIMK1 and perhaps other genes are associated with visual-spatial difficulties, unique behavioral characteristics and other cognitive difficulties. Deletion of the NCF1 gene is related to a decreased risk of developing hypertension. Therefore, the loss of this gene appears to be a protective factor.

EXAMPLE REPORTING OF CHROMOSOME 7q11.23 DELETIONS – WILLIAMS SYNDROME – NORMAL

ISCN Result: ish 7q11.23(ELN/LIMK1)

Interpretation

Fluorescence *in situ* hybridization (FISH) analysis was performed with the 7q11.23 probe (ELN/LIMK1) and control (D7S486, D7S522) probes.

This analysis showed no evidence for a deletion or rearrangement in 10 metaphase cells scored, which is commonly observed in Williams syndrome.

Since FISH showed only normal results, consider microarray analysis to detect smaller deletions or regions not covered by this FISH probe.

EXAMPLE REPORTING OF CHROMOSOME 7q11.23 DELETIONS – WILLIAMS SYNDROME – ABNORMAL

ISCN Result: ish del(7)(q11.23q11.23)(ELN/LIMK1)

Interpretation

Fluorescence *in situ* hybridization (FISH) analysis was performed with the 7q11.23 probe (ELN/LIMK1) and control (D7S486, D7S522) probes.

This analysis showed a deletion of chromosome 7q11.23 region in 10 metaphase cells scored. This result is consistent with the diagnosis of Williams syndrome.

Parental chromosome and FISH studies are recommended to rule out either the presence of a balanced rearrangement or whether this abnormality is *de novo* in origin.

Genetic counseling is recommended.

Chromosome 8q24.1 deletions – Langer–Gideon syndrome

SPECIFIC FEATURES OF LANGER–GIDEON SYNDROME (LGS)

Langer–Gideon syndrome, also called trichorhinophalangeal syndrome type II, is a microdeletion syndrome involving the chromosome 8q24.1 region. Deletions of the LGS critical region include the Exostosin-1 (EXT1) and Zinc finger transcription factor (TRPS1) genes, but there are other genes within this deletion region. LGS is often described as a contiguous gene deletion syndrome due to the loss of several genes.

The main clinical manifestations of this disorder include bone abnormalities and distinctive facial features. Benign bone tumors called exostoses are common, as are short stature, epiphyses of the long bones, sparse scalp hair, a rounded nose, a long, flat area between the nose and philtrum and a thin upper lip. Many cases have been reported with individuals who have some intellectual disability and loose skin in infancy that is usually resolved by adulthood.

The clinical manifestations vary with the genes involved. Deletions and mutations of the EXT1 gene are associated with multiple exostoses; deletions and mutations of the TRPS1 gene are associated with bone and facial abnormalities. The loss of function of additional genes in the LGS region contributes to the varied features of this condition.

Fluorescence *in situ* hybridization analysis will detect deletions that are 100 Kb or greater within the deletion region; however, it is possible that smaller deletions are present and will need molecular analysis for a diagnosis. This is a rare disorder whose incidence has not yet been determined.

EXAMPLE REPORTING OF CHROMOSOME 8q24.1 DELETIONS – LANGER–GIDEON SYNDROME – NORMAL

ISCN Result: ish 8q24.1(EXT1,TRPS1)x2

Interpretation

Fluorescence *in situ* hybridization (FISH) analysis was performed with the 7q11.23 probes (EXT1, TRPS1).

This analysis showed no evidence for a deletion or rearrangement in 10 metaphase cells scored, which is commonly observed in Langer–Gideon syndrome.

Since FISH showed only normal results, consider microarray analysis to detect smaller deletions or regions not covered by this FISH probe.

EXAMPLE REPORTING OF CHROMOSOME 8q24.1 DELETIONS – LANGER–GIDEON SYNDROME – ABNORMAL

ISCN Result: ish del(7)(q11.23q11.23)(EXT1-,TRPS1-)

Interpretation

Fluorescence *in situ* hybridization (FISH) analysis was performed with the 7q11.23 probes (EXT1, TRPS1).

This analysis showed a deletion of chromosome 8q24.1 region in 10 metaphase cells scored. This result is consistent with the diagnosis of Langer–Gideon syndrome.

Parental chromosome and FISH studies are recommended to rule out either the presence of a balanced rearrangement or whether this abnormality is *de novo* in origin.

Genetic counseling is recommended.

Chromosome 15q11.2 deletions

SPECIFIC FEATURES OF 15q11.2 DELETION SYNDROMES

Both Prader–Willi and Angelman syndromes are the result of deletions of the same region of chromosome 15, at band 15q11.2. They are distinct clinical syndromes, manifesting very different phenotypes, but both are the result of the loss of function of different genes within a 4 Mb region whose loci on chromosome 15 are derived from either the paternal or maternal chromosome. Loss of function of the paternal allele of this chromosomal region results in Prader–Willi syndrome and loss of function of the maternal allele results in Angelman syndrome. The loss of function may result from either a deletion that can be detected by FISH analysis or by uniparental disomy in which both homologs of chromosome 15 are inherited from a single parent. Loss of function may also result from a mutation of one of these critical genes, especially when it involves the imprinting center (IC) on chromosome 15q11.2, which is a 100 Kb DNA segment just proximal to the SNRPN gene.

EXAMPLE REPORTING OF CHROMOSOME 15q11.2 DELETIONS – ANGELMAN OR PRADER–WILLI SYNDROMES – NORMAL

ISCN Result: ish 15q11.2q12(SNRPN,D15S10)x2

Interpretation

Fluorescence *in situ* hybridization (FISH) analysis was performed with the 15q11.2-q12 probes (SNRPN, D15S10) and a control PML probe.

This analysis showed no evidence for a deletion or rearrangement in 10 metaphase cells scored, which is commonly observed in Angelman and Prader–Willi syndromes.

Since FISH showed only normal results, consider microarray analysis to detect smaller deletions or regions not covered by these FISH probes.

EXAMPLE REPORTING OF CHROMOSOME 15q11.2 DELETIONS – ANGELMAN OR PRADER–WILLI SYNDROMES – ABNORMAL

ISCN Result: ish del(15)(q11.2q12)(SNRPN-,D15S10-)

Interpretation

Fluorescence *in situ* hybridization (FISH) analysis was performed with the 15q11.2-q12 probes (SNRPN, D15S10) and a PML control probe.

This analysis showed a deletion of chromosome 15q11.2-q12 region in 10 metaphase cells scored. This result is consistent with the diagnosis of either Prader–Willi or Angelman syndrome, depending on parent-of-origin.

Parental chromosome and FISH studies are recommended to rule out either the presence of a balanced rearrangement or whether this abnormality is *de novo* in origin.

Genetic counseling is recommended.

Angelman syndrome

SPECIFIC FEATURES OF ANGELMAN SYNDROME

Angelman syndrome (AS), which is caused by the loss of function of the chromosome 15q11.2 region, is due to loss of the maternal functioning gene, UBE3A. In 70% of patients, a deletion of this region is detectable by FISH analysis. In 20% of patients, the UBE3A gene contains mutations; in 5% of patients, uniparental disomy (UPD) of the paternal allele is present, causing loss of the maternal allele, opposite to the allele inheritance of Prader–Willi syndrome.

The clinical features of Angelman syndrome include severe mental impairment, developmental delay, poor or absent speech, reduced hair and skin pigmentation, facial dysmorphism and microcephaly. Many of the effects of this disorder occur in the brain, which is exemplified by an abnormal electroencephalogram with posterior high-voltage sharp waves. A characteristic behavioral pattern is seen in these patients, including jerky or spasmodic movements and times of inappropriate laughter.

The overall incidence of AS is 1 in 20,000. As with Prader–Willi syndrome, *de novo* mutations are usually sporadic, with a low recurrence risk. However, maternally inherited chromosome rearrangements have been reported in Angelman syndrome.

EXAMPLE REPORTING OF CHROMOSOME 15q11.2 DELETIONS – ANGELMAN SYNDROME – NORMAL

ISCN Result: ish 15q11.2q12(D15S10x2)

Interpretation

Fluorescence *in situ* hybridization (FISH) analysis was performed with the 15q11.2-q12 probe (D15S10) and a control PML probe.

This analysis showed no evidence for a deletion or rearrangement in 10 metaphase cells scored, which is commonly observed in Angelman syndrome.

Since FISH showed only normal results, consider microarray analysis to detect smaller deletions or regions not covered by this FISH probe.

EXAMPLE REPORTING OF CHROMOSOME 15q11.2 DELETIONS – ANGELMAN SYNDROME – ABNORMAL

ISCN Result: ish del(15)(q11.2q12)(D15S10-)

Interpretation

Fluorescence *in situ* hybridization (FISH) analysis was performed with the 15q11.2-q12 probe (D15S10) and a control PML probe.

This analysis showed a deletion of chromosome 15q11.2-q12 region in 10 metaphase cells scored. This result is consistent with the diagnosis of Angelman syndrome.

Maternal chromosome and FISH studies are recommended to rule out either the presence of a balanced rearrangement or whether this abnormality is *de novo* in origin.

Genetic counseling is recommended.

Prader–Willi syndrome

SPECIFIC FEATURES OF PRADER–WILLI SYNDROME

Prader–Willi syndrome (PWS), caused by loss of function of the paternal allele of chromosome 15q11.2, shows an interstitial deletion of 3–4 Mb, including the critical gene SNRPN, in approximately 70–75% of patients. This deletion region is detected by FISH analysis using the SNRPN gene or a combination of genes in the critical region of 15q11.2. FISH probes are generally 150 Kb or greater and thus cannot detect smaller deletions or mutations of the imprinting center, nor can they detect UPD as the cause of loss of function rather than a deletion.

The clinical features of PWS include hypotonia and poor swallowing in newborns, hypogonadism, mental impairment, obesity due to uncontrolled eating habits, prominent forehead, palpebral fissures, and notably small hands and feet.

The overall incidence is 1 in 10,000 and the recurrence risk is very low, especially when a *de novo* deletion is detected. However, 2% of patients have a parentally derived structural chromosomal rearrangement. In 1% of patients, a mutation or microdeletion of the IC is present, which is too small to detect by FISH. The remaining 20–25% of patients have UPD of the maternal chromosome 15, resulting in loss of the paternal chromosome, and loss of function of the paternal SNRPN gene.

EXAMPLE REPORTING OF CHROMOSOME 15q11.2 DELETIONS – PRADER–WILLI SYNDROME – NORMAL

ISCN Result: ish 15q11.2q12(SNRPNx2)

Interpretation

Fluorescence *in situ* hybridization (FISH) analysis was performed with the 15q11.2-q12 probe (SNRPN) and a control PML probe.

This analysis showed no evidence for a deletion or rearrangement in 10 metaphase cells scored, which is commonly observed in Prader–Willi syndrome.

Since FISH showed only normal results, consider microarray analysis to detect smaller deletions or regions not covered by this FISH probe.

EXAMPLE REPORTING OF CHROMOSOME 15q11.2 DELETIONS – PRADER–WILLI SYNDROME – ABNORMAL

ISCN Result: ish del(15)(q11.2q12)(SNRPN-)

Interpretation

Fluorescence *in situ* hybridization (FISH) analysis was performed with the 15q11.2-q12 probe (SNRPN) and a control PML probe.

This analysis showed a deletion of chromosome 15q11.2-q12 region in 10 metaphase cells scored. This result is consistent with the diagnosis of Prader–Willi syndrome.

Paternal chromosome and FISH studies are recommended to rule out either the presence of a balanced rearrangement or whether this abnormality is *de novo* in origin.

Genetic counseling is recommended.

Chromosome 17p11.2 deletions – Smith–Magenis syndrome

SPECIFIC FEATURES OF SMITH–MAGENIS SYNDROME

Smith–Magenis syndrome is a microdeletion syndrome involving the chromosome 17q11.2 region. The phenotype of this syndrome may be variable, but common manifestations include failure to thrive, mental impairment, brachycephaly, microcephaly, prominent forehead, flat and broad mid-face and nasal bridge, strabismus, myopia, cleft palate, malformed ears, prognathism, scoliosis and cryptorchidism. This syndrome also includes behavioral features, including self-mutilation, sleep abnormalities and hyperactivity.

There is a corresponding microduplication syndrome of this region (Potocki–Lupski syndrome), which will be discussed in the microduplication section of this chapter.

EXAMPLE REPORTING OF CHROMOSOME 17p11.2 DELETIONS – SMITH–MAGENIS SYNDROME – NORMAL

ISCN Result: ish 17p11.2(SMSx2)

Interpretation

Fluorescence *in situ* hybridization (FISH) analysis was performed with the 17p11.2 probe (SMS) and a control RARA probe.

This analysis showed no evidence for a deletion or rearrangement in 10 metaphase cells scored, which is commonly observed in Smith–Magenis syndrome.

Since FISH showed only normal results, consider microarray analysis to detect smaller deletions or regions not covered by this FISH probe.

EXAMPLE REPORTING OF CHROMOSOME 17p11.2 DELETIONS – SMITH–MAGENIS SYNDROME – ABNORMAL

ISCN Result: ish del(17)(p11.2p11.2)(SMS-)

Interpretation

Fluorescence *in situ* hybridization (FISH) analysis was performed with the 17p11.2 probe (SMS) and a control RARA probe.

This analysis showed a deletion of chromosome 17p11.2 region in 10 metaphase cells scored. This result is consistent with the diagnosis of Smith–Magenis syndrome.

Parental chromosome and FISH studies are recommended to rule out either the presence of a balanced rearrangement or whether this abnormality is *de novo* in origin.

Genetic counseling is recommended.

Chromosome 17p13.3 deletions – Miller–Dieker syndrome

SPECIFIC FEATURES OF MILLER–DIEKER SYNDROME

Miller–Dieker syndrome (MDS) is characterized by a microdeletion involving chromosome 17p13.3, including the LIS1 (also known as PAFAH1B1) gene and additional telomeric genes. In 80% of patients, there is a deletion that is detectable by FISH analysis. In the remaining 20% of patients, there is an inherited chromosomal rearrangement causing an unbalanced form of the rearrangement, resulting in a deletion or mutation of the LIS1 gene.

The main clinical feature of MDS is lissencephaly caused by loss or mutations of the LIS1 gene, resulting in LIS1-associated lissencephaly/subcortical band heterotopia (SBH). Another term for lissencephaly is a smooth brain with absent gyri (agyria) or abnormally wide gyri (pachygyria). SBH refers to a band of heterotopic gray matter located just beneath the cortex and separated by a thin zone of normal white matter. Lissencephaly and SBH are cortical malformations caused by deficient neuronal migration during embryogenesis.

Other manifestations are variable, but include characteristic facial features of high and prominent forehead, bitemporal hollowing, short nose, upturned nares, prominent upper lip and small jaw. Features also include severe neurological abnormalities, developmental delay, omphalocele and congenital heart defects.

EXAMPLE REPORTING OF CHROMOSOME 17p13.3 DELETIONS – MILLER–DIEKER SYNDROME – NORMAL

ISCN Result: ish 17p13.3(MDCRx2)

Interpretation

Fluorescence *in situ* hybridization (FISH) analysis was performed with the 17p13.3 probe (MDCR) and a control RARA probe.

This analysis showed no evidence for a deletion or rearrangement in 10 metaphase cells scored, which is commonly observed in Miller–Dieker syndrome.

Since FISH showed only normal results, consider microarray analysis to detect smaller deletions or regions not covered by this FISH probe.

EXAMPLE REPORTING OF CHROMOSOME 17p13.3 DELETIONS – MILLER–DIEKER SYNDROME – ABNORMAL

ISCN Result: ish del(17)(p13.3p13.3)(MDCR-)

Interpretation

Fluorescence *in situ* hybridization (FISH) analysis was performed with the 17p13.3 probe (MDCR) and a control RARA probe.

This analysis showed a deletion of chromosome 17p13.3 region in 10 metaphase cells scored. This result is consistent with the diagnosis of Miller–Dieker syndrome.

Parental chromosome and FISH studies are recommended to rule out either the presence of a balanced rearrangement or whether this abnormality is *de novo* in origin.

Genetic counseling is recommended.

Chromosome 20p12.2 deletions – Alagille syndrome

SPECIFIC FEATURES OF ALAGILLE SYNDROME

Alagille syndrome is characterized by a microdeletion involving chromosome 20p12.2, including the JAG1 gene, and may also involve the NOTCH2 gene. Although many patients are now identified with a molecular deletion or mutation of one of these genes, patients may also show a microdeletion by FISH analysis of this region.

The severity of the clinical manifestations of this disorder can vary, even within the same family, ranging from manifestations so mild as to go unnoticed, to severe heart and/or liver disease. One of the major clinical findings in Alagille syndrome is liver damage caused by abnormalities in the bile ducts. Also commonly observed are heart defects, including tetralogy of Fallot. Other features associated with Alagille syndrome are distinctive facial features including a broad, prominent forehead, deep-set eyes and a small, pointed chin. Circulatory and bone problems may also be present, including abnormalities of the blood vessels within the central nervous system and kidneys and an unusual butterfly shape of the bones of the spinal column.

The incidence of this disorder is approximately 1 in 70,000 liveborns. However, this figure is based on diagnoses of liver disease in infants and may be underestimated due to the lack of liver problems in infancy of many individuals. Approximately 30–50% of individuals inherit the mutation or deletion from an affected parent. The remainder results from new mutations in the gene or a *de novo* deletion of genetic material on chromosome 20.

EXAMPLE REPORTING OF CHROMOSOME 20p12.2 DELETIONS – ALAGILLE SYNDROME – NORMAL

ISCN Result: ish 20p12.2(JAG1x2)

Interpretation

Fluorescence *in situ* hybridization (FISH) analysis was performed with the 20p12.2 probe (JAG1) and a control CEP20 probe.

This analysis showed no evidence for a deletion or rearrangement in 10 metaphase cells scored, which is commonly observed in Alagille syndrome.

Since FISH showed only normal results, consider microarray analysis to detect smaller deletions or regions not covered by this FISH probe.

EXAMPLE REPORTING OF CHROMOSOME 20p12.2 DELETIONS – ALAGILLE SYNDROME – ABNORMAL

ISCN Result: ish del(20)(p12.2p12.2)(JAG1-)

Interpretation

Fluorescence *in situ* hybridization (FISH) analysis was performed with the 20p12.2 probe (JAG1) and a control CEP20 probe.

This analysis showed a deletion of chromosome 20p12.2 region in 10 metaphase cells scored. This result is consistent with the diagnosis of Alagille syndrome.

Parental chromosome and FISH studies are recommended to rule out either the presence of a balanced rearrangement or whether this abnormality is *de novo* in origin.

Genetic counseling is recommended.

Chromosome 22q11.2 deletions – DiGeorge and velocardiofacial syndromes (Figure 13.3)

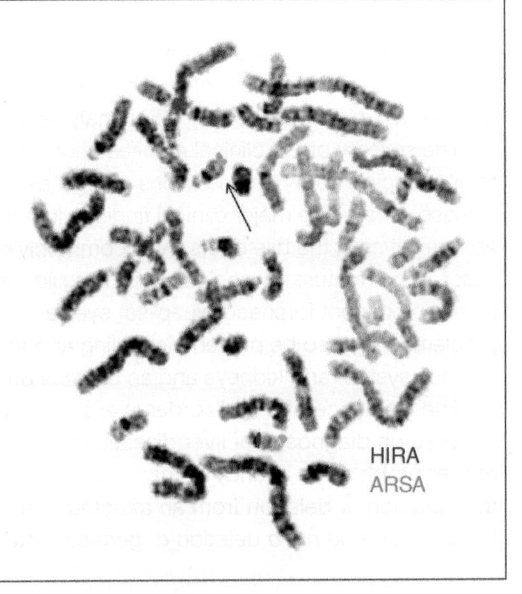

Figure 13.3 Metaphase FISH analysis resulting in velocardiofacial syndrome: ish del(22)(q11.2q11.2)(HIRA-). Courtesy of Lauren Jenkins PhD, Kaiser Permanente Regional Laboratory, Northern California.

SPECIFIC FEATURES OF DIGEORGE (DG)/VELOCARDIOFACIAL (VCF) SYNDROMES

DiGeorge and velocardiofacial syndromes are characterized by a microdeletion of chromosome 22q11.2, including the DiGeorge critical region (DGCR) gene and TUPLE1 gene (also known as HIRA). In 90% of patients, there is a deletion that is detectable by FISH analysis, which involves a minimum critical region of 480–575 Kb but may be as large as a 2 Mb region. Other genes are usually involved in these deletions, including citrate transport protein (CTP) and cathrin heavy chain genes (CLTC), which are involved in the etiology of VCFS. Therefore, this deletion region is described as a contiguous gene deletion syndrome because it results from the loss of several neighboring genes.

Clinical features of the syndromes associated with 22q11.2 deletions vary depending on the region and size of the deletion, and this region has been given the name of CATCH22, due to the various manifestations and disorders associated with it. CATCH22 refers to the phenotypic findings of cardiac defects, thymic hypoplasia, cleft palate and hypocalcemia, which make up the conotruncal anomaly face syndromes.

The incidence of these deletion-associated disorders is approximately 1 in 4000 livebirths, which is the most common of the microdeletion syndromes to date.

EXAMPLE REPORTING OF CHROMOSOME 22q11.2 DELETIONS – DIGEORGE/VELOCARDIOFACIAL SYNDROMES – NORMAL

ISCN Result: ish 22q11.2(HIRAx2)

Interpretation

Fluorescence *in situ* hybridization (FISH) analysis was performed with the 22q11.2 probe (HIRA) and a control ARSA probe.

This analysis showed no evidence for a deletion or rearrangement in 10 metaphase cells scored, which is commonly observed in DiGeorge/velocardiofacial syndromes.

Since FISH showed only normal results, consider microarray analysis to detect smaller deletions or regions not covered by this FISH probe.

EXAMPLE REPORTING OF CHROMOSOME 22q11.2 DELETIONS – DIGEORGE/VELOCARDIOFACIAL SYNDROMES – ABNORMAL

ISCN Result: ish del(22)(q11.2q11.2)(HIRA-)

Interpretation

Fluorescence *in situ* hybridization (FISH) analysis was performed with the 22q11.2 probe (HIRA) and a control ARSA probe.

This analysis showed a deletion of chromosome 22q11.2 region in 10 metaphase cells scored. This result is consistent with the diagnosis of DiGeorge/velocardiofacial syndromes.

Parental chromosome and FISH studies are recommended to rule out either the presence of a balanced rearrangement or whether this abnormality is *de novo* in origin.

Genetic counseling is recommended.

Chromosome 22q13.3 deletion

SPECIFIC FEATURES OF CHROMOSOME 22q13.3 DELETION

The chromosome 22q13.3 deletions are now a recognizable phenotype characterized by hypotonia, absence of speech or speech delay, moderate-to-severe mental impairment and mild dysmorphic features. The probe used to identify this abnormality is a subtelomeric probe of the long arm of chromosome 22, which is distal to the ARSA probe. This probe cocktail also uses the N25 or HIRA probes as controls, thereby allowing FISH analysis of both the VCFS deletion region and this region in the same test mixture. A candidate gene for this syndrome, SHANK3, has been reported and identified by chromosomal microarray analysis.

EXAMPLE REPORTING OF CHROMOSOME 22q13.3 DELETION – NORMAL

ISCN Result: ish 22q13.3(N85A3x2)

Interpretation

Fluorescence *in situ* hybridization (FISH) analysis was performed with the 22q13.3 probe (N85A3) and a control D22S75 probe.

This analysis showed no evidence for a deletion or rearrangement in 10 metaphase cells scored, which is commonly observed in chromosome 22q13.3 deletions.

Since FISH showed only normal results, consider microarray analysis to detect smaller deletions or regions not covered by this FISH probe.

EXAMPLE REPORTING OF CHROMOSOME 22q13.3 DELETION – ABNORMAL

ISCN Result: ish del(22)(q13.3q13.3)(N85A3-)

Interpretation

Fluorescence *in situ* hybridization (FISH) analysis was performed with the 22q13.3 probe (N85A3) and a control D22S75 probe.

This analysis showed a deletion of chromosome 22q13.3 region in 10 metaphase cells scored. This result is consistent with the diagnosis of chromosome 22q13.3 deletion syndrome.

Parental chromosome and FISH studies are recommended to rule out either the presence of a balanced rearrangement or whether this abnormality is *de novo* in origin.

Genetic counseling is recommended.

Chromosome Xp22.3 deletions

SPECIFIC FEATURES OF KALLMANN AND STEROID SULFATASE SYNDROMES

Complex phenotypic abnormalities resulting from the presence of a contiguous gene syndrome involving a deletion of disease genes located in the Xp22.3 region include genes for Kallmann syndrome, steroid sulfatase deficiency, X-linked recessive ichthyosis, chondrodysplasia punctata and ocular albinism. Some of these patients with Xp22.3 contiguous gene deletion syndromes have cytogenetically visible abnormalities, such as large terminal Xp deletions or X/Y translocations, but many have only small deletions that can be discerned by FISH studies and not by standard cytogenetic studies.

Chromosome Xp22.3 deletions – Kallmann syndrome

SPECIFIC FEATURES OF KALLMANN SYNDROME (KS1)

Kallmann syndrome type 1 is characterized by a microdeletion or mutation involving chromosome Xp22.3, including the KAL1gene. Other types of Kallmann syndrome result from mutations in the FGFR1, PROKR2, PROK2, CHD7 and FGF8 genes. KS1 is inherited in an X-linked manner, whereas KS2 (FGFR1), KS3 (PROKR2), KS4 (PROK2), KS5 (CHD7) and KS6 (FGF8) are predominantly inherited in an autosomal dominant manner. KS3 (PROKR2) and KS4 (PROK2) can also be inherited in an autosomal recessive manner. The mode of inheritance is often unclear within families and is likely to be dependent on a mutation of more than one gene. Together, mutations in these six genes account for about 25–35% of all KS. Deletions of KAL1 by FISH analysis are a rare cause of KS; rather, small mutations account for the majority of detectable cases.

The main clinical manifestations of Kallman syndrome type 1 are delayed or absent puberty and an impaired sense of smell. This disorder is a form of hypogonadotropic hypogonadism (HH), affecting the production of hormones that direct sexual development. Due to this disorder being X-linked, the vast majority of affected individuals are male. Males with hypogonadotropic hypogonadism often have micropenis and cryptorchidism. At puberty, most affected individuals do not develop secondary sex characteristics. Affected females usually have amenorrhea and little or no breast development. In some individuals, puberty is incomplete or delayed.

Another characteristic of this syndrome includes hyposmia or anosmia.This feature distinguishes Kallmann syndrome from most other forms of hypogonadotropic hypogonadism, which do not affect the sense of smell. Additional features may include renal agenesis, cleft lip with or without cleft palate, abnormal eye movements, hearing loss and abnormalities of tooth development.

EXAMPLE REPORTING OF CHROMOSOME Xp22.3 DELETIONS – KALLMANN SYNDROME – NORMAL

ISCN Result: ish Xp22.3(KAL1x2)

Interpretation

Fluorescence *in situ* hybridization (FISH) analysis was performed with the Xp22.3 probe (KAL), and a control CEP X probe.

This analysis showed no evidence for a deletion or rearrangement in 10 metaphase cells scored, which is commonly observed in Kallmann syndrome.

Since FISH showed only normal results, consider microarray analysis to detect smaller deletions or regions not covered by this FISH probe.

EXAMPLE REPORTING OF CHROMOSOME Xp22.3 DELETIONS – KALLMANN SYNDROME – ABNORMAL

ISCN Result: ish del(X)(p22.3p22.3)(KAL1-)

Interpretation

Fluorescence *in situ* hybridization (FISH) analysis was performed with the Xp22.3 probe (KAL), and a control CEP X probe.

This analysis showed a deletion of chromosome Xp22.3 region in 10 metaphase cells scored. This result is consistent with the diagnosis of Kallmann syndrome. The Xp deletion may constitute a contiguous gene deletion syndrome, which could include X-linked ichthyosis and mental retardation.

Females carrying the deletion may present with short stature and have an increased risk for male offspring with Kallmann syndrome or other contiguous gene syndrome phenotypic manifestations.

Maternal chromosome and FISH studies are recommended to rule out either the presence of a balanced rearrangement or whether this abnormality is *de novo* in origin.

Genetic counseling is recommended.

Chromosome Xp22.3 deletions – steroid sulfatase deficiency syndrome

SPECIFIC FEATURES OF STEROID SULFATASE DEFICIENCY SYNDROME

Steroid sulfatase deficiency syndrome (STS), also known as X-linked ichthyosis, is characterized by a microdeletion or mutation involving chromosome Xp22.3, including the STS gene. Approximately 85–90% of individuals with STS deficiency have deletions spanning the entire STS gene and flanking markers that can be detected by FISH analysis. Most of these deletions appear to be caused by abnormal recombination due to the presence of low copy-number repeats within the region. Approximately 10% of patients with STS deficiency have point mutations in the STS gene that need to be tested by molecular analysis. The remaining 5–10% have a complex phenotype resulting from the presence of a contiguous gene deletion syndrome involving a deletion of additional disease genes located in the Xp22.3 region.

The main clinical manifestation of STS deficiency is an inborn error of metabolism causing X-linked ichthyosis, a skin disorder inherited as an X-linked trait. This presents with dark scaly skin in a newborn. *In utero*, deficiency of the STS gene causes diminished estrogen biosynthesis by the placenta, often resulting in prolonged labor due to difficulty in cervical effacement. Individuals with STS deficiency also have increased levels of cholesterol sulfate in both plasma and stratum corneum, which is thought to be responsible for the ichthyotic changes. STS deficiency affects 1 in 2000–6000 males.

EXAMPLE REPORTING OF CHROMOSOME Xp22.3 DELETIONS – STEROID SULFATASE DEFICIENCY SYNDROME – NORMAL

ISCN Result: ish Xp22.3(STSx2)

Interpretation

Fluorescence *in situ* hybridization (FISH) analysis was performed with the Xp22.3 probe (STS) and a control CEP X probe.

This analysis showed no evidence for a deletion or rearrangement in 10 metaphase cells scored, which is commonly observed in steroid sulfatase deficiency syndrome.

Since FISH showed only normal results, consider microarray analysis to detect smaller deletions or regions not covered by this FISH probe.

EXAMPLE REPORTING OF CHROMOSOME Xp22.3 DELETIONS – STEROID SULFATASE DEFICIENCY SYNDROME – ABNORMAL

ISCN Result: ish del(X)(p22.3p22.3)(STS-)

Interpretation

Fluorescence *in situ* hybridization (FISH) analysis was performed with the Xp22.3 probe (STS) and a control CEP X probe.

This analysis showed a deletion of chromosome Xp22.3 region in 10 metaphase cells scored. This result is consistent with the diagnosis of steroid sulfatase deficiency syndrome. The Xp deletion may constitute a contiguous gene deletion syndrome, which could include X-linked Kallmann syndrome and mental retardation.

Females carrying the deletion may present with short stature and have an increased risk for male offspring with Kallmann syndrome or other contiguous gene syndrome phenotypic manifestations.

Maternal chromosome and FISH studies are recommended to rule out either the presence of a balanced rearrangement or whether this abnormality is *de novo* in origin.

Genetic counseling is recommended.

Chromosome Xq13.2 deletions – XIST deletions

SPECIFIC FEATURES OF XIST DELETIONS

Deletions of the XIST gene are characterized by a microdeletion involving chromosome Xq13.2. The XIST gene is involved in the inactivation of one X chromosome that occurs early in female embryogenesis. The XIST gene has been mapped to the X inactivation center (XIC), which is responsible for the silencing of one X chromosome in females, leaving only one X chromosome transcriptionally active in all cells, which may result in Turner syndrome or one of its variants. The XIST gene must be present and intact for the gene to be functionally capable of inactivating the X chromosome. The finding of an XIST gene deletion may be the result of loss of just the XIST gene, or loss of part or the whole X chromosome. Therefore, it is suggested that XIST gene deletions detected by FISH analysis be confirmed by standard cytogenetic analysis. If standard cytogenetic analysis was performed and showed loss of the X chromosome with the presence of a small marker chromosome, or a deletion within the long arm of one of the X chromosomes, then FISH analysis of the XIST gene may help clarify the cytogenetic result.

EXAMPLE REPORTING OF CHROMOSOME Xq13.2 DELETIONS – XIST DELETION REGION – NORMAL

ISCN Result: ish Xq13.2(XISTx2)

Interpretation

Fluorescence *in situ* hybridization (FISH) analysis was performed with the Xq13.2 probe (XIST) and a control CEP X probe.

This analysis showed no evidence for a deletion or rearrangement in 10 metaphase cells scored, which is commonly observed in Turner syndrome or other X chromosome disorders.

Since FISH showed only normal results, consider microarray analysis to detect smaller deletions or regions not covered by this FISH probe.

EXAMPLE REPORTING OF CHROMOSOME Xq13.2 DELETIONS – XIST DELETION REGION – ABNORMAL

ISCN Result: ish del(X)(q13.2q13.2)(XIST-)

Interpretation

Fluorescence *in situ* hybridization (FISH) analysis was performed with the Xq13.2 probe (XIST) and a control CEP X probe.

This analysis showed a deletion of chromosome Xq13.2 region in 10 metaphase cells scored. Deletions of the XIST gene are suggestive of Turner syndrome or a variant of Turner syndrome.

Maternal chromosome and FISH studies are recommended to rule out either the presence of a balanced rearrangement or whether this abnormality is *de novo* in origin.

Genetic counseling is recommended.

Chromosome Yp11.3 deletions

SPECIFIC FEATURES OF Yp11.3 DELETIONS

Chromosome Yp11.3 deletions are clinically significant when they include the deletion of male-determining genes, especially the SRY gene (sex-determining region on the Y chromosome). Loss of the SRY gene results in a female phenotype and is critical in determining the sex of an individual at birth as well as at the age of puberty when secondary sexual characteristics emerge. Therefore, in individuals with ambiguous genitalia, males with infertility, XY females and XX males are primary candidates for SRY gene analysis. It is important to correlate FISH analysis with standard cytogenetics in these cases in order to clarify the genotype with associated phenotypic findings.

The SRY gene is responsible for normal male development, although numerous other genes are required for complete sexual development. Thus, SRY gene deletions in a 46,XY male will result in a female phenotype. A 46,XX female with the SRY gene inserted into an X chromosome will result in a male phenotype. Intermediate phenotypic patterns are also possible, including ambiguous genitalia in 46,XX males who are SRY negative. Those who are SRY positive are usually males with azoospermia. 46,XX females with the SRY gene may have ambiguous genitalia, infertility and other manifestations. A 46,XY female that is SRY negative may have other genes with a mutation or deletion, such as the SOX9 gene, and should be evaluated by molecular analysis if the SRY gene is not present.

Also, there is a significant risk of gonadoblastoma in females with the SRY gene present, and reporting this possibility is important for the management of the patient.

EXAMPLE REPORTING OF CHROMOSOME Yp11.3 DELETIONS – Yp11.3 DELETION REGION – NORMAL

ISCN Result: ish Yp11.3(SRYx2)

Interpretation

Fluorescence *in situ* hybridization (FISH) analysis was performed with the Yp11.2 probe (SRY).

This analysis showed no evidence for a deletion or rearrangement in 10 metaphase cells scored. These results confirm the presence of the SRY on the Y chromosome.

Since FISH showed only normal results, consider microarray analysis to detect smaller deletions or regions not covered by this FISH probe.

EXAMPLE REPORTING OF CHROMOSOME Yp11.3 DELETIONS – Yp11.3 DELETION REGION – ABNORMAL

ISCN Result: ish del(Y)(p11.3p11.3)(SRY-)

Interpretation

Fluorescence *in situ* hybridization (FISH) analysis was performed with the Yp11.2 probe (SRY).

This analysis showed a deletion of chromosome Yp11.2 region in 10 metaphase cells scored. Deletions of the SRY gene are consistent with a female phenotype. SRY-negative 46,XX males are at increased risk for ambiguous genitalia and true hermaphroditism.

Paternal chromosome and FISH studies are recommended to rule out either the presence of a balanced rearrangement or whether this abnormality is *de novo* in origin.

Genetic counseling is recommended.

Example reporting of FISH with more than one microdeletion

ISCN RULES FOR REPORTING FISH WITH MORE THAN ONE MICRODELETION

The same rules apply for multiple FISH analyses as for a single probe tested except for the following rules.

Only the clinically significant or informative results within a single chromosome need to be included in the nomenclature, but all probes tested may be designated for clarity.

For example, when both the DiGeorge probe (D22S75) at 22q11.2 and the ARSA probe at 22q13.3 are tested and only the ARSA probe is deleted, then the nomenclature may be written as either: ish del(22)(q13.3q13.3)(ARSA-) or: ish 22q11.2(D22S75x2),del(22)(q13.3q13.3)(ARSA-).

Note that the order of probes is based first on chromosome number, followed by chromosome band, proximal to distal bands, from the short arm to the long arm; therefore, 22q11.2 precedes 22q13.3 in order, regardless of its normal or abnormal status.

When more than one probe is tested and probes are on different chromosomes, then all the probes are listed, whether normal or abnormal, in order of chromosome number, then by band, proximal to distal, and from the short arm to the long arm.

For example, for the DiGeorge/velocardiofacial syndromes and Prader–Willi/Angelman syndromes, the FISH nomenclature would be written as: ish 15q11.2(SNRPNx2,D15S10x2), del(22)(q11.2q11.2)(D22S75-).

EXAMPLE REPORTING OF FISH WITH MORE THAN ONE MICRODELETION

ISCN Result: ish 15q11.2(SNRPNx2,D15S10x2),del(22)(q11.2q11.2)(D22S75-)

Interpretation

Fluorescence *in situ* hybridization (FISH) analysis was performed with the 15q11.2 probe (SNRPN) and 22q11.2 probe (HIRA).

This analysis showed a deletion of chromosome 22q11.2 region in 10 metaphase cells scored. This result is consistent with the diagnosis of DiGeorge/velocardiofacial syndrome.

Analysis with the SNRPN probe showed no evidence for a deletion or rearrangement in 10 metaphase cells scored, which is commonly observed in Prader–Willi syndrome.

Parental chromosome and FISH studies are recommended to rule out either the presence of a balanced rearrangement or whether the chromosome 22 abnormality is *de novo* in origin.

Genetic counseling is recommended.

13.5 Microduplications

Microduplications discussed in this section refer to submicroscopic duplications that are detected by FISH analysis and result in a recognizable clinical disorder. As with microdeletions, many of these duplications are now diagnosed by microarrays, but FISH analysis is still used by many laboratories as a simpler approach to identifying these disorders on a singular basis. Also, as in the case of microdeletions, many of these disorders are contiguous gene syndromes, with more than one gene involved, resulting in variable phenotypic consequences depending on the genes involved in the duplicated region. The mechanisms for many of these gene syndromes are the same as in microdeletions, namely non-allelic homologous recombination, leading to a deletion or duplication of a small chromosomal region. These regions are considered "hot spots," whereby these crossing-over events are non-random, giving rise to specific and distinctive disorders.

As technologies continue to improve the resolution of genomic regions of chromosomes, more microduplication syndromes will be identified, but they may also be diagnosed at the cytogenetic molecular level of microarrays. Other small duplications and regions not detectable by FISH analysis will be discussed in Section 3.

13.5.1 ISCN rules for reporting microduplications

ISCN RULES FOR REPORTING FISH WITH MICRODUPLICATIONS

The same rules apply for microduplications as for microdeletions except for the following.

- Write "dup" instead of "del" where appropriate.
- After the probe name, write "x2" for a normal result and "++" for an abnormal result.

For example, for 22q11.2 duplication syndrome, a normal FISH result would be written as: ish 22q11.2(HIRAx2). An abnormal FISH result would be written as: ish dup(22)(q11.2q11.2)(D22S75++).

13.5.2 Example reporting of microduplications

EXAMPLE REPORTING OF NORMAL RESULTS

ISCN Result: ish __ chromosome(probex2)

Interpretation

Fluorescence *in situ* hybridization (FISH) analysis was performed with the __ chromosome region (probe).

This analysis showed no evidence for a duplication at _ in _ metaphase cells scored, which is commonly observed in ___ syndrome.

Since FISH showed only normal results, consider microarray analysis to detect smaller duplications or regions not covered by this FISH probe.

EXAMPLE REPORTING OF ABNORMAL RESULTS

ISCN Result: ish dup(chromosome)(bands)(probe++)

Interpretation

Fluorescence *in situ* hybridization (FISH) analysis was performed with the __chromosome region probe (probe) and a control __ probe (probe).

This analysis showed a duplication of chromosome __ region in 10 metaphase cells scored. This result is consistent with the diagnosis of __ syndrome.

Parental chromosome and FISH studies are recommended to rule out either the presence of a balanced rearrangement or whether this abnormality is *de novo* in origin.

Genetic counseling is recommended.

13.5.3 Microduplication disorders by chromosome order

Chromosome 12p duplications – Pallister–Killian syndrome (see also i(12p) in Chapter 8)

SPECIFIC FEATURES OF PALLISTER–KILLIAN SYNDROME

Pallister–Killian syndrome is characterized by a duplication of the short arm material of chromosome 12. It is usually seen as a supernumerary chromosome and results in tetrasomy for the whole short arm. This chromosome is generally described as an isochromosome for the short arm of chromosome 12 and is mosaic with a normal cell line. It is important to distinguish this abnormality with an isochromosome of the long arm of chromosome 21, as they look similar cytogenetically. Since i(21q) is most often seen as a trisomy, only one normal chromosome 21 is present along with the isochromosome whereas i(12p) is generally seen with two normal chromosome 12s. It is also prudent to confirm the cytogenetic findings with FISH analysis, as this is the fastest and easiest way to determine the chromosome identity of the abnormal chromosome.

The clinical features of Pallister–Killian syndrome include hypotonia in infancy and early childhood, intellectual disability, sparse hair, abnormal pigmentation and distinctive facial features, including a high, rounded forehead, broad nasal bridge, short nose, widely spaced eyes, low-set ears, wide mouth with a thin upper lip and a large tongue, and cleft palate. Other features include hearing loss, vision impairment, seizures, extra nipples, genital abnormalities, heart defects and skeletal abnormalities. About 40% of affected infants are born with a congenital diaphragmatic hernia. Since this syndrome is seen in the mosaic form, phenotypic features may vary in form and severity based on the cell type and proportion of abnormal cells present.

Also, since i(12p) is always seen in the mosaic form, the mechanism is mostly likely due to the loss of the abnormal isochromosome early in development. If this abnormality is present in all cells, the amount of extra genetic material would be lethal.

The incidence of i(12p) is rare, with approximately 100 cases reported in the literature. This is most likely an understated number of true abnormal cases, in part due to the underreporting of individuals who are undetected due to mild phenotypes when the mosaicism is seen in a small proportion of cells.

EXAMPLE REPORTING OF CHROMOSOME 12p DUPLICATIONS – PALLISTER–KILLIAN SYNDROME – NORMAL

ISCN Result: ish 12p11.1q11(D12Z3x2),12p13(ETV6x2)

Interpretation

Fluorescence *in situ* hybridization (FISH) analysis was performed with the chromosome 12 centromere probe (D12Z3) and the ETV6 gene at chromosome 12p13.

This analysis showed no evidence for a duplication or rearrangement in 10 metaphase cells scored, which is commonly observed in Pallister–Killian syndrome.

Since FISH showed only normal results, consider microarray analysis to detect smaller duplications or regions not covered by this FISH probe.

EXAMPLE REPORTING OF CHROMOSOME 12p DUPLICATIONS – PALLISTER–KILLIAN SYNDROME – ABNORMAL (FIGURE 13.4)

ISCN Result: ish dup(12)(12p11.1q11)(D12Z3+),12p13(ETV6++)

Interpretation

Fluorescence *in situ* hybridization (FISH) analysis was performed with the 12 centromere probe (D12Z3) and the ETV6 gene at chromosome 12p13.

This analysis showed a duplication of the chromosome 12 centromere region and tetrasomy of chromosome 12p13 in 10 metaphase cells scored. This result is consistent with the diagnosis of Pallister–Killian syndrome.

Parental chromosome and FISH studies are recommended to rule out either the presence of a balanced rearrangement or whether this abnormality is *de novo* in origin.

Genetic counseling is recommended.

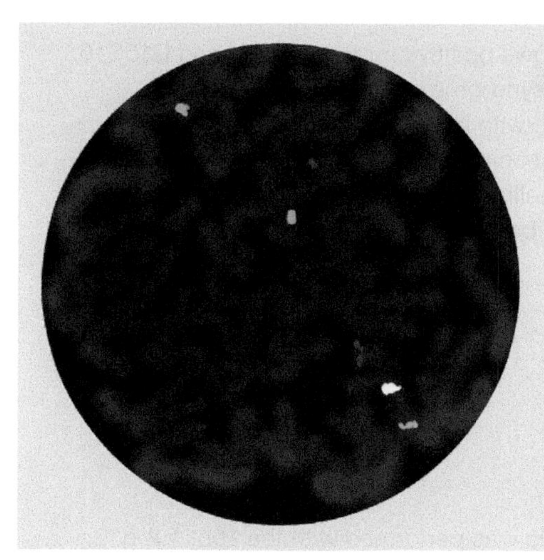

Figure 13.4 Metaphase FISH analysis resulting in Pallister –Killian syndrome, showing two extra 12p signals (ETV6 probe – green) on a supernumerary chromosome (red is 21q22 – AML1 probe): ish + i(12)(p13.2)(ETV6++). Courtesy of Sarah South PhD, ARUP Laboratories.

Chromosome 15q11.2 duplications

SPECIFIC FEATURES OF CHROMOSOME 15q11.2 DUPLICATION

Duplications of chromosome 15q11.2 are usually seen as an extra chromosome. This is the same abnormality as the inv dup 15q11.2 discussed in Chapter 7 under marker chromosomes. Since these are typically seen as bisatellited chromosomes of unknown origin by chromosome analysis, FISH analyses using the chromosome 15q11.2 probes are useful in identifying whether or not the supernumerary chromosome is this syndromic region. The probes that are often used to identify this region are the SNRPN and D15S10 probes, which are used to identify Prader–Willi and Angelman syndromes. Other probes within this region may also be used and are just as effective in diagnosing the presence of extra copies of this chromosomal region.

Fluorescence *in situ* hybridization is useful in distinguishing between extra copies of 15q11.2 as either a supernumerary chromosome or an interstitial duplication within one chromosome 15 homolog. The outcome may result in either trisomy or tetrasomy for the region involved. Generally, a supernumerary 15q marker chromosome results in tetrasomy whereas interstitial duplications are trisomic for the genetic material. In supernumerary chromosomes, the material duplicated encompasses the short arm to a region of the long arm of chromosome 15 of varying lengths. Therefore, this duplication syndrome may actually involve the 15q11.2 to 15q13 region.

Phenotypic variation is extensive with this duplication, depending on the amount of material duplicated as well as the parent of origin. Small supernumerary chromosomes that are shown to be negative for the major probes in the region may have little or no clinical consequence, since little, if any, euchromatic material is present. These small chromosomes are often inherited from a parent, also signifying little to no phenotypic effect.

As the duplication region gets larger, it follows that the risk of euchromatic material gets larger, giving rise to more and more significant clinical manifestations. It also follows that fewer of these chromosomes will be inherited, as the burden of phenotypic consequences increases.

When the chromosome 15q11.2 duplication shows positive for the SNRPN and D15S10 probes, the phenotypic findings in this duplication syndrome generally include mental impairment and behavioral abnormalities consistent with autism spectrum disorders. Interstitial tandem duplications of this region are associated with autistic features when maternally inherited and are less likely when paternally inherited.

More information about detection of autism will be discussed in Section 3.

EXAMPLE REPORTING OF CHROMOSOME 15q11.2 DUPLICATIONS – NORMAL

ISCN Result: ish 15q11.2q12(SNRPN,D15S10)x2

Interpretation

Fluorescence *in situ* hybridization (FISH) analysis was performed with the 15q11.2-q12 probes (SNRPN, D15S10) and a control PML probe.

This analysis showed no evidence for a duplication or rearrangement in 10 metaphase cells scored, which is commonly observed in chromosome 15q11.2 duplication syndrome.

Since FISH showed only normal results, consider microarray analysis to detect smaller duplications or regions not covered by these FISH probes.

EXAMPLE REPORTING OF CHROMOSOME 15q11.2 DUPLICATIONS – ABNORMAL

ISCN Result: ish dup(15)(q11.2q12)(SNRPN++,D15S10++)

Interpretation

Fluorescence *in situ* hybridization (FISH) analysis was performed with the 15q11.2-q12 probes (SNRPN, D15S10) and a control PML probe.

This analysis showed a duplication of chromosome 15q11.2-q12 region in 10 metaphase cells scored. This result is consistent with the diagnosis of chromosome 15q11.2 duplication syndrome.

Parental chromosome and FISH studies are recommended to rule out either the presence of a balanced rearrangement or whether this abnormality is *de novo* in origin.

Genetic counseling is recommended.

EXAMPLE REPORTING OF CHROMOSOME 15q11.2 DUPLICATIONS – ABNORMAL AS A SUPERNUMERARY CHROMOSOME IDENTIFIED BY CHROMOSOME ANALYSIS

ISCN Result: 47,XY,+mar.ish dup(15)(q11.2q12)(SNRPN++,D15S10++)

Interpretation

Chromosome and fluorescence *in situ* hybridization (FISH) analyses (15q11.2-q12 probes SNRPN, D15S10) and a control PML probe were performed on metaphases for each assay.

Chromosome analysis showed an abnormal male chromosome complement with the presence of a small supernumerary bisatellited chromosome in all 20 cells examined. FISH analysis confirmed the presence of a small chromosome with a duplication of chromosome 15q11.2-q12 region in 10 metaphase cells scored. This result is consistent with the diagnosis of chromosome 15q11.2 duplication syndrome.

Parental chromosome and FISH studies are recommended to identify if this is an inherited abnormality or *de novo* in origin.

Genetic counseling is recommended.

Chromosome 17p11.2 duplications – Potocki–Lupski syndrome

SPECIFIC FEATURES OF POTOCKI–LUPSKI SYNDROME (PTLS)

Duplications of chromosome 17p11.2 involve the same genomic region as that of 17p11.2 deletions, consistent with Smith–Magenis syndrome (SMS). PTLS was the first reported reciprocal of a homologous recombination where both non-allelic homologous recombinations resulting in microdeletions and microduplications lead to a contiguous gene syndrome.

Patients with PTLS generally have milder phenotypic manifestations than SMS; however, many features overlap in both syndromes, including mental impairment and features of autistic spectrum disorders. Other unique features of PTLS include hypotonia, sleep apnea, heart defects, attention deficit disorders, short stature and failure to thrive.

The incidence is far more rare than SMS, at approximately 1 in 20,000 livebirths.

EXAMPLE REPORTING OF CHROMOSOME 17p11.2 DUPLICATIONS – POTOCKI–LUPSKI SYNDROME – NORMAL

ISCN Result: ish 17p11.2(RAI1x2)

Interpretation

Fluorescence *in situ* hybridization (FISH) analysis was performed with the 17p11.2 probe (RAI1) and a control RARA probe.

This analysis showed no evidence for a duplication or rearrangement in 10 metaphase cells scored, which is commonly observed in Potocki–Lupski syndrome.

Since FISH showed only normal results, consider microarray analysis to detect smaller duplications or regions not covered by this FISH probe.

EXAMPLE REPORTING OF CHROMOSOME 17p11.2 DUPLICATIONS – POTOCKI–LUPSKI SYNDROME – ABNORMAL

ISCN Result: ish dup(17)(p11.2p11.2)(RAI1++)

Interpretation

Fluorescence *in situ* hybridization (FISH) analysis was performed with the 17p11.2 probe (RAI1) and a control RARA probe.

This analysis showed a duplication of chromosome 17p11.2 region in 10 metaphase cells scored. This result is consistent with the diagnosis of Potocki–Lupski syndrome.

Parental chromosome and FISH studies are recommended to rule out either the presence of a balanced rearrangement or whether this abnormality is *de novo* in origin.

Genetic counseling is recommended.

Chromosome 22q11.2 duplications and cat-eye syndrome

SPECIFIC FEATURES OF CHROMOSOME 22q11.2 DUPLICATION

Duplications of chromosome 22q11.2 involve the same genomic region as that of 22q11.2 deletions, consistent with DGS/VCFS (see p.000). As in chromosome 15q11.2 duplications, chromosome 22q11.2 duplications can be seen as either a supernumerary chromosome (cat-eye syndrome) or as an interstitial tandem duplication of one chromosome 22 homolog.

A small supernumerary bisatellited marker chromosome detected by chromosome analysis, if not derived from chromosome 15, is most likely of chromosome 22 origin, resulting in tetrasomy of this region. FISH analysis of chromosome 15q11.2 and 22q11.2 probes will clarify the chromosome of origin. Generally, the same probe as that used in 22q11.2 deletion syndromes is used to identify a duplication, namely the HIRA probe.

Interstitial tandem duplications of this region form in the same pattern as that of the 17p11.2 deletions/duplications; that is, by non-allelic homologous recombination. Duplications of 22q11.2 vary in size and in the genes involved. These duplications generally include a typically common 3 Mb region, with a 1.5 Mb nested duplication, of distinct low-copy repeats. Smaller microduplications may occur with frequent rearrangements arising from alternative low-copy repeats in the non-allelic homologous recombination mechanism. The majority of these duplications are inherited from a parent with a normal or near-normal phenotype. Therefore, many of these individuals are undiagnosed, unlike the microdeletion syndromes of DGS and VCFS associated with the corresponding deletion.

The common phenotypic manifestations of interstitial duplications of 22q11.2, such as mild but variable features, include mild mental impairment to normal intelligence, learning disabilities, developmental and growth delay and hypotonia.

When duplications of 22q11.2 are seen as a supernumerary chromosome, the resulting phenotype is cat-eye syndrome (also see cat-eye syndrome in Chapter 8). Cat-eye syndrome is characterized by unilateral and bilateral coloboma of the eye, but most other phenotypic features are variable, depending on the size of the duplicated region, even in affected members of the same family. Characteristic features include mild growth delay *in utero*, mild mental deficiency, downward-slanting palpebral fissures, preauricular pits/tags, short stature, skeletal problems such as scoliosis, cardiac defects, micrognathia, hernias, cleft palate, and abnormalities of the kidneys and/or the anal region.

Chromosome 22q11.2 duplications are rare and their incidence is not defined, most likely because of the underdiagnosis of interstitial duplications due to mild to no phenotypic manifestations.

EXAMPLE REPORTING OF CHROMOSOME 22q11.2 DUPLICATIONS – NORMAL

ISCN Result: ish 22q11.2(HIRAx2)

Interpretation

Fluorescence *in situ* hybridization (FISH) analysis was performed with the 22q11.2 probe (HIRA) and a control ARSA probe.

This analysis showed no evidence for a duplication or rearrangement in 10 metaphase cells scored, which is commonly observed in chromosome 22q11.2 duplications.

Since FISH showed only normal results, consider microarray analysis to detect smaller duplications or regions not covered by this FISH probe.

EXAMPLE REPORTING OF CHROMOSOME 22q11.2 DUPLICATIONS – ABNORMAL

ISCN Result: ish dup(22)(q11.2q11.2)(HIRA++)

Interpretation

Fluorescence *in situ* hybridization (FISH) analysis was performed with the 22q11.2 probe (HIRA) and a control ARSA probe.

This analysis showed a duplication of chromosome 22q11.2 region in 10 metaphase cells scored. This result is consistent with the diagnosis of chromosome 22q11.2 duplication syndrome.

Parental chromosome and FISH studies are recommended to rule out either the presence of a balanced rearrangement or whether this abnormality is *de novo* in origin.

Genetic counseling is recommended.

EXAMPLE REPORTING OF CHROMOSOME 22q11.2 DUPLICATIONS – ABNORMAL AS A SUPERNUMERARY CHROMOSOME IDENTIFIED BY CHROMOSOME ANALYSIS (FIGURE 13.5)

ISCN Result: 47,XY,+mar.ish dup(22)(q11.2q12)(HIRA++)

Interpretation

Chromosome analysis and fluorescence *in situ* hybridization (FISH) analysis (22q11.2 probe HIRA) were performed on metaphases for each assay.

Chromosome analysis showed an abnormal male chromosome complement with the presence of a small supernumerary bisatellited chromosome in all 20 cells examined. FISH analysis confirmed the presence of a small chromosome with a duplication of chromosome 22q11.2 region in 10 metaphase cells scored. This is consistent with the diagnosis of cat-eye syndrome.

Parental chromosome and FISH studies are recommended to identify whether this is an inherited abnormality or *de novo* in origin.

Genetic counseling is recommended.

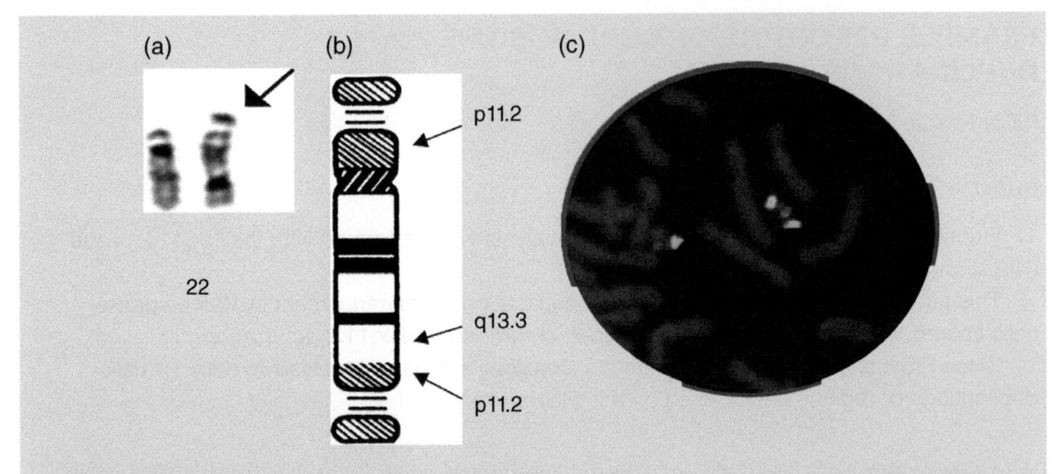

Figure 13.5 Chromosome 22 of a fetus with a bisatellited duplication of 22p11.2 and a deletion of the distal 22q13.3 band. (a) G-band picture depicting a bisatellited duplication of 22p11.2. (b) Ideogram of the bisatellited chromosome 22 duplication. (c) Metaphase FISH picture showing the bisatellited duplication of 22p11.2.

13.6 Fluorescence *in situ* hybridization for chromosome identification

Cytogenetic testing using standard chromosome analysis may contain a chromsosome that is unidentifiable based on chromosome banding alone. Parental chromosome studies may help to identify the marker chromosome if it is inherited, but often these chromosomes are *de novo* in origin.

These "marker" chromosomes may then be further studied by FISH analysis to help clarify the chromosomal origin and possibly the band regions involved. This is useful information, since knowing the origin of the chromosome may help determine the clinical consequence of the extra material present.

One method of FISH for chromosome identification is the "shotgun" approach, in which an educated guess, based on the banding pattern, might give clues to the chromosome of origin. Centromere probes or locus-specific probes of a particular chromosome may be used to help identify the origin of the DNA material. This is certainly true for cases of bisatellited small marker chromosomes that may be of chromosome 15 origin. In bisatellited markers, if the marker is not of chromosome 15 origin, the other acrocentric centromere or locus-specific probes may give the answer to the identity of the chromosome. This is a useful technique, especially if no euchromatic material appears to be present, since FISH probes are usually of centromere, telomere or euchromatic material. Other alpha-satellite FISH probes are generally not specific to particular chromosomes and may not be helpful in identifying the chromosomal origin of the marker.

Another approach for chromosome identity is to use whole chromosome paint probes or 24-color FISH probes that will paint each chromosome a different color and, with a computer program, identify each chromosome by the fluorescent dye intensity. Since each chromosome has a distinct fluorescent dye configuration, markers can be identified by this technique. It may be difficult to pursue this method if the marker is not seen in all cells. The lower the level of mosaicism, the more difficult it is to assess the presence of the marker by FISH analysis. Although this approach is expensive, tedious and difficult to perform with good results, it is a useful tool to employ in these cases.

In cases with multiple chromosome aberrations, 24-color FISH analysis results are often surprising with the high degree of chromosomal rearrangements that exist, many of which are unidentifiable by standard cytogenetics. This discussion will be pursued in greater depth in Chapter 31, where complex karyotypic abnormalities in neoplastic disorders are more common than in constitutional studies.

EXAMPLE REPORTING OF A MARKER OF CHROMOSOME 15 ORIGIN (FIGURE 13.6)

ISCN Result: 47,XY,+mar.ish+del(15)(q11.2q22)(SNRPN++,PML++)

Interpretation

Chromosome and fluorescence *in situ* hybridization (FISH) analyses (15q11.2/SNRPN and 15q22/PML probes) were performed on metaphases for each assay.

Chromosome analysis showed an abnormal male chromosome complement with the presence of a small supernumerary bisatellited chromosome in all 20 cells examined. FISH analysis confirmed the presence of a small chromosome with a duplication of chromosome 15q11.2 to 15q22 region in 10 metaphase cells scored. This finding is consistent with chromosome 15q duplication syndrome.

Parental chromosome and FISH studies are recommended to identify whether this is an inherited abnormality or *de novo* in origin.

Genetic counseling is recommended.

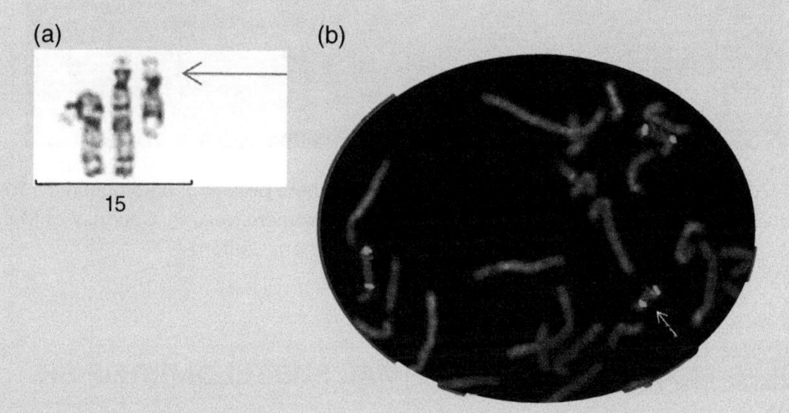

Figure 13.6 Supernumerary marker chromosome of chromosome 15q origin. (a) G-band picture of the marker placed with the chromosome 15 pair. (b) Metaphase FISH analysis showing the gain of a chromosome 15 (*arrow*) with duplication of the SNRPN (*orange*) and PML (*blue*) probes. Courtesy of Mehdi Jamehdor MD. Kaiser Permanente Regional Laboratory, Southern California.

13.7 Subtelomere fluorescence *in situ* hybridization analysis

SPECIFIC FEATURES OF SUBTELOMERE FISH ANALYSIS

Subtelomere FISH analysis is generally performed with all the subtelomeres in one test; that is, as a FISH panel of tests. This allows for all 41 unique chromosome ends to be hybridized and analyzed as one complete panel. The 41 probes consist of all the short and long arm subtelomeric DNA sequences of metacentric and submetacentric chromosomes (chromosomes 1–12, 16–20), which equals 17 chromosomes × 2 probes each (34 probes), plus the long arm subtelomeric DNA sequences of the acrocentric chromosomes (chromosomes 13–15, 21–22), which equals five chromosomes × 1 each (five probes), plus X and Y short arm and long arm shared subtelomere DNA sequences (two probes), which totals to 41 unique probes (Figure 13.7).

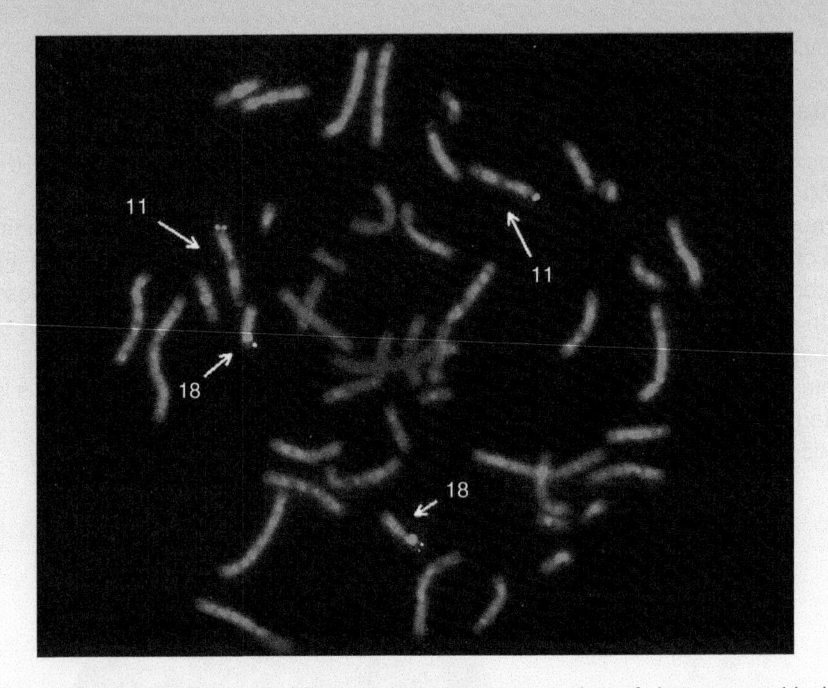

Figure 13.7 Metaphase FISH analysis with example subtelomere probes of chromosome 11p (*orange*) and q (*green*) arms, and chromosome 18p (*aqua*) and 18 centromere (*yellow*). Courtesy of Mehdi Jamehdor MD, Kaiser Permanente Regional Laboratory, Southern California.

ISCN RULES FOR REPORTING NORMAL SUBTELOMERE FISH ANALYSIS

- Subtelomeric FISH nomenclature always begins with “ish subtel” followed, in parentheses, by 41x2.

For example, a normal subtelomere FISH result would be written as: 46,XY.ish subtel(41x2).

EXAMPLE REPORTING OF SUBTELOMERE FISH ANALYSIS – NORMAL

ISCN Result: ish subtel(41x2)

Interpretation

Florescence *in situ* hybridization (FISH) analysis using 41 probes of a subtelomeric panel, with the DNA probes specific to subtelomeric regions of each chromosome, showed normal results in 10 metaphase cells scored for each probe.

Since FISH analysis showed normal results, consider microarray analysis to detect smaller deletions or duplications or abnormalities not covered by these FISH probes.

13.7.1 General results – abnormal

ISCN RULES FOR REPORTING A BALANCED TRANSLOCATION USING SUBTELOMERE FISH ANALYSIS

- Abnormal subtelomere FISH nomenclature only describes the chromosome arm designations of the abnormality present.
- If the abnormality is a balanced translocation of two chromosome subtelomeric regions, then first write "ish" followed, in parentheses, by the two chromosomes involved, separated by a semicolon.
- Then write the arm designation of the first chromosome involved in the translocation, p for the short arm or q for the long arm, followed by "-" since the subtelomere is not present on the chromosome arm that is translocated to the second chromosome.
- Then write a comma, followed by the second chromosome arm (p or q) involved in the translocation of the chromosome that contains an intact arm, followed by "+" since that subtelomere is present on that chromosome arm.
- Next write a semicolon, followed by the second chromosome arm (p or q) involved in the balanced translocation, followed by a "-".
- Next write a comma, followed by the intact arm of the second chromosome, followed by "+".
- Next write the probe designations of each of the subtelomeres, in the order of the chromosome arms of the nomenclature.
- Parentheses enclose the chromosomes involved, the arms involved, and the probes involved.

Note that even though all the chromosomes are being analyzed, only the chromosomes involved in the translocation are written in the nomenclature, although the remaining normal chromosome probes may be mentioned in the interpretation of the report for clarity.

For example, a balanced translocation between the short arm of chromosome 7 and the long arm of chromosome 17 using subtelomere probes would be written as: ish t(7;17)(p22-,q25+;q25-,p22+)(VIJyRM2185-,D17S928+;D17S928-, VIJyRM2185+).

Note that the probe names are based on the manufacturer's description.

EXAMPLE REPORTING OF SUBTELOMERE FISH ANALYSIS – ABNORMAL WITH A BALANCED TRANSLOCATION

ISCN Result: ish t(7;17)(p22-,q25+;q25-,p22+)(VIJyRM2185-,D17S928+;D17S928-, VIJyRM2185+)

Interpretation

Fluorescence *in situ* hybridization (FISH) analysis showed an apparently balanced translocation between the short arm telomeric region of chromosome 7 with the long arm telomeric region of chromosome 17 in 10 metaphase cells scored. All other subtelomere DNA probes showed a normal hybridization pattern in 10 metaphase cells scored for each probe.

Parental chromosome studies are suggested to identify a familial chromosomal rearrangement.

Genetic counseling is recommended.

ISCN RULES FOR REPORTING AN UNBALANCED TRANSLOCATION USING SUBTELOMERE FISH ANALYSIS

Abnormal subtelomere FISH nomenclature only describes the chromosome arm designations of the abnormality present.

- If the abnormality is an unbalanced translocation resulting in a deletion, first write "ish" followed by "der" followed by the chromosome with the deletion, enclosed in parentheses.
- Then write "t" followed by the two chromosomes involved in the unbalanced translocation, separated by a semicolon, and enclosed by parentheses.
- Then write the arm designation of the chromosome with the deletion, followed by "-" since the subtelomere is not present on the chromosome arm that is deleted.
- Then write a comma, followed by the second chromosome arm (p or q) of the chromosome that contains an intact arm, followed by "+" since that subtelomere is present on that chromosome arm.
- Parentheses enclose the chromosomes involved and enclose the arms involved.
- Next write the probe designations of each of the subtelomeres, in the order of the chromosome arms of the nomenclature. Each probe is followed by either "-" or "+" based on whether the probe is absent or present, respectively.
- Parentheses enclose the chromosomes involved, the arms involved and the probes involved. Only the probe names of the abnormal chromosome with the deletion are described.

Note that even though all the chromosomes are being analyzed, only the chromosomes involved in the translocation are written in the nomenclature, although the remaining normal chromosome probes may be mentioned in the interpretation of the report for clarity.

For example, an unbalanced translocation between the short arm of chromosome 7 and the long arm of chromosome 17 that results in a deletion of the subtelomere of chromosome 7p, which is replaced by the subtelomere of chromosome 17q, would be written as: ish der(7) t(7;17)(p22-,q25+)(VIJyRM2185-,D17S928+).

Note that the probe names are based on the manufacturer's description.

EXAMPLE REPORTING OF SUBTELOMERE FISH ANALYSIS – ABNORMAL WITH A DELETION

ISCN Result: ish der(7)t(7;17)(p22-,q25+)(VIJyRM2185-,D17S928+)

Interpretation

Fluorescence *in situ* hybridization (FISH) analysis showed a deletion of the short arm telomeric region of chromosome 7, resulting from an unbalanced translocation between the short arm of chromosome 7 and the long arm of chromosome 17, in 10 metaphase cells scored. All other subtelomere DNA probes showed a normal hybridization pattern in 10 metaphase cells scored for each probe.

Parental chromosome studies are suggested to identify a familial chromosomal rearrangement involving the deletion region. True subtelomeric deletions are associated with abnormal phenotypes and malformations.

Genetic counseling is recommended.

Bibliography

Battaglia A, Carey JC. Wolf-Hirschhorn syndrome and Pitt-Rogers-Danks syndrome. Am J Med Genet 1998; 75(5): 541.

Battaglia A, Hoyme HE, Dallapiccola B, et al. Further delineation of deletion 1p36 syndrome in 60 patients: a recognizable phenotype and common cause of developmental delay and mental retardation. Pediatrics 2008; 121(2): 404–410.

Bergemann AD, Cole F, Hirschhorn K. The etiology of Wolf-Hirschhorn syndrome. Trends Genet 2005; 21(3): 188–195.

Bhattacharjee Y. Friendly faces and unusual minds. Science 2005; 310(5749): 802–804.

Carrasco X, Castillo S, Aravena T, Rothhammer P, Aboitiz F. Williams syndrome: pediatric, neurologic, and cognitive development. Pediatr Neurol 2005; 32(3): 166–172.

Cornish K, Bramble D. Cri du chat syndrome: genotype-phenotype correlations and recommendations for clinical management. Dev Med Child Neurol 2002; 44(7): 494–497.

Gajecka M, Mackay KL, Shaffer LG. Monosomy 1p36 deletion syndrome. Am J Med Genet C Semin Med Genet 2007; 145C(4): 346–356.

Gardner RJM, Sutherland GR. *Chromosome Abnormalities and Genetic Counselling*. Oxford Monographs on Medical Genetics. Oxford University Press, Oxford, 2003.

Gersen S, Keagle M (eds). *Principles of Clinical Cytogenetics*. Humana Press, Totowa, New Jersey, 1999.

Heilstedt HA, Ballif BC, Howard LA, et al. Physical map of 1p36, placement of breakpoints in monosomy 1p36, and clinical characterization of the Syndrome. Am J Hum Genet 2003; 72(5): 1200–1212.

Lahortiga I, Vázquez I, Belloni E, et al. FISH analysis of hematological neoplasias with 1p36 rearrangements allows the definition of a cluster of 2.5 Mb included in the minimal region deleted in 1p36 deletion syndrome. Hum Genet 2005; 116(6): 476–485.

Lee ML, Tsao LY, Wang BT, Lee MH, Chiu IS. Revisit on a distinctive chromosome 1p36 deletion syndrome: report of one case and review of the English literature. Int J Cardiol 2004; 96(3): 477–480.

Medina M, Marinescu RC, Overhauser J, Kosik KS. Hemizygosity of delta-catenin (CTNND2) is associated with severe mental retardation in cri-du-chat syndrome. Genomics 2000; 63(2): 157–164.

Mervis CB, Becerra AM. Language and communicative development in Williams syndrome. Ment Retard Dev Disabil Res Rev 2007; 13(1): 3-15.

Meyer-Lindenberg A, Kohn P, Mervis CB, et al. Neural basis of genetically determined visuospatial construction deficit in Williams syndrome. Neuron 2004; 43(5): 623–631.

Morris CA, Mervis CB. Williams syndrome and related disorders. Annu Rev Genomics Hum Genet 2000; 1: 461–484.

Nieminen P, Kotilainen J, Aalto Y, Knuutila S, Pirinen S, Thesleff I. MSX1 gene is deleted in Wolf-Hirschhorn syndrome patients with oligodontia. J Dent Res 2003; 82(12): 1013–1017.

Potocki L, Bi W, Treadwell-Deering D, et al. Characterization of Potocki-Lupski syndrome (dup(17)(p11.2p11.2)) and delineation of a dosage-sensitive critical interval that can convey an autism phenotype. Am J Hum Genet 2007; 80: 633–649.

Rooney DE, Czepulkowski BH. *Human Cytogenetics. A Practical Approach*. Oxford University Press, New York, 1992.

Shaffer LG, McGowan-Jordan J, Schmid M (eds). *ISCN 2013: An International System for Human Cytogenetic Nomenclature*. Karger Publishers, Unionville, CT, 2013.

Shprintzen RJ. Velo-cardio-facial syndrome: 30 years of study. Dev Disabil Res Rev 2008; 14: 3–10.

Sigafoos J, O'Reilly MF, Lancioni GE. Cri-du-chat. Dev Neurorehabil 2009; 12(3): 119–121.

South ST, Whitby H, Battaglia A, Carey JC, Brothman AR. Comprehensive analysis of Wolf-Hirschhorn syndrome using array CGH indicates a high prevalence of translocations. Eur J Hum Genet 2008; 16(1): 45–52.

Tassabehji M. Williams-Beuren syndrome: a challenge for genotype-phenotype correlations. Hum Mol Genet 2003 Oct 15;12 Spec No 2:R229-37. Epub 2003 Sep 2.

Tobias ES, Connor M, Ferguson Smith M. *Essential Medical Genetics*, 6th edn. Wiley-Blackwell, Oxford, 2011.

Wu Q, Niebuhr E, Yang H, Hansen L. Determination of the 'critical region' for cat-like cry of Cri-du-chat syndrome and analysis of candidate genes by quantitative PCR. Eur J Hum Genet 2005; 13(4): 475–485.

Zhang FR, Aurias A, Delattre O, et al. Mapping of human chromosome 22 by in situ hybridization. Genomics 1990; 7: 319–324.

Zollino M, Lecce R, Fischetto R, et al. Mapping the Wolf-Hirschhorn syndrome phenotype outside the currently accepted WHS critical region and defining a new critical region, WHSCR-2. Am J Hum Genet 2003; 72(3): 590–597.

Commercial FISH probes are available at these websites

www.AbbottMolecular.com

www.Kreatech.com

www.Rainbowscientific.com

www.empiregenomics.com

www.abnova.com

www.cytocell.com

www.exiqon.com

www.cambridgebluegnome.com

CHAPTER 14

Interphase analysis

14.1 Introduction

This chapter describes fluorescence *in situ* hybridization (FISH) for interphase analysis. This methodology can be used to detect aneuploidies and confirm the presence of deletions and duplications as well as other constitutional rearrangements. Since interphase FISH analysis is easy to perform, this technique is useful for many applications. It is fast and does not require the presence of dividing cells, as does standard chromosome analysis or metaphase FISH analysis. This is a great advantage in that it does not require any cell manipulation in the laboratory to stimulate cell division, resulting in a much more rapid result time and also allowing for a greater number of cells to be analyzed. Thus, this technique provides for the best cytogenetic test as a semi-quantitative result, rather than the standard metaphase analysis, which is truly a qualitative test. Interphase FISH analysis also gives a better estimation of mosaicism, since it is a semi-quantitative test with 50–500 cells analyzed per probe, than performing metaphase analysis on typically only 20–50 cells.

Interphase FISH analysis describes the presence or absence of DNA probes that are locus specific, or at the centromere or telomere, rather than describing the composition of DNA as may be done with metaphase analysis. However, interphase FISH analysis is a powerful tool to utilize when the presence or absence of a DNA segment is in question.

Interphase FISH analysis is prevalent in aneuploid detection, which is a very useful tool in prenatal diagnosis, neonatal diagnosis and fetal demises. Since one aspect of prenatal diagnosis is the need for a rapid turn-around time for the management of the fetus, this test has become widespread and the standard of care for many pregnancies. Follow-up chromosome analysis is recommended or even required in many countries, not only to confirm the aneuploid test but also to identify other possible cytogenetic abnormalities that may be present, including balanced and unbalanced rearrangements that are not detectable by common aneuploid tests. The most common aneuploid testing includes DNA probes to identify the presence of trisomy 13, 18 and 21, and the copy number of the sex chromosomes. DNA probes of the other autosomes are generally not included unless it is for fetal demises, in which chromosomes 16 and 22 are typically added, since they are commonly observed in spontaneous abortions.

Cytogenetic Abnormalities: Chromosomal, FISH and Microarray-Based Clinical Reporting, First Edition. Susan Mahler Zneimer.

Another form of aneuploid detection discussed in this chapter is preimplantation genetic diagnosis (PGD), since it has now become a common practice among infertile couples. A summary of the techniques currently employed for diagnosis, with example reporting of aneuploidies observed, will be discussed.

Due to a priority of turn-around time, laboratory testing of interphase FISH analysis is usually reported before metaphase FISH or chromosome analyses are completed; therefore, more than one report is usually generated for a single patient. When separate reports are generated for different tests, a decision must be made on whether or not to incorporate all the results in the final report. Some laboratories view each report as separate and "stand alone" while other labs prefer to provide a comprehensive interpretation of all the cytogenetic findings performed on a single patient. Some laboratories choose to amend their original cytogenetic reports when the metaphase analyses are completed; others generate new reports for each analysis performed.

Interphase FISH analysis may utilize FDA-approved tests (or the equivalent), in which case disclaimers are different from those used with analyte-specific tests (ASRs). Some of the more common disclaimers are described on pages 207–208.

One aspect of interphase analysis is the concept that as a quantitative test, normal and abnormal cut-off values need to be assessed and added to the reports to ensure the accuracy of the analysis. This is unlike metaphase analysis which is qualitative, where the cut-off criteria equal 46,XX or 46,XY. To assess cut-off values for the commonly used aneuploid tests, proper calibration verification and validation of each probe are required, and testing of normal individuals is used to assess the number of abnormal cells observed. Each laboratory would ultimately need to determine its own cut-off value for each probe and test performed. In addition, each laboratory must decide on whether each probe analyzed is a single test or whether a group of probes make up a panel that ultimately consists of a single test. For example, most laboratories today perform aneuploid testing for chromosomes 13, 18, 21, X and Y for prenatal and postnatal diagnoses. These are combined as a single aneuploid test. This has a very different implication for billing and reporting purposes than having each of these probes analyzed and reported singly. Further information on this topic is discussed in Part 2 where many "panels" of probes are used to diagnose specific hematological disorders.

Many laboratories have a disclaimer in their reports stating that interphase FISH for prenatal samples should be performed in conjunction with standard cytogenetics, and that prenatal interphase FISH should not be a stand-alone test in which medical decisions are made solely on these results.

FISH probes may have different DNA composition depending on the manufacturer or laboratory that developed the probe. Therefore, the examples given below for each probe are designed to be generic and not specific to a particular company or laboratory. However, there may be descriptions of probes that appear to be specifically designed, especially FDA-approved probes; this is for the purpose of clarity and not to be misconstrued as the only possible configuration for the particular probe described.

For an example of a FISH ASR assay validation plan, see Appendix 1.

14.2 Example report of interphase analysis

LABORATORY NAME

Cytogenetics Report

Patient Name: Jane Doe	Collection Date: 1/1/2014
Ordering Physician: Dr Smith	Report Date: 1/10/2014
Date of Birth: 1/1/1980	Laboratory Number: A14-000021

Specimen Type: Amniotic fluid
Indication for Study: Rule out chromosomal aneuploidy

Number of cells analyzed by FISH: 200 interphase cells

ISCN Result: nuc ish Xcen(DXZ1x2),Ycen(DYZ3x0),13q14(RB1x2),
18cen(D18Z1x2),21q22.13(D21S259/D21S341/D21S342x2)

Interpretation

Interphase fluorescence *in situ* hybridization (FISH) analysis showed a female chromosome complement with no evidence of a numerical abnormality for chromosomes 13, 18, 21, X, and Y in 200 cells scored for each probe.

Date: January 10, 2014

Susan Zneimer, PhD, FACMG
Cytogenetics Laboratory Director

(Any disclaimer here you may want to provide.)

This test was performed at: Address of Laboratory

Dr Jones, Medical Director
Laboratory License Number: 00001

(Cytogenetic Image)

14.3 Common disclaimers

14.3.1 FISH cut-off comment

This FISH result is considered abnormal based on laboratory validation data for this probe, indicating that>_% abnormal cells is considered a positive result.

14.3.2 Result near the cut-off value

This FISH result is considered abnormal based on laboratory validation data for this probe, indicating that >_% abnormal cells is considered a positive result. However, since the percentage of abnormal cells in this case is near the cut-off value, correlation of this finding with other laboratory and clinical data is strongly recommended.

14.3.3 Disclaimer for undetected abnormalities

This test will not detect numerical abnormalities of other chromosomes, structural abnormalities of any chromosome or low-level mosaicism. Also, genetic variation of the target DNA may lead to errors in the detection of aneuploidy.

14.3.4 Normal cytogenetics reflexing to FISH

Since chromosome analysis on metaphase cells showed normal results, fluorescence *in situ* hybridization (FISH) on interphase cells may prove informative in this case. The preparations from the current specimen can be used for further analysis of chromosome aberrations by FISH with probes to identify a possible abnormality.

14.3.5 FDA-approved comment for Aneuvysion®

Aneuvysion® has been cleared by the FDA as an *in vitro* diagnostic test to be used only in conjunction with a standard cytogenetic analysis. Medical decisions should not be made solely on the basis of this prenatal screening test.

14.3.6 Example disclaimer for FISH in conjunction with standard cytogenetics

This test should be performed in conjunction with standard cytogenetic analysis. It is recommended that medical decisions should not be made solely on the basis of this prenatal test.

14.3.7 More extensive disclaimer for prenatal interphase FISH

This interphase FISH analysis is only designed to detect the common numerical abnormalities for chromosomes 13, 18, 21, X and Y and will not detect approximately 30% of prenatal chromosome abnormalities, which include mosaicism, structural abnormalities and other numerical chromosome abnormalities. Therefore, routine cytogenetic analysis is recommended for the final interpretation.

14.3.8 Interphase FISH disclaimer – neonatal

This interphase FISH analysis is only designed to detect the common numerical abnormalities for chromosomes 13, 18, 21, X and Y and will not detect other chromosome abnormalities, such as mosaicism, structural abnormalities and numerical abnormalities of other chromosomes. Therefore, either routine cytogenetic or genomic microarray analysis is recommended for further characterization of chromosomal imbalances that may be related to this patient's phenotype.

14.3.9 When aneuploidy or triploidy is observed in a prenatal specimen, add this comment

The American College of Medical Genetics recommends that clinical decision making should be based on information from two or three of the following tests: positive FISH results, confirmatory chromosome analysis or consistent clinical information.

14.3.10 For chorionic villus sampling analysis

Cytogenetic analysis performed on chorionic villus sampling (CVS) presumes that the fetal chromosome complement is accurately reflected in the tissue provided. There are rare examples in which the karyotype of the CVS is not consistent with that of the fetus. In addition, contamination of the sample with cells of maternal origin may result in the analysis of maternal rather than fetal chromosomes.

14.4 Reporting normal results

The number of cells scored for interphase FISH analysis may vary from 50 to 200 depending on specimen type. For prenatal studies, 50 cells per probe may be sufficient. For peripheral blood studies, generally 200 cells are scored for each probe.

ISCN RULES FOR REPORTING NORMAL INTERPHASE FISH ANALYSIS

- Interphase FISH nomenclature always begins with "nuc ish" followed by a space.
- Then write the chromosome involved, followed by the chromosome band of the region tested. Centromere probes are usually designated as CEP.
- Then write the probe name, which can be referred to as the clone name or, if unavailable, then use the locus designation utilized by the Genome Database (GDB). If that probe is unavailable, then use the HUGO-approved nomenclature. It is also acceptable to use the GDB nucleotide number in a specific genome build, such as NCBI build 35 (B35).
- The probe name is usually an acronym, so it is written all capitalized and not italicized.
- With normal FISH results, the probe name is followed by the multiplication sign "x" followed by the number of signals observed (typically two signals, one per chromosome).

Note that for normal results, the chromosome and band regions are not surrounded by parentheses, but the probe and signal number are surrounded by parentheses.

For example, an interphase aneuploidy analysis of chromosomes 13, 18, 21, X and Y, a normal FISH result in a female, would be written as: nuc ish Xcen(DXZ1x2),Ycen(DYZ3x0), 13q14(RB1x2),18cen(D18Z1x2),21q22.13(D21S259/D21S341/D21S342x2).

EXAMPLE REPORTING OF GENERAL NORMAL RESULTS

ISCN Result: nuc ish [ ]

Interpretation

Interphase fluorescence *in situ* hybridization (FISH) analysis with the _ probe(s) showed a normal result with no evidence of _ in 200 interphase nuclei scored.

This finding is consistent with the results of the cytogenetics study of this case (Acc. #).

Consider a genomic microarray and uniparental disomy if clinically indicated.

14.4.1 Normal prenatal/neonatal results

EXAMPLE REPORTING OF NORMAL RESULTS – FEMALE (FIGURE 14.1)

ISCN Result: nuc ish Xcen(DXZ1x2),Ycen(DYZ3x0),13q14(RB1x2), 18cen(D18Z1x2),21q22.13(D21S259/D21S341/D21S342x2)

Interpretation

Interphase fluorescence *in situ* hybridization (FISH) analysis showed a female chromosome complement with no evidence of a numerical abnormality for chromosomes 13, 18, 21, X and Y in 200 cells scored for each probe.

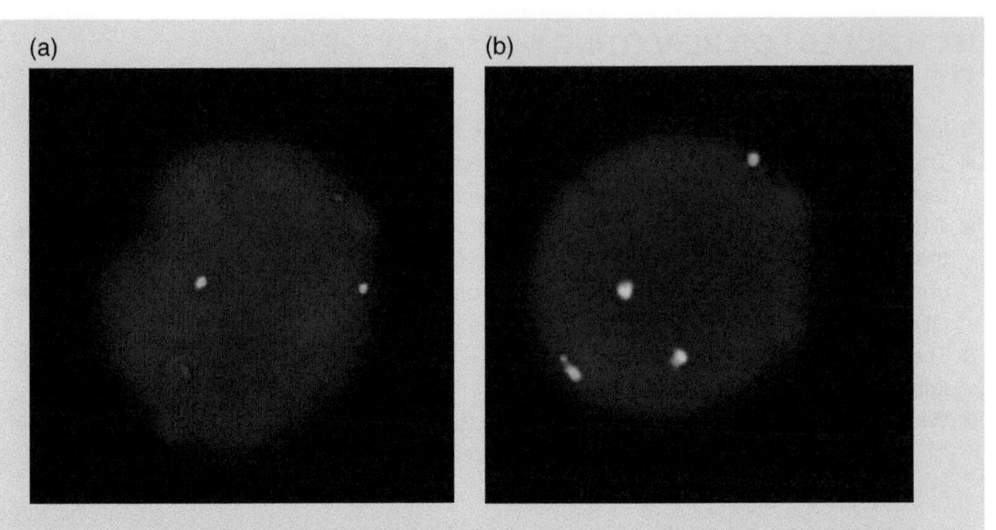

Figure 14.1 Interphase FISH analysis of a normal female showing two signals of chromosomes 13, 18, 21 and X. (a) Red signals = chromosome 13, green signals = chromosome 21. (b) Aqua = chromosome 18, green = X chromosome, no orange signals for the Y chromosome.

EXAMPLE REPORTING OF NORMAL RESULTS – MALE (FIGURE 14.2)

ISCN Result: nuc ish Xcen(DXZ1x1),Ycen(DYZ3x1),13q14(RB1x2), 18cen(D18Z1x2),21q22.13(D21S259/D21S341/D21S342x2)

Interpretation

Interphase fluorescence *in situ* hybridization (FISH) analysis showed a male chromosome complement with no evidence of a numerical abnormality for chromosomes 13, 18, 21, X and Y in 200 cells scored for each probe.

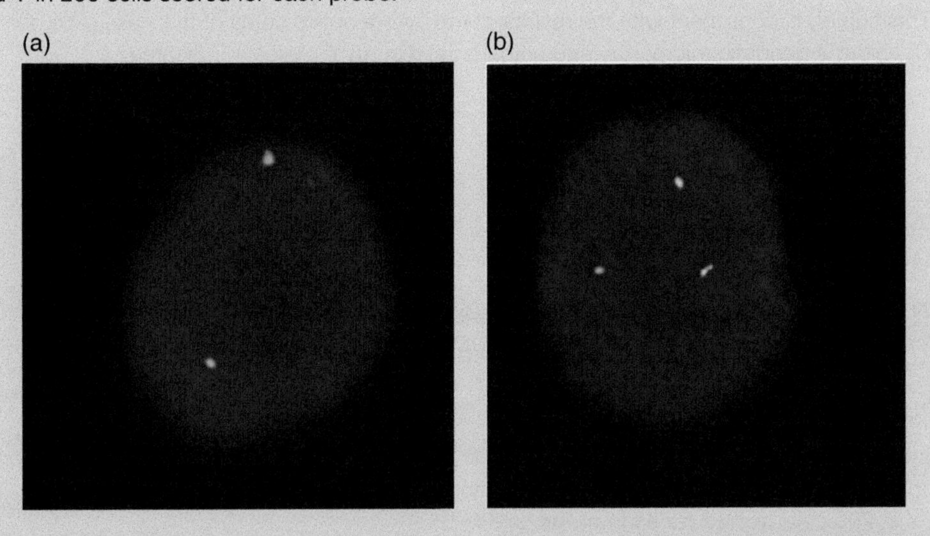

Figure 14.2 Interphase FISH analysis of a normal male showing two signals of chromosomes 13, 18, 21 and one signal for chromosomes X and Y Interphase FISH analysis. (a) Red signals = chromosome 13, green signals = chromosome 21. (b) Aqua = chromosome 18, green = X chromosome, orange = Y chromosome.

14.4.2 Normal fetal demise results

Fetal demise specimens pose a special problem for chromosome analysis. Since, *a priori*, the fetus has expired some time before the procedure in obtaining the specimen for analysis, these cells tend not to be viable in culture. The earlier the fetus demised before obtaining the specimen, the less viable the cells will be in culture, often resulting in no growth of the cells for chromosome analysis. Thus, aneuploid FISH analysis has a great advantage over chromosome analysis in helping to identify the most common abnormalities that cause a fetal demise. As discussed previously with chromosome analysis for fetal demises, aneuploidies are the most common aberration observed and can more easily be accomplished by this type of FISH analysis. The most common probes utilized for this testing include the pre- and postnatal tests, including chromosomes 13, 18, 21, X and Y, plus chromosomes 16 and 22 for the trisomy 16 and trisomy 22 aneuploidies commonly seen in first-trimester spontaneous abortions.

EXAMPLE REPORTING OF NORMAL RESULTS – FEMALE

ISCN Result: nuc ish Xcen(DXZ1x2),Ycen(DYZ3x0),13q14(RB1x2),16cen(D16Z3x2), 18cen(D18Z1x2),21q22.13(D21S259/D21S341/D21S342x2),22q11.2(BCRx2)

Interpretation

Interphase fluorescence *in situ* hybridization (FISH) analysis showed a female chromosome complement with no evidence of a numerical abnormality for chromosomes 13, 16, 18, 21, 22, X and Y in 200 cells scored for each probe.

EXAMPLE REPORTING OF NORMAL RESULTS – MALE

ISCN Result: nuc ish Xcen(DXZ1x1),Ycen(DYZ3x1),13q14(RB1x2),16cen(D16Z3x2),18cen (D18Z1x2),21q22.13q22.2(D21S259,D21S341,D21S342)x2,22q11.2(BCRx2)

Interpretation

Interphase fluorescence *in situ* hybridization (FISH) analysis showed a male chromosome complement with no evidence of a numerical abnormality for chromosomes 13, 16, 18, 21, 22, X and Y in 200 cells scored for each probe.

14.5 Abnormal prenatal/neonatal results

ISCN RULES FOR REPORTING ABNORMAL INTERPHASE FISH ANALYSIS

- Interphase FISH nomenclature always begins with "nuc ish" followed by a space.
- Then write the chromosome involved, followed by the chromosome band of the region tested. Centromere probes are usually designated as CEP.
- Next write the probe name, which can be referred to as the clone name or, if unavailable, then use the locus designation used by the GDB. If that probe is unavailable, then use the HUGO-approved nomenclature. It is also acceptable to use the GDB nucleotide number in a specific genome build, such as NCBI build 35 (B35).

- The probe name is usually an acronym, so it is written all capitalized and not italicized.
- With abnormal FISH results, the probe name is followed by the multiplication sign “x” followed by the number of signals observed (for loss of one signal, then x1; for gain of one signal, then x3).

Note that abnormal results, the chromosome and band regions are not surrounded by parentheses, but the probe and signal number are surrounded by parentheses.

For example, for an interphase aneuploidy analysis of chromosomes 13, 18, 21, X and Y, an abnormal FISH result in a male with trisomy 21 would be written as: nuc ish Xcen(DXZ1x1), Ycen(DYZ3x1),13q14(RB1x2),18cen(D18Z1x2), 21q22.13(D21S259/D21S341/D21S342x3).

EXAMPLE REPORTING OF GENERAL ABNORMAL RESULTS

ISCN Result: nuc ish Xcen(DXZ1x_),Ycen(DYZ3x_),13q14(RB1x_),18cen(D18Z1x_), 21q22.13(D21S259/D21S341/D21S342x_)

Interpretation

Interphase fluorescence *in situ* hybridization (FISH) analysis with the _ probe(s) showed an abnormal result with evidence of _ in 200 interphase nuclei scored. The remaining probes showed a normal two-signal pattern in 200 cells scored for each probe.

This finding is consistent with the diagnosis of _ syndrome.

This test should be performed in conjunction with standard cytogenetic analysis. It is recommended that medical decisions should not be made solely on the basis of this prenatal test.

Genetic counseling is recommended.

14.5.1 Aneuploidies

Trisomy 13

EXAMPLE REPORTING OF TRISOMY 13 IN A FEMALE

ISCN Result: nuc ish Xcen(DXZ1x2),Ycen(DYZ3x0),13q14(RB1x3),18cen(D18Z1x2), 21q22.13(D21S259/D21S341/D21S342x2)

Interpretation

Interphase fluorescence *in situ* hybridization (FISH) showed three hybridization signals for chromosome 13, consistent with a female with trisomy 13 (Patau syndrome). The remaining probes showed a normal two-signal pattern in 200 cells scored for each probe.

This test should be performed in conjunction with standard cytogenetic analysis. It is recommended that medical decisions should not be made solely on the basis of this prenatal test.

Genetic counseling is recommended.

EXAMPLE REPORTING OF TRISOMY 13 IN A MALE

ISCN Result: nuc ish Xcen(DXZ1x1),Ycen(DYZ3x1),13q14(RB1x3),18cen(D18Z1x2), 21q22.13(D21S259/D21S341/D21S342x2)

Interpretation

Interphase fluorescence *in situ* hybridization (FISH) showed three hybridization signals for chromosome 13, consistent with a male with trisomy 13 (Patau syndrome). The remaining probes showed a normal two-signal pattern in 200 cells scored for each probe.

This test should be performed in conjunction with standard cytogenetic analysis. It is recommended that medical decisions should not be made solely on the basis of this prenatal test.

Genetic counseling is recommended.

Trisomy 18

EXAMPLE REPORTING OF TRISOMY 18 IN A FEMALE

ISCN Result: nuc ish Xcen(DXZ1x2),Ycen(DYZ3x0),13q14(RB1x2),18cen(D18Z1x3), 21q22.13(D21S259/D21S341/D21S342x2)

Interpretation

Interphase fluorescence *in situ* hybridization (FISH) showed three hybridization signals for chromosome 18, consistent with a female with trisomy 18 (Edwards syndrome). The remaining probes showed a normal two-signal pattern in 200 cells scored for each probe.

This test should be performed in conjunction with standard cytogenetic analysis. It is recommended that medical decisions should not be made solely on the basis of this prenatal test.

Genetic counseling is recommended.

EXAMPLE REPORTING OF TRISOMY 18 IN A MALE

ISCN Result: nuc ish Xcen(DXZ1x1),Ycen(DYZ3x1),13q14(RB1x2),18cen(D18Z1x3), 21q22.13(D21S259/D21S341/D21S342x2)

Interpretation

Interphase fluorescence *in situ* hybridization (FISH) showed three hybridization signals for chromosome 18, consistent with a male with trisomy 18 (Edwards syndrome). The remaining probes showed a normal two-signal pattern in 200 cells scored for each probe.

This test should be performed in conjunction with standard cytogenetic analysis. It is recommended that medical decisions should not be made solely on the basis of this prenatal test.

Genetic counseling is recommended.

Trisomy 21

EXAMPLE REPORTING OF TRISOMY 21 IN A FEMALE

ISCN Result: nuc ish Xcen(DXZ1x2),Ycen(DYZ3x0),13q14(RB1x2),18cen(D18Z1x2), 21q22.13(D21S259/D21S341/D21S342x3)

Interpretation

Interphase fluorescence *in situ* hybridization (FISH) showed three hybridization signals for chromosome 21, consistent with a female with trisomy 21 (Down syndrome). The remaining probes showed a normal two-signal pattern in 200 cells scored for each probe.

This test should be performed in conjunction with standard cytogenetic analysis. It is recommended that medical decisions should not be made solely on the basis of this prenatal test.

Genetic counseling is recommended.

EXAMPLE REPORTING OF TRISOMY 21 IN A MALE

ISCN Result: nuc ish Xcen(DXZ1x1),Ycen(DYZ3x1),13q14(RB1x2),18cen(D18Z1x2), 21q22.13(D21S259/D21S341/D21S342x3)

Interpretation

Interphase fluorescence *in situ* hybridization (FISH) showed three hybridization signals for chromosome 21, consistent with a male with trisomy 21 (Down syndrome). The remaining probes showed a normal two-signal pattern in 200 cells scored for each probe.

This test should be performed in conjunction with standard cytogenetic analysis. It is recommended that medical decisions should not be made solely on the basis of this prenatal test.

Genetic counseling is recommended.

14.5.2 Mosaic and inconclusive aneuploid results

There are instances when the FISH results show signals consistent with both normal and abnormal patterns due to mosaicism. When mosaicism is in a large proportion for the abnormal signal pattern, the report may be written as true mosaicism as long as the laboratory has cut-off values established for this finding. However, when the abnormal signal pattern is in a low proportion (estimated at 20% or lower), then the mosaicism should be considered inconclusive. The results should be confirmed by standard cytogenetic studies.

EXAMPLE REPORTING OF MOSIAC RESULTS FOR TRISOMY 21

ISCN Result: nuc ish Xcen(DXZ1x2),Ycen(DYZ3x0),13q14(RB1x2),18cen(D18Z1x2), 21q22.13(D21S259/D21S341/D21S342x3)

Interpretation

Interphase fluorescence *in situ* hybridization (FISH) showed three hybridization signals for chromosome 21 in 51% of interphase cells scored, consistent with a female with mosaic trisomy 21 (Down syndrome); however, standard chromosome analysis is necessary to confirm and fully interpret these results. The remaining probes showed a normal two-signal pattern in 200 cells scored for each probe.

This test should be performed in conjunction with standard cytogenetic analysis. It is recommended that medical decisions should not be made solely on the basis of this prenatal test.

Genetic counseling is recommended.

EXAMPLE REPORTING OF INCONCLUSIVE RESULTS

ISCN Result: nuc ish nuc ish Xcen(DXZ1x1),Ycen(DYZ3x1),13q14(RB1x2),18cen(D18Z1x2-3), 21q22.13(D21S259/D21S341/D21S342x2)

Interpretation

Interphase fluorescence *in situ* hybridization (FISH) analysis showed a male chromosome complement with no evidence of a numerical abnormality for chromosomes 13, 16, 21, 22, X and Y in 200 cells scored for each probe.

The results of chromosome 18 analysis were inconclusive. Six cells (3.0%) analyzed showed three signals for chromosome 18, which most likely represents *in vitro* artifact. However, the possibility that this finding represents low-level mosaicism for trisomy 18 cannot be excluded. Standard chromosome analysis is necessary to confirm and fully interpret these results.

This test should be performed in conjunction with standard cytogenetic analysis. It is recommended that medical decisions should not be made solely on the basis of this prenatal test.

Genetic counseling is recommended.

14.5.3 Triploidy

Triploid XXX

EXAMPLE REPORTING OF TRIPLOIDY IN A FEMALE

ISCN Result: nuc ish Xcen(DXZ1x3),Ycen(DYZ3x0),13q14(RB1x3),18cen(D18Z1x3), 21q22.13(D21S259/D21S341/D21S342x3)

Interpretation

Interphase fluorescence *in situ* hybridization (FISH) showed three hybridization signals each for chromosomes 13, 18 and 21, and three signals for the X chromosome in 200 cells scored for each probe, consistent with a female with triploidy.

This test should be performed in conjunction with standard cytogenetic analysis. It is recommended that medical decisions should not be made solely on the basis of this prenatal test.

Genetic counseling is recommended.

Triploid XXY

EXAMPLE REPORTING OF TRIPLOIDY IN A XXY MALE

ISCN Result: nuc ish Xcen (DXZ1x2),Ycen(DYZ3x1),13q14(RB1x3),18cen(D18Z1x3), 21q22.132(D21S259/D21S341/D21S342x3)

Interpretation

Interphase fluorescence *in situ* hybridization (FISH) showed three hybridization signals each for chromosomes 13, 18 and 21, and two signals for X and one signal for Y in 200 cells scored for each probe, consistent with a male with triploidy.

This test should be performed in conjunction with standard cytogenetic analysis. It is recommended that medical decisions should not be made solely on the basis of this prenatal test.

Genetic counseling is recommended.

Triploid XYY

EXAMPLE REPORTING OF TRIPLOIDY IN A XYY MALE

ISCN Result: nuc ish Xcen (DXZ1x1),Ycen(DYZ3x2),13q14(RB1x3),18cen(D18Z1x3), 21q22.132(D21S259/D21S341/D21S342x3)

Interpretation

Interphase fluorescence *in situ* hybridization (FISH) showed three hybridization signals each for chromosomes 13, 18 and 21, and one signal for X and two signals for Y in 200 cells scored for each probe, consistent with a male with triploidy.

This test should be performed in conjunction with standard cytogenetic analysis. It is recommended that medical decisions should not be made solely on the basis of this prenatal test.

Genetic counseling is recommended.

14.5.4 Sex chromosome abnormalities in a prenatal specimen

45,X

EXAMPLE REPORTING OF 45,X (FIGURE 14.3)

ISCN Result: nuc ish Xcen(DXZ1x1),Ycen(DYZ3x0),13q14(RB1x2),18cen(D18Z1x2), 21q22.13(D21S259/D21S341/D21S342x2)

Interpretation

Interphase fluorescence *in situ* hybridization (FISH) showed one hybridization signal for the X chromosome and no signal for the Y chromosome. This result is consistent with a female with monosomy X (Turner syndrome). The remaining probes showed a normal two-signal pattern in 200 cells scored for each probe.

This test should be performed in conjunction with standard cytogenetic analysis. It is recommended that medical decisions should not be made solely on the basis of this prenatal test.

Genetic counseling is recommended.

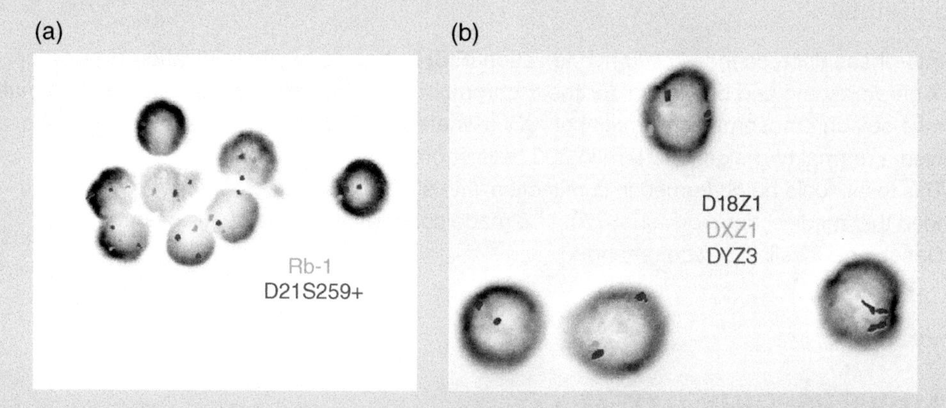

Figure 14.3 Interphase FISH analysis of an abnormal female fetus with 45,X showing two signals of chromosomes 13, 18, 21 and one signal for the X chromosome. (a) Green signals = chromosome 13, red signals = chromosome 21. (b) Blue = chromosome 18, green = X chromosome, no orange signals for the Y chromosome. Courtesy of Lauren Jenkins PhD, Kaiser Permanente Regional Laboratory, Northern California.

47,XXX

EXAMPLE REPORTING OF 47,XXX

ISCN Result: nuc ish Xcen(DXZ1x3),Ycen(DYZ3x0),13q14(RB1x2),18cen(D18Z1x2), 21q22.13(D21S259/D21S341/D21S342x2)

Interpretation

Interphase fluorescence *in situ* hybridization (FISH) showed three hybridization signals for the X chromosome. This result is consistent with a female with the sex chromosome abnormality of XXX (trisomy X syndrome). The remaining probes showed a normal two-signal pattern in 200 cells scored for each probe.

This test should be performed in conjunction with standard cytogenetic analysis. It is recommended that medical decisions should not be made solely on the basis of this prenatal test.

Genetic counseling is recommended.

47,XXY

EXAMPLE REPORTING OF 47,XXY

ISCN Result: nuc ish Xcen(DXZ1x2),Ycen(DYZ3x1),13q14(RB1x2),18cen(D18Z1x2), 21q22.13(D21S259/D21S341/D21S342x2)

Interpretation

Interphase fluorescence *in situ* hybridization (FISH) showed two hybridization signals for the X chromosome and one signal for the Y chromosome. This result is consistent with a male with the sex chromosome abnormality of XXY (Klinefelter syndrome). The remaining probes showed a normal two-signal pattern in 200 cells scored for each probe.

This test should be performed in conjunction with standard cytogenetic analysis. It is recommended that medical decisions should not be made solely on the basis of this prenatal test.

Genetic counseling is recommended.

47,XYY

EXAMPLE REPORTING OF 47,XYY

ISCN Result: nuc ish Xcen(DXZ1x1),Ycen(DYZ3x2),13q14(RB1x2),18cen(D18Z1x2), 21q22.13(D21S259/D21S341/D21S342x2)

Interpretation

Interphase fluorescence *in situ* hybridization (FISH) showed one hybridization signal for the X chromosome and two signals for the Y chromosome. This result is consistent with a male with the sex chromosome abnormality of XYY (XYY syndrome). The remaining probes showed a normal two-signal pattern in 200 cells scored for each probe.

This test should be performed in conjunction with standard cytogenetic analysis. It is recommended that medical decisions should not be made solely on the basis of this prenatal test.

Genetic counseling is recommended.

14.6 Abnormal product of conception FISH abnormalities

14.6.1 Aneuploidies

Trisomy 16

EXAMPLE REPORTING OF TRISOMY 16 IN A MALE PRODUCT OF CONCEPTION

ISCN Result: nuc ish Xcen(DXZ1x1),Ycen(DYZ3x1),13q14(RB1x2),16cen(D16Z3x3),18cen (D18Z1x2),21q22.13(D21S259/D21S341/D21S342x2),22q11.2(BCRx2)

Interpretation

Interphase fluorescence *in situ* hybridization (FISH) showed three hybridization signals for chromosome 16, consistent with a male fetus with trisomy 16. The remaining probes showed a normal two-signal pattern in 200 cells scored for each probe.

Genetic counseling is recommended.

EXAMPLE REPORTING OF TRISOMY 16 IN A FEMALE PRODUCT OF CONCEPTION

ISCN Result: nuc ish Xcen(DXZ1x2),Ycen(DYZ3x0),13q14(RB1x2),16cen(D16Z3x3),18cen (D18Z1x2),21q22.13(D21S259/D21S341/D21S342x2),22q11.2(BCRx2)

Interpretation

Interphase fluorescence *in situ* hybridization (FISH) showed three hybridization signals for chromosome 16, consistent with a female fetus with trisomy 16. The remaining probes showed a normal two-signal pattern in 200 cells scored for each probe.

Genetic counseling is recommended.

Trisomy 22

EXAMPLE REPORTING OF TRISOMY 22 IN A MALE PRODUCT OF CONCEPTION

ISCN Result: nuc ish Xcen(DXZ1x1),Ycen(DYZ3x1),13q14(RB1x2),16cen(D16Z3x2),18cen (D18Z1x2),21q22.13(D21S259/D21S341/D21S342x2),22q11.2(BCRx3)

Interpretation

Interphase fluorescence *in situ* hybridization (FISH) showed three hybridization signals for chromosome 22, consistent with a male fetus with trisomy 22. The remaining probes showed a normal two-signal pattern in 200 cells scored for each probe.

Genetic counseling is recommended.

EXAMPLE REPORTING OF TRISOMY 22 IN A FEMALE PRODUCT OF CONCEPTION

ISCN Result: nuc ish Xcen(DXZ1x2),Ycen(DYZ3x0),13q14(RB1x2),16cen(D16Z3x2), 18cen(D18Z1x2),21q22.13(D21S259/D21S341/D21S342x2),22q11.2(BCRx3)

Interpretation

Interphase fluorescence *in situ* hybridization (FISH) showed three hybridization signals for chromosome 22, consistent with a female fetus with trisomy 22. The remaining probes showed a normal two-signal pattern in 200 cells scored for each probe.

Genetic counseling is recommended.

14.6.2 Triploidy/tetraploidy in a POC specimen

Triploidy

EXAMPLE REPORTING OF TRIPLOIDY IN A FEMALE PRODUCT OF CONCEPTION

ISCN Result: nuc ish Xcen(DXZ1x3),Ycen(DYZ3x0),13q14(RB1x3),16cen(D16Z3x3), 18cen(D18Z1x3), 21q22.2(D21S259x4,D21S341x4,D21S342x3),22q11.2(BCRx3)

Interpretation

Interphase fluorescence *in situ* hybridization (FISH) showed three hybridization signals for chromosomes X, 13, 16, 18, 21 and 22, consistent with a female fetus with triploidy.

Genetic counseling is recommended.

EXAMPLE REPORTING OF TRIPLOIDY IN A MALE XXY PRODUCT OF CONCEPTION

ISCN Result: nuc ish Xcen(DXZ1x2),Ycen(DYZx1),13q14(RB1x3),16cen(D16Z3x3),18cen (D18Z1x3),21q22.2(D21S259x4,D21S341x4,D21S342x3),22q11.2(BCRx3)

Interpretation

Interphase fluorescence *in situ* hybridization (FISH) showed two hybridization signals for chromosome X, one signal for chromosome Y, and three signals for chromosomes 13, 16, 18, 21 and 22, consistent with a male fetus with triploidy.

Genetic counseling is recommended.

Tetraploidy

EXAMPLE REPORTING OF TETRAPLOIDY IN A FEMALE PRODUCT OF CONCEPTION

ISCN Result: nuc ish Xcen(DXZ1x4),Ycen(DYZ3x0),13q14(RB1x4),16cen(D16Z3x4), 18cen(D18Z1x4),21q22.2(D21S259x4,D21S341x4,D21S342x4),22q11.2(BCRx4)

Interpretation

Interphase fluorescence *in situ* hybridization (FISH) showed four hybridization signals for chromosomes X, 13, 16, 18, 21 and 22, consistent with a female fetus with tetraploidy.

Genetic counseling is recommended.

EXAMPLE REPORTING OF TETRAPLOIDY IN A MALE PRODUCT OF CONCEPTION

ISCN Result: nuc ish Xcen(DXZ1x2),Ycen(DYZx2),13q14(RB1x4),16cen(D16Z3x4),18cen (D18Z1x4),21q22.2(D21S259x4,D21S341x4,D21S342x4),22q11.2(BCRx4)

Interpretation

Interphase fluorescence *in situ* hybridization (FISH) showed two hybridization signals for chromosomes X and Y, and four hybridization signals for chromosomes 13, 16, 18, 21 and 22, consistent with a male fetus with tetraploidy.

Genetic counseling is recommended.

14.6.3 Sex chromosome abnormalities

45,X

EXAMPLE REPORTING OF 45,X

ISCN Result: nuc ish Xcen(DXZ1x1),Ycen(DYZ3x0),13q14(RB1x2),16cen(D16Z3x2),
18cen(D18Z1x2),21q22.13(D21S259/D21S341/D21S342x2),22q11.2(BCRx2)

Interpretation

Interphase fluorescence *in situ* hybridization (FISH) showed one hybridization signal for the X chromosome and no signal for the Y chromosome. This result is consistent with a female fetus with monosomy X (Turner syndrome). The remaining probes showed a normal two-signal pattern in 200 cells scored for each probe.

Monosomy X is a common cytogenetic abnormality in products of conception, with over 95% of conceptions with a 45,X resulting in spontaneous pregnancy loss.

Genetic counseling is recommended.

47,XXX

EXAMPLE REPORTING OF 47,XXX

ISCN Result: nuc ish Xcen(DXZ1x3),Ycen(DYZ3x0),13q14(RB1x2),16cen(D16Z3x2),
18cen(D18Z1x2),21q22.13(D21S259/D21S341/D21S342x2),22q11.2(BCRx2)

Interpretation

Interphase fluorescence *in situ* hybridization (FISH) showed three hybridization signals for the X chromosome. This result is consistent with a female fetus with the sex chromosome abnormality of XXX (trisomy X syndrome). The remaining probes showed a normal two-signal pattern in 200 cells scored for each probe.

Genetic counseling is recommended.

47,XXY

EXAMPLE REPORTING OF 47,XYY

ISCN Result: nuc ish Xcen(DXZ1x2),Ycen(DYZ3x1),13q14(RB1x2),16cen(D16Z3x2),
18cen(D18Z1x2),21q22.13(D21S259/D21S341/D21S342x2),22q11.2(BCRx2)

Interpretation

Interphase fluorescence *in situ* hybridization (FISH) showed two hybridization signals for the X chromosome and one signal for the Y chromosome. This result is consistent with a male fetus with the sex chromosome abnormality of XXY (Klinefelter syndrome). The remaining probes showed a normal two-signal pattern in 200 cells scored for each probe.

Genetic counseling is recommended.

47,XYY

EXAMPLE REPORTING OF 47,XXY

ISCN Result: nuc ish Xcen(DXZ1x1),Ycen(DYZ3x2),13q14(RB1x2),16cen(D16Z3x2), 18cen(D18Z1x2),21q22.13(D21S259/D21S341/D21S342x2),22q11.2(BCRx2)

Interpretation

Interphase fluorescence *in situ* hybridization (FISH) showed one hybridization signal for the X chromosome and two signals for the Y chromosome. This result is consistent with a male fetus with the sex chromosome abnormality of XYY (XYY syndrome). The remaining probes showed a normal two-signal pattern in 200 cells scored for each probe.

Genetic counseling is recommended.

14.7 Molar pregnancies

EXAMPLE COMMENT FOR MOLAR PREGNANCIES

Cytogenetic analysis can detect an abnormality only in cases of molar pregnancies due to triploidy. Alternative testing by molecular or immunohistochemical methods is suggested for those cases in which a normal diploid chromosome complement is detected. Molecular testing for molar pregnancy requires both maternal and fetal tissue and can be performed on formalin-fixed, paraffin-embedded tissue specimens. Testing of cultured villous tissue in combination with a maternal blood sample can be performed if the laboratory is notified within one week of the report date.

14.8 Preimplantation genetic diagnosis

Preimplantation genetic diagnosis (PGD) is a clinical prenatal procedure to test for specific genetic conditions very early in pregnancy. It was originally a procedure that was intended to replace standard prenatal diagnosis by amniocentesis or chorionic villus sampling, but has evolved in the last 10 years to primarily provide genetic testing for both fertile couples and infertile couples undergoing *in vitro* fertilization.

Preimplantation genetic diagnosis is a technique that takes cells from human embryos that have been cultured *in vitro*. It is used to test for various genetic conditions, including single gene disorders, X-linked genetic disorders, chromosomal aneuploidies and chromosome translocations. The procedure obtains cells from an early embryo that are taken before implantation into the uterus. Cells are biopsied from either the cleavage stage embryo, when they have reached their third cell division (six or more cells – see Figure 14.4), or at the blastocyst stage (up to 300 cells – see Figure 14.4). Both stages have advantages and disadvantages that affect the viability of the embryo, which must be weighed to determine the best approach for the necessary testing of the fetus. Another approach is to do a polar body biopsy, which has the advantage of not affecting the viability of the embryo at all. However, since polar bodies reflect oocyte DNA content, only testing of the maternal genotype is possible with this approach.

Once cells are obtained for genetic analysis, either polymerase chain reaction (PCR) techniques are used to detect specific single gene disorders and X-linked genes for couples with a family history of a gene defect, or FISH analysis is used for chromosomal translocations and aneuploidies. A newer

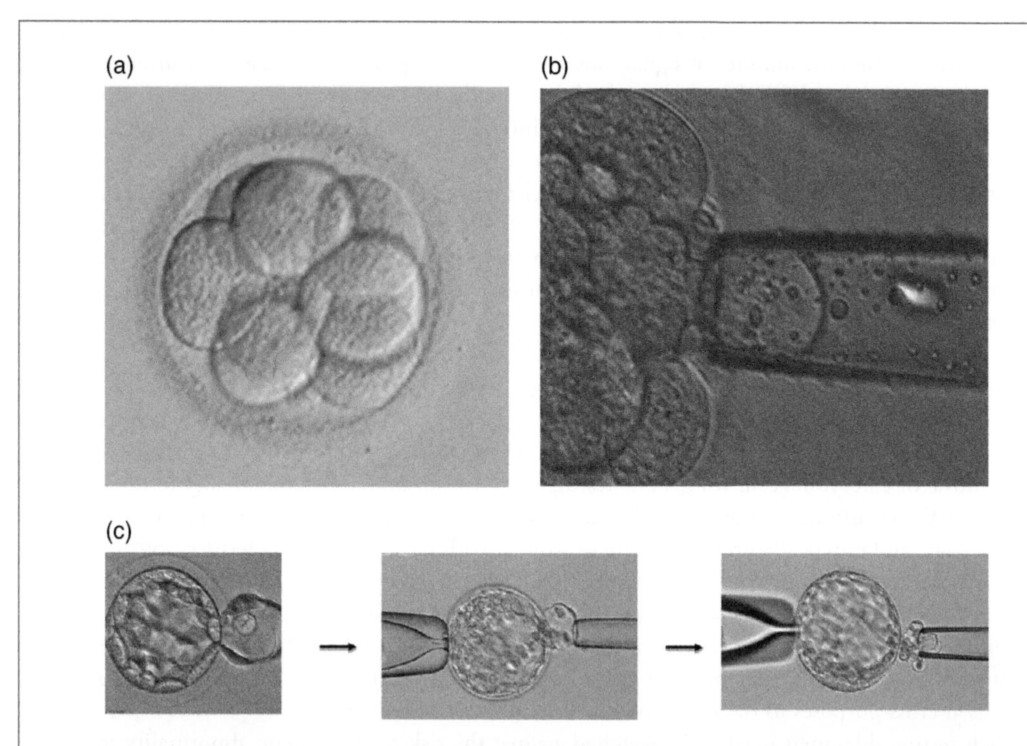

Figure 14.4 Biopsied cells for a PGD study at the cleavage and blastocyst stages. (a, b) Cleavage stage embryo. (c) Blastocyst stage embryo. Courtesy of William Kearns, LabCorp, Inc.

technology currently being used, but at much smaller proportions at present, is comparative genomic hybridization (CGH). This technique is growing in acceptance and will be discussed for this application in Section 3. This section only discusses the FISH application for chromosomal abnormalities and not single gene defects that can only be diagnosed by PCR.

However, the most common indication for PGD currently is for chromosomal aneuploidies. It is estimated that up to 80% of initial embryos are aneuploid for one or more chromosomes. Aneuploidy increases with maternal age, as is well documented for standard prenatal diagnosis. It is also well documented that up to 50–60% of all spontaneous abortions from clinically recognized pregnancies carry an abnormal chromosomal complement. Therefore, aneuploid screening (PGD-AS) is useful for identifying aneuploidies in both women of maternal age and for those individuals with recurrent spontaneous abortions. Since these cohort groups are in such prevalence for PGD testing, FISH analysis has become a widely used screening technique for early embryos. Most recent data suggest that the most commonly used FISH probes greatly enhance the normal implantation rates, and that the misdiagnosis rate is low, at approximately 7%, though accurate rates are still undetermined.

The possibility of a chromosome translocation is another major indication for PGD. Once a translocation is identified in a family, PGD has the advantage of identifying a possible unbalanced form at the earliest stage in pregnancy. If the familial rearrangement is a Robertsonian translocation, then FISH probes anywhere along the long arm of the two acrocentric chromosomes can easily be performed. If the familial rearrangement is a reciprocal translocation involving chromosomes other than the acrocentric chromosomes, then whole chromosome-specific painting probes, alpha-satellite repeat probes, locus-specific probes or subtelomere probes of the chromosomes involved may be performed, as long as the probes are within the translocation region. Unbalanced versus balanced rearrangements may be inferred from the number of signals observed from each probe. If the familial rearrangement is maternal in origin, then PGD from a polar body may be the best source for FISH analysis, since polar bodies are specific for maternal DNA and do not interfere with the viability of the embryo, thereby eliminating any risk of embryo loss due to genetic testing.

Couples who have had a previous child with a chromosome abnormality are also good candidates for PGD, including aneuploidy and translocations. Other common abnormalities, such as microdeletion and microduplication syndromes, may also be tested by PGD. FISH probes for the common deletion and duplication syndromes are easily performed on blastomeres and polar bodies. Since these abnormalities, when seen in a previous pregnancy, carry a risk in further offspring, PGD analysis is a good alternative for early detection.

Some problems exist with PGD testing, raising questions about its validity as a preferred testing methodology for genetic abnormalities. One issue is the efficacy of embryo viability when removing one or more cells of the embryo proper for testing. There has been extensive documentation to show that removal of embryonic cells for genetic testing carries a substantial risk of lethality to the embryo. The ESHRE consortium data on PGD-AS has shown a 28% pregnancy rate for women of advanced maternal age, and only a 7% pregnancy rate in women with recurrent *in vitro* fertilization (IVF) failure. Considering that most women opting for PGD are already at high risk of miscarriages or the inability to conceive, then the risk of PGD is at cross-purposes to the intent of having viable offspring. Also, there is a risk that the testing will not yield complete or accurate results. There might be a need to confirm results with further prenatal testing, again giving rise to a risk of spontaneous abortion, which is at cross-purposes to the intent of a successful viable offspring.

Such testing dilemmas need to be weighed against the risk of the genetic abnormality to determine the appropriate course of action. In some cases, the testing itself might reduce the chances of keeping a viable pregnancy to term to a greater degree than the risk of having a chromosome abnormality.

14.8.1 FISH analysis for preimplantation genetic diagnosis

Interphase FISH analysis is the most common technique for the analysis of chromosome abnormalities with blastomere-derived PGD. The types of FISH probes used for the detection of abnormalities depend on the specificity of the chromosomal abnormalities of interest. For sexing purposes, only the probes for X and Y are needed for diagnosis. For aneuploidies, the aneuploidy screen used for prenatal diagnosis including probes for chromosomes X, Y, 13, 18 and 21, with added chromosomes for clinical implications, such as chromosomes 8 and 9, is the FISH panel of choice. The POC panel, which includes the prenatal panel plus chromosomes 14, 15, 16 and 22, may be used to test for the most common cytogenetic causes of fetal demises and Robertsonian translocations. If there is a family history of a chromosome translocation, then probes specific to the chromosomes involved in the translocation may be used, usually either a locus-specific probe on each of the chromosomes involved or a subtelomere probe for the chromosomes involved.

Example reporting of PGD analysis is similar to standard prenatal interphase FISH analysis (see interphase FISH analysis for prenatal diagnosis). Appropriate disclaimers for ruling out only the probes and/or disorders tested with this technology should be included in these reports, as well as the limitations of testing depending on the type of PGD utilized. An example of reporting PGD analysis is seen below. It may be easier to write a report in a different format than standard cytogenetic analysis since analysis includes many embryonic results, and analysis of each chromosome tested needs to be reported.

See Section 3 for a further discussion about using a microarray technique for PGD analysis.

14.8.2 Regulations of preimplantation genetic diagnosis

The use of PGD is worldwide, and corresponding regulations differ substantially. Each country has different opinions and attitudes regarding assisted reproduction, invasive prenatal testing, testing on human embryos in particular, and the concept of embryo selection. In the United States, there are no specific regulations for PGD as a separate entity of prenatal genetics testing. However, in the United Kingdom, PGD is regulated by the Human Fertilization and Embryology Act 1990. The Human Fertilization and Embryology Authority (HFEA) code of practice states that embryos may only be used for PGD and for research to develop new diagnostic methods under license from the HFEA. The HFEA also states that PGD is only to be used for serious genetic diseases and not for social purposes; that is, for choosing a child that has been designed simply to meet the desires of the parents. This could be inferred to mean choosing an embryo for its sex alone or for compatability of an embryo with a current sibling for possible hematopoetic stem cell transfer, based on HLA typing, or weeding out a heterozygous state of a non-lethal disorder, such as achondroplasia or inherited hearing loss. PGD is allowed in France but not in Austria, Germany or Switzerland. In Germany, severe restrictions only allow prenatal genetic procedures that show a direct benefit to the embryo. Since PGD might result in the loss of the embryo, PGD is not allowed, owing to its past associations with Nazi eugenic policies. Other countries where it is not allowed include Argentina and Taiwan. This list of countries does not include many of those with specific regulations regarding prenatal diagnosis with PGD. However, this limited example gives an idea of the complexity of ethical, social and scientific dilemmas countries face in making decisions regarding this area of laboratory testing.

EXAMPLE DISCLAIMERS FOR PREIMPLANTATION GENETIC DIAGNOSIS

- This test evaluates only the copy numbers of the chromosomes listed in this analysis. Numerical abnormalities of the other chromosomes and structural abnormalities of all the chromosomes cannot be detected. This test does not detect abnormalities at the molecular level.
- The accuracy of this test for aneuploidy is approximately 90–92%. Limitations of PGD testing on a single blastomere include a high incidence of mosaicism in early embryos, estimated to be between 30% and 50%, depending on the stage of development.
- This test was developed and its performance characteristics determined by [Name of laboratory here]. It has not been cleared or approved by the US Food and Drug Administration. The FDA has determined that such clearance or approval is not necessary. Pursuant to the requirements of CLIA '88, this laboratory has established and verified the test's accuracy and precision.

14.8.3 Example reporting of preimplantation genetic diagnosis

EXAMPLE REPORTING OF PREIMPLANTATION GENETIC DIAGNOSIS ANALYSIS

Results

Embryo 1:
Sex chromosome complement: XY
Two signals for chromosomes 8, 9, 13, 15, 16, 18, 21 and 22 were observed.

Embryo 2:
Sex chromosome complement: XY
Two signals for chromosomes 8, 9, 13, 15, 16, 21 and 22 were observed.
One signal for chromosome 18 was observed.

Embryo 3:
Sex chromosome complement: XX
Two signals for chromosomes 8, 9, 13, 15, 16, 18, 21 and 22 were observed.

Embryo 4:
Sex chromosome complement: XY
Two signals for chromosomes 8, 9, 13, 15, 16, 18 and 22 were observed.
Three signals for chromosome 21 were observed.

Embryo 5:
Sex chromosome complement: XY
Two signals for chromosomes 9, 13, 15, 16, 18, 21 and 22 were observed.
Inconclusive results for chromosome 8, a third signal appears overlapped with chromosome 9.

Embryo 6:
Sex chromosome complement: XX
Two signals for chromosomes 8, 9, 15, 16, 18, 21 and 22 were observed.
Three signals for chromosome 13 were observed.

Embryo 7:
No signals observed.

Embryo 8:
Sex chromosome complement: X
Two signals for chromosomes 8, 9 15, 16, 18, 21 and 22 were observed.

Interpretation

Embryo 1: normal male
Embryo 2: abnormal male with monosomy 18
Embryo 3: normal female
Embryo 4: abnormal male with trisomy 21
Embryo 5: male with inconclusive results
Embryo 6: abnormal female with trisomy 13
Embryo 7: no results
Embryo 8: abnormal sex chromosome complement with monosomy X

Comments

Preimplantation genetic diagnosis (PGD) FISH analysis was performed on blastomere biopsies to detect aneuploidies of chromosomes 8, 9, 13, 15, 16, 18 and 21. The panel of DNA probes consists of DXZ1 at the chromosome X centromere, DYZ3 at the chromosome Y centromere, D8Z2 at the chromosome 8 centromere, CEP9 at the chromosome centromere 9, RB1 at chromosome 13q14, D15Z4 at the chromosome 15 centromere, D16Z3 at chromosome 16q11.2, D18Z1 at the chromosome 18 centromere, D21S341/D21S342/D21S339/ERG/D21S338 at chromosome 21q22.13-q22.2, and BCR at chromosome 22q11.2.

Follow-up prenatal diagnosis by CVS or amniocentesis is recommended.

Bibliography

Angell RR, Aitken RJ, van Look PF, Lumsden MA, Templeton AA. Chromosome abnormalities in human embryos after in vitro fertilization. Nature 1983; 303: 336–338.

Baart EB, van Opstal D, Los FJ, Fauser BCJM, Martini E. Fluorescence in situ hybridization analysis of two blastomeres from day 3 frozen-thawed embryos followed by analysis of the remaining embryo on day 5. Hum Reprod 2004; 19(3): 685–693.

Bahçe M, Escudero T, Sandalinas M. Improvements of preimplantation diagnosis of aneuploidy by using microwave-hybridization, cell recycling, and monocolor labeling of probes. Mol Hum Reprod 2000; 6: 849–854.

Braude P, Pickering S, Flinter F, Ogilvie CM. Preimplantation genetic diagnosis. Nature Rev Genet 2002; 3: 941–955.

Cohen J, Grifo JA. Multicentre trial of preimplantation genetic screening reported in the New England Journal of Medicine: an in-depth look at the findings. Reprod BioMed Online 2008; 15(4): 365–366.

Delhanty JD, Griffin DK, Handyside AH, et al. Detection of aneuploidy and chromosomal mosaicism in human embryos during preimplantation sex determination by fluorescent in situ hybridisation (FISH). Hum Mol Genet 1993; 2: 1183–1185.

Findlay I, Corby N, Rutherford A, Quirke P. Comparison of FISH, PRINS, and conventional and fluorescent PCR for single-cell sexing: suitability for preimplantation genetic diagnosis. J Assist Reprod Genet 1998; 15: 258–265.

Findlay I, Urquhart A, Quirke P, Sullivan K, Rutherford AJ, Litford RJ. Simultaneous DNA 'fingerprinting', diagnosis of sex and single-gene defect status from single cells. Mol Hum Reprod 1995; 1(2): 85–93.

Fragouli E, Wells D, Thornhill A, et al. Comparative genomic hybridization analysis of human oocytes and polar bodies. Hum Reprod 2006; 21(9): 2319–2328.

Gardner RJM, Sutherland GR. *Chromosome Abnormalities and Genetic Counselling*. Oxford Monographs on Medical Genetics. Oxford University Press, Oxford, 2003.

Gersen S, Keagle M (eds). *Principles of Clinical Cytogenetics*. Humana Press, Totowa, New Jersey, 1999.

Gosden R. Genetic test may lead to waste of healthy embryos. Nature 2007; 446: 372.

Gunning J. Regulating assisted reproduction technologies. Med Law 2001; 20: 425–433.

Handyside AH, Kontogianni EH, Hardy K, Winston RM. Pregnancies from biopsied human preimplantation embryos sexed by Y-specific DNA amplification. Nature 1990; 344: 768–770.

Handyside AH, Lesko JG, Tarin JJ, Winston RM, Hughes MR. Birth of a normal girl after in vitro fertilization and preimplantation diagnostic testing for cystic fibrosis. N Engl J Med 1992; 327: 905–909.

Hardarson T, Hanson C, Lundin K, et al. Preimplantation genetic screening in women of advanced maternal age caused a decrease in clinical pregnancy rate: a randomized controlled trial. Hum Reprod 2008; 23(12): 2806–2812.

Heng BC, Cao T. Refund fertility-treatment costs for donated embryos. Nature 22006; 443: 26.

Hook EB, Cross PK, Schreinemachers DM. Chromosomal abnormality rates at amniocentesis and in live-born infants. JAMA 1983; (15): 2034–2038.

Hu DG, Webb G, Hussey N. Aneuploidy detection in single cells using DNA array-based comparative genomic hybridization. Mol Hum Reprod 2004; 10: 283–289.

Kamiguchi Y, Rosenbusch B, Sterzik K, Mikamo K. Chromosomal analysis of unfertilized human oocytes prepared by a gradual fixation-air drying method. Hum Genet 1993; 90: 533–541.

Kuliev A, Cieslak J, Ilkevitch Y, Verlinsky Y. Chromosomal abnormalities in a series of 6,733 human oocytes in preimplantation diagnosis for age-related aneuploidies. Reprod Biomed Online 2003; 6: 54–59.

Landwehr C, Montag M, van der Ven K, Weber RG. Rapid comparative genomic hybridization protocol for prenatal diagnosis and its application to aneuploidy screening of human polar bodies. Fertil Steril 2008; 90(3): 488–496.

Márquez CSM, Bahçe M, Alikani M, Munné S. Chromosome abnormalities in 1255 cleavage-stage human embryos. Reprod Biomed Online 2000; 1(1): 17–26.

McArthur SJ, Leigh D, Marshall JT, de Boer KA, Jansen RP. Pregnancies and live births after trophectoderm biopsy and preimplantation genetic testing of human blastocysts. Fertil Steril 2005; 84: 1628–1636.

Menasha J, Levy B, Hirschhorn K, Kardon NB. Incidence and spectrum of chromosome abnormalities in spontaneous abortions: new insights from a 12-year study. Genet Med 2005; 7: 251–263.

Meyer LR, Klipstein S, Hazlett WD, Nasta T, Mangan P, Karande VC. A prospective randomized controlled trial of preimplantation genetic screening in the "good prognosis" patient. Fertil Steril 2008; 91(5): 1731–1738.

Munné S, Chen S, Fisher J, Colls P, et al. Preimplantation genetic diagnosis reduces pregnancy loss in women aged 35 years and older with a history of recurrent miscarriages. Fertil Steril 2005; 84: 331–335.

Munné S, Magli C, Cohen J, et al. Positive outcome after preimplantation diagnosis of aneuploidy in human embryos. Hum Reprod 1999; 14: 2191–2199.

Munné S, Sandalinas M, Escudero T, et al. Improved implantation after preimplantation genetic diagnosis of aneuploidy. Reprod Biomed Online 2003; 7: 91–97.

Munné S, Velilla E, Colls P, et al. Self-correction of chromosomally abnormal embryos in culture and implications for stem cell production. Fertil Steril 2005; 84: 1328–1334.

O'Connor C. Prenatal screen detects fetal abnormalities. Nature Educ 2008; 1: 106.

Ogilvie CM, Braude PR, Scriven PN. Preimplantation genetic diagnosis – an overview. J Histochem Cytochem 2005; 53: 255–260.

Pellestor F, André B, Arnal F, Humeau C, Demaille J. Maternal aging and chromosomal abnormalities: new data drawn from in vitro unfertilized human oocytes. Hum Genet 2003; 112: 195–203.

Renwick PJ, Trusser J, Ostad-Saffari E, et al. Proof of principle and first cases using preimplantation genetic haplotyping a paradigm shift for embryo diagnosis. Reprod Biomed Online 2006; 13: 110–119.

Rooney DE, Czepulkowski BH. *Human Cytogenetics. A Practical Approach*. Oxford University Press, New York, 1992.

Sandalinas M, Marquez C, Munné S. Spectral karyotyping of fresh, non-inseminated oocytes. Mol Hum Reprod 2002; 8: 580–585.

Schoolcraft WB, Katz-Jaffe MG, Stevens J, Rawlins M, Munné S. Preimplantation aneuploidy testing for infertile patients of advanced maternal age: a randomized prospective trial. Fertil Steril 2008; 92(1): 157–162.

Sermon K, van Steirteghem A, Liebaers I. Preimplantation genetic diagnosis. Lancet 2004; 363: 1633–1641.

Shaffer LG, McGowan-Jordan J, Schmid M (eds). *ISCN 2013: An International System for Human Cytogenetic Nomenclature*. Karger Publishers, Unionville, CT, 2013.

Shkumatov A, Kuznyetsov V, Cieslak J, Ilkevitch Y, Verlinsky Y. Obtaining metaphase spreads from single blastomeres for PGD of chromosomal rearrangements. Reprod Biomed Online 2007; 14(4): 498–503.

Staessen C, Platteau P, van Assche E, et al. Comparison of blastocyst transfer with or without preimplantation genetic diagnosis for aneuploidy screening in couples with advanced maternal age: a prospective randomized controlled trial. Hum Reprod 2004; 19: 2849–2858.

Tobias ES, Connor M, Ferguson Smith M. *Essential Medical Genetics*, 6th edn. Wiley-Blackwell, Oxford, 2011.

Veiga A, Calderon G, Santalo J, Barri PN, Egozcue J. Chromosome studies in oocytes and zygotes from an IVF programme. Hum Reprod 1987; 2: 425–430.

Wells D, Delhanty J. Evaluating comparative genomic hybridisation (CGH) as a strategy for preimplantation diagnosis of unbalanced chromosome complements. Eur J Hum Genet 1996; 4(Suppl 1): 125.

Wells D, Fragouli E, Stevens J, Munné S, Schoolcraft W, Katz-Jaffe M. High pregnancy rate after comprehensive chromosomal screening of blastocysts. Fertil Steril 2008; 90: S80.

Commercial FISH probes are available at these websites

www.AbbottMolecular.com

www.Kreatech.com

www.Rainbowscientific.com

www.empiregenomics.com

www.abnova.com

www.cytocell.com

www.exiqon.com

www.cambridgebluegnome.com

CHAPTER 15

Integrated chromosome and FISH analyses

When both chromosome analysis and FISH studies are performed, many laboratories provide separate reports for each study when completed. This may be done because the turn-around times for FISH and chromosome studies are very different when both tests are ordered at the same time the specimen is received. Usually, FISH testing can be completed within 1–3 days whereas chromosome analysis takes an average of 5–10 days for completion, unless it is a STAT case. In these cases, if the clinical indication is for a specific disorder that is performed by FISH analysis, such as a microdeletion or microduplication disorder, then a report with just the FISH analysis can be provided to the clinician in a timely manner if the report is not held for cytogenetic analysis, which may take up to a week longer.

There may also be instances when one study is performed as a reflex based on the finding of the first test. For example, if the chromosomes show a suspicious or subtle abnormality when the clinical indication is non-specific for a disorder, such as developmental delay or multiple congenital anomalies, and chromosome analysis is the only test ordered, FISH analysis may help clarify the chromosome findings. The chromosome findings may be reported when completed, and a separate report is generated for the FISH results. When chromosome and FISH testing are done in different departments of a laboratory, reporting is usually done separately unless the case reviewers are shared between departments. When separate reports are prepared, it is up to a genetic counselor or clinician to coordinate and interpret all the laboratory findings for the patient.

However, laboratories might choose to report both chromosomes and FISH together in one report. If necessary, the first test performed is reported separately. When the second test is completed, then an amended or integrated report can be generated. The advantage of integrated reporting of all tests is to provide a comprehensive result and interpretation for the clinician, to relay to the patient. Clinicians who are not geneticists most likely rely on the laboratory report to be comprehensive so as not to misinterpret the findings. This chapter gives examples of integrated chromosome and FISH reports with different normal and abnormal results. The ISCN rules are given for both types of testing, as are comments in reports that combine the results and interpretation with both types of testing. Further integrated reports with chromosomes, FISH and microarray analyses will be given at the end of Section 3.

Cytogenetic Abnormalities: Chromosomal, FISH and Microarray-Based Clinical Reporting, First Edition. Susan Mahler Zneimer.

15.1 ISCN rules and reporting normal results by chromosomes and FISH

ISCN RULES FOR REPORTING CHROMOSOME AND NORMAL FISH MICRODELETION/MICRODUPLICATION TESTING

- When describing metaphase FISH with cytogenetics, first write the chromosome designation, i.e. modal chromosome number, followed by the sex designation, followed by any visible chromosome abnormality (for details, see Section 1).
- After the chromosome designation, write a period (.), followed by the FISH nomenclature.
- Metaphase FISH nomenclature always begins with "ish" and interphase FISH always begins with "nuc ish" followed by a space.
- Then write the chromosome involved, followed by the band of the region tested.
- Then write the probe name, which can be referred to as the clone name or, if unavailable, then use the locus designation used by the Genome Database (GDB). If that probe is unavailable, then use the HUGO-approved nomenclature. It is also acceptable to use the GDB nucleotide number in a specific genome build, such as NCBI build 35 (B35).
- The probe name is usually an acronym, so it is written as all capitalized and not italicized.
- With normal FISH results, the probe name is followed by the multiplication sign "x" followed by the number of signals observed (typically two signals, one per chromosome).

Note that for normal results, the chromosome and band region are not surrounded by parentheses, but the probe and signal number are surrounded by parentheses.

Also note that even though a control probe may be present in the analysis, only the disease gene is written in the nomenclature.

For example, for the DiGeorge/velocardialfacial syndromes, a normal FISH result and a normal chromosome study with the gene name would be written as: 46,XY.ish 22q11.2(HIRAx2).nuc ish 22q11.2(HIRAx2).

EXAMPLE REPORTING OF NORMAL CHROMOSOMES AND FISH ANALYSIS FOR CHROMOSOME 22q11.2 – DIGEORGE/ VELOCARDIOFACIAL SYNDROMES

ISCN Result: 46,XY.ish 22q11.2(HIRAx2),nuc ish 22q11.2(HIRAx2)

Interpretation

Chromosome analysis and fluorescence *in situ* hybridization (FISH) analyses on both metaphase and interphase cells were performed with the 22q11.2 probe (HIRA) and a control ARSA probe.

Chromosome analysis showed a normal male chromosome complement in all cells examined, and FISH analyses showed no evidence for a deletion or rearrangement in 10 metaphase cells and 100 interphase cells scored, which is commonly observed in DiGeorge/velocardiofacial syndromes.

Since both chromosomes and FISH analyses show only normal results, consider microarray analysis to detect smaller deletions or regions not covered by these assays.

15.2 ISCN rules and reporting abnormal chromosomes and FISH

ISCN RULES FOR REPORTING ABNORMAL CYTOGENETICS AND FISH MICRODELETION/MICRODUPLICATION TESTING

- When describing metaphase FISH with cytogenetics, first write the chromosome designation, i.e. modal chromosome number, followed by the sex designation, followed by any visible chromosome abnormality (for details, see Section 1).
- After the chromosome designation, write a period (.), followed by the FISH nomenclature.
- Metaphase FISH nomenclature always begins with "ish" and interphase FISH always begins with "nuc ish" followed by a space.
- Then write "del" or "dup" followed by the chromosome involved, surrounded by parentheses.
- Then write the chromosome band of the region tested, surrounded by parentheses.
- The band region deleted/duplicated is usually a single interstitial band, so the nomenclature needs to have the band written twice (without a semicolon) to reflect that it is not terminal.
- Next write the probe name, which can be referred to as the clone name or, if unavailable, use the locus designation used by the GDB. If that probe is unavailable, then use the HUGO-approved nomenclature. It is also acceptable to use the GDB nucleotide number in a specific genome build, such as NCBI build 35 (B35).
- The probe name is usually an acronym, so it is written as all capitalized and not italicized.
- With abnormal metaphase FISH results, the probe name is followed by a minus sign "–" for a deletion and "+" sign for a duplication. With abnormal interphase FISH results, the probe name is followed by the multiplication sign "x" followed by the number of signals observed (for loss of one signal, then x1, for gain of one signal, then x3).
- The probe name, minus/plus sign and signal number are all surrounded by parentheses.

Note that for abnormal results, the chromosome, band region, the probe and signal number are surrounded by parentheses.

Also note that even though a control probe may be present in the analysis, only the disease gene is written in the nomenclature, although the control probe may be mentioned in the interpretation of the report for clarity.

For example, for chromosome 22 duplication syndrome, an abnormal FISH result and a normal chromosome study using the GDB D-number would be written as: 46,XY.ish dup(22)(q11.2q11.2)(HIRA+).nuc ish dup(22)(q11.2q11.2)(HIRAx3).

EXAMPLE REPORTING OF NORMAL CHROMOSOMES AND ABNORMAL FISH ANALYSES FOR CHROMOSOME 22q11.2 DUPLICATION SYNDROME (FIGURE 15.1)

ISCN Result: 46,XY.ish dup(22)(q11.2q11.2)(HIRA+),nuc ish dup(22)(q11.2q11.2)(HIRAx3).

Interpretation

Chromosome analysis and fluorescence *in situ* hybridization (FISH) analysis on both metaphase and interphase cells were performed with the 22q11.2 probe (HIRA) and a control ARSA probe.

Chromosome analysis showed a normal male chromosome complement in all cells examined. However, FISH analysis showed a duplication of chromosome 22q11.2 region in 10 metaphase cells and 100 interphase cells scored. These FISH results are consistent with the diagnosis of chromosome 22q11.2 duplication syndrome.

Parental chromosome and FISH studies are recommended to rule out either the presence of a balanced rearrangement or whether this abnormality is *de novo* in origin.

Genetic counseling is recommended.

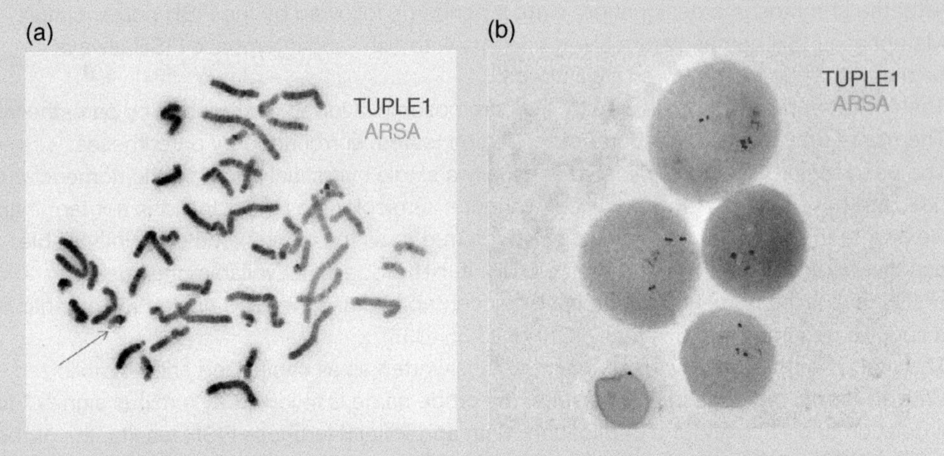

Figure 15.1 Duplication of the HIRA/TUPLE1 probe resulting in chromosome 22q11.2 duplication syndrome. (a) Metaphase cell showing a duplication of the HIRA/TUPLE1 probe of chromosome 22 in red depicted by a more intense, wider and brighter signal than the other chromosome 22. (b) Interphase cells with an abnormal signal pattern of three copies of HIRA/TUPLE1 (*red*) and two copies of ARSA probe (*green*). Courtesy of Lauren Jenkins PhD, Kaiser Permanente Regional Laboratory, Northern California.

EXAMPLE REPORTING OF ABNORMAL CHROMOSOME AND FISH ANALYSIS FOR CHROMOSOME 4p15-16 DELETIONS – WOLF–HIRSCHHORN SYNDROME (FIGURE 15.2)

ISCN Result: 46,XX,del(4)(p15.1).ish del(4)(p16.3)(WHSC1-).nuc ish del(4)(p16.3)(WHSC1x1)

Interpretation

Chromosome analysis and fluorescence *in situ* hybridization (FISH) analysis on both metaphase and interphase cells were performed with the 4p16.3 probe (WHSC1) and a control CEP4 probe.

Chromosome analysis showed an abnormal female chromosome complement in all cells examined with a terminal deletion of the distal short arm of chromosome 4 from band p15.1 to the terminus.

FISH analysis confirmed a deletion of chromosome 4p16.3 region in 10 metaphase cells and 100 interphase cells scored. Both the chromosome and FISH results are consistent with the diagnosis of Wolf–Hirschhorn syndrome.

Parental chromosome and FISH studies are recommended to rule out either the presence of a balanced rearrangement or whether this abnormality is *de novo* in origin.

Genetic counseling is recommended.

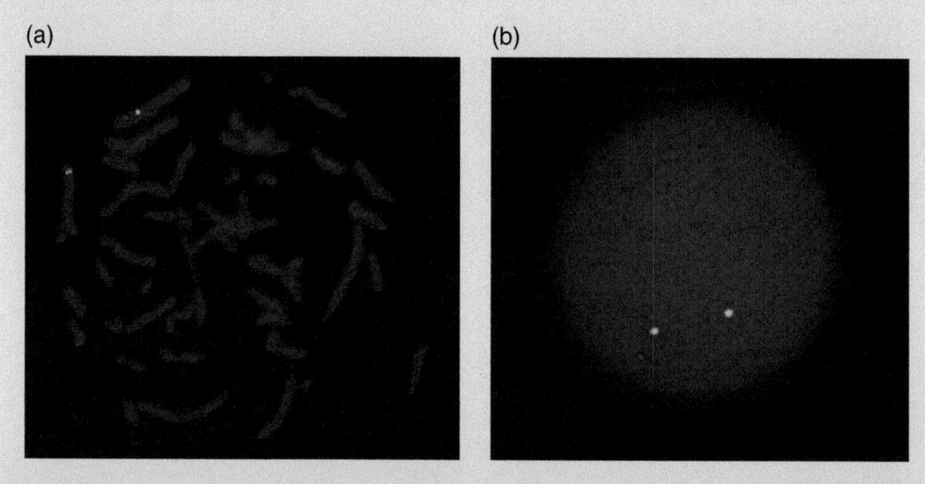

Figure 15.2 Deletion of the WHS critical region resulting in Wolf–Hirschhorn syndrome. (a) Metaphase cell showing a deletion of the WHS probe (*red*). (b) Interphase cell confirming the deletion with one red signal.

EXAMPLE REPORTING OF NEONATAL ABNORMAL FEMALE CHROMOSOME AND INTERPHASE FISH RESULTS WITH TRISOMY 21 (FIGURE 15.3)

ISCN Result: 47,XX,+21.nuc ish Xcen(DXZ1x2),Ycen(DYZ3x0),13q14(RB1x2),18cen(D18Z1x2), 21q22.13(D21S259/D21S341/D21S342x3)

Interpretation

Chromosome analysis showed an abnormal female chromosome complement with the gain of chromosome 21 (trisomy 21) in all cells examined. Corresponding interphase FISH analysis showed three hybridization signals for chromosome 21, consistent with trisomy 21 (Down syndrome).

The remaining probes showed a normal two-signal pattern in 200 cells scored for each probe.

Genetic counseling is recommended.

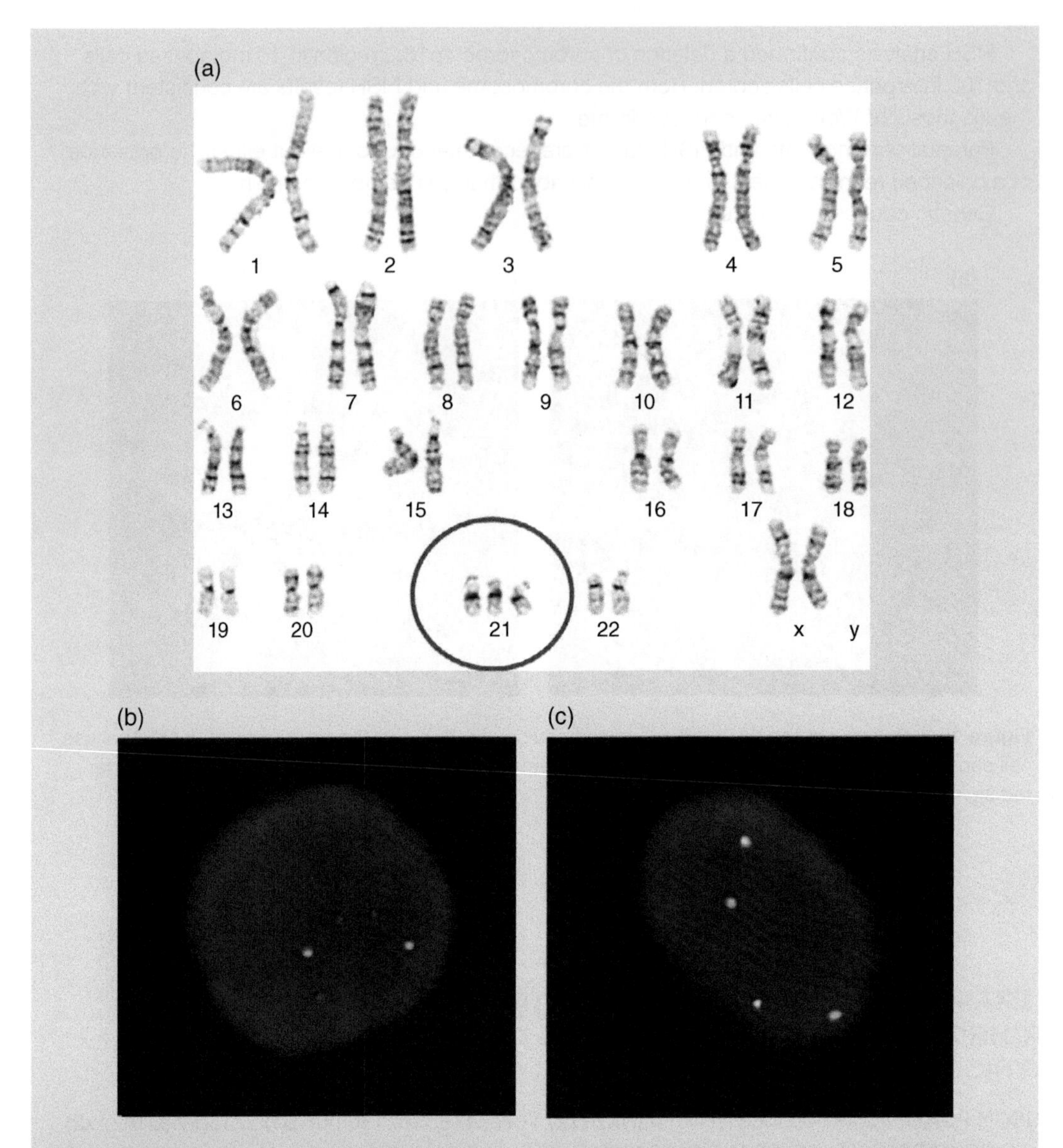

Figure 15.3 Chromosome and interphase FISH analysis in a female with trisomy 21. (a) Chromosome analysis showing three copies of chromosome 21 (*circled*). (b) Interphase FISH analysis showing two copies of chromosome 13 (*green*) and three copies of chromosome 21 (*red*). (c) Interphase FISH analysis showing two copies of chromosome 18 (*aqua*) and two copies of the X chromosome (*green*), and no copy of the Y chromosome (*red*).

15.3 ISCN rules and reporting of chromosomes and subtelomere FISH

ISCN RULES FOR REPORTING NORMAL CHROMOSOME AND SUBTELOMERE FISH ANALYSES

- When describing metaphase FISH with cytogenetics, first write the chromosome designation, i.e. modal chromosome number, followed by the sex designation, followed by any visible chromosome abnormality (for details, see Section 1).
- After the chromosome designation, write a period (.), followed by the FISH nomenclature.
- Subtelomeric FISH nomenclature always begins with "ish subtel" followed, in parentheses, by 41x2.

For example, for a normal FISH result and a normal chromosome study, the result would be written as: 46,XY.ish subtel(41x2).

EXAMPLE REPORTING OF CHROMOSOMES AND SUBTELOMERE FISH ANALYSIS – NORMAL

ISCN Result: 46,XY.ish subtel(41x2)

Interpretation

Chromosome analysis and fluorescence *in situ* hybridization (FISH) analysis with 41 probes of the subtelomere panel, corresponding to DNA probes specific to subtelomeric regions of each chromosome, were performed on metaphase cells from peripheral blood.

Chromosome analysis showed a normal male chromosome complement in all cells examined. Corresponding FISH analysis showed normal results in 10 metaphase cells scored for each probe examined.

Since both chromosome and FISH analyses showed normal results, consider microarray analysis to detect smaller deletions or duplications, or abnormalities not covered by these FISH probes.

ISCN RULES FOR REPORTING ABNORMAL CHROMOSOME AND SUBTELOMERE FISH ANALYSIS

- When describing metaphase FISH with cytogenetics, first write the chromosome designation, i.e. modal chromosome number, followed by the sex designation, followed by any visible chromosome abnormality (for details, see Section 1).
- After the chromosome designation, write a period (.), followed by the FISH nomenclature.
- Abnormal subtelomere FISH nomenclature only describes the chromosome arm designations of the abnormality present.
- If the abnormality is an unbalanced translocation resulting in a deletion, first write "ish" followed by der, followed, in parentheses, by the chromosome with the deletion.
- Then write "t" followed, in parentheses, by the two chromosomes involved in the unbalanced translocation, separated by a semicolon.

- Then write the arm designation of the chromosome with the deletion, followed by "–" since the subtelomere is not present on the chromosome arm that is deleted.
- Then write a comma, followed by the second chromosome arm (p or q) of the chromosome that contains an intact arm, followed by "+" since that subtelomere is present on that chromosome arm.
- Parentheses enclose the chromosomes involved and enclose the arms involved.
- Next write the probe designations of each of the subtelomeres, in the order of the chromosome arms of the nomenclature. Each probe is followed by either "–" or "+" based on whether the probe is absent or present, respectively.
- Parentheses enclose the chromosomes involved, the arms involved and the probes involved. Only the probe names of the abnormal chromosome with the deletion is described.

Note that even though all the chromosomes are being analyzed, only the chromosomes involved in the translocation are written in the nomenclature, although the remaining normal chromosome probes may be mentioned in the interpretation of the report for clarity.

For example, for an unbalanced translocation between the short arm of chromosome 7 and the short arm of chromosome 22, resulting in pseudodicentric chromosome and a deletion of the subtelomere of chromosome 7p, the result would be written as: ish der(7) t(7;17)(p22–,q25+)(VIJyRM2185–,D17S928+).

Note that the probe names are based on the manufacturer's description.

Often a combination of probes is used to help clarify the origin of an unidentifiable chromosome. One example with different probes was used to identify a pseudodicentric chromosome involving chromosome identification in a patient. In this case, other banding techniques, including centromere banding (C-banding) and silver staining (NOR banding), were used to confirm the presence of specific chromosomal regions.

EXAMPLE REPORTING OF A PSEUDODICENTRIC CHROMOSOME WITH SUBTELOMERE FISH AND OTHER BANDING TECHNIQUES (FIGURE 15.4)

ISCN Result: 45,XY,der(22)psu dic(22;7)(p13;p22.3)del(7)(p11.2p15.1).ish der(22)t(22;7)(q13+,p22–;p22–,q13+)(MS607;ACR+; VIJyRM2185–)

Interpretation

Chromosome analysis showed an abnormal male chromosome complement with a pseudodicentric chromosome involving chromosomes 7 and 22 in all cells examined. The abnormal chromosome is the result of the fusion of chromosome 7 at the short arm telomere to the short arm of chromosome 22, wherein this chromosome also contains a deletion of the proximal short arm of chromosome 7.

Fluorescence *in situ* hybridization (FISH) analysis, using subtelomere probes of chromosomes 7 and 22, confirmed the derivative chromosome, resulting in the presence of the chromosome 7 short arm telomere translocated to the pseudodicentric chromosome 22, in 10 metaphase cells scored.

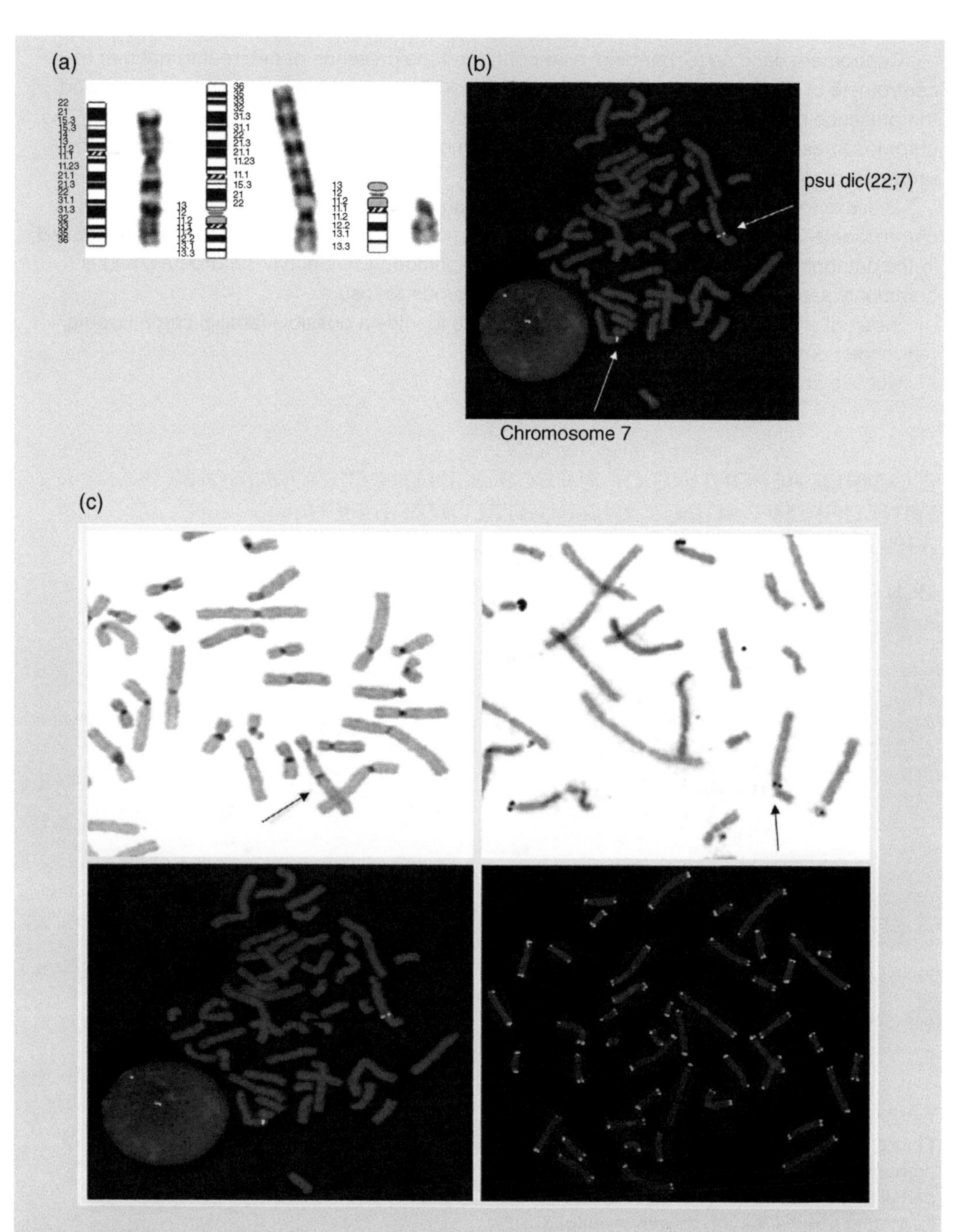

Figure 15.4 Partial karyotype showing a psu dic(22;7) with ideogram (a) and metaphase FISH picture (b) depicting subtelomere probe D7S589 of the short arm of chromosome 7. C-banding, AgNOR banding, subtelomere and telomere banding (c) shows the dicentric chromosome and that all the telomeres are present. These findings result in the karyotype designation: 45,XY,der(22)psu dic(22;7)(p13;p22.3)del(7)(p11.2p15.1).

Centromere banding (C-banding) also confirmed the presence of heterochromatin at the centromere of the pseudodicentric chromosome. NOR banding (silver staining) also showed the presence of a nuclear organizer region, typically seen on short arm acrocentric chromosomes, seen on the pseudodicentric chromosome. Telomere banding confirmed the presence of telomeres on all the chromosomes.

These cytogenetic and FISH findings result in an interstitial deletion of the short arm of chromosome 7. Loss of genetic material of this region of chromosome 7p has been implicated in the pathophysiology of craniosynostosis and cephalopolysyndactyly syndromes, and is commonly seen in patients with agenesis of the corpus callosum.

Parental chromosome studies are suggested to identify a possible familial chromosomal rearrangement.

Genetic counseling is recommended.

EXAMPLE REPORTING OF A RECOMBINANT CHROMOSOME WITH CHROMOSOMES AND SUBTELOMERE FISH (FIGURE 15.5)

ISCN Result: 46,XY,rec(8)inv(8)(p23.1q22)mat.ish inv(8)(p23.1q22)(p23.3)(RP8-1011+)

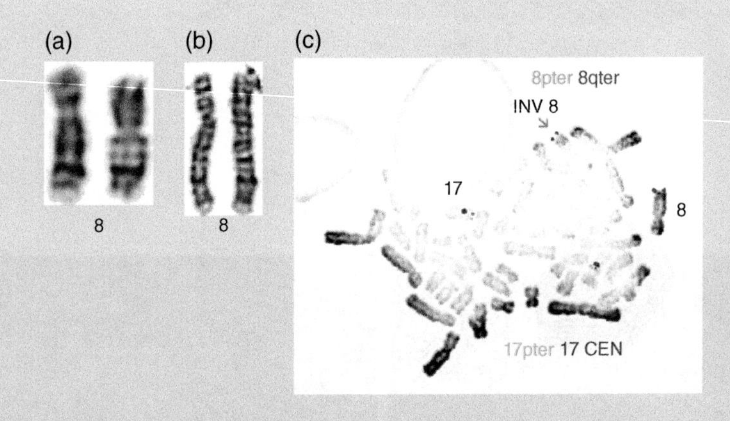

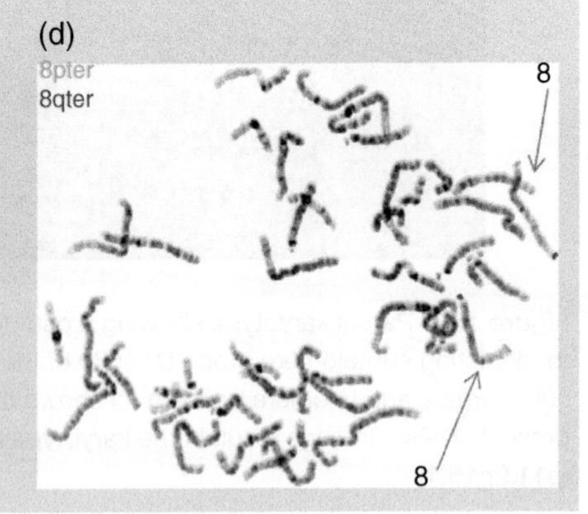

Figure 15.5 Recombinant chromosome 8 in a proband, maternally inherited. (a) Proband's partial karyotype with rec(8)inv(8)(p23.1q22)mat. (b) Mother's partial karyotype with inv(8)(p23.1q22). (c) Subtelomere 8p and 8q with gain of 8pter signal within 8q region in the proband. (d) Normal subtelomere 8p and 8q of mother. Courtesy of Lauren Jenkins PhD, Kaiser Permanente Regional Laboratory, Northern California.

Interpretation

Chromosome analysis revealed an abnormal male chromosome complement with a chromosome 8 showing a recombination event due to a maternally derived inversion of chromosome 8. This finding results in the gain of chromosome 8p23.1-pter.

FISH analysis confirmed a duplication of the subtelomere 8p region proximal to the subtelomere of 8p. FISH analysis also confirmed a normal 8p and 8q signal pattern of the mother's metaphase cells.

Genetic counseling is recommended.

Bibliography

Gardner RJM, Sutherland GR. *Chromosome Abnormalities and Genetic Counselling*. Oxford Monographs on Medical Genetics. Oxford University Press, Oxford, 2003.

Gersen S, Keagle M (eds). *Principles of Clinical Cytogenetics*. Humana Press, Totowa, New Jersey, 1999.

Rooney DE, Czepulkowski BH. *Human Cytogenetics. A Practical Approach*. Oxford University Press, New York, 1992.

Shaffer LG, McGowan-Jordan J, Schmid M (eds). *ISCN* 2013: *An International System for Human Cytogenetic Nomenclature*. Karger Publishers, Unionville, CT, 2013.

Smith AC, Spuhler K, Williams TM, McConnell T, Sujansky E, Robinson A. Genetic risk for recombinant 8 syndrome and the transmission rate of balanced inversion 8 in the Hispanic population of the southwestern United States. Am J Hum Genet 1987; 41(6): 1083–1103.

Tobias ES, Connor M, Ferguson Smith M. *Essential Medical Genetics*, 6th edn. Wiley-Blackwell, Oxford, 2011.

Zneimer SM, Cotter P, Stewart SD. Telomere-telomere fusion of chromosomes 7 and 22 with an interstitial deletion of chromosome 7p11.2-p15.1: phenotypic consequences and possible mechanisms. Clin Genet 2000; 58: 129–133.

Commercial FISH probes are available at these websites

www.AbbottMolecular.com

www.Kreatech.com

www.Rainbowscientific.com

www.empiregenomics.com

www.abnova.com

www.cytocell.com

www.exiqon.com

www.cambridgebluegnome.com

Section 3 Chromosomal Microarray Analysis (CMA)

CHAPTER 16

Bacterial artificial chromosome, oligoarray and single nucleotide polymorphism array methodologies for analysis

16.1 Introduction

Chromosomal microarray (CMA) technologies have opened a new wave of genetic clinical testing for a group of disorders that have so far been poorly diagnosed. Patients with developmental delay (DD), mental impairment (MI), multiple congenital anomalies (MCA) and dysmorphic features show highly variable phenotypic manifestations that may be present in any number of genetic disorders. These indications for testing may now be evaluated by microarray analysis as a better method for the diagnosis of an underlying genetic defect associated with these manifestations.

Developed in the 1990s, chromosomal microarray technology uses genome-wide analyses of cDNA clones to profile gene expression as well as identifying the genetic composition of individuals. Microarrays consist of DNA sequences that vary in genomic coverage from bacterial artificial chromosome (BAC) clones, which yield coverage at ~1 Mb genomic intervals, to oligonucleotides at ~100 Kb intervals throughout the genome, to single nucleotide polymorphism (SNP) coverage at ~30–35 Kb intervals.

For BAC and oligoarrays, control DNA sequences are labeled with a specific color dye (e.g. red) and are fixed to a microscope slide at precise locations on a microarray grid to which a patient's sample of DNA is hybridized, and then labeled with a different fluorescent colored dye (e.g. green). After hybridization, washing and drying steps, fluorescent labels are detected on a laser scanner and analyzed with specific software that will show the intensity of signals present (Figures 16.1, 16.2). By this

Cytogenetic Abnormalities: Chromosomal, FISH and Microarray-Based Clinical Reporting, First Edition. Susan Mahler Zneimer.

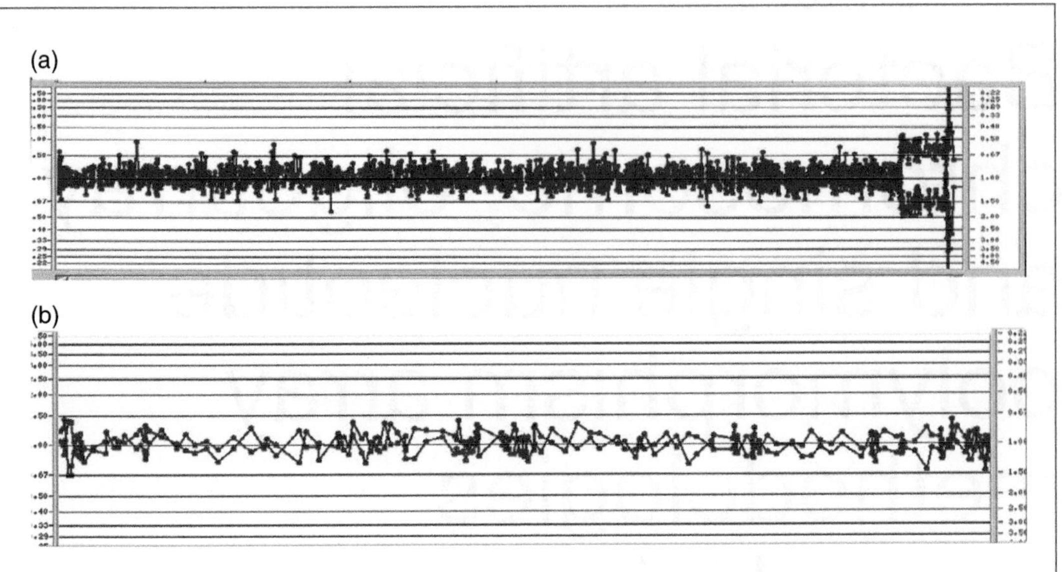

Figure 16.1 Normal female BAC array with a male reference. (a) Whole genome view of all the chromosomes. (b) Normal array of chromosome 1. Courtesy of Karine Hovanes PhD, CombiMatrix, Inc.

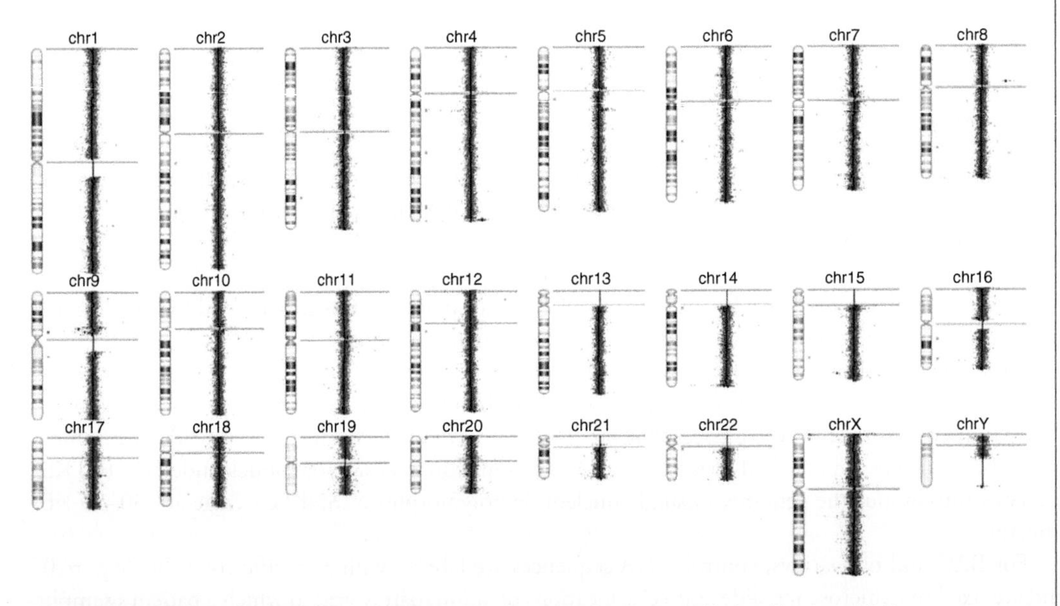

Figure 16.2 Normal male oligoarray – whole genome view. Courtesy of Karine Hovanes PhD, CombiMatrix, Inc.

technique of comparative genomic hybridization, differences in copy number between a patient's DNA sample and control samples will be detected by identifying the ratio of red to green signals. For example, a larger red ratio in a patient may indicate a copy number gain, and a lower red ratio may indicate a deletion. In microarrays, only gains and losses of DNA are identified, and chromosomal abnormalities that are truly balanced will appear normal.

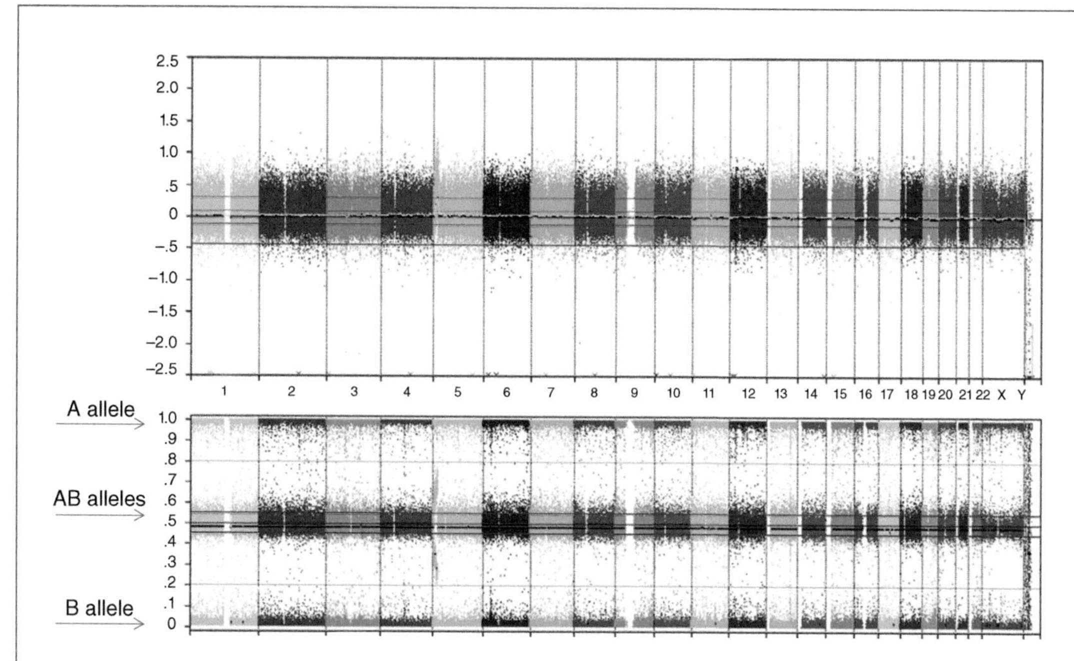

Figure 16.3 Normal female SNP array showing at the top: normalized total intensity = log R ratio, and at the bottom: the allelic intensity ratios = A and B allele frequencies. Courtesy of Karine Hovanes PhD, CombiMatrix, Inc.

For SNP arrays, no controls are used. Rather, labeled patient DNA is hybridized to the array, and copy number changes are seen as variations in probe intensity compared to a pool of normal individuals, where deletions will show only a single allele and duplications show three alleles. SNP arrays will have two types of data displayed for each individual tested. One graph will be the log R ratio, which is a metric that normalizes the total signal intensity for copy number changes. The second graph will show the allelic intensity ratios for the alleles A and B.

Therefore, in the bottom graph of a SNP array, there will be three bands. The top band represents the A allele frequency, the bottom band represents the B allele frequency, and the middle band represents the heterozygote AB allele frequency. Each identified region has an expected R, which is equal to the sum of normalized red and green signal intensities (Figure 16.3), such that:

$$\text{Log R ratio} = \log_2 \times \frac{R_{obs}}{R_{exp}}$$

Figure 16.4 shows an example of two genomic gains corresponding to the regions marked with a blue line as well as a purple line, which denotes an allelic imbalance. Note that for gains, in the bottom graph, the allelic imbalance will show a spread of two bands in the heterozygote AB allele band. Therefore, a true duplication will show both purple and blue lines, in which both graphs depict the gain. The red line depicts a deletion.

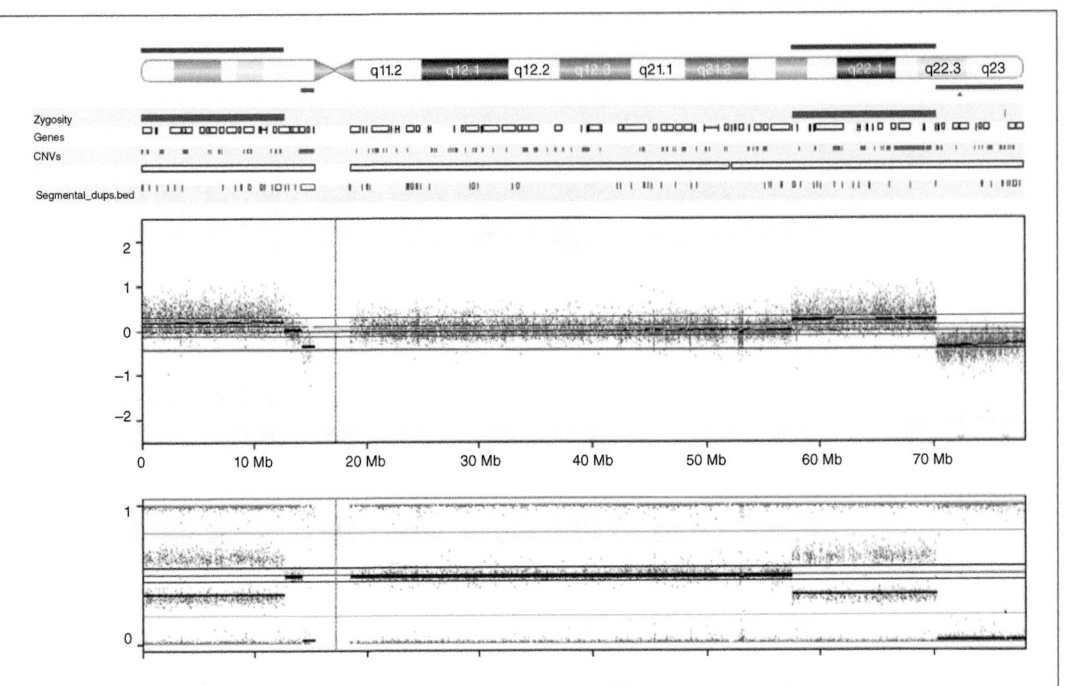

Figure 16.4 SNP array in which the blue line above the ideogram indicates a gain, the red line below the ideogram indicates a loss, and the purple line (zygosity) indicates an allelic imbalance, with a gain or mosaic gain. Corresponding gains and losses are in the upper track, and the allele patterns are in the bottom track. Courtesy of Karine Hovanes PhD, CombiMatrix, Inc.

Figure 16.5 shows a deletion indicated by the red line whereas a yellow line depicts loss of heterozygosity. Therefore, a true deletion will need to have both red and yellow lines, in which case a loss is seen in both graphs.

With all these techniques, a patient's genome is evaluated to identify gains and losses of genetic material that are typically too small to detect by standard chromosome analysis and are genome wide. In contrast, the targeted FISH probes identify only very specific regions of the genome.

Laboratories performing microarray analysis have many obstacles to overcome before reporting test results, which are unlike any seen in other cytogenetics testing. As a laboratory developed test, specific performance specifications must be met. These specifications follow CLIA, CAP and other national standards that adhere primarily to the following laboratory guidelines:

- accuracy
- sensitivity
- specificity
- reproducibility
- reportable range
- interfering substances
- limit of detection.

Although these are common parameters used for specific FISH assays (see Appendix 1), using the whole genome as a single test magnifies the performance parameters extensively. This book will not delve into the laboratory technique used for each of the types of arrays currently available; rather, only

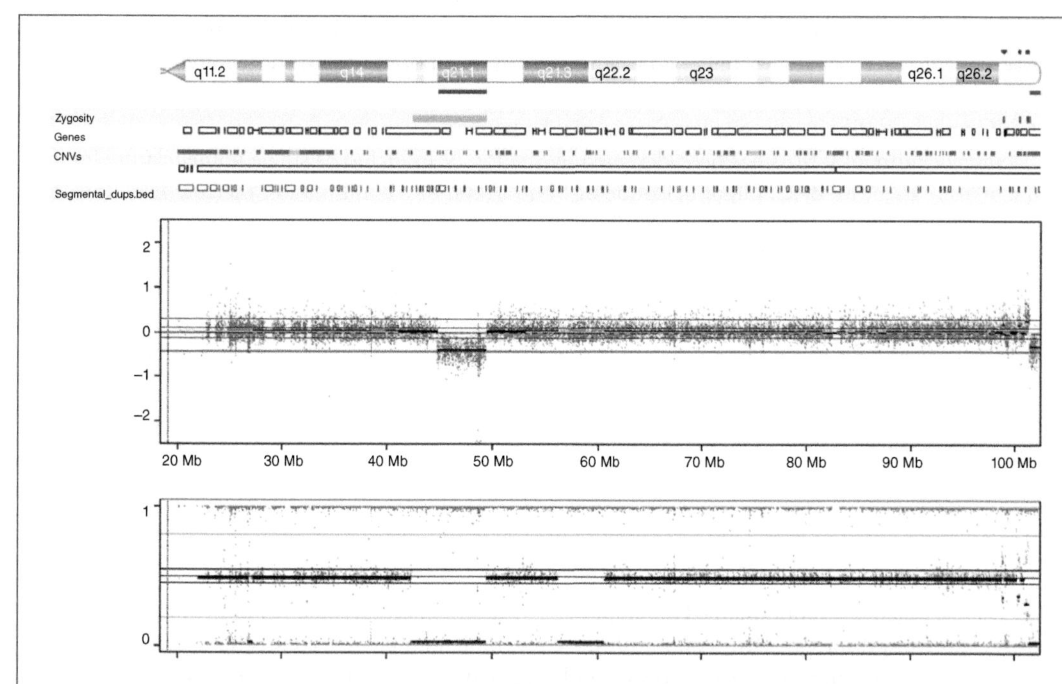

Figure 16.5 SNP array with a deletion denoted by the red line below the ideogram and loss of heterozygosity (LOH) denoted by the yellow line (zygosity). Courtesy of Karine Hovanes PhD, CombiMatrix, Inc.

a summary of its intent will be discussed. The aim of this book is to assume that the laboratory techniques are being performed correctly, and it is the interpretation and reporting of the results that will be emphasized. There are currently no written standards for reporting microarray data, and each laboratory needs to address only a few details in a clinical report to fulfill any, if not all, requirements by most regulatory agencies. This book will attempt to give relevant clinical information in example reports to be used as a possible model for laboratories to follow, which may help standardize the approach for reporting microarray data in a clinical study.

As much as microarrays are proving to be a valuable tool in genetics testing, there are limitations to array testing. Arrays cannot currently detect balanced rearrangements, such as translocations or inversions, or detect whether aneuploidies are derived from a translocation event or from the gain or loss of a single chromosome. Some aneuploidies may also be missed, such as a sex chromosome gain, if the incorrect gender control is used. Some gains, such as unidentifiable marker chromosomes, may also be missed depending on the size and DNA composition and array coverage of the chromosomal region present. Centromeric and telomeric chromosomal regions may be especially difficult to identify true pathogenic abnormalities. Mosaicism may also be difficult to assess when observed lower than the detection limit, which varies depending on the type of array platform. The detection of abnormalities also depends on the coverage of genomic nucleotides, including the length and space between DNA probes used throughout the genome. Also, the algorithm used to determine the threshold of gains and losses may vary between laboratories. All these criteria make this type of genetic testing quite variable among laboratories, and results may differ accordingly.

Interpretation of arrays may also be difficult to access. For example, the presence of a copy number change (CNC) in an individual may not have a phenotypic consequence. That is, many of the CNCs are benign, and are either present in a significant number of normal individuals or the CNC is not the underlying defect of a patient's phenotype. Therefore, it is important to utilize genomic databases that document normal DNA variants in the interpretation of CMA data and research the genes involved in a CNC to help clarify its clinical significance for the patient.

Nomenclature of CMAs has been designed by the ISCN to include a single nomenclature for any type of array used, by BAC, oligonucleotide or SNP arrays, or a combination thereof. Therefore, the type of array is not included in the nomenclature, but may be described in the interpretation. Two ISCN systems have been devised: one with a detailed description that includes the abnormal nucleotides as well as the bordering normal nucleotides, and one with a short description that includes only the abnormal nucleotides. Adding the specific human genome build (e.g. hg 18, hg 19) is optional for abnormal cases and is generally not included in normal cases. It is also optional to describe results in a format other than as a string of abnormalities when the results are complex, such as in a listing or a table. Rules for the detailed and short versions will be given. However, only the short version will be used in example reports.

16.2 Clinical utility of chromosomal microarray analysis

The clinical diagnosis of unexplained developmental delay, intellectual disability or mental impairment, autism spectrum disorders (ASD) and multiple congenital anomalies accounts for a large percentage of cytogenetics testing. These disorders combined have an incidence of approximately 3% in the general population, in which ASDs alone affect approximately 1 in 150 individuals. The International Standard of Cytogenomics Arrays (ISCA) consortium reviewed 33 published studies on CMA, including over 21,000 patients, and found a 15–20% abnormality rate for these diagnoses over the 3% detection rate for standard chromosome analysis alone (excluding Down syndrome and other recognizable chromosomal syndromes). Thus, the ISCA and other genetic professional organizations, including the American College of Medical Genetics (ACMG), have now recommended CMA as a first-tier clinical diagnostic test for these indications. This decision has led to the ability of decreasing costs for genetic testing by other means, such as standard chromosome analysis and FISH analysis in 80% of these patients.

The ISCA and others have also addressed descriptive categories of abnormalities for the interpretation of results; however, the reporting of results is far from standardized. Consequently, this section attempts to use the ISCA guidelines and other published works in conjunction with the best information from different laboratories to set a standard approach to writing CMA reports structured into a single style that can be used for any type of array and for any specimen type.

The format used here is intended to be comprehensive in the nomenclature and the interpretation of an abnormality observed, with the exclusion of genes involved, and without a listing of possible non-clinically relevant copy number changes that might be present in a particular specimen. References for all the abnormalities described in a report are listed in the back of the chapter.

Other comments that are commonly used in reporting CMAs are also discussed in this section. Since, at the time of this writing, the amount and type of information included in reports vary to a great degree from lab to lab, these reports are just examples to be modified where needed and updated with time when human genome builds continue to be modified. This section is divided by the type of abnormality observed: normal, abnormal, variants of unknown significance (VOUS) and benign CNCs.

This chapter describes the ISCN rules and example reports of normal results, while Chapter 17 describes abnormal results, inherited and *de novo* abnormalities and other common comments often included in reports. Chapter 18 gives example reports of pathogenic abnormalities commonly diagnosed by CMAs, and Chapter 19 gives example reports integrating standard chromosomes, FISH and array analyses. These examples describe various types of abnormalities, using these analyses as possible prototypes for any individual study, which can be used as a model for reporting similar findings in any laboratory. Since methodologies vary extensively and are modified constantly with time, these examples may not be all-inclusive, but are meant as the beginning of possibly standardizing the integrative reporting process for these types of tests.

As in any CNC that is either pathogenic or VOUS, any number of genes may be associated with various syndromes or phenotypic manifestations. Therefore, only a few disorders are described here, since the list of disease associations could be extensive. Also, any number of examples of CNCs could be included here; consequently, only the commonly reported pathogenic and VOUS CNCs are described.

Determining whether a CNC is pathogenic or benign may be problematic in some instances. The ISCA has developed five categories to define a CNC: pathogenic, likely pathogenic, variant of unknown significance, likely benign and benign. Some guidelines exist to help classify specific changes into a pathogenic category versus a benign category.

16.3 Guidelines for classification states

16.3.1 Guidelines for a likely pathogenic state

- CNC change that differs from both parents
- CNC that is present in an affected parent or relative
- CNC overlaps with a genomic region consistent with a syndrome
- CNC overlaps with a genomic region consistent with known symptoms (e.g. mental retardation, autism, developmental delay)
- CNC containing OMIM morbid genes located in the OMIM database
- CNC is a homozygous deletion
- CNC is within a gene that is known to convey clinical significance (e.g. PMP22, DAX1)
- large gains or losses encompassing numerous genes (>1 Mb genomic region)
- deletions that overlap autosomal recessive genes that may unmask a recessive allele associated with a syndrome

16.3.2 Guidelines for a likely benign state

- CNC inherited from a normal parent or relative
- CNC overlaps with segmental duplications or polymorphic genomic regions
- Regions that contain no genes
- Heterozygous duplications with no known OMIM morbid genes
- Known regions in the genome listed as benign in ISCA or other databases, based on large patient samples

16.3.3 Guidelines for a variant of unknown significance (VOUS) state

- Rare CNC with unknown inheritance
- CNC in which the region is not associated with genetic syndromes or symptoms of disease
- Deletions that overlap autosomal recessive genes that may unmask a recessive allele but are not associated with a syndrome
- *De novo* CNC with no OMIM morbid genes or no genes associated with disease

- Novel CNC not listed in any clinically relevant database
- Rare CNCs on the X chromosome observed in a male with maternal inheritance, but with no or limited information in any database

16.4 ISCN rules and reporting of normal results

ISCN RULES FOR NORMAL CHROMOSOMAL MICROARRAY RESULTS

- First write "arr" to describe any array analysis.
- If the results are for a normal female, then write, all in parentheses, "1–22" followed by a comma, followed by "X".
- Since autosomes and the X chromosome are both in two copies, then write "x2".

For example, a normal female is written as: arr(1–22,X)x2.

- If the results are for a normal male, then write "1–22" enclosed in parentheses, followed by "x2", since only the autosomes are in two copies each, followed by a comma.
- Then write "XY" enclosed in parenthesis, followed by "x1" since the X and Y chromosomes are only in one copy each.

For example, a normal male is written as: arr(1–22)x2,(XY)x1.

Note that autosomes are described before sex chromosomes.

Also note that the interpretation may include the genome build, the type of platform used, the genome resolution, and whether or not the DNA along the entire genome was used.

EXAMPLE REPORTING OF A NORMAL MALE MICROARRAY RESULT

ISCN Result: arr(1–22)x2,(XY)x1

Interpretation

Chromosomal microarray analysis revealed a normal male chromosome complement. No pathogenic copy number changes were observed.

Benign copy number changes that may be present in this patient are not listed in this report.

EXAMPLE REPORTING OF A NORMAL FEMALE MICROARRAY RESULT

ISCN Result: arr(1–22,X)x2

Interpretation

Chromosomal microarray analysis revealed a normal female chromosome complement. No pathogenic copy number changes were observed.

Benign copy number changes that may be present in this patient are not listed in this report.

EXAMPLE REPORT FORMATTING OF A NORMAL MICROARRAY RESULT

Laboratory Name
Cytogenetics Report

Patient Name: Jane Doe
Ordering Physician: Dr Smith
Date of Birth: 1/1/2000

Collection Date: 1/1/2014
Report Date: 1 /10/2014
Laboratory Number: A14-000021

Specimen Type: Peripheral Blood
Indication for Study: Autism Spectrum Disorder
Array Platform: Post Oligo180K

ISCN Result: arr(1–22,X)x2

Interpretation

Chromosomal microarray analysis revealed a normal female chromosome complement. No pathogenic copy number changes were observed.

Benign copy number changes that may be present in this patient are not listed in this report.

Date: January 10, 2014

Susan Zneimer, PhD, FACMG
Cytogenetics Laboratory Director

(Any disclaimer here you may want to provide.)

This test was performed at: Address of Laboratory

Dr Jones, Medical Director
Laboratory License Number: 00001

(Cytogenetic Image)

16.5 Comments, disclaimers and recommendations

- Clinical correlation, genetic counseling and parental studies are recommended.
- Normal findings do not rule out the diagnosis of any known disorders. This test may not detect genetic abnormalities below the resolution of this assay. Clinical implications of some copy number changes may not be currently known.
- Copy number variants that currently have no known clinical significance are not listed in this report. These variants may include those that overlap known polymorphic regions, contain no genes, or contain genes that are not associated with disorders in the Online Mendelian Inheritance in Man (OMIM) database.
- There was no growth on the cultures that were set up for microarray analysis; therefore, cells were not available for DNA extraction. Consequently, chromosomal microarray analysis could not be performed.
- No copy number changes involving chromosome __ were detected, suggesting that the chromosomal rearrangement is likely balanced. However, position effects on gene expression resulting from structural rearrangements are not detectable by chromosomal microarray analysis.

- Approximately 10% of patients with autism spectrum disorders have clinically significant abnormalities, which can be detected by chromosomal microarray analysis. Since this patient has an indication of autism and showed a normal chromosome complement by standard karyotypic analysis, additional testing by chromosomal microarray analysis is suggested.
- Approximately 11–15% of patients referred for developmental delay/mental impairment (DD/MI) have clinically significant abnormalities, which can be detected by chromosomal microarray analysis. Since this patient has an indication of developmental delay and showed a normal chromosome complement by standard karyotypic analysis, additional testing by chromosomal microarray analysis (CMA) is suggested. CMA testing provides the highest diagnostic yield for patients being referred for DD/MI; therefore, it may be considered as a first-tier test for future patients with these diagnoses.
- Chromosomal microarray analysis revealed an abnormal copy number change in this patient. However, this copy number variant has been observed in multiple individuals with no phenotypic consequence. Therefore, this variant is considered to have no clinical significance.
- Benign copy number changes that may be present in this patient are not listed in this report.
- Chromosomal microarray analysis (CMA) detects deletions, duplications, aneuploidies and greater than 20% mosaicism. This CMA test will not detect balanced rearrangements (including translocations and inversions), triploidy, small genomic gains or losses, or DNA point mutations.

16.5.1 CLIA '88 disclaimer for non-FDA-approved tests

This test was developed and its performance characteristics determined by [name of laboratory here] as required by CLIA '88 regulations. It has not been cleared or approved for specific uses by the US Food and Drug Administration. This laboratory has established and verified the test's accuracy and precision.

16.5.2 Other related non-FDA-approved disclaimers

- This test was developed and its performance characteristics determined by the [name of laboratory here] as required by the CLIA '88 regulations. It has not been cleared or approved by the US Food and Drug Administration (FDA). The FDA has determined that such clearance or approval is not necessary. These results are provided for informational purposes only, and should be interpreted only in the context of established procedures and/or diagnostic criteria.
- This test uses analyte-specific reagents (ASRs), whose performance characteristics were determined by [name of laboratory here]. It has not been cleared or approved by the US Food and Drug Administration. This test should be regarded as investigational or for research purposes only.

16.5.3 Example comments describing array technology in a report

BAC array

Chromosomal microarray analysis (CMA) uses comparative genomic hybridization (array CGH) of over 3000 unique large-insert clones that cover coding and non-coding human genome sequences from the UCSC hg18 human genome (NCBI build 36, March 2006). DNA copy number in the patient sample is compared with a reference diploid DNA sample, utilizing DNA probes corresponding to genomic loci encompassing the whole genome on all 22 autosomes and the sex chromosomes. If applicable, confirmation is performed by FISH using probes targeted to the region delineated by the BAC array CGH result. Results from the patient sample are compared to a normal reference sample and to parental samples.

Oligoarray

Chromosomal microarray analysis using a 180K oligonucleotide CGH array was performed on this extracted DNA sample referenced to a normal male/female control DNA sample with the [add company name here] designed by the International Standard Cytogenomic Array (ISCA) Consortium and modified by [product name here if appropriate]. This test utilized DNA probes corresponding to genomic loci encompassing the whole genome on all 22 autosomes and the sex chromosomes. This test utilizes approximately 60,000 oligonucleotide probes with enhanced coverage within known microdeletion and microduplication syndromes. Resolution averages approximately 5–10 Kb for targeted regions and 75 Kb for the remaining regions throughout the genome.

SNP array

Chromosomal microarray analysis (CMA) was performed using a [name of product here] single nucleotide polymorphism (SNP) array. This array contains over 800,000 SNP markers covering both coding and non-coding human genome sequences throughout the genome. The median spatial resolution between probes is 1 Kb within gene-rich regions and 5 Kb for the remaining regions of the genome. Extracted DNA was evaluated for copy number changes involving greater than 15 probes and regions of homozygosity of greater than 5 Mb. Genomic imbalances are reported using UCSC Human Genome Build 19 (NCBI build 37, Feb 2009). Mosaicism for partial or whole chromosome aneuploidy is reported when present at or above the detection threshold of 15%.

Bibliography

Aston E, Whitby H, Maxwell T, et al. Comparison of targeted and whole genome analysis of postnatal specimens using a commercially available array based comparative genomic hybridisation (aCGH) microarray platform. J Med Genet 2008; 45: 268–274.

Baldwin EL, Lee JY, Blake DM, et al. Enhanced detection of clinically relevant genomic imbalances using a targeted plus whole genome oligonucleotide microarray. Genet Med 2008; 10: 415–429.

Cheung SW, Shaw CA, Scott DA, et al. Microarray-based CGH detects chromosomal mosaicism not revealed by conventional cytogenetics. Am J Med Genet A 2007; 143A: 1679–1686.

Conlin LK, Thiel BD, Bonnemann CG, et al. Mechanisms of mosaicism, chimerism and uniparental disomy identified by single nucleotide polymorphism array analysis. Hum Mol Genet 2010; 19: 1263–1275.

Conrad DF, Pinto D, Redon R, et al. Origins and functional impact of copy number variation in the human genome. Nature 2010; 464: 704–712.

Cooper GM, Zerr T, Kidd JM, Eichler EE, Nickerson DA. Systematic assessment of copy number variant detection via genome-wide SNP genotyping. Nat Genet 2008; 40: 1199–1203.

Firth HV, Richards SM, Bevan AP, et al. DECIPHER: Database of Chromosomal Imbalance and Phenotype in Humans using Ensembl Resources. Am J Hum Genet 2009; 84: 524–533.

Hoyer J, Dreweke A, Becker C, et al. Molecular karyotyping in patients with mental retardation using 100K single-nucleotide polymorphism arrays. J Med Genet 2007; 44: 629–636.

Itsara A, Cooper GM, Baker C, et al. Population analysis of large copy number variants and hotspots of human genetic disease. Am J Hum Genet 2009; 84: 148–161.

Lee C, Iafrate AJ, Brothman AR. Copy number variations and clinical cytogenetic diagnosis of constitutional disorders. Nat Genet 2007; 39: S48–S54.

Manning M, Hudgins L. Use of array-based technology in the practice of medical genetics. Genet Med 2007; 9: 650–653.

Manning M, Hudgins L, Array-based technology and recommendations for utilization in medical genetic practice for detection of chromosomal abnormalities. Genet Med 2010; 10: 742–745.

Miller DT, Adam MP, Aradhya S, et al. Consensus statement: Chromosomal microarray is a first-tier clinical diagnostic test for individuals with developmental disabilities or congenital anomalies. Am J Hum Genet 2010; 86: 749–764.

Poot M, Hochstenbach R. A three-step workflow procedure for the interpretation of array-based comparative genome hybridization results in patients with idiopathic mental retardation and congenital anomalies. Genet Med 2010; 12: 478–485.

Rauch A, Ruschendorf F, Huang J, et al. Molecular karyotyping using an SNP array for genomewide genotyping. J Med Genet 2004; 41: 916–922.

Sambrook J, Russell DW. *Molecular Cloning. A Laboratory Manual.* Cold Spring Harbor Laboratory Press, Woodbury, NY, 2001.

Scott SA, Cohen N, Brandt T, et al. Detection of low-level mosaicism and placental mosaicism by oligonucleotide array comparative genomic hybridization. Genet Med 2010; 12: 85–92.

Shaffer L, Beaudet A, Brothman A, et al. Microarray analysis for constitutional cytogenetic abnormalities. Genet Med 2007; 9: 654–662.

Shaffer LG, McGowan-Jordan J, Schmid M (eds). *ISCN 2013*: *An International System for Human Cytogenetic Nomenclature.* Karger Publishers, Unionville, CT, 2013.

Shen Y, Irons M, Miller DT, et al. Development of a focused oligonucleotide-array comparative genomic hybridization chip for clinical diagnosis of genomic imbalance. Clin Chem 2007; 53: 2051–2059.

Strachen T, Read A. *Human Molecular Genetics*, 4th edn. Garland Science Press, New York, 2011.

Theisen A, Rosenfeld JA, Farrell SA, et al. aCGH detects partial tetrasomy of 12p in blood from Pallister-Killian syndrome cases without invasive skin biopsy. Am J Med Genet A 2009; 149A: 914–918.

Tsuchiya KD, Shaffer LG, Aradhya S, et al. Variability in interpreting and reporting copy number changes detected by array-based technology in clinical laboratories. Genet Med 2009; 11: 866–873.

Microarray database resources

Database of Genomic Variants (http://projects.tcag.ca/variation/)

Online Mendelian Inheritance in Man (www.ncbi.nlm.nih.gov/omim/)

DECIPHER (www.sanger.ac.uk/research/areas/)

dbVar – database of Structural Variation (www.ncbi.nlm.nih.gov/dbvar).

dbGaP – database of Genotypes and Phenotypes (www.ncbi.nlm.nih.gov/gap).

UCSC Genome Bioinformatics (http://genome.ucsc.edu/cgi-bin/hgGateway)

Ensembl (http://uswest.ensembl.org/Homo_sapiens/Gene/Summary)

International Standards for Cytogenomics Arrays Consortium (www.iscaconsortium.org/)

Wellcome Trust Sanger Institute (www.sanger.ac.uk/genetics/CGP/cosmic/)

CHAPTER 17

Microarray abnormal results

17.1 Reporting of abnormal results

ISCN RULES FOR CHROMOSOMAL MICROARRAYS – ABNORMAL RESULTS – COMMON RULES

- For abnormal results, only the abnormal chromosomes detected are described. Sex chromosomes are listed first, then autosomes by ascending order (lowest chromosome number first), without regard to copy number change.
- Only the chromosome band designation encompassing the copy number change (CNC) is described, and the band order and nucleotide order are always from pter to qter for each chromosome arm and abnormality, respectively.
- Single nucleotides are listed individually, separated by a comma, and a segment of DNA (BAC or oligonucleotide) is listed from start nucleotide to end nucleotide, separated by a dash.
- First write "arr".
- If the genome build is to be added, next write "hg", followed by the build, e.g. 19, enclosed in brackets. For example: arr[hg19].

EXAMPLE REPORTING OF AN ABNORMAL MALE MICROARRAY RESULT

ISCN Result: arr[hg19] 16p11.2(28,692,936-29,230,338)x1

Interpretation

Chromosomal microarray analysis (CMA) revealed an abnormal male chromosome complement, including the loss of 537 Kb of chromosome 16p11.2. This deletion is termed the "distal 16p11.2 deletion". Phenotypic findings with this deletion include developmental delay,

Cytogenetic Abnormalities: Chromosomal, FISH and Microarray-Based Clinical Reporting, First Edition. Susan Mahler Zneimer.

learning disability, behavioral problems, dysmorphology and obesity with incomplete penetrance.

Benign copy number changes that may be present in this patient are not listed in this report. Genetic counseling and parental CMA studies are recommended.

17.2 Loss or gain of a single chromosome

Whole chromosome gains and losses are easily detectable by all three array methodologies (bacterial artificial chromosome (BAC), oligoarray and single nucleotide polymorphism (SNP)). However, the identification appears visually different for these three techniques (Figure 17.1). The ISCN shows differing ways to describe aneuploidies, by either the short version or the long version and with or without the human genome build. See below for example reports for the various versions of reporting.

ISCN RULES FOR CHROMOSOMAL MICROARRAYS – ABNORMAL RESULTS – LOSS OR GAIN OF A SINGLE CHROMOSOME – SHORT VERSION

- With the **short version**, no nucleotides are described. Only the chromosome is described, within parentheses, followed by the number of copies present.

For example, for gain of an X chromosome in a female, the nomenclature is written as: arr(X)x3. Loss of a single X chromosome is written as: arr(X)x1.

For example, trisomy 21 and 22 would be written as: arr(21)x3,(22)x3.

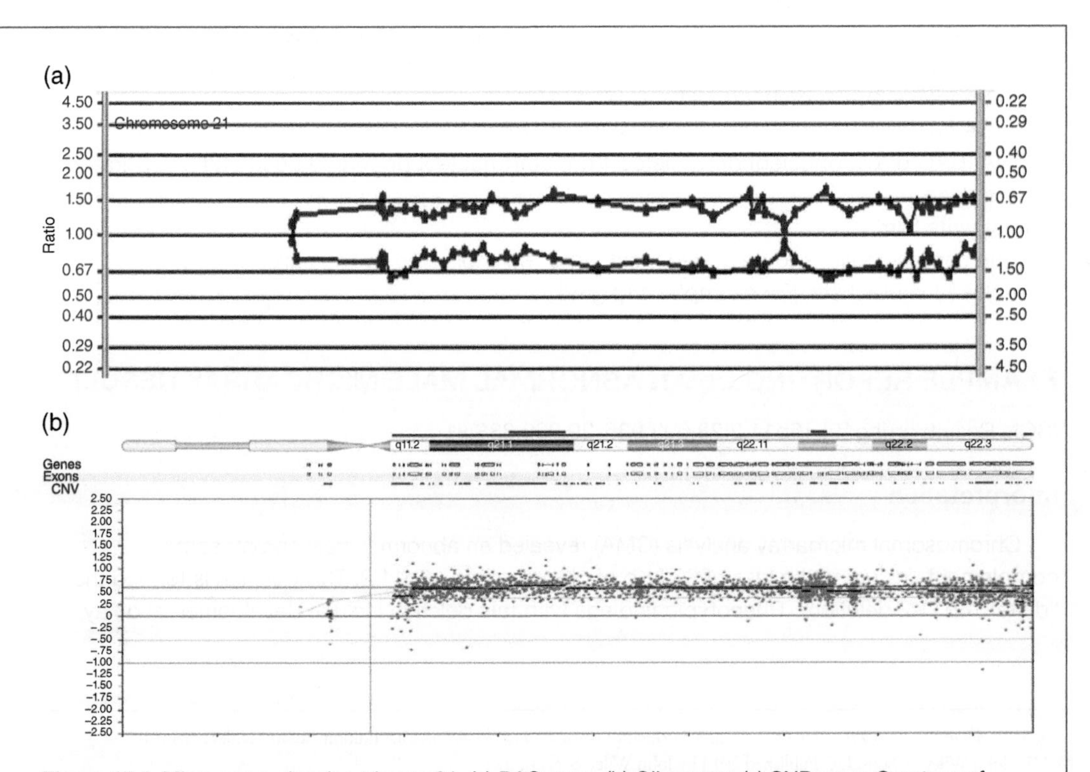

Figure 17.1 Microarrays showing trisomy 21. (a) BAC array. (b) Oligoarray. (c) SNP array. Courtesy of Karine Hovanes PhD, CombiMatrix, Inc.

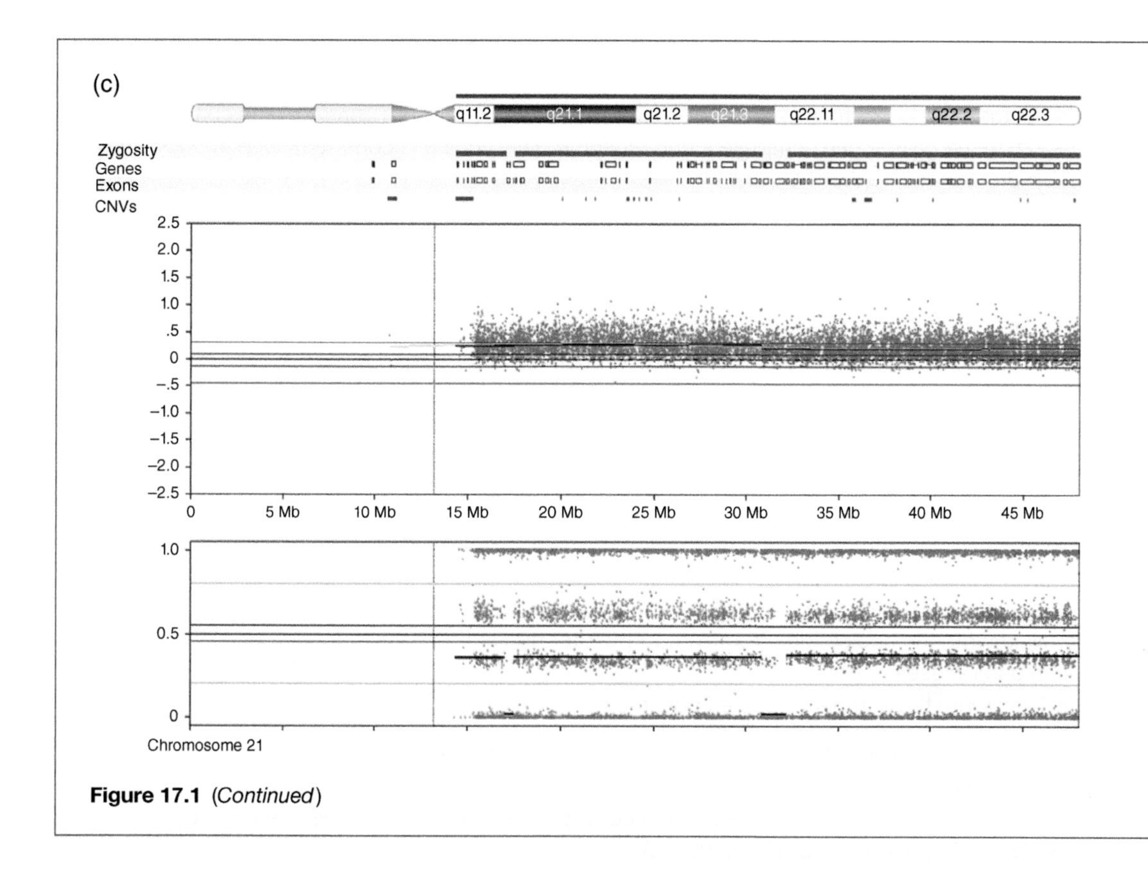

Figure 17.1 (*Continued*)

EXAMPLE REPORTING OF AN ABNORMAL FEMALE MICROARRAY RESULT WITH GAIN OF ONE X CHROMOSOME (FIGURE 17.2)

ISCN Result: arr(X)x3

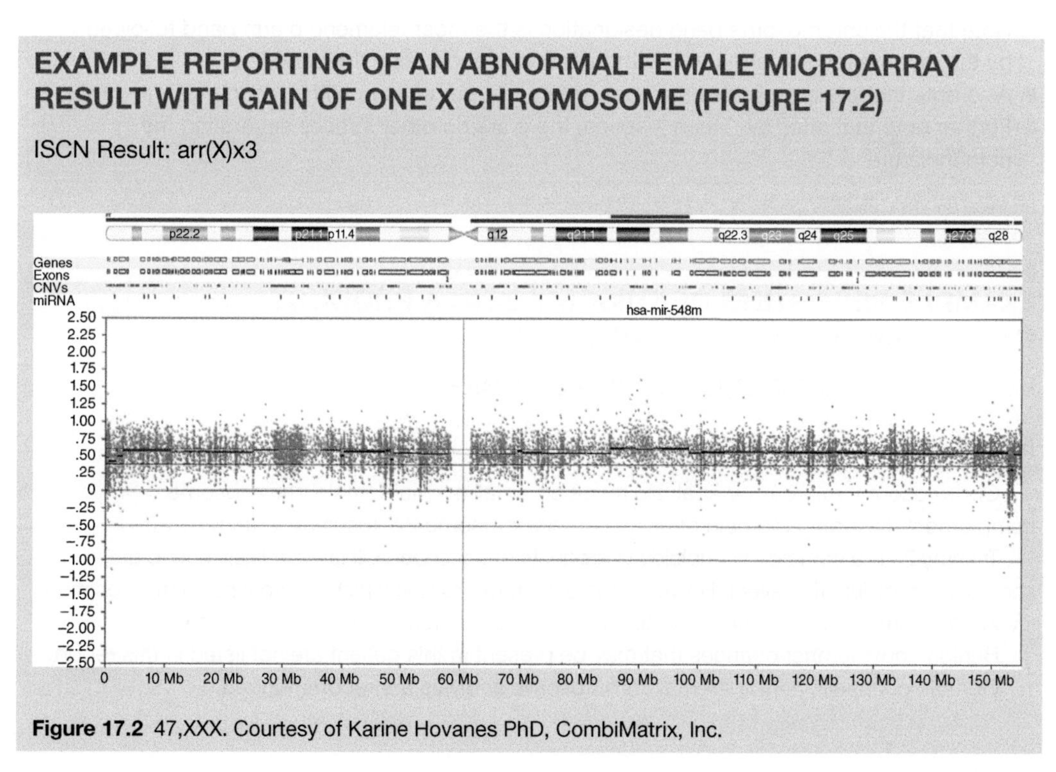

Figure 17.2 47,XXX. Courtesy of Karine Hovanes PhD, CombiMatrix, Inc.

Interpretation

Chromosomal microarray analysis revealed an abnormal female chromosome complement, including the gain of one complete X chromosome. This finding is consistent with 47,XXX – triple X syndrome.

Individuals with 47,XXX chromosome complement have variable phenotypic abnormalities, but generally have normal pubertal development.

Benign copy number changes that may be present in this patient are not listed in this report.

Genetic counseling is recommended.

ISCN RULES FOR CHROMOSOMAL MICROARRAYS – ABNORMAL RESULTS – LOSS OR GAIN OF A SINGLE CHROMOSOME – LONG VERSION

- With the **long version**, after "arr" and a space, write the chromosome. Then the band designation is described, from the order of pter to qter.
- Next write the nucleotides from zero to the qter nucleotide base pair number, all enclosed in parentheses.
- Then write the number of copies observed.

For example, for gain of an X chromosome, the nomenclature is written as: arr Xp22.3q28(1–247,249,719)x3.

For example, trisomy 21 and 22 would be written as: arr 21p13q22.3(9,769,099-46,940,056)x3,22p13q22.3(15,761,719-49,572,994)x3.

- Note that the chromosome band designation is the most telomeric p arm band followed by the most terminal q arm band without a space separating the bands.
- Also note that the nucleotides may be written with commas or without commas.
- Further note that after "arr" then a space, there are no other spaces separating the nomenclature.

EXAMPLE REPORTING OF TRISOMY 22 IN A PRODUCT OF CONCEPTION (POC) (FIGURE 17.3)

ISCN Result: arr 22p13q22.3(15,761,719-49,572,994)x3

Interpretation

Chromosomal microarray analysis revealed an abnormal female chromosome complement, including the gain of all loci on chromosome 22, resulting in trisomy 22.

Trisomy 22 is a common aneuploidy in products of conception. It is most likely due to a sporadic non-disjunction event. However, there are rare cases in which a parent has a translocation involving chromosome 22, which significantly increases the recurrence risk for trisomy.

Benign copy number changes that may be present in this patient are not listed in this report.

Genetic counseling and parental chromosome analysis are recommended.

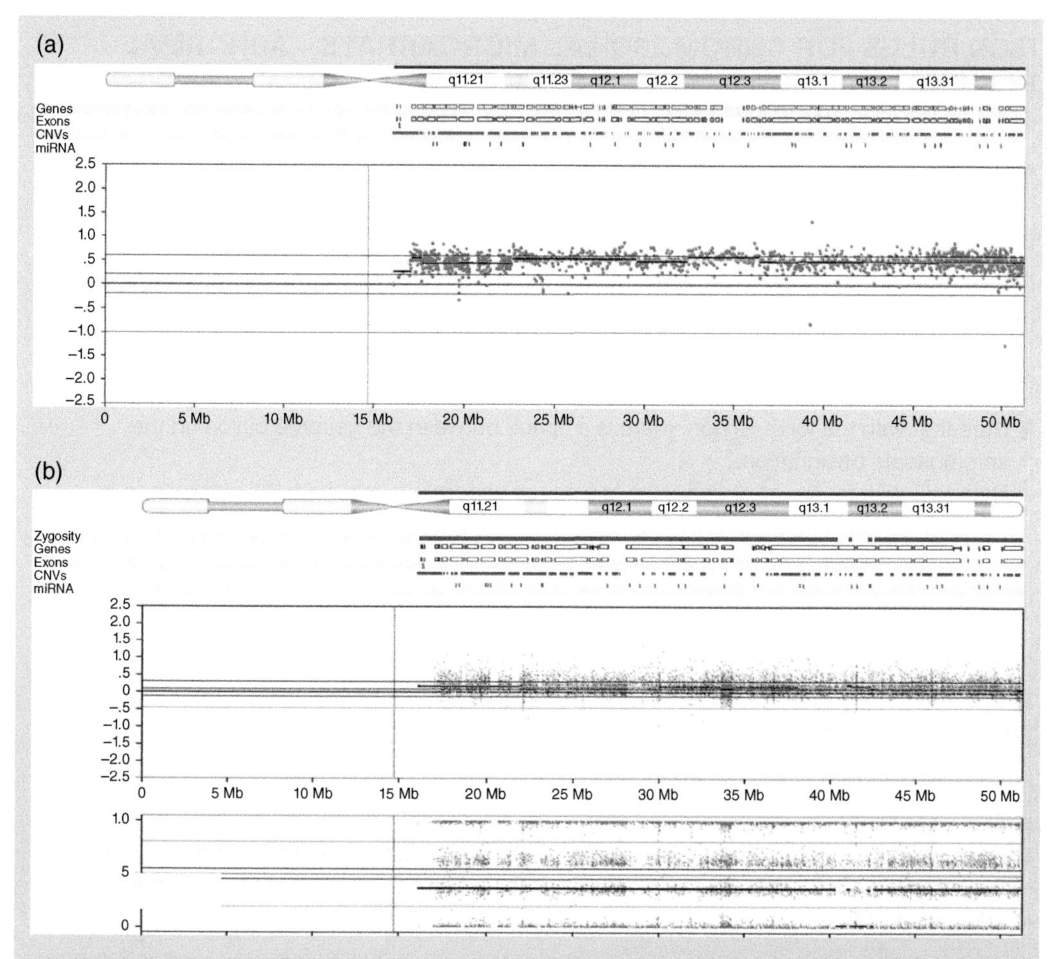

Figure 17.3 Microarray showing trisomy 22. (a) Oligoarray. (b) SNP array. Courtesy of Karine Hovanes PhD, CombiMatrix, Inc.

EXAMPLE REPORTING OF DOUBLE TRISOMY IN A POC

ISCN Result: arr 21p13q22.3(9,769,099-46,940,056)x3,22p13q22.3(15,761,719-49,572,994)x3

Interpretation

Chromosomal microarray analysis revealed an abnormal female chromosome complement, including the gain of all loci on chromosomes 21 and 22, resulting in trisomy 21 and trisomy 22.

While double trisomy represents a relatively rare finding, numerous cases have been reported in products of conception. They are most likely due to sporadic non-disjunction events. However, there are rare cases in which a parent has a translocation involving chromosome 21 and 22, which significantly increases the recurrence risk for trisomy.

Benign copy number changes that may be present in this patient are not listed in this report.

Genetic counseling and parental chromosome analysis are recommended.

ISCN RULES FOR CHROMOSOMAL MICROARRAYS – ABNORMAL RESULTS – LOSS OR GAIN OF A SINGLE CHROMOSOME – ADDING THE HUMAN GENOME BUILD

- With the human genome build in the nomenclature, after "arr" add a bracket, then the hg build number, followed by a bracket.
- Next write the short or long version of the abnormality without a space.

For example, for the short version with the genome build, the nomenclature is written as: arr[hg19](X)x3.

For example, for the long version with the genome build, the nomenclature is written as: arr[hg19] Xp22.3q28(1–247,249,719)x3.

- Note that with the long version, there is a space between the genome build and the chromosome designation.
- Note that with the long version there is a space separating "arr" with the chromosome number, but not with parentheses or brackets.

17.3 Loss or gain of a whole chromosome complement

ISCN RULES FOR CHROMOSOMAL MICROARRAYS – ABNORMAL RESULTS – LOSS OR GAIN OF A WHOLE CHROMOSOME COMPLEMENT – TRIPLOIDY

- First write "arr".
- For a female who would have three copies of all the autosomes and the X chromosome, write the autosomes 1–22, followed by a comma, followed by "X", all enclosed in parentheses.
- Then write "x3".

For example, a female triploid chromosome complement would be written as: arr(1–22,X)x3.

- For a male who would have three copies of all the autosomes, only 1–2 copies of the X chromosome and 1–2 copies of the Y chromosome, write the autosomes 1–22, enclosed in parentheses, followed by "x3".
- Then write the X chromosome in parentheses, followed by the copy number, followed by a comma.
- Then write the Y chromosome in parentheses, followed by the copy number.

For example, a male triploid chromosome complement would be written as: arr(1–22)x3,(X)x2,(Y)x1.

EXAMPLE REPORTING OF XXX TRIPLOIDY (FIGURE 17.4)

ISCN Result: arr(1–22,X)x3

Interpretation

Chromosomal microarray analysis revealed an abnormal female chromosome complement, including the gain of a whole chromosome complement with the presence of three X chromosomes. This finding results in a female fetus with triploidy.

Benign copy number changes that may be present in this patient are not listed in this report.

Genetic counseling is recommended.

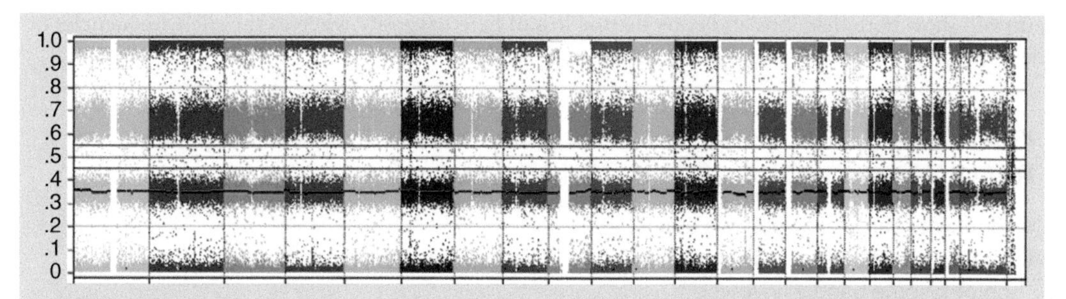

Figure 17.4 SNP array showing XXX triploidy. Note the heterozygous AB allele pattern in the middle of the graph, showing gain of each chromosome. Courtesy of Karine Hovanes PhD, CombiMatrix, Inc.

17.4 Microdeletions

Figure 17.5 shows images of a deletion of 22q11.2 by both oligo and SNP arrays. With a SNP array, if the R_{obs} is less than the R_{exp}, then there will be a decreased signal intensity resulting in a deletion or loss of copy number. However, a deletion needs to be confirmed with both data sets, since the log R

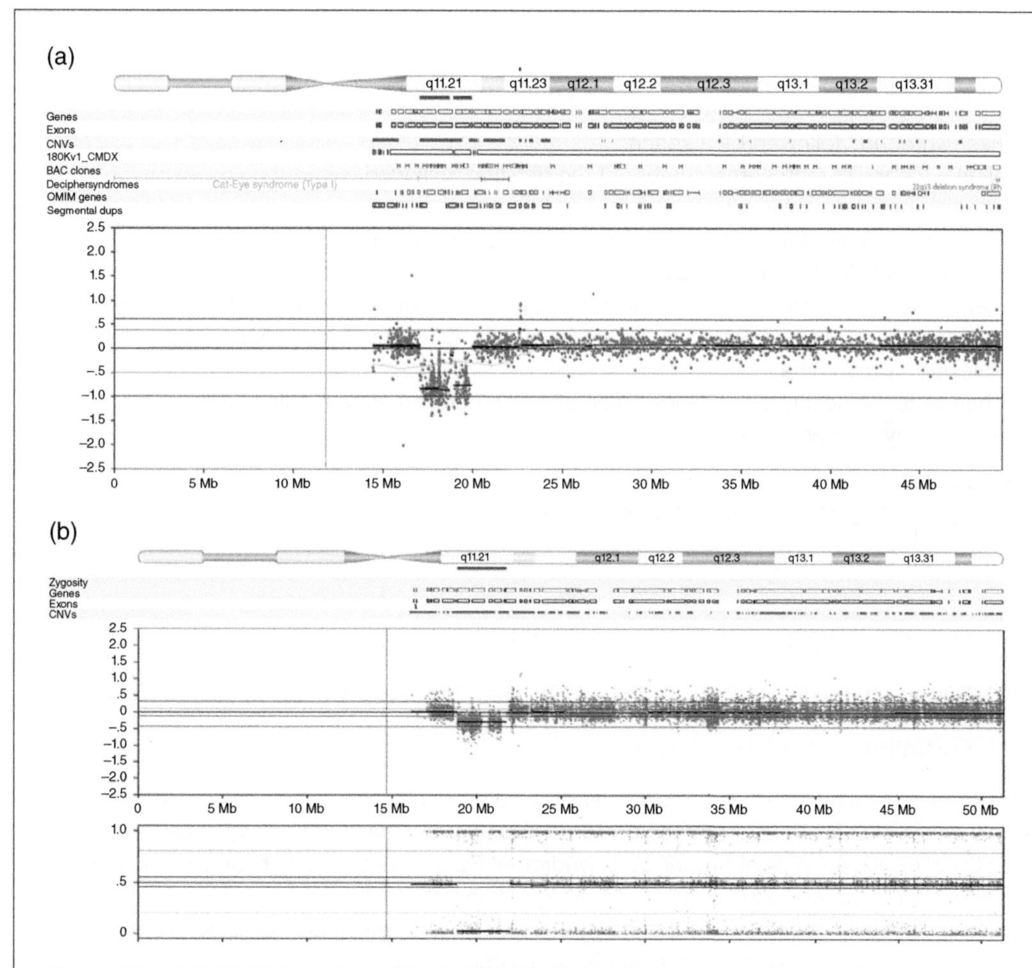

Figure 17.5 del(22)(q11.2q11.2) resulting in DGS/VCFS. (a) Oligoarray. (b) SNP array. Courtesy of Karine Hovanes PhD, CombiMatrix, Inc.

ratio will show both deletions and copy neutral loss of heterozygosity whereas the allele frequencies in the lower graph will show the copy number loss.

ISCN RULES FOR CHROMOSOMAL MICROARRAYS – ABNORMAL RESULTS – MICRODELETION – SHORT VERSION

- First write "arr" followed by a space.
- Next write the chromosome number of the deletion, followed (without a space) by the chromosome band(s) involved.
- Next write the start nucleotide of the deleted region, followed by a dash, followed by the end nucleotide of the deletion, enclosed in parentheses.
- Then write "x1" to designate a heterozygous deletion and "x0" to designate a homozygous deletion.

For example, a chromosome 15q11.2-q13.1 microdeletion would be written as: arr 15q11 .2q13.1(23,289,365-28,542,401)x1.

ISCN RULES FOR CHROMOSOMAL MICROARRAYS – ABNORMAL RESULTS – MICRODELETION – LONG VERSION

- First write "arr" followed by a space.
- Next write the chromosome number of the deletion, followed (without a space) by the chromosome band(s) involved.
- Next write in parenthesis the most proximal nucleotide identified before the deletion begins, followed by "x2", followed by a comma.
- Next write the start nucleotide of the deleted region, followed by a dash, followed by the end nucleotide of the deletion, followed by "x1" or x0" for the copy number of the deletion, followed by a comma.
- Then write the next nucleotide identified after the deletion ended, followed by "x2", followed by a parenthesis.

For example, a chromosome 15q11.2-q13.1 microdeletion would be written as: arr 15q11.2q1 3.1(23,002,343x2,23,289,365-28,542,401x1,28,964,213x2).

EXAMPLE REPORTING OF A MICRODELETION

ISCN Result: ish del(15)(q11.2q13.1)(RP11-373J1-).arr 15q11.2q13.1(23,289,365-28,542,401)x1

Interpretation

Chromosomal microarray analysis revealed an abnormal female chromosome complement with the loss of 5.25 Mb of chromosome 15q11.2-q13.1. Deletions of this region are consistent with the clinical diagnosis of Prader–Willi syndrome. Fluorescence *in situ* hybridization (FISH) analysis confirmed the 15q11.2-q13.1 deletion seen in this study.

Benign copy number changes that may be present in this patient are not listed in this report. Genetic counseling and parental studies are recommended.

17.5 Microduplications

Figure 17.6 shows a microduplication by SNP array.

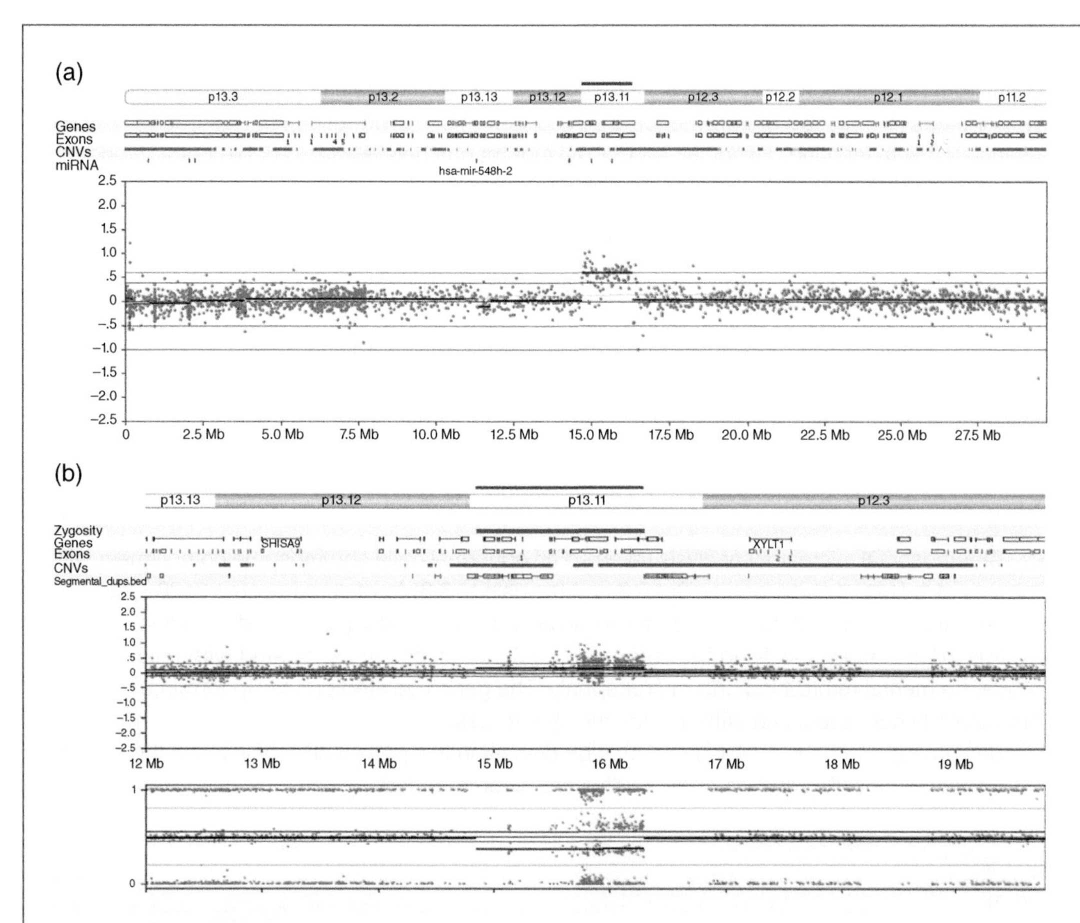

Figure 17.6 Duplication of chromosome 16 by both (a) oligo and (b) SNP arrays. Courtesy of Karine Hovanes PhD, CombiMatrix, Inc.

ISCN RULES FOR CHROMOSOMAL MICROARRAYS – ABNORMAL RESULTS – MICRODUPLICATION – SHORT VERSION

- First write "arr", followed by a space.
- Next write the chromosome number of the duplication, followed (without a space) by the chromosome band(s) involved.
- Next write the start nucleotide of the duplicated region, followed by a dash, followed by the end nucleotide of the duplication, enclosed in parentheses.
- Then write "x3" to designate gain of one copy and "x4" to designate gain of two copies.

For example, chromosome 15q11.2-q13.1 microduplication would be written as: arr 15q11.2q13.1(23,289,365-28,542,401)x3.

EXAMPLE REPORTING OF A MICRODUPLICATION

ISCN Result: nuc ish 22q11.23(RP11-124F9x3).arr 22q11.23(23,664,331-25,018,378)x3

Interpretation

Chromosomal microarray analysis revealed an abnormal female chromosome complement with the gain of 1.35 Mb of chromosome 22q11.23. This microduplication has been associated with learning difficulties and/or developmental delays with variable phenotypes and may be inherited from a phenotypically normal parent. Fluorescence *in situ* hybridization (FISH) analysis confirmed the 22q11.23 duplication seen in this study.

Benign copy number changes that may be present in this patient are not listed in this report.

Genetic counseling and parental studies are recommended.

EXAMPLE REPORTING OF AN ARRAY WITH BOTH A MICRODELETION AND MICRODUPLICATION (FIGURE 17.7)

ISCN Result: arr 3p26.3(0–1,480,687)x1,3q28q29(189,446,245-198,022,430)x3

Interpretation

Chromosomal microarray analysis revealed an abnormal male chromosome complement with the loss of 1.4 Mb of chromosome 3p26.3 and the gain of 8.3 Mb of chromosome 3q28-q29. These results are consistent with a 3p terminal microdeletion and 3q terminal duplication.

Microduplications involving the chromosome 3q29 region are associated with mild to moderate mental retardation and microcephaly. The genes deleted on the 3p terminal region are currently not associated with any known syndromes.

Benign copy number changes that may be present in this patient are not listed in this report.

Genetic counseling and parental studies are recommended.

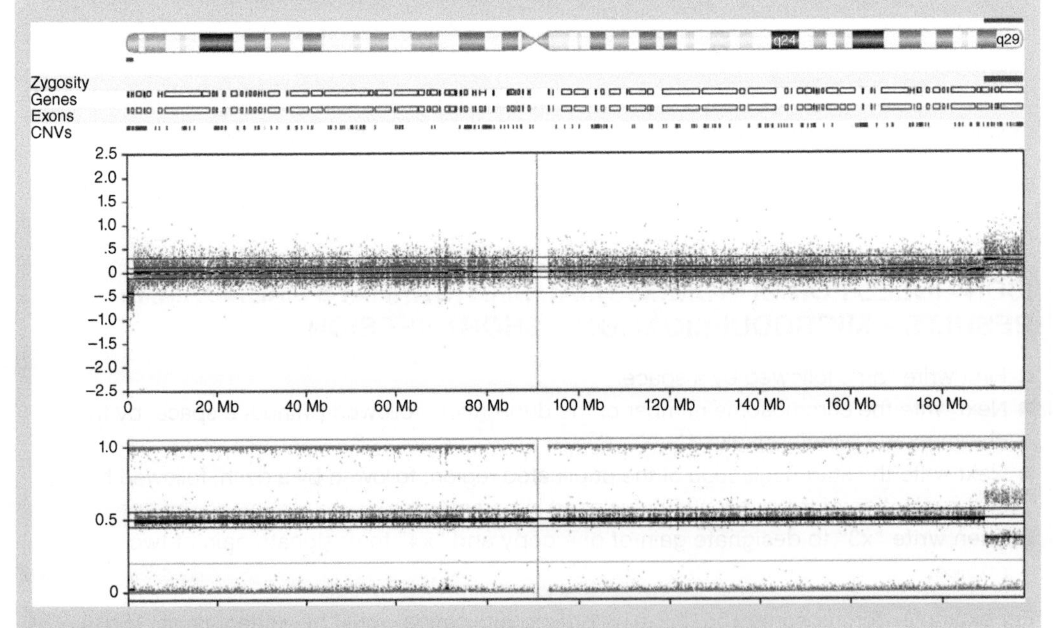

Figure 17.7 Microduplication (gain of chromosome 3q28-q29) and microdeletion (loss of chromosome 3p26.3) using a SNP array. Courtesy of Karine Hovanes PhD, CombiMatrix, Inc.

17.6 Derivative chromosomes

ISCN RULES FOR CHROMOSOMAL MICROARRAYS – ABNORMAL RESULTS WITH TWO CHROMOSOME ABNORMALITIES – A DERIVATIVE CHROMOSOME

- When two or more abnormalities are present in a single patient's analysis, the order of the listed abnormalities is by chromosome number in ascending order, regardless of the type of abnormality present.

For example, for a patient who is identified with both a 14q deletion and a 17q duplication, the nomenclature is written as: arr 14q32.33(106,710,826-107,349,540)x1,17q24.2q25.3 (66,296,272-81,195,210)x3.

EXAMPLE REPORTING OF TWO ABNORMALITIES IN A CHORIONIC VILLUS SAMPLING (CVS) CULTURE WITH AN ABNORMAL ULTRASOUND AND ABNORMAL CHROMOSOME ANALYSIS (FIGURE 17.8)

ISCN Result: 46,XX,add(14)(q32.2).arr 14q32.33(106,710,826-107,349,540)x1, 17q24.2q25.3(66,296,272-81,195,210)x3

Interpretation

Chromosomal microarray analysis revealed an abnormal female chromosome complement with the loss of 840 Kb of chromosome 14q32.33-qter and the gain of 14.9 Mb of chromosome 17q24.2-qter. This result is consistent with a 14q32.33 terminal deletion and a 17q24.2-q25.3 terminal duplication.

The mother is reported to have a balanced translocation between the terminal long arms of chromosomes 14 and 17 [46,XX,t(14;17)(q32.3;q23.3)]. This microarray result is consistent with an unbalanced translocation due to a derivative chromosome 14 that is monosomic for 14q32.33-qter and trisomic for 17q24.2-ter. This result is also consistent with the fetal karyotype, which was previously reported as 46,XX,add(14)(q32.2), and is presently reported as 46,XX,der(14)t(14;17)(q32.3;q23.3).

Benign copy number changes that may be present in this patient are not listed in this report.

Genetic counseling is recommended.

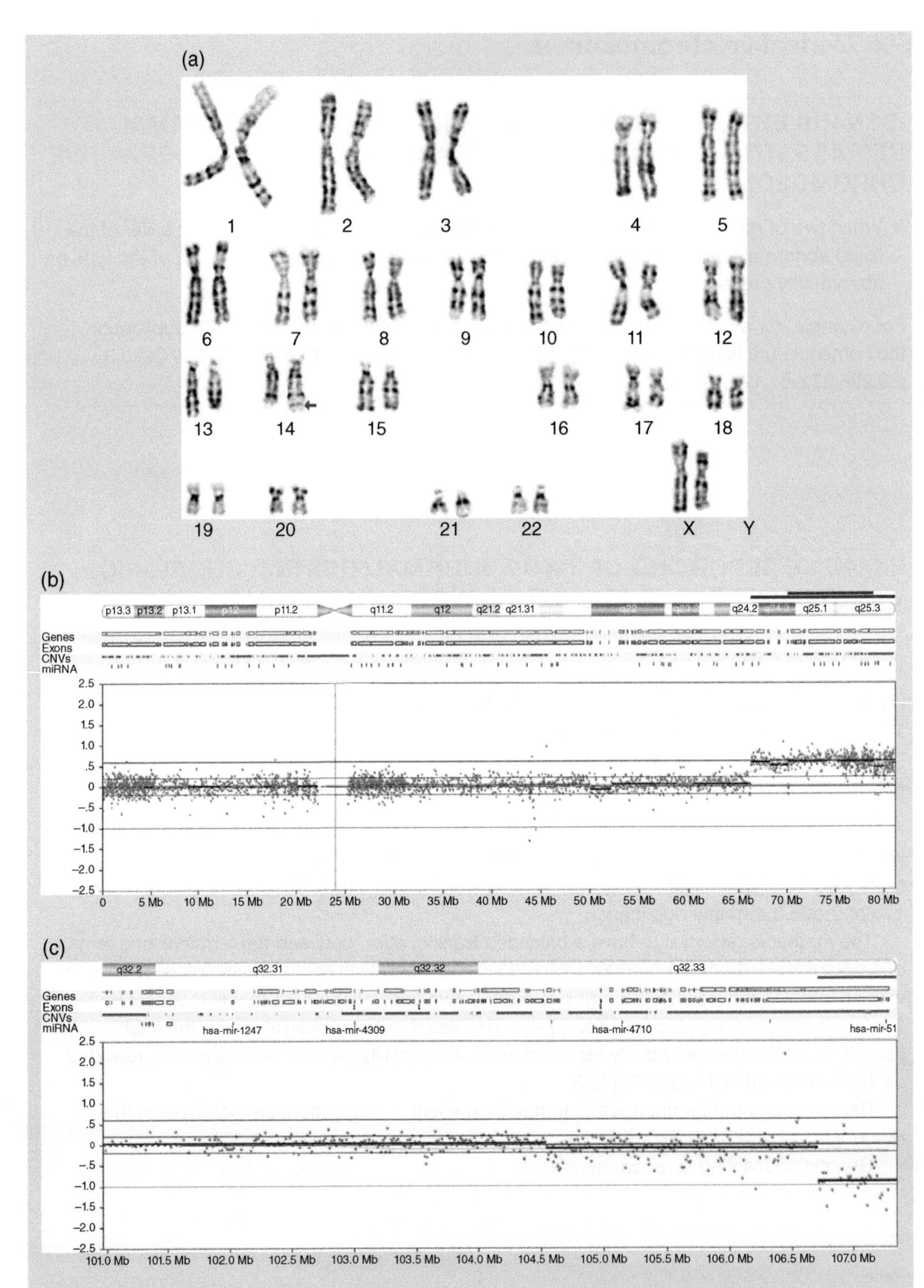

Figure 17.8 An unbalanced translocation detected by chromosomes and array. (a) Karyotype - 46,XX,add(14)(q32.2). (b) Array results showing chromosome 17q duplication. (c) Array results showing chromosome 14q deletion. ISCN: arr 14q32.33(106,710,826-107,349,540)x1,17q24.2q25.3 (66,296,272-81,195,210)x3. Courtesy of Karine Hovanes PhD, CombiMatrix, Inc.

17.7 Variants of unknown significance

EXAMPLE REPORTING OF A VARIANT OF UNKNOWN SIGNIFICANCE (VOUS) (FIGURE 17.9)

ISCN Result: arr 2p25.3(1,722,995-1,855,067)x3

Interpretation

Chromosomal microarray analysis revealed a male with a variant of unknown significance (VOUS), including the gain of chromosome 2p25.3, partially overlapping the MYT1L and PXDN genes.

This region has not been reported as a normal copy number variation in the population, nor has it been associated with known syndromes. However, a recent report showed an association of a duplication involving the MYT1L and PXDN genes with autism. Another report describes MYT1L as a candidate gene for intellectual disability in patients with 2p25.3 deletions; however, the clinical significance of this duplication is not known.

Benign copy number changes that may be present in this patient are not listed in this report.

Genetic counseling and parental CMA studies are recommended.

Genes involved in the gain of 2p25.3: MYT1L (partial), PXDN (partial).

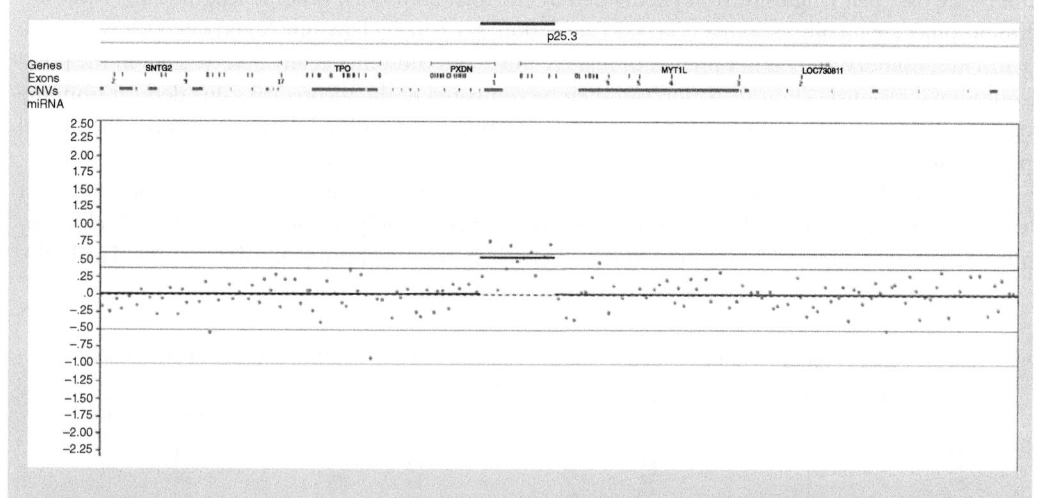

Figure 17.9 Array results with a variant of unknown significance (VOUS) of chromosome 2 duplication: arr 2p25.3(1,722,995-1,855,067)x3. Courtesy of Karine Hovanes PhD, CombiMatrix, Inc.

17.8 Uniparental disomy/loss of heterozygosity/regions of homozygosity

Loss of heterozygosity (LOH) or regions of homozygosity (ROH) results from either uniparental disomy (discussed in Chapter 12) or a deletion of a whole chromosome or within a chromosome. Uniparental disomy (UPD) is defined as the inheritance of chromosome homologs or segments of both homologs from a single parent and loss of the corresponding homolog from the other parent. UPD may be detected by SNP arrays in which both parents carry the same allele at any one single nucleotide. Generally, only large contiguous segments of homozygosity on a single chromosome, equaling >5 Mb, are observed and indicative of UPD and possibly reported using a SNP array.

There are three main mechanisms causing UPD: trisomy rescue, monosomy rescue and postfertilization mitotic non-disjunction. Any of these three mechanisms may lead to UPD in any tissue type and may also lead to mosaicism for ROH. Since UPD of a single chromosome is seen predominantly in the mosaic form, SNP array analysis is a good approach to detecting low levels of mosaicism that other cytogenetic techniques cannot match. For example, SNP arrays provide a greater sensitivity for the detection of mosaicism than chromosome analysis. Furthermore, SNP analysis can provide evidence of long stretches of ROH that may be due to UPD. This is important in copy neutral mosaicism, which will not be detected by chromosome analysis or by comparative genomic oligo or BAC arrays. UPD seen within a chromosome is also termed segmental UPD, caused by mitotic recombination, which results in copy number neutral changes. This type of UPD is a postzygotic event that may occur at different times during development and can also be detected by SNP arrays.

When long stretches of homozygosity occur throughout the genome, it is more indicative of regions identical by descent (IBD). These ROH regions originate from a common ancestor and may indicate consanguinity. The greater the region in the genome that contains ROH, the greater the relationship of an individual to that of the parent(s). The relative proportion of ROH that is IBD in offspring of related parents is usually identified by the coefficient of inbreeding (F), and can be estimated by SNP arrays. For example, children of first-cousin parents will have 1/16th (6.25%) of the genome IBD. To estimate the level of consanguinity when large areas of homozygosity involving many chromosomes exist, the sum of the base pair lengths of ROH is divided by the total autosomal genome length (2881 Mb). Sex chromosomes are usually excluded from the calculation, since males have one X and one Y chromosome each. Thus, homozygosity is not present in male X and Y chromosomes at any SNP (except in the pseudoautosomal regions). This calculation can then be compared to the theoretical value derived from the coefficient of inbreeding for any trio (parent and offspring) relationship. However, this is a theoretical approach for IBD estimation, and empirically significant variations may occur.

Reporting ROH that result from UPD or consanguinity is variable, and each laboratory must develop thresholds for rates that are to be reported. Generally, only regions >5 Mb on each chromosome are reported. Figure 17.10 shows ROH of 11.8% in regions throughout the genome. For UPD,

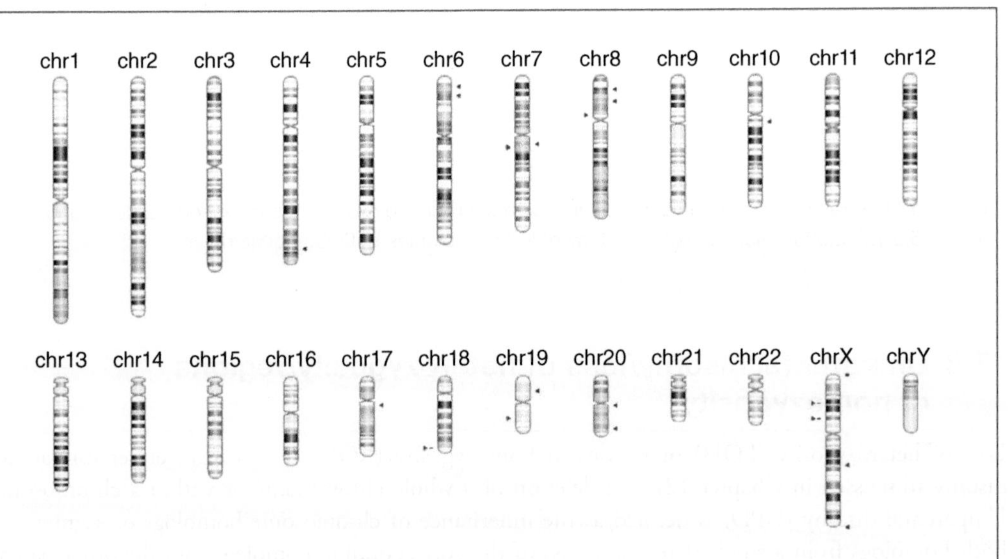

Figure 17.10 SNP array showing 11.8% total ROH with regions designated in yellow throughout the genome. Courtesy of Karine Hovanes PhD, CombiMatrix, Inc.

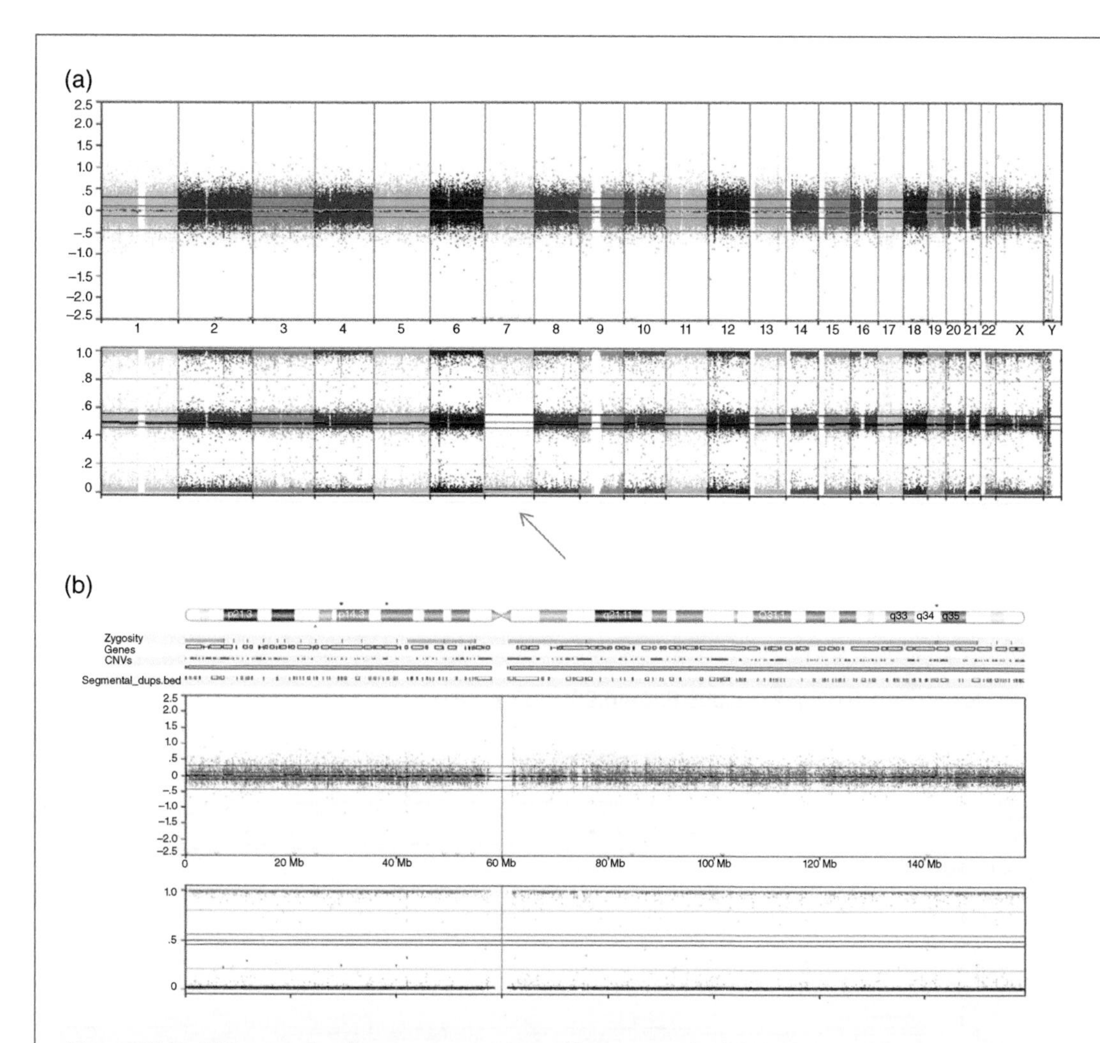

Figure 17.11 SNP array showing copy neutral loss of heterozygosity (LOH) of chromosome 7. (a) Whole genome view with arrow pointing to chromosome 7. (b) UPD7/LOH of chromosome 7 depicted in the lower graph. Note the loss of heterozygote alleles AB in the middle band of the lower graph. Courtesy of Karine Hovanes PhD, CombiMatrix, Inc.

generally >20% Mb on a single chromosome will be reportable as segmental UPD. However, due to the possibility of meiotic recombination, hetero-UPD will not easily be detected.

Figure 17.11 shows a case with ROH/LOH of the whole chromosome 7, consistent with UPD7, and Figure 17.12 shows segmental UPD of chromosome 15q.

17.9 Mosaicism

Mosaicism cannot be detected by BAC or oligoarrays easily, but can be more easily depicted by SNP arrays. However, even with SNP arrays, the graph showing the log R ratio appears normal whereas the allele signal intensity will show either a gain or loss, but with a different amplitude than a full gain or loss.

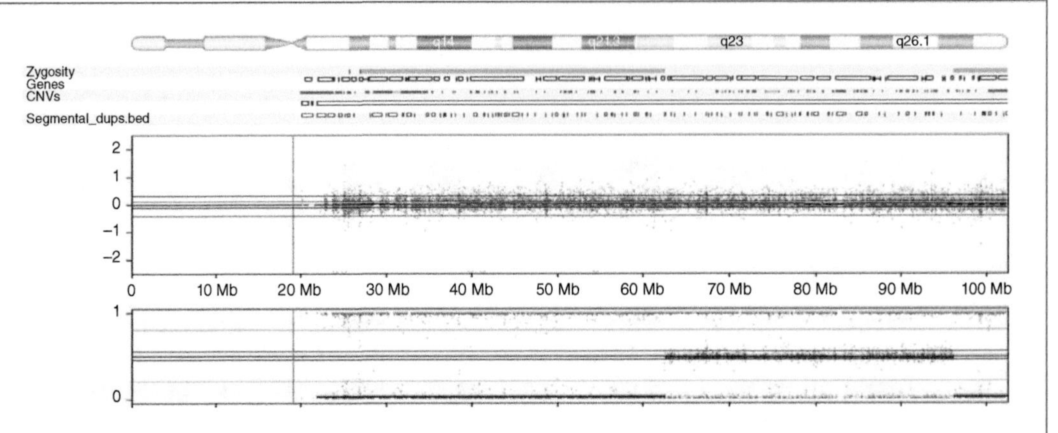

Figure 17.12 SNP array showing segemental UPD of chromosome 15q, in the regions corresponding to the yellow line in the lower graph. Courtesy of Karine Hovanes PhD, CombiMatrix, Inc.

EXAMPLE REPORTING OF MOSAIC TRISOMY 21 (FIGURE 17.13)

ISCN Result: arr 21p13q22.3(9,896,630-46,905,339)x2~3

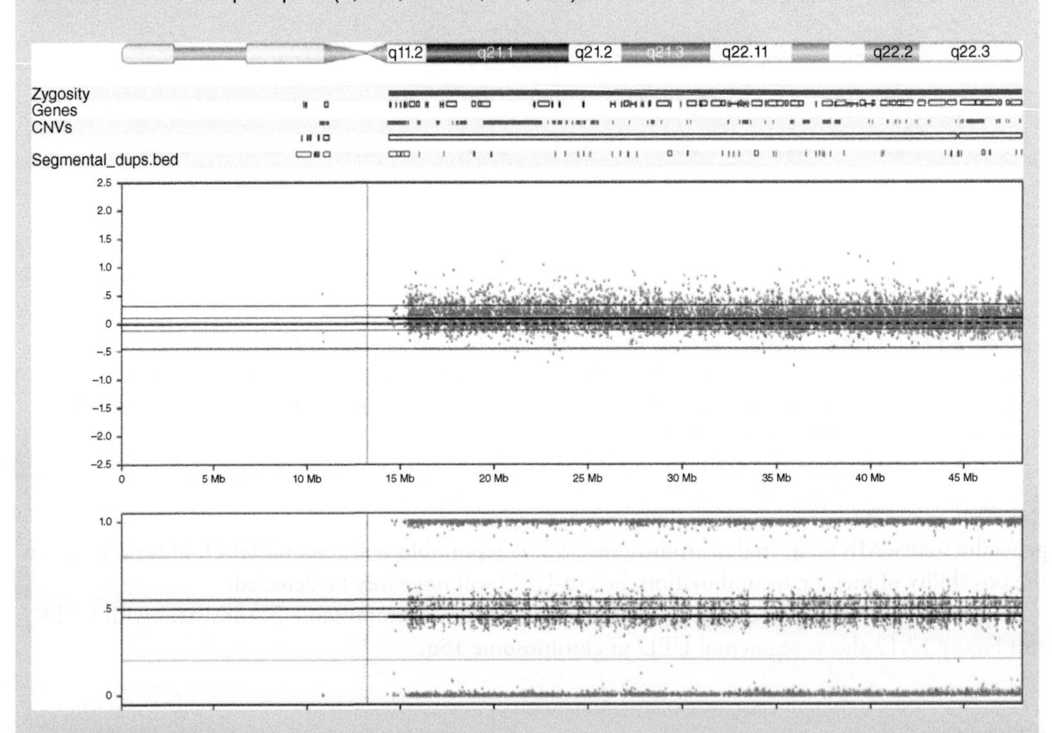

Figure 17.13 SNP array showing mosaic trisomy 21. The upper graph shows a normal pattern, but the purple line depicts a gain, shown in the lower graph, with a smaller amplitude of allele change as in full gains. FISH estimated this mosaicism at approximately 20%. Courtesy of Karine Hovanes PhD, CombiMatrix, Inc.

Interpretation

Chromosomal microarray analysis revealed an abnormal female chromosome complement, with a mosaic copy number gain of all loci on chromosome 21, resulting in mosaic Down syndrome. Fluorescence *in situ* hybridization (FISH) analysis confirmed trisomy 21 in approximately 20% of interphase nuclei scored. Patients with low-level mosaic trisomy 21 present with fewer phenotypic manifestations and higher IQs than patients with full trisomy.

Benign copy number changes that may be present in this patient are not listed in this report.

Genetic counseling and parental chromosome analysis are recommended.

17.10 Common comments in abnormal reports

ADDING PARENTAL STUDIES

- Chromosomal microarray results in this study did not show the copy number change observed in this patient's son/daughter.
- Chromosomal microarray results in this study showed the copy number change that was present in this patient's daughter/son. Clinical correlation and genetic counseling are recommended
- Chromosomal microarray results in this patient's mother/father also showed this copy number change; therefore, this copy number change is considered maternally/paternally inherited.
- Chromosomal microarray results in both the mother and father of this patient showed no abnormality for this region; therefore, this is considered a *de novo* event.
- This derivative chromosome appears to have resulted from a balanced translocation observed in the mother/father, which showed a translocation between chromosomes __ and __ .
- Chromosomal microarray results in both the mother and father of this patient showed no abnormality for this region; therefore, this is considered a *de novo* event. However, a parental balanced structural rearrangement resulting in an unbalanced genomic abnormality in offspring cannot be excluded.
- Metaphase FISH analysis was performed to determine if this abnormality was inherited from a parent. Results in both parental studies showed no abnormality involving chromosomes __ and __, suggesting that the abnormality observed in this patient is a *de novo* event. Standard chromosome analysis on both parents is recommended to help identify a rearrangement undetectable by FISH analysis.

MARKER CHROMOSOMES NOT DETECTED BY ARRAY

- Chromosomal microarray results showed no copy number changes representing the mosaic marker chromosome which was detected by standard chromosome analysis. This result is consistent with either the presence of a supernumerary marker chromosome that does not contain euchromatic material or the level of mosaicism is below the level of detection of chromosomal microarray analysis. Genetic counseling is recommended.
- Chromosomal microarray analysis (CMA) showed no copy number gain of genomic material representing the marker chromosome. Therefore, the supernumerary marker chromosome observed by standard chromosome analysis is likely composed of repetitive genomic sequences and not euchromatic material that is undetectable by CMA. Genetic counseling is recommended.

OTHER COMMENTS FOR ABNORMAL REPORTS

- Genes involved in the gain of__ include ___.
- The absence of the gain of the Y chromosome with chromosomal microarray analysis (CMA), which would suggest mosaic XYY/XY, is likely due to the low level of mosaicism below the threshold of detection by CMA.
- Neutral copy number genomic changes, and position effects resulting from the previously reported __ chromosome inversion/translocation, are not detectable by chromosomal microarray analysis.
- Copy number changes on the X chromosome may be phenotypically neutral in females due to X inactivation. Therefore, the effect of this copy number change in a female is currently not known.
- Copy number variants are observed at multiple sites in the genomes of normal individuals. There are public databases of such variants; however, the abnormality detected in this patient was not observed in any database. It may reflect a new copy number variant, or it may be associated with the phenotype of this patient. Nevertheless, the clinical significance is currently not known.

17.11 Microarrays with concurrent FISH studies and/or chromosome studies

ISCN RULES FOR CHROMOSOMAL MICROARRAYS – ABNORMAL RESULTS WITH CONFIRMATION FISH ANALYSIS – SHORT VERSION

- First write the FISH results (see Section 2 for nomenclature rules).
- Then write the array results, as described above.

For example, a chromosome 15q11.2-q13.1 microdeletion would be written as: ish del(15)(q11.2q13.1)(RP11-373J1-).arr 15q11.2q13.1(23,289,365-28,542,401)x1.

ISCN RULES FOR CHROMOSOMAL MICROARRAYS – ABNORMAL RESULTS WITH CHROMOSOME ANALYSIS – SHORT VERSION

- First write the chromosome result (see Section 1 for nomenclature rules).
- Then write the array results, as described above.

For example, chromosome 15q11.2-q13.1 microduplication would be written as: 47,XY,+mar.arr 15q11.2q13.1(23,289,365-28,542,401)x3.

17.12 Microarrays with concurrent parental studies

ISCN RULES FOR CHROMOSOMAL MICROARRAYS – ABNORMAL RESULTS WITH PARENTAL STUDIES

- First write the array results, as described above.
- If the array results end in the copy number, add a space, then write the parental origin of the abnormality ("pat" or "mat", for paternal or maternal inheritance, respectively).
- If the results end in a parenthesis, write pat or mat without a space.
- If the results are *de novo* in origin, known by normal parental studies, write "dn" instead of "pat" or "mat", with the same rules for spacing after the copy number or parenthesis.

For example, a chromosome 15q11.2-q13.1 microdeletion would be written in the short version as: arr 15q11.2q13.1(23,289,365-28,542,401)x1 pat.

The nomenclature for the long version would be written as: arr 15q11.2q13.1(23,002, 343x2,23,289,365-28,542,401x1,28,964,213x2)pat.

Note: If the parental studies were performed by standard chromosome analysis or FISH analysis, then the "pat", "mat" or "dn" would be written after the chromosomes or FISH nomenclature. That is, the parental origin is written only after the nomenclature of the analysis performed.

17.13 Preimplantation genetic diagnosis testing

Reporting of preimplantation genetic diagnosis (PGD) of aneuploid screening using microarray analysis is similar to FISH analysis. However, the methodology used is vastly different, usually utilizing BAC, oligo or SNP arrays to determine chromosome gains and losses of the total chromosome complement of each single cell of the embryo of a fetus. With microarray testing, all 24 chromosomes may be analyzed in a single test, unlike FISH analysis which usually only analyzes the major chromosomes of interest. As in FISH analysis, limitations of this test include mosaicism, since only single cells are analyzed, as well as non-complete chromosomal aneuploidy including single gene disorders and partial chromosomal gains and losses.

Uniparental disomy may also be identified by SNP microarrays and can be classified as either heterodisomy (homologous chromosomes inherited from a single parent) or isodisomy (identical chromosomes inherited from a single parent). However, UPD may not be indistinguishable from monosomy of a given chromosome.

EXAMPLE REPORTING OF PGD EMBRYOS – DAY 3 CLEAVAGE STAGE EMBRYOS

Results

Embryo 1: 46,XXX,−8
Embryo 2: 23,X
Embryo 3: 47,XX,+5
Embryo 4: 46,XY: normal male
Embryo 5: 48,XX,+12,+14
Embryo 6: 46,XX: normal female
Embryo 7: No signals observed
Embryo 8: 45,X,−1,+10

Interpretation

Embryos with normal results include:

Embryo 4: normal male
Embryo 6: normal female

Embryos with aneuploidies include:

Embryo 1: abnormal female with monosomy 8 and trisomy for the X chromosome
Embryo 2: abnormal fetus with a haploid chromosome complement and a single X chromosome

Embryo 3: abnormal female with trisomy 5
Embryo 5: abnormal female with trisomy 12 and trisomy 14
Embryo 7: no results
Embryo 8: abnormal female with monosomy for chromosomes X and 1, and trisomy 10

Follow-up prenatal diagnosis by CVS or amniocentesis is recommended.

17.14 Non-invasive prenatal testing

Non-invasive prenatal testing (NIPT) obtains diagnostic information about a pregnancy without risk of an invasive procedure to the fetus. Currently, diagnosis of cytogenetic disorders or other genetic testing *in utero* is performed using invasive procedures, including amniocentesis or CVS, which carry a slight risk for miscarriage. In the 1990s, researchers discovered cell-free fetal DNA (cffDNA) within the circulation of pregnant women, which requires only a simple blood draw to diagnose genetic disorders early in pregnancy (beginning at 8–10 weeks' gestation).

However, there are challenges to non-invasive prenatal testing, including the acquisition of cffDNA that is present in maternal circulation. Since the amount of fetal cells in the maternal peripheral blood is very small, this process is very difficult, as is the testing of very small quantities of cffDNA.

Cell-free fetal DNA only stays in the maternal blood for hours before being degraded. Therefore, all cffDNA is representative of the current pregnancy. This is in contrast to intact fetal cells that remain in maternal circulating blood for years after a pregnancy, which does not guarantee that these cells analyzed represent the current fetus. However, it is thought that only 10–15% of cffDNA circulating in maternal blood is of fetal origin. Therefore, it is critical to quantitate the differences between DNA markers from the mother and fetus. Laboratories performing this testing need to establish clear standards for knowing the difference between fetal DNA and maternal DNA to ensure accuracy of testing. Figure 17.14 shows SNP array allele patterns differentiating cell-free maternal versus fetal DNA.

Currently, NIPT is performed to detect specific fetal aneuploidies, including trisomy 21, 18, 13 and the sex chromosomes and triploidy. Abnormalities involving only regions within a chromosome are generally not detected. However, certain microdeletions are now being tested in addition to whole chromosome aneuploidies. The detection rate for trisomy 18 and 21 is considered to be 99%. However,

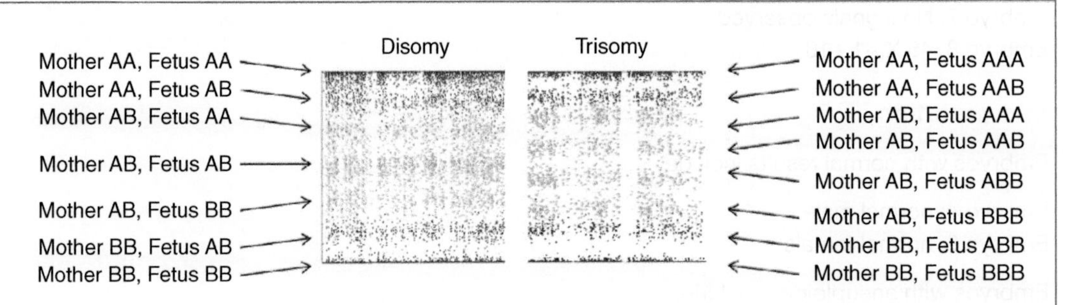

Figure 17.14 SNP array showing the allele patterns from cell free DNA with non-invasive prenatal testing. Courtesy of Zachary Demko PhD, Natera, Inc.

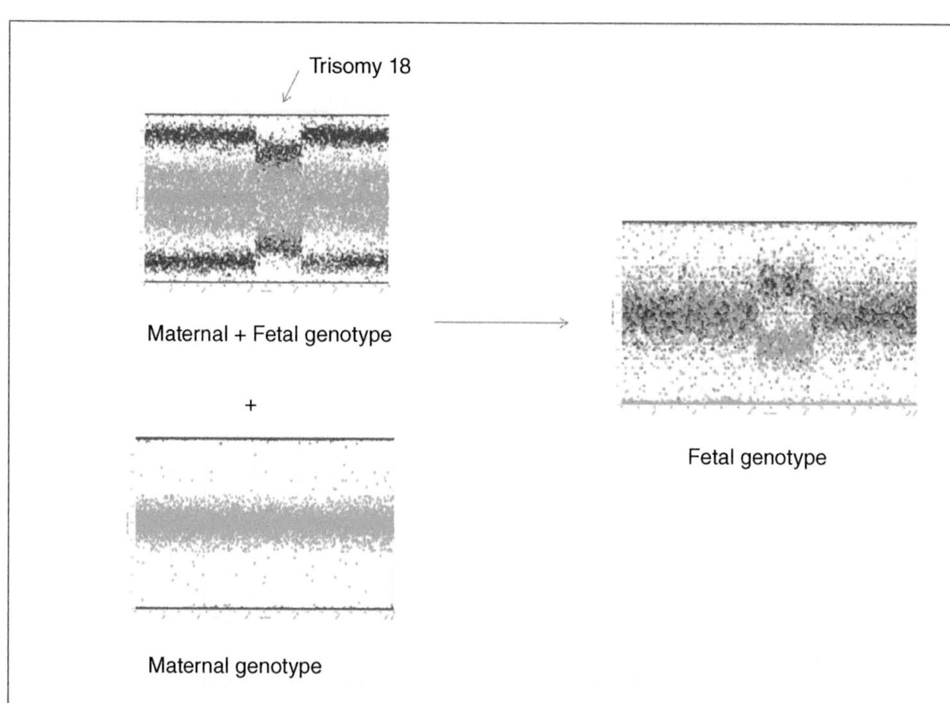

Figure 17.15 SNP array depicting specifically analyzed chromosomes, with an arrow pointing to trisomy 18 seen in the scatter plot. Courtesy of Zachary Demko PhD, Natera, Inc.

there is a false-negative rate of 1% in this screening test. The false-positive rate is approximately 0.2% for trisomy 18 and 21 (Figure 17.15). The detection rate for trisomy 13 is lower, within a range of 79–92%, and the false-positive rate is higher than 1%, due in part to technical issues and the low incidence of trisomy 13. Corresponding data for sex chromosome aneuploidies have not yet been published.

Reporting of NIPT is not uniform, with some laboratories reporting results as positive and negative for a trisomy while other laboratories report a quantitative risk for trisomies. Laboratories should be reporting the cut-off values for normal versus abnormal results, and where the patient results fall within the range of normal versus abnormal.

Although the detection rate reaches 99%, due to the screening nature of this test with both false-positive and false-negative results possible, it is recommended that all positive results be followed up with confirmatory testing by either a CVS or amniocentesis. Since trisomies may arise due to a translocation as well as chromosomal malsegregation, confirmatory testing will be able to identify the cause of the trisomy, which has important risk implications for future pregnancies of the parents and possibly other family members.

However, NIPT has proven to be a valuable tool for prenatal testing in general and is gaining importance as a prenatal screening test. The American College of Obstetrics and Gynecology has stated that NIPT should be an informed patient choice after pretest counseling, but should not at this time be a routine prenatal laboratory assessment or offered to low-risk women or women with multiple gestations. The International Society of Prenatal Diagnosis states that NIPT can be helpful as a screening test for women who are at high risk for trisomy 21, with suitable genetic counseling. It is generally agreed among professional groups that women need to have genetic counseling for all possible risks and outcome possibilities with NIPT, that a negative cffDNA test result does not ensure an unaffected fetus, and that a patient with a positive result should be offered an invasive prenatal diagnostic test for confirmation of test results.

Due to the limitations of obtaining sufficient cell-free fetal DNA, it is possible for NIPT specimens to yield poor or no results. A range of 0.5–7.0% of specimens, depending on the laboratory, will yield no results. Other limitations of NIPT include testing for twins or other multiple gestations, validation studies for low-risk pregnancies, egg donor pregnancies, surrogacy or women who have previously received a bone marrow transplant. Also, maternal sex chromosome abnormalities may interfere with the sex chromosome findings of the fetus, giving no results for the X and Y chromosomes. This test may also show ROH, also referred to as loss of heterozygosity (LOH). Regions of ROH/LOH are generally not reported due to the lack of confidence about their clinical significance in the fetus. When ROH/LOH is seen segmentally, and not as a whole chromosome, it often gives no result for that chromosome, as it will interfere with the statistical algorithm for distinguishing aneuploidy. When ROH/LOH is seen throughout the whole chromosome, it is consistent with uniparental disomy, but may be unreportable if the confidence that it is solely in the fetus is not substantiated.

EXAMPLE NIPT DISCLAIMER COMMENTS

- NIPT could not be performed. This test is not currently available for multiple gestations, pregnancy by egg donor or surrogacy, or bone marrow transplant recipients.
- NIPT could not be performed. The total fetal cell fraction was below the reportable cut-off for testing. Submission of a new sample is necessary for further testing.
- NIPT could not be performed. The sample failed to meet the quality control measures for this test; therefore, results are not obtainable. Submission of a new sample is necessary for further testing.
- NIPT revealed a low risk for trisomy ___. However, results for chromosome __ were not possible, due to DNA findings outside the current reportable range of the test. Follow-up testing is recommended. The possible causes of these findings include the presence of regions of homozygosity, mosaicism or maternal chromosome mosaicism.

EXAMPLE REPORTING OF NORMAL RESULTS

Results

Chromosome X: risk score of X chromosome aneuploidy is <1/10,000 (0.01%)
Chromosome 13: risk score of trisomy 13 is <1/10,000 (0.01%)
Chromosome 18: risk score of trisomy 18 is <1/10,000 (0.01%)
Chromosome 21: risk score of trisomy 21 is <1/10,000 (0.01%)

Interpretation

Chromosomal microarray analysis of chromosomes X, 13, 18 and 21 revealed a low risk (less than 0.01% probability) for trisomy 13, 18 and 21 and aneuploidy for the X chromosome.

These findings are based on 99% sensitivity and specificity confidence intervals. Genetic counseling is recommended.

EXAMPLE REPORTING OF ABNORMAL RESULTS SUGGESTING TRISOMY 21

Results

Chromosome X: risk score of monosomy/trisomy X is <1/10,000 (0.01%)
Chromosome 13: risk score of trisomy 13 is <1/10,000 (0.01%)
Chromosome 18: risk score of trisomy 18 is <1/10,000 (0.01%)
Chromosome 21: risk score of trisomy 21 is >99/100 (99.0%)

Interpretation

Chromosomal microarray analysis of chromosomes X, 13, 18 and 21 revealed a high risk of trisomy 21 with a greater than 99.0% probability.

The remaining results revealed a low risk for trisomy 13 and 18, and aneuploidy for the X chromosome (less than 0.01% probability).

These findings are based on 99% sensitivity and specificity confidence intervals.

Follow-up prenatal diagnosis by CVS or amniocentesis and genetic counseling are recommended.

Bibliography

American College of Obstetricians and Gynecologists. ACOG Practice Bulletin No. 77: Screening for Fetal Chromosomal Abnormalities. Obstet Gynecol 2007; 109: 217–227.

American College of Obstetricians and Gynecologists. ACOG Committee Opinion No. 545, December 2012. www.acog.org/Resources And Publications/ Committee Opinions/Committee on Genetics/Noninvasive Prenatal Testing for Fetal Aneuploidy.

Aradhya S, Cherry AM. Array-based comparative genomic hybridization: clinical contexts for targeted and whole-genome designs, Genet Med 2007; 9: 553–559.

Ashoor G, Poon L, Syngelaki A, et al. Fetal fraction in maternal plasma cell-free 374 DNA at 11–13 weeks' gestation: effect of maternal and fetal factors. Fetal Diagn Ther 2012; 375: 237–243.

Aston E, Whitby H, Maxwell T, et al. Comparison of targeted and whole genome analysis of postnatal specimens using a commercially available array based comparative genomic hybridisation (aCGH) microarray platform. J Med Genet 2008; 5: 268–274.

Bachmann-Gagescu R, Mefford H, Cowan C, et al. Recurrent 200-kb deletions of 16p11.2 that include the SH2B1 gene are associated with developmental delay and obesity. Genet Med 2010; 12(10): 641–647.

Baldwin E, Lee J, Blake D, et al. Enhanced detection of clinically relevant genomic imbalances using a targeted plus whole genome oligonucleotide microarray. Genet Med 2008; 10: 415–429.

Barge-Schaapveld D, Maas S, Polstra A, et al. The atypical 16p11.2 deletion: a not so atypical microdeletion syndrome? Am J Med Genet Part A 2011; 155: 1066–1072.

Bianchi DW, Platt LK, Goldberg JD, et al. Genome-wide fetal aneuploidy detection by maternal plasma DNA sequencing. Obstet Gynecol 2012; 119: 1–12.

Burgess AW. How many red flags does it take? Am J Nurs 2007; 107: 28–31.

Canick JA, Kloza EM, Lambert-Messerlian GM, et al. DNA sequencing of maternal plasma to identify Down syndrome and other trisomies in multiple gestations. Prenat Diagn 2012; 32(8): 730–734.

Carothers AD, Rudan I, Kolcic I, et al. Estimating human inbreeding coefficients: comparison of genealogical and marker heterozygosity approaches. Ann Hum Genet 2006; 70(Pt 5): 666–676.

Cassidy S, Driscoll D. Prader-Willi syndrome. Eur J Hum Genet 2009; 17(1): 3–13.

Centers for Disease Control and Prevention, American Society for Reproductive Medicine. *2006 Assisted Reproductive Technology Success Rates: National Summary and Fertility Clinic Reports.* US Department of Health and Human Services, Centers for Disease Control and Prevention, Atlanta, GA, 2008.

Chiu RW, Akolekar R, Zheng YWL, et al. Non-invasive prenatal assessment of 297 trisomy 21 by multiplexed maternal plasma DNA sequencing: large scale validity study. BMJ 2011; 342: c7401.

Conlin LK, Kaur M, Izumi K, et al. Utility of SNP arrays in detecting, quantifying, and determining meiotic origin of tetrasomy 12p in blood from individuals with Pallister–Killian syndrome. Am J Med Genet Part A 2012; 158A: 3046–3053.

Conlin LK, Thiel BD, Bonnemann CG, et al. Mechanisms of mosaicism, chimerism and uniparental disomy identified by single nucleotide polymorphism array analysis. Hum Mol Genet 2010; 19: 1263–1275.

Coppinger J, McDonald-McGinn D, Zackai E, et. al. Identification of familial and de novo microduplications of 22q11.21–q11.23 distal to the 22q11.21 microdeletion syndrome region. Hum Mol Genet 2009; 18(8): 1377–1383.

De Vries BB, Pfundt R, Leisink M, et al. Diagnostic genome profiling in mental retardation. Am J Hum Genet 2005; 77: 606–616.

Ding C, Chiu RW, Lau TK, et al. MS analysis of single-nucleotide differences in circulating nucleic acids: application to noninvasive prenatal diagnosis. Proc Natl Acad Sci USA 2004; 101(29): 10762–10767.

Ehrich M, Deciu C, Zwiefelhofer T, et al. Noninvasive detection of fetal trisomy 21 by 301 sequencing of DNA in maternal blood: a study in a clinical setting. Am J Obstet Gynecol 2011; 204(3): 205.e1–11.

Finning KM, Martin PG, Soothill PW, Avent ND. Prediction of fetal D status from maternal plasma: introduction of a new noninvasive fetal RHD genotyping service. Transfusion 2002; 42(8): 1079–1085.

Garfield SS, Armstrong SA. Clinical and cost consequences of incorporating a novel non-invasive prenatal test into the diagnostic pathway for fetal trisomies. J Manag Care Med 2012; 15: 34–41.

Gautier E, Benachi A, Giovazngrandi Y, et al. Fetal RhD genotyping by maternal serum analysis: a two-year experience. Am J Obstet Gynecol 2005; 192(3): 666–669.

Hamamy H, Antonarakis SE, Cavalli-Sforza LL, et al. Consanguineous marriages, pearls and perils: Geneva International Consanguinity Workshop Report. Genet Med 2011; 13: 841–847.

Hartl DL, Clark AG. *Principles of Population Genetics.* Sinauer Associates, Sunderland, MA, 1997.

Hochstenbach R, Ploos van Amstel HK. Poot M. Microarray-based genome investigation: molecular karyotyping or segmental aneuploidy profiling? Eur J Hum Genet 2006; 14: 262–265.

Hoyer J, Dreweke A, Becker C, et al. Molecular karyotyping in patients with mental retardation using 100K single-nucleotide polymorphism arrays. J Med Genet 2007; 44: 629–636.

Izumi K, Santani AB, Deardorff MA, et al. Mosaic maternal uniparental disomy of chromosome 15 in Prader–Willi syndrome: utility of genome-wide SNP array. Am J Med Genet Part A 2013; 161A: 166–171.

Jones KL. *Smith's Recognizable Patterns of Human Malformation*, 6th edn. Saunders, London, 2006.

Kearney HM, Kearney JB, Conlin LK. Diagnostic implications of excessive homozygosity detected by SNP-based microarrays: consanguinity, uniparental disomy, and recessive single-gene mutations. Clin Lab Med 2011; 31: 595–613.

Leon E, Zou Y, Milunsky J. Mosaic Down syndrome in a patient with low-level mosaicism detected by microarray. Am J Med Genet Part A 2010; 152A: 3154–3156.

Lo YM, Chiu RWK. Prenatal diagnosis: progress through plasma nucleic acids. Nature Rev Genet 2006; 8: 71–77.

Lo YM, Corbetta N, Chamberlain PF, et al. Presence of fetal DNA in maternal plasma and serum. Lancet 1997; 350(9076): 485–487.

Manning M, Hudgins L. Use of array-based technology in the practice of medical genetics. Genet Med 2007; 9: 650–653.

Manning M, Hudgins L. Array-based technology and recommendations for utilization in medical genetics practice for detection of chromosomal abnormalities for the Professional Practice and Guidelines Committee. Genet Med 2010; 12: 742–745.

McQuillan R, Leutenegger AL, Abdel-Rahman R, et al. Runs of homozygosity in European populations. Am J Hum Genet 2008; 83: 359–372.

Meyer KJ, Axelsen M, Sheffield V, et al. Germline mosaic transmission of a novel duplication of PXDN and MYT1L to two male half-siblings with autism. Psychiatr Genet 2012; 22: 137–140.

Micale M, Insko J, Ebrahim S, et. al. Double trisomy revisited – a multicenter experience. Prenat Diagn 2010; 30: 173–176.

Michels E, de Preter K, van Roy N, Speleman F. Detection of DNA copy number alterations in cancer by array comparative genomic hybridization. Genet Med 2007; 9: 574–584.

Miller DT, Adam MP, Aradhya S, et al. Consensus statement: Chromosomal microarray is a first-tier clinical diagnostic test for individuals with developmental disabilities or congenital anomalies. Am J Hum Genet 2010; 86: 749–764.

Moeschler JB. Genetic evaluation of intellectual disabilities. Semin Pediatr Neurol 2008; 15: 2–9.

Nicolaides KH. Screening for fetal aneuploidies at 11 to 13 weeks. Prenat Diagn 2011; 31: 7–15.

Nicolaides KH, Syngelaki A, Ashoor G, et al. Noninvasive prenatal testing for fetal trisomies in a routinely screened first-trimester population. Am J Obstet Gynecol 2012; 207: 374.e1–6.

Norton ME, Brar H, Weiss J, et al. Non-invasive chromosomal evaluation (NICE) 321 study: results of a multicenter prospective cohort study for detection of fetal trisomy 21 322 and trisomy 18. Am J Obstet Gynecol 2012; 207: 137.e1–8.

O'Connor, C. Prenatal screen detects fetal abnormalities. Nature Educ 2008; 1: 106.

Palomaki GE, Deciu C, Kloza EM, et al. DNA sequencing of maternal plasma 346 reliably identifies trisomy 18 and trisomy 13 as well as Down syndrome: an international 347 collaborative study. Genet Med 2012; 14: 296–305.

Pang AW, MacDonald JR, Pinto D, et al. Towards a comprehensive structural variation map of an individual human genome. Genome Biol 2010; 11: R52.

Papenhausen P, Schwartz S, Risheg H, et al. UPD detection using homozygosity profiling with aSNP genotyping microarray. Am J Med Genet Part A 2011; 155: 757–768.

Pergament E. Controversies and challenges of array comparative genomic hybridization in prenatal genetic diagnosis, Genet Med 2007; 9: 596–599.

Poot M, Hochstenbach R. A three-step workflow procedure for the interpretation of array-based comparative genome hybridization results in patients with idiopathic mental retardation and congenital anomalies. Genet Med 2010; 12: 478–485.

Rauch A, Ruschendorf F, Huang J, et al. Molecular karyotyping using an SNP array for genomewide genotyping. J Med Genet 2004; 41: 916–922.

Reddy KS. Double trisomy in spontaneous abortions. Hum Genet 1997; 101(3): 339–345.

Rehder CW, David KL, Hirsch B, Toriello HV, Wilson CM, Kearney HM. American College of Medical Genetics and Genomics: standards and guidelines for documenting suspected consanguinity as an incidental finding of genomic testing. Genet Med 2013; 15(2): 150–152.

Rodriguez-Revenga L, Mila M, Rosenberg C, Lamb A, Lee C. Structural variation in the human genome: the impact of copy number variants on clinical diagnosis. Genet Med 2007; 9: 600–606.

Rosenfeld JA, Ballif BC, Torchia BS, et al. Copy number variations associated with autism spectrum disorders contribute to a spectrum of neurodevelopmental disorders. Genet Med 2010; 12: 694–702.

Sampson M, Coughlin CR 2nd, Kaplan P, et al. Evidence for a recurrent microdeletion at chromosome 16p11.2 associated with congenital anomalies of the kidney and urinary tract (CAKUT) and Hirschsprung disease. Am J Med Genet A 2010; 152A(10): 2618–2622.

Schaaf CP, Scott DA, Wiszniewska J, Beaudet AL. Identification of incestuous parental relationships by SNP-based DNA microarrays. Lancet 2011; 377: 555–556.

Shaffer LG, Beaudet AL, Brothman AR, et al., for the Working Group of the Laboratory Quality Assurance Committee of the American College of Medical Genetics. Microarray analysis for constitutional cytogenetic abnormalities. Genet Med 2007; 9: 654–662.

Shaffer LG, McGowan-Jordan J, Schmid M (eds). *ISCN 2013: An International System for Human Cytogenetic Nomenclature*. Karger Publishers, Unionville, CT, 2013.

Shaikh TH. Oligonucleotide arrays for high-resolution analysis of copy number alteration in mental retardation/multiple congenital anomalies. Genet Med 2007; 9: 617–625.

Stevens SJ, van Ravenswaaij-Arts CM, Janssen JW, et al. MYT1L is a candidate gene for intellectual disability in patients with 2p25.3 (2pter) deletions. Am J Med Genet Part A 2011; 1552739–2745.

Tabor HK, Cho MK. Ethical implications of array comparative genomic hybridization in complex phenotypes: points to consider in research, Genet Med 2007; 9: 626–631.

Thorland EC, Gonzales PR, Gliem TJ, Wiktor AE, Ketterling RP. Comprehensive validation of array comparative genomic hybridization platforms: how much is enough? Genet Med 2007; 9: 632–641.

Tsuchiya KD, Shaffer LG, Aradhya S, et al. Variability in interpreting and reporting copy number changes detected by array-based technology in clinical laboratories. Genet Med 2009; 11: 866–873.

Van Buggenhout G, Fryns J. Angelman syndrome. Eur J Hum Genet 2009; 17(11): 1367–1373.

Van Vooren S, Coessens B, de Moor B, Moreau Y, Vermeesch JR. Array comparative genomic hybridization and computational genome annotation in constitutional cytogenetics: suggesting candidate genes for novel submicroscopic chromosomal imbalance syndromes. Genet Med 2007; 9: 642–649.

Vermeesch JR, Rauch A. Reply to Hochstenbach et al. 'Molecular karyotyping'. Eur J Hum Genet 2006; 14: 1063–1064.

Vermeesch JR, Fiegler H, de Leeuw N, et al. Guidelines for molecular karyotyping in constitutional genetic diagnosis. Eur J Hum Genet 2007; 15: 1105–1114.

Zimmermann B, Hill M, Gemelos G, et al. Noninvasive prenatal aneuploidy testing of 354 chromosomes 13, 18, 21, X, and Y using targeted sequencing of polymorphic loci. Prenat Diagn 2012; 32: 1233–1241.

Microarray database resources

Database of Genomic Variants (http://projects.tcag.ca/variation/)

Online Mendelian Inheritance in Man (www.ncbi.nlm.nih.gov/omim/)

DECIPHER (www.sanger.ac.uk/research/areas/)

dbVar – database of Structural Variation (www.ncbi.nlm.nih.gov/dbvar)

dbGaP – database of Genotypes and Phenotypes (www.ncbi.nlm.nih.gov/gap)

UCSC Genome Bioinformatics (http://genome.ucsc.edu/cgi-bin/hgGateway)

Ensembl (http://uswest.ensembl.org/Homo_sapiens/Gene/Summary)

International Standards for Cytogenomics Arrays Consortium (www.iscaconsortium.org/)

Wellcome Trust Sanger Institute (www.sanger.ac.uk/genetics/CGP/cosmic/)

CHAPTER 18

Pathogenic chromosomal microarray copy number changes by chromosome order

18.1 Chromosome 1

18.1.1 1p terminal deletion

EXAMPLE REPORTING OF 1p TERMINAL DELETION

ISCN Result: arr[hg19]1p36(__)x1

Interpretation

Chromosomal microarray analysis revealed an abnormal female/male chromosome complement with the loss of __ Mb of chromosome 1p36.

Deletion of the chromosome 1p36 region is the most common terminal deletion syndrome. Clinical features include microcephaly, brachycephaly, developmental delay, hypotonia and seizures.

Benign copy number changes that may be present in this patient are not listed in this report.

Genetic counseling and parental chromosomal microarray analysis are recommended.

Cytogenetic Abnormalities: Chromosomal, FISH and Microarray-Based Clinical Reporting, First Edition. Susan Mahler Zneimer.

18.1.2 1q21.1 deletion

EXAMPLE REPORTING OF 1q TAR DELETION (FIGURE 18.1)

ISCN Result: arr[hg19] 1q21.1(145,618,328-145,764,483)x1

Interpretation

Chromosomal microarray analysis revealed an abnormal female/male chromosome complement with the loss of 0.14 Mb of chromosome 1q21.1.

Deletions of this region are clinically associated with thrombocytopenia absent radius (TAR) syndrome. However, the haploinsufficiency of the deleted region is not sufficient to cause TAR syndrome. It is proposed that the presence of an additional modifier is required for phenotypic manifestations.

Benign copy number changes that may be present in this patient are not listed in this report.

Genetic counseling and parental chromosomal microarray analysis are recommended.

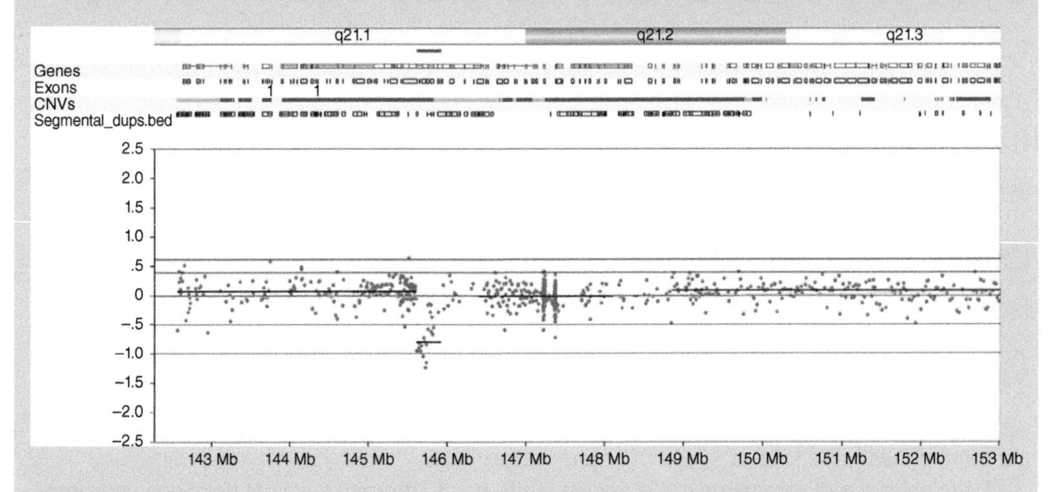

Figure 18.1 Oligoarray showing chromosome 1q21.1 deletion. Courtesy of Karine Hovanes PhD, CombiMatrix, Inc.

18.1.3 1q21.1 deletion/duplication

EXAMPLE REPORTING OF 1q21.1 DELETION/DUPLICATION

ISCN Result: arr[hg19] 1q21.1(145,773,406-149,190,276)x1/3

Interpretation

Chromosomal microarray analysis revealed an abnormal female/male chromosome complement with the loss/gain of 3.4 Mb of chromosome 1q21.1.

Abnormalities of this region are known to show copy number variation in normal individuals. However, reciprocal microdeletions and microduplications involving this region are also clinically associated with a range of developmental disorders including abnormal head size, mental retardation, autism spectrum disorder and congenital heart defects.

This copy number change may be inherited from a parent or may be a *de novo* event. A parent with the microdeletion may show a normal phenotype or an abnormal phenotype that is similar to but usually less severe than that of his/her child.

Benign copy number changes that may be present in this patient are not listed in this report.

Genetic counseling and parental chromosomal microarray analysis are recommended.

18.2 Chromosome 2

18.2.1 2p16.3 deletion/duplication

EXAMPLE REPORTING OF 2p16.3 DELETION/DUPLICATION

ISCN Result: arr[hg19] 2p16.3( )x1/3

Interpretation

Chromosomal microarray analysis revealed an abnormal female/male chromosome complement with the loss/gain of __ Mb of chromosome 2p16.3 overlapping the NRXN1 gene.

Segments of this region are known to show copy number variation in normal individuals. However, disruption of the neurexin 1 (NRXN1) gene is clinically associated with autism spectrum disorder, developmental delay and speech delay.

Benign copy number changes that may be present in this patient are not listed in this report.

Genetic counseling and parental chromosomal microarray analysis are recommended.

18.2.2 2q13 deletion

EXAMPLE REPORTING OF 2q13 DELETION

ISCN Result: arr[hg19] 2q13( )x1

Interpretation

Chromosomal microarray analysis revealed an abnormal female/male chromosome complement with loss of __ Mb of chromosome 2q13.

This deletion is clinically associated with increased risk for developmental delay and cranial facial dysmorphism.

Benign copy number changes that may be present in this patient are not listed in this report.

Genetic counseling and parental chromosomal microarray analysis are recommended.

18.2.3 2q24.3 deletion

EXAMPLE REPORTING OF 2q24.3 DELETION

ISCN Result: arr[hg19] 2q24.3( )x1

Interpretation

Chromosomal microarray analysis revealed an abnormal female/male chromosome complement with loss of __ Mb of chromosome 2q24.3.

This deletion is clinically associated with increased risk for low birthweight, postnatal growth restriction, mental impairment, dysmorphic features and seizures. The haploinsufficiency of SCN1A and SCN2A is likely to contribute to epilepsy in these patients.

Benign copy number changes that may be present in this patient are not listed in this report.

Genetic counseling and parental chromosomal microarray analysis are recommended.

18.2.4 2q31.1 deletion

EXAMPLE REPORTING OF 2q31.1 DELETION

ISCN Result: arr[hg19] 2q31.1(171,189,504-173,866,207)x1

Interpretation

Chromosomal microarray analysis revealed an abnormal female/male chromosome complement with the loss of __ Mb of chromosome 2q31.1.

This deletion is clinically associated with increased risk for variable phenotypic manifestations including developmental delay and dysmorphic features.

Benign copy number changes that may be present in this patient are not listed in this report.

Genetic counseling and parental chromosomal microarray analysis are recommended.

18.2.5 2q31.1-q32.2 deletion

EXAMPLE REPORTING OF 2q31.1-q32.2 DELETION

ISCN Result: arr[hg19] 2q31.1q32.2(176,414,548-191,013,728)x1

Interpretation

Chromosomal microarray analysis revealed an abnormal female/male chromosome complement with the loss of 15 Mb of chromosome 2q31.1-2q32.2.

This deletion is clinically associated with increased risk for multiple dysmorphisms, developmental delay, mental impairment and behavioral disturbances.

Benign copy number changes that may be present in this patient are not listed in this report.

Genetic counseling and parental chromosomal microarray analysis are recommended.

18.2.6 2q33.1 deletion

EXAMPLE REPORTING OF 2q33.1 DELETION

ISCN Result: arr[hg19] 2q33.1( )x1

Interpretation

Chromosomal microarray analysis revealed an abnormal female/male chromosome complement with the loss of __ Mb of chromosome 2q33.1.

This deletion is clinically associated with increased risk for significant learning difficulties, growth retardation, dysmorphic features, thin and sparse hair and feeding difficulties.

Benign copy number changes that may be present in this patient are not listed in this report.

Genetic counseling and parental chromosomal microarray analysis are recommended.

18.2.7 2q37 terminal deletion

EXAMPLE REPORTING OF 2q37 TERMINAL DELETION

ISCN Result: arr[hg19] 2q37( )x1

Interpretation

Chromosomal microarray analysis revealed an abnormal female/male chromosome complement with the loss of __ Mb of chromosome 2q37.

This terminal deletion is clinically associated with increased risk for congenital anomalies and mental impairment.

Benign copy number changes that may be present in this patient are not listed in this report.

Genetic counseling and parental chromosomal microarray analysis are recommended.

18.3 Chromosome 3

18.3.1 3p deletion

EXAMPLE REPORTING OF 3p DELETION

ISCN Result: arr[hg19] 3p( )x1

Interpretation

Chromosomal microarray analysis revealed an abnormal female/male chromosome complement with the loss of __ Mb of chromosome 3p region.

This region is known to show copy number variation in normal individuals and is also clinically associated with autism spectrum disorder and developmental delay.

Benign copy number changes that may be present in this patient are not listed in this report.

Genetic counseling and parental chromosomal microarray analysis are recommended.

18.3.2 3q29 deletion

EXAMPLE REPORTING OF 3q29 DELETION

ISCN Result: arr[hg19] 3q29(195,673,381-197,358,134)x1

Interpretation

Chromosomal microarray analysis revealed an abnormal female/male chromosome complement with the loss of 1.6 Mb of chromosome 3q29.

This 1.6 Mb deletion is associated with a clinical phenotype with variable features including mild to moderate mental retardation and mild dysmorphic facial features.

Benign copy number changes that may be present in this patient are not listed in this report.

Genetic counseling and parental chromosomal microarray analysis are recommended.

18.4 Chromosome 4

18.4.1 4p16.3 deletion

EXAMPLE REPORTING OF 4p16.3 DELETION (FIGURE 18.2)

ISCN Result: arr[hg19] 4p16.3(1,180,884-2,521,606)x1

Interpretation

Chromosomal microarray analysis revealed an abnormal female/male chromosome complement with the loss of 1.34 Mb of chromosome 4p16.3.

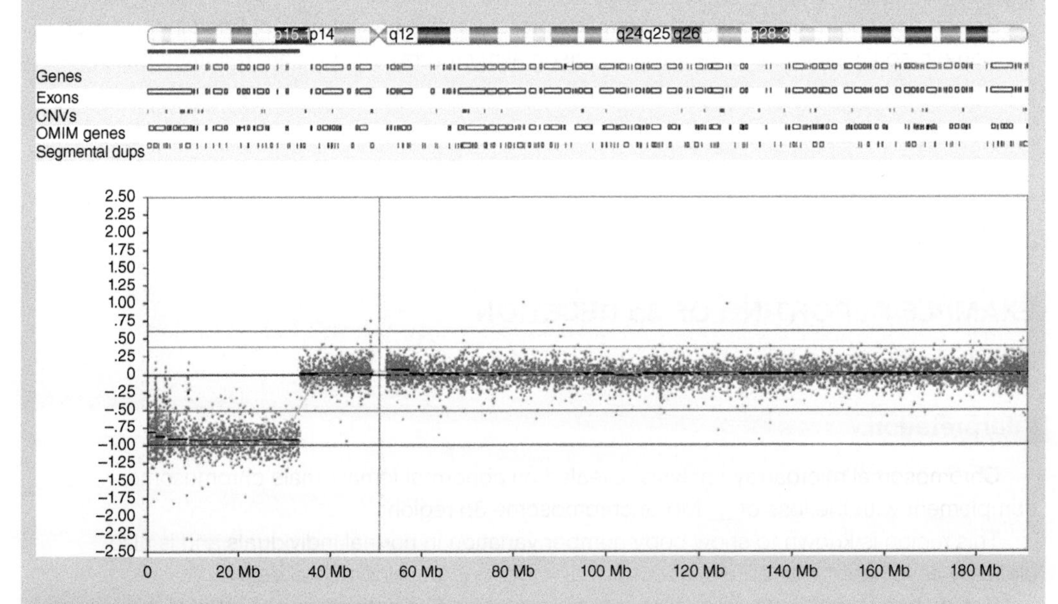

Figure 18.2 Oligoarray showing chromosome 4p16.3-p14 deletion resulting in Wolf–Hirschhorn syndrome. Courtesy of Karine Hovanes PhD, CombiMatrix, Inc.

This deletion is clinically associated with Wolf-Hirschhorn Syndrome, which is characterized by low birthweight and postnatal failure to thrive, hypotonia, microcephaly and developmental delay. Variable clinical phenotypes, depending on the size of the deletion, have been reported.

Benign copy number changes that may be present in this patient are not listed in this report.

Genetic counseling and parental chromosomal microarray analysis are recommended.

18.5 Chromosome 5

18.5.1 5p terminal deletion

EXAMPLE REPORTING OF 5p TERMINAL DELETION

ISCN Result: arr[hg19] 5p( )x1

Interpretation

Chromosomal microarray analysis revealed an abnormal female/male chromosome complement with the loss of __ Mb of chromosome 5p region.

This result likely represents a female/male with a 5p terminal deletion. Terminal deletions of chromosome 5p are clinically associated with cri-du-chat syndrome, but with phenotypic variation depending on the region deleted.

Benign copy number changes that may be present in this patient are not listed in this report.

Genetic counseling and parental chromosomal microarray analysis are recommended.

18.5.2 5q23.2 duplication

EXAMPLE REPORTING OF 5q23.2 DUPLICATION

ISCN Result: arr[hg19] 5q23.2( )x3

Interpretation

Chromosomal microarray analysis revealed an abnormal female/male chromosome complement with the gain of __ Mb of chromosome 5q23.2.

This duplication region includes the LMNB1 gene and is clinically associated with adult-onset, autosomal dominant leukodystrophy (ADLD).

Benign copy number changes that may be present in this patient are not listed in this report.

Genetic counseling and parental chromosomal microarray analysis are recommended.

18.5.3 5q35.3 deletion

EXAMPLE REPORTING OF 5q35.3 DELETION

ISCN Result: arr[hg19] 5q35.3(175,130,402-177,456,545)x1

Interpretation

Chromosomal microarray analysis revealed an abnormal female/male chromosome complement with the loss of 2.3 Mb of chromosome 5q35.3.

This deletion region includes the NSD1 gene and is clinically associated with Sotos syndrome.

Benign copy number changes that may be present in this patient are not listed in this report.

Genetic counseling and parental chromosomal microarray analysis are recommended.

18.6 Chromosome 7

18.6.1 7q11.22 deletion

EXAMPLE REPORTING OF 7q11.22 DELETION

ISCN Result: arr[hg19] 7q11.22( )x1

Interpretation

Chromosomal microarray analysis revealed an abnormal female/male chromosome complement with the loss of 2.3 Mb of chromosome 7q11.22.

This deletion partially overlaps the AUTS2 gene. Mutations and disruption of AUTS2 gene are clinically associated with mental impairment.

Benign copy number changes that may be present in this patient are not listed in this report.

Genetic counseling and parental chromosomal microarray analysis are recommended.

18.6.2 7q11.23 deletion

EXAMPLE REPORTING OF 7q11.23 DELETION (FIGURE 18.3)

ISCN Result: arr[hg19] 7q11.23(72,709,362-74,157,590)x1

Interpretation

Chromosomal microarray analysis revealed an abnormal female/male chromosome complement with the loss of 1.4 Mb of chromosome 7q11.23.

Deletions of this region are clinically associated with Williams syndrome.

Benign copy number changes that may be present in this patient are not listed in this report.

Genetic counseling and parental chromosomal microarray analysis are recommended.

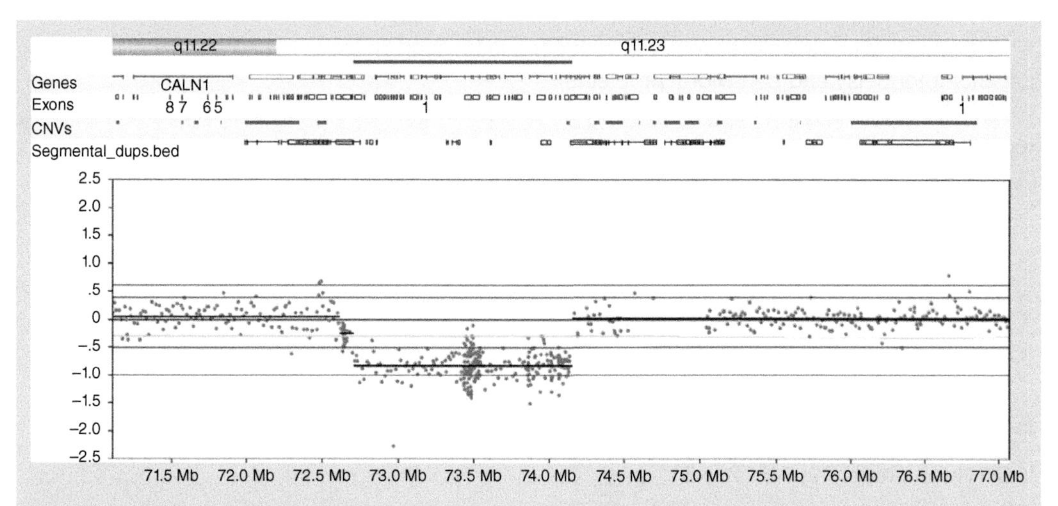

Figure 18.3 Oligoarray showing chromosome 7q11.23 deletion resulting in Williams syndrome. Courtesy of Karine Hovanes PhD, CombiMatrix, Inc.

18.6.3 7q11.23 duplication

EXAMPLE REPORTING OF 7q11.23 DUPLICATION

ISCN Result: arr[hg19] 7q11.23(72,722,437-74,157,591)x3

Interpretation

Chromosomal microarray analysis revealed an abnormal female/male chromosome complement with the gain of 1.4 Mb of chromosome 7q11.23.

This is a duplication of the Williams syndrome deletion region and is associated with variable developmental, psychomotor and language delay in the absence of marked dysmorphic features.

Benign copy number changes that may be present in this patient are not listed in this report.

Genetic counseling and parental chromosomal microarray analysis are recommended.

18.7 Chromosome 8

18.7.1 8p23.1 deletion

EXAMPLE REPORTING OF 8p23.1 DELETION

ISCN Result: arr[hg19] 8p23.1(7,135,980-11,976,572)x1

Interpretation

Chromosomal microarray analysis revealed an abnormal female/male chromosome complement with the loss of 4.8 Mb of chromosome 8p23.1.

This deletion is clinically associated with congenital heart defects, microcephaly, psychomotor delay and behavioral problems.

Benign copy number changes that may be present in this patient are not listed in this report.

Genetic counseling and parental chromosomal microarray analysis are recommended.

18.8 Chromosome 14

18.8.1 14q12 duplication

EXAMPLE REPORTING OF 14q12 DUPLICATION

ISCN Result: arr[hg19] 14q12( )x3

Interpretation

Chromosomal microarray analysis revealed an abnormal female/male chromosome complement with the gain of __ Mb of chromosome 14q12.

This duplication includes the FOXG1 gene and is clinically associated with developmental epilepsy, mental impairment and severe speech delay.

Benign copy number changes that may be present in this patient are not listed in this report.

Genetic counseling and parental chromosomal microarray analysis are recommended.

18.9 Chromosome 15

18.9.1 15q11.2 deletion

EXAMPLE REPORTING OF 15q11.2 DELETION

ISCN Result: arr[hg19] 15q11.2( )x1

Interpretation

Chromosomal microarray analysis revealed an abnormal female/male chromosome complement with the loss of __ Mb of chromosome 15q11.2.

Loss of four highly conserved genes (CYFIP1, NIPA1, NIPA2 and TUBGCP5) in this region (between BP1 and BP2 of Prader–Willi critical region) is associated with a microdeletion syndrome with clinical features including mental impairment, behavioral disturbances, autistic features, obsessive-compulsive behavior, attention deficit hyperactivity disorder (ADHD) and dysmorphic features with incomplete penetrance.

Benign copy number changes that may be present in this patient are not listed in this report.

Genetic counseling and parental chromosomal microarray analysis are recommended.

18.9.2 15q11.2-q13.1 deletion

EXAMPLE REPORTING OF 15q11.2-q13.1 DELETION (FIGURE 18.4)

ISCN Result: arr[hg19] 15q11.2q13.1(23,691,284-28,602,810)x1

Interpretation

Chromosomal microarray analysis revealed an abnormal female/male chromosome complement with the loss of 5.0 Mb of chromosome 15q11.2-q13.1.

Deletions of this region are clinically associated with Prader–Willi and Angelman syndromes. The referring indication for this analysis is seizures, which are most frequently associated with Angelman syndrome. However, if clinical correlation with the full phenotypic picture does not clearly distinguish between the two syndromes, methylation analysis may be considered for further clarification.

Benign copy number changes that may be present in this patient are not listed in this report.

Genetic counseling and parental chromosomal microarray analysis are recommended.

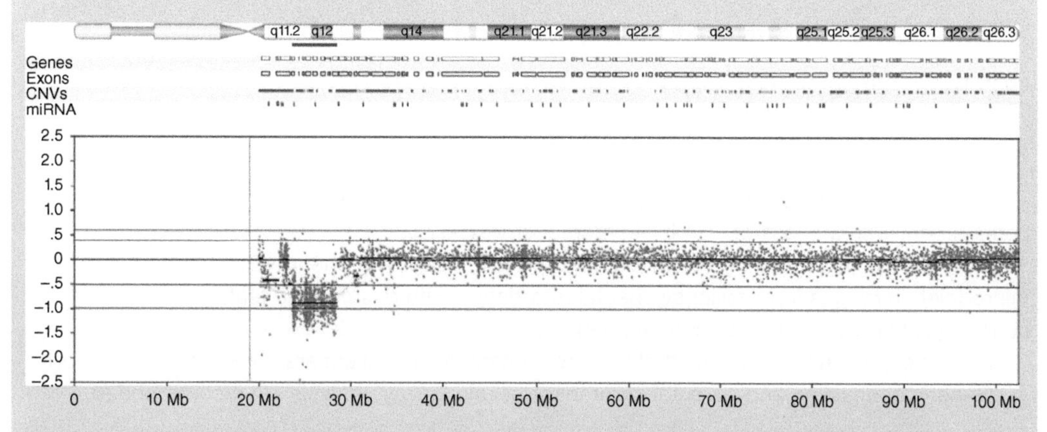

Figure 18.4 Oligoarray showing chromosome 15q11.2q13.1 deletion resulting in Prader–Willi and Angelman syndromes. Courtesy of Karine Hovanes PhD, CombiMatrix, Inc.

18.9.3 15q11.2-q13.1 duplication

EXAMPLE REPORTING OF 15q11.2-q13.1 DUPLICATION

ISCN Result: arr[hg19] 15q11.2q13.1(23,564,854-28,542,402)x3

Interpretation

Chromosomal microarray analysis revealed an abnormal female/male chromosome complement with the gain of 5.0 Mb of chromosome 15q11.2-q13.1.

Maternally derived duplications of this region are clinically associated with developmental delay, behavioral problems and autism.

Benign copy number changes that may be present in this patient are not listed in this report.

Genetic counseling and parental chromosomal microarray analysis are recommended.

18.9.4 15q13.2-q13.3 deletion

EXAMPLE REPORTING OF 15q13.2-q13.3 DELETION

ISCN Result: arr[hg19] 15q13.2q13.3(30,350,488-32,928,400)x1

Interpretation

Chromosomal microarray analysis revealed an abnormal female/male chromosome complement with the loss of 2.5 Mb of chromosome 15q13.2-q13.3.

Chromosome 15q13 microdeletions have highly variable intra- and interfamilial phenotypes including intellectual disability, cardiac malformations, seizures and autism.

Benign copy number changes that may be present in this patient are not listed in this report.

Genetic counseling and parental chromosomal microarray analysis are recommended.

18.9.5 15q13.3 deletion

EXAMPLE REPORTING OF 15q13.3 DELETION

ISCN Result: arr[hg19] 15q13.3( )x1

Interpretation

Chromosomal microarray analysis revealed an abnormal female/male chromosome complement with the loss of __ Mb of chromosome 15q13.3.

This 15q13.3 microdeletion includes the CHRNA7 gene. The phenotypic spectrum of this microdeletion ranges from epilepsy, mental retardation with dysmorphic features, neuropsychiatric anomalies to normal phenotype.

Benign copy number changes that may be present in this patient are not listed in this report.

Genetic counseling and parental chromosomal microarray analysis are recommended.

18.10 Chromosome 16

18.10.1 16p13.11 deletion/duplication

EXAMPLE REPORTING OF 16p13.11 DELETION/DUPLICATION

ISCN Result: arr[hg19] 16p13.11(14,901,892-16,586,261)x1/3

Interpretation

Chromosomal microarray analysis revealed an abnormal female/male chromosome complement with the loss/gain of 1.7 Mb of chromosome 16p13.11.

This deletion/duplication is considered a risk factor for neurocognitive disease and multiple congenital anomalies.

Patients carrying duplications of this region have varied clinical features including behavioral abnormalities, cognitive impairment, congenital heart defects and skeletal manifestations with incomplete penetrance.

Benign copy number changes that may be present in this patient are not listed in this report.

Genetic counseling and parental chromosomal microarray analysis are recommended.

18.10.2 16p12.1 deletion

EXAMPLE REPORTING OF 16p12.1 DELETION

ISCN Result: arr[hg19] 16p12.1(21,894,465-22,512,952)x1

Interpretation

Chromosomal microarray analysis revealed an abnormal female/male chromosome complement with the loss of 0.6 Mb of chromosome 16p12.1.

This 16p12.1 microdeletion is clinically associated with developmental delay with a highly variable phenotype. The deletion is often inherited from a carrier parent with the proband having an additional large copy number variant.

Benign copy number changes that may be present in this patient are not listed in this report.

Genetic counseling and parental chromosomal microarray analysis are recommended.

18.10.3 16p11.2 distal deletion

EXAMPLE REPORTING OF 16p11.2 DISTAL DELETION

ISCN Result: arr[hg19] 16p11.2(28,822,798-29,102,395)x1

Interpretation

Chromosomal microarray analysis revealed an abnormal female/male chromosome complement with the loss of 0.3 Mb of chromosome 16p11.2.

This 16p11.2 distal microdeletion is clinically associated with developmental delay, learning disability, behavioral problems, dysmorphology and obesity, kidney and urinary tract anomalies with incomplete penetrance.

Benign copy number changes that may be present in this patient are not listed in this report.

Genetic counseling and parental chromosomal microarray analysis are recommended.

18.10.4 16p11.2 deletion/duplication

EXAMPLE REPORTING OF 16p11.2 DELETION/DUPLICATION

ISCN Result: arr[hg19] 16p11.2(29,324,809-30,274,073)x1/3

Interpretation

Chromosomal microarray analysis revealed an abnormal female/male chromosome complement with the loss/gain of 0.9 Mb of chromosome 16p11.2.

Copy number changes involving this 16p11.2 region are reported to carry substantial susceptibility to speech and developmental delay, autism, behavior abnormalities and obesity.

Benign copy number changes that may be present in this patient are not listed in this report.

Genetic counseling and parental chromosomal microarray analysis are recommended.

18.11 Chromosome 17

18.11.1 17p13.3 deletion

EXAMPLE REPORTING OF 17p13.3 DELETION

ISCN Result: arr[hg19] 17p13.3(2,563,963-2,810,508)x1

Interpretation

Chromosomal microarray analysis revealed an abnormal female/male chromosome complement with the loss of 3.0 Mb of chromosome 17p13.3.

This deletion includes the LIS1 (PAFAH1B1) gene. Deletions and mutations of the LIS1 gene are clinically associated with lissencephaly and Miller–Dieker syndrome.

Benign copy number changes that may be present in this patient are not listed in this report.

Genetic counseling and parental chromosomal microarray analysis are recommended.

18.11.2 17p13.3 microdeletion

EXAMPLE REPORTING OF 17p13.3 MICRODELETION

ISCN Result: arr[hg19] 17p13.3( )x1

Interpretation

Chromosomal microarray analysis revealed an abnormal female/male chromosome complement with the loss of 258 Kb of chromosome 17p13.3.

This deletion encompasses the critical region of 17p13.3 in patients with developmental delay and dysmorphic features.

Benign copy number changes that may be present in this patient are not listed in this report.

Genetic counseling and parental chromosomal microarray analysis are recommended.

18.11.3 17p12 deletion

EXAMPLE REPORTING OF 17p12 DELETION

ISCN Result: arr[hg19] 17p12(14,092,685-15,490,348)x1

Interpretation

Chromosomal microarray analysis revealed an abnormal female/male chromosome complement with the loss of 1.5 Mb of chromosome 17p12.

Deletion of this region is clinically associated with hereditary neuropathy with a liability to pressure palsies (HNPP), an adult-onset autosomal dominant condition characterized by episodic recurrent pressure palsies at points of nerve entrapment.

Benign copy number changes that may be present in this patient are not listed in this report.

Genetic counseling and parental chromosomal microarray analysis are recommended.

18.11.4 17p12 duplication

EXAMPLE REPORTING OF 17p12 DUPLICATION

ISCN Result: arr[hg19] 17p12(14,092,685-15,490,348)x3

Interpretation

Chromosomal microarray analysis revealed an abnormal female/male chromosome complement with the gain of 1.5 Mb of chromosome 17p12.

This microduplication is clinically associated with Charcot–Marie–Tooth neuropathy type 1 syndrome.

Benign copy number changes that may be present in this patient are not listed in this report.

Genetic counseling and parental chromosomal microarray analysis are recommended.

18.11.5 17p11.2 deletion

EXAMPLE REPORTING OF 17p11.2 DELETION

ISCN Result: arr[hg19] 17p11.2(16,596,953-20,363,910)x1

Interpretation

Chromosomal microarray analysis revealed an abnormal female/male chromosome complement with the loss of 3.7 Mb of chromosome 17p11.2 region.

This deletion is clinically associated with Smith–Magenis syndrome.

Benign copy number changes that may be present in this patient are not listed in this report.

Genetic counseling and parental chromosomal microarray analysis are recommended.

18.11.6 17p11.2 duplication

EXAMPLE REPORTING OF 17p11.2 DUPLICATION

ISCN Result: arr[hg19] 17p11.2(16,596,953-19,808,039)x3

Interpretation

Chromosomal microarray analysis revealed an abnormal female/male chromosome complement with the gain of 3.2 Mb of chromosome 17p11.2 region.

This deletion is clinically associated with Potocki–Lupski syndrome.

Benign copy number changes that may be present in this patient are not listed in this report.

Genetic counseling and parental chromosomal microarray analysis are recommended.

18.11.7 17q12 deletion

EXAMPLE REPORTING OF 17q12 DELETION

ISCN Result: arr[hg19] 17q12(34,803,845-36,263,636)x1

Interpretation

Chromosomal microarray analysis revealed an abnormal female/male chromosome complement with the loss of 1.4 Mb of chromosome 17q12.

This deletion is clinically associated with Mayer–Rokitansky–Kuster–Hauser syndrome, renal disease, diabetes and epilepsy and an associated high risk of autism and schizophrenia.

Benign copy number changes that may be present in this patient are not listed in this report.

Genetic counseling and parental chromosomal microarray analysis are recommended.

18.11.8 17q12 duplication

EXAMPLE REPORTING OF 17q12 DUPLICATION

ISCN Result: arr[hg19] 17q12(34,803,845-36,263,636)x3

Interpretation

Chromosomal microarray analysis revealed an abnormal female/male chromosome complement with the gain of 1.4 Mb of chromosome 17q12 region.

Duplications of this region are clinically associated with intellectual disabilities or cognitive impairment and some of them are also associated with seizures and behavioral problems. The microduplication is mostly inherited from a parent and is also seen in unaffected individuals.

Benign copy number changes that may be present in this patient are not listed in this report.

Genetic counseling and parental chromosomal microarray analysis are recommended.

18.11.9 17q21.31 deletion

EXAMPLE REPORTING OF 17q21.31 DELETION

ISCN Result: arr[hg19] 17q21.31(43,516,471-44,364,715)x1

Interpretation

Chromosomal microarray analysis revealed an abnormal female/male chromosome complement with the loss of 0.8 Mb of chromosome 17q21.31.

This microdeletion is clinically associated with oromotor dyspraxia, hypotonia, learning disability and congenital anomalies.

Benign copy number changes that may be present in this patient are not listed in this report.

Genetic counseling and parental chromosomal microarray analysis are recommended.

18.11.10 17q23.1-q23.2 deletion

EXAMPLE REPORTING OF 17q23.1-q23.2 DELETION

ISCN Result: arr[hg19] 17q23.1q23.2(58,466,304-60,326,070)x1

Interpretation

Chromosomal microarray analysis revealed an abnormal female/male chromosome complement with the loss of 1.8 Mb of chromosome 17q23.1-q23.2.

Deletions of this region are clinically associated with mild to moderate developmental delay, microcephaly, postnatal growth retardation and heart defects, and hand, foot and limb abnormalities.

Benign copy number changes that may be present in this patient are not listed in this report.

Genetic counseling and parental chromosomal microarray analysis are recommended.

18.11.11 17q23.1-q23.2 duplication

EXAMPLE REPORTING OF 17q23.1-q23.2 DUPLICATION

ISCN Result: arr[hg19] 17q23.1q23.2(58,466,304-60,326,070)x3

Interpretation

Chromosomal microarray analysis revealed an abnormal female/male chromosome complement with the gain of 1.9 Mb of chromosome 17q23.1-q23.2.

This duplication is clinically associated with microcephaly, growth retardation, hypotonia, short webbed neck, congenital heart defects, agenesis of the corpus callosum and abnormalities of the iris.

Benign copy number changes that may be present in this patient are not listed in this report.

Genetic counseling and parental chromosomal microarray analysis are recommended.

18.12 Chromosome 19

18.12.1 19p13.3 deletion

EXAMPLE REPORTING OF 19p13.3 DELETION

ISCN Result: arr[hg19] 19p13.3(3,415,096-4,735,468)x1

Interpretation

Chromosomal microarray analysis revealed an abnormal female/male chromosome complement with the loss of 1.8 Mb of chromosome 19p13.3.

Deletions of this region are clinically associated with cutis aplasia, dysmorphic features, multiple congenital anomalies and mild mental retardation.

Benign copy number changes that may be present in this patient are not listed in this report.

Genetic counseling and parental chromosomal microarray analysis are recommended.

18.12.2 19p13.13 deletion/duplication

EXAMPLE REPORTING OF 19p13.13 DELETION/DUPLICATION

ISCN Result: arr[hg19] 19p13.13(12,932,474-13,243,643)x1/3

Interpretation

Chromosomal microarray analysis revealed an abnormal female/male chromosome complement with the loss/gain of 0.83 Mb of chromosome 19p13.13.

This is a described microdeletion/microduplication syndrome of 19p13.13. Patients with 19p13.13 deletions/duplications present with distinctive manifestations including developmental disabilities, overgrowth, macrocephaly and ophthalmological and gastrointestinal findings.

Benign copy number changes that may be present in this patient are not listed in this report.

Genetic counseling and parental chromosomal microarray analysis are recommended.

18.12.3 19p13.13 microdeletion

EXAMPLE REPORTING OF 19p13.13 MICRODELETION

ISCN Result: arr[hg19] 19p13.13(14,240,000-14,640,000)x1

Interpretation

Chromosomal microarray analysis revealed an abnormal female/male chromosome complement with the loss of 0.4 Mb of chromosome 19p13.13.

Deletions of this region are clinically associated with mental impairment, psychomotor and language delay, hearing impairment and dysmorphic features.

Benign copy number changes that may be present in this patient are not listed in this report.

Genetic counseling and parental chromosomal microarray analysis are recommended.

18.13 Chromosome 22

18.13.1 22q11.21 microduplication

EXAMPLE REPORTING OF 22q11.21 MICRODUPLICATION

ISCN Result: arr[hg19] 22q11.21( )x3

Interpretation

Chromosomal microarray analysis revealed an abnormal female/male chromosome complement with the gain of __ Mb of chromosome 22q11.21.

Microduplications of 22q11.21 are clinically associated with variable phenotypes including developmental delay, learning disabilities, seizures and subtle facial dysmorphism. This duplication has also been detected in apparently normal parents of affected probands.

Benign copy number changes that may be present in this patient are not listed in this report.

Genetic counseling and parental chromosomal microarray analysis are recommended.

18.13.2 22q11.21 deletion

EXAMPLE REPORTING OF 22q11.21 DELETION (FIGURE 18.5)

ISCN Result: arr[hg19] 22q11.21(18,765,346-21,628,163)x1

Interpretation

Chromosomal microarray analysis revealed an abnormal female/male chromosome complement with the loss of 28 Mb of chromosome 22q11.21.

Deletions of this region are clinically associated with DiGeorge and velocardiofacial syndromes.

Benign copy number changes that may be present in this patient are not listed in this report.

Genetic counseling and parental chromosomal microarray analysis are recommended.

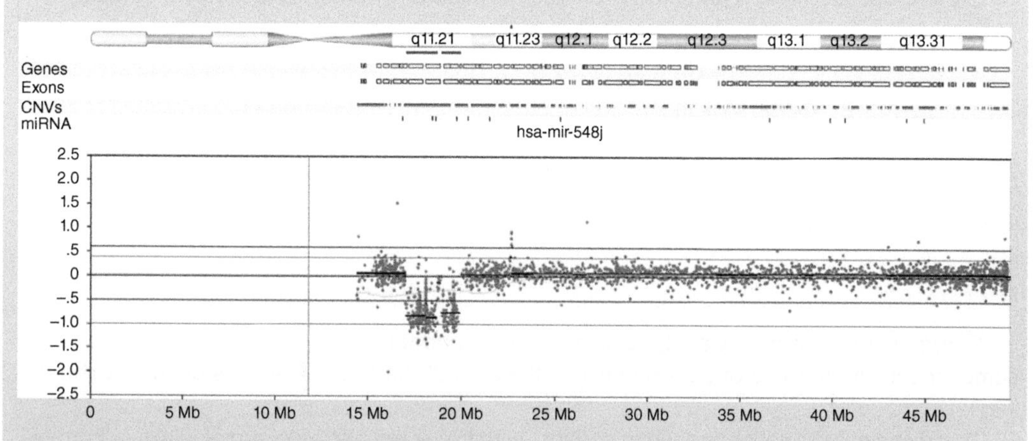

Figure 18.5 Oligoarray showing chromosome 22q11.21 deletion resulting in velocardiofacial syndrome. Courtesy of Karine Hovanes PhD, CombiMatrix, Inc.

18.13.3 22q11.21 microdeletion

EXAMPLE REPORTING OF 22q11.21 MICRODELETION

ISCN Result: arr[hg19] 22q11.21(21,073,593-21,522,853)x1

Interpretation

Chromosomal microarray analysis revealed an abnormal female/male chromosome complement with the loss of 0.5 Mb between the LCR-B and LCR-D regions of chromosome 22q11.21.

This loss represents an atypical 22q11.21 deletion seen in patients with DiGeorge syndrome, which does not encompass the TBX1 gene.

Benign copy number changes that may be present in this patient are not listed in this report.

Genetic counseling and parental chromosomal microarray analysis are recommended.

18.13.4 22q11.21 discontinuous deletion

EXAMPLE REPORTING OF 22q11.21 DISCONTINUOUS DELETION

ISCN Result: arr[hg19] 22q11.21( )x1

Interpretation

Chromosomal microarray analysis revealed an abnormal female/male chromosome complement with a discontinuous loss of 3.0 Mb between the low-copy number repeats A and D (LCR-A and LCR-D) of chromosome 22q11.21.

Deletions of this region are clinically associated with DiGeorge and velocardiofacial syndromes.

Benign copy number changes that may be present in this patient are not listed in this report.

Genetic counseling and parental chromosomal microarray analysis are recommended.

18.13.5 22q11.21 distal deletion

EXAMPLE REPORTING OF 22q11.21 DISTAL DELETION

ISCN Result: arr[hg19] 22q11.21( )x1

Interpretation

Chromosomal microarray analysis revealed an abnormal female/male chromosome complement with the loss of 3.0 Mb between the LCR-3b and LCR-4 regions of chromosome 22q11.21.

This distal microdeletion is seen in patients with DiGeorge and velocardiofacial syndromes. The distal deletion is associated with significant developmental delay, atypical congenital heart defects and mild dysmorphism.

Benign copy number changes that may be present in this patient are not listed in this report.

Genetic counseling and parental chromosomal microarray analysis are recommended.

18.13.6 22q11.1-q11.23 deletion/duplication

EXAMPLE REPORTING OF 22q11.1-q11.23 DELETION/DUPLICATION

ISCN Result: arr[hg19] 22q11.1q11.23( )x1/3

Interpretation

Chromosomal microarray analysis revealed an abnormal female/male chromosome complement with the loss/gain of __ Mb of the terminal region of chromosome 22q11.1-q11.23.

This region contains several segmental duplications which lead to deletions and duplications of varying sizes. Variable phenotypic manifestations have been reported with copy number changes within this region.

Benign copy number changes that may be present in this patient are not listed in this report.

Genetic counseling and parental chromosomal microarray analysis are recommended.

18.13.7 22q11.1-q11.23 duplication

EXAMPLE REPORTING OF 22q11.1-q11.23 DUPLICATION

ISCN Result: arr[hg19] 22q11.1q11.23( )x3

Interpretation

Chromosomal microarray analysis revealed an abnormal female/male chromosome complement with the gain of __ Mb of chromosome 22q11.1-q11.23.

Duplication of the proximal region of 22q is clinically associated with variable phenotypes from apparently normal intelligence to mental impairment, learning disabilities, delayed psychomotor development, growth retardation and/or hypotonia. A few cases of patients with cat-eye syndrome and trisomy of 22q11.2 region have also been reported.

Benign copy number changes that may be present in this patient are not listed in this report.

Genetic counseling and parental chromosomal microarray analysis are recommended.

18.13.8 22q11.21-q11.23 deletion

EXAMPLE REPORTING OF 22q11.21-q11.23 DELETION

ISCN Result: arr[hg19] 22q11.21q11.23(21,863,961-24,067,072)x1

Interpretation

Chromosomal microarray analysis revealed an abnormal female/male chromosome complement with the loss of 2.1 Mb of chromosome 22q11.21-q11.23.

Microdeletions of this region are clinically associated with prenatal and postnatal growth restriction, learning difficulties and/or developmental delays. Some patients also present with skeletal abnormalities and a truncus arteriosus.

Benign copy number changes that may be present in this patient are not listed in this report.

Genetic counseling and parental chromosomal microarray analysis are recommended.

18.13.9 22q11.23 duplication

EXAMPLE REPORTING OF 22q11.23 DUPLICATION

ISCN Result: arr[hg19] 22q11.23(23,682,241-25,018,378)x3

Interpretation

Chromosomal microarray analysis revealed an abnormal female/male chromosome complement with the gain of 1.3 Mb of chromosome 22q11.23.

Microduplications of this region are clinically associated with learning difficulties and developmental delay with variable phenotypes and may be inherited from a phenotypically normal parent.

Benign copy number changes that may be present in this patient are not listed in this report.

Genetic counseling and parental chromosomal microarray analysis are recommended.

18.13.10 22q12.3-q13.33 duplication

EXAMPLE REPORTING OF 22q12.3-q13.33 DUPLICATION

ISCN Result: arr[hg19] 22q12.3q13.33(37,488,339-51,063,624)x3

Interpretation

Chromosomal microarray analysis revealed an abnormal female/male chromosome complement with the gain of 13.5 Mb of chromosome 22q12.3-q13.33.

This duplication is clinically associated with developmental delay, mental impairment, microcephaly and mild facial dysmorphism.

Benign copy number changes that may be present in this patient are not listed in this report.

Genetic counseling and parental chromosomal microarray analysis are recommended.

18.13.11 22q13.33 terminal deletion

EXAMPLE REPORTING OF 22q13.33 TERMINAL DELETION

ISCN Result: arr[hg19] 22q13.33( )x1

Interpretation

Chromosomal microarray analysis revealed an abnormal female/male chromosome complement with the loss of __ Mb of chromosome 22q13.33.

Deletions of this region include the SHANK3 gene and are clinically associated with neurological syndromes.

Benign copy number changes that may be present in this patient are not listed in this report.

Genetic counseling and parental chromosomal microarray analysis are recommended.

18.14 Chromosome X

18.14.1 Xp22.33 deletion/duplication

EXAMPLE REPORTING OF Xp22.33 DELETION/DUPLICATION

ISCN Result: arr[hg19] Xp22.33( )x1/3

Interpretation

Chromosomal microarray analysis revealed an abnormal female/male chromosome complement with the loss/gain of __ Mb of chromosome Xp22.33.

This region is known to show copy number variation in normal individuals. However, duplication of the SHOX gene is clinically associated with variable phenotypes which are not well described. Mutations and deletions of the SHOX gene are associated with SHOX-related haploinsufficiency disorders, which range from Leri–Weill dyschondrosteosis to SHOX-related short stature.

Benign copy number changes that may be present in this patient are not listed in this report.

Genetic counseling and parental chromosomal microarray analysis are recommended.

18.14.2 Xp22.33 deletion

EXAMPLE REPORTING OF Xp22.33 DELETION

ISCN Result: arr[hg19] Xp22.33(6,454,730-8,142,343)x1

Interpretation

Chromosomal microarray analysis revealed an abnormal female/male chromosome complement with the loss of 1.64 Mb of chromosome Xp22.33, including the STS gene.

Deletions of this region are clinically associated with X-linked ichthyosis and variable phenotypes ranging from normal intellect to mental retardation.

Benign copy number changes that may be present in this patient are not listed in this report.

Genetic counseling and parental chromosomal microarray analysis are recommended.

18.14.3 Xp21.1 deletion

EXAMPLE REPORTING OF Xp21.1 DELETION

ISCN Result: arr[hg19] Xp21.1( )x1

Interpretation

Chromosomal microarray analysis revealed an abnormal female/male chromosome complement with the loss of __ Mb of chromosome Xp21.1.

Mutations and deletions of this region are clinically associated with Becker muscular dystrophy, Duchenne muscular dystrophy (DMD) and DMD-associated dilated cardiomyopathy.

Benign copy number changes that may be present in this patient are not listed in this report.

Genetic counseling and parental chromosomal microarray analysis are recommended.

18.14.4 Xq28 duplication

EXAMPLE REPORTING OF Xq28 DUPLICATION

ISCN Result: arr[hg19] Xq28(153,178,367-153,625,736)x3

Interpretation

Chromosomal microarray analysis revealed an abnormal female/male chromosome complement with the gain of 0.4 Mb of chromosome Xq28.

This duplication includes the MECP2 gene and is clinically associated with a neurodevelopmental disorder in males characterized by infantile hypotonia, mental impairment, poor speech development and progressive spasticity.

Benign copy number changes that may be present in this patient are not listed in this report.

Genetic counseling and parental chromosomal microarray analysis are recommended.

18.14.5 Xq28 microduplication

EXAMPLE REPORTING OF Xq28 MICRODUPLICATION

ISCN Result: arr[hg19] Xq28(153,655,695-153,690,983)x3

Interpretation

Chromosomal microarray analysis revealed an abnormal female/male chromosome complement with the gain of 0.035 Mb of chromosome Xq28.

This duplication includes the GDI1 gene and is clinically associated with a phenotypic spectrum ranging from moderate to severe mental impairment, epilepsy and brain anomalies. The copy numbers of the GDI1 gene correlate with the severity of the phenotype.

Benign copy number changes that may be present in this patient are not listed in this report.

Genetic counseling and parental chromosomal microarray analysis are recommended.

Bibliography

Milunsky A, Milunsky J. *Genetic Disorders and the Fetus*, 6th edn. Wiley-Blackwell, Ames, IA, 2010.

Abdelmoity AT, LePichon JB, Nyp SS, Soden SE, Daniel CA, Yu S. 15q11.2 proximal imbalances associated with a diverse array of neuropsychiatric disorders and mild dysmorphic features. J Dev Behav Pediatr 2012; 33(7): 570–576.

Aldred MA, Sanford R, Thomas N, et al. Molecular analysis of 20 patients with 2q37.3 monosomy: definition of minimum deletion intervals for key phenotypes. J Med Genet 2004; 41: 433–439.

Al-Kateb H, Hahn A, Gastier-Foster JM, et al. Molecular characterization of a novel, de novo, cryptic interstitial deletion on 19p13.3 in a child with a cutis aplasia and multiple congenital anomalies. Am J Med Genet Part A 2010; 152A: 3148–3153.

Antonacci F, Kidd JM, Marques-Bonet T, et al. A large and complex structural polymorphism at 16p12.1 underlies microdeletion disease risk. Nature Genet 2010; 42: 745–751.

Auvin S, Holder-Espinasse M, Lamblin MD, et al. Array-CGH detection of a de novo 0.7-Mb deletion in 19p13.13 including CACNA1A associated with mental retardation and epilepsy with infantile spasms. Epilepsia 2009; 50(11): 2501–2503.

Bachmann-Gagescu R, Mefford H, Cowan C, et al. Recurrent 200-kb deletions of 16p11.2 that include the SH2B1 gene are associated with developmental delay and obesity. Genet Med 2010; 12(10): 641–647.

Balasubramanian M, Smith K, Basel-Vanagaite L, et al. Case series: 2q33.1 microdeletion syndrome – further delineation of the phenotype. J Med Genet 2011; 48(5): 290–298.

Ballarati L, Cereda A, Caselli R, et al. Genotype-phenotype correlations in a new case of 8p23.1 deletion and review of the literature. Eur J Med Genet 2011; 54(1): 55–59.

Ballif B, Theisen A, Coppinger J, et al. Expanding the clinical phenotype of the 3q29 microdeletion syndrome and characterization of the reciprocal microduplication. Mol Cytogenet 2008; 1: 8.

Ballif B, Theisen A, Rosenfeld J, et al. dentification of a recurrent microdeletion at 17q23.1q23.2 flanked by segmental duplications associated with heart defects and limb abnormalities. Am J Hum Genet 2010; 86: 454–461.

Barge-Schaapveld D, Maas S, Polstra A, et al. The atypical 16p11.2 deletion: a not so atypical microdeletion syndrome? Am J Med Genet Part A 2011; 155: 1066–1072.

Bartsch O, Gebauer K, Lechno S, et al. Four unrelated patients with Lubs X-linked mental retardation syndrome and different Xq28 duplications. Am J Med Genet Part A 2010; 152A: 305–312.

Bellanné-Chantelot C, Clauin S, Chauveau D, et al. Large genomic rearrangements in the hepatocyte nuclear factor-1β (TCF2) gene are the most frequent cause of maturity-onset diabetes of the young type 5. Diabetes 2005; 54: 3126–3132.

Ben-Shachar S, Ou Z, Shaw C, et al. 22q11.2 distal deletion: a recurrent genomic disorder distinct from DiGeorge syndrome and velocardiofacial syndrome. Am J Hum Genet 2008; 82: 214–221.

Bittel D, Yu S, Newkirk H, et al. Refining the 22q11.2 deletion breakpoints in DiGeorge syndrome by aCGH. Cytogenet Genome Res 2009; 124: 113–120.

Bonaglia M, Marelli S, Novara F, et al. Genotype–phenotype relationship in three cases with overlapping 19p13.12 microdeletions. Eur J Hum Genet 2010; 18: 1302–1309.

Brandt T, Desai K, Grodberg D, et al. Complex autism spectrum disorder in a patient with a 17q12 microduplication. Am J Med Genet Part A 2012; 158A: 1170–1177.

Brunetti-Pierri N, Berg J, Scaglia F, et al. Recurrent reciprocal 1q21.1 deletions and duplications associated with microcephaly or macrocephaly and developmental and behavioral abnormalities. Nature Genet 2008; 40: 1466–1471.

Brunetti-Pierri N, Paciorkowski AR, Ciccone R, et al. Duplications of FOXG1 in 14q12 are associated with developmental epilepsy, mental retardation, and severe speech impairment. Eur J Hum Genet 2011; 19: 102–107.

Bruno D, Anderlid B, Lindstrand A, et al. Further molecular and clinical delineation of co-locating 17p13.3 microdeletions and microduplications that show distinctive phenotypes. J Med Genet 2010; 47: 299.

Burnside RD, Pasion R, Mikhail FM, et al. Microdeletion/microduplication of proximal 15q11.2 between BP1 and BP2: a susceptibility region for neurological dysfunction including developmental and language delay. Hum Genet 2011; 130: 517–528.

Cardoso C, Leventer RJ, Ward HL, et al. Refinement of a 400-kb critical region allows genotypic differentiation between isolated lissencephaly, Miller-Dieker syndrome, and other phenotypes secondary to deletions of 17p13.3. Am J Hum Genet 2003; 72: 918–930.

Cassidy SB, Driscoll DJ. Prader-Willi syndrome. Eur J Hum Genet 2009; 17(1): 3–13.

Cocchella A, Malacarne M, Forzano F, et al. The refinement of the critical region for the 2q31.2q32.3 deletion syndrome indicates candidate genes for mental retardation and speech impairment. Am J Med Genet B Neuropsychiatr Genet 2010; 153B: 1342–1346.

Coppinger J, McDonald-McGinn D, Zackai E, et al. Identification of familial and de novo microduplications of 22q11.21–q11.23 distal to the 22q11.21 microdeletion syndrome region. Hum Mol Genet 2009; 18: 1377–1383.

Cottrell C, Bir N, Varga E, et al. Contactin 4 as an autism susceptibility locus. Autism Res 2011; 4(3): 189–199.

Dhar SU, del Gaudio D, German JR, et al. 22q13.3 deletion syndrome: clinical and molecular analysis using array CGH. Am J Med Genet A 2010; 152A(3): 573–581.

Dolan M, Mendelsohn N, Pierpont M, et al. A novel microdeletion/microduplication syndrome of 19p13.13. Genet Med 2010; 12: 8.

Doornbos M, Sikkema-Raddatz B, Ruijvenkamp C, et al. Nine patients with a microdeletion 15q11.2 between breakpoints 1 and 2 of the Prader–Willi critical region, possibly associated with behavioural disturbances. Eur J Med Genet 2009;52: 108–115.

Elsea SH, Girirajan S. Smith–Magenis syndrome. Eur J Hum Genet 2008; 16: 412–421.

Feenstra I, Koolen DA, van der Pas J, et al. Cryptic duplication of the distal segment of 22q due to a translocation (21;22): three case reports and a review of the literature. Eur J Med Genet 2006; 49(5): 384–395.

Fernandez T, Morgan T, Davis N, et al. Disruption of contactin 4 (CNTN4) results in developmental delay and other features of 3p deletion syndrome. Am J Hum Genet 2004; 74: 1286–1293.

Gajecka M, Mackay KL, Shaffer LG. Monosomy 1p36 deletion syndrome. Am J Med Genet C Semin Med Genet 2007; 145C(4): 346–356.

Garcia-Miñaur S, Fantes J, Murray R, et al. A novel atypical 22q11.2 distal deletion in father and son. J Med Genet 2002; 39: e62.

Hannes F, Sharp A, Mefford H, et al. Recurrent reciprocal deletions and duplications of 16p13.11: the deletion is a risk factor for MR/MCA while the duplication may be a rare benign variant. J Med Genet 2009; 46: 223–232.

Hazan C, Orlow SJ, Schaffer JV. X-linked recessive ichthyosis. Dermatol Online J 2005; 11(4): 12.

Heilstedt HA, Ballif BC, Howard LA, et al. Physical map of 1p36, placement of breakpoints in monosomy 1p36, and clinical characterization of the syndrome. Clin Genet 2003; 64: 310–316.

Houlden H, Reilly MR. Molecular genetics of autosomal-dominant demyelinating Charcot-Marie-Tooth disease. Neuromol Med 2006; 8(1-2): 43–62.

Huang XL, Zou YS, Maher TA, et al. A de novo balanced translocation breakpoint truncating the autism susceptibility candidate 2 (AUTS2) gene in a patient with autism. Am J Med Genet A 2010; 152A(8): 2112–2114.

Ingason A, Kirov G, Giegling I, et al. Maternally derived microduplications at 15q11-q13: implication of imprinted genes in psychotic illness. Am J Psychiatry 2011; 168: 408–417.

Kalscheuer V, FitzPatrick D, Tommerup N, et al. Mutations in autism susceptibility candidate 2 (AUTS2) in patients with mental retardation. Hum Genet 2007; 121: 501–509.

Kim HG, Kishikawa S, Higgins AW, et al. Disruption of neurexin 1 associated with autism spectrum disorder. Am J Hum Genet 2008; 82: 199–207.

Kitsiou-Tzeli S, Sismani C, Ioannides M, et al. Array-CGH analysis and clinical description of 2q37.3 de novo subtelomeric deletion. Eur J Med Genet 2007; 50: 73–78.

Jones KL. *Smith's Recognizable Patterns of Human Malformation*, 6th edn. Saunders, London, 2006.

Kurahashi H, Tsuda E, Kohama R, et al. Another critical region for deletion of 22q11: a study of 100 patients. Am J Med Genet 1997; 72: 180–185.

Maas NM, van Buggenhout G, Hannes F, et al. Genotype–phenotype correlation in 21 patients with Wolf–Hirschhorn syndrome using high resolution array comparative genome hybridisation (CGH). J Med Genet 2008; 45: 71-80.

Masurel-Paulet A, Andrieux J, Callier P, et al. Delineation of 15q13.3 microdeletions. Clin Genet 2010; 78(2): 149–161.

Mefford HC, Cooper GM, Zerr T, et al. A method for rapid, targeted CNV genotyping identifies rare variants associated with neurocognitive disease. Genome Res 2009; 19: 1579–1585.

Mefford HC, Sharp A, Baker C, et al. Recurrent rearrangements of chromosome 1q21.1 and variable pediatric phenotypes. N Engl J Med 2009; 359: 16.

Mefford HC, Clauin S, Sharp A, et al. Recurrent reciprocal genomic rearrangements of 17q12 are associated with renal disease, diabetes, and epilepsy. Am J Hum Genet 2007; 81: 1057–1069.

Meins M, Burfeind P, Motsch S, et al. Partial trisomy of chromosome 22 resulting from an interstitial duplication of 22q11.2 in a child with typical cat eye syndrome. J Med Genet 2003; 40: e62.

Merla G, Brunetti-Pierri N, Micale L, Fusco C. Copy number variants at Williams-Beuren syndrome 7q11.23 region. Hum Genet 2010; 128(1): 3–26.

Mitter D, delle Chiaie B, Lüdecke H, et al. Genotype–phenotype correlation in eight new patients with a deletion encompassing 2q31.1. Am J Med Genet Part A 2010; 152A: 1213–1224.

Moreno-De-Luca D, Mulle J, Kaminsky E, et al. Deletion 17q12 is a recurrent copy number variant that confers high risk of autism and schizophrenia. Am J Hum Genet 2010; 87: 618–630.

Murthy SK, Nygren AOH, El Shakankiry HM, et al. Detection of a novel familial deletion of four genes between BP1 and BP2 of the Prader-Willi/Angelman syndrome critical region by oligo-array CGH in a child with neurological disorder and speech impairment. Cytogenet Genome Res 2007; 116: 135–140.

Nagamani et. al. Clinical spectrum associated with recurrent genomic rearrangements in chromosome 17q12. Eur J Hum Genet 2010; 18: 278–284.

Nagamani SC, Erez A, Bader P, et al. Phenotypic manifestations of copy number variation in chromosome 16p13.11. Eur J Hum Genet 2011; 19: 280–286.

Orellana C, Bernabeu J, Monfort S, et al. Duplication of the Williams–Beuren critical region: case report and further delineation of the phenotypic spectrum. J Med Genet 2008; 45: 187–189.

Ou Z, Berg JS, Yonath H, et al. Microduplications of 22q11.2 are frequently inherited and are associated with variable phenotypes. Genet Med 2008; 10: 267–277.

Padiath QS, Saigoh K, Schiffmann R, et al. Lamin B1 duplications cause autosomal dominant leukodystrophy. Nature Genet 2006; 38(10): 1114–1123.

Paskulin GA, Zen PR, Rosa RF, et al. Report of a child with a complete de novo 17p duplication localized to the terminal region of the long arm of chromosome 17. Am J Med Genet A 2007; 143A(12): 1366–1370.

Peeters H, Vermeesch J, Fryns JP. A cryptic duplication 22q13.31 to qter leads to a distinct phenotype with mental retardation, microcephaly and mild facial dysmorphism. Genet Couns 2008; 19(4): 365–371.

Piard J, Philippe C, Marvier M, et al. Clinical and molecular characterization of a large family with an interstitial 15q11q13 duplication. Am J Med Genet A 2010; 152A(8): 1933–1941.

Ramocki MB, Bartnik M, Szafranski P, et al. Recurrent distal 7q11.23 deletion including HIP1 and YWHAG identified in patients with intellectual disabilities, epilepsy, and neurobehavioral problems. Am J Hum Genet 2010; 87: 857–865.

Rauch A, Zink S, Zweier C, et al. Systematic assessment of atypical deletions reveals genotype–phenotype correlation in 22q11.2. J Med Genet 2005; 42: 871–876.

Roohi J, Montagna C, Tegay D, et al. Disruption of contactin 4 in three subjects with autism spectrum disorder. J Med Genet 2009; 46: 176-182.

Roos L, Brondum Nielsen K, et al. A duplication encompassing the SHOX gene and the downstream evolutionarily conserved sequences. Am J Med Genet Part A 2009; 149A: 2900–2901.

Rosenfeld J, Coppinger J, Bejjani B, et al. Speech delays and behavioral problems are the predominant features in individuals with developmental delays and 16p11.2 microdeletions and microduplications. J Neurodevelop Disord 2010; 2: 26–38.

Ross JL, Scott C Jr, Marttila P, et al. Phenotypes associated with SHOX deficiency. J Clin Endocrinol Metab 2001; 86: 5674–5680.

Shaffer LG, McGowan-Jordan J, Schmid M (eds). *ISCN 2013: An International System for Human Cytogenetic Nomenclature*. Karger Publishers, Unionville, CT, 2013.

Sharkey FH, Morrison N, Murray R, et al. 17q21.31 microdeletion syndrome: further expanding the clinical phenotype. Cytogenet Genome Res 2009; 127(1): 61–66.

Sharp AJ, Mefford HC, Li K, et al. A recurrent 15q13.3 microdeletion syndrome associated with mental retardation and seizures. Nature Genet 2008; 40: 322–328.

Shimojima K, Imai K, Yamamoto T. A de novo 22q11.22q11.23 interchromosomal tandem duplication in a boy with developmental delay, hyperactivity, and epilepsy. Am J Med Genet A 2010; 152A(11): 2820–2826.

Shinawi M, Schaaf CP, Bhatt SS, et al. A small recurrent deletion within 15q13.3 is associated with a range of neurodevelopmental phenotypes. Nature Genet 2009; 41(12): 1269–1271.

Siggberg L, Olsén P, Näntö-Salonen K, Knuutila S. 19p13.3 aberrations are associated with dysmorphic features and deviant psychomotor development. Cytogenet Genome Res 2011; 132: 8–15.

Simon EW, Haas-Givler B, Finucane B. A longitudinal follow-up study of autistic symptoms in children and adults with duplications of 15q11-13. Am J Med Genet B Neuropsychiatr Genet 2010; 153B(2): 463–467.

Soemedi R, Topf A, Wilson I, et al. Phenotype-specific effect of chromosome 1q21.1 rearrangements and GJA5 duplications in 2436 congenital heart disease patients and 6760 controls. Hum Mol Genet 2012; 21: 1513–1520.

Takatsuki S, Nakamura R, Haga Y, et al. Severe pulmonary emphysema in a girl with interstitial deletion of 2q24.2q24.3 including ITGB6. Am J Med Genet A 2010; 152A(4): 1020–1025.

Tatton-Brown K, Douglas J, Coleman K, et al. Genotype-phenotype associations in Sotos syndrome: an analysis of 266 individuals with NSD1 aberrations. Am J Hum Genet 2005; 77: 193–204.

Van Bon B, Mefford H, Menten B, et al. Further delineation of the 15q13 microdeletion and duplication syndromes: a clinical spectrum varying from non-pathogenic to a severe outcome. J Med Genet 2009; 46(8): 511–523.

Van Buggenhout G, Fryns JP. Angelman syndrome. Eur J Hum Genet 2009; 17(11): 1367–1373.

Vandewalle J, van Esch H, Govaerts K, et al. Dosage-dependent severity of the phenotype in patients with mental retardation due to a recurrent copy-number gain at Xq28 mediated by an unusual recombination. Am J Hum Genet 2009; 85(6): 809–822.

Walters RG, Jacquemont S, Valsesia A, et al. A new highly penetrant form of obesity due to deletions on chromosome 16p11.2. Nature 2010; 463(7281): 671–675.

Weiss LA, Shen Y, Korn JM, et al. Association between microdeletion and microduplication at 16p11.2 and sutism. N Engl J Med 2008; 358: 667–675.

Wentzel C, Fernstrom M, Ohrner Y, et al. Clinical variability of the 22q11.2 duplication syndrome. Eur J Med Genet 2008; 51: 501–510.

Wilson H, Wong A, Shaw S, et al. Molecular characterisation of the 22q13 deletion syndrome supports the role of haploinsufficiency of SHANK3/PROSAP2 in the major neurological symptoms. J Med Genet 2003; 40: 575–584.

Yu HE, Hawash K, Picker J, et al. A recurrent 1.71 Mb genomic imbalance at 2q13 increases the risk of developmental delay and dysmorphism. Clin Genet 2012; 81: 257–264.

Yu S, Cox K, Friend K, et al. Familial 22q11.2 duplication: a three-generation family with a 3-Mb duplication and a familial 1.5-Mb duplication. Clin Genet 2008: 73: 160–164.

Zhang X, Snijders A, Segraves R, et al. High-resolution mapping of genotype-phenotype relationships in cri du chat syndrome using array comparative genomic hybridization. Am J Hum Genet 2005; 76: 312–326.

Zhang F, Seeman P, Liu P, et al. Mechanisms for nonrecurrent genomic rearrangements associated with CMT1A or HNPP: rare CNVs as a cause for missing heritability. Am J Hum Genet 2010; 86(6): 892–903.

Microarray database resources

GeneReviews (www.ncbi.nlm.nih.gov)

Database of Genomic Variants (http://projects.tcag.ca/variation/)

Online Mendelian Inheritance in Man (www.ncbi.nlm.nih.gov/omim/)

DECIPHER (http:www.sanger.ac.uk/research/areas/)

dbVar – database of Structural Variation (www.ncbi.nlm.nih.gov/dbvar)

dbGaP– database of Genotypes and Phenotypes (www.ncbi.nlm.nih.gov/gap)

UCSC Genome Bioinformatics (http://genome.ucsc.edu/cgi-bin/hgGateway)

Ensembl (http://uswest.ensembl.org/Homo_sapiens/Gene/Summary)

International Standards for Cytogenomics Arrays Consortium (www.iscaconsortium.org/)

Wellcome Trust Sanger Institute (www.sanger.ac.uk/genetics/CGP/cosmic/)

CHAPTER 19

Integrated reports with cytogenetics, FISH and microarrays

19.1 Reporting of a deletion

EXAMPLE REPORTING OF A PRENATAL DELETION (FIGURE 19.1)

ISCN Result: 46,XY,del(18)(q21.2q23). ish del(18)(q21.2q23)(RP11-958N13-).arr 18q21.2q23 (50,506,593-74,395,578)x1

Interpretation

Chromosome analysis on cultured cells from amniotic fluid revealed an abnormal male chromosome complement with an interstitial deletion within the long arm of chromosome 18, from bands 18q21.2-q23

Chromosomal microarray analysis confirmed the loss of 23.88 Mb of the 18q21.2-q23 region. Fluorescence *in situ* hybridization (FISH) analysis also confirmed the 18q21.2q23 deletion seen in this study.

It is possible that this abnormality is the result of a balanced rearrangement inherited from a parent. Therefore, parental chromosome analyses are suggested to look for an inherited rearrangement.

Chromosome 18q deletions show clinical manifestations including microcephaly, short stature, congenital aural atresia, cleft palate with or without cleft lip, foot deformities, white matter alterations and delayed myelination.

Benign copy number changes that may be present in this patient are not listed in this report.

Clinical correlation and genetic counseling are recommended.

Cytogenetic Abnormalities: Chromosomal, FISH and Microarray-Based Clinical Reporting, First Edition. Susan Mahler Zneimer.

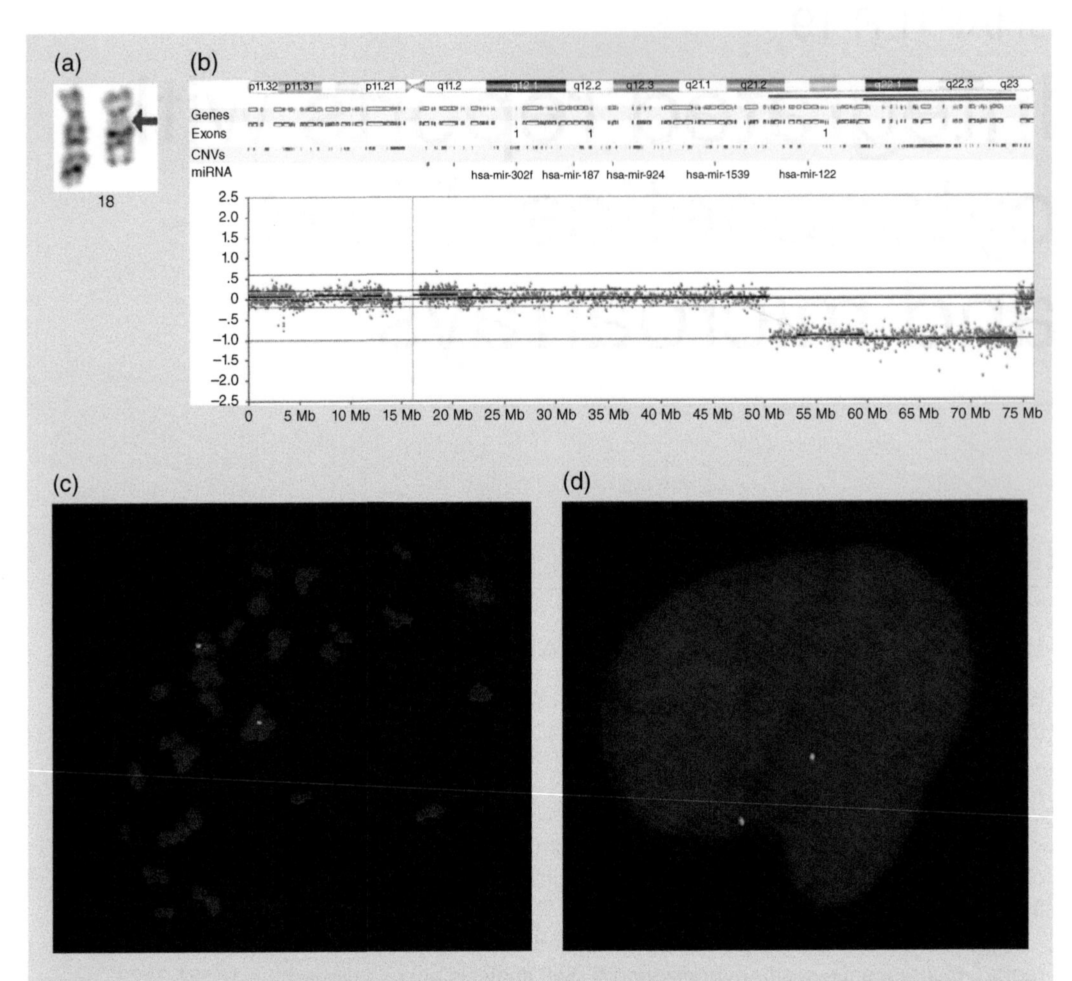

Figure 19.1 Chromosome analysis of a deletion of part of the long arm of chromosome 18. (a) Partial karyotype showing del(18)(q21.2q23). (b) Oligoarray showing an interstitial deletion of 23.88 Mb of the 18q21.2-q23 region. (c) Metaphase FISH results: ish del(18)(q21.2q23)(RP11-958N13-). (d) Interphase FISH results confirming the chromosome 18q deletion. Courtesy of Karine Hovanes PhD, CombiMatrix, Inc.

19.2 Reporting of a supernumerary chromosome

EXAMPLE REPORTING OF PALLISTER–KILLIAN SYNDROME (FIGURE 19.2)

ISCN Result: 47,XY,+i(12)p10)[12/50] ish i(12)(p10)(RP11-642E18+)[2/30].nuc ish 12p11.1(RP11-642E18x4)[9/200] dn.arr 12p13.33q11(0–35,400,000)x2~4

Interpretation

Cytogenetic analysis revealed an abnormal male chromosome complement in 12 of 50 metaphase cells with the gain of a chromosome containing two copies of chromosome 12 short arm material, resulting in mosaicism for four copies of chromosome 12p. These findings are consistent with the diagnosis of Pallister–Killian syndrome.

Chromosomal microarray analysis confirmed mosaicism for the gain of 35.4 Mb of chromosome 12p13.33-q11. Fluorescence *in situ* hybridization (FISH) analysis also confirmed the 12p13.33-q11duplication.

Clinical features of Pallister–Killian syndrome include profound mental retardation, seizures, streaks of hypo- or hyperpigmentation, short neck and facial anomalies.

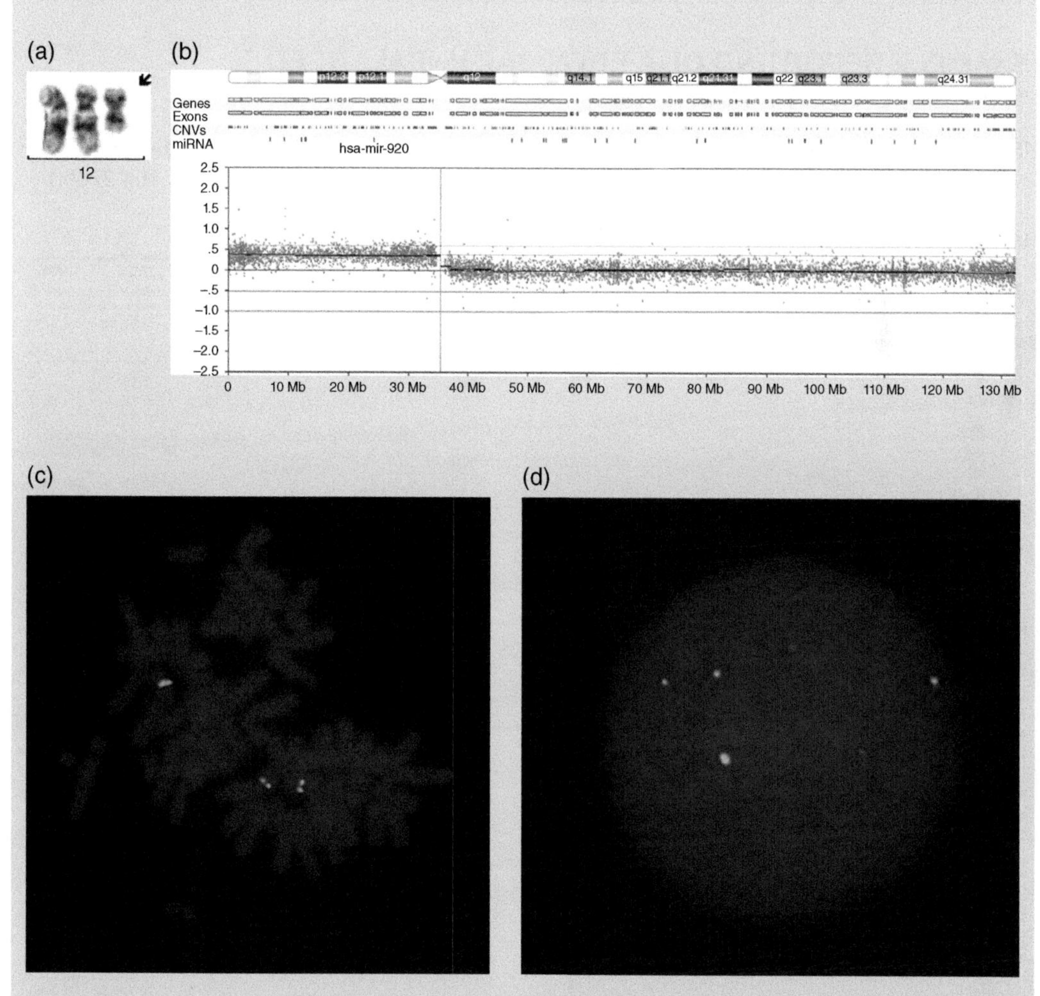

Figure 19.2 Chromosome analysis resulting in tetrasomy 12p – Pallister–Killian syndrome. (a) Partial karyotype showing the gain of chromosome 12p: i(12)(p10). Courtesy of Sarah South PhD, ARUP Laboratories. (b) Oligoarray showing the gain of the short arm of chromosome 12. Courtesy of Karine Hovanes PhD, CombiMatrix, Inc. (c) Metaphase FISH and (d) interphase FISH analysis confirming the gain of two copies of chromosome 12p, where green represents the centromere of chromosome 12 and red represents a custom probe on the long arm of chromosome 12. Courtesy of Karine Hovanes PhD, CombiMatrix, Inc.

Gain of chromosome 12p was not observed in the samples provided from the patient's parents by FISH analysis, suggesting a *de novo* event in the child. However, low-level mosaicism cannot be excluded.

Benign copy number changes that may be present in this patient are not listed in this report.

Genetic counseling is recommended.

19.3 Reporting of an unbalanced translocation – deletion/duplication

EXAMPLE REPORTING OF A PRENATAL UNBALANCED TRANSLOCATION (FIGURE 19.3)

ISCN Result: 46,XY,der(4)t(1;4)(q25;q35).ish der(4)t(1;4)(q25.3;q35.2)(RP11-974M21+;RP11-521G19-)mat.arr 1q25.3q44(184,315,011-249,250,621)x3,4q35.2(190,458,786-191,154,276)x1

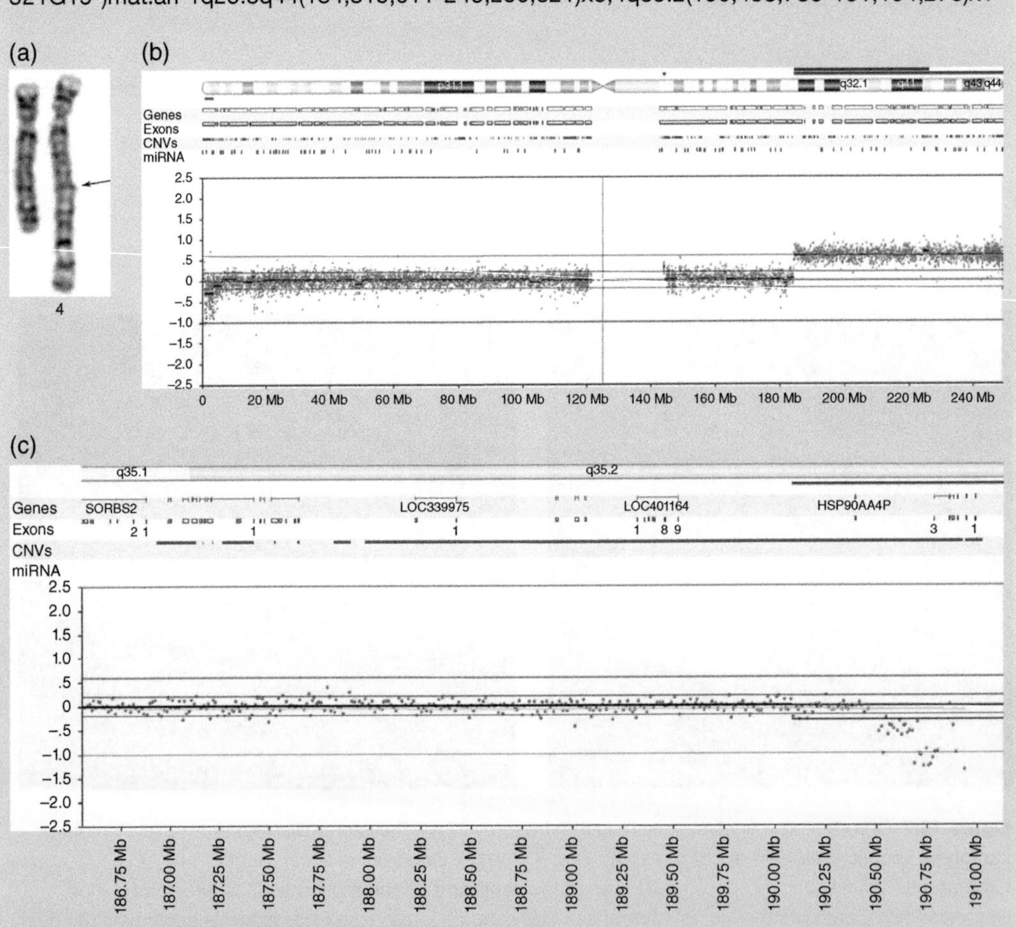

Figure 19.3 Cytogenetic analysis of an unbalanced translocation in a child involving chromosomes 1 and 4 inherited from the mother. (a) Partial karyotype showing the derivative chromosome 4 with added material from chromosome 1: der(4)t(1;4)(q25;q35)mat. (b) Oligoarray showing the gain of chromosome 1q material: arr 1q25.3q44(184,315,011-249,250,621)x3. (c) Oligoarray showing the loss of chromosome 4q material: arr 4q35.2(190,458,786-191,154,276)x1.

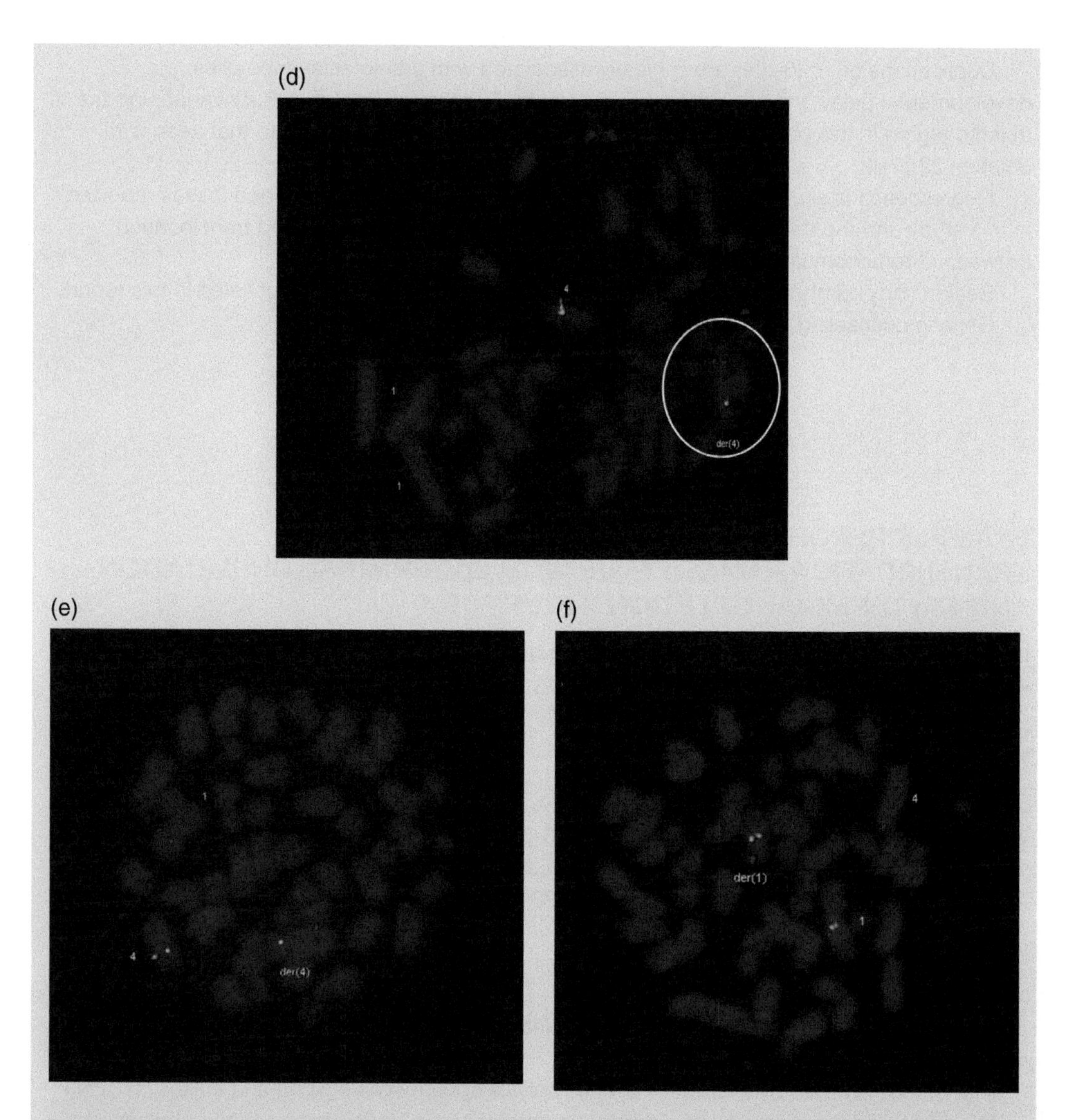

Figure 19.3 (*Continued*) (d) Metaphase FISH showing two normal chromosome 1 s (*red signal*), one normal chromosome 4 (*green signal*), and the derivative chromosome with both red and green signals (*circled*): ish der(4)t(1;4)(q25.3;q35.2)(RP11-974M21+;RP11-521G19-)mat. (e,f) Mother's FISH analysis showing red and green signals on both the translocated chromosomes 1 and 4 resulting in t(1;4)(q25;q35). Courtesy of Karine Hovanes PhD, CombiMatrix, Inc.

Interpretation

Chromosome analysis revealed an abnormal male chromosome complement with a chromosome 1q25.3-qter terminal duplication and chromosome 4q35.2-4qter terminal deletion.

Chromosomal microarray analysis confirmed the gain of 65 Mb of chromosome 1q25.3-qter and loss of ~0.7 Mb of chromosome region 4q34.2-qter. This abnormality corresponds to a der(4)t(1;4)(q25;q35) that is an unbalanced version of a t(1;4) translocation, which was maternally inherited.

Duplications of 1q25-qter are clinically associated with craniofacial anomalies, developmental delay and congenital heart defects. Deletion of 4q34.1-q35.2 overlapping the deleted region in this patient has been reported with a phenotype similar to that seen with deletion 22q11.2.

Fluorescence *in situ* hybridization (FISH) analysis confirmed the gains and losses reported in this study and the derivative chromosome 4 derived from an unbalanced translocation between chromosomes 1q and 4q.

Benign copy number changes that may be present in this patient are not listed in this report.

Genetic counseling are recommended.

EXAMPLE REPORTING OF A PRENATAL UNBALANCED TRANSLOCATION – MALE FETUS WITH 8p TERMINAL DUPLICATION AND 18q TERMINAL DELETION (FIGURE 19.4)

ISCN Result: ish der(18)t(8;18)(p23.1;q23)(RP11-44L22+;RP11-90L3-).arr 8p23.3p23.1(0–7,938,904) x3,8p23.1p22(10,838,833-12,785,036)x3,18q23(73,249,895-76,117,153)x1

Interpretation

Chromosomal microarray analysis showed an abnormal male chromosome complement with the gain of 8.0 Mb on chromosome 8p23.3-p23.1, and the loss of 2.8 Mb on chromosome 18q23. This result is consistent with a male fetus with a chromosomal rearrangement resulting in trisomy 8p23–>pter and monosomy of 18q23–>qter.

The results obtained from this analysis suggest an unbalanced derivative chromosome 18q containing genomic material from the terminal region of chromosome 8p. A duplication less than 2.0 Mb on 8p23.1 was also observed. This duplication may be a familial polymorphism or generated as a result of the unbalanced rearrangement between the chromosomes 8 and 18.

Fluorescence *in situ* hybridization (FISH) analysis confirmed the gains and losses reported by this study and the presence of a derivative chromosome 18 derived from an unbalanced translocation between chromosomes 8p and 18q, which is consistent with the additional material on the 18q reported by the cytogenetic study.

Benign copy number changes that may be present in this patient are not listed in this report.

Correlation with ultrasound findings, genetic counseling and parental testing are recommended.

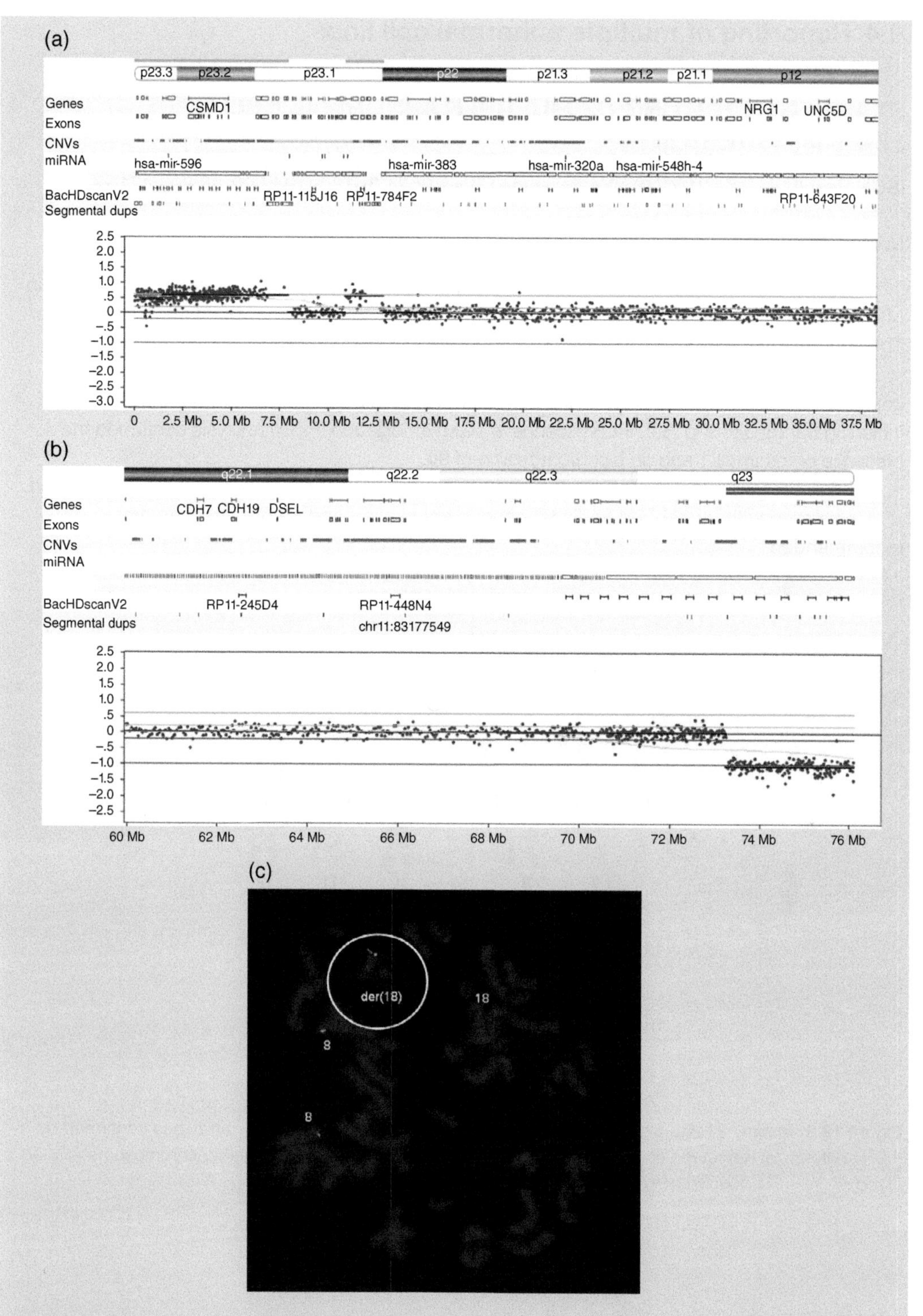

Figure 19.4 Male fetus with oligoarrays showing a chromosome 8p terminal duplication (a) and 18q terminal deletion (b). (c) Metaphase FISH analysis with green signals on chromosome 8p23 and red signals on chromosome 18q12.1 with the derivative chromosome 18 circled. Courtesy of Karine Hovanes PhD, CombiMatrix, Inc.

19.4 Reporting of multiple abnormal cell lines

EXAMPLE REPORTING OF MULTIPLE ABNORMAL CELL LINES (FIGURE 19.5)

ISCN Result: 47,XY,+21[7]/48,XY+21,+mar[13].arr 9p24.3p13.2(194,090-38,805,471)x4

Interpretation

Chromosome analysis revealed an abnormal male chromosome complement with the gain of chromosome 21 in all 20 cells examined, consistent with the diagnosis of Down syndrome. Thirteen of these cells also contained a small supernumerary marker chromosome.

Chromosomal microarray analysis confirmed trisomy 21, plus mosaicism for the gain of two copies of a portion of the short arm of chromosome 9, resulting in an isochromosome of 9p including bands p24.3-p13.2. Fluorescence *in situ* hybridization (FISH) analysis confirmed the presence of trisomy 21 and an isochromosome of 9p.

Benign copy number changes that may be present in this patient are not listed in this report.

Correlation with ultrasound findings, genetic counseling and parental testing are recommended.

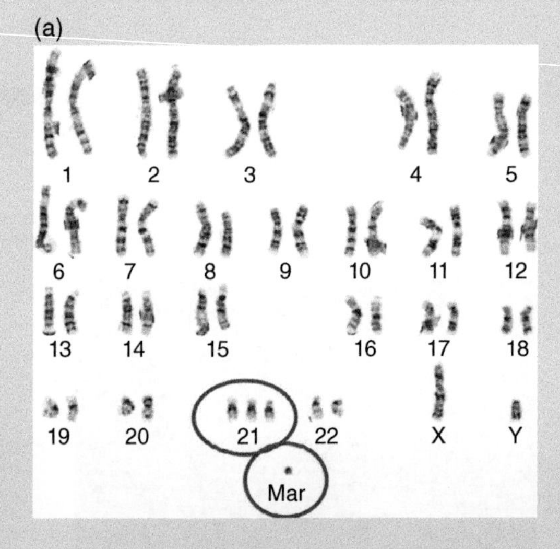

Figure 19.5 Trisomy 21 and a supernumerary marker chromosome identified by array and confirmed by FISH analysis. (a) Karyotype showing trisomy 21 (*circled in blue*) plus a small marker chromosome (*circled in red*).

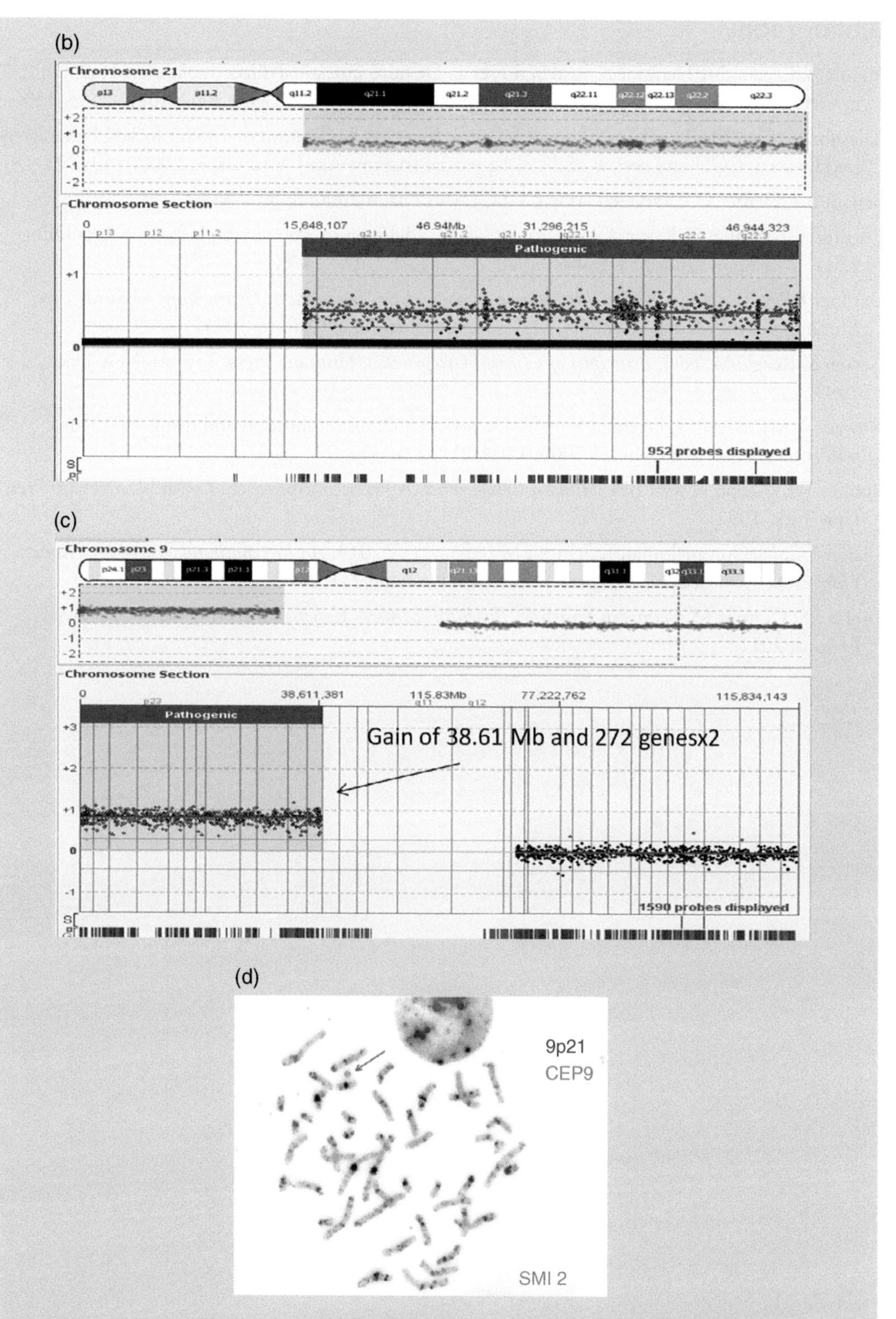

Figure 19.5 (*Continued*) (b) Oligoarray showing the gain one copy of chromosome 21. (c) Oligoarray showing the gain of two copies of chromosome 9p:arr9p24.3p13.2(194,090-38,805,471)x4. (d) Metaphase FISH analysis showing the identification of the extra marker chromosome with two copies of chromosome 9p, resulting in isochromosome 9p. Courtesy of Lauren Jenkins PhD, Kaiser Permanente, Northern California.

Bibliography

Aboura A, Coulomb-L'Herminé A, Audibert F, et al. De novo interstitial direct duplication 1(q23.1q31.1) in a fetus with Pierre Robin sequence and camptodactyly. Am J Med Genet 2002; 108(2): 153–159.

Balasubramanian M, Barber J, Collinson M, et al. Inverted duplication of 1q32.1 to 1q44 characterized by array CGH and review of distal 1q partial trisomy. Am J Med Genet 2008; 149: 793–797.

Bejjani BA, Shaffer LG. Targeted array CGH. J Mol Diagn 2006; 8: 537–539.

Cuturilo G, Menten B, Krstic A, et al. 4q34.1-q35.2 deletion in a boy with phenotype resembling 22q11.2 deletion syndrome. Eur J Pediatr 2011; 170: 1465–1470.

Gardner RJM, Sutherland GR. *Chromosome Abnormalities and Genetic Counselling.* Oxford Monographs on Medical Genetics. Oxford University Press, Oxford, 2003.

Gersen S, Keagle M (eds). *Principles of Clinical Cytogenetics.* Humana Press, Totowa, New Jersey, 1999.

Nowaczyk MJ, Bayani J, Freeman V, et al. De novo 1q32q44 duplication and distal 1q trisomy syndrome. Am J Med Genet A 2003; 120A(2): 229–233.

Rooney DE, Czepulkowski BH. *Human Cytogenetics.* A Practical Approach. Oxford University Press, New York, 1992.

Shaffer LG, McGowan-Jordan J, Schmid M (eds). *ISCN* 2013: *An International System for Human Cytogenetic Nomenclature.* Karger Publishers, Unionville, CT, 2013.

Tobias ES, Connor M, Ferguson Smith M. *Essential Medical Genetics*, 6th edn. Wiley-Blackwell, Oxford, 2011.

Part 2
Acquired Abnormalities in Hematological and Tumor Malignancies

Section 1
Chromosome Analysis

CHAPTER 20

Introduction

Chromosome analysis for hematological and solid tumor malignancies has been performed for approximately 40 years. The first cytogenetic abnormality described was the observance of the Philadelphia chromosome associated with chronic myeloid leukemia, which we now know is derived from a translocation between the long arms of chromosomes 9 and 22, resulting in a BCR/ABL gene rearrangement. The paradigm of a specifically acquired cytogenetic change that is associated with a particular form of cancer is currently an accepted theory in genetics and cancer biology, but it is also accompanied by many complications. To date, there are over 200 subtypes of cancer diseases and over 600 known genes associated with these diseases to date, yet there is no single paradigm which describes the association of genetic changes with cancer even though we know that cancer is genetically based. Generally, all cancer diseases have an underlying genetic defect; however, this concept is too simplistic to explain all the known types of cancer diseases. Cancer is generally a cascade of events for which genetic changes develop in the tissue of disease, followed by either or both environmental cues and/or other genetic changes or a predisposition of genetic factors, giving rise to a precursor disease state that may become active disease, which may lead to metastasis.

Laboratory studies of cancer relies on many biological technologies in addition to the standard pathological morphology assessment of tissue. Critical to the diagnosis and prognosis of disease subtypes is immunohistochemical staining, flow cytometry, chromosomal, FISH, microarray analyses, and molecular genetic analysis of regions within the genome that give rise to the classification of disease.

This part of the book (Part 2) describes the cytogenetic component of cancer disorders. It is divided into three sections, as in the constitutional part of the book, including chromosome abnormalities associated with hematological and solid tumor disorders, FISH analyses as stand-alone tests in disease diagnosis and prognosis, and microarray analysis to identify regions in the genome that are associated with specific cancer disorders. Part 2 will also give examples of clinical reports of abnormalities within each technology used for testing, as well as integrated reporting utilizing two or all three of the methodologies for testing patients with an acquired neoplastic disease.

Cytogenetic Abnormalities: Chromosomal, FISH and Microarray-Based Clinical Reporting, First Edition. Susan Mahler Zneimer.

Additionally, Part 2 gives examples of abnormalities seen in both hematologic disorders and solid tumors, as well as within the subtypes of hematological disorders. Hematological disorders are divided into myeloid disorders and lymphoid disorders. It also gives an overview of the classification of cancer entities based on the World Health Organization (WHO) system. It then presents some details on how specimens are specifically set up with particular culture conditions in cytogenetic laboratories that are important in optimizing the identification of abnormalities by the origin of the diseased cell type.

Part 2 also gives examples of report writing, for hematological and tumor malignancies, including normal results, and ISCN rules for describing chromosome abnormalities with acquired changes seen in cancer versus the rules for describing constitutional disorders. Examples of reporting non-random, recurrent chromosome abnormalities are described next, followed by recurrent abnormalities using FISH analysis, and lastly, disease entities that utilize microarrays for the diagnosis and prognosis of specific cancer disorders.

Report writing for cancer diseases has been quite variable among cytogenetic laboratories, ranging from detailed descriptions of each cytogenetic abnormality observed, whether it be a clonal abnormality or a single cell aberration that might or might not be associated with disease, to a brief summary of the chromosomes involved in either a numerical or structural rearrangement, including only a sentence to relate these changes to a disease entity. Due to the vast range of detail in report writing in cancer disorders, Part 2 will give example reports that laboratories can use as possible "macros" which attempt to include the best of all possible reports into a detailed but somewhat condensed approach to describing chromosomal abnormalities. The correct ISCN nomenclature will also be given as a "result" followed by a somewhat detailed summary of the chromosomal findings in the "interpretation" in a readable form for clinicians, geneticists and subspecialty medical professionals in order to understand the relationship between the cytogenetic findings and a cancer disease.

Although many laboratories include references of clinical significance in each report, the references for related information will be included at the end of each chapter rather than in each example report, solely for reasons of efficiency and space.

Example report formatting for hematological malignancies

Normal results

LABORATORY NAME

Cytogenetics Report

Patient Name: John Doe
Ordering Physician: Dr Smith
Date of Birth: 1/1/1950

Collection Date: 1/1/2014
Report Date: 1/10/2014
Laboratory Number: A14-000021

Specimen Type: Bone marrow
Indication for Study: Pancytopenia

Cells counted: 20
Cells analyzed: 20
Number of cells karyotyped: 2
GTG-Band resolution: 400–450

ISCN Result: 46,XY[20]

Interpretation

Cytogenetic analysis revealed a normal male chromosome complement with no evidence of a clonal abnormality in all cells examined.

Date: January 10, 2014

Susan Zneimer, PhD, FACMG
Cytogenetics Laboratory Director

(Any disclaimer here you may want to provide.)

This test was performed at: Address of Laboratory

Dr Jones, Medical Director
Laboratory License Number: 00001

(Cytogenetic Image)

Abnormal results in all cells examined

LABORATORY NAME

Cytogenetics Report

Patient Name: Jane Doe
Ordering Physician: Dr Smith
Date of Birth: 1/1/1950

Collection Date: 1/1/2014
Report Date: 1/10/2014
Laboratory Number: A14-000021

Specimen Type: Bone marrow
Indication for Study: Chronic myeloid leukemia

Cells counted: 20
Cells analyzed: 20
Number of cells karyotyped: 4
GTG-Band resolution: 400–450

ISCN Result: 46,XY,t(9;22)(q34;q11.2)[20]

Interpretation

Cytogenetic analysis revealed an abnormal male chromosome complement in all cells examined with a translocation between the long arms of chromosomes 9 and 22, resulting in a Philadelphia chromosome. These findings are consistent with a diagnosis of chronic myeloid leukemia.

FISH analysis for the BCR/ABL gene rearrangement is suggested to monitor this abnormal cell line.

Date: January 10, 2014

Susan Zneimer, PhD, FACMG
Cytogenetics Laboratory Director

(Any disclaimer here you may want to provide.)

This test was performed at: Address of Laboratory

Dr Jones, Medical Director
Laboratory License Number: 00001

(Cytogenetic Image)

20.1 Description of World Health Organization classification for hematological malignancies

The WHO classification system for tumors and hematological malignancies is on its fourth edition at the time of this writing. For the first time, it describes genetic abnormalities as an integral part of its disease stratification. This book uses the WHO classification only as a means to clearly describe known cancer entities with cytogenetic abnormalities, and will not give a detailed description of the morphology or clinical description of each cancer subtype. Rather, it is a means to integrate the known, recurrent or non-random cytogenetic abnormalities with the disease entity.

20.2 Description of different tumor types with significant cytogenetic abnormalities

There are a vast number of solid tumors causing cancer, many of which are generally not studied by cytogenetic laboratories due to the difficulty in establishing cell cultures for cell division, the lack of specific genetic changes associated with specific disease entities and the lack of efficiency of testing without a known clinical utility once results are obtained. The culture initiation, maintenance and harvesting of cells can be overwhelming for both financial and efficiency reasons to develop protocols for each type of tumor for most laboratories to attempt tumor cytogenetics. Many laboratories will specialize in specific tumor types, especially if they are involved in research of those particular tumors. Also, the success rate among laboratories for rare tumor culturing is usually poor, making this a difficult process for the laboratory to perform. The tumor tests that give the best results in the laboratory are those that are performed in greater quantity, to allow for improved experience with time, and those that grow and divide well in culture, for example, fast-growing, spontaneously dividing tumors such as pediatric tumors.

Tumors may be studied from paraffin-embedded specimens for FISH and microarray analysis, forgoing chromosome analysis, which allows for analysis of the genome from previously obtained specimens and those specimens that have specific genomic regions which may give prognostic information. This helps determine treatment decisions for patients, such as those with non-small cell lung cancer (NSCLC), breast cancer and gastric cancer, to name a few.

For chromosome analysis, some culture set-up considerations and microscope analyses are discussed that are useful for reporting and interpreting results. This section is not intended to be used for obtaining protocols for the laboratory in performing chromosome analysis. Rather, it sets out a rationale for when and why to set up cultures of specific tumor types in order to best obtain results. Standard operating procedures (SOPs) have been published (see references) and have appeared in online resources; however, there is definitely a lack of information on how to write and interpret results from various tumors that will allow physicians and geneticists to use this information to treat patients.

Only tumors with a fair degree of incidence in the population and with recurrent, non-random abnormalities that are seen in a majority of tumor types are discussed, and only when there is relevant clinical information to be used for patient care. Therefore, most tumor types are not reviewed here.

The vast number of other tumors that are seen with far less frequency or with sporadic cytogenetic changes will not be reviewed here. Therefore, the tumor types discussed here for chromosome analysis include mainly pediatric tumors. Tumors that are studied by FISH and microarray analyses include lung, breast, bladder and other tumors with genetic changes that result in specific treatment protocols.

20.3 Set-up and analysis of specific cultures for optimal results

20.3.1 Specimen type and culture set-up

Cytogenetic studies for the detection of genetic abnormalities in neoplastic disorders are essential in obtaining diagnostic and prognostic information about the disease involved. Cancer is genetic, and identifying the underlying genetic defect that gives rise to disease is an important step in treating and hopefully curing cancer. Cytogenetic studies are divided into the same three categories of testing, as with constitutional studies, based on methodology techniques. These are chromosome, FISH and microarray analyses. All three methodologies are used based on their different advantages in the detection of acquired genetic changes in those cells causing a hematological or tumor disease.

Obtaining the correct specimen type for the laboratory to perform genetics testing is essential for the best possible results. For hematological disorders, bone marrow is the specimen of choice, since peripheral blood often yields only normal results due to the overwhelming number of normal cells versus neoplastic cells in circulation. Peripheral blood may be used for lymphoproliferative disorders since many of the genetically abnormal cells are mature and quiescent and may be present in enough numbers in circulation. For chromosome analysis, cells need to be cultured in order to undergo cell division and be analyzed at the metaphase stage of mitosis. Therefore, obtaining enough abnormal cells that are actively undergoing cell division is critical for proper analysis. For solid tumors, sending tissue with known tumor region included is critical since, as with cytogenetic analysis, a mixture of normal and abnormal cells will be used for analysis, and the abnormal cells need to be present in a large enough proportion to be observed.

Knowing the morphology of cell types is important for analyzing the correct cells involved in disease. For example, acute lymphoblastic leukemia has a distinct "blast look" as it is the large, puffy cells that are more likely to contain the abnormalities, rather than the smaller, discrete lymphocytes that are likely normal cells dividing in culture. Also, myeloid cells that are indicative of chronic myeloid leukemia often have a "polymorphonucleated (PMN) look" where there are three lobes joined as one cell, which is more than likely the type of cell to use for FISH analysis, rather than the discrete round cells that are most likely normal cells in circulation.

Another aspect of cytogenetic analysis that is critical for a successful result is having a clear indication for testing or a presumptive diagnosis of the disease in question for each patient. For chromosome analysis in particular, specimens are set up in culture that will identify specific abnormalities based on the type of culture initiated. These cultures are set up based on whether myeloid or lymphoid cells are involved in disease and whether blast cells are present or not. Lymphoid and myeloid cells, and acute versus chronic disease states, are each set up differently in the laboratory to enhance the possibility of analyzing the acquired abnormal cells present in the specimen. Also, the type of specimen received will greatly influence the number of possible abnormal cells present. Therefore, it is essential for laboratory testing of genetic abnormalities by cytogenetics to have the most detailed information at the time the specimen is obtained to ensure the best possible results.

As an example, for lymphoid hematological diseases, a specific procedure is used to set up acute lymphoblastic leukemias (ALL), since these cells are rapidly spontaneously dividing and will die out within one day of being obtained for genetic testing. Therefore, only cultures set up for

24 hours or less will likely give rise to the abnormal cells for analysis. It is imperative that patients with a presumptive diagnosis of ALL be sent to the genetics laboratory quickly in order to be set up with as many short-term cultures as possible. Usually a same-day culture and an overnight to 24-hour culture are used to analyze these cells. Chronic lymphocytic leukemia, however, is set up with specific T- or B-cell mitogens to stimulate the ordinarily quiescent cells to divide in culture. This may require 3–5-day cultures to obtain the greatest number of abnormal cells to divide and analyze.

For myeloid disorders, no mitogens are added to cultures, since only actively spontaneously dividing cells are representative of the cells underlying disease. Therefore, bone marrow is the specimen of choice, whether the presumptive disease is acute or chronic in nature.

Due to the high probability that a mixture of normal and abnormal cells is present in the marrow, there is often a mosaic pattern of genetic results. That is, a normal cell population and one or more abnormal cell populations are present in any specimen. In order to maximize the analysis of abnormal cells, sending either bone marrow for hematological disorders or a clearly identified tumor specimen is critical for success in identifying the genetic abnormalities that may be present in the cancerous cells.

Therefore, the success of cytogenetic analysis first rests with the clinician obtaining the proper specimen, the amount of specimen, and enough clinical information to give the laboratory the proper tools for initiating the correct cultures as well as analyzing the possible abnormal cells for the most accurate genetic information giving rise to disease. The laboratory must then know which cultures to set up based on the indication for testing and which cells to analyze that best represent the abnormal cells. For an example set-up procedure for each indication type, see Table 20.1.

Cultures initiated for the spectrum of solid tumor types vary depending on the growth pattern of each tumor. One critical factor in cytogenetic analysis of tumors is obtaining the tumor section in a large enough quantity, in contrast to the region of non-tumor tissue. That is, if the segment of tissue received is primarily non-tumor in origin but rather peripheral tissue surrounding the tumor region, then much of what will be analyzed by cytogenetics will not be tumor material. For chromosome analysis, when tissue is set up in the laboratory, growth media are used to optimize cell growth and division. However, many tumors are quiescent, in which case normal tissue will overgrow the tumor cells and the wrong cell type will be analyzed, giving a normal chromosome result and misleading the clinician to believe that the results are normal. Therefore, when appropriate, a comment in

Table 20.1 Typical set-up cultures for chromosome analysis for hematological malignancies

	Acute lymphoblastic leukemia	Acute myeloid leukemia/ myeloproliferative disorders	Chronic lymphoproliferative disorders	Plasma cell disorders
Cultures	Direct unstimulated	24 hour unstimulated	24 hour unstimulated	24 hour unstimulated
	Overnight unstimulated	48 hour unstimulated	72 hour stimulated with T/B-cell mitogens	72 hour stimulated with B-cell mitogens
	24 hour unstimulated		96 hour stimulated with T/B-cell mitogens	96 hour stimulated with B-cell mitogens

cytogenetic reports regarding the possibility that non-tumor cells were analyzed, which may not be representative of the tumor region, is advisable.

Adding mitogens to help initiate cell growth is complicated as most tumors will divide spontaneously at unsynchronized rates, making it difficult to anticipate when to harvest cells during cell division for chromosome analysis. When mitogens are added to culture media, this may in fact be enhancing the growth and division of normal cells, exacerbating the dilemma of knowing what cell type is being analyzed. Since tumors have various cell growth rates, it is often difficult to get dividing cells in enough quantity at any one time to identify enough cells for chromosome analysis. Thus, constant monitoring of cell growth in culture is critical in order to know the appropriate time to harvest cells. Each tumor type must be monitored separately, as each tumor of origin differs in its cell cycle time. No specifics on cell culturing or harvesting are described here, as there are published protocols for tumor cell growth. However, it is important to note that absence of growth or normal results may not reflect the tumor viability or true genomic characteristics, respectively.

20.3.2 Methodologies for testing

The advantage of performing chromosome analysis on neoplastic cells is to be able to see all the possible genetic changes that are microscopically visible. Although there are a multitude of non-random, recurrent genetic abnormalities associated with different hematological and solid tumor diseases, many genetic abnormalities do not recur and are often seen in advanced disease states. Secondary cytogenetic changes are present in cancer cells that are not easily identified by any means other than chromosome analysis. However, one limitation of chromosome studies is that it requires actively dividing cells to see the intact chromosomes in the metaphase stage of cell division in order to identify rearrangements and aneuploidy that may be present in the abnormal cells. This mitotic yield is often difficult to obtain and requires days to analyze the cells before a study can be completed for reporting.

Fluorescence *in situ* hybridization analysis offers an advantage over chromosome studies in that it is performed on interphase cells, i.e. those that do not need to go through cell division for analysis. Since many hematological and tumor specimens are quiescent in culture and do not spontaneously divide, metaphase chromosome analysis may only yield normally dividing cells and not the population of cells resulting in disease. Interphase analysis, although predominantly performed on a mixture of abnormal cancerous cells and normal cells in the tissues being studied, usually has enough neoplastic cells present to identify some abnormally cytogenetic cells causing disease. FISH also offers the advantage of being a semi-quantitative test, in which 200 or more cells may be analyzed per specimen to obtain a percentage of abnormal cells. These initial results can then be used to monitor differences in frequency in follow-up studies. This allows an opportunity to follow trends in a patient's disease for better prognosis and treatment strategies. One limitation of FISH analysis is that it is a targeted genetic analysis, as it only tests specific genetic rearrangements or aneuploidy that is thought to be significant for a specific disease entity, rather than the whole genome approach utilized in chromosome studies.

Currently, various FISH panels are being used to target specific recurring abnormalities seen in different hematological malignancies, including myelodysplasias, acute myeloid and lymphoid leukemias, chronic lymphocytic leukemia, non-Hodgkin lymphomas and multiple myeloma. These panels of tests are easily performed as a single test, making ordering the test simple, and they are fairly rapid in turn-around time compared to chromosome analysis.

Microarray analysis offers a great advantage in genetic testing for hematological disorders and some tumors in that it gives a whole genome analysis of a patient, much like chromosome analysis. It also analyzes the genome at a much more precise level of DNA than does chromosome analysis. Chromosome analysis on neoplastic cells is often at a 300–400 band level, translating to a 1–5 million base pair level of detection of abnormalities. That is, any genetic change less than 1–5 million base pairs will not be detected by chromosome analysis. Microarray analysis can offer detection from hundreds of base pair genetic changes to single nucleotide changes. Similar to cytogenetic analysis, having complex chromosomal changes in a cell may pose a difficulty in interpreting results. As often occurs in complex karyotypic abnormalities, the aberrations present are not recurring changes, but may be unique changes not clearly associated with any disease. Unique changes are much more prevalent in microarray analysis, due to the possible changes that occur within small genetic regions, compared to the 1–5 million base pair detection limit of chromosome analysis. Therefore, interpreting microarray analysis may be difficult at best in specific cases and, at worst, "educated guesses" for the interpretation of genetic changes.

Another disadvantage of microarray analysis is that generally only unbalanced rearrangements are detectable, not balanced rearrangements. This allows for the detection of aneuploidies and small gains and losses of chromosome regions, but does not diagnose the multitude of recurrent abnormalities associated with many diseases, such as the balanced translocations juxtaposing two genes resulting in a gene fusion product. New technologies are emerging that attempt to identify balanced translocations by microarray analyses, but these are not yet ready for clinical use.

Therefore, all three methodologies of cytogenetics offer advantages and disadvantages of testing, giving rise to the question of which tests to order for each disease. It has become common practice to order at least both chromosome and FISH analyses for the majority of hematological diseases and offer microarray analysis only for those hematological diseases that give abnormal results more often in a panel of tests, such as lymphoproliferative disorders. Microarray analysis is also useful for ERBB2 (Her2) testing and other solid tumors that have specific recurring abnormalities with multiple variations of the genes known to cause disease. FISH analysis is very useful for formalin-fixed paraffin-embedded tissue that may be treated and performed with specific probes that are clearly recurring in specific diseases.

For tumor analysis, either fresh tissue or paraffin-embedded tissue may be analyzed, depending on the viability of cell growth and the ability to obtain cells for a complete genomic analysis, versus the need for knowing a single to a few specific genetic changes that can be performed by targeted FISH or microarray DNA probes. Certainly for chromosome analysis, a full 20-cell microscopic analysis is the optimum for chromosome changes. However, many laboratories now will perform FISH analysis on tumors, so that cell culture is not required when specific FISH probes are available for the identification of pertinent genetic changes related to disease diagnosis, prognosis and treatment strategies. Microarray analysis is becoming more widely used for tumors as cell culturing is difficult, and many tumors have specific genetic changes that are available as DNA probe regions that are easily accommodated by a genomic microarray platform. All three methodologies are discussed in their specific sections with tumor types that are typically studied with these approaches.

20.4 Nomenclature rules for normal and simple abnormal results

The ISCN rules for neoplastic syndromes vary from constitutional nomenclature in a few ways. One difference is in the description of abnormal cell lines. As in constitutional studies, when both normal and abnormal cell lines are present, the abnormal cell line(s) is always written first and the normal cell line is always written last.

One important distinction for reporting neoplastic cytogenetic results is the inclusion of the number of cells enclosed in brackets, even when only a single cell line is observed.

EXAMPLE REPORTING OF NORMAL RESULTS

ISCN Result: 46,XY[20]

Interpretation

Cytogenetic analysis revealed a normal male chromosome complement with no evidence of a clonal abnormality in all cells examined.

As in constitutional cell lines, a specific cell line, or clone, is defined as having derived from a single progenitor cell. Once a clone has been identified, it may act as a progenitor cell for other cells, giving rise to subclones. Since the evolution of clonal abnormalities is common in neoplasia and complex karyotypic abnormalities are the norm and not a rarity, the description of different cell lines is more precisely defined in this section.

As in constitutional studies, the definition of a cell line is the presence of two or more cells with the same chromosomal gain or structural abnormality and three cells with the same chromosomal loss. Therefore, a single abnormal cell with gains or a structural change is considered non-clonal, and two cells with the same chromosomal loss are non-clonal. Non-clonal abnormalities should not be reported in the ISCN nomenclature. However, laboratories will often mention non-clonal abnormalities in a report for future reference or to suggest further studies with FISH analysis to confirm a possible low-level, abnormal clonal population. One exception to reporting single cell abnormalities is when that single cell represents a previously identified abnormal cell line; therefore, the current single cell represents persistence of that previously identified clone.

ISCN RULES FOR REPORTING ONE ABNORMAL CELL LINE AND ONE NORMAL CELL LINE

- First write the modal chromosome number of the abnormal cell line, followed by a comma, followed by the sex designation, separated by a comma.
- Then write the clonal abnormalities present, using the same nomenclature as described in the constitutional section.
- Then write the number of cells in the abnormal cell line, enclosed in brackets.
- Next write a slash "/" to separate the next cell line present.
- The normal cell line is always written last in the string of cell lines and is always followed by the number of cells present, enclosed by brackets.

For example, a male with one cell line with a (9;22) translocation as the sole anomaly seen in 16 cells and a normal cell line seen in four cells would be written as: 46,XY,t(9;22)(q34;q11.2)[16]/46,XY[4].

EXAMPLE REPORTING OF ONE ABNORMAL AND ONE NORMAL CELL LINE

ISCN Result: 46,XY,t(9;22)(q34;q11.2)[16]/46,XY[4]

Interpretation

Cytogenetic analysis revealed an abnormal male chromosome complement in 16 of 20 cells analyzed with a translocation between the long arms of chromosomes 9 and 22, resulting in a Philadelphia chromosome, as the sole anomaly. The remaining four cells showed a normal male chromosome complement.

Translocation (9;22) results in the fusion of the ABL1 gene at 9q34 and the BCR gene at 22q11.2, and is consistent with the diagnosis of chronic myeloid leukemia (CML).

FISH analysis for the t(9;22) is suggested to monitor this abnormal cell line.

EXAMPLE REPORTING OF CELLS WITH NON-CLONAL ABNORMALITIES

ISCN Result: 46,XY[18]

Non-clonal abnormalities:
47,XY,+11[1]
46,XY,t(2;3)(q10;q10)[1]

Interpretation

Cytogenetic analysis revealed a normal male chromosome complement in 18 of 20 cells analyzed. There was no significant clonal chromosomal abnormality detected.

However, two cells contained differing, non-repetitive abnormalities as described in the genetic nomenclature. No clinical significance can be ascribed to these singular findings at the present time.

One exception for reporting single cell abnormalities is when specific abnormalities are observed in an initial cytogenetic study and the follow-up study shows the same aberrations. It is then appropriate to include the single abnormal cell in the ISCN nomenclature.

EXAMPLE REPORTING OF A PRESUMED STEMLINE

ISCN Result: Presumed stemline: 46,XY,t(9;22)(q34;q11.2)[1]/46,XY[19]

Intepretation

Cytogenetic analysis revealed an abnormal male chromosome complement in 1 of 20 cells analyzed with a translocation between the long arms of chromosomes 9 and 22, resulting in a Philadelphia chromosome, as the sole anomaly. The remaining 19 cells showed a normal male chromosome complement.

This translocation was observed in the previous specimen of this patient; therefore, this one abnormal cell is consistent with persistence of chronic myeloid leukemia.

FISH analysis for the t(9;22) is suggested to confirm and monitor this abnormal cell line.

20.5 Common report comments for hematological malignancies

20.5.1 Disclaimer comments

- The conventional cytogenetic results reported here may not identify a low percent mosaicism or a small (subtle) structural aberration.

- The standard cytogenetic methodology utilized in this analysis may not detect small rearrangements, low-level mosaicism, microdeletions or other submicroscopic genetic changes.

20.5.2 Cultures with growth factors

- The majority of mitotic cells analyzed were cultured in the presence of myeloid growth factors. Thus, the normal karyotype observed most likely represents the cytogenetic profile of myeloid, not lymphoid, precursors.

20.5.3 Stimulated cultures only

- Multiple cultures were processed but all metaphase cells were analyzed from only B/T-cell stimulated cultures. Normal results from these types of cultures may represent the constitutional karyotype of this patient.

20.5.4 Non-recurrent abnormalities

- These complex karyotypic changes generally support a diagnosis of acute leukemia; however, the abnormalities observed are not characteristic of any leukemia subtype. The complexity of these structural changes is generally associated with a poor prognosis.
- This abnormality has not been reported as a recurrent cytogenetic abnormality in hematological malignancies. However, it does represent an acquired clonal change which can be followed in subsequent studies for cytogenetic persistence, relapse or remission of disease.

20.5.5 Cancelled studies

- Conventional cytogenetic studies on this specimen were cancelled by the physician's request.
- Conventional cytogenetics analysis for this peripheral blood specimen was cancelled because the concurrent bone marrow specimen was sufficient for analysis. Please refer to bone marrow study ( ) for cytogenetic results.

20.5.6 Abnormal results with unclear clinical significance

- The significance of these findings is not clear. Clinical follow-up and correlation with other studies is suggested.
- Although the abnormalities are not diagnostic for a particular type of hematological malignancy, the number and complexity of abnormalities present are generally associated with a poor prognosis.

20.5.7 Follow-up studies

- This patient's disease course can be monitored in future studies by conventional cytogenetics and FISH analyses to detect the ___ clonal abnormality(ies).
- These findings are suspicious, but not definitive, of an abnormal clonal population. Cytogenetic studies at follow-up are suggested to monitor for development of an overt clone.

20.5.8 Incomplete studies

- Only * mitotic cells were available for cytogenetic analysis from this bone marrow specimen, precluding a full 20-cell study.

- This specimen failed to yield sufficient mitotic cells for a complete 20-cell analysis.
- The specimen was __ days old upon receipt; therefore, the sample quality was suboptimal for chromosome analysis and thus yielded an incomplete study.
- Suboptimal chromosome banding and morphology precluded definitive characterization of the structural rearrangement(s) observed.
- A repeat specimen, when clinically appropriate, may provide additional cytogenetic information.
- Only __ metaphase cells were available for analysis. This poor mitotic yield may have prohibited the detection of an abnormal clonal population.

20.5.9 Random loss of chromosomes

- The observed chromosome losses were non-repetitive, consistent with random *in vitro* artifact.

20.5.10 Presence of unidentifiable additional material or unidentifiable marker chromosomes

- Supernumerary marker chromosomes have been reported as either a constitutional cytogenetic abnormality or an acquired cancer cytogenetic aberration.
- In constitutional cases, mosaicism with a normal cell line is found in a majority of cases. A PHA-stimulated peripheral blood lymphocyte cytogenetics study is suggested to rule out the possibility of a constitutional abnormality.

20.5.11 Specific abnormalities not detected

- Specifically, there was no evidence for the * abnormality in any of the * cells examined.
- Specifically, there was no evidence for a Philadelphia chromosome or any other structural aberration commonly associated with a myeloproliferative disorder.

20.5.12 Polyploidy

- The number of polyploid cells detected in this study are significantly above the average control range of 0–3% reported for normal bone marrow specimens.
- One tetraploid cell was observed; this finding is within normal limits for a bone marrow specimen.

20.5.13 Preliminary results

- A preliminary report of these findings was discussed with Dr * on */*/*. The final report concurs with the preliminary report.

20.5.14 Abnormal results correlated with prognosis

- The multiple numerical and complex structural abnormalities and the evolution of at least __ sidelines from the complex stemline are generally associated with a poor prognosis.
- These results are consistent with cytogenetic relapse of disease post chemotherapy.
- These results are consistent with cytogenetic relapse of disease post bone marrow transplantation.

20.5.15 Constitutional trisomy 21 (Down syndrome)

- Patients with constitutional trisomy 21 (Down syndrome) have an increased risk of developing acute leukemia, occurring at a rate 10–20-fold higher than the general population. The peak ages of onset are in the newborn period and again at 3–6 years of age.

20.5.16 Normal karyotypes with abnormal immunophenotyping results

Acute lymphoblastic leukemia

- Cytogenetic analysis revealed no clonal numerical or structural abnormalities. However, the concurrent immunophenotype is consistent with acute lymphoblastic leukemia (ALL). The absence of a cytogenetic anomaly is not inconsistent with the diagnosis of ALL, as up to 33% of morphologically defined cases of adult ALL and 20% of childhood ALL will have a normal karyotype at diagnosis.
- If clinically indicated, consider analysis of the ALL FISH panel to identify the presence of abnormalities commonly observed in leukemic cells that did not divide in culture.

Chronic lymphocytic leukemia

- Cytogenetic analysis revealed no clonal numerical or structural abnormalities. However, the concurrent immunophenotype is consistent with B-cell chronic lymphocytic leukemia (CLL). Since the cells involved in lymphoproliferative disorders have a very low mitotic rate, they may not be reflected in the metaphase population available for analysis.
- If clinically indicated, consider analysis of the CLL FISH panel to identify the presence of abnormalities commonly observed in leukemic cells that did not divide in culture.

Chronic lymphoproliferative disorders

- Cytogenetic analysis revealed no clonal numerical or structural abnormalities. However, the concurrent immunophenotype revealed a monoclonal B-cell population. Since the cells involved in lymphoproliferative disorders have a very low mitotic rate, they may not be reflected in the metaphase population available for analysis.

Myelodysplasia

- Cytogenetic analysis revealed no clonal numerical or structural abnormalities. However, the concurrent immunophenotype is suggestive of a myelodysplastic disorder (MDS). The absence of a cytogenetic anomaly is not inconsistent with the diagnosis of MDS, as approximately 50% of primary and less than 15% of secondary morphologically defined cases of MDS will have a normal karyotype at diagnosis.
- If clinically indicated, consider analysis of the MDS FISH panel to identify the presence of abnormalities commonly observed in leukemic cells that did not divide in culture.

Acute myeloid leukemia

- Cytogenetic analysis revealed no clonal numerical or structural abnormalities. However, the concurrent immunophenotype is consistent with acute myeloid leukemia (AML). The absence of a cytogenetic anomaly is not inconsistent with the diagnosis of AML, as up to 33% of morphologically defined cases of AML will have a normal karyotype at diagnosis.

- If clinically indicated, consider analysis of the AML FISH panel to identify the presence of abnormalities commonly observed in leukemic cells that did not divide in culture.

Myeloproliferative disorders

- Cytogenetic analysis revealed no clonal numerical or structural abnormalities. However, the concurrent flow cytometric analysis is suggestive of a myeloproliferative disorder.
- BCR/ABL gene rearrangement studies by FISH or molecular genetic analysis may be informative, as approximately 5% of CML cases will demonstrate BCR/ABL fusion only by molecular studies.
- Molecular testing for a JAK2 mutation implicated in the pathogenesis of myeloproliferative disorders (polycythemia vera (PV), myelofibrosis (MF), essential thrombocythemia (ET)) may provide additional information. More than 80% of patients with PV, and 40% of patients with ET and MF, have been shown to demonstrate this mutation.

Refractory anemia with excess blasts

- Cytogenetic analysis revealed no clonal numerical or structural abnormalities. However, the concurrent immunophenotype is consistent with refractory anemia with excess blasts (RAEB). The absence of a cytogenetic anomaly does not preclude the diagnosis of RAEB, as 50–70% of morphologically defined cases of RAEB will have a normal karyotype at diagnosis.

Multiple myeloma

- Cytogenetic analysis revealed no clonal numerical or structural abnormalities. However, the concurrent immunophenotype revealed a monoclonal plasma cell population, which is consistent with a diagnosis of multiple myeloma (MM). Since plasma cells usually have a low growth and mitotic rate, standard cytogenetic analysis is often uninformative, showing only normal karyotypic results. In contrast, targeted plasma FISH analysis detects chromosomal abnormalities associated with plasma cell disorders in the majority of cases with active disease.
- If clinically indicated, consider analysis of the MM FISH panel to look for the presence of abnormalities commonly observed in leukemic cells for MM that did not divide in culture.

20.5.17 Presence of a constitutional abnormality

- All cells examined revealed a translocation between chromosomes __ and __. This rearrangement is not associated with any particular hematological disorder, and it may represent a constitutional rearrangement. A peripheral blood sample for routine chromosome analysis is suggested in order to clarify if this abnormality is constitutional or an acquired genetic change associated with disease.
- Small marker chromosomes can be constitutional; therefore, a disease association cannot be confirmed. A peripheral blood sample for routine chromosome analysis is suggested in order to clarify if this abnormality is constitutional or an acquired genetic change associated with disease.

20.5.18 Adding FISH or other testing

- Residual material from this sample is available for FISH studies, if requested.
- If clinically indicated, consider analysis of __ FISH to look for the presence of abnormalities commonly observed in __ cells that did not divide in culture.

20.5.19 Persistence of disease

- These results were observed in the previous analysis from this patient; therefore, these current findings are consistent with persistence of disease.

20.5.20 Composite karyotypes

- There was some variation from cell to cell with the abnormalities present; therefore, a composite of all the clonal abnormalities is described in the nomenclature.

20.5.21 Example reporting after mismatched bone marrow transplant

DONOR SEX CHROMOSOME COMPLEMENT IDENTIFIED IN ALL CELLS

ISCN Result: //46,XX[20] Female (donor) karyotype

Interpretation

Cytogenetic analysis revealed a normal female (donor) sex chromosome complement with no evidence of a clonal abnormality in all cells examined.

CHIMERIC SEX CHROMOSOME COMPLEMENT

ISCN Result: 46,XY[6]//46,XX[14]

Interpretation

Cytogenetic analysis revealed six of 20 cells analyzed with a normal male (host) chromosome complement. The remaining 14 cells showed a normal female (donor) chromosome complement. These findings are consistent with a chimeric engraftment status of this specimen.

20.5.22 Other bone marrow transplant comments

- Specifically, there was no evidence of ___(male/female) cells or ___abnormality seen in a previous, pretransplantation analysis.
- These results are consistent with cytogenetic relapse of disease post bone marrow transplantation.
- Cytogenetic analysis showed no clonal numerical or structural abnormalities. However, the sex chromosome complement of the metaphase cells examined is inconsistent with the patient's gender. These cells are therefore interpreted as being of donor origin consistent with the provided clinical history.

20.5.23 Chromosome instability comments

- Marked chromosome instability, expressed as non-clonal chromosomal abnormalities, was observed. No clinical significance can be ascribed to these singular findings at the present time.
- The clinical significance of the observed chromosome instability is uncertain, as this could be the result of previous cytotoxic exposure or could indicate cellular instability preceding development of an overt clone. Cytogenetic studies at follow-up are suggested to monitor for development of an overt clonal population.
- The related sidelines are evidence of clonal evolution of disease. Secondary mutations related to increased chromosomal instability often accompany histological transformation into higher-grade disease.
- Chromosomal fragile sites are inherited as simple co-dominant traits and are prone to breakage upon replication stress. The molecular basis underlying their fragility and their significance in cancer risk, development and treatment are largely unknown.

Bibliography

Alyea EP, Kim HT, Ho V, et al. Impact of conditioning regimen intensity on outcome of allogeneic hematopoietic cell transplantation for advanced acute myelogenous leukemia and myelodysplastic syndrome. Biol Blood Marrow Transplant 2006; 12(10): 1047–1055.

American College of Medical Genetics. *Guidelines in Clinical Cytogenetics. Section E10 Methods in Fluorescence In situ Hybridization – Interphase/Nuclear Fluorescence In Situ Hybridization.* American College of Medical Genetics, Bethesda, MD. www.acmg.net/StaticContent/SGs/Section_E.html, accessed 28 January 2014.

Atlas of Genetics and Cytogenetics in Oncology and Haematology. http://AtlasGeneticsOncology.org/Anomalies, accessed 28 January 2014.

Bishop MR, Pavletic SZ. Hematopoietic stem cell transplantation. In: Abeloff MD, Armitage JO, Niederhuber JE, Kastan MB, McKena WG (eds) *Clinical Oncology*, 4th edn. Elsevier Churchill Livingstone, Philadelphia, 2008.

Dewald GW, Brockman SR, Paternoster SF, et al. Chromosome anomalies detected by interphase fluorescence in situ hybridization: correlation with significant biological features of B-cell chronic lymphocytic leukaemia. Br J Haematol 2003; 121: 287–295.

Elad S, Zadik Y, Zeevi I, et al. Oral cancer in patients after hematopoietic stem-cell transplantation: long-term follow-up suggests an increased risk for recurrence. Transplantation 2010; 90(11): 1243–1244.

Geigl JB, Obenauf AC, Schwarzbraun T, Speicher MR. Defining 'chromosomal instability'. Trends Genet 2008; 24(2): 64–69.

Gersen S, Keagle M (eds). *Principles of Clinical Cytogenetics*. Humana Press, Totowa, New Jersey, 1999.

Gratwohl A, Baldomero H, Aljurf M, et al. Hematopoietic stem cell transplantation: a global perspective. JAMA 2010; 303(16): 1617–1624.

Halter J, Kodera Y, Urbano-Ipizua A, et al. Severe events in donors after allogeneic hematopoietic stem cell donation. Haematologica 2009; 94(1): 94–101.

Heim S, Mitelman F. *Cancer Cytogenetics*, 2nd edn. Wiley-Liss, New York, 1995.

Jennings CD, Foon KA. Recent advances in flow cytometry: application to the diagnosis of hematologic malignancy. Blood 1997; 90: 2863–2892.

Kussick SJ, Kalnoski M, Braziel RM, Wood BL. Prominent clonal B-cell populations identified by flow cytometry in histologically reactive lymphoid proliferations. Am J Clin Pathol 2004; 121: 464–472.

Li S, Eshleman JR, Borowitz MJ. Lack of surface immunoglobulin light chain expression by flow cytometric immunophenotyping can help diagnose peripheral B-cell lymphoma. Am J Clin Pathol 2002; 118: 229–234.

Rawstron AC, Villamor N, Ritgen M, et al. International standardized approach for flow cytometric residual disease monitoring in chronic lymphocytic leukaemia. Leukemia 2007; 21: 956–964.

Remstein ED, Dogan A, Einerson RR, et al. The incidence and anatomic site specificity of chromosomal translocations in primary extranodal marginal zone B-cell lymphoma of mucosaassociated lymphoid tissue (MALT lymphoma) in North America. Am J Surg Pathol 2006; 30: 1546–1553.

Rooney DE, Czepulkowski BH. *Human Cytogenetics. A Practical Approach*. Oxford University Press, New York, 1992.

Shaffer LG, McGowan-Jordan J, Schmid M (eds). *ISCN 2013: An International System for Human Cytogenetic Nomenclature*. Karger Publishers, Unionville, CT, 2013.

Stetler-Stevenson M, Davis B, Wood B, Braylan R. 2006 Bethesda International Consensus Conference on Flow Cytometry Immunophenotyping of Hematolymphoid Neoplasia. Cytomet B Clin Cytom 2007; 72B: S3.

Swerdlow SH, Campo E, Harris NL, et al. (eds). *WHO Classification of Tumours of Haematopoietic and Lymphoid Tissues*. IARC Press, Lyon, 2008.

Vose JM, Pavletic SZ. Hematopoietic stem cell transplantation. In: Goldman L, Schafer AI (eds) *Cecil Medicine*, 24th edn. Elsevier Saunders, Philadelphia, 2011.

Wood B. 9-color and 10-color flow cytometry in the clinical laboratory. Arch Pathol Lab Med 2006; 130: 680–690.

Wood BL, Arroz M, Barnett D, et al. 2006 Bethesda International Consensus recommendations on the immunophneotypic analysis of hematolymphoid neoplasia by flow cytometry: optimal reagents and reporting for the flow cytometric diagnosis of hematopoietic neoplasia. Cytomet B Clin Cytom 2007; 72B: S14–S22.

Wright EG. Inherited and inducible chromosomal instability: a fragile bridge between genome integrity mechanisms and tumourigenesis. J Pathol 1999; 187(1): 19–27.

CHAPTER 21

Results with constitutional or other non-neoplastic abnormalities

21.1 Possible constitutional abnormalities observed

When studying neoplastic cells, a constitutional abnormality may be detected. In this case, it is important to distinguish whether the abnormality present is truly constitutional or an acquired abnormality associated with disease. To make this distinction, a peripheral blood specimen studied for routine chromosome abnormalities is necessary. However, this is not as critical when the abnormality is a common variant or a well-described balanced rearrangement commonly seen in the population, such as Robertsonian translocations or sex chromosome aneuploidy. Even a balanced translocation that is not a recurrent change described in hematological malignancies, and is seen in all cells examined in a bone marrow or unstimulated peripheral blood specimen, will not distinguish between an acquired and a constitutional chromosomal change. Only when the change is a non-random abnormality associated with disease, such as t(9;22) or t(8;21) that clearly are pathenogenic of a hematological malignancy, should it be interpreted as an acquired change associated with disease.

A listing of the common constitutional chromosome changes is given below.

21.1.1 Common polymorphic variants

The common polymorphic variants that can be excluded as an acquired chromosomal change associated with a hematological malignancy include:

- Inv(9)(p12q13)
- Inv(2)p12q12)
- Inv(Y)(p12q13)
- any other variant listed in the ISCN book as a common variant.

EXAMPLE COMMENT REPORTING A COMMON POLYMORPHIC VARIANT

Cytogenetic analysis revealed no evidence of an acquired clonal abnormality. However, all cells contained a pericentric inversion of chromosome 9 around the centromere. This inversion of chromosome 9 is considered to be a normal variant in the population with no known clinical significance.

Cytogenetic Abnormalities: Chromosomal, FISH and Microarray-Based Clinical Reporting, First Edition. Susan Mahler Zneimer.

21.1.2 Sex chromosome aneuploidies

Sex chromosome aneuploidies are commonly observed among aging females and males in the population. It is common to find these abnormalities as part of a hematological malignancy as well, but they are usually observed as one of many karyotypic abnormalities. Therefore, when observing a sex chromosome aneuploidy as the sole aberration, a comment in the report is useful so as to not mislead the physician in interpreting the abnormality as definitive of an association with disease.

47,XXY/47,XXX

Gain of an X chromosome in males or females as the sole anomaly is most likely constitutional in origin. In a male, 47,XXY is consistent with a clinical diagnosis of Klinefelter syndrome; in a female, 47,XXX is consistent with triple X syndrome. An extra X chromosome, when seen with other chromosome changes, is a common secondary change in hematological malignancies and most likely associated with disease. However, when gain of the X chromosome is seen in all cells, while other chromosome abnormalities exist in only a subset of cells, then the 47,XXY/47,XXX karyotype may be interpreted as constitutional, provisionally, with confirmation by a routine chromosome peripheral blood study.

EXAMPLE REPORTING OF 47,XXY AS THE SOLE ANOMALY

ISCN Result: 47,XXY?c[20]

Interpretation

Cytogenetic analysis revealed an abnormal male chromosome complement with the gain of an X chromosome as the sole anomaly in all cells analyzed. Gain of an X chromosome in a male as the sole anomaly is most likely constitutional in origin and consistent with a clinical diagnosis of Klinefelter syndrome.

Consider sending a peripheral blood specimen for a routine chromosome study in order to clarify if this is a constitutional cytogenetic change or an acquired change associated with a hematological disorder.

EXAMPLE REPORTING OF 47,XXY WITH OTHER CYTOGENETIC CHANGES

ISCN Result: 47,XXY?c,t(9;22)(q34;q11.2)[12]/47,XXY?c[8]

Interpretation

Cytogenetic analysis revealed an abnormal male chromosome complement with a translocation between the long arms of chromosomes 9 and 22, resulting in a Philadelphia (Ph) chromosome in 12 of 20 metaphase cells analyzed. This result is consistent with a diagnosis of chronic myeloid leukemia.

All cells analyzed contained the gain of an X chromosome. Gain of an X chromosome in a male as the sole anomaly is most likely constitutional in origin and consistent with a clinical diagnosis of Klinefelter syndrome.

Since the 47,XXY cells were seen in all cells examined, consider sending a peripheral blood specimen for a routine chromosome study in order to clarify if the gain of the X chromosome is a constitutional cytogenetic change or an acquired change associated with a hematological disorder.

21.1.3 Trisomy 21

EXAMPLE REPORTING OF CONSTITUTIONAL TRISOMY 21 FOR A HEMATOLOGICAL MALIGNANCY

ISCN Result: 47,XY,+21c[20]

Interpretation

Chromosome analysis revealed an abnormal male chromosome complement with an extra copy (trisomy) of chromosome 21 as the sole anomaly in all cells examined. This trisomy 21 is known to be a constitutional abnormality (by history) in this patient and not an acquired cytogenetic change.

Patients with constitutional trisomy 21 (Down syndrome) have an increased risk of developing acute leukemia, with a 10–20-fold higher risk than the general population, and with the peak age of onset in the newborn period and again at 3–6 years of age.

21.2 Age-related abnormalities

21.2.1 Loss of the Y chromosome in males

EXAMPLE REPORTING OF LOSS OF THE Y CHROMOSOME IN A MINORITY OF CELLS

ISCN Result: 45,X,-Y[6]/46,XY[14]

Interpretation

Cytogenetic analysis revealed an abnormal male chromosome complement in 6 of 20 (30%) cells analyzed with the loss of the Y chromosome as the sole anomaly. The remaining 14 cells showed a normal male chromosome complement.

Loss of the Y chromosome as the sole abnormality has been reported in both myeloid and lymphoproliferative disorders. However, loss of the Y chromosome is also common in hematopoietic cells of older males and is generally considered to be an age-related phenomenon with no known clinical significance when seen in less than 75% of cells examined.

EXAMPLE REPORTING OF LOSS OF THE Y CHROMOSOME IN A MAJORITY OF CELLS

ISCN Result: 45,X,-Y[19]/46,XY[1]

Interpretation

Cytogenetic analysis revealed an abnormal male chromosome complement in 19 of 20 (95%) cells analyzed with the loss of the Y chromosome as the sole anomaly. The remaining cell showed a normal male chromosome complement.

Loss of the Y chromosome is common in elderly males and is generally an age-related phenomenon without clinical significance. However, when observed in such a high proportion of cells (>75% of cells), a malignant association cannot be excluded.

Loss of the Y chromosome is most frequently observed in myeloproliferative disorders, myelodysplastic syndromes and acute myeloid leukemia. Correlation with other clinical and histopathological findings is suggested.

21.2.2 Aneuploidy of the X chromosome in females

Aneuploidy of the X chromosome, either loss or gain, in 10% or less in females is thought to be an age-related phenomenon and cannot be assumed to be an acquired change associated with hematological disease. It also cannot be assumed to be a constitutional disorder such as Turner syndrome or triple X syndrome, when the number of cells identified is small. Therefore, a comment is usually written in reports for these instances that explain the various possibilities of X chromosome aneuploidy and their probabilities seen in the hematopoietic system in an adult female.

EXAMPLE REPORTING OF LOSS OF THE X CHROMOSOME IN FEMALES

ISCN Result: 45,X[3]/46,XX[17]

Interpretation

Cytogenetic analysis revealed an abnormal female chromosome complement in 3 of 20 cells analyzed with the loss of the X chromosome as the sole anomaly. The remaining 17 cells showed a normal female chromosome complement.

The clinical significance of the X chromosome aneuploidy is uncertain, but most likely represents an age-related phenomenon or a constitutional mosaicism, rather than a neoplastic-related change. A standard stimulated peripheral blood chromosome study is suggested to determine the constitutional karyotype for this individual.

EXAMPLE REPORTING OF GAIN OF THE X CHROMOSOME IN FEMALES

ISCN Result: 47,XX,+X?c[3]/46,XX[17]

Interpretation

Cytogenetic analysis revealed an abnormal female chromosome complement in 3 of 20 cells analyzed with the gain of the X chromosome as the sole anomaly. The remaining 17 cells showed a normal female chromosome complement.

The clinical significance of the X chromosome aneuploidy is uncertain, but most likely represents an age-related phenomenon or a constitutional mosaicism, rather than a neoplastic-related change.

A standard stimulated peripheral blood chromosome study is suggested to determine the constitutional karyotype for this individual.

21.2.3 Trisomy 15 with loss of the Y chromosome in males

Loss of the Y chromosome as the sole anomaly has been discussed above as an age-related phenomenon. It is also observed in conjunction with trisomy 15. There are reports of both trisomy 15 and loss of Y seen together in hematopoietic cells that have been observed in normal aging males, not as an acquired change related to a hematological malignancy. Therefore, it is important to include a comment in a report explaining the possibility of these two abnormalities.

EXAMPLE REPORTING FOR TRISOMY 15 WITH LOSS OF THE Y CHROMOSOME

ISCN Result: 46,X,-Y,+15[4]/46,XY[16]

Interpretation

Cytogenetic analysis revealed an abnormal male chromosome complement in 4 of 20 cells analyzed with the loss of the Y chromosome and gain of chromosome 15. The remaining 16 cells showed a normal male chromosome complement.

Trisomy 15, as the sole anomaly, is a rare finding in hematological disorders and has been reported in association with myelodysplasias. However, most patients with trisomy 15 are elderly males, suggesting that the presence of trisomy 15, when seen as a minor clone, may reflect an age-related phenomenon, particularly in association with concurrent loss of the Y chromosome.

Cytogenetic studies at follow-up are suggested.

EXAMPLE REPORTING FOR TRISOMY 15 IN A FEMALE

ISCN Result: 47,XX,+15[4]/46,XX[16]

Interpretation

Cytogenetic analysis revealed an abnormal female chromosome complement in 4 of 20 cells analyzed with the gain of chromosome 15 as the sole anomaly. The remaining 16 cells showed a normal female chromosome complement.

Trisomy 15, as the sole anomaly, is a rare finding in hematological disorders and has been reported in association with myelodysplasias. However, it has also been reported as an age-related phenomenon without evidence of a hematological disease.

Cytogenetic studies at follow-up are suggested.

21.3 Non-clonal aberrations

Non-clonal aberrations refer to a single cell gain, structural abnormality or two cells with the loss of the same chromosome and occur frequently in the hematopoietic system, especially in bone marrow. Since bone marrow is constantly regenerating, these cells may undergo non-disjunction, yielding single cell abnormalities, or they may be the result of an *in vitro* artifact. Nevertheless, the possibility also exists that a single abnormal cell represents low-level mosaicism for an abnormal cell line. Since the mechanistic origin of the non-clonal abnormality cannot be established with a chromosome study, all three possibilities exist.

There is a variety of ways in which these non-clonal abnormalities may be reported (or not reported). Some laboratories will report each and every non-clonal change seen, either in the ISCN result or in the interpretation of the report, or both. Other labs never report non-clonals but some labs will report non-clonals if they are significant, recurrent changes in hematological or tumor diseases.

There is no standard for writing these types of abnormalities in reports, and there are no guidelines for discussing them in reports. Laboratories that describe non-clonals do so because when patients are repeatedly studied, some of the abnormalities detected reappear as clonal changes. Other labs can see the chromosome results of previous studies easily and may not need to report non-clonals, but wait until they become clonal since they can look for previous non-clonals in subsequent studies.

As there is no right or wrong way to write these cells, much depends on the method of coordinating previous studies with current analyses to ensure that previous non-clonals are consciously looked for in future studies in order not to miss a possible cell line. Certainly, a single abnormal cell of clinical significance is a good start to reporting non-clonal abnormalities.

Laboratories also vary on how they report single cell losses. Generally, these abnormalities would be considered random loss and only if two abnormal cells signify a possible abnormal cell line will it be mentioned in a report. However, the definition of a true abnormal cell line requires three cells with the loss of the same chromosome. Still, some labs do report single cell losses, such as monosomy 7. There is a risk, however, of overstating these abnormalities if the clinician fails to realize that these cells do not represent an abnormal cell line and may, in fact, be an innocuous finding.

EXAMPLE REPORTING OF NORMAL RESULTS WITH EXTRA CELLS ANALYZED FOR A SINGLE CELL ABNORMALITY

ISCN Result: 46,XY[20]

Interpretation

Cytogenetic analysis revealed a normal male chromosome complement with no evidence of a clonal abnormality in all cells examined. However, a single cell containing ___ was present.

An additional 20 cells were analyzed looking for evidence of a small abnormal clonal population, but no additional abnormal cells were detected.

Although the single abnormal cell does not define an abnormal cell line, it may represent a low-level abnormal clonal population. FISH analysis is suggested to help clarify the presence of an abnormal clonal cell line.

EXAMPLE REPORTING OF NORMAL RESULTS WITH A SINGLE CELL WITH A SUSPICIOUS AND SIGNIFICANT ABNORMALITY [inv(16)]

ISCN Result: 46,XY[20]

Interpretation

Cytogenetic analysis revealed a normal male chromosome complement with no evidence of a clonal abnormality in all cells examined.

However, one cell was suspicious for an inversion of chromosome 16. Without the presence of other abnormal cells or without positive FISH results, this abnormality cannot be confirmed. Inversion 16 is the same abnormality seen in a previous cytogenetic study of this patient (Accession #_, dated _). Therefore, this single cell may represent the previous stemline abnormality.

Clinical-pathologic correlation is suggested.

EXAMPLE REPORTING OF NORMAL RESULTS WITH TWO CELLS WITH A SUSPICIOUS AND SIGNIFICANT ABNORMALITY [MONOSOMY 7]

ISCN Result: 46,XY[20]

Interpretation

Cytogenetic analysis revealed a normal male chromosome complement with no evidence of a clonal abnormality in all cells examined.

However, two cells revealed the loss of chromosome 7. Twenty additional cells were scored for this abnormality; however, further cells with monosomy 7 were not seen. Though the two cells with monosomy 7 fail to qualify as an abnormal clone, they may represent a low-frequency clonal population. FISH analysis is suggested to help clarify the presence of an abnormal clonal cell line.

Clinical-pathologic correlation is suggested.

21.3.1 One cell abnormality comment

- One metaphase cell showed (abnormality description/trisomy/monosomy, translocation/ deletion/ duplication/add, etc.). Twenty additional cells were scored for (abnormality), but no other cells with (abbreviated description of abnormality: del(5q), trisomy 8, t(8;14), etc.) were detected. This single abnormal cell may represent a low-frequency abnormal clone. (Abnormality: Del(5q),..) is associated with ___ disorders.
- Follow-up FISH studies on interphase cells are recommended to detect a low-frequency emerging/ residual abnormal cell line.

21.3.2 Comments for the presence of non-clonal abnormalities

- _____was observed as a non-clonal aberration in this study.
- The multiple complex aberrations, observed as both clonal and non-clonal abnormalities, are indicative of highly disturbed hematopoiesis.
- In addition to the multiple clonal abnormalities, numerous non-clonal aberrations were observed, indicating chromosomal instability and the potential transition for further clonal evolution.
- __ cells contained differing, non-repetitive abnormalities as described in the ISCN nomenclature. No clinical significance can be ascribed to these singular findings at the present time.
- One cell contained a __; however, this finding was non-repetitive in the __ cell analysis. No clinical significance can be ascribed to this singular finding at the present time.
- One cell contained a __; however, this finding was non-repetitive in the __ cell analysis. Although this is a singular finding in the current study, this aberration is noteworthy due to the association

of __ with __. Cytogenetic studies at follow-up are suggested to monitor for development of an overt clonal population.

- This single abnormal cell is interpreted as an *in vitro* artifact but may represent an emerging abnormal clonal population. The presence of this cell has been noted to correlate with future studies of this patient.

21.4 No growth and poor growth

Cases with no growth in culture yielding no analyzable metaphases, or very poor growth leading to few metaphases to analyze, are seen in a portion of specimens of any cell type. An acceptable range of no growth cases should be established for each laboratory for each specimen type, and is usually part of a quality assurance (QA) or quality improvement (QI) program for the laboratory. There is no standard establishing a growth requirement for specimen types, but an acceptable range may be established based on other laboratories' findings or empirical data from one's own laboratory through time. Generally, for bone marrow specimens, where enough marrow is received in the laboratory within 48 hours of collection, <5% of cases may yield no growth. Peripheral blood specimens usually have a higher no-growth rate based on the indication for testing, the cell count and/or blast cell count, and whether cells require a mitogen stimulant or are cultured without a mitogen stimulant. Solid tumor growth is very dependent on culturing with the correct medium, its cell type, whether the tumor that is collected prevails or the specimen is viable when received in the laboratory.

If the origin of the tumor or hematological malignancy in question is mainly from quiescent cells, then harvesting enough spontaneously dividing cells or even mitogen-stimulated cells to divide at any given time may be difficult. Therefore, with tumors and some hematological malignancies, it may be difficult to estimate an acceptable no-growth range, but experience in the laboratory is a major factor.

The definition of poor growth needs to be established by each laboratory, as it may imply less than five cells analyzed in one laboratory, whereas any number under 20 fully analyzed cells may define poor growth in another laboratory. Reporting cases as an incomplete study varies greatly among laboratories, since there is no standard of reporting. Since laboratories need to state how many cells were analyzed on the report for each study, it is evident if less than a standard 20-cell study was performed. However, the reports vary in their interpretation of these results.

Poor or no growth may also be due to previous cytotoxic exposure during treatment for disease and this, if known, should be written in the report to explain the lack of results. The same may be said for specimens known to have a long transport time to the laboratory or if a specimen is too small to be optimally grown in the laboratory.

Another problem exists when a specimen is received in the laboratory without a clear presumptive diagnosis or symptoms of disease. Since chronic lymphoproliferative diseases need mitogen stimulation in cultures for optimal results, a lack of clinical information may mean that the wrong culture is initiated in the laboratory, giving rise to poor or no growth or the wrong culture being analyzed for the disorder in question.

Also, if a peripheral blood specimen is received with an indication of a lymphoblastic leukemia or myeloid disorder, when too few blasts are in the circulation, then there is an *a priori* risk of not having enough dividing abnormal cells to analyze. The same is true for tumors that are sent for chromosome analysis, when much of the tissue is derived from non-malignant or non-pertinent tumor region. Obtaining the correct specimen type/portion is essential for optimal results for each disease entity.

21.4.1 No growth comments

- No metaphase cells were apparent from this bone marrow specimen, precluding cytogenetic analysis. The corresponding bone marrow aspirate morphology was aparticulate and hemodilute (per pathology report).
- No metaphase cells were apparent from this unstimulated peripheral blood specimen. These results are consistent with the lack of spontaneously dividing mitotic cells in the circulation. Follow-up studies on a bone marrow specimen may prove more informative.
- These results are consistent with cytotoxic effects related to recent chemotherapy (day *).
- These results may be related to environmental issues including, but not limited to, delay in specimen receipt, inadequate protection from extreme temperature exposure during transport, or contamination resulting from non-aseptic specimen collection or processing.

21.4.2 Poor growth comment

- All cultures established from this peripheral blood specimen yielded poor growth. Despite the multiple culture and harvest attempts, an insufficient number of metaphases were available for analysis. A bone marrow specimen, when clinically appropriate, may provide additional cytogenetic information.

Bibliography

American College of Medical Genetics. *Guidelines in Clinical Cytogenetics. Section E10 Methods in Fluorescence In situ Hybridization – Interphase/Nuclear Fluorescence In Situ Hybridization.* American College of Medical Genetics, Bethesda, MD. www.acmg.net/StaticContent/SGs/Section_E.html, accessed 28 January 2014.

Atlas of Genetics and Cytogenetics in Oncology and Haematology. http://AtlasGeneticsOncology.org/Anomalies, accessed 28 January 2014.

Gersen S, Keagle M (eds). *Principles of Clinical Cytogenetics.* Humana Press, Totowa, New Jersey, 1999.

Hanson CA, Steensma D, Hodnefield J, et al. Isolated trisomy 15: a clonal chromosome abnormality in bone marrow with doubtful hematologic significance. Am J Clin Pathol 2008; 129: 478–485.

Heim S, Mitelman F. *Cancer Cytogenetics*, 2nd edn. Wiley-Liss, New York, 1995.

Morel F, Le Bris MJ, Herry A, et al. Trisomy 15 as the sole abnormality in myelodysplastic syndromes: case report and review of the literature. Leuk Lymphoma 2003; 44: 549–551.

Natelson EA. Myelodysplasia with isolated trisomy 15: a 15-year follow-up without specific therapy. Am J Med Sci 2006; 331: 157–158.

Rooney DE, Czepulkowski BH. *Human Cytogenetics. A Practical Approach.* Oxford University Press, New York, 1992.

Sashida G, Tauchi T, Kimura Y, et al. Trisomy 15 as a single autosomal abnormality in a patient with unclassifiable myelodysplastic syndromes [letter]. Cancer Genet Cytogenet 2001; 127: 91–92.

Shaffer LG, McGowan-Jordan J, Schmid M (eds). *ISCN 2013: An International System for Human Cytogenetic Nomenclature.* Karger Publishers, Unionville, CT, 2013.

Sinclair EJ, Potter AM, Watmore AE, et al. Trisomy 15 associated with loss of the Y chromosome in bone marrow: a possible new aging effect. Cancer Genet Cytogenet 1998; 105: 20–23.

Smith A, Watson N, Sharma P. Frequency of trisomy 15 and loss of the Y chromosome in adult leukemia. Cancer Genet Cytogenet 1999; 114: 108–111.

Spurbeck JL, Carlson RO, Allen JE, et al. Culturing and robotic harvesting of bone marrow, lymph nodes, peripheral blood, fibroblasts, and solid tumors with in situ techniques. Cancer Genet Cytogenet 1988; 32: 59–66.

Swerdlow SH, Campo E, Harris NL, et al. (eds). *WHO Classification of Tumours of Haematopoietic and Lymphoid Tissues*. IARC Press, Lyon, 2008.

Vadlamani I, Ma E, Brink DS, et al. Trisomy 15 in a case of pediatric hemangiopericytoma and review of the literature. Cancer Genet Cytogenet 2002; 138: 116–119.

Wiktor A, Rybicki BA, Piao ZS, et al. Clinical significance of Y chromosome loss in hematologic disease. Genes Chromosomes Cancer 2000; 27: 11–16.

CHAPTER 22

Cytogenetic abnormalities in myeloid disorders

22.1 Introduction to myeloid disorders

The WHO classification includes a number of subtypes of myeloid diseases based on the underlying genetic defect. Therefore, it is recommended that for myeloid neoplasms, a chromosome analysis be performed at the time of the initial evaluation of a patient to establish the cytogenetic composition and, at subsequent regular intervals, to detect any genetic evolution. Further genetic tests should be performed based on the initial cytogenetic results as well as clinical, morphological and immunophenotyping studies. Fluorescence *in situ* hybridization (FISH) and/or polymerase chain reaction (PCR) testing for specific genetic abnormalities may be necessary to rule out a low frequency of an abnormal cell line that may not be observed in the initial chromosome study, or to confirm an abnormal clonal population seen by chromosome analysis for a baseline frequency of abnormal cells at diagnosis. Also, specific genetic mutations or rearrangements may be submicroscopic; therefore, only FISH, chromosomal microarray analysis (CMA) or molecular studies will identify an abnormal clonal population. The three main types of myeloid neoplasms discussed here, based on their underlying genetic defects, are the myelodysplastic syndromes (MDS), acute myeloid leukemia (AML) and myeloproliferative neoplasms (MPN).

The myelodysplastic syndromes are usually characterized by the proliferation of hematopoietic cells, leading to cytopenia and dysplasia of one or more of the myeloid cell lineages. MDS has an increased risk of developing AML, and apoptosis is increased in the diseased cells. MDS is defined as having <20% blasts in the peripheral blood and bone marrow. Approximately 50% of MDS cases will show chromosome abnormalities with specific genetic defects. These often define different subtypes of MDS. See Table 22.1 for a list of the specific genetic abnormalities associated with MDS. Some chromosome abnormalities observed in MDS, when seen as the sole anomaly, do not give definitive evidence of disease in the absence of morphological criteria and should be noted in the report, including -Y, +8 and del(20q).

Acute myeloid leukemia is defined as having >20% of myeloid blasts in peripheral blood and bone marrow of one or more of the myeloid cell lineages. Patients with <20% of blasts may be defined as having AML with the presence of an underlying genetic defect associated with AML, including t(8;21), inv(16), t(16;16), or t(15;17). Therefore, the new WHO classification system describes each of these genetic defects as individual subtypes of AML. These genetic rearrangements generally involve chromosomal translocations including transcription factors that have distinct characteristic clinical,

Cytogenetic Abnormalities: Chromosomal, FISH and Microarray-Based Clinical Reporting, First Edition. Susan Mahler Zneimer.

Table 22.1 Recurrent cytogenetic abnormalities in myelodysplastic disorders (MDS) at the time of diagnosis (modified from the WHO classification system)

Abnormality	Frequency in MDS at diagnosis (%)	Frequency in MDS after therapy
Del(5q)/-5	10	40
Del(7q)/-7	10	50
Del(20q)	5–8	
-Y	5	
i(17q)/t(17p)	3–5	
Del(13q)/-13	3	
Del(11q)	3	
Del(12p)/t(12p)	3	
Del(9q)	1–2	
Idic(X)(q13)	1–2	
T(1;3)	1	
T(2;11)	1	
Inv(3)/t(3;3)	1	
T(3;21)	-	2
T(6;9)	1	
T(11;16)	-	3
+8	10	

morphological and immunophenotypic features. Often, secondary genetic abnormalities exist with these rearrangements, including further cytogenetic changes or gene mutations. Common gene mutations include the FLT3, KIT or NPM1 genes, which promote signal transduction pathways, resulting in the proliferation of the abnormal cells or loss of genes or chromosomal regions, such as TP53/17p13.1. These secondary changes generally lead to progression of disease and worsening prognosis. Therefore, cytogenetic analysis gives necessary information of the underlying defect causing disease and an understanding of the pathogenesis of the neoplastic disorder, prognosis and specificity of therapy to target the genetic defect. See Box 22.1 for specific genetic abnormalities associated with AML.

EXAMPLE REPORTING OF NORMAL RESULTS WITH A NEW DIAGNOSIS OF AML

ISCN Result: 46,XY[20]

Interpretation

Cytogenetic analysis revealed a normal male chromosome complement with no evidence of a clonal abnormality in all cells examined.

Normal chromosome results with an indication of acute myeloid leukemia (AML) may benefit from specific genetic molecular testing. FLT3-ITD (internal tandem duplication) represents one of the most frequent genetic changes in AML with an estimated occurrence of 20–40%. FLT3-ITD mutations are associated with a less favorable outcome in cytogenetically normal AML cases.

Also, mutations of the NPM1 and CEBPA genes are often observed in AML with a normal karyotype. NPM1 mutations are observed in approximately one-third of adult AML cases, and CEBPA mutations are observed in approximately 15–18% of AML cases with normal karyotypes. NPM1 and CEBPA mutations are generally associated with a favorable prognosis in the absence of FLT3-ITD mutations.

If clinically indicated, consider NPM1, CEBPA and FLT3 mutation analysis to look for the presence of abnormalities commonly observed in acute myeloid leukemia with normal chromosome analysis.

BOX 22.1 RECURRENT CYTOGENETIC ABNORMALITIES IN ACUTE MYELOID LEUKEMIA AT THE TIME OF DIAGNOSIS

t(1;22)(p13;q13)
inv(3)(q21q26.2)
t(3;3)(q21;q26.2)
t(6;9)(p23;q34)
t(8;21)(q22;q22)
t(9;11)(p22;q23)
t(15;17)(q24;q21)
inv(16)(p13.1q22)
t(16;16)(p13.1;q22)

The MPN disorders are characterized by the proliferation and hypercellularity of one or more of the myeloid cell lineages. Genetic evolution in these disorders is common, leading to disease progression. This can be evidenced by increased blasts in the bone marrow. A level of 10–19% of blasts indicates accelerated disease, and >20% of blasts is considered blast crisis. One WHO classification subtype of MPN is the presence of the Philadelphia chromosome, evidenced by a t(9;22) and/or BCR/ABL gene rearrangement, consistent with the diagnosis of CML. Absence of the Philadelphia chromosome is consistent with other MPN subtypes, which often contain other genetic abnormalities, either chromosomal or gene-level mutations. They generally involve genes that encode tyrosine kinase activity, such as the JAK2 gene, which is commonly seen in polycythemia vera, primary myelofibrosis and essential thrombocythemia. Also in this classification category are the rearrangements of PDGFRA and B, and FGFR1 rearrangements associated with hypereosinophilia. See Box 22.2 for a list of recurrent abnormalities associated with MPNs.

BOX 22.2 RECURRENT CYTOGENETIC ABNORMALITIES IN MYELOPROLIFERATIVE NEOPLASMS AT THE TIME OF DIAGNOSIS

Trisomy 1q
Del 5q
Del 7q
Abnormalities of 8p11
Trisomy 8
Abnormalities of 9p
t(9;22)(q34;q11.2)
Trisomy 9
Del(12p)
Del 13q
Del 17p
i(17q)
Del 20q

22.2 Individual myeloid abnormalities by chromosome order

22.2.1 Chromosome 1

Translocation (1;7)(q10;p10) (Figure 22.1)

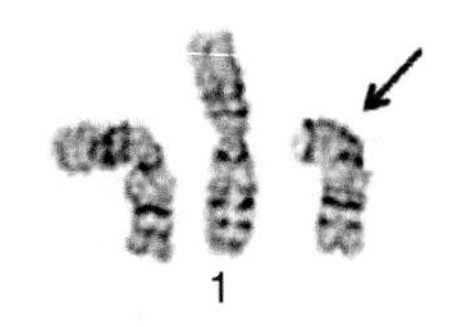

Figure 22.1 Partial karyotype showing translocation (1;7)(q10;p10) with two normal chromosome 1s and only one normal chromosome 7 (not shown).

SPECIFIC FEATURES

Translocation (1;7)(q10;p10) is a recurrent abnormality in MDS, MPN and AML. This translocation is most often seen as an unbalanced rearrangement in which there is a net gain of the long arm of chromosome 1 (1q trisomy) and loss of the long arm of chromosome 7 (7q deletion). This abnormality is seen in approximately 2% of MDS cases, in 0.5% of AML cases and less frequently in MPN cases. T(1;7), when seen in MDS, often evolves into AML. Those cases diagnosed as AML frequently have a preceding MDS, and approximately 50% of cases are seen as a secondary change due to previous cytotoxic exposure. This abnormality is mainly seen in adults with a median age of 60 years, with a male predominance. The result of the unbalanced form of this rearrangement, resulting in trisomy 1q and del(7q), can be detected by cytogenetics and FISH analysis. This translocation is seen as the sole karyotypic abnormality in approximately 50% of cases. Hyperdiploidy, including trisomy 8, 9 and 21 are common secondary changes with this abnormality. T(1;7) is associated with a poor prognosis, with the median survival of only 11 months.

EXAMPLE REPORTING OF t(1;7)

ISCN Result: 46,XY,+1,t(1;7)(q10;p10)[20]

Interpretation

Chromosome analysis revealed an abnormal male chromosome complement with the gain of chromosome 1, and an unbalanced whole arm translocation between the long arm of chromosome 1 and the short arm of chromosome 7 in all cells examined.

This unbalanced (1;7) translocation results in the loss of the long arm of chromosome 7 and gain of the long arm of chromosome 1 and is a recurrent, non-random abnormality observed in acute myeloid leukemia, which is often preceded by myelodysplasia. It is also often seen in secondary disease with prior radiation or chemotherapeutic treatment. This rearrangement is generally associated with a poor prognosis.

FISH analysis for chromosomes 1q and 7q is suggested to confirm and monitor this abnormal cell line.

Translocation (1;22)(p13;q13)

SPECIFIC FEATURES

Translocation (1;22)(p13;q13) is a recurrent abnormality in AML and is a distinct genetic category by the WHO classification. T(1;22) involves the maturation of the megakaryocyte lineage, similar to acute megakaryocytic leukemia. This is a fairly rare abnormality in AML, which is seen in <1% of cases and most often in infants and young children (less than 3 years of age), with a predominance of females. It is seen as a *de novo* event with thrombocytopenia and often with an elevated white blood count. This abnormality results in the fusion of the RBM15 gene at 1p13 and the MKL1 gene at 22q13, and it can be detected by cytogenetics and FISH analysis. This translocation is usually the sole karyotypic abnormality, without common secondary changes. With appropriate treatment, the prognosis is fairly good with long disease-free survival.

EXAMPLE REPORTING OF t(1;22)

ISCN Result: 46,XY,t(1;22)(p13;q13)[20]

Interpretation

Chromosome analysis revealed an abnormal male chromosome complement with a translocation between the short arm of chromosome 1 and the long arm of chromosome 22 in all cells examined.

This translocation is a rare but recurrent, non-random abnormality observed in acute myeloid leukemia and results in the fusion of the RBM15 gene at 1p13 and the MKL1 gene at 22q13. This rearrangement is generally associated with a fairly good prognosis.

FISH analysis for chromosomes 1p and 22q is suggested to confirm and monitor this abnormal cell line.

22.2.2 Chromosome 3

Inv(3)(q21q26.2) and t(3;3)(q21;q26.2)

SPECIFIC FEATURES

Chromosome 3 inversions and translocations involving the breakpoints q21 and q26 are recurrent abnormalities in AML and are within a distinct genetic category by the WHO classification. Inv(3) and t(3;3) are seen both as a *de novo* abnormality and as a development from prior MDS. They are seen in 1–2% of all AML cases, most often in adults, and equally among males and females. Inv(3) and t(3;3) are often associated with normal or elevated platelet count, atypical megakaryocyte morphology with mono- or bilobed nuclei and are seen in multilineage dysplasia. The example report below shows an inv(3)(q21q26.2), but t(3;3)(q21;q26.2) is also encountered. The genes involved include RPN1 at 3q21 and EVI1 at 3q26.2, and the abnormalities result in the fusion of these two genes and can be detected by cytogenetics and FISH analysis. Common secondary changes associated with these abnormalities include monosomy 7, seen in approximately 50% of cases, and del(5q). Complex karyotypic abnormalities are also common. Chromosome 3 inversions and translocations are generally associated with a poor prognosis and aggressive disease.

EXAMPLE REPORTING OF inv(3) (FIGURE 22.2)

ISCN Result: 46,XY,inv(3)(q21q26.2)[20]

Interpretation

Chromosome analysis revealed an abnormal male chromosome complement with a paracentric inversion of the long arm of chromosome 3 in all cells examined.

Chromosomal rearrangements of the long arm of chromosome 3 are recurrent, non-random abnormalities in myeloid disorders. Overexpression of the EVI1 gene at chromosome 3q26 appears to be a consistent feature of chromosome 3q21q26 syndrome, and is associated with acute myeloid leukemia (AML) and myelodysplastic syndromes (MDS).

Acute leukemias characterized by 3q abnormalities are often associated with disturbed thrombocytopoiesis, elevated platelet counts and aberrant overexpression of EVI1, a putative transcription factor. Leukemias with 3q aberrations are generally non-responsive to standard chemotherapeutic regimens and are associated with a poor prognosis.

FISH analysis for the EVI1 gene is suggested to confirm and monitor this abnormal cell line.

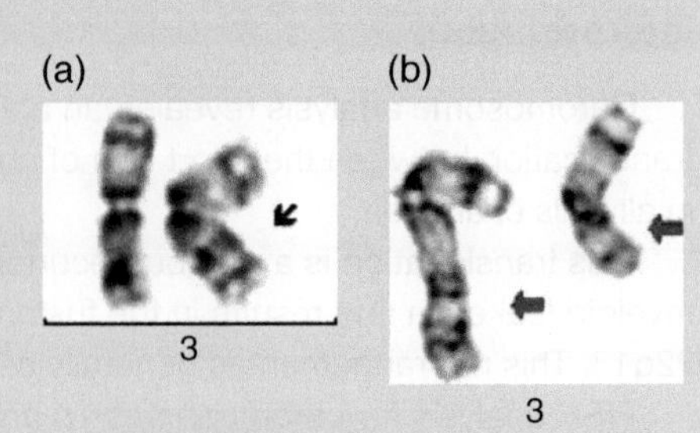

Figure 22.2 Inversion and translocation of chromosome 3. (a) Paracentric inversion of chromosome 3 – inv(3)(q21q26.2). Courtesy of Sarah South PhD, ARUP Laboratories. (b) Translocation (3;3) – t(3;3)(q21;q26.2).

EXAMPLE REPORTING OF inv(3) WITH MONOSOMY 7

ISCN Result: 46,XY,inv(3)(q21q26.2),-7[20]

Interpretation

Chromosome analysis revealed an abnormal male chromosome complement with a paracentric inversion of the long arm of chromosome 3 and loss of chromosome 7 in all cells examined.

Chromosomal rearrangements of the long arm of chromosome 3 are recurrent, non-random abnormalities in myeloid disorders. Overexpression of the EVI1 gene at chromosome 3q26 appears to be a consistent feature of chromosome 3q21q26 syndrome, and is associated with acute myeloid leukemia (AML) and myelodysplastic syndromes (MDS).

Acute leukemias characterized by 3q abnormalities are often associated with disturbed thrombocytopoiesis, elevated platelet counts and aberrant overexpression of EVI1, a putative transcription factor. Leukemias with 3q aberrations are generally non-responsive to standard chemotherapeutic regimens and are associated with a poorer prognosis.

Monosomy 7 is a recurrent, non-random abnormality commonly observed in the myelodysplastic syndromes and acute myeloid leukemia, occurring in both *de novo* and secondary disease prior to exposure of chemical mutagens or chemotherapy treatments with alkylating agents. Monosomy 7 is often associated with chromosome 3q aberrations and is also associated with a poor prognosis.

FISH analyses for the EVI1 gene and chromosome 7 are suggested to confirm and monitor this abnormal cell line.

22.2.3 Chromosome 4

Trisomy 4

SPECIFIC FEATURES

Trisomy 4 as the sole anomaly is a rare chromosomal abnormality associated with a specific subtype of AML with myelomonocytic morphology. It may be seen in conjunction with double minute chromosomes as well as with t(8;21). It has been found in both primary and treatment-related disease. Trisomy 4 with the presence of c-Kit mutations is generally associated with rapid disease progression.

EXAMPLE REPORTING OF TRISOMY 4

ISCN Result: 47,XY,+4[20]

Interpretation

Chromosome analysis revealed an abnormal male chromosome complement with the gain of chromosome 4 as the sole anomaly in all cells examined.

Trisomy 4 as the sole anomaly is a rare chromosomal abnormality associated with a specific subtype of acute myeloid leukemia (AML) with myelomonocytic morphology.

FISH analysis for chromosome 4 is suggested to confirm and monitor this abnormal cell line.

22.2.4 Chromosome 5

Del(5q) (Figure 22.3)

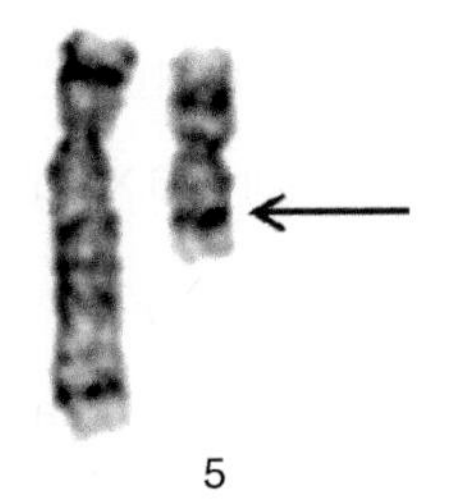

Figure 22.3 Partial karyotype showing del(5)(q13q33).

SPECIFIC FEATURES

Deletion within the long arm of chromosome 5 is a recurrent abnormality in MDS and AML and is described in both AML and MDS categories by the WHO classification. When seen as the sole anomaly, it is considered "isolated 5q- syndrome" and is characterized by anemia with or without thrombocytosis. It occurs in adults, with a median age of 67, and predominantly in females. The size of the deletion varies, but the critical region contains the 5q31-q33 bands. When seen as the sole anomaly, the prognosis is generally good with transformation to AML seen in <10% of cases. When del(5q) is seen with other karyotypic abnormalities or excess blasts, it is classified in the MDS/MPN WHO classification and the prognosis is usually worse than in isolated 5q- syndrome. The gene(s) responsible for this disorder is unknown presently, though there are candidate genes, including EGR1 and CTNNA1, which are presumed to be tumor suppressor genes within the 5q31-q33 region. Del(5q) can be detected by cytogenetics and FISH analysis.

EXAMPLE REPORTING OF DELETION OF CHROMOSOME 5q

ISCN Result: 46,XY,del(5)(q13q33)[20]

Interpretation

Chromosome analysis revealed an abnormal male chromosome complement with a deletion within the long arm of chromosome 5 as the sole anomaly in all cells examined.

Deletion of the long arm of chromosome 5 is a recurrent, non-random finding in myeloid disorders, including myelodysplastic syndromes (MDS) and acute myeloid leukemia (AML). In MDS, as a sole anomaly, del(5q) is particularly frequent in refractory anemia (RA), being described as "5q- syndrome" with characteristic clinical-hematological features. The majority of patients with the syndrome are elderly females with macrocytic anemia resistant to therapy. The clinical course is generally mild, and transformation to AML is relatively rare when the del(5q) is the sole anomaly at diagnosis.

FISH analysis for chromosome 5q is suggested to confirm and monitor this abnormal cell line.

22.2.5 Chromosome 6

Translocation (6;9)(p23;q34) (Figure 22.4)

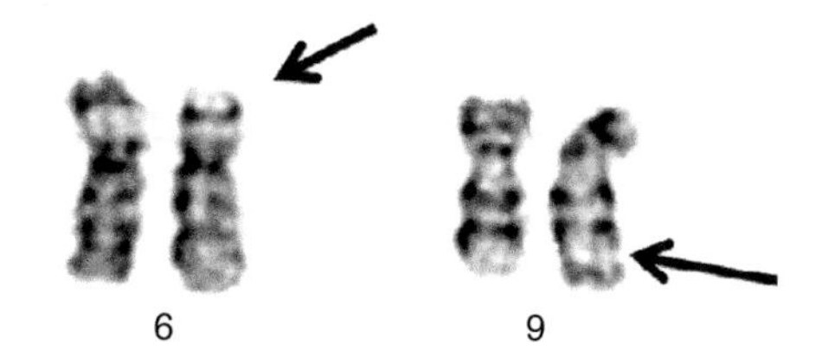

Figure 22.4 Partial karyotype showing translocation (6;9)(p23;q34).

SPECIFIC FEATURES

Translocation (6;9) is a recurrent abnormality associated with AML and is a distinct genetic category in the WHO classification. It is characterized by anemia and thrombocytopenia, often with pancytopenia and multilineage dysplasia with or without monocytic features. T(6;9) is seen in approximately 1–2% of AML cases in children with a median age of 13 years, as well as adults with a median age of 35 years. T(6;9) results in the fusion of the DEK gene on 6p23 and the NUP214 gene on 9q34, which is detectable by cytogenetics and FISH analyses. T(6;9) is seen most often as the sole anomaly, but may also be seen with complex karyotypic changes as well as with FLT3-ITD mutations. This abnormality is generally associated with a poor prognosis in both children and adults. When t(6;9) is seen with <20% of blasts, patients should be monitored closely for the development of overt AML.

EXAMPLE REPORTING OF t(6;9)

ISCN Result: 46,XY,t(6;9)(p23;q34)[20]

Interpretation

Chromosome analysis revealed an abnormal male chromosome complement with a translocation between the short arm of chromosome 6 and the long arm of chromosome 9 as the sole anomaly in all cells examined.

The t(6;9) translocation, which results in the fusion of the DEK gene at 6p23 with the NUP214 gene at 9q34, is a recurrent, non-random abnormality observed in acute myeloid leukemia and myelodysplastic syndromes and is often associated with absolute basophilia. This translocation is commonly associated with FLT3 internal tandem duplications. The t(6;9) is predominantly associated with younger adults and carries an unfavorable prognosis with standard chemotherapy.

FISH analysis for this rearrangement is suggested to confirm and monitor this abnormal cell line.

22.2.6 Chromosome 7

Del(7q)/monosomy 7 (Figure 22.5)

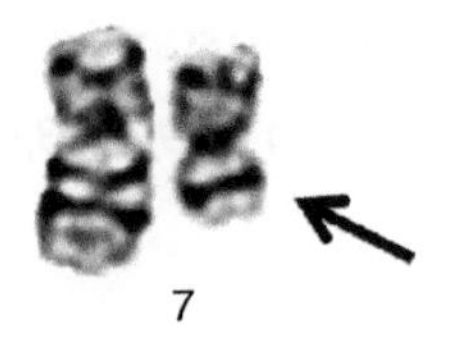

Figure 22.5 Partial karyotype showing del(7)(q31q36). Courtesy of Sarah South PhD, ARUP Laboratories.

SPECIFIC FEATURES

Deletion within the long arm of chromosome 7 is a recurrent abnormality in MDS and AML and is described in both AML and MDS categories by the WHO classification. The main breakpoints in the deletion comprise two common minimal regions, one including 7q22 and another including 7q32-34. Del(7q)/monosomy 7 occurs in *de novo* cases, secondary to treatments with alkylating agents, and in patients with predisposing leukemia syndromes, incuding Fanconi anemia and Kostman syndrome. Although del(7q) and monosomy 7 are seen as the sole anomaly in 75% of MDS cases and in 32% of AML cases, additional anomalies may also occur, including del(5q), monosomy 5 and trisomy 8.

Translocation t(1;7)(q10;p10) may also be present and results in del(7q) and may be considered a variant del(7q), while other unbalanced translocations involving 7q occur that have the consequence of partial monosomy 7 of the 7q22 to 7q34 bands.

Monosomy 7 is also seen often in juvenile chronic myeloid leukemia (JCML). It is the most frequent abnormality seen in childhood myeloid disorders, found in 30% of childhood MDS and in 4% of childhood AML. Young patients with monosomy 7 due to exposure to alkylating agents usually have an MDS phase preceding acute leukemia with multilineage bone marrow dysplasia. Prognosis in children with chromosome 7 abnormalities is poor, with survival <2 years. However, the European Working Group on MDS in childhood noted a superior survival for children with MDS having monosomy 7 as the sole anomaly compared to those with MDS that have other anomalies. Survival was not good in children with AML with monosomy 7.

In adults, del(7q) and monosomy 7 in MDS are seen in 30% of refractory anemia with excess blasts (RAEB)/RAEB-T, 20% of chronic myelomonocytic leukemia (CMML), and only 5% of RA and, in AML, usually with the FAB subtype M4 or M6. In MDS, del(7q) and monosomy 7 are usually secondary events contributing to disease, rather than being the primary cytogenetic change. In addition, monosomy 7 is seen as the most frequent abnormality in secondary myeloid disorders, found in 51% of cases, whereas del(7q) was found in 7% and partial monosomy 7, as a result of an unbalanced translocation, in 8% of secondary cases. The proportion of adults with monosomy 7 myeloid disorder grows dramatically after 60 years of age. A monosomy 7 myeloid disorder is characterized by infectious susceptibility, quick aggravation and treatment resistance. In adults, monosomy 7 is classified as a poor prognostic indicator by the International Prognostic Scoring System (IPSS), with an 82% relapse rate within 1 year, and 7-year event-free survival is only 6%. Monosomy 7 postallogenic bone marrow transplantation is also associated with a poor prognosis.

EXAMPLE REPORTING OF DELETION OF CHROMOSOME 7q

ISCN Result: 46,XY,del(7)(q31q36)[20]

Interpretation

Chromosome analysis revealed an abnormal male chromosome complement with a deletion within the long arm of chromosome 7 as the sole anomaly in all cells examined.

Deletion of 7q is a recurrent, non-random abnormality observed in both *de novo* and secondary myelodysplastic syndromes and acute myeloid leukemia, and is generally associated with a poor prognosis.

FISH analysis for chromosome 7q is suggested to confirm and monitor this abnormal cell line.

EXAMPLE REPORTING OF MONOSOMY 7

ISCN Result: 45,XY,-7[20]

Interpretation

Chromosome analysis revealed an abnormal male chromosome complement with loss of chromosome 7 as the sole anomaly in all cells examined.

Monosomy 7 is a recurrent, non-random abnormality commonly observed in myelodysplastic syndromes and acute myeloid leukemia, occurring in both *de novo* and secondary disease prior to exposure of chemical mutagens or chemotherapy treatments with alkylating agents. Monosomy 7 is generally associated with a poor prognosis.

FISH analysis for chromosome 7 is suggested to confirm and monitor this abnormal cell line.

For cases with loss of chromosome 7 and the presence of an unidentifiable marker chromosome, the marker chromosome is most likely of chromosome 7 origin. FISH analysis of the chromosome 7q probe is suggested to clarify the findings. Without FISH analysis, reports may be written as -7,+mar, or with a derivative chromosome 7, in which a possible deletion of both the short and long arms is present. If FISH analysis confirms a chromosome 7 deletion, the der(7) nomenclature is appropriate and a comment regarding the results of the FISH in the report will fully explain the chromosomal findings (see Chapter 31 for an example report).

EXAMPLE REPORTING OF MONOSOMY 7 WITH A MARKER CHROMOSOME

ISCN Result: 46,XY,-7,+mar[20] **or** 46,XY,der(7)del(7)(?p11.2)del(7)(?q11.2q36)[20]

Interpretation

Chromosome analysis revealed an abnormal male chromosome complement with loss of chromosome 7 and the presence of a small unidentifiable marker chromosome in all cells examined.

The presence of a small marker chromosome with the loss of chromosome 7 is most likely of chromosome 7 origin, resulting in deletions within both the short and long arms. FISH analysis for chromosome 7 is suggested to confirm and monitor this abnormal cell line.

Monosomy 7/del(7q) are recurrent, non-random abnormalities commonly observed in the myelodysplastic syndromes and acute myeloid leukemia, occurring in both *de novo* and secondary disease. Monosomy 7 is generally associated with a poor prognosis.

i(7)(p10) or idic(7)(q11.2) (Figure 22.6)

Figure 22.6 Partial karyotype showing i(7)(p10).

SPECIFIC FEATURES

Isodicentric 7q or isochromosome 7p is a rare chromosomal abnormality in AML; however, it does result in the loss of the long arm of chromosome 7 as a consequence of recombination and centromeric duplication. Isochromosome 7p is seen in elderly patients with AML, often as a sole abnormality. This abnormality has shown a positive response to induction chemotherapy, unlike other forms of del(7q), possibly due to the genes mapped to the short arm of chromosome 7 that are known to have protective oncogenic potential, including EGFR, PDGFA, MDR1, T-cell invasion and metastasis-inducing locus, and an aryl hydrocarbon receptor (AHR).

EXAMPLE REPORTING OF ISODICENTRIC 7q11.2

ISCN Result: 46,XY,idic(7)(q11.2) [20]

Interpretation

Chromosome analysis revealed an abnormal male chromosome complement with an isodicentric chromosome 7, resulting in a deletion of most of the long arm of chromosome 7, as the sole anomaly in all cells examined.

Deletion of 7q is a recurrent, non-random abnormality observed in both *de novo* and secondary myelodysplastic syndromes and acute myeloid leukemia, and is generally associated with a poor prognosis. However, recent reports have shown a favorable outcome in patients with idic(7)(q11.2), possibly due to the protective nature of the genes located on chromosome 7p.

FISH analysis for chromosome 7q is suggested to confirm and monitor this abnormal cell line.

22.2.7 Chromosome 8

Translocation (8;16)(p11;p13)

SPECIFIC FEATURES

Translocation (8;16)(p11;p13) is a recurrent abnormality in AML, consistent with the FAB subtypes M4, M5a and M5b, with involvement of granulomonocytic maturation, the presence of disseminated intravascular coagulation in many cases, and erythrophagocytosis.

This is a fairly rare abnormality in AML, seen in <1% of cases and most often in children, including infants and young adults, equally among females and males. It is seen as a *de novo* event in most cases, but up to 20% of cases have been reported as therapy related. This abnormality results in the fusion of the MOZ (monocytic leukemia zinc finger) gene at 8p11.2 and the CBP (CREB binding protein) gene at 16p13.3. FISH probes are not readily available but it is detectable by cytogenetics analysis. This translocation is usually the sole karyotypic abnormality in half the cases, with the other half showing trisomy 8 and complex abnormalities. This rearrangement is generally associated with a poor prognosis, with remission obtained in only half of the cases, in which infections and bleeding are common and survival is often <1 year.

EXAMPLE REPORTING OF t(8;16)

ISCN Result: 46,XY,t(8;16)(p11;p13)[20]

Interpretation

Chromosome analysis revealed an abnormal male chromosome complement with a translocation between the short arms of chromosomes 8 and 16 as the sole anomaly in all cells examined.

The t(8;16) translocation is a recurrent, non-random abnormality observed in acute myeloid leukemia and is often associated with disseminated intravascular coagulation. This translocation is generally associated with a poor prognosis.

Translocation (8;21)(q22;q22) (Figure 22.7)

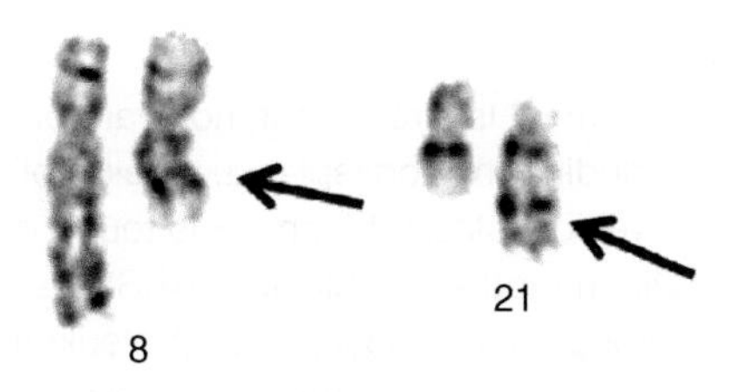

Figure 22.7 Partial karyotype showing translocation (8;21)(q22;q22).

SPECIFIC FEATURES

Translocation (8;21) is a recurrent abnormality associated with AML and is a distinct genetic category in the WHO classification. It is characterized by maturation of the neutrophilic lineage and is often present with tumors, such as myeloid sarcomas. T(8;21) is seen in approximately 5–10% of AML cases, most often in young patients. T(8;21) results in the fusion of the RUNX1T1 (ETO) gene on 8q22 with the RUNX1 (AML1) gene on 21q22, which is detectable by cytogenetics and FISH analysis. T(8;21) is seen often (70%) with additional karyotypic changes, including loss of a sex chromosome and del(9q). In addition, KRAS and NRAS mutations are present in approximately 30% of childhood cases, and 20–25% of all t(8;21) translocations also have KIT mutations. T(8;21) AML is generally associated with a good prognosis and good response to therapy. However, the presence of KIT mutations may have an adverse prognosis.

EXAMPLE REPORTING OF t(8;21)

ISCN Result: 46,XY,t(8;21)(q22;q22)[20]

Interpretation

Chromosome analysis revealed an abnormal male chromosome complement with a translocation between the long arms of chromosomes 8 and 21 as the sole anomaly in all cells examined.

The t(8;21) translocation, which results in the fusion of the RUNX1T1 (ETO) gene at 8q22 with the RUNX1 (AML) gene at 21q22, is a recurrent, non-random abnormality observed in acute myeloid leukemia, occurring in 15–18% of all FAB-classified cases. This translocation is generally associated with a good prognosis.

The c-KIT mutation, occurring in 20–40% of AML with t(8;21), confers an increased risk of relapse and a lower overall survival. Consider c-KIT mutation molecular analysis, which generally can be performed on the remaining sample.

FISH analysis for this rearrangement is suggested to confirm and monitor this abnormal cell line.

Trisomy 8

SPECIFIC FEATURES

Trisomy 8 is a recurrent, non-random abnormality seen primarily in myeloid disorders, including myelodysplasia, myeloproliferative neoplasms and acute myeloid leukemia. In myelodysplasia, trisomy 8 is found in 15–20% of all cases and in 30% of refractory anemia with ringed sideroblasts (RARS) cases in particular. Patients with trisomy 8 generally proceed through a myelodysplastic preleukemic phase prior to development of overt leukemia. When seen as the sole anomaly, it is not considered definitive evidence for MDS in the absence of morphological criteria, based on the WHO classification.

Trisomy 8 is also seen in CML as a major additional secondary change with the t(9;22). Trisomy 8 may arise after interferon and/or imatinib treatment but its clinical significance is unknown. Trisomy 8 is also found in other chronic myeloproliferative diseases, including 20% of polycythemia vera (PV) and 10% of ideopathic myelofibrosis.

Trisomy 8 is generally seen as the sole anomaly in 55–65% of myelodysplastic cases and found with other anomalies in the remaining cases, often with del(5q), monosomy 5, t(1;7)(q10;p10) and del(20q). Progression from MDS towards AML occurs in about half of the cases when trisomy 8 is seen as the sole anomaly.

Trisomy 8 is found in 10–15% of AML and 10% of treatment-related AML. It is present in each FAB subgroup, and many cases may arise from a preceding MDS. Trisomy 8 as the sole anomaly in AML is seen in 40% of cases, with the remaining cases seen with other cytogenetic changes including del(5q), del(7q), monosomy 5 and 7, often associated with complex karyotypic anomalies. The prognosis of AML in adults with trisomy 8 is generally associated with an intermediate to poor prognosis. Interestingly, individuals with constitutional trisomy 8 syndrome have an increased risk of developing leukemia.

EXAMPLE REPORTING OF TRISOMY 8

ISCN Result: 47,XY,+8[20]

Interpretation

Chromosome analysis revealed an abnormal male chromosome complement with the gain of chromosome 8 as the sole anomaly in all cells examined.

Trisomy 8 is a recurrent, non-random abnormality seen primarily in myeloid disorders, including myelodysplasia, myeloproliferative neoplasms and acute myeloid leukemia. In myelodysplasia, trisomy 8 is found in 20% of all cases and in 30% of RARS cases in particular. Patients with trisomy 8 generally proceed through a myelodysplastic, preleukemic phase prior to development of overt leukemia. When seen as the sole anomaly, it is not considered definitive evidence for MDS in the absence of morphological criteria, based on the WHO classification.

FISH analysis for chromosome 8 is suggested to confirm and monitor this abnormal cell line.

22.2.8 Chromosome 9

Del(9q) (Figure 22.8)

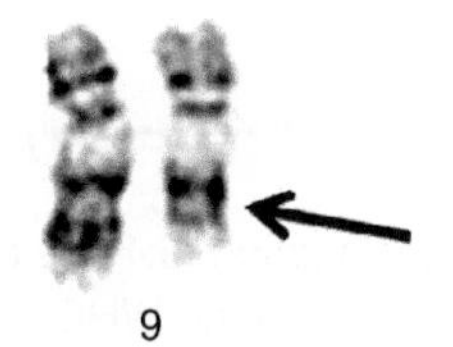

Figure 22.8 Partial karyotype showing del(9)(q21).

SPECIFIC FEATURES

Del(9q) is seen predominantly in AML and rarely observed in MDS or MPNs. It is seen in biphenotypic leukemias as well, and in both adults and children. As mentioned previously, del(9q) is a common secondary change in AML with t(8;21). Del(9q) involves a variable deleted chromosomal region, but usually includes the bands 9q21-22. Del(9q) is usually not present at diagnosis of AML, but appears as a secondary change at relapse. It is usually not described in association with other recurrent primary changes but unbalanced translocations involving 9q may be considered as del(9q) variants.

EXAMPLE REPORTING OF DELETION OF CHROMOSOME 9q

ISCN Result: 46,XY,del(9)(q21)[20]

Interpretation

Chromosome analysis revealed an abnormal male chromosome complement with a deletion within the long arm of chromosome 9 as the sole anomaly in all cells examined.

Deletion of chromosome 9q is a recurrent, non-random abnormality observed primarily in acute myeloid leukemia, but may also be seen in mixed phenotypic acute leukemia. As the sole aberration, it is associated with a characteristic phenotype with single, long, slender Auer rods, vacuolation of the granulocytic lineage and erythroid dysplasia and with an intermediate prognosis.

Translocation (9;11)(p22;q23) (Figure 22.9)

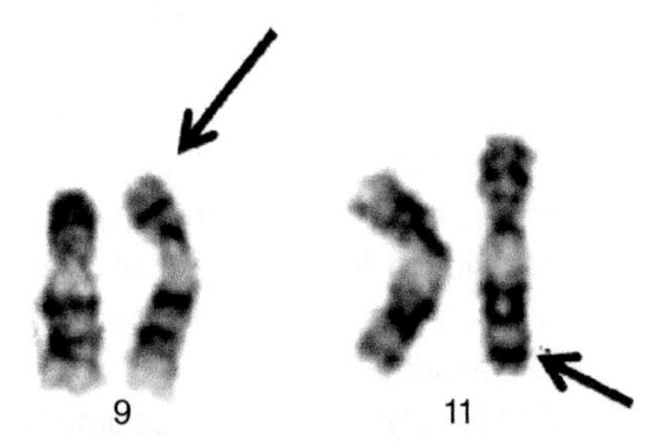

Figure 22.9 Partial karyotype showing translocation (9;11)(p22;q23).

SPECIFIC FEATURES

Translocation (9;11) is a recurrent abnormality associated with AML and is a distinct genetic category in the WHO classification. It is characterized by maturation of the monocytic lineage, often presents with disseminated intravascular coagulation, and may also have extramedullary myeloid (monocytic) sarcomas and various tissue infiltration. T(9;11) is generally seen in conjunction with monocytic and myelomonocytic leukemias, with or without maturation. It is seen predominantly in children, i.e. in 9–12% of pediatric cases, and in 2% of adult AML.

T(9;11) results in the fusion of the MLLT3 (AF9) gene on 9p22 with the MLL gene on 11q23, which is detectable by cytogenetics and FISH analysis. T(9;11) is seen often with additional karyotypic changes, most commonly with trisomy 8. This abnormality is generally associated with an intermediate prognosis in both children and adults. When t(9;11) is seen with <20% of blasts, patients should be monitored closely for the development of overt AML.

EXAMPLE REPORTING OF t(9;11)

ISCN Result: 46,XY,t(9;11)(p22;q23)[20]

Interpretation

Chromosome analysis revealed an abnormal male chromosome complement with a translocation between the short arm of chromosome 9 and the long arm of chromosome 11 as the sole anomaly in all cells examined.

The t(9;11) translocation, which results in the fusion of the MLLT3 gene at 9p22 with the MLL gene at 11q23, is a recurrent, non-random abnormality observed in acute myeloid leukemia. This translocation is generally associated with an intermediate to poor prognosis.

FISH analysis for this rearrangement is suggested to confirm and monitor this abnormal cell line.

EXAMPLE REPORTING OF t(9;11) AS A THERAPY-RELATED ABNORMALITY

ISCN Result: 46,XY,t(9;11)(p22;q23)[20]

Interpretation

Chromosome analysis revealed an abnormal male chromosome complement with a translocation between the short arm of chromosome 9 and the long arm of chromosome 11 as the sole anomaly in all cells examined.

Translocation t(9;11) leukemias are generally associated with a monocytic phenotype and may occur in either *de novo* or therapy-related AML post treatment with anti-topoisomerase II drugs. The prognosis for this 11q23/MLL-positive leukemia may not be as poor as other 11q23/MLL leukemias in *de novo* cases; however, in secondary or therapy-related disease, the prognosis with standard therapy is poor.

FISH analysis for this rearrangement is suggested to confirm and monitor this abnormal cell line.

Translocation (9;22)(q34;q11.2) (Figure 22.10)

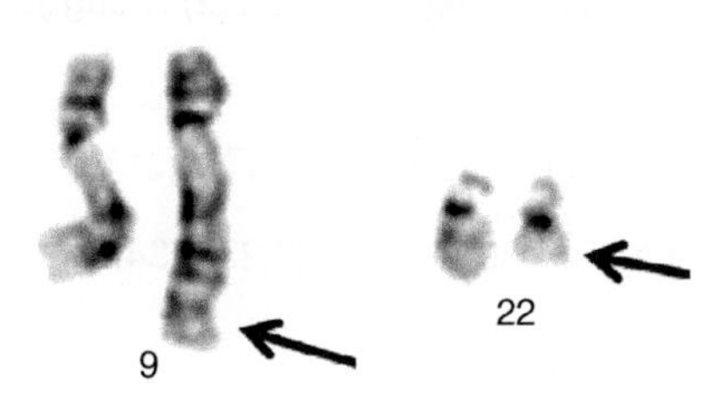

Figure 22.10 Partial karyotype showing translocation (9;22)(q34;q11.2).

SPECIFIC FEATURES

Translocation (9;22) is a recurrent abnormality associated with CML and is a distinct genetic category in the WHO classification. Generally, all CML cases have a t(9;22), at least at the molecular level, with the presence of a BCR/ABL gene rearrangement. However, some (9;22) translocations are found in AML and acute lymphoblastic leukemia (ALL). T(9;22) is found in myeloid progenitor cells and in B-lymphocytic progenitor cells. The age of onset of CML with t(9;22) varies from 30 to 60 years. The chronic phase lasts approximately 3 years, followed by an accelerated phase with a blast count of <15% and a blast crisis phase with a blast cell count of over 30%. The median survival in CML is 4 years with conventional therapy (hydroxyurea, busulfan), and 6 years with alpha-interferon therapy. Allogenic bone marrow transplantation may cure the patient.

In CML, the t(9;22) persists during remission but with acute leukemia cases, the t(9;22) disappears. Additional anomalies are seen in 10% of CML at diagnosis or may appear during the course of the disease. Even though the additional anomalies do not indicate the imminence of blast crisis, these additional anomalies emerge frequently at the time of acute transformation.

The most common secondary changes in CML with t(9;22) include an extra der(22) or Philadelphia chromosome, trisomy 8, 17, 19 and 21, i(17q), loss of chromosomes Y, 7 and 17, and a t(3;21)(q26;q22).

Variant translocations of the t(9;22) are often seen, including 3–4 chromosomes involved in a translocation, as well as variant chromosome partners with 9q34 or 22q11.2, which are found in 5–10% of cases. However, the 9q34-3'ABL gene region always joins the 22q11-5' BCR region in true CML.

EXAMPLE REPORTING OF t(9;22) WITH A NON-SPECIFIC INDICATION OR "PANCYTOPENIA"

ISCN Result: 46,XY,t(9;22)(q34;q11.2)[20]

Interpretation

Chromosome analysis revealed an abnormal male chromosome complement with a translocation between the long arms of chromosomes 9 and 22, resulting in the Philadelphia (Ph) chromosome, as the sole anomaly in all cells examined.

Translocation (9;22) results in the fusion of the ABL1 gene at 9q34 and the BCR gene at 22q11. Translocation (9;22) is observed in approximately 95% of chronic myeloid leukemia (CML) and 15–20% of acute lymphoblastic leukemia (ALL). It is seen in approximately 25% of adult B-ALL and 2–4% of childhood ALL. In both children and adults, t(9;22) is generally associated with a poor prognosis among patients with ALL.

FISH analysis for t(9;22) is suggested to confirm and monitor this abnormal cell line.

EXAMPLE REPORTING OF t(9;22) WITH AN INDICATION OF CML

ISCN Result: 46,XY,t(9;22)(q34;q11.2)[20]

Interpretation

Chromosome analysis revealed an abnormal male chromosome complement with a translocation between the long arms of chromosomes 9 and 22, resulting in the Philadelphia (Ph) chromosome, as the sole anomaly in all cells examined.

The t(9;22) translocation, which results in the fusion of the ABL1 gene at 9q34 and the BCR gene at 22q11.2, is consistent with a diagnosis of chronic myeloid leukemia in chronic phase. Additional clonal cytogenetic aberrations, commonly seen in association with clonal evolution of disease, were not observed.

FISH analysis for t(9;22) is suggested to confirm and monitor this abnormal cell line.

EXAMPLE REPORTING OF t(9;22) WITH AN INDICATION OF CML WITH SECONDARY CHANGES

ISCN Result: 48,XY,+8,t(9;22)(q34;q11.2),+der(22)t(9;22)[12]/46,XY[8]

Interpretation

Chromosome analysis revealed an abnormal male chromosome complement in 12 of 20 cells with the following clonal changes:

- a translocation between the long arms of chromosomes 9 and 22, resulting in the Philadelphia (Ph) chromosome
- a second copy of the Ph chromosome
- and gain of chromosome 8.

The remaining eight cells showed a normal male chromosome complement.

The t(9;22) translocation, which results in the fusion of the ABL1 gene at 9q34 and the BCR gene at 22q11, is consistent with a diagnosis of chronic myeloid leukemia (CML). Additional clonal cytogenetic aberrations were observed, including a second Ph chromosome and trisomy 8. These secondary changes are commonly observed in association with impending blast crisis of CML.

FISH analysis for t(9;22) is suggested to confirm and monitor this abnormal cell line.

EXAMPLE REPORTING OF A COMPLEX t(1;9;22;17) (FIGURE 22.11)

ISCN Result: 46,XY,t(1;9;22;17)(p34.2;q34;q11.2;p13)[20]

Interpretation

Chromosome analysis revealed an abnormal male chromosome complement with a four-way translocation between the short arm of chromosome 1, the long arms of chromosomes 9 and 22 and the short arm of chromosome 17, resulting in a Philadelphia (Ph) chromosome, in all cells examined.

In 5–10% of all cases, the Philadelphia chromosome arises through a variant complex translocation involving other chromosomes in addition to chromosomes 9 and 22. This variant t(9;22) results in the fusion of the ABL1 gene at 9q34 and the BCR gene at 22q11.2, and is consistent with a diagnosis of chronic myeloid leukemia in chronic phase. The additional chromosomes involved in the translocation do not alter the diagnosis or prognosis from that of the standard t(9;22) when seen as the sole anomaly.

FISH analysis for the t(9;22) is suggested to confirm and monitor this abnormal cell line.

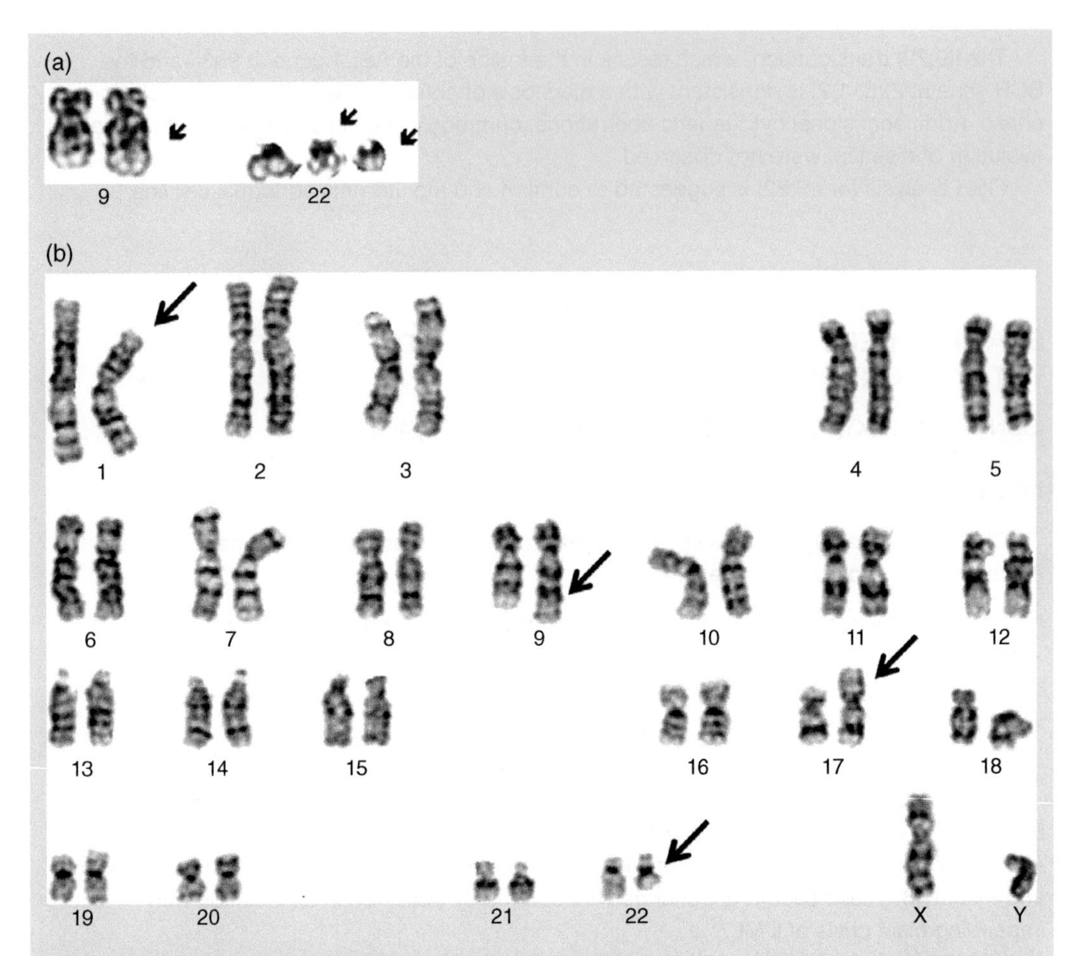

Figure 22.11 Variant abnormalities associated with t(9;22). (a) The presence of a second Philadelphia (Ph) chromosome: 47,XX,t(9;22)(q34;q11.2),+der(22)t(9;22). (b) A four-way translocation: 46,XY,t(1;9;22;17).

OTHER COMMENTS WITH SECONDARY CHANGES IN CML

- The 3;21 translocation is a recurrent, non-random abnormality associated with disease progression in chronic myeloid leukemia.
- In 5–10% of all cases, the Philadelphia chromosome arises through a variant complex translocation involving other chromosomes in addition to chromosomes 9 and 22.
- Numerical abnormalities involving chromosomes Y, 7, 17, 19 or 21 are common in the minor route of cytogenetic evolution during acceleration of disease in chronic myeloid leukemia (CML).
- During disease progression, 75–80% of CML cases develop additional chromosomal abnormalities. These secondary changes may precede hematological and clinical manifestations of more malignant disease by several months and, thus, may serve as valuable prognostic indicators.
- Disease progression in CML is associated with the acquisition of secondary cytogenetic aberrations; on average, nearly three additional cytogenetic abnormalities are present in blast crisis.

COMMENTS ON PHILADELPHIA CHROMOSOME-NEGATIVE CML

- Although a t(9;22), Philadelphia chromosome was not detected in this analysis, approximately 5% of patients with chronic myeloid leukemia are t(9;22) negative but are BCR/ABL positive by FISH or reverse transcription PCR (RT-PCR), resulting in a "cryptic" BCR/ABL gene rearrangement. If clinically indicated, FISH analysis using a BCR/ABL probe could further evaluate the presence of this gene rearrangement.
- During therapy with imatinib, some patients (approximately 2–17%) with chronic myeloid leukemia develop chromosomal abnormalities that are Philadelphia chromosome negative. These aberrations are frequently transient and their clinical consequence is unclear; however, high-risk MDS and AML after imatinib treatment occurs in a small subset (0.1%), suggesting the need for continued monitoring.
- The ___ abnormality observed in ___ cells represents a new Ph-negative clone in this patient. Recent reports describe new Ph-negative clones in chronic myeloid leukemia patients treated with imatinib mesylate. The clinical impact of these new Ph-negative clones is unclear.
- Cytogenetic studies at follow-up are suggested.

Trisomy 9

SPECIFIC FEATURES

Trisomy 9 is seen in a large spectrum of myeloid malignancies, including MDS, AML and MPN. There is a strong correlation with MPNs and especially with PV. Trisomy 9 is seen in approximately 2% of all MPNs, in 7% of PV cases, and in a small percentage (<1%) of all AML and MDS cases. It may be seen as the sole anomaly or occur with other abnormalities, including del(20q), trisomy 8 and del(13q). Trisomy 9 may represent a gain-of-function mechanism with respect to the JAK2 gene on 9p24 coding for the JAK2 kinase. It is generally associated with an intermediate prognosis as the sole aberration or with the addition of trisomy 8. Trisomy 9 seen with complex karyotypic abnormalities is generally associated with a poor prognosis.

EXAMPLE REPORTING OF TRISOMY 9

ISCN Result: 47,XY,+9[20]

Interpretation

Chromosome analysis revealed an abnormal male chromosome complement with the gain of chromosome 9 as the sole anomaly in all cells examined.

Trisomy 9 as the sole abnormality is a recurrent, non-random abnormality seen primarily in myeloid disorders including BCR/ABL1-negative chronic myeloproliferative disorders, such as polycythemia vera, chronic idiopathic myelofibrosis and essential thrombocythemia as well as in myelodysplasia and acute myeloid leukemia.

FISH analysis for chromosome 9 is suggested to confirm and monitor this abnormal cell line.

22.2.9 Chromosome 11

Del(11)(q23) (Figure 22.12)

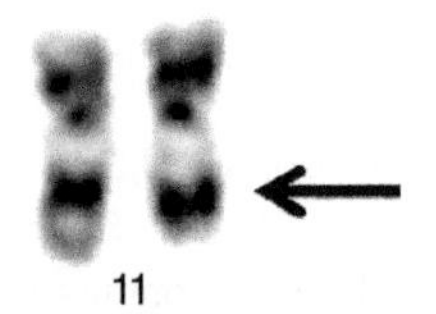

Figure 22.12 Partial karyotype showing del(11)(q23).

SPECIFIC FEATURES

Chromosome 11q23 abnormalities, including deletions, are seen in both *de novo* and therapy-related myeloid and lymphoblastic leukemias and myelodysplasias. Approximately 5–10% of 11q23 rearrangements are found in treatment-related disease, mainly post treatment with anti-topoisomerase II or intercalating topoisomerase II inhibitors, as well as after treatments with alkylating agents and/or radiotherapy.

Molecular studies have shown that MLL is rearranged more frequently than is revealed by conventional cytogenetic studies. A partial tandem duplication of the MLL gene has been reported in the majority of adult patients whose leukemic blast cells have trisomy 11 as well as in some cases with normal karyotypes.

There is a strong association between AML M4/M5 and 11q23 deletions and translocations involving 11q23. Approximately 50% of AML cases are seen in FAB subtype M5a and 20% in M4. In MDS, cases are most often refractory anemia or refractory anemia with excess blasts.

Two clinical subgroups of AML-M5 patients have a high frequency of 11q23 aberrations: one group is AML in infants with MLL rearrangement in about 50% of cases; the other group is "secondary leukemia" (sAML), potentially after treatment with DNA topoisomerase II inhibitors. In general, the translocations in these leukemias are the same as those occurring in *de novo* leukemia, i.e. t(9;11), t(11;19). Prognosis is very poor in general, yet variable according to the translocation, phenotype, age of onset and whether the leukemia is *de novo* or treatment related.

EXAMPLE REPORTING OF CHROMOSOME 11q23 DELETIONS

ISCN Result: 46,XY,del(11)(q23)[20]

Interpretation

Chromosome analysis revealed an abnormal male chromosome complement with a deletion within the long arm of chromosome 11 as the sole anomaly in all cells examined.

Deletions of the 11q23 (MLL gene) region are observed in *de novo* and therapy-related acute myeloid leukemia (AML). These rearrangements are generally associated with a poor prognosis.

FISH analysis for the MLL gene is suggested to confirm and monitor this abnormal cell line.

Del(11)(p11.2p13) and del(11)(q14q23) (Figure 22.13)

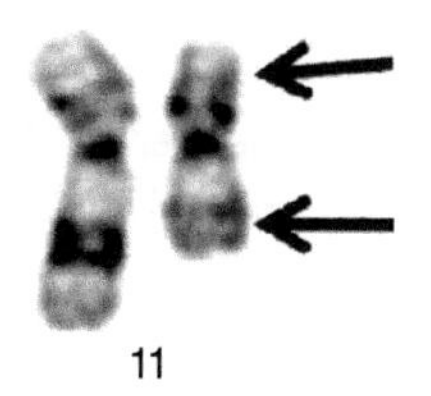

Figure 22.13 Partial karyotype showing both del(11)(p11.2p13) and del(11)(q14q23) on the same chromosome 11.

EXAMPLE REPORTING OF CHROMOSOME 11q23 DELETIONS

ISCN Result: 46,XX,der(11)del(11)(p11.2p13)del(11)(q14q23)[20]

Interpretation

Chromosome analysis revealed an abnormal female chromosome complement with deletions within both the short and long arms of chromosome 11 in all cells examined.

Deletions of the 11q23 (MLL gene) region are observed in *de novo* and therapy-related acute myeloid leukemia (AML). These rearrangements are often associated with an unfavorable prognosis.

FISH analysis for the MLL gene is suggested to confirm and monitor this abnormal cell line.

Translocation (11;17)(q13;q21)

SPECIFIC FEATURES

Translocation (11;17) is a recurrent abnormality associated with AML and is a distinct genetic category in the WHO classification. T(11;17) is a variant of the t(15;17), which is characteristic of acute promyelocytic leukemia (APL), with characteristic abnormal promyelocytes and frequently associated with disseminated intravascular coagulation (DIC). T(11;17) results in the fusion of either the ZBTB16 (PLZF) gene on 11q23 with the RARA gene on 17q12 or the NUMA1 gene at 11q13. The t(11;17) gene rearrangement with the RARA gene responds to all trans-retinoic receptor acid (ATRA) therapy, though there are morphological differences to the standard t(15;17). Therefore, it is recommended that with a presumptive diagnosis of APL showing normal chromosomes, FISH analysis with the RARA probe be performed. The standard translocation (15;17) is discussed below.

EXAMPLE REPORTING OF A VARIANT RARA GENE REARRANGEMENT

ISCN Result: 46,XY,t(11;17)(q13;q21)[20]

Interpretation

Chromosome analysis revealed an abnormal male chromosome complement with a translocation between the long arms of chromosomes 11 and 17 as the sole anomaly in all cells examined.

Translocation (11;17) is a variant of the t(15;17), which is characteristic of acute promyelocytic leukemia (APL). T(11;17) results in the fusion of the NUMA1 gene at 11q13 with the RARA gene on 17q12. This gene rearrangement involving the RARA gene responds to all trans-retinoic receptor acid (ATRA) therapy, though there are morphological differences to the standard t(15;17).

FISH analysis for the RARA gene is suggested to confirm and monitor this abnormal cell line.

Translocation (11;19)(q23;p13.1) - associated with AML (Figure 22.14)

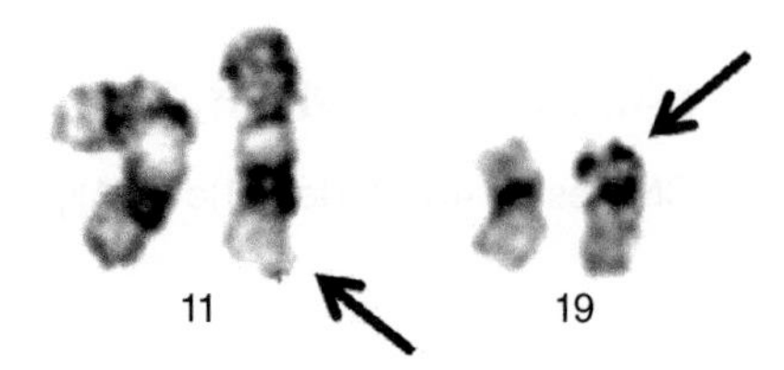

Figure 22.14 Partial karyotype showing translocation (11;19)(q23;p13.1).

SPECIFIC FEATURES

Two different translocations involving chromosomes 11 and 19 correlate with two clinical disease entities and are recurrent, non-random abnormalities in AML. Both translocations involve the MLL gene region at 11q23, but differ in the chromosome 19p13 breakpoints. These recurrent translocations involve the 19p13.1 and 19p13.3 breakpoints. The t(11;19)(q23;p13.3) is seen in both AML and lymphoid disorders and is discussed in Chapter 23. The t(11;19)(q23;p13.1) translocation is seen mainly in the FAB subtypes M4 and M5, and is often a therapy-related abnormality. This abnormality is also seen as a biphenotypic rearrangement, as are many 11q23/MLL abnormalities. This abnormality is seen in both children and adults, and often as the sole anomaly, and is generally associated with a very poor prognosis.

EXAMPLE REPORTING OF t(11;19)(q23;p13.1)

ISCN Result: 46,XY,t(11;19)(q23;p13.1)[20]

Interpretation

Chromosome analysis revealed an abnormal male chromosome complement with a translocation between the long arm of chromosome 11 and the short arm of chromosome 19 as the sole anomaly in all cells examined.

This translocation results in the fusion of the MLL gene at 11q23 with the ELL gene at 19p13.1 and is consistent with a diagnosis of acute myeloid leukemia (AML). AML with this translocation is generally associated with a poor prognosis.

FISH analysis for the 11q23/MLL gene is suggested to confirm and monitor this abnormal cell line.

Translocation (v;11)(V;q23)

SPECIFIC FEATURES

Variant rearrangements of the 11q23 region involving the MLL locus are recurrent abnormalities associated with AML and are in a distinct genetic category in the WHO classification. Over 50 translocation partners fused with the MLL gene are reported in the literature. They are seen in both pediatric and adult AML cases, but with different disease entities. For example, the t(4;11)/MLL/MLLT2 (AF4) gene rearrangement results predominantly in acute lymphoblastic leukemia whereas the t(9;11)/MLL/MLLT3 (AF9) gene rearrangement results primarily in AML, thereby making MLL rearrangements characteristic of biphenotypic leukemia. For a list of MLL translocation partners, see Table 22.2.

Table 22.2 Most common 11q23 rearrangements

11q23 translocations	Breakpoints of partner chromosome	Gene identified in partner chromosome
t(1;11)	1q21	AF1* (EPS15, MLLT5)
t(4;11)	4q21	AF4 (MLLT2)
t(6;11)	6q27	AF6 (MLLT4)
t(9;11)	9p21.3	AF9 (LTG9, MLLT3)
t(10;11)	10p11.2	AF10 (ABI1, E3B1)
t(11;17)	17q21	AF17 (MLLT6)
t(11;19)	19p13.1	ELL
t(11;19)	19p13.3	ELN

*AF, acute lymphoblastic leukemia (ALL) fused gene.

EXAMPLE REPORTING OF t(10;11)

ISCN Result: 46,XY,t(10;11)(p11.2;q23)[20]

Interpretation

Chromosome analysis revealed an abnormal male chromosome complement with a translocation between the short arm of chromosome 10 and the long arm of chromosome 11 as the sole anomaly in all cells examined.

This translocation is a rare but recurrent, non-random abnormality observed in acute myeloid leukemia with monocytic involvement (FAB subtype M4 and M5). This translocation results in fusion between the MLL gene on 11q23 and the ABI1 gene located on 10p11.2. However, the possibility that the breakpoint on chromosome 10 is p12, with involvement of the AF10 gene, cannot be excluded.

FISH analysis for the 11q23/MLL gene is suggested to confirm and monitor this abnormal cell line.

Trisomy 11

SPECIFIC FEATURES

Trisomy 11, when seen as the sole anomaly, is observed mainly in MDS and AML, most prominently in FAB subtypes M1, M2 and M4. It is also observed in therapy-related AML and in MDS evolving towards AML. Trisomy 11 has been reported in 1% of AML and 1% of MDS cases, with an equal male-to-female ratio. This is mainly an adult-onset disease with the median age of 60 years. Prognosis is usually intermediate to poor.

EXAMPLE REPORTING OF TRISOMY 11

ISCN Result: 47,XY,+11[20]

Interpretation

Chromosome analysis revealed an abnormal male chromosome complement with the gain of chromosome 11 as the sole anomaly in all cells examined.

Trisomy 11 as the sole abnormality is a recurrent, non-random abnormality seen primarily in acute myeloid leukemia and myelodysplasia, and is generally associated with an intermediate to poor prognosis.

FISH analysis for chromosome 11 is suggested to confirm and monitor this abnormal cell line.

22.2.10 Chromosome 12

Add(12p)

SPECIFIC FEATURES

Rearrangements of chromosome 12p are recurrent abnormalities associated with a broad spectrum of hematological malignancies including acute leukemias, MDS and MPNs. These abnormalities are seen in approximately 5% of AML and MDS cases and are also seen in approximately 10% of secondary leukemias after prior mutagenic exposure (10%). Chromosome 12p aberrations are seen as the sole anomaly in 20% of cases. When seen with other karyotypic changes, numerical and structural rearrangements of chromosomes 5 and 7 are most frequently observed.

Additional material on chromosome 12p is seen frequently, which can be considered as an unbalanced translocation whose partner chromosome is unidentified. When translocations involve chromosome 12p, they are mostly associated with loss of 12p material.

Chromosome 12p abnormalities are generally associated with a poor prognosis, especially when seen in conjunction with other cytogenetic aberrations. However, different clinical courses are defined based on the amount of 12p material deleted. For example, small deletions have a better prognosis in general and tend not to have additional chromosomal rearrangements. A minimal interstitial deletion region involves the ETV6 and CDKN1B genes.

EXAMPLE REPORTING OF CHROMOSOME 12p REARRANGEMENTS

ISCN Result: 46,XY,add(12)(p13)

Interpretation

Chromosome analysis revealed an abnormal male chromosome complement with added material on the short arm of chromosome 12 as the sole anomaly in all cells examined.

Rearrangements of the short arm of chromosome 12 are recurrent, non-random abnormalities in myelodysplasia (MDS) and acute myeloid leukemia (AML).

The 12p13 breakpoint involves a putative transcription factor gene, ETV6 (TEL), which has been identified as a fusion gene partner with a variety of chromosome partners in myeloid as well as lymphoid leukemias.

FISH analysis for the ETV6 gene is suggested to confirm and monitor this abnormal cell line.

22.2.11 Chromosome 13

Del(13q) (Figure 22.15)

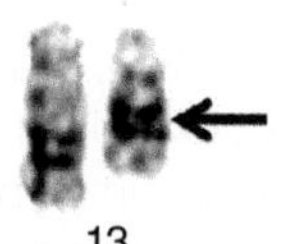

Figure 22.15 Partial karyotype showing del(13)(q12q14).

SPECIFIC FEATURES

Deletions within the long arm of chromosome 13 are recurrent, non-random abnormalities associated with both myeloid and lymphoid disorders including AML, MDS and MPNs, mostly in polycythemia vera and idiopathic myelofibrosis. When seen in myeloid disorders, other deletions including chromosomes 5, 7 and 20 are often observed. It is also a common secondary change post therapy.

Deletion of 13q is also a common aberration found in an accelerated phase of CML with persistent or relapsed disease following bone marrow transplant. Chromosome 13 rearrangements are composed of both translocations and deletions, where a common region of deletion is identified at bands 13q12-q14.

The deletion has been described as interstitial in most cases, with breakpoints at q13-q21 (seen most frequently), q13-q22, q14-q22 or q12-q21. Loss of 13q12-q32 appears to be prevalent in MPD and MDS, and loss of the 13q21 band is more frequent in AML. Translocations involving chromosome 13 may occur, sometimes with cryptic microdeletions; however, apart from the t(12;13)(p12;q14), all are seen sporadically.

EXAMPLE REPORTING OF DELETION OF CHROMOSOME 13q

ISCN Result: 46,XY,del(13)(q12q14)[20]

Interpretation

Chromosome analysis revealed an abnormal male chromosome complement with a deletion within the long arm of chromosome 13 as the sole anomaly in all cells examined.

Deletion of 13q is a recurrent, non-random abnormality in myeloid disorders, including myelodysplasia, acute myeloid leukemia and myeloproliferative disorders, and also in the accelerated phase of chronic myeloid leukemia. Del(13q) is also a recurrent finding in lymphoproliferative disorders. Therefore, clinical-pathological correlation of results is suggested.

FISH analysis for chromosome 13q is suggested to confirm and monitor this abnormal cell line.

22.2.12 Chromosome 14

Trisomy 14

SPECIFIC FEATURES

Trisomy 14, when seen as the sole anomaly, is primarily observed in myeloid disorders, including MDS, AML and MPNs. It is mainly seen in adults with a median age of 60–65 years, and the sex ratio is slightly higher in males. This abnormality is seen in all FAB subtypes of MDS, and in atypical CML cases with dysplastic features and in which non-lobulated megakaryocytes are often found.

EXAMPLE REPORTING OF TRISOMY 14

ISCN Result: 47,XY,+14[20]

Interpretation

Chromosome analysis revealed an abnormal male chromosome complement with the gain of chromosome 14 as the sole anomaly in all cells examined.

Trisomy 14 as a sole abnormality is a recurrent, non-random abnormality seen primarily in myelodysplasia, acute myeloid leukemia and atypical chronic myeloid leukemia.

FISH analysis for chromosome 14 is suggested to confirm and monitor this abnormal cell line.

22.2.13 Chromosome 15

Translocation (15;17)(q24;q21) (Figure 22.16)

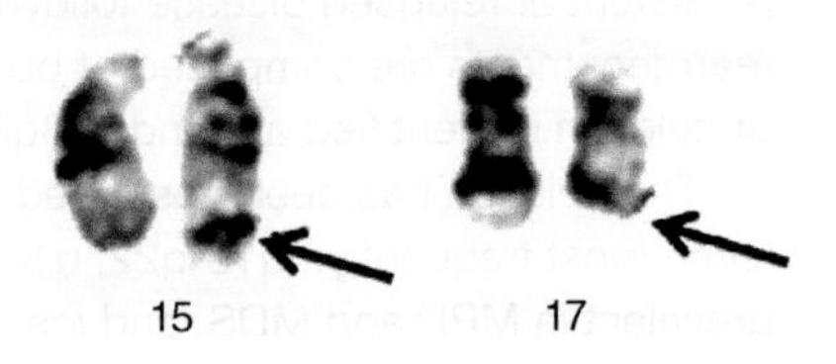

Figure 22.16 Partial karyotype showing translocation (15;17)(q24;q21).

SPECIFIC FEATURES

Translocation (15;17) is a recurrent abnormality associated with AML and is a distinct genetic category in the WHO classification. T(15;17) is pathonomonic for acute promyelocytic leukemia, with characteristic abnormal promyelocytes and frequently associated with disseminated intravascular coagulation (DIC). T(15;17) is seen in 5–8% of AML cases, most of which are middle-aged adults, though it is seen in adults of any age. T(15;17) results in the fusion of the PML gene on 15q24 with the RARA gene on 17q12, which is detectable by cytogenetics and FISH analyses. T(15;17) with the PML/RARA gene rearrangement responds well to ATRA therapy and, when diagnosed early in disease will have a better outcome than without this treatment. T(15;17) is seen often (40%) with additional karyotypic changes, most commonly trisomy 8. Mutations of the FLT3 gene are also seen in approximately 35–40% of APL patients.

Acute promyelocytic leukemia with variant RARA translocations are common, including complex translocations or masked RARA translocations without a visible chromosome abnormality. Variant translocations of t(15;17) are also common. The most common variant translocation is the t(11;17), which is discussed above. Also reported as a recurrent translocation is a t(5;17)(q35;q12), which also involves the RARA locus and responds well with ATRA therapy. Therefore, it is recommended that with a presumptive diagnosis of APL showing normal chromosomes, FISH analysis with the RARA probe be performed.

T(15;17) APL is generally associated with a good prognosis when treated with ATRA therapy; however, prognosis with the presence of FLT3 mutations has not been well described.

EXAMPLE REPORTING OF t(15;17)(q24;q21)

ISCN Result: 46,XY,t(15;17)(q24;q21)[20]

Interpretation

Chromosome analysis revealed an abnormal male chromosome complement with a translocation between the long arms of chromosomes 15 and 17 as the sole anomaly in all cells examined.

This translocation is a recurrent, non-random abnormality which results in the fusion of the PML gene at 15q24 with the RARA gene at 17q21 and is consistent with a diagnosis of acute promyelocytic leukemia (APL). T(15;17) APL is generally associated with a good prognosis when treated with ATRA therapy.

FISH analysis for the RARA gene is suggested to confirm and monitor this abnormal cell line.

22.2.14 Chromosome 16

Inv(16)(p13.1q22) (Figure 22.17)

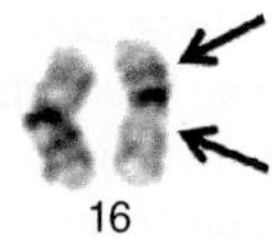

Figure 22.17 Partial karyotype showing inv(16)(p13.1q22).

SPECIFIC FEATURES

Inversions of chromosome 16 and translocation (16;16) are recurrent abnormalities associated with AML and are a distinct genetic category in the WHO classification. Both abnormalities are characterized by maturation of the monocytic and granulocytic lineages and are often present with abnormal eosinophils in the marrow. They may present with myeloid sarcomas at diagnosis or at relapse. Inv(16) and t(16;16) are seen in approximately 5–8% of AML cases and most often in younger adult patients. The inv(16) has also been observed as a secondary, therapy-related leukemia in conjunction with refractory anemia with excess blasts.

Inv(16) and t(16;16) result in the fusion of the CBFB gene on 16q22 with the MYH11 gene on 16p13.1, which is detectable by cytogenetics and FISH analyses. However, both of these abnormalities are subtle cytogenetic changes and may be missed. Consequently, if chromosome studies are suboptimal or normal, FISH analysis is suggested to rule out a chromosome 16 abnormality. Inv(16) and t(16;16) are seen with additional karyotypic changes in approximately 40% of cases, most commonly with trisomy 8, 21 or 22, or del(7q).

Trisomy 22 is a common secondary change with inv(16) and is rarely seen with other AML abnormalities. Therefore, if trisomy 22 is observed by chromosome analysis, attention to chromosome 16 abnormalities is suggested. Inv(16) is also associated with c-KIT mutations, which occur in approximately 30% of cases.

This abnormality is generally associated with a good prognosis even with trisomy 22 as a secondary change. However, older patients have a decreased survival rate, as do patients with c-KIT mutations.

EXAMPLE REPORTING OF CHROMOSOME 16q REARRANGEMENTS

ISCN Result: 46,XY,inv(16)(p13.1q22)[20]

Interpretation

Chromosome analysis revealed an abnormal male chromosome complement with a pericentric inversion of chromosome 16 as the sole anomaly in all cells examined.

Inversion 16 is a recurrent, non-random abnormality associated with acute myeloid leukemia FAB-M4 or FAB-M2, and is often associated with eosinophilia. The inversion creates a novel fusion gene product involving the myosin heavy chain gene, MYH11 at 16p13.1, and the CBFB transcription factor gene at 16q22. The inv(16) has also been observed as a secondary, therapy-related leukemia in conjunction with refractory anemia with excess blasts.

A favorable response to therapy with a prolonged remission duration after high-dose cytarabine has been reported for inv(16) leukemia. The c-KIT mutation, occurring in 20–40% of AML with inv(16), carries an increased risk of relapse and a lower overall survival; therefore, c-KIT mutation analysis is suggested.

FISH analysis for the CBFB gene is suggested to confirm and monitor this abnormal cell line.

22.2.15 Chromosome 17

Del(17p)

SPECIFIC FEATURES

Deletions of chromosome 17p, usually involving theTP53 gene at 17p13.1, are recurrent abnormalities associated with AML and MDS with a particular form of morphological dysgranulopoiesis. These morphological abnormalities involve neutrophilic as well as eosinophilic and basophilic lineages. These are seen mainly in refractory anemia with excess blasts in MDS, and often in AML FAB subtypes of M2 or M6, and can include multilineage involvement.

Approximately 30% of myeloid cases with 17p deletions are therapy related, occurring after a lymphoid neoplasm or a solid tumor treated by chemotherapy with an alkylating agent, or after essential thrombocythemia or polycythemia vera treated by hydroxyurea alone or associated with other drugs.

Chromosome 17p deletions may result from monosomy 17, isochromosome 17q or an unbalanced translocation between 17p and another chromosome. This abnormality is mainly seen in adults with a mean age of over 60 years and equally among females and males. Chromosome 17p deletions are generally associated with a very poor prognosis with a median survival of only 4 months.

EXAMPLE REPORTING OF DELETION OF CHROMOSOME 17p (FIGURE 22.18)

ISCN Result: 46,XY,del(17)(p11.2p13)[20]

Interpretation

Chromosome analysis revealed an abnormal male chromosome complement with a deletion within the short arm of chromosome 17 as the sole anomaly in all cells examined.

Deletion of 17p is a non-random, recurrent abnormality observed in both *de novo* and secondary myelodysplasia and acute myeloid leukemia. Deletions of chromosome 17p, resulting in loss of the TP53 gene, are generally associated with a poor prognosis.

FISH analysis for the TP53 gene is suggested to confirm and monitor this abnormal cell line.

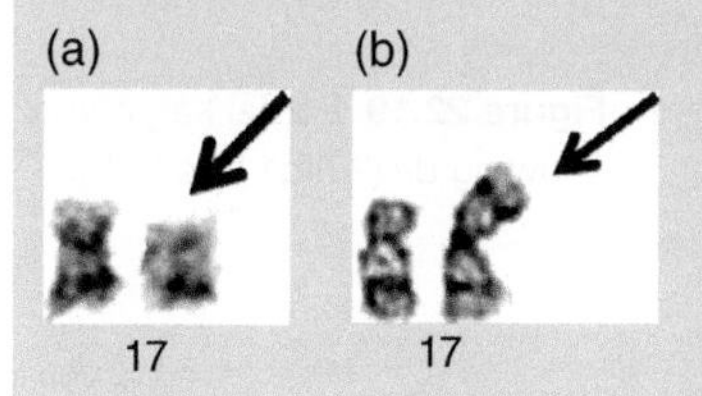

Figure 22.18 Chromosome 17p abnormalities. (a) del(17)(p11.2). (b) 46,XY,i(17)(q10).

i(17q)

SPECIFIC FEATURES

Isochromosome of the long arm of chromosome 17 is a recurrent abnormality associated with AML. It results in loss of the short arm (17p) and duplication of the long arm (17q). I(17q) is usually observed with other chromosomal abnormalities in myeloid disorders, including CML,

AML, MDS and MPN, as well as lymphoid disorders and solid tumors. In CML, i(17q) is a frequent secondary anomaly, either solely in 10% of cases or with other additional anomalies, in particular with trisomy 8.

It is possible that i(17q), when seen as the sole abnormality, is a distinctive clinical-pathological entity with a high risk of leukemic progression. However, a subset may present as *de novo* AML. The distinctive morphological features include multilineage dysplasia and concurrent myeloproliferative features. Since i(17q) usually occurs at the time of blast transformation, it is generally associated with an aggressive clinical course and poor prognosis.

EXAMPLE REPORTING OF ISOCHROMOSOME 17q (FIGURE 22.18b)

ISCN Result: 46,XY,i(17)(q10)[20]

Interpretation

Chromosome analysis revealed an abnormal male chromosome complement with an isochromosome of the long arm of chromosome 17, resulting in loss of the short arm and the TP53 gene, as the sole anomaly in all cells examined.

Isochromosome 17q often occurs in patients with dysplastic bone marrow and is seen less frequently in myeloproliferative disorders. The tumor suppressor TP53 gene, mapped to 17p13.1, is deleted and generally, patients progress rapidly to acute leukemia and are resistant to chemotherapy with an associated shorter survival duration after transformation.

FISH analysis for the TP53 gene is suggested to confirm and monitor this abnormal cell line.

22.2.16 Chromosome 20

Del(20)(q11.2q13.3) (Figure 22.19)

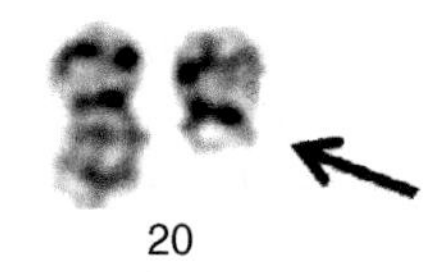

Figure 22.19 Partial karyotype showing del(20)(q11.2q13.3).

SPECIFIC FEATURES

Deletions of the long arm of chromosome 20 are seen in a large spectrum of hematological malignancies, including MDS, AML and MPNs including polycythemia vera and chronic neutrophilic leukemia. Del(20q) appears as a primary cytogenetic abnormality in disease, occurring in pluripotent hematopoietic stem cells. However, the pathogenic mechanism by which del(20q) alters the hematopoietic stem cells in hematological disorders remains unknown. Del(20q) is frequently associated with other cytogenetic abnormalities, including del(5q), trisomy 8, trisomy 21, and deletions or translocations involving the long arm of chromosome 13.

When seen in MDS as the sole anomaly, del(20q) is generally associated with a good prognosis regarding survival and potential for AML evolution, as defined by the IPSS. However, in *de novo* acute leukemia, there is generally a poor response to treatment and a reduced survival time. In MPNs, the presence of del(20q) does not appear to adversely affect survival. When seen as the sole anomaly, it is not considered definitive evidence for MDS in the absence of morphological criteria, based on the WHO classification.

EXAMPLE REPORTING OF DELETION OF CHROMOSOME 20q

ISCN Result: 46,XY,del(20)(q11.2q13.3)[20]

Interpretation

Chromosome analysis revealed an abnormal male chromosome complement with a deletion within the long arm of chromosome 20 as the sole anomaly in all cells examined.

Deletion of 20q is a recurrent, non-random abnormality observed in a large spectrum of hematological malignancies, including myelodysplasia (MDS), acute myeloid leukemia and polycythemia vera. In MDS, del(20q) as the sole abnormality is associated with a favorable prognosis. However, del(20q) as a sole abnormality is not definitive evidence of MDS in the absence of morphological criteria.

FISH analysis for chromosome 20q is suggested to confirm and monitor this abnormal cell line.

ider(20)(q10)del(20)(q11.2q13.1) (Figure 22.20)

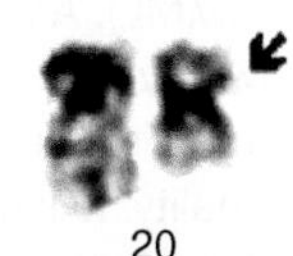

Figure 22.20 Partial karyotype showing ider(20)(q10)del(20)(q11.2q13.1). Courtesy of Sarah South PhD, ARUP Laboratories.

SPECIFIC FEATURES

When monosomy of chromosome 20 with small metacentric marker chromosome is observed, it is most likely an isoderivative of chromosome 20. The ider(20)(q10)del(20)(q11.2q13.1) is a variant of del(20)(q11.2q13.1). The ider(20q) can be both monocentric or dicentric, with the proximal breakpoint located at band 20q11.2. The commonly deleted region includes the short arm of chromosome 20 and a large region on the long arm of chromosome 20, spanning from 20q11.2 to 20q13.1. Additional anomalies may be seen with the ider(20q), including del(20q), detected by standard cytogenetics or by FISH, two copies of the ider(20q) chromosome, monosomy 7 and complex chromosomal abnormalities. Prognosis of patients with ider(20q) seems to be poor compared to those with del(20q), but is unclear due to the small number of cases reported.

EXAMPLE REPORTING OF ISODERIVATIVE CHROMOSOME 20q

ISCN Result: 46,XY, 46,XY,ider(20)(q10)del(20)(q11.2q13.1)[20]

Interpretation

Chromosome analysis revealed an abnormal male chromosome complement with an isoderivative chromosome 20, resulting in a deletion within the long arm of chromosome 20 as the sole anomaly in all cells examined.

Ider(20q) is a variant of the del(20q) and is a recurrent, non-random abnormality observed in a large spectrum of hematological malignancies including myelodysplasia (MDS), acute myeloid leukemia and polycythemia vera. Prognosis of patients with ider(20q) seems to be poor compared to patients with del(20q), but is unclear due to the small number of cases reported.

FISH analysis for chromosome 20q is suggested to confirm and monitor this abnormal cell line.

22.2.17 Chromosome 21

Trisomy 21

SPECIFIC FEATURES

Trisomy 21 as an acquired abnormality is a recurrent, non-random aberration associated with myeloid disorders, including MDS and AML. Trisomy 21 is not seen in any specific FAB subtype, but is possibly seen in a higher incidence in monocytic phenotypes (ANLL-M4 and -M5, and CMML). AML type M7 with acquired trisomy 21 is rare whereas AML type M7 is frequent in Down syndrome with AML disease. Trisomy 21 is the second most frequently acquired trisomy, after trisomy 8, in adult-onset myeloid disorders and is rarely observed as the sole abnormality. It is more frequently associated with chromosome 5 and 7 abnormalities, trisomy 8 and translocations (8;21), (15;17) and inv(16).

Gains of chromosome 21, including tetrasomy or pentasomy 21, as well as single or multiple copies of a structurally rearranged chromosome 21, such as i(21q), psu dic(21q) or r(21), are common. Amplified chromosome 21 as a homogeneous staining region (HSR) is also observed.

EXAMPLE REPORTING OF TRISOMY 21

ISCN Result: 47,XY,+21[20]

Interpretation

Chromosome analysis revealed an abnormal male chromosome complement with the gain of chromosome 21 as the sole anomaly in all cells examined.

Trisomy 21 is a recurrent, non-random abnormality in myelodysplasia, acute myeloid leukemia and acute lymphoblastic leukemia. When it occurs as the sole cytogenetic abnormality, trisomy 21 is generally associated with a poor prognosis.

Since trisomy 21 is also seen as a constitutional genetic disorder (Down syndrome), a peripheral blood specimen for routine chromosome analysis is suggested if a clinical indication of Down syndrome is suspected.

FISH analysis for chromosome 21 is suggested to confirm and monitor this abnormal cell line.

22.2.18 Chromosome X

idic(X)(q13) (Figure 22.21)

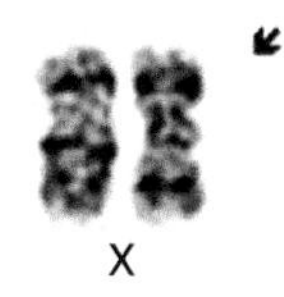

Figure 22.21 Partial karyotype showing idic(X)(q13).

SPECIFIC FEATURES

Isodicentric chromosome Xq13 is a recurrent abnormality associated with myeloid disorders, including MDS, MPNs and AML. It is seen in AML subtypes M1, M2 and M4, often with preceding MDS, and in MDS often with RARS in which an early progenitor cell is involved. It has only been reported in female patients, with one normal X chromosome and one X chromosome that involves the duplication of Xpter-Xq13 and a deletion of Xq13-Xqter. The breakpoint is located within a 450 kb region proximal from the XIST gene and contains an inverted repeat segment of DNA; however, both centromeres appear to be active.

This abnormality appears similar to the i(Xp) abnormality. However, since it is clear that two centromeres and a small part of the long arm of the X chromosome are present, seen by the split of the centromeres, the appropriate ISCN designation is idic(X)(q13), rather than i(X)(p10).

EXAMPLE REPORTING OF idic(X)(q13)

ISCN Result: 46,XX,idic(X)(q13)[20]

Interpretation

Chromosome analysis revealed an abnormal female chromosome complement with an isodicentric X chromosome of the long arm as the sole anomaly in all cells examined.

Isodicentric Xq13 rearrangements result in the loss of part of the long arm (Xq13-Xqter) and gain of the short arm (Xpter-Xq13), and are recurrent, non-random abnormalities in myeloid disorders, including myelodysplasia, myeloproliferative neoplasms and acute myeloid leukemia.

Bibliography

American College of Medical Genetics. *Guidelines in Clinical Cytogenetics. Section E10 Methods in Fluorescence In situ Hybridization – Interphase/Nuclear Fluorescence In Situ Hybridization.* American College of Medical Genetics, Bethesda, MD. www.acmg.net/StaticContent/SGs/Section_E.html, accessed 28 January 2014.

Atlas of Genetics and Cytogenetics in Oncology and Haematology. http://AtlasGeneticsOncology.org/Anomalies, accessed 28 January 2014.

Bennett JM. A comparative review of classification systems in myelodysplastic syndromes (MDS). Semin Oncol 2005; 32(4 suppl 5): S3–S10.

Buijs A, Sherr S, van Baal S, et al. Translocation (12;22) (p13;q11) in myeloproliferative disorders results in fusion of the ETS-like TEL gene on 12p13 to the MN1 gene on 22q11. Oncogene 1995; 10: 1511–1519.

Cherian S, Bagg A. The genetics of the myelodysplastic syndromes: classical cytogenetics and recent molecular insights. Hematology 2006; 11: 1–13.

Bloomfield C, Lawrence D, Byrd J, et al. Frequency of prolonged remission duration after high-dose cytarabine intensification in acute myeloid leukemia varies by cytogenetic subtype. Cancer Res 1998; 58:4173-4179, 1998.

Dierlamm J, Michaux L, Criel A, et al. Isodicentric (X)(q13) in haematological malignancies: presentation of five new cases, application of fluorescence in situ hybridization (FISH) and review of the literature. Br J Haematol 1995; 91(4): 885–891.

Gersen S, Keagle M (eds). Principles of Clinical Cytogenetics. Humana Press, Totowa, New Jersey, 1999.

Greenberg P, Cox C, LeBeau MM, et al. International scoring system for evaluating prognosis in myelodysplastic syndromes. Blood 1997; 89: 2079–2088.

Heaney ML, Golde DW. Medical progress: myelodysplasia. N Engl J Med 1999; 340: 1649–1660.

Heim S, Mitelman F. *Cancer Cytogenetics*, 2nd edn. Wiley-Liss, New York, 1995.

Hellman A, Zlotorynski E, Scherer SW, et al. A role for common fragile site induction in amplification of human oncogenes. Cancer Cell 2002; 1: 89–97.

Hori T, Takahashi E, Murata M. Nature of distamycin A-inducible fragile sites. Cancer Genet Cytogenet 1988; 34: 189–194.

Pennings J, van de Locht L, Jansen J, et al. Degradable dU-based DNA template as a standard in real-time PCR quantitation. Leukemia 1997; 11: 2217–2219.

Kovitz C, Kantarjian H, Garcia-Manero G, Abruzzo LV, Cortes J. Myelodysplastic syndromes and acute leukemia developing after imatinib mesylate therapy for chronic myeloid leukemia. Blood 2006; 108: 2811–2813.

Kuffel DG, Schultz CG, Ash RC, Dewald GW. Normal cytogenetic values for bone marrow based on studies of bone marrow transplant donors. Cancer Genet Cytogenet 1991; 55: 39–48.

Melchert M, Kale V, List A. The role of lenalidomide in the treatment of patients with chromosome 5q deletion and other myelodysplastic syndromes. Curr Opin Hematol 2007; 14(2): 123–129.

Morerio C, Rosanda C, Rapella A, Micalizzi C, Panarello C. Is t(10;11)(p11.2;q23) involving MLL and ABI-1 genes associated with congenital acute monocytic leukemia? Cancer Genet Cytogenet 2002; 139: 57–59.

Murata M, Takahashi E, Minamihisamatsu M, et al. Heritable rare fragile sites in patients with leukemia and other hematologic disorders. Cancer Genet Cytogenet 1988; 31: 95–103.

Ogawa S, Kurokawa M, Tanaka T, et al. Structurally altered Evi-1 protein generated in the 3q21q26 syndrome. Oncogene 1996; 13(1): 183–191.

Ogawa S, Mitani K, Kurokawa M, et al. Abnormal expression of Evi-1 gene in human leukemias. Hum Cell 1996; 9: 323–332.

Peniket A, Wainscoat J, Side L, et al. Del (9q) AML: clinical and cytological characteristics and prognostic implications. Br J Haematol 2005; 129: 210–220.

Plantier I, Lai JL, Wattel E, Bauters F, Fenaux P. Inv(16) may be one of the only 'favorable' factors in acute myeloid leukemia: a report on 19 cases with prolonged follow-up. Leukemia Res 1994; 18: 885–888.

Reiter E, Greinix H, Rabitsch W, et al. Low curative potential of bone marrow transplantation for highly aggressive acute myelogenous leukemia with inversioin inv (3)(q21q26) or homologous translocation t(3;3) (q21;q26). Ann Hematol 2000; 79: 374–377.

Rooney DE, Czepulkowski BH. *Human Cytogenetics. A Practical Approach*. Oxford University Press, New York, 1992.

Sakai H, Karasawa M, Okamoto K, et al. Leukemic transformation in three patients with polycythemia vera. Analysis of the clinicopathological features and N-ras gene mutation. J Med 1996; 27(3-4): 183–191.

Shaffer LG, McGowan-Jordan J, Schmid M (eds). *ISCN 2013: An International System for Human Cytogenetic Nomenclature*. Karger Publishers, Unionville, CT, 2013.

Smith SR, Rowe D. Trisomy 15 in hematological malignancies: six cases and review of the literature. Cancer Genet Cytogenet 1996; 89: 27–30.

Sole F, Espinet B, Sanz GF, et al. Incidence, characterization and prognostic significance of chromosomal abnormalities in 640 patients with primary myelodysplastic syndromes. Br J Haematol 2000; 108: 346–356.

Steensma DP, Bennett JM. The myelodysplastic syndromes: diagnosis and treatment. Mayo Clin Proc 2006; 81: 104–130.

Steensma DP, List AF. Genetic testing in the myelodysplastic syndromes: molecular insights into hematologic diversity. Mayo Clin Proc 2005; 80: 681–698.

Steensma DP, Tefferi A. The myelodysplastic syndrome(s): a perspective and review highlighting current controversies. Leuk Res 2003; 27: 95–120.

Swerdlow SH, Campo E, Harris NL, et al. (eds). *WHO Classification of Tumours of Haematopoietic and Lymphoid Tissues*. IARC Press, Lyon, 2008.

Warburton D. De novo balanced chromosome rearrangements and extra marker chromosomes identified at prenatal diagnosis: clinical significance and distribution of breakpoints. Am J Hum Genet 1991; 49:995–1013.

Wiktor A, Rybicki BA, Piao ZS, et al. Clinical significance of Y chromosome loss in hematologic disease. Genes Chromosomes Cancer 2000; 27: 11–16.

Wlodarska I, Mecucci C, Baens M, Marynen P, van den Berghe H. ETV6 gene rearrangements in hematopoietic malignant disorders. Leuk Lymphoma 1996; 23: 287–295.

Wlodarska I, Mecucci C, Marynen P, et al. TEL gene is involved in myelodysplastic syndromes with either the typical t(5;12)(q33;p13) translocation or its variant t(10;12)(q24;p13). Blood 1995; 85: 2848–2852.

CHAPTER 23

Cytogenetic abnormalities in lymphoid disorders

23.1 Introduction to lymphoid disorders

In the WHO classification, lymphoid disorders are defined as clonal abnormalities of immature and mature B-cells, T -cells or natural killer (NK) cells, including acute lymphoblastic leukemia, chronic lymphocytic leukemia, non-Hodgkin lymphoma, Hodgkin lymphoma and the plasma cell disorders. Many of these neoplasms correspond to the degree of cell maturation and are subtyped accordingly. However, other disorders do not follow this model, for example, hairy cell leukemia.

From the genetic perspective, certain lymphoid neoplasms have specific chromosomal changes that are involved in the underlying etiology of the disease process. Therefore, chromosome and fluorescence *in situ* hybridization (FISH) analyses are very useful in the classification of disease and are prognostic indicators of disease outcomes. For example, t(11;14) is seen predominantly in mantle cell lymphoma (MCL), t(11;18) in mucosa-associated lymphoid tissue (MALT) lymphoma, t(14;18) in follicular lymphoma (FL) and t(8;14) in Burkitt lymphoma. Although these rearrangements are not exclusively associated with one specific disease entity, often differences in the translocation breakpoints exist, giving rise to variation in protein products and consequently the disease phenotype. For example, the very common chromosomal rearrangement involving the immunoglobulin heavy chain (IGH) at chromosome 14q32 acts as a promoter that enhances a protooncogene that becomes overexpressed. One such rearrangement, the t(14;18) seen in FL, results in the BCL2 gene on chromosome 18q22 being fused with the IGH gene on chromosome 14q32, giving rise to overexpression of BCL2, which stops apoptosis in the germinal center B-cells. Most of the cytogenetic abnormalities are described in B-cell neoplasms, with very few T-cell chromosomal anomalies. The exception to this is anaplastic large cell lymphoma, a T-cell disorder which is defined by rearrangements of the ALK gene at 5q23.

Acute lymphoblastic leukemia (ALL) and lymphoma are defined as neoplasms of precursor lymphoblasts that involve the bone marrow and, with lymphomas, involve nodal or extranodal sites. Cytogenetic abnormalities are seen in the majority of B-cell ALL, often with defined specific disease subgroups as well as their prognoses, such as hyper- or hypodiploidy, t(9;22), t(4;11) and

Cytogenetic Abnormalities: Chromosomal, FISH and Microarray-Based Clinical Reporting, First Edition. Susan Mahler Zneimer.

t(12;21). However, there are numerous chromosomal abnormalities that are recurrent but non-specific to a subtype of disease and are seen in many lymphoid disorders, such as del(6q), del(9p) and del(12p).

For chromosomal analysis of lymphoid disorders, the type of culture initiated in the laboratory for each specimen is very important in optimizing the detection of abnormal cells. Mitogen-stimulated cultures for the detection of mature B-cell lymphoproliferative disorders (LPDs) are commonly used to stimulate the otherwise quiescent abnormal cells that do not spontaneously divide in culture. For LPDs, most standard chromosome studies yield a normal result. Therefore, FISH analysis is important as a conjunctive test to help identify recurrent abnormalities present in these disorders. Since FISH analysis is performed on interphase cells that are not undergoing active mitosis, it does not require cell culturing for metaphase analysis. This is a good approach to detect many abnormalities rather than chromosome analysis alone.

For ALL, only unstimulated cultures are used, and those need to be 24 hours or less in culture before the spontaneously dividing cells become inviable. The chromosomes from these cells are generally from blast cells that have a very poor morphology and look puffy and indistinct under the microscope. However, these cells are the ones most likely to show chromosome abnormalities in comparison to the morphologically better appearing cells, which are more likely to be normal lymphocytes and not abnormal blast cells.

Plasma cell disorders (PCD) are difficult to study with standard chromosome analysis, as the number of plasma cells in bone marrow or the peripheral blood is very small, usually only 1–2% of the total cells present in circulation. Therefore, the vast majority of PCD indications will yield normal chromosomal results unless in advanced disease state, which often has complex karyotypic abnormalities. Consequently, FISH analysis should be performed in conjunction with standard cytogenetic analysis. However, to maximize the detection of abnormalities in PCDs by FISH, it is recommended to sort the plasma cells from the whole bone marrow specimen in order to analyze only the plasma cell portion of the marrow cells. Cell sorting can be achieved by many methods, including semi-automated approaches that take only hours in which to collect enough cells for analysis. Unfortunately, sorting plasma cells for chromosome analysis is currently not possible. Therefore, FISH analysis is essential for the diagnosis of abnormalities in these disorders.

This chapter will describe abnormalities seen in lymphoid disorders, first with ploidy level abnormalities, then single chromosome abnormalities by chromosome order.

23.2 Hyperdiploidy and hypodiploidy

23.2.1 Hyperdiploidy and hypodiploidy in multiple myeloma (MM)

SPECIFIC FEATURES

Aneuploidy is detected in 67–90% of MM cases, with either hyperdiploidy or hypodiploidy. Hyperdiploidy is primarily seen with gains of odd-numbered chromosomes, including 3, 5, 7, 9, 11, 15, 19 and 21. Hyperdiploidy is usually seen without structural chromosome changes and is generally associated with a good prognosis. The exception to the gain of an odd chromosome is chromosome 13, which is generally seen as a loss (monosomy 13) or a deletion [(del(13q))]. Hypodiploidy, as well as hypotetraploidy due to endoreduplication of a prior hypodiploid karyotype, most often occurs with complex structural rearrangements, including chromosome 14q32 translocations, and del(13q)/monosomy 13, and is generally associated with a more aggressive disease.

EXAMPLE REPORTING OF HYPERDIPLOIDY IN MULTIPLE MYELOMA

ISCN Result: 50,XX,+7,+9,+15,+21[4]/46,XX[16]

Interpretation

Chromosome analysis revealed an abnormal female hyperdiploid chromosome complement in 4 of 20 cells examined with the gain of chromosomes 7, 9, 15 and 21. The remaining 16 cells showed a normal female chromosome complement.

Hyperdiploidy is detected in approximately 40–50% of patients with a diagnosis of multiple myeloma (MM), with the gain of chromosomes 3, 7, 9, 11 and 15 most frequently involved. In general, hyperdiploid multiple myeloma is associated with kappa immunoglobulin subtype, symptomatic bone disease and better survival compared to non-hyperdiploid MM.

Fluorescence *in situ* hybridization (FISH) analysis for the MM panel of probes is suggested to look for the presence of commonly observed cytogenetic abnormalities that are not easily detectable in actively dividing cells by standard cytogenetics.

23.2.2 Hyperdiploidy and hypodiploidy in acute lymphoblastic leukemia (ALL)

SPECIFIC FEATURES

Hyperdiploidy in ALL is most commonly seen in children, detected in 20–30% of childhood ALL, and is rarely seen in adults (5% of adult ALL). High hyperdiploidy (51–65 chromosomes) is associated with FAB type L1 or L2. The prognosis for children with hyperdiploidy is good and high hyperdiploidy even better with long-term survival in 70–80% of patients. In adults, the prognosis is not as good as with children, with an event-free survival of 59% at 3 years. The good prognosis in children is thought to be related to leukemic cell sensitivity to a number of antileukemic drugs and the propensity of cells to respond to apoptosis. The presence of trisomies 4 and 10 specifically is associated with a good prognosis. High hyperdiploidy tends to show a pattern of chromosome gain with extra copies of chromosomes 4, 6, 10, 14, 18 and 21. The gain of chromosome 21 is often seen as tetrasomy.

Non-random translocations, including t(9;22), t(1;19), t(4;11) and t(12;21), are present in approximately half of high hyperdiploid cases as well as the duplication of 1q and the deletion of 6q. Generally, rearrangements are considered the primary cytogenetic change and hyperdiploidy is probably a secondary event. Therefore, ALL should be classified according to the translocation present rather than the ploidy group in order to assign the correct prognostic implications.

Hypodiploidy is a rare type of ALL. Near-haploid karyotypes tend to be seen in children or teenagers and are generally associated with a poor prognosis, although long survival of patients has been reported. Near haploidy is usually defined as the presence of less than 30 chromosomes with a typical abnormal clone containing 23–28 chromosomes. The pattern of chromosome loss in near haploidy is not random with a preferential retention of two copies of chromosomes 6, 8, 10, 14, 18, 21 and the sex chromosomes.

It is also common to see a normal diploid cell line with an abnormal cell line with double the near-haploid number of chromosomes resulting in hyperdiploidy. Thus, with hyperdiploidy, the chromosomes must be examined closely, since hyperdiploidy resulting from double hypodiploidy is distinct from the general hyperdiploidy that is associated with a good prognosis. The hyperdiploid cells arising from double hypodiploidy tend to have 2 or 4 copies of chromosomes, unlike typical over-50 chromosome hyperdiploid ALL with 3 copies of chromosomes. It is important to distinguish these gains, as near haploidy defines a rare type of childhood ALL associated with a very poor prognosis.

EXAMPLE REPORTING OF HYPERDIPLOIDY IN ACUTE LYMPHOBLASTIC LEUKEMIA

ISCN Result: 54,XX,+4,+6,+10,+11,+14,+18,+21,+21[6]/46,XX[14]

Interpretation

Chromosome analysis revealed an abnormal female hyperdiploid chromosome complement in 6 of 20 cells examined with the gain of one copy of chromosomes 4, 6, 10, 11, 14 and 18, and the gain of two copies of chromosome 21. The remaining 14 cells showed a normal female chromosome complement.

High hyperdiploidy comprises the most common cytogenetic subgroup in acute lymphoblastic leukemia (ALL), accounting for approximately 25% of cases, with a median modal chromosome number approximating 55 chromosomes. Although any chromosome can be involved in trisomies, the most frequently gained chromosomes include X, 4, 6, 10, 14, 18 and 21, with tetrasomy 21 also common. Hyperdiploidy is generally associated with a good prognosis in children with ALL.

Fluorescence *in situ* hybridization (FISH) analysis for the ALL panel of probes is suggested to look for the presence of other commonly observed cytogenetic abnormalities that are not easily detectable in actively dividing cells by standard cytogenetics.

23.3 Individual lymphoid abnormalities by chromosome order

23.3.1 Chromosome 1

Translocation (1;19)(q23;p13.3) (Figure 23.1)

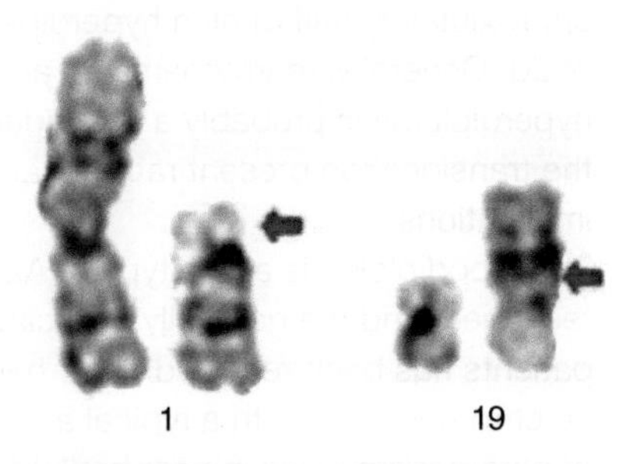

Figure 23.1 Partial karyotype showing translocation (1;19)(q23;p13.3).

SPECIFIC FEATURES

Translocation (1;19)(q23;p13.3) is a recurrent abnormality associated with acute lymphoblastic leukemia/lymphoma and is a distinct genetic category in the WHO classification. T(1;19) is seen in approximately 6% of B-ALL cases and most often in children. It is also seen in a minority of adult cases. T(1;19) results in the fusion of the PBX1 gene on 1q23 with the E2A (TCF3) gene on 19p13.3, and is detectable by cytogenetic and FISH analyses. This fusion gene rearrangement may be seen as a balanced translocation, and as an unbalanced translocation with two normal chromosome 1 s and loss of the derivative chromosome 1. This results in partial chromosome 1 gain and loss of chromosome 19p13.3 to the terminus. This rearrangement is generally associated with a poor prognosis, but more intensive therapy has recently improved survival rates.

EXAMPLE REPORTING OF t(1;19)(q23;p13.3) – BALANCED FORM

ISCN Result: 46,XY,t(1;19)(q23;p13.3)[20]

Interpretation

Chromosome analysis revealed an abnormal male chromosome complement with a translocation between the long arm of chromosome 1 and the short arm of chromosome 19 as the sole anomaly in all cells examined.

This translocation is a recurrent, non-random abnormality observed in B-cell acute lymphoblastic leukemia, and is generally associated with a poor prognosis.

Fluorescence *in situ* hybridization (FISH) analysis for the ALL panels of probes is suggested to look for the presence of other commonly observed cytogenetic abnormalities that are not easily detectable in actively dividing cells by standard cytogenetics.

EXAMPLE REPORTING OF t(1;19)(q23;p13.3) – UNBALANCED FORM

ISCN Result: 46,XY,der(19)t(1;19)(q23;p13.3)[20]

Interpretation

Chromosome analysis revealed an abnormal male chromosome complement with an unbalanced translocation between the long arm of chromosome 1 and the short arm of chromosome 19, resulting in trisomy for chromosome 1q23-qter and monosomy 19p13.3 to pter as the sole anomaly in all cells examined.

This translocation is a recurrent, non-random abnormality observed in B-cell acute lymphoblastic leukemia, and is generally associated with a poor prognosis.

Fluorescence *in situ* hybridization (FISH) analysis for the ALL panels of probes is suggested to look for the presence of other commonly observed cytogenetic abnormalities that are not easily detectable in actively dividing cells by standard cytogenetics.

Trisomy (1q) (Figure 23.2)

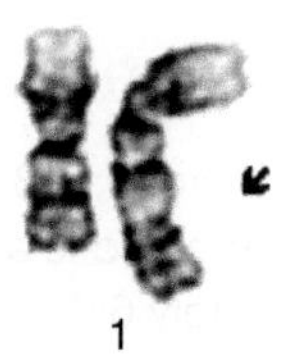

Figure 23.2 Trisomy 1q resulting from a duplication of part of the long arm of chromosome 1: inv dup(1)(q32q21). Courtesy of Sarah South PhD, ARUP Laboratories.

SPECIFIC FEATURES

Abnormalities resulting in trisomy 1q are recurrent abnormalities associated most commonly with multiple myeloma (MM) and are a distinct genetic category in the WHO classification. Rearrangements of chromosome 1q are one of the most frequent structural abnormalities in MM, seen in 16–26% of abnormal cases, but always as a secondary change. This is due to such factors as cytotoxic drugs, ionizing radiation and oncogenic viruses, which are suspected to induce decondensation of pericentric heterochromatin that forms triradials, giving rise to extra copies of chromosome 1q. This abnormality is seen as an adult-onset disorder, with a mean age of 62 years, where the majority of cases (71%) are males. It is generally associated with tumor progression and advanced disease.

Trisomy 1q results in duplication of all or part of chromosome 1q from isochromosomes or translocations, some of which are jumping translocations. Chromosome 1q may be involved in translocations with telomeres from different chromosome partners, most commonly with chromosomes 8pter, 9pter, 12qter, 13pter, 15pter, 17qter, 19pter, 19qter, 21pter and 22pter. Also seen are whole-arm centromere-to-centromere translocations, for example between 16p and 1q.

EXAMPLE REPORTING OF TRISOMY 1q FROM AN INVERTED DUPLICATION

ISCN Result: 46,XY, inv dup(1)(q32q21)[20]

Interpretation

Chromosome analysis revealed an abnormal male chromosome complement with duplicated material on the long arm of chromosome 1, resulting in trisomy for chromosome 1q21-q23, as the sole anomaly in all cells examined.

This rearrangement, resulting in partial trisomy 1q, is a recurrent, non-random abnormality observed in multiple myeloma (MM) and is generally associated with a poor prognosis.

Fluorescence *in situ* hybridization (FISH) analysis for the MM panels of probes is suggested to look for the presence of other commonly observed cytogenetic abnormalities that are not easily detectable in actively dividing cells by standard cytogenetics.

23.3.2 Chromosome 2

Translocation (2;5)(p23;q35)

SPECIFIC FEATURES

Translocation (2;5) is a recurrent abnormality associated with anaplastic large cell lymphoma (ALCL) and is a distinct genetic category in the WHO classification. This rearrangement is characterized by anaplastic lymphoma kinase (ALK)-positive lymphoma with distinctive lymphoid cell morphology. ALCL lacking an ALK rearrangement is considered a different category from ALK-positive disease. ALK-positive ALCL is seen in approximately 3% of adult non-Hodgkin lymphomas and 10–20% of childhood lymphomas and has a male predominance.

T(2;5) results in the fusion of the ALK gene on 2p23 with the NPM gene on 5q35, which is detectable by cytogenetic and FISH analyses and seen in approximately 84% of ALK-positive cases. However, multiple genes are seen with ALK translocations. Therefore, FISH analysis with the ALK gene is suggested to identify any of the possible gene rearrangements. ALK-positive rearrangements are often seen with additional karyotypic changes, including monosomy 4, del(11q), del(13q) and trisomy for chromosomes 7, 17p and 17q. There is no known prognostic difference between t(2;5) and other variant ALK-positive translocations. It is generally associated with a good prognosis compared with ALK-negative ALCL.

EXAMPLE REPORTING OF t(2;5)(p23;q35)

ISCN Result: 46,XY,t(2;5)(p23;q35)[20]

Interpretation

Chromosome analysis revealed an abnormal male chromosome complement with a translocation between the short arm of chromosome 2 and the long arm of chromosome 5 as the sole anomaly in all cells examined.

T(2;5) is associated with anaplastic large cell lymphoma (ALCL), found in approximately 45–60% of patients with this disease, with 83% of pediatric patients and 31% of adult patients having the translocation or variant. The translocation juxtaposes the ALK (anaplastic lymphoma kinase) gene at 2p23 with the NPM (nucleoplasmin) gene (5q35), creating a new fusion gene and gene product. Variants of the translocation have also been described. ALK-positive ALCL patients have a good prognosis compared to ALK-negative ALCL patients. ALK rearrangements have also been observed in large B-cell lymphomas.

Fluorescence *in situ* hybridization (FISH) analysis for the ALK gene is suggested to confirm and monitor this abnormal cell line.

23.3.3 Chromosome 3

Rearrangements of 3q27

SPECIFIC FEATURES

Chromosome 3q27 rearrangements are recurrent, non-random abnormalities in B-cell non-Hodgkin lymphomas (NHL), including diffuse large B-cell lymphomas (DLBCL), follicular lymphomas and marginal zone B-cell lymphomas, and may be seen at a lower incidence in multiple myeloma and chronic lymphocytic leukemia (CLL). The chromosome 3q27 region involved in a rearrangement most often involves the BCL6 gene.

In follicular cell lymphoma, 3q27 translocation partners typically involve immunoglobulin genes, such as IGK at 2p11, IGH at 14q32 and IGL at 22q11.2. Common translocations associated with a 3q27 abnormality include t(14;18)(q32;q21), t(8;14)(q24;q32) and t(11;14)(q24;q32).

Translocations of 3q27 with other chromosome partners typically include 2p12 (IgK), 4p13 (RHOH), 6p22 (H1F1), 6p21.2 (PIM-1), 7p12 (ZNFN1A1), 11q23 (OBF1), 13q14 (LCP1), 14q32 (IgH), 16p13 (MHC2TA), 18p11.2 (EIF4A2) and 22q11 (IGL).

EXAMPLE REPORTING OF CHROMOSOME 3q27 REARRANGEMENTS

ISCN Result: 46,XY,t(3;18)(q27;p11.2)[20]

Interpretation

Chromosome analysis revealed an abnormal male chromosome complement with a translocation between the long arm of chromosome 3 and the short arm of chromosome 18 as the sole anomaly in all cells examined.

Translocations involving the BCL6 gene at 3q27 are observed in approximately 30% of diffuse large B-cell lymphomas and 5–15% of follicular lymphomas, most commonly in grade 3B cases.

Fluorescence *in situ* hybridization (FISH) analysis for the BCL6 probe is suggested to confirm and monitor this abnormal cell line.

23.3.4 Chromosome 4

Translocation (4;11)(q21;q23) (Figure 23.3)

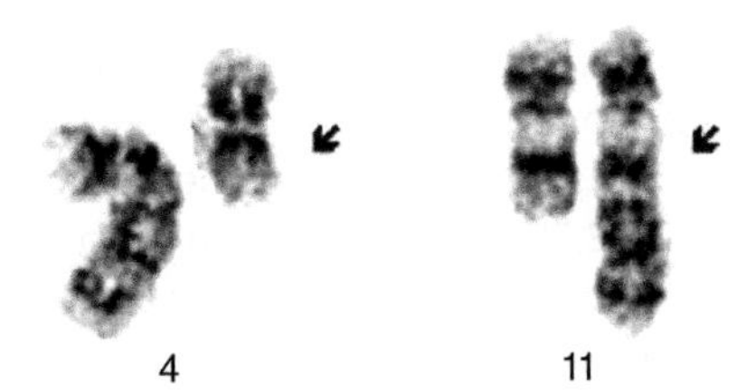

Figure 23.3 Partial karyotype showing translocation (4;11)(q21;q23). Courtesy of Sarah South PhD, ARUP Laboratories.

SPECIFIC FEATURES

The t(4;11) is a recurrent, non-random rearrangement seen predominantly in B-ALL (L1 or L2), biphenotypic acute leukemias and treatment-related leukemias (secondary to epipodophyllotoxin, an anti-topoisomerase drug). This abnormality is common in children, including infants. When seen in children <1–2 years of age, it is considered a congenital leukemia. Half of the cases reported are in children under 4 years of age and 33% in children under 1 year of age with a predominance of females (2:1 ratio). T(4;11) results in the fusion of the AF4 gene on 4q21 with the MLL gene on 11q23, which is detectable by cytogenetic and FISH analyses.

Bone marrow transplantation is highly indicated for these cases, as the prognosis is very poor. Although clinical remission occurs, it is often followed by prompt relapse, and the median survival is only 7 months in adult cases and 9 months in children. T(4;11) is often associated with hyperploidy and i(7q).

EXAMPLE REPORTING OF t(4;11)(q21;q23)

ISCN Result: 46,XY,t(4;11)(q21;q23)[20]

Interpretation

Chromosome analysis revealed an abnormal male chromosome complement with a translocation between the long arms of chromosomes 4 and 11 as the sole anomaly in all cells examined.

Translocation (4;11)(q21;q23) results in the fusion of the MLL gene at 11q23 with the AF4 gene at 4q21 and is a recurrent, non-random rearrangement associated with B-cell acute lymphoblastic leukemia (B-cell ALL). Leukemic cells express an early pre-B (CD10-negative) immunophenotype and frequently co-express myeloid markers. T(4;11) is generally associated with a poor prognosis. Infants with MLL translocations, especially those <6 months of age, have a particularly poor prognosis.

Fluorescence *in situ* hybridization (FISH) analysis for the MLL gene is suggested to confirm and monitor this abnormal cell line.

Translocation (4;14)(p16;q32)

SPECIFIC FEATURES

This t(4;14) is a recurrent, non-random abnormality seen in MM, plasmacytomas and monoclonal gammopathy of unknown significance (MGUS). T(4;14) results in a rearrangement of the FGFR3 gene on 4p16 with the IGH gene on 14q32, which is not easily detectable by cytogenetics and is recommended to be studied by FISH analysis. This IGH/FGFR3 gene rearrangement combines the promoter regions of both partner genes, resulting in overexpression and activation of the FGFR3 gene, which provides an oncogenic signal-enhancing cell proliferation of abnormal cells.

In MM, this IGH rearrangement is associated with a poor prognosis, even in patients treated with high-dose chemotherapy. Additional anomalies include hypodiploidy, del(13q) and monosomy 13.

EXAMPLE REPORTING OF t(4;14)(p16;q32)

ISCN Result: 46,XY,t(4;14)(p16;q32)[20]

Interpretation

Chromosome analysis revealed an abnormal male chromosome complement with a translocation between the short arm of chromosome 4 and the long arm of chromosome 14 as the sole anomaly in all cells examined.

Translocation (4;14) results in the fusion of the FGFR3 gene at 4p16 with the IGH gene at 14q32 and is a recurrent, non-random rearrangement found in multiple myeloma. It is generally associated with a poor prognosis.

Fluorescence *in situ* hybridization (FISH) analysis for the FGFR3/IGH fusion probe is suggested to confirm and monitor this abnormal cell line. Also, FISH analysis for the MM panel of probes is suggested to look for the presence of other commonly observed cytogenetic abnormalities not easily detectable in actively dividing cells by standard cytogenetics.

23.3.5 Chromosome 6

Del(6)(q21) (Figure 23.4)

Figure 23.4 Partial karyotype showing del(6)(q21).

SPECIFIC FEATURES

Deletions of the long arm of chromosome 6 are recurrent, non-random abnormalities seen in many lymphoid malignancies, including ALL, CLL, prolymphocytic leukemia (PLL), MM and NHL. In childhood B- and T-cell ALL, del(6q) is the most common recurrent cytogenetic abnormality detected. It is seen in approximately 5–15% of B-cell malignancies by chromosome analysis and 30% by FISH analysis. In T-cell ALL, it is seen in both children and adults (but less frequently) in 10–20% of cases, and often associated with 14q11 and del(9p) abnormalities. FISH is suggested to identify 6q deletions, since many are small interstitial deletions that are submicroscopic.

Del(6q) is seen as the sole anomaly in approximately 30% of cases, but is also associated with other abnormalities including del(9p), del(12p), t(11;14), t(1;19), t(9;22) and t(12;21).

In MM, del(6q) is detected in 15% of cases involving the 6q15-6q21 band, suggesting that the critical region might contain a recessive tumor suppressor gene.

EXAMPLE REPORTING OF DELETION OF CHROMOSOME 6q

ISCN Result: 46,XY,del(6)(q21)[20]

Interpretation

Chromosome analysis revealed an abnormal male chromosome complement with a deletion within the long arm of chromosome 6 as the sole anomaly in all cells examined.

Deletions of 6q are recurrent, non-random abnormalities observed in lymphoid disorders, including acute and chronic lymphocytic leukemia (CLL), non-Hodgkin lymphoma, multiple myeloma and prolymphocytic leukemia. CLL with del(6q) is characterized by a high incidence of atypical morphology, a classic immunophenotype with CD38 positivity and an intermediate incidence of IGVH somatic hypermutation. Del(6q) is generally considered to be in an intermediate risk category.

Fluorescence *in situ* hybridization (FISH) analysis for the CLL/MM panel of probes is suggested to look for the presence of other commonly observed cytogenetic abnormalities not easily detectable in actively dividing cells by standard cytogenetics.

Trisomy 6

Trisomy 6 is often reported when seen as part of hyperdiploidy in ALL.

EXAMPLE REPORTING OF TRISOMY 6

ISCN Result: 47,XY,+6[20]

Interpretation

Chromosome analysis revealed an abnormal male chromosome complement with the gain of chromosome 6 as the sole anomaly in all cells examined.

For pediatric acute lymphoblastic leukemia (ALL) cases in particular, the presence of an extra copy of chromosome 6 has been reported to be strongly associated with a favorable outcome, as is the combination of trisomy 4 and trisomy 10.

Fluorescence *in situ* hybridization (FISH) analysis for the ALL panel of probes is suggested to look for the presence of other commonly observed cytogenetic abnormalities not easily detectable in actively dividing cells by standard cytogenetics.

23.3.6 Chromosome 7

Del(7q)

SPECIFIC FEATURES

Deletions of chromosome 7q are recurrent abnormalities associated with lymphomas, generally seen in splenic marginal zone lymphomas (SMZL), and are a distinct genetic category in the WHO classification. SMZL is characterized by lymphocytes surrounding the splenic germinal centers of the follicle mantle merging with the marginal zone of the larger cells. Although SMZL is rare, seen in less than 2% of lymphoid neoplasms, it has a distinct morphology involving the spleen and splenic hilar lymph nodes. This abnormality is seen in older patients (over 50 years) and equally among females and males. SMZL patients generally present with splenomegaly, thrombocytopenia and anemia. Over half of these cases have IGH heavy and light chain gene rearrangements and hypermutations. However, deletion of chromosome 7q21-32 is seen in approximately 40% of cases. This is generally an indolent disease. However, the presence of del(7q) is associated with a poor prognosis, and there is a risk of transforming to large B-cell lymphoma.

EXAMPLE REPORTING OF DELETION OF CHROMOSOME 7q

ISCN Result: 46,XY,del(7)(q21q32)[20]

Interpretation

Chromosome analysis revealed an abnormal male chromosome complement with a deletion within the long arm of chromosome 7 as the sole anomaly in all cells examined.

Deletions of 7q are recurrent, non-random abnormalities observed in B-cell low-grade lymphoid malignancies, in particular splenic marginal zone lymphoma (SMZL). In SMZL, the presence of a 7q deletion and unmutated IGHV genes may be associated with a poor outcome.

Fluorescence *in situ* hybridization (FISH) analysis for the non-Hodgkin lymphoma (NHL) panel of probes is suggested to look for the presence of other commonly observed cytogenetic abnormalities not easily detectable in actively dividing cells by standard cytogenetics.

23.3.7 Chromosome 8

Translocation (8;14)(q24;q32) (Figure 23.5)

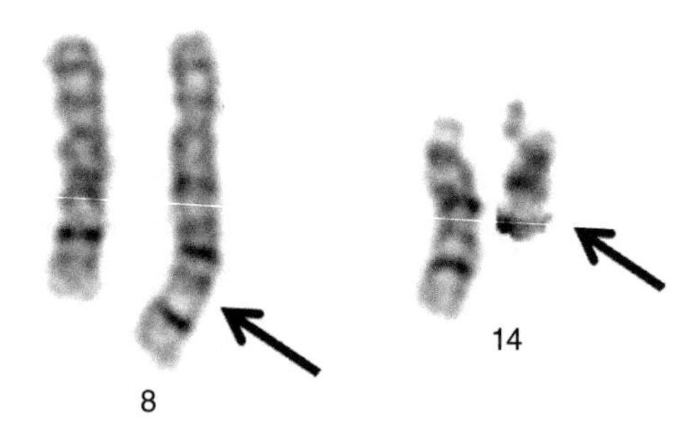

Figure 23.5 Partial karyotype showing translocation (8;14)(q24;q32).

SPECIFIC FEATURES

Translocation (8;14) is observed in both B-cell ALL and NHL, especially in Burkitt lymphoma. This translocation is present in both the endemic African Burkitt lymphoma and in the non-endemic tumor type seen in other parts of the world. It is seen primarily in ALL L3 morphology, according to the FAB classification, and only very occasionally in L1 or L2.

T(8;14) results in the fusion of the MYC gene at 8q24 with the immunoglobulin heavy chain gene (IgH) at 14q32. Variant translocations of the t(8;14) include t(2;8) and t(8 ;22). High-quality metaphases are required to detect the t(8;14) and t(8;22); therefore, MYC FISH analysis is suggested when clinically appropriate. Three-way rearrangements and translocations of submicroscopic chromosome fragments have also been described. In all of these translocations, the oncogene C-MYC is juxtaposed with the immunoglobulin heavy chain locus IGH (14q32), the kappa light chain locus IGK (2p12) or the lambda light chain locus IGL (22q11). T(8;14) often is seen in conjunction with trisomy 1q and rearrangements of 13q34. This abnormality is associated with an aggressive disease and a poor prognosis.

EXAMPLE REPORTING OF t(8;14)(q24;q32)

ISCN Result: 46,XY,t(8;14)(q24;q32)[20]

Interpretation

Chromosome analysis revealed an abnormal male chromosome complement with a translocation between the long arms of chromosomes 8 and 14 as the sole anomaly in all cells examined.

Translocation (8;14) results in the fusion of the MYC gene at 8q24 with the immunoglobulin heavy chain gene (IgH) at 14q32. Translocation (8;14) is a characteristic abnormality found in approximately 80% of Burkitt lymphoma, although it may be seen in diffuse large B-cell lymphoma and in transformed lower-grade lymphomas. This abnormality is generally associated with an aggressive disease and a poor prognosis.

Fluorescence *in situ* hybridization (FISH) analysis for the MYC/IGH fusion probe is suggested to confirm and monitor this abnormal cell line. Also, FISH analysis for the NHL panel of probes is suggested to look for the presence of other commonly observed cytogenetic abnormalities not easily detectable in actively dividing cells by standard cytogenetics.

23.3.8 Chromosome 9

Rearrangements of (9p)

SPECIFIC FEATURES

Chromosome 9p abnormalities occur in approximately 10% of childhood and adult ALL and are generally associated with a poor prognosis in B-cell ALL. Rearrangements of chromosome 9p include i(9q), balanced translocations with 9p breakpoints and dicentric chromosomes.

Deletion of 9p21 results in loss of the CDKN2A gene, also known as p16, and is observed in approximately 10–30% of acute ALL cases. Deletion of the p16 gene is associated with a poor prognosis in pediatric B-lineage, but not in adult or T-lineage ALL. Chromosome 9p abnormalities most often occur with other abnormalities, rather than as a sole anomaly, particularly including chromosome 12p rearrangements and del(6q).

EXAMPLE REPORTING OF DELETION OF CHROMOSOME 9p

ISCN Result: 46,XY,del(9)(p21)[20]

Interpretation

Chromosome analysis revealed an abnormal male chromosome complement with a deletion within the short arm of chromosome 9 as the sole anomaly in all cells examined.

Rearrangements involving 9p are recurrent, non-random abnormalities in acute lymphoblastic leukemia (ALL) and lymphoblastic lymphoma (LBL). Deletion of 9p21 results in loss of the CDKN2A gene, also known as p16, and is observed in approximately 10–30% of acute ALL cases. Deletion of the p16 gene is associated with a poor prognosis in pediatric B-lineage, but not in adult or T-lineage ALL.

Fluorescence *in situ* hybridization (FISH) analysis for the CDKN2A gene is suggested to confirm and monitor this abnormal cell line. Also, FISH analysis for the ALL panel of probes is suggested to look for the presence of other commonly observed cytogenetic abnormalities not easily detectable in actively dividing cells by standard cytogenetics.

Translocation (9;22)(q34;q11.2)

SPECIFIC FEATURES

Translocation (9;22) is seen in approximately 20% of adult ALL and 2–5% of childhood ALL, accompanied frequently with central nervous system (CNS) involvement. Translocation (9;22) results in the fusion of the ABL1 gene at 9q34 and the BCR gene at 22q11.2. When seen in ALL, the prognosis is very poor, especially in lymphoid-committed progenitor cases. Therefore, bone marrow transplant is indicated for these cases. T(9;22) disappears during remission, in contrast to CML cases when treated with conventional therapies. Additional anomalies are found in 50–80% of cases, particularly+der(22), monosomy 7, del(7q), trisomy 8 and hyperdiploidy.

EXAMPLE REPORTING OF t(9;22)(q34;q11.2) IN ACUTE LYMPHOBLASTIC LEUKEMIA

ISCN Result: 46,XY,t(9;22)(q34;q11.2)[20]

Interpretation

Chromosome analysis revealed an abnormal male chromosome complement with a translocation between the long arms of chromosomes 9 and 22, resulting in a Philadelphia chromosome as the sole anomaly in all cells examined.

Translocation (9;22) results in the fusion of the ABL1 gene at 9q34 and the BCR gene at 22q11.2. It is observed in approximately 20% of B-cell acute lymphoblastic leukemia (B-ALL) cases and 4% of cases of childhood B-ALL. In both children and adults, B-ALL/LBL with t(9;22) is generally associated with a very poor prognosis.

Fluorescence *in situ* hybridization (FISH) analysis for the BCR/ABL fusion probe is suggested to confirm and monitor this abnormal cell line. Also, FISH analysis for the ALL panel of probes is suggested to look for the presence of other commonly observed cytogenetic abnormalities not easily detectable in actively dividing cells by standard cytogenetics.

23.3.9 Chromosome 11

Del(11)(q22) (Figure 23.6)

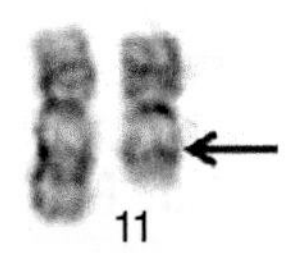

Figure 23.6 Partial karyotype showing del(11)(q22).

SPECIFIC FEATURES

Deletion of chromosome 11q22-23 results in the loss of the ATM gene and is observed in up to 25% of NHL cases, with the highest incidence occurring in MCL, CLL and DLBCL. Del(11q22-23) is generally seen as a secondary change, not a primary finding. These deletions are often submicroscopic; therefore, FISH analysis is suggested in cases where an ATM deletion is suspected. In CLL, del(11q22-23) is generally associated with the presence of bulky abdominal/mediastinal lymphadenopathy, shorter treatment-free survival and relatively aggressive disease.

EXAMPLE REPORTING OF DELETION OF CHROMOSOME 11q

ISCN Result: 46,XY,del(11)(q22)[20]

Interpretation

Chromosome analysis revealed an abnormal male chromosome complement with a deletion within the long arm of chromosome 11 as the sole anomaly in all cells examined.

Deletions of 11q22-23, resulting in the loss of the ATM gene, are observed in up to 25% of non-Hodgkin lymphoma (NHL) cases, with the highest incidence occurring in mantle cell lymphoma (MCL), chronic lymphocytic leukemia (CLL) and diffuse large B-cell lymphoma (DLBCL). In CLL, it is generally associated with the presence of bulky abdominal/mediastinal lymphadenopathy, shorter treatment-free survival and relatively aggressive disease.

Fluorescence *in situ* hybridization (FISH) analysis for the NHL/CLL panel of probes is suggested to look for the presence of other commonly observed cytogenetic abnormalities not easily detectable in actively dividing cells by standard cytogenetics.

Translocation (v;11q23)

SPECIFIC FEATURES

Chromosome 11q23 abnormalities, involving MLL gene rearrangements with variable (v) chromosome partners, have an early pre-B-cell immunophenotype, and may express myeloid antigens but not CD10. MLL gene rearrangements are not typically seen with hyperdiploidy, and when seen in infants, it is considered a clinically distinct entity from that diagnosed in older children. Infant ALL is clinically aggressive and strongly associated with a poor prognosis, and there is evidence from molecular studies that 11q23 abnormalities in infants with ALL occur *in utero*. The MLL gene has an important role in normal hematopoietic growth and differentiation. Since MLL abnormalities can occur very early in hematopoietic stem cell development, *in utero* exposure to natural or synthetic substances that inhibit topoisomerase II (e.g. genistein, catechins, flavonoids) may result in acute leukemia.

The incidence of 11q23/MLL abnormalities among infants with ALL is from 60% to 80% whereas the incidence in children older than 1 year is from 4.5% to 5.7%. Prognosis is variable, depending on age of onset and the type of abnormality detected. However, infants with MLL translocations, especially those less than 6 months of age, have a particularly poor prognosis.

EXAMPLE REPORTING OF t(v;11q23) IN ACUTE LYMPHOBLASTIC LEUKEMIA

ISCN Result: 46,XY, t(v;11q23) [20]

Interpretation

Chromosome analysis revealed an abnormal male chromosome complement with a translocation between the ___ arm of chromosome __ and the long arm of chromosome 11 at band q23 as the sole anomaly in all cells examined.

Patients with B-cell acute lymphoblastic leukemia (B-cell ALL) that contain a translocation involving 11q23 region usually involve the MLL gene, which can fuse with a large number of chromosome partners. Infants with MLL translocations, especially those less than 6 months of age, have a particularly poor prognosis.

Fluorescence *in situ* hybridization (FISH) analysis for the MLL probe is suggested to confirm and monitor this abnormal cell line.

Translocation (11;14)(q13;q32) (Figure 23.7)

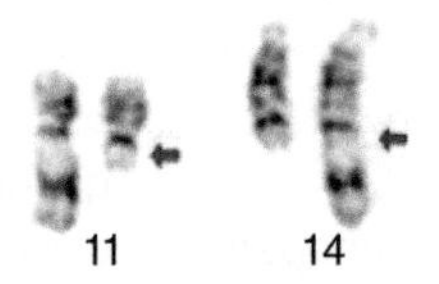

Figure 23.7 Partial karyotype showing translocation (11;14)(q13;q32).

SPECIFIC FEATURES

Translocation (11;14) is seen in approximately 50–70% of mantle cell lymphoma and 10–20% of B-prolymphocytic leukemia, but can also be observed in plasma cell leukemia, splenic lymphoma with villous lymphocytes, chronic lymphocytic leukemia and multiple myeloma. Translocation (11;14) results in the fusion of the BCL1 gene at 11q13 and the IGH gene at 14q32, which may be observed by cytogenetics, but most often by FISH analysis. Lymphomas with t(11;14) are seen most often as part of complex karyotypic abnormalities, particularly with trisomy for chromosomes 3, 7, 12 and 18, monosomy 8, and with del(9p), del(1p), del(6q), del(7q), del(13q) and del(17p).

In MM, t(11;14) is a balanced rearrangement in most cases, but can also occur as -14 or +der(14)t(11;14). T(11;14) is seen most often with other abnormalities, as in lymphomas, particularly with monosomy 13, del(13q) and structural anomalies of chromosome 1. T(11;14) is generally seen in advanced stages of disease with a median survival of 3–4 years.

EXAMPLE REPORTING OF t(11;14)(q13;q32)

ISCN Result: 46,XY,t(11;14)(q13;q32)[20]

Interpretation

Chromosome analysis revealed an abnormal male chromosome complement with a translocation between the long arms of chromosomes 11 and 14 as the sole anomaly in all cells examined.

Translocation (11;14) results in the fusion of the BCL1 (CCND1) gene at 11q13 with the immunoglobulin heavy chain gene (IgH) at 14q32 and is a recurrent, non-random rearrangement found predominantly in mantle cell lymphoma, although it is also observed in B-prolymphocytic leukemia, plasma cell leukemias and multiple myeloma. In plasma cell myelomas, t(11;14) is associated with a favorable prognosis in the absence of other poor prognostic markers.

Fluorescence *in situ* hybridization (FISH) analysis for the BCL1/IGH fusion probe is suggested to confirm and monitor this abnormal cell line. Also, FISH analysis for the MM panel of probes is suggested to look for the presence of other commonly observed cytogenetic abnormalities not easily detectable in actively dividing cells by standard cytogenetics.

Translocation (11;18)(q21;q21)

SPECIFIC FEATURES

Translocation (11;18) is found primarily in extranodal marginal zone B-cell lymphomas of mucosa-associated lymphoid tissue (MZBCL) or MALT-type lymphoma. It is also seen, though less frequently, in gastric MALT-type lymphomas in which t(11;18) is a clonal marker for resistance to *Helicobacter pylori* eradication therapy and antigen independent growth. Translocation (11;18) results in the fusion of the API2 gene at 11q21 and the MALT1 gene at 18q21.

EXAMPLE REPORTING OF t(11;18)(q21;q21)

ISCN Result: 46,XY,t(11;18)(q21;q21)[20]

Interpretation

Chromosome analysis revealed an abnormal male chromosome complement with a translocation between the long arms of chromosomes 11 and 18 as the sole anomaly in all cells examined.

T(11;18) is a recurrent, non-random rearrangement associated with mucosa-associated lymphoid tissue (MALT) lymphoma. It fuses the amino terminal of the API2 gene to the carboxyl terminal of the MALT1 gene and generates a chimeric fusion product.

The translocation is also frequently detected in pulmonary and gastric MALT lymphomas and at moderate frequencies in those from the conjunctiva and orbit. In gastric MALT lymphoma, t(11;18) is significantly associated with infection by CagA-positive strains of *Helicobacter pylori*.

Fluorescence *in situ* hybridization (FISH) analysis for the MALT probe is suggested to confirm and monitor this abnormal cell line.

Translocation (11;19)(q23;p13.3) – associated with lymphoid disorders (Figure 23.8)

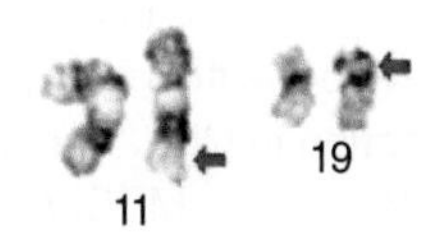

Figure 23.8 Partial karyotype showing translocation (11;19)(q23;p13.3).

SPECIFIC FEATURES

Translocation (11;19) results in the fusion of the MLL gene at 11q23 and the ENL gene at 19p13.3. For further information regarding t(11;19), see Chapter 22.

EXAMPLE REPORTING OF t(11;19)(q23;p13.3)

ISCN Result: 46,XY, t(11;19)(q23;p13.3)[20]

Interpretation

Chromosome analysis revealed an abnormal male chromosome complement with a translocation between the long arm of chromosome 11 and the short arm of chromosome 19 as the sole anomaly in all cells examined.

Translocation (11;19)(q23;p13.3) results in the fusion of the MLL gene at 11q23 with the ENL gene at 19p13.3 and is observed in acute myeloid leukemia (AML) and B- and T-cell lymphoblastic leukemia/lymphoblastic lymphoma (B-ALL/LBL).

Fluorescence *in situ* hybridization (FISH) analysis for the MLL probe is suggested to confirm and monitor this abnormal cell line.

23.3.10 Chromosome 12

Del(12)(p11.2p13) (Figure 23.9)

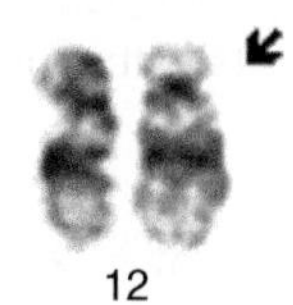

Figure 23.9 Partial karyotype showing del(12)(p11.2p13).

SPECIFIC FEATURES

Del(12p) is seen in approximately 10–15% of pediatric ALL cases and 5% of adult ALL. It can result from a number of various aberrations including balanced translocations with 12p breakpoints, deletions, additional material replacing part of the 12p, monosomy 12 and unbalanced translocations resulting in der(12)t(v;12). Del(12p) is seen as the sole abnormality in less than 20% of cases and is often seen with complex karyotypic abnormalities, particularly including del(6q), del(13q), monosomy 13 and trisomy 21.

Del(12p) or 12p abnormalities are observed in about 1–2% of CLL cases. Atypical morphology and immunophenotypic abnormalities may be present in cases with 12p rearrangements. Frequently there is disease progression. Transformation of CLL to PLL has been observed in cases with 12p abnormalities. The breakpoint most commonly seen in 12p abnormalities is 12p13. Translocation (12;13) is the most common recurring 12p translocation in CLL. Translocation (12;13) has also been found in patients with CML in transformation, MDS, AML and both B- and T-cell ALL. The other recurring translocation in CLL involves chromosomes 12 and 17 with a derivative chromosome 12 resulting from an unbalanced translocation (12;17). This translocation is observed in patients with acute leukemia at diagnosis or relapse and is generally associated with a poor prognosis.

EXAMPLE REPORTING OF DELETION OF CHROMOSOME 12p

ISCN Result: 46,XY,del(12)(p11.2p13)[20]

Interpretation

Chromosome analysis revealed an abnormal male chromosome complement with a deletion within the long arm of chromosome 12 as the sole anomaly in all cells examined.

Deletions of 12p are recurrent, non-random abnormalities observed in pediatric acute lymphoblastic leukemia (ALL) and chronic lymphocytic leukemia (CLL)

Fluorescence *in situ* hybridization (FISH) analysis for the ALL/CLL panel of probes is suggested to look for the presence of other commonly observed cytogenetic abnormalities not easily detectable in actively dividing cells by standard cytogenetics.

Translocation (12;21)(p13;q22)

SPECIFIC FEATURES

Translocation (12;21) is a recurrent abnormality associated with acute lymphoblastic leukemia and is a distinct genetic category in the WHO classification. This rearrangement is characterized by lymphoblasts of the B-cell lineage. T(12;21) is seen in approximately 25% of childhood ALL cases and most often in young children, but generally not in infants, and decreases in frequency in older children and is rare in adulthood.

Translocation (12;21) results in the fusion of the ETV6 gene on 12p13 with the RUNX1 gene on 21q22, which is generally not detectable by cytogenetics. Consequently, FISH analysis is required for detection. This rearrangement is thought to occur early in disease, as it is seen by the presence of blood spots in infants who developed disease years later. Therefore, it is postulated that this rearrangement is necessary but not sufficient for the initiation of disease. This abnormality is associated with a good prognosis, with over 90% of children going into complete remission. Relapse, if it occurs, is generally later than other types of ALL. However, when t(12;21) is present in older children, or those with a high white blood count, the prognosis is not as favorable.

EXAMPLE REPORTING OF t(12;21)(p13;q22)

ISCN Result: 46,XY,t(12;21)(p13;q22)[20]

Interpretation

Chromosome analysis revealed an abnormal male chromosome complement with a translocation between the short arm of chromosome 12 and the long arm of chromosome 21 as the sole anomaly in all cells examined.

Translocation (12;21) results in the fusion of the ETV6 (TEL) gene at 12p13 with the RUNX1 (AML) gene at 21q22, and is a recurrent, non-random rearrangement found in 25% of childhood B-precursor acute lymphoblastic leukemia (ALL). T(12;21) is associated with a good prognosis in childhood ALL.

Fluorescence *in situ* hybridization (FISH) analysis for the ETV6/RUNX1 fusion probe is suggested to confirm and monitor this abnormal cell line.

Trisomy 12

SPECIFIC FEATURES

Trisomy 12 is the most common cytogenetic change in CLL, but is also seen in other subtypes of B-cell lymphoproliferative disorders where it is most often a primary cytogenetic change. Trisomy 12 is detected in approximately 33% of CLL by chromosome analysis and in up to 54% of cases by FISH analysis. Trisomy 12, when seen as the sole anomaly, is associated with an intermediate prognosis. When seen in conjunction with add(14q) abnormalities, trisomy 12 is generally associated with shorter survival time compared to patients with 13q abnormalities and normal karyotypes.

Trisomy 12 is also seen in B-cell non-Hodgkin lymphoma, including prolymphocytic leukemia, hairy cell leukemia, splenic lymphoma with villous lymphocytes (SLVL), Waldenstrom macroglobulinemia, follicular lymphoma, MCL and DLBCL.

EXAMPLE REPORTING OF TRISOMY 12

ISCN Result: 47,XY,+12[20]

Interpretation

Chromosome analysis revealed an abnormal male chromosome complement with the gain of chromosome 12 as the sole anomaly in all cells examined.

Trisomy 12 is the most frequently reported chromosomal aberration seen in B-cell chronic lymphocytic leukemia (CLL), occurring in nearly one-third of all cytogenetically abnormal cases. It is also observed in other subtypes of B-cell lymphoproliferative disorders. Trisomy 12 in CLL is generally associated with an intermediate prognosis.

Fluorescence *in situ* hybridization (FISH) analysis for the CLL panel of probes is suggested to look for the presence of other commonly observed cytogenetic abnormalities not easily detectable in actively dividing cells by standard cytogenetics.

23.3.11 Chromosome 13

Del(13q)

SPECIFIC FEATURES

Del(13q) is seen in a spectrum of B-cell chronic lymphoproliferative disorders, including CLL, atypical CLL, splenic marginal zone B-cell lymphoma and MCL. It is seen in approximately 10–15% of all CLL cases. The majority of MCLs show peripheral blood (PB) involvement at diagnosis or at disease evolution and, in MCL, are often seen with the t(11;14)(q13;q32).

The frequency of del(13q) as the sole anomaly is approximately 8–10% by chromosome analyses in CLL, MM and splenic marginal zone B-cell lymphomas, but with FISH analysis the detection rate increases to approximately 40%. Additional chromosome anomalies most often include trisomy 12, del(6q), del(11q), add(14q) and del(17p).

In MM, del(13q) and monosomy 13 appear as one of the main prognostic factors with a significantly lower event-free survival, overall survival and complete remission duration, either in standard-dose or high-dose therapy protocols. Del(13q) is usually seen as a secondary event, but may occur early in the evolution of MM since it is observed in patients with MGUS.

In NHL, del(13q) is generally associated with the presence of splenomegaly, PB dissemination, lower probability of attaining complete remission and a short survival. However, in CLL it is generally associated with a good prognosis, with the longest median survival in comparison to other cytogenetic abnormalities.

The minimal common region of the deletion is in band 13q14.3, involving the RB1 gene, and loci D13S319 and D13S272, which are approximately 100 kb distal from RB1. A subset of cases shows loss of both 13q34 and 13q14 regions with retention of band 13q21. A critical region of approximately 350 kb was identified at 13q14 with DLEU1, DLEU2 and RFP2 genes.

EXAMPLE REPORTING OF DELETION OF CHROMOSOME 13q

ISCN Result: 46,XY,del(13)(q12q32)[20]

Interpretation

Chromosome analysis revealed an abnormal male chromosome complement with a deletion within the long arm of chromosome 13 as the sole anomaly in all cells examined.

Deletions of 13q are recurrent, non-random abnormalities in lymphoid malignancies, including multiple myeloma, non-Hodgkin lymphoma, and are the most common cytogenetic abnormalities in chronic lymphocytic leukemia (CLL). They result in the loss of the coding region for two micro-RNAs and, when they are the sole defect in CLL, are associated with a good prognosis. In multiple myeloma (MM), deletions of 13q are associated with a less favorable prognosis.

Fluorescence *in situ* hybridization (FISH) analysis for the CLL/NHL/MM panels of probes is suggested to look for the presence of other commonly observed cytogenetic abnormalities not easily detectable in actively dividing cells by standard cytogenetics.

23.3.12 Chromosome 14

Add(14)(q32)

SPECIFIC FEATURES

The human IGH heavy-chain locus is located on chromosome 14 at band 14q32.33. The IGH gene is involved in many abnormalities, including deletions, balanced and unbalanced rearrangements of various chromosome partners. Rearrangements involving the IGH gene are observed primarily in lymphoproliferative disorders, including B-cell NHL, CLL and MM. Paracentric inversions of the long arm of chromosome 14 are recurrent abnormalities in T-cell prolymphocytic leukemia, usually involving the breakpoint 14q11.2 which is the locus for the T-cell receptor alpha gene.

EXAMPLE REPORTING OF add(14)(q32) IN B-CELLS

ISCN Result: 46,XY,add(14)(q32)[20]

Interpretation

Chromosome analysis revealed an abnormal male chromosome complement with added material of unknown origin on the long arm of chromosome 14 as the sole anomaly in all cells examined.

Rearrangements involving the immunoglobulin heavy chain (IGH) on chromosome 14q32 are observed in B-cell non-Hodgkin lymphoma (NHL), chronic lymphocytic leukemia (CLL) and multiple myeloma (MM).

Fluorescence *in situ* hybridization (FISH) analysis for the IGH probe is suggested to confirm and monitor this abnormal cell line. Also, FISH analysis for the NHL/CLL/MM panel of probes is suggested to look for the presence of other commonly observed cytogenetic abnormalities not easily detectable in actively dividing cells by standard cytogenetics.

EXAMPLE REPORTING OF inv(14) IN T-CELLS

ISCN Result: 46,XY,inv(14)(q11.2q32)[20]

Interpretation

Chromosome analysis revealed an abnormal male chromosome complement with a paracentric inversion of the long arm of chromosome 14 as the sole anomaly in all cells examined.

Paracentric inversions of the long arm of chromosome 14 are recurrent, non-random abnormalities observed predominantly in T-cell prolymphocytic leukemia. The breakpoint 14q11.2 is the locus for the T-cell receptor alpha gene.

Translocation (14;18)(q32;q21) (Figure 23.10)

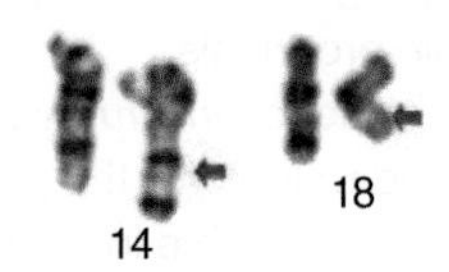

Figure 23.10 Partial karyotype showing translocation (14;18)(q32;q21).

SPECIFIC FEATURES

Translocation (14;18) results in the fusion of the IGH gene at 14q32 and the BCL2 gene at 18q21. It is found in 80–90% of follicular lymphomas, 30% of DLBCL and rarely in other lymphoproliferative disorders.

Although t(14;18) is well recognized by cytogenetics, it is most often identified by FISH analysis. However, since it is often seen with other karyotypic abnormalities, cytogenetics is useful in determining the complexity of abnormalities present. The most common abnormalities associated with t(14;18) include trisomy for chromosomes X, 5, 7, 8, 12, 18 and 21 as well as del(6q), +der(18)t(14;18) and t(14;18)(q24;q32).

Translocation (14;18) resulting in an IGH/MALT1 gene rearrangement is seen in MALT lymphomas and is cytogenetically indistinguishable from t(14;18) that results in an IGH/BCL2 rearrangement. Therefore, FISH analysis with gene-specific probes is necessary to distinguish between these two different rearrangements. Small cleaved cell follicular lymphomas have a median survival of 10 years or more; however, large cell lymphomas have a worse prognosis with t(14;18).

EXAMPLE REPORTING OF t(14;18)(q32;q21)

ISCN Result: 46,XY,t(14;18)(q32;q21)[20]

Interpretation

Chromosome analysis revealed an abnormal male chromosome complement with a translocation between the long arms of chromosomes 14 and 18 as the sole anomaly in all cells examined.

Translocation (14;18) results in the fusion of the immunoglobulin heavy chain gene (IGH) at 14q32 with the BCL2 gene at 18q21, and is a recurrent, non-random rearrangement found in 80–90% of follicular lymphoma cases and in approximately 30% of large B-cell non-Hodgkin lymphoma cases.

Fluorescence *in situ* hybridization (FISH) analysis for the BCL2/IGH fusion probe is suggested to confirm and monitor this abnormal cell line. Also, FISH analysis for the NHL panel of probes is suggested to look for the presence of other commonly observed cytogenetic abnormalities not easily detectable in actively dividing cells by standard cytogenetics.

23.3.13 Chromosome 17

Del(17p)

SPECIFIC FEATURES

Chromosome 17p deletions which involve the TP53 gene at 17p13.1 are seen in a large spectrum of hematological disorders, including both myeloid and lymphoid lineages. In lymphoid malignancies, they occur in all subtypes of lymphoproliferative disorders and are usually a secondary change, not a primary abnormality. The deleted segment may vary in size and many cases are submicroscopic deletions, which require FISH analysis for detection. Other abnormalities which result in del(17p) include unbalanced 17p translocations, dicentric rearrangements and isochromosome of the 17q, all resulting in TP53 gene loss. In general, deletion of the TP53 gene is associated with a poor prognosis. In B-CLL, in particular, it is considered one of the most important independent prognostic indicators associated with poor survival.

EXAMPLE REPORTING OF DELETION OF CHROMOSOME 17p

ISCN Result: 46,XY,del(17)(p12)[20]

Interpretation

Chromosome analysis revealed an abnormal male chromosome complement with a deletion within the long arm of chromosome 17 as the sole anomaly in all cells examined.

Deletion of the long arm of chromosome 17, resulting in the loss of the TP53 gene, is a recurrent, non-random finding in both lymphoid and myeloid malignancies. In lymphoid malignancies, del(17p) is commonly observed in chronic lymphoproliferative disorders, non-Hodgkin lymphoma and multiple myeloma. In general, deletion of the TP53 gene is associated with a poor prognosis. In B-CLL, in particular, it is considered one of the most important independent prognostic indicators associated with poor survival.

Fluorescence *in situ* hybridization (FISH) analysis for the CLL/NHL/MM panels of probes is suggested to look for the presence of other commonly observed cytogenetic abnormalities not easily detectable in actively dividing cells by standard cytogenetics.

23.3.14 Chromosome 18

Add(18)(q21)

SPECIFIC FEATURES

Rearrangements involving the MALT1 gene at 18q21 have been observed in extranodal marginal zone lymphoma of mucosa-associated lymphoid tissue (MALT lymphoma). T(11;18)(q21;q21) is mainly detected in pulmonary and gastric tumors while t(14;18)(q32;q21) has been associated with ocular adnexae/orbit and salivary gland lesions. However, other rearrangements, including unidentifiable additional material on chromosome 18q21, are not uncommon.

EXAMPLE REPORTING OF add(18)(q21)

ISCN Result: 46,XY,add(18)(q21)[20]

Interpretation

Chromosome analysis revealed an abnormal male chromosome complement with added material of unknown origin on the long arm of chromosome 18 as the sole anomaly in all cells examined.

Translocations involving the MALT1 gene at 18q21 have been observed in extranodal marginal zone lymphoma of mucosa-associated lymphoid tissue (MALT lymphoma). Translocation (11;18)(q21;q21) is mainly detected in pulmonary and gastric tumors while translocation (14;18)(q32;q21) has been associated with ocular adnexae/orbit and salivary gland lesions.

Fluorescence *in situ* hybridization (FISH) analysis with the MALT probe is suggested to confirm and monitor this abnormal cell line.

Trisomy 18

SPECIFIC FEATURES

Trisomy 18 is observed in most lymphoproliferative disorders. When seen as the sole anomaly, it is also non-specific. Trisomy 18 is common in hyperdiploid ALL with more than 50 chromosomes, which also shows trisomy for chromosomes X, 4, 6, 10, 14 and 21 (as well as tetrasomy 21), and may have trisomy 17 or an isochromosome 17q. However, trisomy 18 is not common in hyperdiploid ALL that has fewer than 50 chromosomes or with structural abnormalities associated with ALL. The prognosis appears to be neutral to favorable in a karyotype with high hyperdiploidy. There is some evidence of a poor prognosis when trisomy 18 is seen as the sole anomaly.

In MM, trisomy 18 is seen in approximately 10% of hyperdiploid cases, usually with multiple trisomies, tetrasomies and structural abnormalities. The most common structural anomalies that appear with trisomy 18 are chromosome 1 and 14q32 abnormalities, most of which include t(11;14).

In CLL, trisomy 18 is not common. When observed, it usually presents as the sole abnormality or in combination with trisomy for chromosomes 12, 18 and 19.

In Hodgkin disease, trisomy 18 has been reported in a few, quite complex near-triploid karyotypes and in hyperdiploid karyotypes, which include trisomies for chromosomes 2, 7, 12 and 21. The Hodgkin disease cases with trisomy 18 have included both the mixed cellularity and the nodular sclerosis types.

In non-Hodgkin lymphoma, trisomy 18 is observed in approximately 15–33% of cases, including DLBCL, FL and MZBCL. Trisomy 18 may be less frequent in other NHL subtypes. Trisomy 18 is strongly associated with t(14;18) and other 14q32 abnormalities, and may represent a derivative 18 with the gain of the der(18) of the unbalanced form of the t(14;18). In DLBCL, trisomy 18 is usually seen in hyperdiploid cases with the presence of structural rearrangements, including t(14;18), +der(18)t(14;18), t(8;14), other 14q32 rearrangements and i(6)(p10) resulting in loss of 6q. In FL, the general cytogenetic pattern of cases with trisomy 18 is similar to that of DLBCL. In MZBCL, trisomy 18 is seen in association with trisomy 3.

EXAMPLE REPORTING OF TRISOMY 18

ISCN Result: 47,XY,+18[20]

Interpretation

Chromosome analysis revealed an abnormal male chromosome complement with the gain of chromosome 18 as the sole anomaly in all cells examined.

Trisomy 18 is a recurrent, non-random abnormality observed primarily in lymphoproliferative disorders including non-Hodgkin lymphomas (NHL), multiple myeloma (MM) and chronic lymphocytic leukemia (CLL). Of the NHLs, trisomy 18 is seen most commonly in diffuse large B-cell lymphoma (DLBCL), follicular lymphoma (FL) and marginal zone B-cell lymphoma (MZBCL).

Fluorescence *in situ* hybridization (FISH) analysis for the NHL/MM/CLL panel of probes is suggested to look for the presence of other commonly observed cytogenetic abnormalities not easily detectable in actively dividing cells by standard cytogenetics.

23.3.15 Chromosome 21

Amplified(21q)

SPECIFIC FEATURES

Amplification of chromosome 21 including the AML1 (RUNX1 gene) region is a cytogenetic subgroup of ALL, which is characterized by the presence of multiple copies of the RUNX1 gene on a duplicated chromosome 21q. Currently, FISH with probes of the RUNX1 gene is the only reliable method of detection. However, there are cytogenetic abnormalities showing the gain of a region on chromosome 21 that results in RUNX1 gene amplification.

The banding pattern of the abnormal chromosome 21 is highly variable and may be present as a metacentric, acrocentric or ring chromosome. Therefore, if a karyotype with -21 and+mar is observed, FISH for the RUNX1 gene is suggested to identify whether the marker is of chromosome 21 origin. The duplicated 21q is most often seen with complex karyotypic abnormalities with no recurrent secondary abnormalities identified.

The current published definitions of amplification of chromosome 21q include only cases in which the abnormality has been visualized in metaphases and three or more RUNX1 signals are seen on a single abnormal chromosome 21 or, if only FISH is performed, cases which show interphase cells with five or more RUNX1 signals.

The estimated incidence of amp21q in childhood ALL is approximately 1–3%. The majority of patients tend to be older children or adolescents, with nearly three-quarters of reported patients between 6 and 14 years old. Amp21q is generally associated with a poor prognosis in childhood ALL.

EXAMPLE REPORTING OF add(21)(q21)

ISCN Result: 46,XY,add(21)(q21)[20]

Interpretation

Chromosome analysis revealed an abnormal male chromosome complement with added material of unknown origin on the long arm of chromosome 21 as the sole anomaly in all cells examined.

The added material on chromosome 21 may be of chromosome 21 origin, resulting in amplification of chromosome 21. Chromosome 21q amplification usually includes the RUNX1 (AML1) gene, and is a recurrent, non-random abnormality in acute lymphoblastic leukemia and generally associated with a poor prognosis.

Fluorescence *in situ* hybridization (FISH) analysis for the RUNX1 probe is suggested to confirm and monitor this abnormal cell line.

Trisomy 21

SPECIFIC FEATURES

Trisomy 21 is a frequent aneuploidy observed in both adult and childhood ALL. Its overall incidence is approximately 15% of cases. As the sole acquired clonal abnormality, trisomy 21 accounts for 2% of pediatric and less than 1% of adult ALL cases. In childhood ALL, trisomy 21 is very common in hyperdiploid cases.

In childhood ALL, trisomy 21 is also often associated with t(12;21)(p13;q22), chromosome 6q abnormalities, t(1;19)(q23;p13), t(4;11)(q21;q23) and chromosome 14q abnormalities. In adults, +21 is associated most frequently with t(9;22)(q34;q11.2). When trisomy 21 is seen as the sole abnormality, it is generally associated with a good prognosis.

EXAMPLE REPORTING OF TRISOMY 21

ISCN Result: 47,XY,+21[20]

Interpretation

Chromosome analysis revealed an abnormal male chromosome complement with the gain of chromosome 21 as the sole anomaly in all cells examined.

Trisomy 21 is the most frequent aneuploidy observed in both childhood and adult B-cell lymphoblastic leukemia (B-cell ALL) and lymphoblastic lymphoma (LBL). It is observed in approximately 15% of B-cell ALL/LBL cases, usually in conjunction with other abnormalities.

Trisomy 21 has a favorable prognosis as a sole abnormality or in children with a hyperdiploid chromosome count without other unfavorable structural rearrangements.

The presence of constitutional trisomy 21 (Down syndrome) should be excluded.

Fluorescence *in situ* hybridization (FISH) analysis for the ALL panel of probes is suggested to look for the presence of other commonly observed cytogenetic abnormalities not easily detectable in actively dividing cells by standard cytogenetics.

Bibliography

American College of Medical Genetics. *Guidelines in Clinical Cytogenetics. Section E10 Methods in Fluorescence In situ Hybridization – Interphase/Nuclear Fluorescence In Situ Hybridization*. American College of Medical Genetics, Bethesda, MD. www.acmg.net/StaticContent/SGs/Section_E.html, accessed 28 January 2014.

Atlas of Genetics and Cytogenetics in Oncology and Haematology. http://AtlasGeneticsOncology.org/Anomalies, accessed 28 January 2014.

Dewald GW, Jenkins RB. Cytogenetic and molecular genetic studies of patients with monoclonal gammopathies. In: Wiernik P, Canello G, Kyle R, Schiffer C (eds) *Neoplastic Diseases of Blood*, 2nd edn. Churchill Livingstone, New York, 1991, pp. 427–438.

Dewald GW, Therneau T, Larson D, et al. Relationship of patient survival and chromosome anomalies detected in metaphase and/or interphase cells at diagnosis of myeloma. Blood 2005; 106(10): 3553–3558.

Dohner H, Stilgenbauer S, Benner A, et al. Genomic aberrations and survival in chronic lymphocytic leukemia. N Engl J Med 2000; 343: 1910–1916.

Fonseca R, Barlogie B, Bataille R, et al. Genetics and cytogenetics of multiple myeloma: a workshop report. Cancer Res 2004; 64: 1546–1558.

Gersen S, Keagle M (eds). *Principles of Clinical Cytogenetics*. Humana Press, Totowa, New Jersey, 1999.

Gollin SM. Mechanisms leading to nonrandom, nonhomologous chromosomal translocations in leukemia. Semin Cancer Biol 2007; 17(1): 74–79.

Heim S, Mitelman F. *Cancer Cytogenetics*, 2nd edn. Wiley-Liss, New York, 1995.

Janz S, Potter M, Rabkin CS. Lymphoma- and leukemia-associated chromosomal translocations in healthy individuals. Genes Chromosomes Cancer 2003; 36: 211–223.

Korsmeyer SJ. Chromosomal translocations in lymphoid malignancies reveal novel proto-oncogenes. Annu Rev Immunol 1992; 10: 785–807.

McClure RF, Remstein ED, Macon WR, et al. Adult B-cell lymphomas with Burkitt-like morphology are phenotypically and genotypically heterogeneous with aggressive clinical behavior. Am J Surg Pathol 2005; 29: 1652–1660.

Mohamed AN, Bentley G, Bonnett M, et al. Chromosomes aberrations in a series of 120 multiple myeloma cases with abnormal karyotypes. Am J Hematol 2007; 82(12): 1080–1087.

Pierre RV, Dewald GW, Banks PM. Cytogenetic studies in malignant lymphoma: possible role in staging studies. Cancer Genet Cytogenet 1980; 1: 257–261.

Rooney DE, Czepulkowski BH. *Human Cytogenetics. A Practical Approach*. Oxford University Press, New York, 1992.

Shaffer LG, McGowan-Jordan J, Schmid M (eds). *ISCN 2013: An International System for Human Cytogenetic Nomenclature*. Karger Publishers, Unionville, CT, 2013.

Siebert R. Mature B- and T-cell neoplasms and Hodgkin lymphoma. In: Heim S, Mitelman F (eds) *Cancer Cytogenetics*, 3rd ed. Wiley, Hoboken, NJ, 2009, pp. 297–374.

Swerdlow SH, Campo E, Harris NL, et al. (eds). *WHO Classification of Tumours of Haematopoietic and Lymphoid Tissues*. IARC Press, Lyon, 2008.

CHAPTER 24

Common biphenotypic abnormalities and secondary changes

24.1 Translocation (4;11)(q21;q23)

SPECIFIC FEATURES

Translocation (4;11) is a recurrent, non-random finding seen in both myeloid and lymphoid cell lineages, including acute lymphoblastic and acute myeloid leukemias and mixed phenotypic acute leukemia (MPAL). T(4;11) is generally associated with a poor prognosis.

EXAMPLE REPORTING OF t(4;11)(q21;q23)

ISCN Result: 46,XY,t(4;11)(q21;q23)[4]/46,XY[16]

Interpretation

Chromosome analysis revealed an abnormal male chromosome complement in 4 of 20 cells examined with a translocation between the long arms of chromosomes 4 and 11 as the sole anomaly. The remaining 16 cells showed a normal male chromosome complement.

Translocation (4;11) is a recurrent, non-random finding in hematological malignancies, primarily observed in B-cell acute lymphoblastic leukemia (B-cell ALL), but it has also been reported in mixed phenotypic acute leukemia (MPAL) and acute myeloid leukemia (AML). This rearrangement is generally associated with a poor prognosis.

Fluorescence *in situ* hybridization (FISH) analysis for the MLL probe is suggested to confirm and monitor this abnormal cell line.

Cytogenetic Abnormalities: Chromosomal, FISH and Microarray-Based Clinical Reporting, First Edition. Susan Mahler Zneimer.

24.2 Del(9q)

SPECIFIC FEATURES

Deletions of chromosome 9q are recurrent, non-random findings seen in a broad spectrum of hematological malignancies, including chronic myeloid and lymphoid disorders, acute leukemia and lymphomas. Del(9q) is generally associated with a poor prognosis.

EXAMPLE REPORTING OF del(9q)

ISCN Result: 46,XY,del(9)(q13q33)[7]/46,XY[13]

Interpretation

Chromosome analysis revealed an abnormal male chromosome complement in 7 of 20 cells examined with a deletion within the long arm of chromosome 9 as the sole anomaly. The remaining 13 cells showed a normal male chromosome complement.

Deletions of the long arm of chromosome 9 are recurrent abnormalities in hematological disorders, being observed in chronic myeloid and lymphoid disorders, acute leukemia and lymphoma, and generally associated with a poor prognosis.

24.3 Translocation (11;19)(q23;p13.3)

SPECIFIC FEATURES

This translocation results in the fusion of the MLL gene at 11q23 with the ENL gene at 19p13.3 and is observed in both myeloid and lymphoid leukemias, including AML and B-cell ALL. However, it is rare in T-cell lymphoblastic leukemias (T-cell ALL) and therapy-related leukemias. T(11;19) is generally associated with a very poor prognosis except in the rare T-cell cases, which have been reported to show longer survival rates.

EXAMPLE REPORTING OF t(11;19)(q23;p13.3)

ISCN Result: 46,XY,t(11;19)(q23;p13.3)[4]/46,XY[16]

Interpretation

Chromosome analysis revealed an abnormal male chromosome complement in 4 of 20 cells examined with a translocation between the long arm of chromosome 11 and the short arm of chromosome 19 as the sole anomaly. The remaining 16 cells showed a normal male chromosome complement.

This translocation results in the fusion of the MLL gene at 11q23 with the ENL gene at 19p13.3. It is observed in both myeloid and lymphoid leukemias, including acute myeloid leukemia (AML) and B-cell lymphoblastic leukemia (B-cell ALL). It is rare in T-cell lymphoblastic leukemia (T-cell ALL) and therapy-related leukemia. T(11;19) is generally associated with a very poor prognosis except in the rare T-cell cases, which have been reported to show longer survival rates.

Fluorescence *in situ* hybridization (FISH) analysis for the MLL probe is suggested to confirm and monitor this abnormal cell line.

24.4 Del(12)(p11.2p13)

SPECIFIC FEATURES

Deletions of chromosome 12p are recurrent, non-random findings seen in a broad spectrum of hematological malignancies, including acute lymphoblastic and acute myeloid leukemias, myelodysplastic syndromes, chronic myeloproliferative syndromes and non-Hodgkin lymphomas. Del(12p) is generally associated with a poor prognosis.

EXAMPLE REPORTING OF del(12)(p11.2p13)

ISCN Result: 46,XY,del(12)(p11.2p13)[4]/46,XY[16]

Interpretation

Chromosome analysis revealed an abnormal male chromosome complement in 4 of 20 cells examined with a deletion within the short arm of chromosome 12 as the sole anomaly. The remaining 16 cells showed a normal male chromosome complement.

Deletions of chromosome 12p are recurrent, non-random findings seen in a broad spectrum of hematological malignancies, including acute lymphoblastic and acute myeloid leukemias, myelodysplastic syndromes, chronic myeloproliferative syndromes and non-Hodgkin lymphoma. Del(12p) is generally associated with a poor prognosis.

Since a minimal interstitial deletion region usually involves the ETV6 and CDKN1B genes, fluorescence *in situ* hybridization (FISH) analysis for one of these probes is suggested to confirm and monitor this abnormal cell line.

24.5 Trisomy 15

SPECIFIC FEATURES

Trisomy 15, as the sole anomaly, is a rare finding in hematological disorders but has been reported in association with myelodysplasias and lymphoid malignancies. However, most patients with trisomy 15 when seen as the sole anomaly are elderly males, suggesting that the presence of trisomy 15 may reflect an age-related phenomenon. For trisomy 15 seen in conjunction with loss of the Y chromosome, see Chapter 21.

EXAMPLE REPORTING OF TRISOMY 15

ISCN Result: 47,XY,+15[7]/46,XY[13]

Interpretation

Chromosome analysis revealed an abnormal male chromosome complement in 7 of 20 cells examined with an extra copy (trisomy) of chromosome 15 as the sole anomaly. The remaining 13 cells showed a normal male chromosome complement.

Trisomy 15, as a sole cytogenetic anomaly, is a rare but recurrent finding in hematological disorders. It has been reported in both myeloid and lymphoid malignancies although it is most frequently observed in myelodysplasia.

However, trisomy 15 has also been reported in patients free of hematological malignancy. Clinical-pathological correlation of results is suggested.

24.6 i(17q)

SPECIFIC FEATURES

Isochromosome of the long arm of chromosome 17 results in a 17p deletion. This abnormality is a multilineage disorder, mainly seen in adults with a mean age of over 60 years and seen equally among females and males. Chromosome 17p deletions are generally associated with a very poor prognosis with a median survival of only 4 months.

EXAMPLE REPORTING OF CHROMOSOME i(17)(q10) REARRANGEMENTS

ISCN Result: 46,XY, i(17)(q10)[20]

Interpretation

Chromosome analysis revealed an abnormal male chromosome complement with an isochromosome of the long arm of chromosome 17 as the sole anomaly in all cells examined.

Isochromosome 17q results in the gain of the long arm and loss of the short arm of chromosome 17 (including the TP53 gene). Loss of the TP53 gene is a recurrent, non-random finding in both lymphoid and myeloid malignancies.

In lymphoid malignancies, del(17p) is commonly observed in chronic lymphoproliferative diseases, non-Hodgkin lymphoma and multiple myeloma. In particular, chronic lymphocytic leukemia (CLL) patients with chromosome 17p deletions have more rapid disease progression and the shortest median treatment-free interval when compared to other genetically defined CLL risk groups.

In myeloid malignancies, when seen as the sole abnormality, i(17q) is a distinctive hematological entity that is observed as both a *de novo* and secondary cytogenetic change with a high risk for leukemic progression. These neoplasms have shown distinctive morphological features, including multilineage dysplasia and concurrent myeloproliferative features. I(17q) usually occurs at the time of blast transformation and is associated with an aggressive clinical course and poor prognosis.

Fluorescence *in situ* hybridization (FISH) analysis for the TP53 gene is suggested to confirm and monitor this abnormal cell line.

Bibliography

American College of Medical Genetics. *Guidelines in Clinical Cytogenetics. Section E10 Methods in Fluorescence In situ Hybridization – Interphase/Nuclear Fluorescence In Situ Hybridization.* American College of Medical Genetics, Bethesda, MD. www.acmg.net/StaticContent/SGs/Section_E.html, accessed 28 January 2014.

Atlas of Genetics and Cytogenetics in Oncology and Haematology. http://AtlasGeneticsOncology.org/Anomalies, accessed 28 January 2014.

Béné MC, PorwitA. Acute leukemias of ambiguous lineage. Semin Diagn Pathol 2012; 29(1): 12–18.

Borowitz MJ, Rubnitz J, Nash M, Pullen DJ, Camitta B. Surface antigen phenotype can predict TEL-AML1 rearrangement in childhood B-precursor ALL: a Pediatric Oncology Group study. Leukemia 1998; 121764–1770.

Carbonell F, Swansbury J, Min T, et al. Cytogenetic findings in acute biphenotypic leukaemia. Leukemia 1996; 10: 1283–1287.

Chang WJ, Santana-Dávila R, van Wier SA, et al. Prognostic factors for hyperdiploid-myeloma: effects of chromosome 13 deletions and IgH translocations. Leukemia 2006; 20(5): 807–813.

Cuneo A, Rigolin GM, Bigoni R, et al. Chronic lymphocytic leukemia with 6q- shows distinct hematological features and intermediate prognosis. Leukemia 2004; 18: 476–483.

Dewald GW, Brockman SR, Paternoster SF, et al. Chromosome anomalies detected by interphase fluorescence in situ hybridization: correlation with significant biological features of B-cell chronic lymphocytic leukaemia. Br J Haematol 2003; 121: 287–295.

Dickinson JD, Gilmore J, Iqbal J, et al. 11q22.3 deletion in B-chronic lymphocytic leukemia is specifically associated with bulky lymphadenopathy and ZAP-70 expression but not reduced expression of adhesion/cell surface receptor molecules. Leuk Lymphoma 2006; 47(2): 231–244.

Döhner H, Stilgenbauer S, Benner A, et al. Genomic aberrations and survival in chronic lymphocytic leukemia. N Engl J Med 2000; 343: 1910–1916.

Drexler HG, Gignac SM, von Wasielewski R, Werner M, Dirks WG. Pathobiology of NPM-ALK and variant fusion genes in anaplastic large cell lymphoma and other lymphomas. Leukemia 2000; 14(9): 1533–1559.

Gersen S, Keagle M (eds). *Principles of Clinical Cytogenetics.* Humana Press, Totowa, New Jersey, 1999.

Grimwade D, Walker H, Oliver F, et al. The importance of diagnostic cytogenetics on outcome in AML: analysis of 1,612 patients entered into the MRC AML 10 trial. The Medical Research Council Adult and Children's Leukaemia Working Parties. Blood 1998; 92(7): 2322–2333.

Harris MB, Shuster JJ, Carroll A, et al. Trisomy of leukemic cell chromosomes 4 and 10 identifies children with B-progenitor cell acute lymphoblastic leukemia with a very low risk of treatment failure: a Pediatric Oncology Group study. Blood 1992; 79: 3316–3324.

Heim S, Mitelman F. *Cancer Cytogenetics*, 2nd edn. Wiley-Liss, New York, 1995.

Jackson JF, Boyett J, Pullen J, et al. Favorable prognosis associated with hyperdiploidy in children with acute lymphocytic leukemia correlates with extra chromosome 6. A Pediatric Oncology Group study. Cancer 1990; 66: 1183–1189.

Matutes E, Morilla R, Farahat N, et al. Definition of acute biphenotypic leukemia. Haematologica 1997; 82(1): 64–6.

Rooney DE, Czepulkowski BH. *Human Cytogenetics. A Practical Approach.* Oxford University Press, New York, 1992.

Schoch C, Haase D, Haferlach T, et al. Fifty-one patients with acute myeloid leukemia and translocation t(8;21)(q22;q22): an additional deletion in 9q is an adverse prognostic factor. Leukemia 1996; 10: 1288–1295.

Shaffer LG, McGowan-Jordan J, Schmid M (eds). *ISCN 2013: An International System for Human Cytogenetic Nomenclature*. Karger Publishers, Unionville, CT, 2013.

Smith SR, Rowe D. Trisomy 15 in hematological malignancies: six cases and review of the literature. Cancer Genet Cytogenet 1996; 89: 27–30.

Swerdlow SH, Campo E, Harris NL, et al. (eds). *WHO Classification of Tumours of Haematopoietic and Lymphoid Tissues*. IARC Press, Lyon, 2008.

Ye H, Liu H, Attygalle A, et al. Variable frequencies of t(11;18)(q21;q21) in MALT lymphomas of different sites: significant association with CagA strains of H pylori in gastric MALT lymphoma. Blood 2003; 102(3): 1012–1018.

Zhang YM, Wu DP, Sun AN, et al. Study on the clinical characteristics of 32 patients with mixed phenotype acute leukemia. Zhonghua Xue Ye Xue Za Zhi 2011; 32(1): 12–16.

CHAPTER 25

Reporting complex abnormalities and multiple cell lines

One major component of reporting neoplastic cytogenetic results in complex cases is the order of cell lines present. Nomenclature is based on the complexity of abnormalities present and not on the number of cells per clone. Therefore, writing multiple cell lines begins with clones containing the simplest abnormalities followed by clones with increasing complexity of the aberrations present.

Reporting neoplastic cell lines is described using various terms. One term is the mainline, which refers to the cell line composed of the most number of cells compared to the other cell lines. However, the mainline does not address the complexity of abnormalities present. More than one mainline may be present in a specimen if each clone has an equally large number of abnormal cells.

The stemline refers to the abnormal cell line that has the most basic abnormalities. The stemline usually indicates the progenitor of all the cell lines. To identify an abnormal cell line as the stemline, it must contain the common abnormalities present in all the abnormal clones, although it is possible for a cell line to have lost one or more of the stemline abnormalities as part of clonal evolution. Thus, there are instances in which one may not reliably ascertain whether a cell line is the stemline and whether it originated before the other cell lines.

Once the stemline is identified, however, cell lines that contain the stemline abnormalities with the presence of other abnormalities are considered sidelines or subclones. Sidelines or subclones differ from one another in their abnormalities. When sidelines are related by continually adding new abnormalities to each clone, it is described as clonal evolution. When each sideline evolves from the stemline with distinct abnormalities, it is considered clonal divergence. Therefore, the sidelines are a description of their relation of abnormalities to that of the stemline. When more than one sideline is described, each evolving from a previous sideline or the stemline, they are called sideline 1, sideline 2, etc.

When describing stemlines and subclones, the number of cells comprising each cell line does not influence the placement of that clone in the ISCN nomenclature order. Only the complexity of abnormalities present describes the order of subclones. When writing the ISCN nomenclature, specific rules and abbreviations apply, which are described below. One note is that the ISCN accepts two abbreviations for describing subclones of a stemline: “sl” or “idem” (with a description below). However, “sl” is appropriate in most cases.

Cytogenetic Abnormalities: Chromosomal, FISH and Microarray-Based Clinical Reporting, First Edition. Susan Mahler Zneimer.

25.1 Stemline and sideline abnormalities

ISCN RULES FOR DESCRIBING STEMLINE AND SIDELINE ABNORMALITIES

- First write the cell line that represents the stemline; that is, the cell line that has the least complexity of abnormalities present or has the abnormality(ies) in common with all cell lines.
- Then write the number of cells enclosed in brackets, followed by a slash.
- Next write the first sideline, which is the cell line that is the least complex of the sidelines, followed by sidelines in increasing complexity of abnormalities present.
- When writing the first sideline, first write the modal chromosome number, followed by "sl", followed by the abnormalities present. The term "idem" may be used with one sideline present rather than "sl" but "idem" is not used with multiple sidelines. With the term "sl", do not write the sex designation, as it is implied from the stemline nomenclature.
- Then write the abnormalities present, followed by the number of cells in the sideline enclosed in brackets.
- Next write the second sideline. If the second sideline has evolved from the first sideline, then write the modal chromosome number, followed by "sdl1", referring to changes from the first sideline.
- Then write the abnormalities present, followed by the number of cells in the sideline enclosed in brackets.
- If the second sideline deviates from the stemline and not the first sideline, write the modal chromosome number, followed by "sl", referring to changes from the stemline.
- Then write the abnormalities present, followed by the number of cells in the sideline enclosed in brackets.
- Note that cell lines are written in order of the complexity of abnormalities present, not the number of cells present in each cell line.
- Also note that when an abnormality with breakpoints is present in the stemline and is also seen in a subsequent cell line, the breakpoints do not need to be repeated in these cell lines.
- Also note that there are no nomenclature rules for designating which cell line should precede the other if the complexity of the abnormalities is comparable. For example, if each cell line contains a single abnormality, then there is no designation for which cell line is written first.

For example, a stemline and one sideline would be written as:

46,XY,t(9;22)(q34;q11.2)[5]/47,sl,i(17)(q10)[15].

A stemline and two related sidelines would be written as:

46,XY,t(9;22)(q34;q11.2)[5]/47,sl,i(17)(q10)[4]/48,sdl1,+der(22)t(9;22)[11].

A stemline and two sidelines derived separately from the stemline would be written as:

46,XY,t(9;22)(q34;q11.2)[5]/47,sl,i(17)(q10)[8]/47,sl,+der(22)t(9;22)[7].

The presence of a stemline and related sidelines that have acquired new abnormalities from each subsequent sideline represents clonal evolution, since the stemline has changed or evolved with time to include other abnormalities in a sequential order of genetic changes.

An example of *clonal evolution* is:

Stemline: 46,XX,t(9;22)(q34;q11.2)[5]/
Sideline 1: 47,sl,+8[15]

Or with any number of sidelines that progress from the previous sideline, such as:

Stemline: 46,XX,t(9;22)(q34;q11.2)[5]/
Sideline 1: 47,sl,+8[7]/
Sideline 2: 47,sdl1,i(17)(q10)[4]/
Sideline 3: 48,sdl2,+der(22)t(9;22)[4]

When a stemline shows two sidelines that contain distinct abnormalities from one another, this represents clonal divergence.

An example of *clonal divergence* is:

Stemline: 46,XX,t(9;22)(q34;q11.2)[5]/
Sideline 1: 47,sl,+8[7]
Sideline 2: 47,sl,i(17)(q10)[4]/
Sideline 3: 48,sl,+der(22)t(9;22)[4]

The presence of multiple sidelines that are all derived from the stemline can then be described as showing both clonal evolution and clonal divergence, since all the sidelines have evolved from the stemline and all the sidelines have diverged from one another.

Since there are no ISCN rules for describing more than one sideline, except by order of increasing complexity, one can choose the order of sidelines if the complexity of abnormalities seen in each sideline is comparable. For instance, in the example above, with a stemline that contains a t(9;22) as the sole anomaly, when sidelines show a newly acquired single abnormality, such as trisomy 8, i(17q) or +der(22) as sole anomalies, one can write the sidelines in any order. Choosing the order of these sidelines may be based on the number of cells seen in each sideline, the chromosome number involved in the abnormality or the clinical significance or predominance of the abnormality seen in the disease entity.

EXAMPLE REPORTING OF TWO RELATED ABNORMAL CELL LINES (STEMLINE AND SIDELINE)

ISCN Result: 46,XY,t(9;22)(q34;q11.2)[5]/46,sl,i(17)(q10)[15]

Interpretation

Cytogenetic analysis revealed a male chromosome complement with two related abnormal cell lines in all 20 cells examined. The first cell line (stemline), seen in five cells, showed a translocation between the long arms of chromosomes 9 and 22, resulting in a Philadelphia chromosome, as the sole anomaly.

The second cell line (sideline), seen in 15 cells, showed the stemline abnormality plus an isochromosome of the long arm of chromosome 17, resulting in loss of the TP53 gene.

The t(9;22) translocation results in the fusion of the ABL1 gene at 9q34 and the BCR gene at 22q11, and is consistent with a diagnosis of chronic myeloid leukemia (CML). An additional clonal cytogenetic aberration, including isochromosome of 17q, is a recurrent secondary change in CML and is commonly observed in association with impending blast crisis.

Fluorescence *in situ* hybridization (FISH) analysis for the t(9;22) is suggested to monitor this abnormal cell line.

EXAMPLE REPORTING OF THREE RELATED ABNORMAL CELL LINES (STEMLINE AND TWO RELATED SIDELINES DERIVED FROM ONE ANOTHER)

ISCN Result: 46,XY,t(9;22)(q34;q11.2)[5]/46,sl,i(17)(q10)[4]/47,sdl1,+der(22)t(9;22)[11]

Interpretation

Cytogenetic analysis revealed a male chromosome complement with three related abnormal cell lines in all 20 cells examined. The first cell line (stemline), seen in five cells, showed a translocation between the long arms of chromosomes 9 and 22, resulting in a Philadelphia chromosome as the sole anomaly.

The second cell line (sideline 1), seen in four cells, showed the stemline abnormality plus an isochromosome of the long arm of chromosome 17, resulting in loss of the TP53 gene. The third cell line (sideline 2), seen in 11 cells, showed the sideline 1 abnormalities plus an extra copy of the abnormal chromosome 22 (extra Philadelphia chromosome) derived from the t(9;22).

The t(9;22) translocation results in the fusion of the ABL1 gene at 9q34 and the BCR gene at 22q11.2, and is consistent with a diagnosis of chronic myeloid leukemia (CML). Additional clonal cytogenetic aberrations, including an isochromosome of 17q and an extra Philadelphia chromosome, are recurrent secondary changes in CML and are commonly observed in association with impending blast crisis.

Fluorescence *in situ* hybridization (FISH) analysis for the t(9;22) is suggested to monitor this abnormal cell line.

EXAMPLE REPORTING OF THREE RELATED ABNORMAL CELL LINES (STEMLINE AND TWO SIDELINES DERIVED FROM THE STEMLINE)

ISCN Result: 46,XY,t(9;22)(q34;q11.2)[5]/46,sl,i(17)(q10)[8]/47,sl,+der(22)t(9;22)[7].

Interpretation

Cytogenetic analysis revealed a male chromosome complement with three related abnormal cell lines in all 20 cells examined. The first cell line (stemline), seen in five cells, showed a translocation between the long arms of chromosomes 9 and 22, resulting in a Philadelphia chromosome as the sole anomaly.

The second cell line (sideline 1), seen in eight cells, showed the stemline abnormality plus an isochromosome of the long arm of chromosome 17, resulting in loss of the TP53 gene. The third cell line (sideline 2), seen in seven cells, showed the stemline abnormality plus an extra copy of the abnormal chromosome 22 (extra Philadelphia chromosome) derived from the t(9;22).

The t(9;22) translocation results in the fusion of the ABL1 gene at 9q34 and the BCR gene at 22q11.2, and is consistent with a diagnosis of chronic myeloid leukemia (CML). Additional clonal cytogenetic aberrations, including an isochromosome of 17q and an extra Philadelphia chromosome, are recurrent secondary changes in CML and are commonly observed in association with impending blast crisis.

Fluorescence *in situ* hybridization (FISH) analysis for the t(9;22) is suggested to monitor this abnormal cell line.

There are no ISCN rules for describing the sideline order when the sidelines are derived from the stemline or other sidelines. For example, with the clones described below:

46,XX,t(9;22)(q34;q11.2)[5]/
47,XX,t(9;22)(q34;q11.2),+8[5]/
47,XX,t(9;22)(q34;q11.2),+der(22)t(9;22)[4]
47,XX, t(9;22)(q34;q11.2),i(17)(q10),+der(22)t(9;22)[6]

The stemline is clearly 46,XX,t(9;22)(q34;q11.2)[5], since the t(9;22) is the least complex karyotype and contains the same abnormality in common with all the cell lines. Then, there are two sidelines derived from the stemline: 47,XX,t(9;22)(q34;q11.2)+8[5] and 47,XX,t(9;22)(q34;q11.2),+der(22)t(9;22)[4]. These clones each contain a single, newly acquired abnormality from the stemline and are distinct abnormalities from each other, representing clonal divergence.

There is also the clone 47,XX,t(9;22)(q34;q11.2),i(17)(q10),+der(22)t(9;22)[6], which has evolved from a sideline, not directly from the stemline. Thus, there are two sidelines derived from the stemline, and a sideline derived from another sideline. Writing the order of the sidelines can be confusing and is best explained in the interpretation of the report. One suggested approach is to describe the stemline followed by all the related sidelines, and then write the second sideline derived from the stemline last. This avoids the confusion of writing sl, sdl1, sdl2, without knowing which cell line is derived from which.

For example, one approach to describing the following four cell lines is as follows:

46,XX,t(9;22)(q34;q11.2)[5]/
47,XX,t(9;22)(q34;q11.2),+8[5]/
47,XX,t(9;22)(q34;q11.2),+der(22)t(9;22)[4]
47,XX, t(9;22)(q34;q11.2),i(17)(q10),+der(22)t(9;22)[6]

These cell lines could be written as:

46,XX,t(9;22)(q34;q11.2)[5]/
47,sdl1,+der(22)t(9;22)[4]/
47,sdl2,i(17)(q10)[6]/
47,sl,+8[5]

If one wrote the 47,sl,+8[6] cell line before the other sidelines, then it would not be clear that 47,sdl1,+der(22)t(9;22)[4] and 47,sdl2,i(17)(q10)[6] are derived from the +8 clone or the stemline.

The description of more than one abnormal clone not related to each other is referred to as unrelated clones. These are cell lines that do not have any common abnormalities present in each cell line. Unrelated clones are written in the ISCN in order of the number of cells present in each clone, regardless of the complexity of abnormalities present.

25.2 Unrelated abnormal clones

ISCN RULES FOR DESCRIBING UNRELATED ABNORMAL CLONES

- With the presence of abnormal and normal cell lines, always write the normal cell line last.
- First write the cell line with the most cells observed. If both or all cell lines contain the same number of cells, then the order is as follows: first are clones with sex chromosome abnormalities, then clones are ordered by autosome abnormalities present, from lowest to highest chromosome number.
- Then write the second abnormal cell line.
- Separate each cell line with a slash.
- Write the normal cell line last.

For example, two unrelated clones, one with del(20q) seen in 11 cells, and one with del(11q) seen in 4 cells, would be written as: 46,XY,del(20)(q11.2q13.1)[11]/46,XY,del(11)(q13q23)[4]/46,XY[5]. If both these abnormal clones were seen in equal numbers, then it would be written as: 46,XY,del(11)(q13q23)[8]/46,XY,del(20)(q11.2q13.1)[8]/46,XY[4].

EXAMPLE REPORTING OF TWO UNRELATED CLONES

ISCN Result: 46,XY,del(20)(q11.2q13.1)[11]/46,XY,del(11)(q13q23)[4]/46,XY[5]

Interpretation

Cytogenetics analysis revealed a male chromosome complement with two unrelated abnormal cell lines. The first cell line, seen in 11 of 20 cells examined, showed a deletion within the long arm of chromosome 20 as the sole anomaly. The second cell line, seen in 4 of 20 cells examined, showed a deletion within the long arm of chromosome 11 as the sole anomaly. The remaining five cells showed a normal male chromosome complement.

Deletions of the long arm of chromosome 20 are associated with myeloid disorders in general, including myelodysplasias, myeloproliferative neoplasms and acute myeloid leukemia. Deletions of the MLL gene at 11q23 are observed in both *de novo* and therapy-related acute myeloid leukemia.

Fluorescence *in situ* hybridization (FISH) analysis for the del(20q) and 11q23 probes is suggested to monitor these abnormal cell lines.

One important aspect of neoplastic cell lines is the presence of genetic heterogeneity among the abnormalities present in a single clone. That is, clonal abnormalities may be present because two or more cells contain the same gain or structural abnormality, or three or more cells contain the same chromosomal loss. However, with complex karyotypic changes, there may be some variation from cell to cell in the abnormalities present. Therefore, a composite of all the clonal changes is listed in the ISCN nomenclature. This is referred to as a composite karyotype. In this case, there are two ways in which the composition of the karyotypic designation may be written. Composite karyotypes may be written with each abnormality in their normal order and, at the end, a [cp] with the total number of cells described in that clone or, after each abnormality, a bracket with the number of cells of each abnormality may be described followed at the end of the abnormalities with [cp] and the total number of cells in that clone. It may be useful to designate how many cells contain each abnormality if each of the abnormalities varies greatly from cell to cell.

25.3 Composite karyotypes

ISCN RULES FOR DESCRIBING COMPOSITE KARYOTYPES

- First write the modal chromosome number or, if it varies between cells, then write the range of chromosomes observed from all the cells that are to be included in the composite karyotype.
- Next write the sex chromosome designation.
- If the clonal abnormalities present include some cells with a sex chromosome aberration, representing either a gain or loss or a rearrangement, while other cells show normal sex chromosomes, then write the normal sex designation, followed by a comma, followed by the sex chromosome aberration(s).

For example, in a female patient, if some cells contain two normal X chromosomes and some cells show the loss of one X chromosome while other cells gain an X chromosome, then write: 45~47,XX,-X,+X,...

- Then write the clonal abnormalities seen in order by chromosome. If some cells contain one particular abnormality of a chromosome while other cells contain a different abnormality of the same chromosome, then write both abnormalities in the nomenclature, following the rules of alphabetic order.

For example, in the female patient above, if some cells contain added material on the long arm of chromosome 1 while other cells show a translocation between chromosomes 1 and 3, then write: 45~47,XX,-X,+X,add(1)(q13),t(1;3)...

- At the end of writing all the abnormalities present, write "cp", the designation for a composite karyotype, followed by the number of cells observed, enclosed in brackets.

For example: 45~47,XX,add(3)(q12),del(4)(q25),-5,der(7)t(7;15)(q11.2;q11.2),+8,+14,-15,inv(17)(q21q25),add(18)(q23),+20,psu dic(20;21)(q11.2;p11.2)x2,+21[cp17]/46,XX[3].

One may choose to add in brackets the number of cells showing each abnormality, in order to more precisely describe the abnormalities seen. For example, in the karyotype above, the nomenclature could be written as: 45~47,XX,add(3)(q12)[4],del(4)(q25)[12],-5[16],der(7)t(7;15)(q11.2;q11.2)[3],+8[15],+14[5],-15[8],inv(17)(q21q25)[12],add(18)(q23)[6],+20[4],psu dic(20;21)(q11.2;p11.2)x2[8],+21[3][cp17]/ 46,XX[3].

Only clonal abnormalities are included in the nomenclature, not single cell abnormalities, nor chromosome losses seen in one or two cells.

If any of the abnormalities are seen in all cells examined, then this may also be written as a stemline, followed by sidelines, or a composite sideline.

EXAMPLE REPORT OF A COMPOSITE KARYOTPE (FIGURE 25.1)

ISCN Result: 45~47,XX,add(3)(q12),del(4)(q25),-5,der(7)t(7;15)(q11.2;q11.2),+8,+14,-15,inv(17)(q21q25),add(18)(q23),+20,psu dic(20;21)(q11.2;p11.2)x2,+21[cp17]/46,XX[3]

Interpretation

Cytogenetic analysis revealed an abnormal female chromosome complement in 17 of 20 metaphase cells examined with the following clonal abnormalities:

- loss of one copy of chromosomes 5 and 15
- gain of one copy of chromosomes 8, 14, 20 and 21

- additional material of unidentified origin on the long arms of chromosomes 3 and 18
- a deletion within the long arm of chromosome 4
- an unbalanced translocation between chromosomes 7 and 15, resulting in loss of 7q
- an inversion of the long arm of chromosome 17
- and two copies of a pseudodicentric chromosome (20;21), arising from a rearrangement involving the long arm of chromosome 20 and the short arm of chromosome 21.

There was some variation from cell to cell in the abnormalities present; therefore, a composite of all the abnormalities is described in the nomenclature.

The remaining three cells showed a normal female chromosome complement.

Monosomy 5, deletion of 7q and trisomy 8 are recurrent abnormalities in myelodysplastic syndromes (MDS) and acute myeloid leukemia (AML). Patients with complex cytogenetic abnormalities involving chromosomes 5 and 7 generally have a poor prognosis.

Fluorescence *in situ* hybridization (FISH) analysis for the chromosome 5, 7 and 8 probes is suggested to monitor this abnormal cell line.

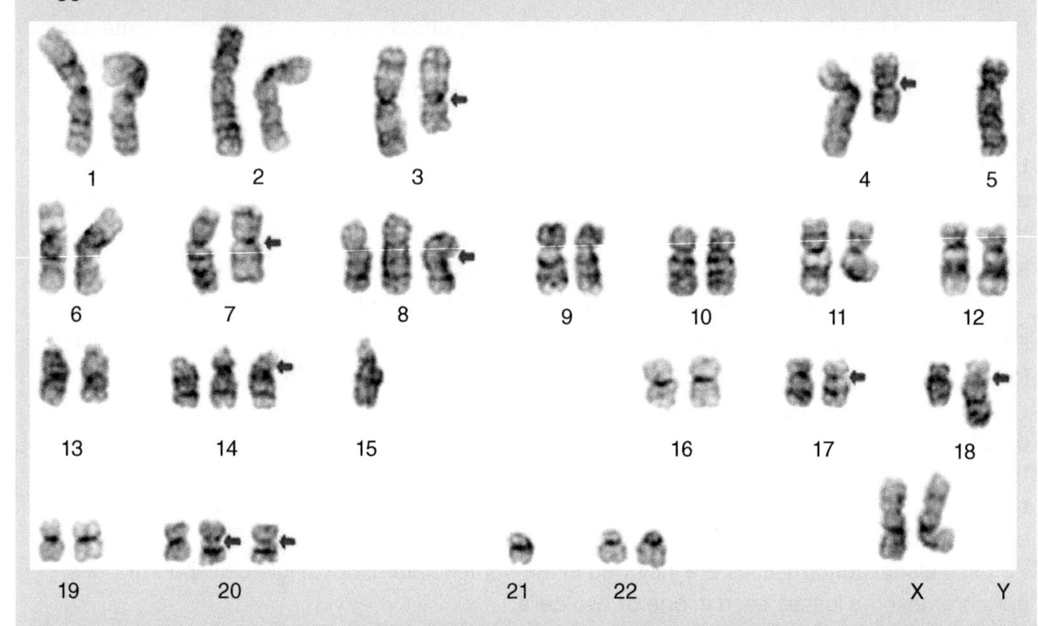

Figure 25.1 Karyotype with the ISCN designation: 45~47,XX,add(3)(q12),del(4)(q25),-5,der(7)t(7;15)(q11.2;q11.2),+8,+14,-15,inv(17)(q21q25),add(18)(q23),+20,psu dic(20;21)(q11.2;p11.2)x2,+21[cp17]/46,XX[3], representing a composite karyotype. Note that in this cell, all the abnormalities are present, except there is only one copy of chromosome 21, without the clonal gain.

25.4 Double minute chromosomes

ISCN RULES FOR DESCRIBING DOUBLE MINUTE CHROMOSOMES (FIGURE 25.2)

- First write the modal chromosome number, excluding the double minute chromosomes.
- Next write the sex chromosome designation.
- If the double minute chromosomes are present with other abnormalities, write the other chromosomal abnormalities as described previously.

- Double minute chromosomes go last in the string of abnormalities present, after markers and ring chromosomes.
- Unlike markers and rings, which have a plus (+) sign included, double minute chromosomes do not. Also, the presence of double minute chromosomes is not included in the modal chromosome number.
- After the last abnormality, followed by a comma, write the number of double minute chromosomes, followed by "dmin".
- If there is a range of double minutes within the abnormal cell line, put the range of double minutes, then "dmin".
- Add the total number of cells in brackets.

For example, in a female patient with five cells containing a deletion within the long arm of chromosome 5 plus seven double minute chromosomes in each of these cells, then write: 46,XX,del(5)(q13q33),7dmin[5].

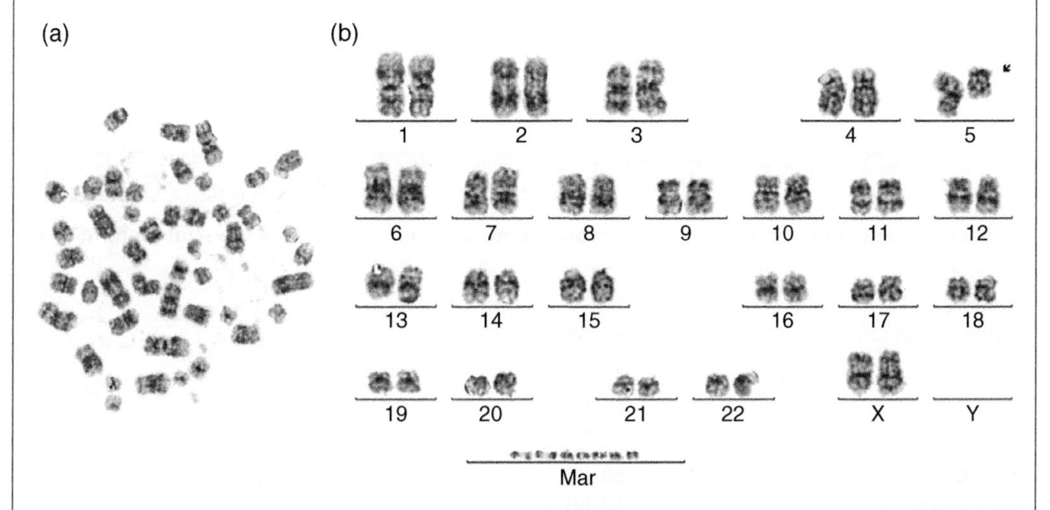

Figure 25.2 Example of a metaphase cell and its corresponding karyotype with the presence of double minute chromosomes. (a) Metaphase with double minutes and del(5q). (b) Karyotype with double minutes and del(5q): 46,XX,del(5)(q13q33),7dmin. Courtesy of Sarah South PhD, ARUP Laboratories.

EXAMPLE REPORT WITH DOUBLE MINUTE CHROMOSOMES

ISCN Result: 46,XX,del(5)(q13q33),7dmin[5]/46,XX[15]

Interpretation

Cytogenetic analysis revealed an abnormal female chromosome complement in 5 of 20 metaphase cells examined with a deletion within the long arm of chromosome 5 and the presence of seven double minute chromosomes.

The remaining 15 cells showed a normal female chromosome complement.

Deletion of 5q is a recurrent abnormality in myelodysplastic syndromes (MDS) and acute myeloid leukemia (AML). Double minute chromosomes are most often seen in acute myeloid leukemia. The chromosomal origin of the double minutes is not known; however, these chromosomes do represent amplification of part of the genome. The most common amplified region of double minutes is the MYC gene/8q24, MLL gene/11q23 or AML1 gene/21q22.

Fluorescence *in situ* hybridization (FISH) analysis for the chromosome 5 probe (and other probes to possibly identify the origin of the double minute chromosomes) is suggested to monitor this abnormal cell line.

25.5 Modal ploidy numbers

Another aspect of writing results of neoplastic studies is the designation of the modal ploidy number. The modal ploidy number refers to the range of chromosome numbers from haploid, diploid, triploid to tetraploid. Table 25.1 shows the description of modal chromosome numbers (modified from the ISCN 2013). The abbreviation may be written in the ISCN nomenclature after the modal number to help clarify the ploidy level. This becomes important when describing and interpreting the number of each chromosome pair that follows the modal number. For example, if the modal number has 92 chromosomes, the ploidy level is tetraploid and all the chromosomes following are considered as 4n. Any variation from 4n of each homolog would be written in the ISCN nomenclature, including either or both clonal loss or gain.

Table 25.1 Description of modal chromosome number (modified from Shaffer et al. 2013)

Modal number description	Chromosome number	Abbreviation for ISCN
Haploidy	≤34	n
Hypohaploid	<23	
Hyperhaploid	24–34	
Diploidy	35–57	2n
Hypodiploid	35–45	
Hyperdiploid	47–57	
Triploidy	58–80	3n
Hypotriploid	58–68	
Hypertriploid	70–80	
Tetraploidy	81–103	4n
Hypotetraploid	81–91	
Hypertetraploid	93–103	
Pentaploidy	104–126	5n
Hypopentaploid	104–114	
Hyperpentaploid	116–126	
Hexaploidy	127–149	6n
Hypohexaploid	127–137	
Hyperhexaploid	139–149	
Heptaploidy	150–172	7n
Hypoheptaploid	150–160	
Hyperheptaploid	162–172	
Octaploidy	173–195	8n
Hypooctaploid	173–183	
Hyperoctaploid	185–195	

ISCN RULES FOR DESCRIBING MODAL CHROMOSOME NUMBER

- First write the most common chromosome modal number in the cell line.
- If the modal chromosome number varies within the cell population, write the range of chromosome number separated by a "~" symbol.
- When the chromosome number is diploid, then continue writing the karyotypic designation as in any other abnormal karyotype. For example: 45~47,XY,+8,+17, -20[cp20].
- If the chromosome number is haploid, triploid, tetraploid, etc., then write the ploidy number, 1n, 3n, 4n, respectively, enclosed by arrow symbols, "< and >", before and after the chromosome number.
- Finish writing the karyotypic abnormalities that are in addition to the ploidy number.

For example: 68~69<3n>,XXY,+8,+17,-20. This karyotype describes a triploid chromosome complement, with three copies of each chromosome, with the exception of having four copies of chromosomes 8 and 17 and two copies of chromosome 20.

EXAMPLE REPORTING OF PLOIDY LEVEL CHROMOSOMAL CHANGES (FIGURE 25.3)

ISCN Result: 68~69<3n>,XX,-X,+2,-4,-5,+6,der(7;9)(q10;p10),+8,+9,der(9;15)(p10;q10),i(9)(p10),-12,+13,-14,der(15;21)(q10;q10),-16,del(16)(q22),-17,add(17)(q23),+18,+19,del(19)(p12),+21,i(21)(q10)[cp12]/46,XX[8]

Interpretation

Cytogenetic analysis revealed a near-triploid abnormal female chromosome complement with complex chromosomal abnormalities in 12 of 20 metaphase cells examined, including the following clonal abnormalities:

- loss from 3n of chromosomes X, 4, 5, 12, 14, 16 and 17
- gain from 3n of chromosomes 2, 6, 8, 9, 13, 18, 19 and 21
- whole arm translocations (derivative chromosomes) involving chromosomes 7 and 9, 9 and 15, and 15 and 21
- an isochromosome of the short arm of chromosome 9 and the long arm of chromosome 21
- a deletion of part of the long arm of chromosome 16 and part of the short arm of chromosome 19
- and added material on the long arm of chromosome 17.

There was some variation from cell to cell in the abnormalities present; therefore, a composite of all the abnormalities is listed in the nomenclature.

The remaining eight cells showed a normal female chromosome complement.

Gains of chromosomes 8, 9, 19 and 21, del(16q), and rearrangements involving 19p have been observed in myeloid malignancies including myelodysplastic syndromes (MDS) and acute myeloid leukemia (AML). The complexity of the abnormalities suggests more advanced disease and a poor prognosis.

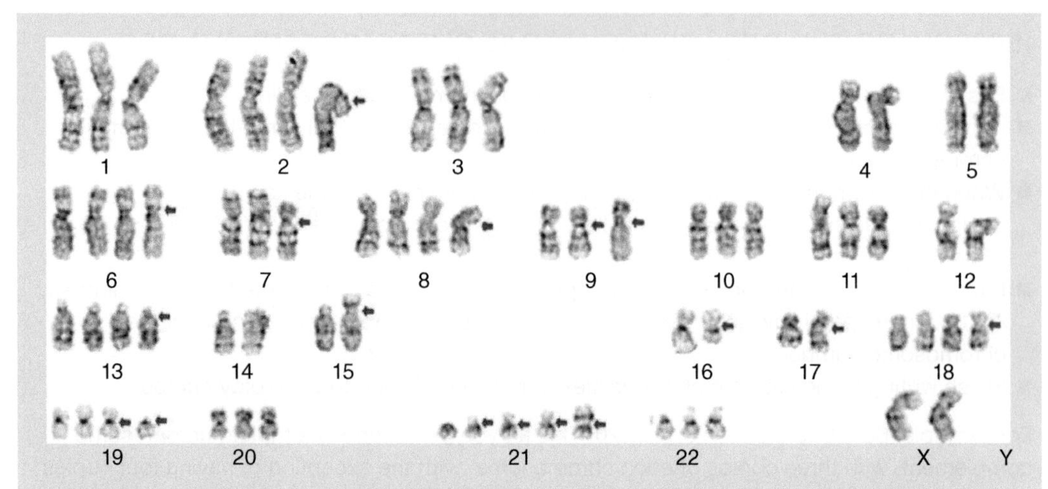

Figure 25.3 Triploid chromosome complement with complex karyotypic abnormalities: 68~69<3n>,XX,-X,+2,-4,-5,+6,der(7;9)(q10;p10),+8,+9,der(9;15)(p10;q10),i(9)(p10),-12,+13,-14,der(15;21)(q10;q10),-16,del(16)(q22),-17,add(17)(q23),+18,+19,del(19)(p12), +21,+21,i(21)(q10)[cp12]. Note that the gain of chromosome 9 and the i(9p) chromosome are missing from this cell, but are clonal abnormalities; therefore, a composite karyotype is written.

25.6 Multiple abnormal cell lines indicative of clonal evolution

EXAMPLE REPORTING OF MULTIPLE ABNORMAL CELL LINES INDICATIVE OF CLONAL EVOLUTION (FIGURE 25.4)

ISCN Result

47,X,t(X;5)(q28;q13),+8[4]/
47,sl,add(2)(q23)[7]/
47,sdl1,i(17)(q10)[5]/
47,sl,del(7)(q22q34)[4]

Interpretation

Cytogenetic analysis revealed an abnormal female chromosome complement with four related abnormal cell lines. The first cell line (stemline), seen in four cells, showed a translocation between the long arms of chromosomes X and 5 and gain of chromosome 8.

The second cell line (sideline 1), seen in seven cells, showed the stemline abnormalities plus added material on the long arm of chromosome 2.

The third cell line (sideline 2), seen in five cells, showed the sideline 1 abnormalities plus an isochromosome of the long arm of chromosome 17, resulting in a deletion of 17p and the TP53 gene region.

The fourth cell line (sideline 3), seen in four cells, showed the stemline abnormalities plus a deletion within the long arm of chromosome 7.

Deletions of 7q and 17p and trisomy 8 are recurrent abnormalities in both *de novo* and secondary myelodysplastic syndromes (MDS) and acute myeloid leukemia (AML). Deletions of 7q and 17p are generally associated with a poor prognosis.

Fluorescence *in situ* hybridization (FISH) analysis for the chromosomes 7, 8 and 17 is suggested to monitor these abnormal cell lines.

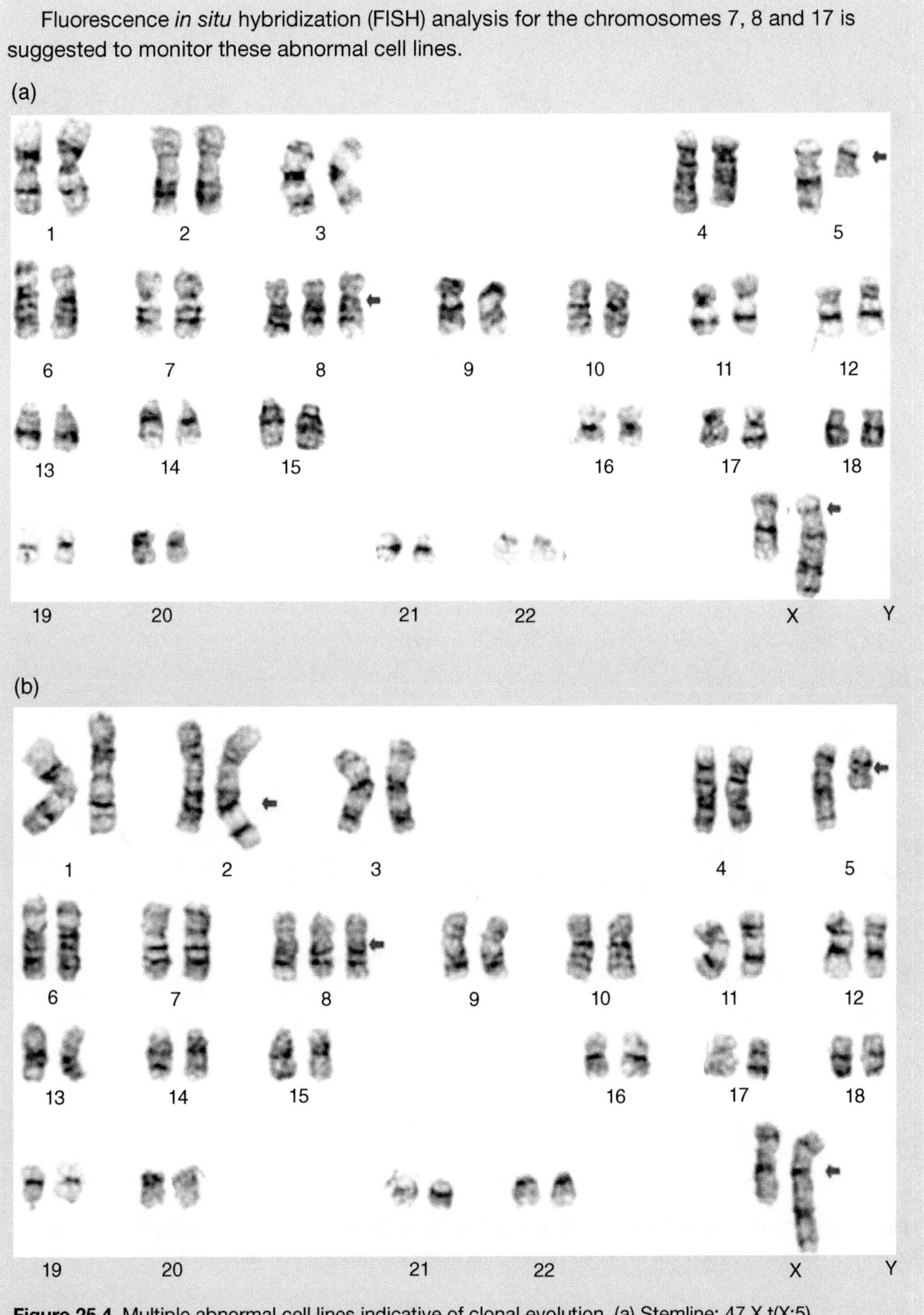

Figure 25.4 Multiple abnormal cell lines indicative of clonal evolution. (a) Stemline: 47,X,t(X;5)(q28;q13),+8. (b) Sideline 1: 47,sl,add(2)(q23).

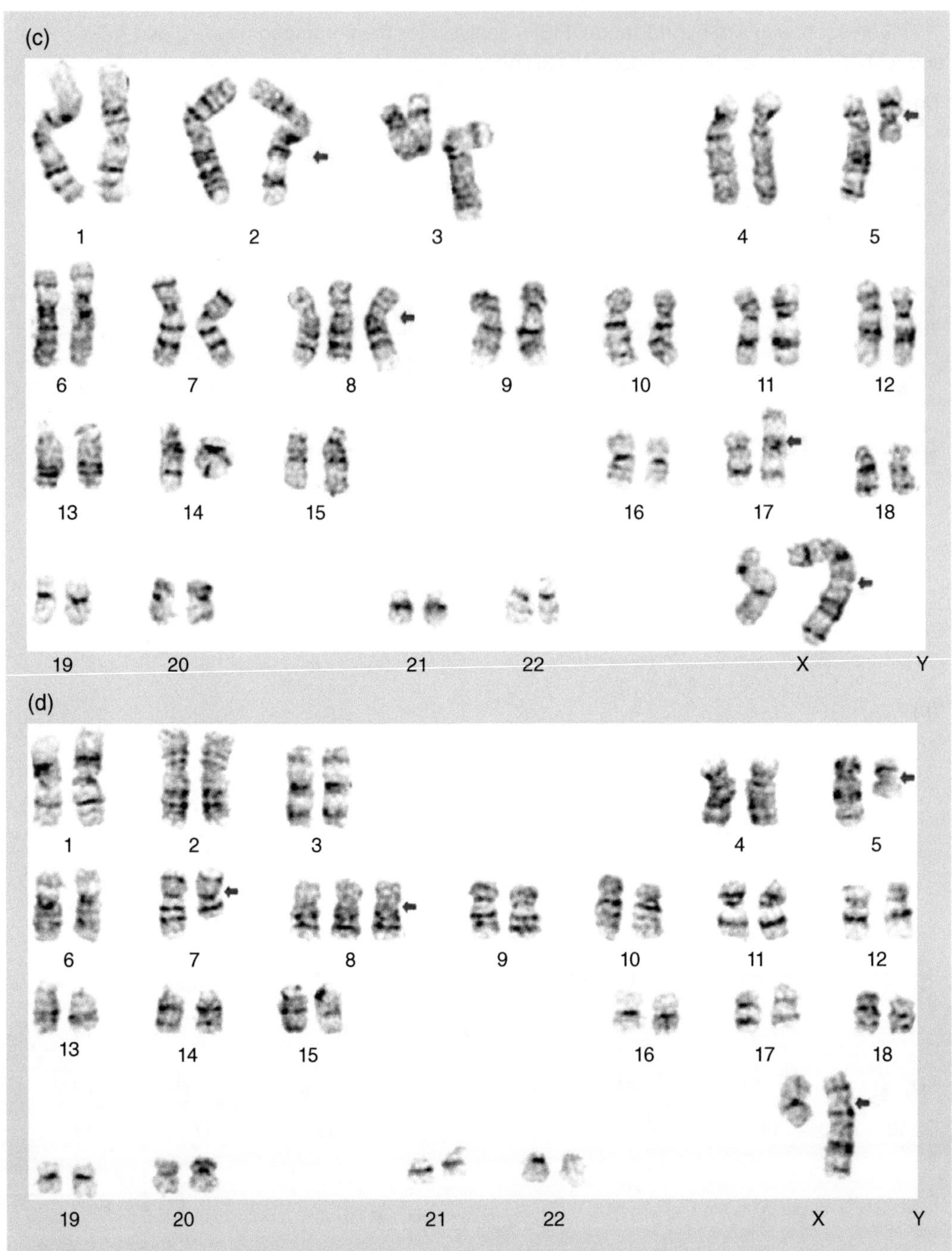

Figure 25.4 (*Continued*) (c) Sideline 2: 47,sdl1,i(17)(q10). (d) Sideline 3: 47,sl,del(7)(q22q34). Note that sidelines 1 and 3 evolved from the stemline and that sideline 2 evolved from sideline 1.

Bibliography

American College of Medical Genetics. *Guidelines in Clinical Cytogenetics. Section E10 Methods in Fluorescence In situ Hybridization – Interphase/Nuclear Fluorescence In Situ Hybridization.* American College of Medical Genetics, Bethesda, MD. www.acmg.net/StaticContent/SGs/Section_E.html, accessed 28 January 2014.

Atlas of Genetics and Cytogenetics in Oncology and Haematology. http://AtlasGeneticsOncology.org/Anomalies, accessed 28 January 2014.

Barker PE. Double minutes in human tumor cells. Cancer Genet Cytogenet 1982; 5(1): 81–94.

Baskin F, Rosenberg RN, Dev V. Correlation of double-minute chromosomes with unstable multidrug cross-resistance in uptake mutants of neuroblastoma cells. Proc Natl Acad Sci USA 1981; 78(6): 3654–3658.

Gersen S, Keagle M (eds). *Principles of Clinical Cytogenetics.* Humana Press, Totowa, New Jersey, 1999.

Greaves M, Maley CG. Clonal evolution in cancer. Nature 2012; 481: 306–313.

Heim S, Mitelman F. *Cancer Cytogenetics,* 2nd edn. Wiley-Liss, New York, 1995.

Masters J, Keeley B, Gay H, Attardi G. Variable content of double minute chromosomes is not correlated with degree of phenotype instability in methotrexate-resistant human cell lines. Mol Cell Biol 1982; 2(5): 498–507.

Merlo LM, Pepper JW, Reid BJ, Maley CC. Cancer as an evolutionary and ecological process. Nature Rev Cancer 6 2006; 12: 924–935.

Rooney DE, Czepulkowski BH. *Human Cytogenetics. A Practical Approach.* Oxford University Press, New York, 1992.

Sait SN, Qadir MU, Conroy JM, Matsui S, Nowak NJ, Baer MR. Double minute chromosomes in acute myeloid leukemia and myelodysplastic syndrome: identification of new amplification regions by fluorescence in situ hybridization and spectral karyotyping. Genes Chromosomes Cancer. 2002; 34(1): 42–47.

Shaffer LG, McGowan-Jordan J, Schmid M (eds). *ISCN 2013: An International System for Human Cytogenetic Nomenclature.* Karger Publishers, Unionville, CT, 2013.

Swerdlow SH, Campo E, Harris NL, et al. (eds). *WHO Classification of Tumours of Haematopoietic and Lymphoid Tissues.* IARC Press, Lyon, 2008.

Tian T, Olson S, Whitacre JM, Harding, A. The origins of cancer robustness and evolvability. Integrat Biol2 2011; 3(1): 17–30.

CHAPTER 26

Breakage disorders

26.1 Ataxia telangiectasia

SPECIFIC FEATURES

Ataxia telangiectasia (AT) is an autosomal recessive disorder with a frequency of approximately 1 in 300,000 newborns, with heterozygotes estimated at 1% of the general population. AT is a chromosome instability syndrome characterized by cerebellar degeneration, immunodeficiency and an increased risk of cancer disorders. This is an early childhood disease, often diagnosed during the second year of life with progressive cerebellar ataxia, oculomotor apraxia, dysarthria, and dystonia leading to muscular atrophy, telangiectasia, combined immunodeficiency, growth retardation, hypogonadism and occasionally diabetes mellitus.

The risk of cancers in AT patients is over 100 times that of the normal population, consisting mainly of T- and B-cell malignancies, carcinomas of the skin, ovary, breast and stomach. Treatments are complicated by radiation and chemosensitivity of these patients due to the lack of DNA repair mechanisms.

Cytogenetic abnormalities include spontaneous chromatid and chromosome breaks, triradials, quadriradials and telomeric associations. Chromosome analysis often shows inv(7)(p14q35), t(14;14)(q11;q32), inv(14q) and non-clonal rearrangements involving 2p12, 7p14, 7q35, 14q11, 14q32 and 22q11, usually involving recombination between immunoglobulin genes. Clonal rearrangements in T-cell acute lymphoblastic leukemia (ALL) and T-cell prolymphocytic leukemia in AT patients are often complex, with recurrent abnormalities including t(14;14)(q11;q32) and t(X;14)(q28;q11).

Cytogenetic Abnormalities: Chromosomal, FISH and Microarray-Based Clinical Reporting, First Edition. Susan Mahler Zneimer.

EXAMPLE REPORTING OF ATAXIA TELANGIECTASIA

ISCN Result: 46,XY[20]

Chromosome Instability Studies:

Spontaneous abnormalities of chromosomes 7 and 14 = 8.0% Abnormal result
Normal range = 0–0.01%
Induced irradiated abnormalities of chromosomes 7 and 14 = 60.0% Abnormal result
Normal range = 0–0.1%

Interpretation

Cytogenetics analysis revealed a normal male chromosome complement in all 20 cells analyzed with a routine cytogenetic study.

Chromosome studies with a PHA-mitogen stimulated lymphocyte (unstressed) culture showed four cells (8.0%) with spontaneous abnormalities of chromosomes 7 and 14 in 50 cells analyzed. Spontaneous abnormalities of chromosome 7 and 14 in ataxia telangiectasia (AT) patients range from 1% to 10%.

Also, irradiation in the G_2 phase of the cell cycle is a diagnostic test for AT diagnosis. A PHA-mitogen stimulated lymphocyte culture stressed with 0.5 and 1.0 Gy X-rays for 4 hours showed 30 cells (60%) with abnormalities of chromosomes 7 and 14. The irradiation stress test in AT patients usually shows abnormalities in greater than 10% of cells analyzed.

These results are consistent with a diagnosis of ataxia telangiectasia. Although the irradiation test is highly specific for the diagnosis of AT, it does not identify heterozygous carriers.

26.2 Bloom syndrome

SPECIFIC FEATURES

Bloom syndrome is an autosomal recessive disorder with a frequency of approximately 2 in 10,000 newborns in the Ashkenazi Jewish population and in Japanese populations. It is much rarer in the general population. The clinical manifestations are variable, but generally include dwarfism and intrauterine growth retardation, hyper- and hypopigmented areas of the skin, sun-sensitive telangiectatic erythema, microcephaly, dolichocephaly, narrow face, prominent nose and/or ears, characteristic high-pitched voice, immune deficiency, myocardiopathy, hypogonadism in male patients and hypertriglyceridemia.

Approximately 50% of patients have at least one cancer, with mean age at cancer onset of 25 years. The most common malignancies include ALL and AML, lymphomas, carcinomas and benign tumors. Prognosis is very poor in these patients with death occurring by a mean age of 24 years.

Chromosomal/chromatid breaks are common findings with triradial and quadriradials, in particular, along with symmetrical quadriradial configuration involving homologous chromosomes. These findings are specific to Bloom syndrome and may be due to a mitotic crossing-over errors.

The diagnosis includes a highly elevated spontaneous sister chromatid exchange (SCE) rate, with 90 SCEs per cell, which is more than 10 times the normal rate. A minor population of low SCE cells exists in some people, which suggests a recombination event between maternal and paternal alleles, giving rise to a wild-type functional gene. Cytogenetic analysis will not detect heterozygote status.

EXAMPLE REPORTING OF SISTER CHROMATID EXCHANGE FOR BLOOM SYNDROME

ISCN Result: 46,XY[20]

Sister Chromatid Exchange (SCE) Studies:

Spontaneous SCE = 45 SCEs/cell Abnormal result
Normal range = 0–5 SCEs/cell
SCE with BrdU Incorporation Stress Test = 80 SCEs/cell Abnormal result
Normal range = 0–10 SCEs/cell

Interpretation

Cytogenetics analysis revealed a normal male chromosome complement in all 20 cells analyzed with a routine cytogenetic study.

Chromosome instability studies with a PHA-mitogen stimulated lymphocyte (unstressed) culture showed 45 SCEs per cell. The spontaneous chromosome breakage in Bloom syndrome patients usually shows over 10 SCEs per cell.

Also, BrdU incorporation in cultures is a diagnostic test for Bloom syndrome patients. A PHA-mitogen stimulated lymphocyte culture stressed with 10 µg/mL of BrdU showed 80 SCEs per cell. The BrdU stress test in Bloom syndrome patients is over 60 SCEs per cell.

These results are consistent with a diagnosis of Bloom syndrome. Although the BrdU test is highly specific for the diagnosis of Bloom syndrome, it does not identify heterozygous carriers.

26.3 Fanconi anemia

SPECIFIC FEATURES

Fanconi anemia (FA) is a chromosome instability syndrome with progressive bone marrow failure and an increased risk of cancers. Common manifestations include growth retardation, skin abnormalities (hyperpigmentation and café-au-lait spots), radial axis defects (absent or hypoplastic thumb or radius) and less frequently, renal anomalies, hypogonadism, mental impairment, heart defects and diabetes mellitus. FA is seen most often in children, with a mean age of onset at 8 years.

There is a 15,000-fold increased risk of MDS and AML in FA patients, with a mean age of developing a hematological malignancy at 13–15 years old. There is also a higher risk of developing hepatocarcinoma or squamous cell carcinoma. The prognosis of FA is poor with a mean age of death at 16 years. Most patients die from bone marrow aplasia, including hemorrhaging or sepsis, or from other malignancies. There may be a certain degree of clinical variation of symptoms even within a given family.

Characteristic cytogenetic findings in FA include spontaneous chromatid and/or chromosome breaks, triradials and quadriradials. There is hypersensitivity to the clastogenic effect of DNA cross-linking agents, such as diepoxybutane, mitomycin C or mechlorethamine hydrochloride, which are used for diagnosis.

Recurrent, non-random chromosomal aberrations in FA patients who have developed MDS or AML include monosomy 5, del(5q), monosomy 7 and del(7q). Reports in the literature state that approximately 10–25% of FA patients show two cell populations, one that is sensitive to the clastogenic effects of DNA cross-linking agents and one that is resistant, consistent with somatic revertant mosaicism, which may result in inconclusive results.

EXAMPLE REPORTING OF FANCONI ANEMIA

ISCN Result: 46,XY[20]

Chromosome Instability Studies:

Spontaneous Chromosome Breakage = 0.20 breaks/cell   Abnormal result
Normal range = 0–0.12 breaks/cell
DEB Stress Test = 12.1 breaks/cell   Abnormal result
Normal range = 0–0.18 breaks/cell

Interpretation

Cytogenetics analysis revealed a normal male chromosome complement in all 20 cells analyzed with a routine cytogenetic study.

Chromosome instability studies with a PHA-mitogen stimulated lymphocyte (unstressed) culture showed 0.20 spontaneous breaks per cell. The spontaneous chromosome breakage in Fanconi anemia (FA) patients ranges from 0.02 to 0.85 breaks per cell.

Also, diepoxybutane (DEB) testing is diagnostic for FA in patients with subtle features. A PHA-mitogen stimulated lymphocyte culture stressed with 0.1 µg/mL of DEB showed 12.1 breaks per cell. The DEB stress test in FA patients ranges from 1.10 to 23.9 breaks per cell.

These results are consistent with a diagnosis of Fanconi anemia. Although the DEB test is highly specific for the diagnosis of FA, it does not distinguish between heterozygous carriers or define the specific FA complementation group.

26.4 Nijmegen syndrome

SPECIFIC FEATURES

The clinical manifestations of Nijmegen syndrome include microcephaly with a “bird-like” face with prominent mid-face, long nose and receding mandible, immunodeficiency, growth and mental retardation, craniofacial dysmorphology, ovarian failure, lymphoreticular malignancies and chromosomal instability. This is an autosomal recessive disorder that has a predisposition to B-cell lymphoid malignancies and radiosensitivity.

Structural chromosome aberrations are common, with rearrangements between chromosomes 7 and 14, at bands 7p13, 7q35, 14q11 and 14q32, in which the most frequent rearrangement is the inv(7)(p13q35). These regions contain immunoglobulin and T-cell receptor genes.

EXAMPLE REPORTING OF NIJMEGEN SYNDROME

ISCN Result: 46,XY[20]

Chromosome Instability Studies:

Spontaneous Abnormalities of Chromosomes 7 and 14 = 8.0% Abnormal result
Normal range = 0–0.01%
Induced Irradiated Abnormalities of chromosomes 7 and 14 = 60.0% Abnormal result
Normal range = 0–0.1%

Interpretation

Cytogenetics analysis revealed a normal male chromosome complement in all 20 cells analyzed with a routine cytogenetic study.

Chromosome studies with a PHA-mitogen stimulated lymphocyte (unstressed) culture showed four cells (8.0%) with spontaneous abnormalities of chromosomes 7 and 14 in 50 cells analyzed. The spontaneous abnormalities of chromosomes 7 and 14 in Nijmegen syndrome patients range from 1% to 10% in 50 cells analyzed with these abnormalities.

Also, irradiation in the G_2 phase of the cell cycle is a diagnostic test for Nijmegen syndrome. A PHA-mitogen stimulated lymphocyte culture stressed with 0.5 and 1.0 Gy X-rays for 4 hours showed 30 cells (60%) with abnormalities of chromosomes 7 and 14. The irradiation stress test in Nijmegen syndrome patients usually shows abnormalities in greater than 10% of cells analyzed.

These results are consistent with a diagnosis of Nijmegen syndrome. Although the irradiation test is highly specific for this diagnosis, it does not identify heterozygous carriers.

Bibliography

Ababou M, Dumaire V, Lacluse Y, Amor-Guret M. Bloom's syndrome protein response to ultraviolet-C radiation and hydroxyurea-mediated DNA synthesis inhibition. Oncogene 2002; 21(13): 2079–2088.

American College of Medical Genetics. *Guidelines in Clinical Cytogenetics. Section E10 Methods in Fluorescence In situ Hybridization – Interphase/Nuclear Fluorescence In Situ Hybridization*. American College of Medical Genetics, Bethesda, MD. www.acmg.net/StaticContent/SGs/Section_E.html, accessed 28 January 2014.

Atlas of Genetics and Cytogenetics in Oncology and Haematology. http://AtlasGeneticsOncology.org/Anomalies, accessed 28 January 2014.

Auerbach A. Diagnosis of Fanconi anemia by diepoxybutane analysis. Curr Protoc Hum Genet 2003; Chapter 8: Unit 8.7.

Auerbach AD, Allen RG. Leukemia and preleukemia in Fanconi anemia patients: a review of the literature and report of the International Fanconi Anemia Registry. Cancer Genet Cytogenet 1991; 51: 1–12.

Callin E, Samper E, Ramirez MJ, et al. Breaks at telomeres and TRF2-independent end fusions in Fanconi anemia. Hum Mol Genet 2002; 11(4): 439–444.

Gaymes TJ, North PS, Brady N, Hickson ID, Mufti GJ, Rassool FV. Increased error-prone non homologous DNA end-joining – a proposed mechanism of chromosomal instability in Bloom's syndrome. Oncogene 2002;21(16): 2525–2533.

Gersen S, Keagle M (eds). *Principles of Clinical Cytogenetics*. Humana Press, Totowa, New Jersey, 1999.

Grompe M, d'Andrea A. Fanconi anemia and DNA repair. Hum Mol Genet 2001; 10(20): 2253–2259.

Heim S, Mitelman F. *Cancer Cytogenetics*, 2nd edn. Wiley-Liss, New York, 1995.

Langland G, Elliott J, Li Y, Creaney J, Dixon K, Groden J. The BLM helicase is necessary for normal DNA double-strand break repair. Cancer Res 2002; 62(10): 2766–2770.

Maser RS, Zinkel R, Petrini JH. An alternative mode of translation permits production of a variant NBS1 protein from the common Nijmegen breakage syndrome allele. Nature Genet 2001; 27(4): 417–421.

Medhurst AL, Huber PA, Waisfisz Q, de Winter JP, Mathew CG. Direct interactions of the five known Fanconi anemia proteins suggest a common functional pathway. Hum Mol Genet 2001; 10(4): 423–429.

Opresko PL, von Kobbe C, Laine JP, Harrigan J, Hickson ID, Bohr VA. Telomere-binding protein TRF2 binds to and stimulates the Werner and Bloom syndrome helicases. J Biol Chem 2002; 277(43): 41110–41119.

Qiao F, Moss A, Kupfer GM. Fanconi anemia proteins localize to chromatin and the nuclear matrix in a DNA damage- and cell cycle-regulated manner. J Biol Chem 2001; 276(26): 23391–23396.

Rooney DE, Czepulkowski BH. *Human Cytogenetics. A Practical Approach*. Oxford University Press, New York, 1992.

Shaffer LG, McGowan-Jordan J, Schmid M (eds). *ISCN 2013: An International System for Human Cytogenetic Nomenclature*. Karger Publishers, Unionville, CT, 2013.

Stavropoulos DJ, Bradshaw PS, Li X, et al. The Bloom syndrome helicase BLM interacts with TRF2 in ALT cells and promotes telomeric DNA synthesis. Hum Mol Genet 2002; 11(25): 3135–3144.

Swerdlow SH, Campo E, Harris NL, et al. (eds). *WHO Classification of Tumours of Haematopoietic and Lymphoid Tissues*. IARC Press, Lyon, 2008.

van Engelen BG, Hiel JA, Gabreis FJ, van den Heuvel LP, van Gent DC, Weemaes CM. Decreased immunoglobulin class switching in Nijmegen Breakage syndrome due to the DNA repair defect. Hum Immunol 2001; 62(12): 1324–1327.

Varon R, Reis A, Henze G, von Einsiedel HG, Sperling K, Seeger K. Mutations in the Nijmegen Breakage Syndrome gene (NBS1) in childhood acute lymphoblastic leukemia (ALL). Cancer Res 2001; 61(9): 3570–3572.

Yamashita T, Nakahata T. Current knowledge on the pathophysiology of Fanconi anemia: from genes to phenotypes. Int J Hematol 2001; 74(1): 33–41.

CHAPTER 27

Cytogenetic abnormalities in solid tumors

27.1 Clear cell sarcoma

SPECIFIC FEATURES

Clear cell sarcomas are soft tissue tumors, manifested as a slow-growing mass causing pain or tenderness particularly in the extremities, and most commonly in the foot and ankle. They are rare tumors seen primarily in young adults. Generally, clear cell sarcomas are associated with a poor prognosis, with a long-term survival rate of only 40–50%. This tumor type shows a recurrent, non-random abnormality that generally does not occur in other tumors, namely translocation (12;22)(q13;q12). This translocation, seen in both balanced and unbalanced forms, has been described in the majority of reported clear cell sarcoma cases. It is the result of the EWSR1 gene at 22q12 fusing with the ATF1 gene at 12q13. Although t(12;22) has been reported as the sole chromosomal aberration in clear cell sarcomas, most cases show additional cytogenetic anomalies, including +7, +8 along with numerical and structural abnormalities of chromosome 22.

EXAMPLE REPORTING OF CLEAR CELL SARCOMAS

ISCN Result: 46,XY,-1,+2,+8,+8,-12,-14,-16,der(22)t(12;22)(q13;q12),+22[11]/46,XY[9]

Interpretation

Chromosome analysis revealed an abnormal male chromosome complement in 11 of 20 cells examined with the following clonal changes:

- gain of chromosomes 2, 8 and 22, and loss of chromosomes 1, 12, 14 and 16
- and an unbalanced translocation between the long arms of chromosomes 12 and 22.

Cytogenetic Abnormalities: Chromosomal, FISH and Microarray-Based Clinical Reporting, First Edition. Susan Mahler Zneimer.

Translocation (12;22) is a recurrent, non-random abnormality associated with clear cell sarcomas, seen in approximately 90% of cases reported. This characteristic genetic anomaly is very rarely detected in other tumors. T(12;22) is often seen in conjunction with the gain of chromosomes 8 and 22.

Fluorescence *in situ* hybridization (FISH) analysis for the EWSR1/ATF1 fusion gene rearrangement is suggested to confirm and monitor this abnormal cell line.

27.2 Chondrosarcoma

SPECIFIC FEATURES

Chondrosarcomas are rare malignant tumors of soft tissue origin, seen more often in males than females. Cytogenetic studies show a recurrent abnormality, t(9;22)(q22;q12), resulting in the fusion of the EWSR1 gene on chromosome 22q22 with the NR4A3 gene on chromosome 9q22. Variant translocations have also been reported, such as t(9;17)(q22;q11) and t(9;15)(q22;q21). The t(9;22)(q22;q12) is seen most commonly in extraskeletal myxoid chondrosarcomas. Other abnormalities that may be present include trisomy 7, seen primarily in malignant cartilaginous tumors, and 9p12-22 abnormalities more commonly seen in central chondrosarcomas. Deletions of chromosome 13q are also seen and are generally associated with progression of disease and metastasis, regardless of tumor grade or size.

EXAMPLE REPORTING OF CHRONDROSARCOMAS

ISCN Result: 46,X,add(Y)(q12),t(8;16)(q24.3;p11.2),t(9;22)(q22;q12),del(13)(q14q32)[20]

Interpretation

Chromosome analysis revealed an abnormal male chromosome complement in all 20 cells examined with the following clonal changes:

- a translocation between the long arms of chromosomes 9 and 22
- a translocation between the long arm of chromosome 8 and the short arm of chromosome 16
- added material on the long arm of chromosome Y
- and a deletion within the long arm of chromosome 13.

Translocation (9;22) is an abnormality resulting in the fusion of the EWRS1 gene on chromosome 22 with the NR4A3 gene on chromosome 9 and is a non-random abnormality associated primarily with extraskeletal myxoid chondrosarcomas. Deletions of chromosome 13q are also seen in chondrosarcomas and are generally associated with progression of disease and metastasis, regardless of tumor grade or size.

27.3 Ewing sarcoma

SPECIFIC FEATURES

Ewing sarcomas are small, round cell tumors derived from neural crest cells. They represent 5–15% of malignant bone and soft tissue tumors. The majority of cases of Ewing tumors are seen in individuals younger than 35 years of age, with a median age of 20 years. The prognosis for these tumors is mainly determined by the presence of metastasis at the time of diagnosis, where the 5-year survival rate is 10–35% in patients with metastasis and 54–74% for patients with localized disease at the time of diagnosis.

Ewing sarcomas show a recurrent, non-random abnormality in approximately 90% of cases with t(11;22)(q24;q12). This translocation results in the fusion of the EWSR1 gene at chromosome 22q12 with the FLI1 gene at 11q24, leading to a chimeric protein product. Variant translocations have also been reported in a minority of cases with t(21;22)(q12;q12) with the ERG gene and t(7;22)(p22;q12) with the ETV1 gene. Additional anomalies mainly consist in chromosome gains, including trisomy for chromosomes 1q, 2, 5, 7, 8, 9 and 12, and an unbalanced t(1q;16q).

EXAMPLE REPORTING OF t(11;22)(q24;q12)

ISCN Result: 46,XY,t(11;22)(q24;q12)[11]/46,XY[9]

Interpretation

Chromosome analysis revealed an abnormal male chromosome complement in 11 of 20 cells examined with a translocation between the long arms of chromosomes 11 and 22 as the sole anomaly. The remaining nine cells showed a normal male chromosome complement.

Rearrangements involving the 22q12 region are most often observed in Ewing tumors, including Ewing sarcoma and peripheral primitive neuroectodermal tumors (PNET).

Fluorescence *in situ* hybridization (FISH) analysis for the EWSR1 gene is suggested to confirm and monitor this abnormal cell line.

27.4 Liposarcoma

SPECIFIC FEATURES

Liposarcomas are lipogenic tumors subclassified into four main histological groups, including well-differentiated liposarcomas, myxoid round cell liposarcomas, pleomorphic liposarcomas and undifferentiated liposarcomas. Myxoid round cell liposarcomas show a specific chromosomal translocation, namely t(12;16)(q13;p11). This translocation results in the fusion of the ATF1 gene at 16p11 with the CHOP gene at 12q13.

There are very few cases of pleomorphic liposarcomas reported in the literature that have been cytogenetically characterized. However, they invariably show complex karyotypic changes and extensive intratumor genetic heterogeneity, characterized by cell-to-cell variation in the abnormalities present and/or multiple cell lines present.

Well-differentiated or atypical lipomatous tumors are characterized by the presence of large ring chromosomes, often supernumerary.

EXAMPLE REPORTING OF LIPOSARCOMAS

ISCN Result: 44~47,X,-Y,-2,del(2)(p21p23),add(8)(q24),add(11)(p13),t(12;16)(q13;p11), add(15)(q24),+r,+1~3mar[cp20]

Interpretation

Chromosome analysis revealed an abnormal male chromosome complement in all 20 cells examined with the following clonal changes:

- loss of chromosomes Y and 2
- a deletion within the short arm of chromosome 2
- added material on the short arm of chromosome 11, and on the long arms of chromosomes 8 and 15
- a translocation between the long arm of chromosome 12 and the short arm of chromosome 16
- the presence of one large supernumerary ring chromosome
- the gain of 1–3 marker chromosomes.

There was some variation from cell to cell in the abnormalities present; therefore, a composite of all the abnormalities is listed in the nomenclature.

Translocation (12;16) is a recurrent, non-random abnormality associated with myxoid round cell liposarcomas. However, the presence of large supernumerary ring chromosomes, as seen here, is more commonly observed in well-differentiated or atypical lipomatous tumors. Pleomorphic liposarcomas usually show complex abnormalities with no characteristic changes identified but with extensive genetic heterogeneity, as seen here. Therefore, clinical-pathological correlation of results is suggested.

27.5 Neuroblastoma

SPECIFIC FEATURES

Peripheral neuroblastic tumors are derived from developing neuronal cells of the sympathetic nervous system, are considered aggressive tumors and are found mostly in infants and young children. The majority of patients (60%) show metastatic disease at diagnosis. Neuroblastomas have distinct cytogenetic abnormalities that predict the clinical phenotype. Two main groups are identified according to the tumor cell ploidy: near-triploid (and pentaploid/hexaploid) chromosome complement, occurring in approximately 55% of cases, and near-diploid (and tetraploid) chromosome complement, occurring in the remaining 45% of cases.

Near-triploid (pentaploid, hexaploid) neuroblastomas show numerical gains and losses of chromosomes, most frequently chromosomes 7 and 17, and are generally associated with a good prognosis. However, structural abnormalities may be present, including N-MYC amplification, deletions of 1p36.3 and 11q, and gains of 17q, and are generally associated with a poor prognosis.

Diploid/tetraploid tumors are generally associated with a very poor prognosis and often include unbalanced structural rearrangements. One such rearrangement is N-MYC amplification, which is predominantly seen as double minute chromosomes (dmins). N-MYC amplification is often seen with 1p deletion and 17q gain. Cases without N-MYC amplification will often show 11q deletions, 17q gain and 3p deletions. Identification of N-MYC amplification, as well as deletions of 11q and diploidy, is an important criterion for defining clinical subgroups.

Linkage analysis using 10 families with neuroblastoma mapped the hereditary neuroblastoma locus at 16p12-p13. Furthermore, heterozygote germline alterations in PHOX2B have recently been identified in patients with familial neuroblastoma. However, these germline aberrations are only rarely associated with the onset of neuroblastomas.

EXAMPLE REPORTING OF NEUROBLASTOMAS

ISCN Result: 56,XY,+2,+3,+7,+9,+10,+17,+20,+20,+2mar,10~24dmin[20]

Interpretation

Chromosome analysis revealed an abnormal male chromosome complement in all 20 cells examined with a near-triploid chromosome complement showing the gain of numerous chromosomes, the gain of two unidentifiable marker chromosomes and the gain of 10–24 double minute chromosomes.

Triploidy and the presence of double minute chromosomes, predominantly consisting of N-MYC amplification, are recurrent, non-random abnormalities seen in neuroblastomas, and generally associated with a poor prognosis.

Fluorescence *in situ* hybridization (FISH) analysis for the N-MYC gene is suggested to confirm and monitor this abnormal cell line.

27.6 Rhabdomyosarcoma

SPECIFIC FEATURES

Rhabdomyosarcomas are mesenchymal tumors related to the skeletal muscle lineage. Rhabdomyosarcomas are the most common soft tissue sarcomas observed in children. The two main histopathological subtypes of rhabdomyosarcomas are embryonal and alveolar. The majority of cases (70–80%) are embryonal and usually occur in young children, with a median age of 6.5 years. Alveolar types are seen in the remaining cases, and occur more often in older children and young adults, with a median age of 12 years. They are also associated with a poorer prognosis.

Alveolar tumors show recurrent translocations, t(2;13)(q35;q14) and t(1;13)(p36;q14), both involving the FOXO1 gene at 13q14. Embryonal tumors usually contain chromosome gains and losses, in particular, loss of chromosome 11p15.5.

EXAMPLE REPORTING OF RHABDOMYOSARCOMA

ISCN Result: 46,XY,t(2;13)(q35;q14)[9]/46,XY[11]

Interpretation

Chromosome analysis revealed an abnormal male chromosome complement in 9 of 20 cells examined with a translocation between the long arms of chromosomes 2 and 13 as the sole anomaly. The remaining 11 cells showed a normal male chromosome complement.

Translocation (2;13) is a recurrent, non-random rearrangement observed in alveolar rhabdomyosarcomas.

Fluorescence *in situ* hybridization (FISH) analysis for the FOXO1 gene is suggested to confirm and monitor this abnormal cell line.

27.7 Synovial sarcoma

SPECIFIC FEATURES

Synovial sarcomas are rare and account for 5–8% of all soft tissue cancers, and are seen most often in adolescents and young adults. They occur primarily in the lower extremities, and are generally associated with a poor prognosis, especially in poorly differentiated small cell neoplasms.

The primary cytogenetic abnormality associated with synovial sarcomas is t(X;18)(p11.2;q11.2), which is found in over 80% of cases regardless of the histological type. T(X;18) appears to be specific to synovial sarcomas and is very rarely detected in other tumor types. This translocation is detectable by metaphase and FISH analyses and is characterized by the fusion of the SYT gene on chromosome Xp11.2 with the SSX1 gene on chromosome 18q11.2. Other structural anomalies are found in 50% of cases and tumors may be hypodiploid, pseudodiploid, hyperdiploid or near tetraploid. There are a few variant translocations reported, including t(5;18) and t(X;7).

EXAMPLE REPORTING OF t(X;18)(p11.2;q11.2)

ISCN Result: 46,Y,t(X;18)(p11.2;q11.2)[20]

Interpretation

Chromosome analysis revealed an abnormal male chromosome complement with a translocation between the short arm of chromosome X and the long arm of chromosome 18 as the sole anomaly in all cells examined.

Translocation(X;18)(p11.2;q11.2) is a recurrent, non-random abnormality seen in 80–90% of synovial sarcomas. This characteristic translocation is very rarely detected in other tumors.

Fluorescence *in situ* hybridization (FISH) analysis for the SYT/SSX1 fusion gene rearrangement is suggested to confirm and monitor this abnormal cell line.

27.8 Wilms tumor

SPECIFIC FEATURES

Wilms tumors (WT) are also called nephroblastomas and are the most common abdominal malignancy in children. The tumor cells are thought to be derived from pluripotent embryonic renal precursor cells.

Approximately 10% of cases are associated with genetic syndromes, including WAGR syndrome (Wilms tumor, aniridia, genitourinary abnormalities and mental retardation), Denys–Drash syndrome (mesangial sclerosis, male pseudohermaphrodism and Wilms tumors), Beckwith–Wiedemann syndrome (exomphalos, macroglossia and gigantism) and Simpson–Golabi–Behmel syndrome (overgrowth, mental impairment, craniofacial anomalies). The majority of Wilms tumors, however, are sporadic. Sporadic forms are usually unilateral and constitute the majority of cases whereas bilateral or multifocal cases account for 10% of cases, possibly due to a germline mutation.

Wilms tumors are characterized by extensive genetic heterogeneity with complex genetic changes, including trisomy for chromosomes 6, 7, 8, 12, 13 and 18, del(11p) and del(16q). Abnormalities involving del(17p) and der(16)t(1;16) are considered to be associated with tumor progression. Translocations involving chromosomes with the Wilms tumor 1 gene (WT1) at 11p13 have been associated with desmoplastic small round cell tumors and are associated with a poor prognosis.

EXAMPLE REPORTING OF WILMS TUMOR

ISCN Result: 50,XY,+6,+8,del(11)(p13),+12,+13,der(16)t(1;16)(p13;q11.2),del(17)(q11.2)[20]

Interpretation

Chromosome analysis revealed an abnormal male chromosome complement in all cells examined with the following clonal changes:

- hyperdiploidy with the gain of chromosomes 6, 8, 12 and 13
- deletions within the short arms of chromosomes 11 and 17
- and an unbalanced translocation between the short arm of chromosome 1 and the long arm of chromosome 16.

Hyperdiploidy, especially with the gains of chromosomes 6, 8, 12 and 13, del(11p) and unbalanced t(1;16), are recurrent, non-random abnormalities observed in Wilms tumors.

Bibliography

American College of Medical Genetics. *Guidelines in Clinical Cytogenetics. Section E10 Methods in Fluorescence In situ Hybridization – Interphase/Nuclear Fluorescence In Situ Hybridization*. American College of Medical Genetics, Bethesda, MD. www.acmg.net/StaticContent/SGs/Section_E.html, accessed 28 January 2014.

Antonescu CR, Tschernyavsky SJ, Woodruff JM, Jungbluth AA, Brennan MF, Ladanyi M. Molecular diagnosis of clear cell sarcoma: detection of EWS-ATF1 and MITF-M transcripts and histopathological and ultrastructural analysis of 12 cases. J Mol Diagn 2002; 4(1): 44–52.

Atlas of Genetics and Cytogenetics in Oncology and Haematology. http://AtlasGeneticsOncology.org/Anomalies, accessed 28 January 2014.

Betts DR, Cohen N, Leibundgut KE, et al. Characterization of karyotypic events and evolution in neuroblastoma. Pediatric Blood Cancer 2005; 44(2): 147–157.

Bovée JV, Sciot R, Cin PD, et al. Chromosome 9 alterations and trisomy 22 in central chondrosarcoma: a cytogenetic and DNA flow cytometric analysis of chondrosarcoma subtypes. Diagn Mol Pathol 2001; 10 (4): 228–235.

Bown N, Cotterill SJ, Roberts P, et al. Cytogenetic abnormalities and clinical outcome in Wilms tumor: a study by the U.K. Cancer Cytogenetics Group and the U.K. Children's Cancer Study Group. Med Pediatr Oncol 2002; 38(1):11–21.

Folpe AL, Goldblum JR, Rubin BP, et al. Morphologic and immunophenotypic diversity in Ewing family tumors: a study of 66 genetically confirmed cases. Am J Surg Pathol 2005; 29(8): 1025–1033.

Fujimura Y, Siddique H, Lee L, Rao VN, Reddy ES. EWS-ATF-1 chimeric protein in soft tissue clear cell sarcoma associates with CREB-binding protein and interferes with p53-mediated trans-activation function. Oncogene 2001; 20(46): 6653–6659.

Gersen S, Keagle M (eds). *Principles of Clinical Cytogenetics.* Humana Press, Totowa, New Jersey, 1999.

Grosfeld JL. Risk-based management: current concepts of treating malignant solid tumors of childhood. J Am Coll Surg 1999; 189(4): 407–425.

Heim S, Mitelman F. *Cancer Cytogenetics*, 2nd edn. Wiley-Liss, New York, 1995.

Hopyan S, Gokgoz N, Poon R, et al. A mutant PTH/PTHrP type I receptor in enchondromatosis. Nature Genet 2002; 30(3): 306–310.

Limon J, Mrozek K, Mandahl N, et al. Cytogenetics of synovial sarcoma: presentation of ten new cases and review of the literature. Genes Chromosomes Cancer 1991; 3(5): 338–345.

Macarenco RS, Erickson-Johnson M, Wang X, Jenkins RB, Nascimento AG, Oliveira AM. Cytogenetic and molecular cytogenetic findings in dedifferentiated liposarcoma with neural-like whorling pattern and metaplastic bone formation. Cancer Genet Cytogenet 2007 ; 172(2): 147–150.

Mandahl N, Gustafson P, Mertens F, et al. Cytogenetic aberrations and their prognostic impact in chondrosarcoma. Genes Chromosomes Cancer 2002; 33(2): 188–200.

Murphy WM, Grignon DJ, Perlman EJ. Wilms tumor. In: *AFIP Atlas of Tumor Pathology: Tumors of the Kidney, Bladder, and Related Urinary Structures*, Series IV. ARP Press, Washington, DC, 2005, pp.10–38.

Nishio J, Iwasaki H, Ishiguro M, et al. Establishment of a novel human dedifferentiated liposarcoma cell line, FU-DDLS-1: conventional and molecular cytogenetic characterization. Int J Oncol 2003; 22(3): 535–542.

Panagopoulos I, Mertens F, D'biec-Rychter M, et al. Molecular genetic characterization of the EWS/ATF1 fusion gene in clear cell sarcoma of tendons and aponeuroses. Int J Cancer 2002; 99(4): 560–567.

Rooney DE, Czepulkowski BH. *Human Cytogenetics. A Practical Approach.* Oxford University Press, New York, 1992.

Saboorian MH, Ashfaq R, Vandersteenhoven JJ, Schneider NR. Cytogenetics as an adjunct in establishing a definitive diagnosis of synovial sarcoma by fine-needle aspiration. Cancer 1997; 81(3): 187–92.

Shaffer LG, McGowan-Jordan J, Schmid M (eds). *ISCN 2013: An International System for Human Cytogenetic Nomenclature.* Karger Publishers, Unionville, CT, 2013.

Shannon BA, Murch A, Cohen RJ. Primary renal synovial sarcoma confirmed by cytogenetic analysis: a lesion distinct from sarcomatoid renal cell carcinoma. Arch Pathol Lab Med 2005; 129(2): 238–240.

Swerdlow SH, Campo E, Harris NL, et al. (eds). *WHO Classification of Tumours of Haematopoietic and Lymphoid Tissues.* IARC Press, Lyon, 2008.

Unni KK, Inwards CY, Bridge JA, Kindblom L-G, Wold LE. Ewing's sarcoma. In: *AFIP Atlas of Tumor Pathology: Tumors of the Kidney, Bladder, and Related Urinary Structures*, Series IV. ARP Press, Washington, DC, 2005, pp.209–222.

Walterhouse D, Watson A. Optimal management strategies for rhabdomyosarcoma in children. Paediatr Drugs 2007; 9(6): 391–400.

Section 2
Fluorescence *In Situ* Hybridization (FISH) Analysis

CHAPTER 28

Introduction to FISH analysis for hematological disorders and solid tumors

Fluorescence *in situ* hybridization (FISH) allows for the accurate localization and quantification of fluorescent signals in the cell directly. This technique uses DNA probes that hybridize to repetitive regions of chromosomes or to unique sequences within the genome. Double-stranded target DNA of the chromosomes is denatured and permitted to re-anneal in the presence of a high concentration of denatured probe DNA. Appropriate temperatures and appropriate concentrations of salt and formamide enable the annealing of the probe to its exact complementary sequence on the chromosomes. Following hybridization, unbound and non-specifically bound probe is removed by a series of stringency washes. Nuclear DNA is counterstained with DAPI for visualization. FISH probes vary in their configuration, depending on the desired result.

The most common types of FISH probes used to study neoplastic disorders consist of centromere probes, locus-specific probes, fusion probes and breakapart probes. Centromere probes are useful in determining the copy number of specific chromosomes for the detection of aneuploidies and hyperdiploidy, which have non-random copy number changes for specific disorders. Locus-specific probes are useful for determining the presence and absence of certain DNA sequences known to be deleted, gained or amplified in disorders. Fusion probes are used to determine the presence of a rearrangement of specific chromosome regions, including inversions and translocations, known to be involved in diseases. Breakapart probes are useful for significant gene regions known to be involved in disease, but that have rearrangements with more than one chromosome, such that it is more informative to know that the significant gene is rearranged, but the partner chromosome is not critical to the diagnosis or prognosis. For example, the IGH gene is involved in many hematological malignancies, and has numerous chromosome partners. Certain partners are associated with specific disorders, such as t(11;14), seen predominantly in mantle cell lymphomas as well as other lymphomas; therefore, a

Cytogenetic Abnormalities: Chromosomal, FISH and Microarray-Based Clinical Reporting, First Edition. Susan Mahler Zneimer.

CCND1/IGH fusion probe may be most informative. However, if a presumptive diagnosis is not clear, or if fusion probes are uninformative, then an IGH breakapart probe may be more useful in establishing the presence/absence of any IGH gene rearrangement (for more details on IGH gene rearrangements, see Chapter 30).

Current molecular methods used to identify changes in the DNA for the diagnosis of malignancies include flow cytometry, polymerase chain reaction (PCR), standard cytogenetics and FISH. Flow cytometry measures the DNA ploidy from tumor cells, but lacks the sensitivity to detect specific genetic aberrations and/or gains or loss of chromosomes. PCR is used to detect well-characterized chromosome translocations, but cannot detect whole or partial chromosome gains or deletions. Standard cytogenetics is still the best method used to detect translocations, deletions, aneuploidy, marker chromosomes and non-random chromosome abnormalities. However, cytogenetic studies require actively dividing cells and are also limited by low mitotic index, poor morphology and low-level mosaicism for abnormal cell lines. FISH analysis offers a combination of specificity of genetic changes known to be associated with disease with an accurate, rapid and fairly inexpensive approach to identifying the most common recurrent genetic changes in cancer.

Fluorescence *in situ* hybridization probes may have different DNA composition depending on the manufacturer or laboratory that developed the probe. Therefore, the examples given below for each probe are designed to be generic and not specific to a particular company or laboratory. However, there may be descriptions of probes that appear to be specifically designed but this is for the purpose of clarity and not to be misconstrued as the only possible configuration for the particular probe described.

For an example of an analyte-specific reagent FISH assay validation plan, see Appendix 1.

28.1 General results

28.1.1 Expected signal patterns

Metaphase and interphase FISH nomenclature that follows the same rules as constitutional abnormalities will not be discussed here. This section emphasizes interphase FISH analyses that have various types of probe configurations and the use of multiple probes in a single test (panels of probes). FISH results may be described with breakpoints (long version) or without breakpoints (short version), and either the long or the short version is acceptable ISCN nomenclature. This section describes FISH analyses with breakpoints to provide more clarity of the probes performed. However, for certain probe configurations, such as amplification of DNA sequences, the short version will be used. If more than one probe is used for a specific chromosomal region, all the names of the probes are described in the nomenclature.

ISCN RULES FOR THE SHORT VERSION OF INTERPHASE FISH NOMENCLATURE

- First write “nuc ish” for describing interphase FISH analysis.
- Next, in parentheses, write the locus designation, followed by “x”, followed by the number of signals present.
- Then place the number of cells analyzed enclosed by brackets (for neoplastic studies).

For example: nuc ish(TP53x2)[200].

> **ISCN RULES FOR THE LONG VERSION OF INTERPHASE FISH NOMENCLATURE**
>
> - First write "nuc ish", then leave a space, and then write the chromosome number.
> - Next write the chromosomal band designation.
> - Then, in parentheses, write the locus designation, followed by "x", followed by the number of signals present.
> - Then place the number of cells analyzed enclosed by brackets (for neoplastic studies).
>
> For example, nuc ish 17p11.2(TP53x2)[200].

Deletion probes, e.g. del(5q) (Figure 28.1)

In a normal cell hybridized with dual color deletion probes, e.g. EGR1/D5S23 and D5S721, using orange and green fluorochromes, the expected FISH pattern is two orange and two green (2O2G) signals. In an abnormal cell containing a deletion, e.g. del(5q31), where the locus of interest is in orange, then a one orange and two green (1O2G) signal pattern will be observed. With loss of the chromosome (e.g. monosomy 5), a one orange and one green (1O1G) signal pattern will be observed.

Centromere probes, e.g. trisomy 21 (Figure 28.2)

In a normal cell hybridized with a single color centromere probe, e.g. chromosome 21 (CEP 21), using an orange fluorochrome, the expected FISH pattern is two orange (2O) signals. In an abnormal cell containing loss of chromosome 21, a one orange (1O) signal pattern will be observed. In an abnormal cell with the gain of chromosome 21, a three orange (3O) signal pattern will be observed.

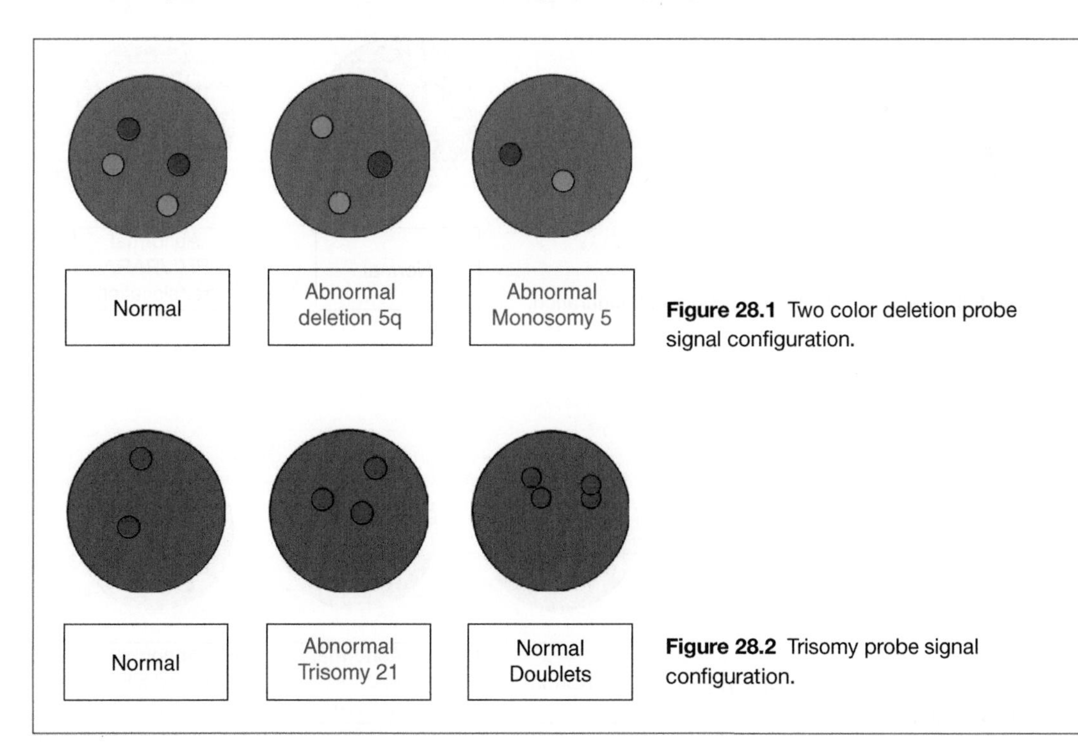

Figure 28.1 Two color deletion probe signal configuration.

Figure 28.2 Trisomy probe signal configuration.

Some normal nuclei may appear to have a slight separation of the orange and green signals due to decondensation of the chromatin. It should be noted that due to the highly repetitive DNA present at the centromere, non-specific hybridization can occur if the washing step is not stringent enough. Doublets may occur due to duplication of the chromatin in G2 and can result in what might appear as two separate signals, when in fact the signals are from the nucleus of a cell preparing for division (doubling its chromatin). If the distance between the two signals is less than the diameter of an average single signal, then the two signals should be scored as one.

Dual color, single fusion probes, e.g. PML/RARA (Figure 28.3)

In a normal cell hybridized with a dual color, single fusion probe with orange and green fluorochromes, e.g. the PML/RARA, the expected FISH pattern is two orange and two green (2O2G) signals. In an abnormal cell containing a PML/RARA fusion, a one green (RARA), one orange (PML) and one closely adjacent or fused green/orange (yellow) signal pattern (1O1G1F) is observed.

Dual color, dual fusion probes, e.g. BCR/ABL (Figure 28.4)

In a normal cell hybridized with a dual color, dual fusion probe with orange and green fluorochromes, e.g. the BCR/ABL, the expected FISH pattern is two orange and two green (2O2G) signals. In an abnormal cell containing a BCR/ABL fusion, one green and one orange signal for the normal chromosomes 9 and 22 and two orange/green (yellow) fusion signals, one each from the derivative chromosomes 9 and 22 (1O1G2F), will be observed. In some instances, deletions may occur 3' of the BCR breakpoint and/or 5' of the ABL breakpoint, resulting in either an extra signal (extra orange or green) or a single fusion pattern.

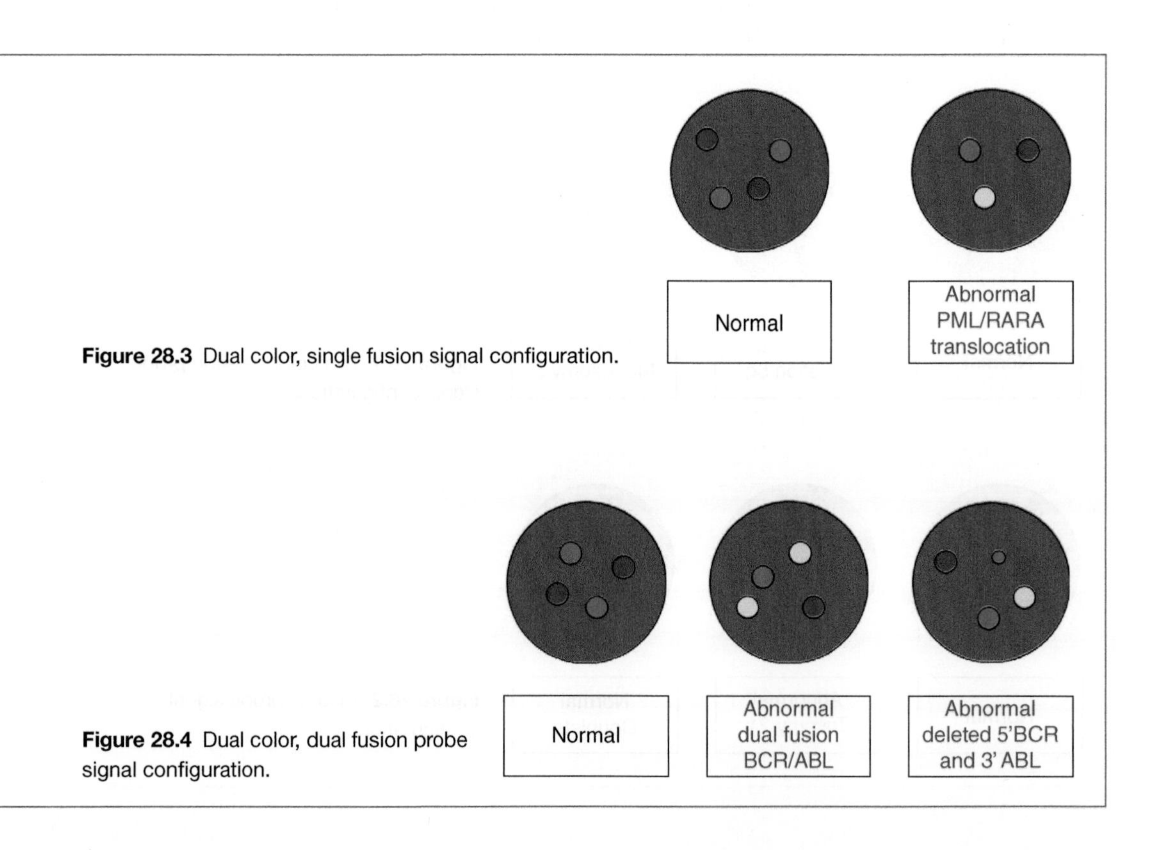

Figure 28.3 Dual color, single fusion signal configuration.

Figure 28.4 Dual color, dual fusion probe signal configuration.

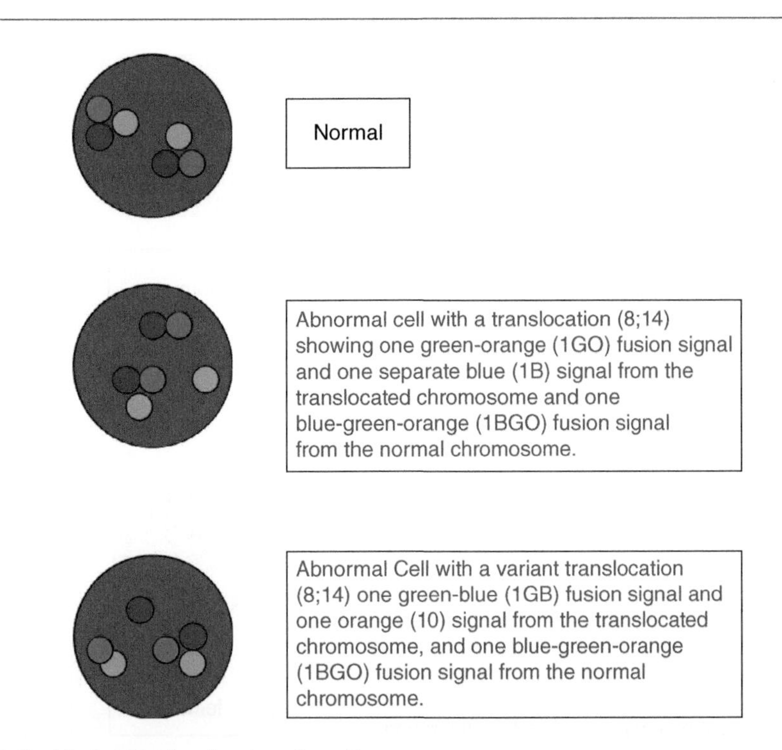

Normal
Abnormal cell with a translocation (8;14) showing one green-orange (1GO) fusion signal and one separate blue (1B) signal from the translocated chromosome and one blue-green-orange (1BGO) fusion signal from the normal chromosome.
Abnormal Cell with a variant translocation (8;14) one green-blue (1GB) fusion signal and one orange (1O) signal from the translocated chromosome, and one blue-green-orange (1BGO) fusion signal from the normal chromosome.

Figure 28.5 Dual fusion tricolor signal configuration.

Tricolor dual fusion probes, e.g. IGH/MYC/CEP8 (Figure 28.5)

In a normal cell hybridized with a tricolor dual fusion probe with blue, orange and green fluorochromes, e.g. IGH/MYC and CEP 8 probes (MYC-orange, IGH-green and CEP8-blue), the expected FISH pattern is two blue, two orange and two green (2A2O2G) signals. In an abnormal cell containing a reciprocal (8;14) translocation (MYC/IGH), one orange/green fusion, one orange/green/blue fusion, and one single blue signal (1O1G2F2A) will be observed. If the cell contains a variant t(8;14) with the breakpoint far from the 5' end of MYC, a fusion on the der(8) may not be visible or may be very weak, as little or no orange probe target would remain on the der(8).

Breakapart probes, e.g. IGH (Figure 28.6)

In a normal cell hybridized with a dual color, breakapart probe with orange and green fluorochromes, e.g. the IGH probe, the expected FISH pattern is two orange/green or yellow (2F) fusion signals. In an abnormal cell with a translocation breakpoint within the gap region between the 5' and 3' ends of the probe, one orange, one green, and one fusion yellow signal (1O1G1F) is observed. Some normal nuclei may appear to have a slight separation of the orange and green signals due to a gap between the hybridization targets of the two probes.

X/Y probes (Figure 28.7)

In cells hybridized with the CEP X/Y probes with the X chromosome labeled in orange and the Y chromosome labeled in green, the expected FISH pattern for a normal male cell is one orange and one green (1O1G) signal. In a normal female cell, two orange (2O) signals will be observed.

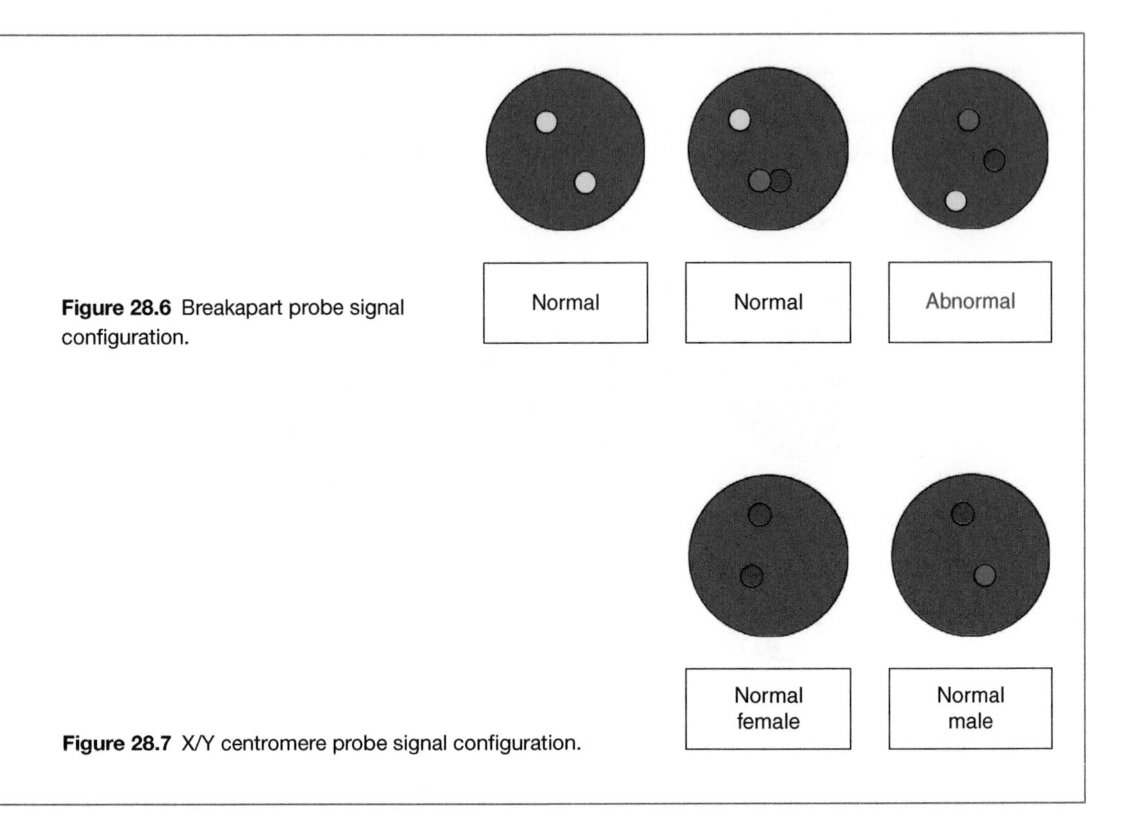

Figure 28.6 Breakapart probe signal configuration.

Figure 28.7 X/Y centromere probe signal configuration.

28.1.2 General reporting comments

EXAMPLE REPORTING OF A NORMAL RESULT

ISCN Result: ish or nuc ish

Interpretation

Fluorescence *in situ* hybridization (FISH) analysis with the _ probe showed a normal result with no evidence of an abnormality in 10 metaphase cells and 200 interphase nuclei scored.

EXAMPLE REPORTING OF AN ABNORMAL RESULT

ISCN Result: ish or nuc ish

Interpretation

Fluorescence *in situ* hybridization (FISH) analysis with the _ probe showed an abnormal result with evidence of _ abnormality in 10 metaphase cells and in _ /200 (%) interphase nuclei scored.

This finding is a recurrent, non-random abnormality observed in _disorder(s).

Follow-up studies to monitor for this abnormal cell line by fluorescence *in situ* hybridization (FISH) analysis with _ probe in addition to standard chromosome studies are suggested.

28.1.3 Common Disclaimers

FISH results near the cut-off value

- These FISH findings are considered to be abnormal based on our laboratory validation data for this probe, which indicate that greater than _% abnormal cells is considered a positive result. However, as the percentage of abnormal cells in this case is close to our normal cut-off value, correlation of this finding with other laboratory and clinical data is strongly recommended.

Insufficient cell sorting for plasma cell disorders

- Sorting of CD138+ cells was attempted, but insufficient cellularity of the sorted sample was obtained. Therefore, this FISH analysis was performed on unsorted bone marrow cells.

Non-FDA-approved testing comment

- This test was developed and its performance characteristics determined by [laboratory name] as required by the CLIA '88 regulations. It has not been cleared or approved by the US Food and Drug Administration (FDA). The FDA has determined that such clearance or approval is not necessary.

Reflex FISH comments

- Follow-up studies to monitor for this abnormal cell line by FISH analysis with _ probe in addition to standard chromosome studies are suggested.
- Since chromosome analysis on metaphase cells showed *normal results*, fluorescence *in situ* hybridization (FISH) on interphase cells may provide further information. The preparations from this specimen can be used for further FISH analysis to detect recurrent chromosome aberrations and gene rearrangements for this clinical indication.
- Since chromosome analysis on metaphase cells showed *no growth*, fluorescence *in situ* hybridization (FISH) on interphase cells may provide further information. The preparations from this specimen can be used for further FISH analysis to detect recurrent chromosome aberrations and gene rearrangements for this clinical indication.
- Since chromosome analysis on metaphase cells showed a single cell abnormality that has been shown to be associated with a neoplastic process, fluorescence *in situ* hybridization (FISH) on interphase cells may provide further information. The preparations from this specimen can be used for further FISH analysis to detect recurrent chromosome aberrations and gene rearrangements for this clinical indication.
- Since chromosome analysis showed both female and male cells, fluorescence *in situ* hybridization (FISH) on interphase cells may provide further information in the evaluation of this sex-mismatched bone marrow transplant case. The preparations from this specimen can be used for further FISH analysis to detect the presence of donor versus host cells using X- and Y-specific FISH probes.

Comment with concurrent chromosome analysis

- These findings were confirmed by the cytogenetic result (case number) showing _ in _% of metaphase cells analyzed.

28.2 Bone marrow transplantation results

Fluorescence *in situ* hybridization for CEP X/Y is a mixture of two fluorochrome colors, such as orange-labeled CEP X probe and green-labeled CEP Y probes. This dual color probe hybridizes to the alpha satellite sequences in the centromere region of chromosome X and the satellite III DNA at chromosome Yq12, and is used to identify the X and Y sex chromosomes.

These probes are used as an adjunct to standard chromosome analysis to evaluate engraftment success in recipients of sex-mismatched bone marrow transplantation (BMT) by evaluating the proportion of XX and XY sex gonosomes. Following transplantation, an assessment of the proportion of cells belonging to the donor and to the recipient can be used to evaluate engraftment, detect the presence of clonal neoplasms and determine disease recurrence. Bone marrow transplantation is a critical therapeutic strategy in the management of hematological malignancies. Thus, this FISH test offers timely results that enhance patient care.

EXAMPLE REPORTING OF BMT WITH A MALE PATIENT, FEMALE DONOR

ISCN Result: nuc ish Xcen(DXZ1x2),Ycen(DYZ3x0)[200]

Interpretation

Fluorescence *in situ* hybridization (FISH) analysis showed an XX (female donor) population in 200/200 (100%) cells scored. No recipient (male) cells were observed.

These findings are consistent with a female donor sex chromosome complement post bone marrow transplantation.

EXAMPLE REPORTING OF BMT WITH A FEMALE PATIENT, MALE DONOR

ISCN Result: nuc ish Xcen(DXZ1x1),Ycen(DYZ3x1)[200]

Interpretation

Fluorescence *in situ* hybridization (FISH) analysis showed an XY (male donor) population in 200/200 (100%) cells scored. No recipient (female) cells were observed.

These findings are consistent with a male donor sex chromosome complement post bone marrow transplantation.

EXAMPLE REPORTING OF CHIMERIC RESULTS POST BMT WITH A MALE PATIENT, FEMALE DONOR

ISCN Result: nuc ish Xcen(DXZ1x2),Ycen(DYZ3x0)[120]/Xcen(DXZ1x1),Ycen(DYZ3x1)[80]

60% Donor (XX)

40% Recipient (XY)

Interpretation

Fluorescence *in situ* hybridization (FISH) analysis showed an XX (female donor) population in 120/200 (60%) cells scored, and an XY (recipient) population in 80/200 (40%) cells scored.

These findings are consistent with a chimeric sex chromosome complement post bone marrow transplantation.

EXAMPLE REPORTING OF CHIMERIC RESULTS POST BMT WITH A FEMALE PATIENT, MALE DONOR

ISCN Result: nuc ish Xcen(DXZ1x1),Ycen(DYZ3x1)[180]/Xcen(DXZ1x2),Ycen(DYZ3x0)[20]
90% Donor (XY)
10% Recipient (XX)

Interpretation

Fluorescence *in situ* hybridization (FISH) analysis showed an XY (male donor) population in 180/200 (90%) cells scored, and an XX (recipient) population in 20/200 (10%) cells scored.

These findings are consistent with a chimeric sex chromosome complement post bone marrow transplantation.

EXAMPLE REPORTING OF CHIMERIC RESULTS WITH INDETERMINATE CELLS POST BMT WITH A MALE PATIENT, FEMALE DONOR

ISCN Result: nuc ish Xcen(DXZ1x2),Ycen(DYZ3x0)[195]/Xcen(DXZ1x1),Ycen(DYZ3x0)[5]
97.5% Donor (XX)
0% Recipient (XY)
2.5% Indeterminate

Interpretation

Fluorescence *in situ* hybridization (FISH) analysis showed an XX (female donor) population in 195/200 (97.5%) cells scored, and no XY (recipient) cells scored.

In 5/200 (2.5%) cells, only a single X signal and no Y signal was observed. The clinical significance of this finding is uncertain, as it cannot be determined if this represents loss of an X or Y signal.

Bibliography

Albertson D, Collins C, McCormick F, Gray J. Chromosome aberrations in solid tumors. Nature Genet 2003; 34: 369–376.

American College of Medical Genetics. *Guidelines in Clinical Cytogenetics. Section E10 Methods in Fluorescence In situ Hybridization – Interphase/Nuclear Fluorescence In Situ Hybridization*. American College of Medical Genetics, Bethesda, MD. www.acmg.net/StaticContent/SGs/Section_E.html, accessed 28 January 2014.

Atlas of Genetics and Cytogenetics in Oncology and Haematology. http://AtlasGeneticsOncology.org/Anomalies, accessed 28 January 2014.

Bishop R. Applications of fluorescence in situ hybridization (FISH) in detecting genetic aberrations of medical significance. Biosci Horizons 2010; 3(1): 85–95.

Gersen S, Keagle M (eds). *Principles of Clinical Cytogenetics*. Humana Press, Totowa, New Jersey, 1999.

Gozzetti A, Le Beau MM. Fluorescence in situ hybridization: uses and limitations. Semin Hematol 2000; 37: 320–333.

Heim S, Mitelman F. *Cancer Cytogenetics*, 2nd edn. Wiley-Liss, New York, 1995.

Kearney L. Molecular cytogenetics. Best Pract Res Clin Haematol 2001; 14: 645–668.

Lobo I. Chromosome abnormalities and cancer cytogenetics. Nature Educ 2008; 1(1): 68.

O'Connor C. Fluorescence in situ hybridization (FISH). Nature Educ 2008; 1(1): 171.

Rooney DE, Czepulkowski BH. *Human Cytogenetics. A Practical Approach*. Oxford University Press, New York, 1992.

Shaffer LG, McGowan-Jordan J, Schmid M (eds). *ISCN 2013: An International System for Human Cytogenetic Nomenclature*. Karger Publishers, Unionville, CT, 2013.

Speicher MR, Carter NP. The new cytogenetics: blurring the boundaries with molecular biology. Nature Rev Genet 2005; 6: 782–792.

Swerdlow SH, Campo E, Harris NL, et al. (eds). *WHO Classification of Tumours of Haematopoietic and Lymphoid Tissues*. IARC Press, Lyon, 2008.

Commercial FISH probes are available at these websites

www.AbbottMolecular.com

www.Kreatech.com

www.Rainbowscientific.com

www.empiregenomics.com

www.abnova.com

www.cytocell.com

www.exiqon.com

www.cambridgebluegnome.com

CHAPTER 29

Recurrent FISH abnormalities in myeloid disorders

29.1 Individual abnormalities in myeloid disorders by chromosome order

29.1.1 Chromosome 3

Rearrangements of 3q21-q26.2

SPECIFIC FEATURES

The EVI1 gene/t(3;3)/inv(3) can be viewed as either a dual color or triple color probe that is used to detect inversions of chromosome 3 involving the EVI1 gene region at 3q26 as well as translocation (3;3).

Overexpression of the EVI1 gene appears to be a consistent feature of 3q21q26 abnormalities, and is a recurrent, non-random abnormality in acute myeloid leukemia (AML) and myelodysplastic syndromes (MDS). Specific chromosomal aberrations involving both 3q21 and 3q26.2 are seen, including inv(3)(q21q26.2) and t(3;3)(q21;q26.2). EVI1, which maps to the q26.2 region of chromosome 3, encodes a zinc finger protein overly expressed in leukemic cells of 2–5% of AML and MDS cases.

Normal signal pattern for chromosome 3
nuc ish 3q26.2(EVI1x2)[200]

Abnormal signal pattern for 3q26 rearrangement
nuc ish 3q26.2(EVI1x2)(5’EVI1 sep 3’EVI1x1)[ ]

Cytogenetic Abnormalities: Chromosomal, FISH and Microarray-Based Clinical Reporting, First Edition. Susan Mahler Zneimer.

EXAMPLE REPORTING OF 3q26 REARRANGEMENT – TRIPLE COLOR, DUAL FUSION PROBE (FIGURE 29.1)

ISCN Result: nuc ish 3q26.2(EVI1x2)(5'EVI1 sep 3'EVI1x1)[156/200]

Interpretation

Fluorescence *in situ* hybridization (FISH) analysis with the EVI1 probe showed evidence of a 3q26 rearrangement involving the EVI1 locus in 156 of 200 (78%) interphase nuclei scored. The partner of this rearrangement is not revealed by this FISH analysis.

Overexpression of the EVI1 gene appears to be a consistent feature of a 3q21q26 rearrangement, which is a recurrent, non-random abnormality in acute myeloid leukemia (AML) and myelodysplastic syndromes (MDS). Specific chromosomal aberrations involving both 3q21 and 3q26.2 are seen, including inv(3)(q21q26.2) and t(3;3)(q21q26.2). EVI1, which maps to the q26.2 region of chromosome 3, encodes a zinc finger protein overly expressed in leukemic cells of 2–5% of AML and MDS cases.

Leukemias with 3q aberrations are generally non-responsive to standard chemotherapeutic regimens and are associated with a poor prognosis.

Follow-up studies to monitor for this abnormality by FISH analysis in addition to standard chromosome studies are suggested.

29.1.2 Chromosome 4

Deletion of 4q12

SPECIFIC FEATURES

The CHIC2, FIP1L1, PDGFRA dual or triple color probe is used to detect deletions of chromosome 4q12. The fluorochromes for each gene in this probe set are situated along the chromosome 4q12 region to detect specific gene deletions.

Deletions of the CHIC2 gene are seen predominantly in patients with idiopathic hypereosinophilic syndrome (HES) and chronic eosinophilic leukemia (CEL). A deletion will result in the fusion of FIP1L1 with PDGFRA (platelet-derived growth factor alpha), while deleting the CHIC2 gene region, which produces a tyrosine kinase that will deregulate hematopoietic cells.

Normal signal pattern for chromosome 4
nuc ish 4q12(CHIC2,FIP1L1,PDGFRA)x2[200]

Abnormal signal pattern for chromosome 4 deletion
nuc ish 4q12(CHIC2x1),(FIP1L1x2),(PDGFRAx2)(FIP1L1 con PDGFRAx1)[ ]

EXAMPLE REPORTING OF 4q12 DELETION – TRIPLE COLOR DELETION PROBE WITH INDICATION OF EOSINOPHILIA (FIGURE 29.1)

ISCN Result: nuc ish 4q12(CHIC2x1),(FIP1L1x2),(PDGFRAx2)(FIP1L1 con PDGFRAx1) [156/200]

Interpretation

Fluorescence *in situ* hybridization (FISH) analysis with the PDGFRA probe showed evidence of a deletion of the CHIC2 locus, resulting in fusion of FIP1L1 with PDGFRA in 156 of 200 (78%) interphase nuclei scored.

This finding is consistent with a specific group of myeloproliferative and lymphoid neoplasms, which result from the formation of a fusion gene encoding an aberrant tyrosine kinase with rearrangements involving PDGFRA, PDGFRB or FGFR1.

In PDGFRA-related disease, cases often present as chronic eosinophilic leukemia/ hypereosinophilic syndrome (CEL/HES), but can also present as acute myeloid leukemia or precursor T-cell lymphoblastic lymphoma (T-cell LBL).

Follow-up studies to monitor for this abnormality by FISH analysis are suggested as this abnormality is not visible by standard chromosome analysis.

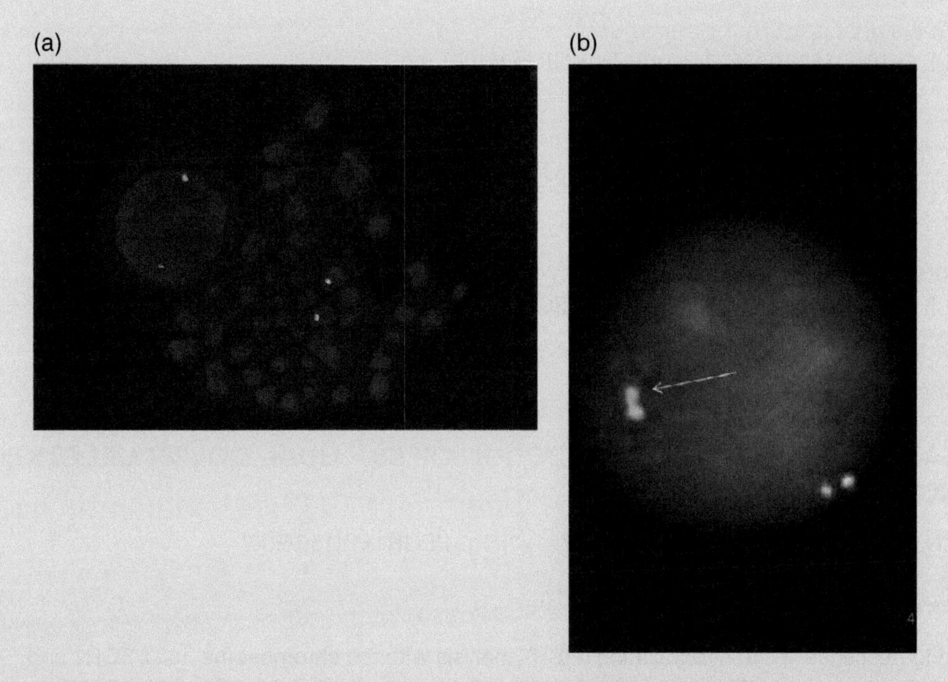

Figure 29.1 Tricolor deletion probe. (a) Normal – showing the two green, orange, aqua fusion signals. (b) Abnormal – with an arrow showing one green, aqua fusion signal with the orange signal deleted, corresponding to an interstitial deletion of the CHIC2 locus, resulting in the fusion of FIP1L1 with PDGFRA gene regions. Courtesy of Mehdi Jamehdor MD, Kaiser Permanente Regional Laboratory, Southern California.

29.1.3 Chromosome 5

Monosomy 5/deletion of 5q

SPECIFIC FEATURES

The EGR1/D5S23, D5S721 dual color probe is used to detect deletions of chromosome 5q31 containing the EGR1 locus as well as monosomy 5. The 5p15.31 region is often used as an internal control for chromosome 5 to determine loss of the whole chromosome (monosomy 5) versus del(5q).

Deletions of chromosome 5 are the most common structural rearrangement observed in both *de novo* and therapy-related MDS and AML. In MDS, it is observed particularly in refractory anemia whereas in AML it does not occur preferentially in any particular FAB subtype and can result from genotoxic exposure.

A subset of patients with a deletion of chromosome 5 is classified as "del(5q) syndrome" which generally occurs in elderly women with refractory macrocytic anemia, normal to elevated platelets, and showing morphologically characteristic megakaryocytic abnormalities. The clinical course for these patients is usually mild, and transformation to AML is fairly rare.

Normal signal pattern for chromosome 5
nuc ish 5p12(D5S23/D5S721x2),5q31(EGR1x2[200]

Abnormal signal pattern for del(5q)
nuc ish 5p15.2(D5S23/D5S721x2),5q31(EGR1x1)[ ]

Abnormal signal pattern of monosomy 5
nuc ish 5p15.2(D5S23/D5S721x1),5q31(EGR1x1)[ ]

EXAMPLE REPORTING OF DELETION OF 5q – DUAL COLOR DELETION PROBE (FIGURE 29.2)

ISCN Result: nuc ish 5p12(D5S23/D5S721x2),5q31(EGR1x1)[156/200]

Interpretation

Fluorescence *in situ* hybridization (FISH) analysis with the chromosome 5q31/EGR1 and 5p15.2 probes showed evidence of a deletion of chromosome 5q31 in 156 of 200 (78%) interphase nuclei scored.

Del(5q) is a recurrent, non-random abnormality observed in myeloid malignancies, including myelodysplasia, myeloproliferative disorders and acute myeloid leukemia.

Follow-up studies to monitor for this abnormality by FISH analysis in addition to standard chromosome studies are suggested.

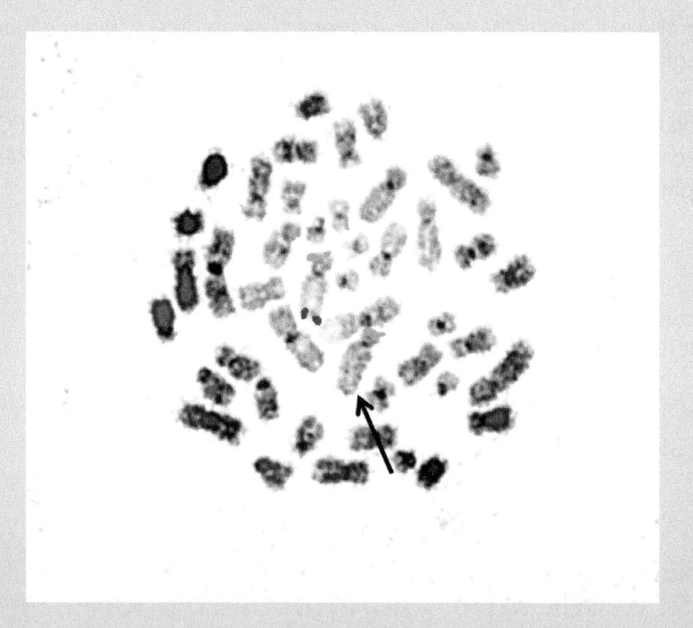

Figure 29.2 Dual color deletion probe with the arrow pointing to a deletion of the red signal corresponding to a 5q31/EGR1 gene. Courtesy of Lauren Jenkins PhD, Kaiser Permanente Regional Laboratory, Northern California.

EXAMPLE REPORTING OF MONOSOMY 5 – DUAL COLOR DELETION PROBE

ISCN Result: nuc ish 5p12(D5S23/D5S721x1),5q31(EGR1x1)[156/200]

Interpretation

Fluorescence *in situ* hybridization (FISH) analysis with the chromosome 5q31/EGR1 and 5p15.2 probes showed evidence of monosomy 5 in 156 of 200 (78%) interphase nuclei scored.

Monosomy 5 is a recurrent, non-random abnormality observed in myeloid malignancies, including myelodysplasia, myeloproliferative disorders and acute myeloid leukemia.

Follow-up studies to monitor for this abnormality by FISH analysis in addition to standard chromosome studies are suggested.

Rearrangements of 5q33

SPECIFIC FEATURES

The PDGFRB probe is a dual color breakapart probe used to detect translocations involving the PDGFRB region at 5q33. This finding is consistent with a specific group of myeloproliferative and lymphoid neoplasms, which results in a fusion gene encoding an aberrant tyrosine kinase with rearrangements involving PDGFRA, PDGFRB or FGFR1. In PDGFRB-related disease, cases often present as atypical chronic myeloid leukemia (aCML) or chronic eosinophilic leukemia/hypereosinophilic syndrome (CEL/HES).

Normal signal pattern for chromosome 5
nuc ish 5q33(PDGFRBx2)[200]

Abnormal signal pattern for a 5q33 rearrangement
nuc ish 5q33(PDGFRBx2)(5’PDGFRB sep 3’PDGFRBx1)[ ]

EXAMPLE REPORTING OF 5q33 REARRANGEMENT – DUAL COLOR BREAKAPART PROBE WITH INDICATION OF EOSINOPHILIA (FIGURE 29.3)

ISCN Result: nuc ish 5q33(PDGFRBx2)(5’PDGFRB sep 3’PDGFRBx1)[156/200]

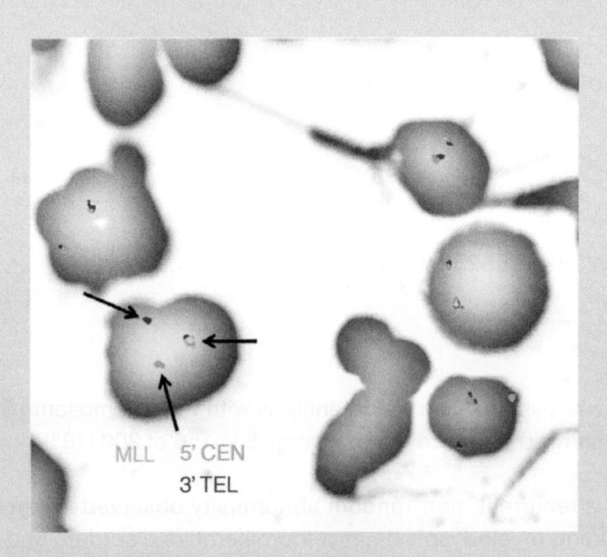

Figure 29.3 Dual color breakapart probe. Interphase cells with normal and abnormal cells. Normal cells show two fusion chromosomes while the abnormal cell (*arrows*) shows one fusion signal, one green signal and one red signal, corresponding to one chromosome split, resulting in a rearrangement. Courtesy of Lauren Jenkins PhD, Kaiser Permanente Regional Laboratory, Northern California.

Interpretation

Fluorescence *in situ* hybridization (FISH) analysis with the PDGFRB probe showed evidence of a 5q33 rearrangement in 156 of 200 (78%) interphase nuclei scored. The partner of this rearrangement is not revealed by this FISH analysis.

This finding is consistent with a specific group of myeloproliferative and lymphoid neoplasms, which result in a fusion gene encoding an aberrant tyrosine kinase with rearrangements involving PDGFRA, PDGFRB or FGFR1.

In PDGFRB-related disease, cases often present as atypical chronic myeloid leukemia (aCML) or chronic eosinophilic leukemia/hypereosinophilic syndrome (CEL/HES).

Follow-up studies to monitor for this abnormality by FISH analysis are suggested, as this abnormality is not visible by standard chromosome analysis.

29.1.4 Chromosome 6

Translocation (6;9)(p23;q34)

SPECIFIC FEATURES

The t(6;9)(p22;q34) chromosomal translocation is found in a subset of patients with AML and constitutes 0.5–4% of all AML cases. This translocation results in a fusion between the nuclear phosphoprotein DEK and the nucleoporin NUP214 which is seen as a dual color probe. Patients with t(6;9) AML generally have a poor prognosis. Diagnosis is important because these patients may benefit from early allogeneic stem cell transplant.

Normal signal pattern for chromosome 6 and 9
nuc ish 6p23(DEKx2),9q34(NUP214x2)[200]

Abnormal signal pattern for a (6;9) rearrangement
nuc ish 6p23(DEKx3),9q34(NUP214x3),(DEK con NUP214x2)[ ]

EXAMPLE REPORTING OF t(6;9) – DUAL COLOR, DUAL FUSION PROBE (FIGURE 29.4)

ISCN Result: nuc ish 6p23(DEKx3),9q34(NUP214x3),(DEK con NUP214x2)[156/200]

Interpretation

Fluorescence *in situ* hybridization (FISH) analysis with the DEK/NUP214 probes showed an abnormal result with evidence of a 6;9 translocation in 156 of 200 (78%) interphase nuclei scored.

Translocation (6;9)(p23;q34) results in the fusion of the DEK gene at 6p23 with the NUP214 gene at 9q34, and is a recurrent, non-random abnormality in acute myeloid leukemia (AML) and generally associated with a poor prognosis. Patients with t(6;9) and less than 20% blasts should be carefully monitored for emergence of overt AML.

Follow-up studies to monitor for this abnormality by FISH analysis in addition to standard chromosome studies are suggested.

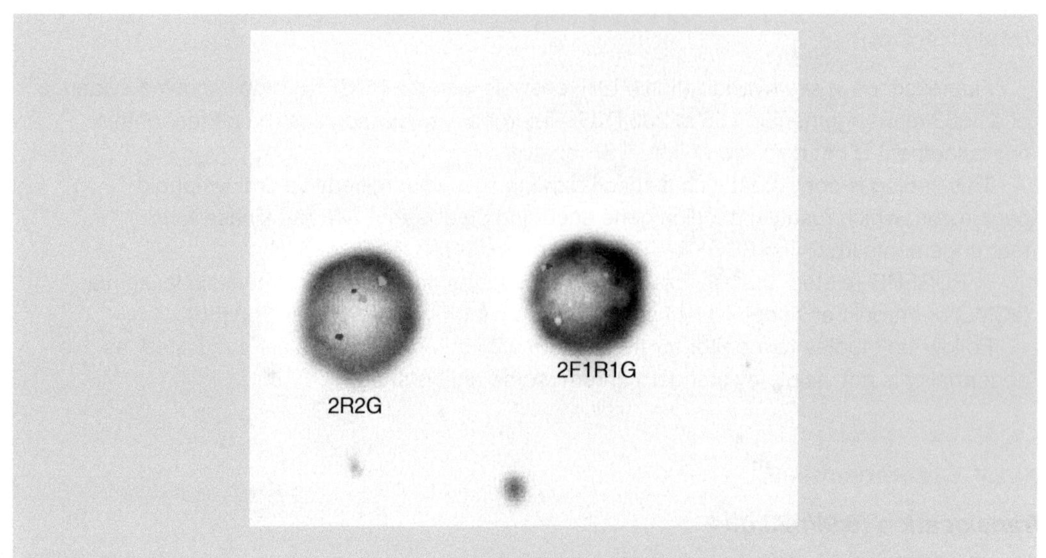

Figure 29.4 Dual color, dual fusion probe. The cell on the left shows two red and two green signals corresponding to the two normal chromosomes, while the cell on the right shows two fusion signals, one each from the derivative chromosomes, one red and one green signal, each corresponding to the normal chromosomes (2F1R1G). Courtesy of Lauren Jenkins PhD, Kaiser Permanente Regional Laboratory, Northern California.

29.1.5 Chromosome 7

Monosomy 7/deletion of 7q

SPECIFIC FEATURES

The D7S486 dual color probe is used to detect deletions of 7q31 containing the D7S486 locus as well as monosomy 7. The CEP7 or 7q22 region is often used as an internal control for chromosome 7 to determine loss of the whole chromosome (monosomy 7) versus del(7q).

Deletions of the long arm of chromosome 7 and monosomy 7 are observed in both *de novo* and secondary MDS and AML. In MDS, deletion/monosomy 7 is most common in refractory anemia with excess blasts (RAEB)/RAEB-T, followed by chronic myelomonocytic leukemia (CMML) and refractory anemia (RA). In AML, it is more frequently associated with the FAB types M4 and M5 but is not specific to these subtypes. Both del(7q) and monosomy 7 are generally associated with a poor prognosis.

Normal signal pattern for chromosome 7
nuc ish 7cen(D7Z1x2),7q31(D7S486x2)[200]

Abnormal signal pattern for del(7q)
nuc ish 7cen(D7Z1x2),7q31(D7S486x1)[ ]

Abnormal signal pattern of monosomy 7
nuc ish 7cen(D7Z1x1),7q31(D7S486x1)[ ]

EXAMPLE REPORTING OF MONOSOMY 7 – DUAL COLOR, DELETION PROBE

ISCN Result: nuc ish 7cen(D7Z1x1),7q31(D7S486x1)[156/200]

Interpretation

Fluorescence *in situ* hybridization (FISH) analysis with the chromosome 7q31 and chromosome 7 centromere probes showed evidence of monosomy 7 in 156 of 200 (78%) interphase nuclei scored.

Monosomy 7 is a recurrent, non-random abnormality observed in myeloid malignancies, including myelodysplasia, myeloproliferative disorders and acute myeloid leukemia.

Follow-up studies to monitor for this abnormality by FISH analysis in addition to standard chromosome studies are suggested.

EXAMPLE REPORTING OF DELETION OF 7q – DUAL COLOR, DELETION PROBE

ISCN Result: nuc ish 7cen(D7Z1x2),7q31(D7S486x1)[156/200]

Interpretation

Fluorescence *in situ* hybridization (FISH) analysis with the chromosome 7q31 and 7 centromere probes showed evidence of a deletion of chromosome 7q31 in 156 of 200 (78%) interphase nuclei scored.

Del(7q) is a recurrent, non-random abnormality observed in myeloid malignancies, including myelodysplasia, myeloproliferative disorders and acute myeloid leukemia.

Follow-up studies to monitor for this abnormality by FISH analysis in addition to standard chromosome studies are suggested.

29.1.6 Chromosome 8

Trisomy 8

SPECIFIC FEATURES

The trisomy 8 FISH assay uses either a centromeric probe or locus-specific probe to detect copy number changes of chromosome 8. Trisomy 8 is the most frequent numerical aberration seen in MDS and AML. It occurs in 20–30% of MDS cases and 10–15% of all AML cases. Trisomy 8 appears to be relatively specific for myeloid disorders, as it is very rare in the lymphocytic leukemias. Trisomy 8 in AML is often preceded by MDS and is not restricted to any particular FAB group, but is seen as a primary change most often in M1 (AML), followed by M4 (acute myelomonocytic leukemia – AMML) and M5 (AML).

Normal signal pattern for chromosome 8
nuc ish 8cen(D8Z2x2)[200]

Abnormal signal pattern for trisomy 8
nuc ish 8cen(D8Z2x3)[ ]

EXAMPLE REPORTING OF TRISOMY 8 – CENTROMERIC PROBE (FIGURE 29.5)

ISCN Result: nuc ish 8cen(D8Z2x3)[156/200]

Interpretation

Fluorescence *in situ* hybridization (FISH) analysis with the chromosome 8 centromeric probe (D8Z2) showed evidence of trisomy 8 in 156 of 200 (78%) interphase nuclei scored.

Trisomy 8 is a recurrent, non-random abnormality in myeloid disorders, including myelodysplasia, acute myeloid leukemia and myeloproliferative neoplasms.

Follow-up studies to monitor for this abnormality by FISH analysis in addition to standard chromosome studies are suggested.

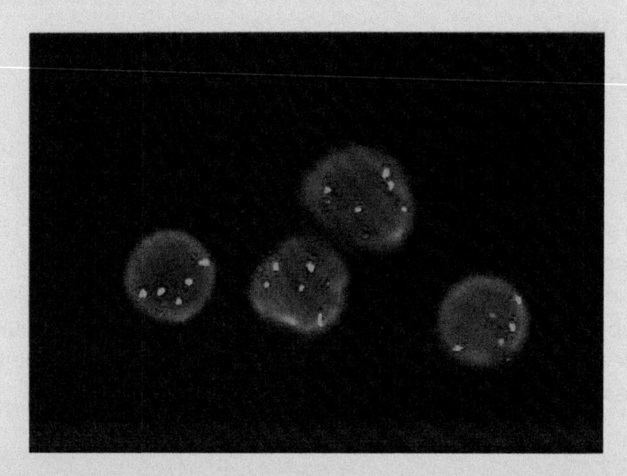

Figure 29.5 Three color trisomy probes with cells showing two red and blue signals corresponding to disomy, and three green signals resulting in trisomy 8. Courtesy of Mehdi Jamehdor MD, Kaiser Permanente Regional Laboratory, Southern California.

Rearrangements of 8p12

SPECIFIC FEATURES

Rearrangements of chromosome 8p12 use a dual color breakapart probe for the detection of myeloid and lymphoid neoplasms with FGFR1 rearrangements, also known as 8p11-12 myeloproliferative syndrome. Myeloid and lymphoid neoplasms associated with rearrangement of FGFR1 are hematologically and genetically heterogeneous, but share common features. The most frequent presentation is seen as a myeloproliferative neoplasm or as a T-cell lymphoblastic lymphoma with eosinophilia. Less often, they present as acute myeloid leukemia. Although FGFR1 encodes a tyrosine kinase, there is, as yet, no recognized effective targeted treatment for this group of neoplasms. The presence of an FGFR1 rearrangement is also important since only a trial of midostaurin or an allogeneic transplant offers any current treatment prospects.

Normal signal pattern for chromosome 8p12
nuc ish 8p12(FGFR1x2)[200]

Abnormal signal pattern for chromosome 8p12 rearrangement
nuc ish 8p12(FGFR1x2)(5'FGFR1 sep 3'FGFR1x1)[ ]

EXAMPLE REPORTING OF 8p12 REARRANGEMENT – DUAL COLOR, BREAKAPART PROBE WITH INDICATION OF EOSINOPHILIA

ISCN Result: nuc ish 8p12(FGFR1x2)(5'FGFR1 sep 3'FGFR1x1)[156/200]

Interpretation

Fluorescence *in situ* hybridization (FISH) analysis with the FGFR1 probe showed evidence of an 8p12 rearrangement involving the FGFR1 locus in 156 of 200 (78%) interphase nuclei scored. The partner of this rearrangement is not revealed by this FISH analysis.

FGFR1 rearrangements often present as a myeloproliferative neoplasm or in transformation as acute myeloid leukemia (AML), T-cell or B-cell lineage lymphoblastic lymphoma/leukemia or mixed phenotype acute leukemia.

Follow-up studies to monitor for this abnormality by FISH analysis are suggested, as this abnormality is not visible by standard chromosome analysis.

Translocation (8;21)(q22;q22)

SPECIFIC FEATURES

The RUNX1T1/RUNX1 (ETO/AML1) dual color, dual fusion probe is designed to detect translocations of the ETO gene locus on chromosome 8q22 with the AML1 gene locus on chromosome 21q22. The 8q22 breakpoint is within the putative zinc finger DNA binding gene ETO (RUNX1T1). The 21q22 breakpoint is within the AML1 (RUNX1) gene. The 8;21 translocation fusion product results in a novel chimeric gene. Translocation (8;21) is most commonly associated with AML (FAB subtype M2).

Normal signal pattern for chromosomes 8 and 21
nuc ish 8q22(RUNX1T1x2),21q22(RUNX1x2)[200]

Abnormal signal pattern for a t(8;21)rearrangement
nuc ish 8q22(RUNX1T1x3),21q22(RUNX1x3),(RUNX1T1 con RUNX1x2)[ ]

EXAMPLE REPORTING OF t(8;21) REARRANGEMENT – DUAL COLOR, DUAL FUSION PROBE

ISCN Result: nuc ish 8q22(RUNX1T1x3),21q22(RUNX1x3),(RUNX1T1 con RUNX1x2)[156/200]

Interpretation

Fluorescence *in situ* hybridization (FISH) analysis with the RUNX1/RUNX1T1, also known as AML1/ETO, probes showed evidence of an 8;21 translocation in 156 of 200 (78%) interphase nuclei scored.

This rearrangement is a recurrent, non-random abnormality consistent with the diagnosis of acute myeloid leukemia (AML) and is generally associated with a favorable prognosis.

Follow-up studies to monitor for this abnormality by FISH analysis in addition to standard chromosome studies are suggested.

29.1.7 Chromosome 9

Translocation (9;22)(q34;q11.2)

SPECIFIC FEATURES

The BCR/ABL dual or triple color probe is used to detect the (9;22) translocation associated with CML. The BCR/ABL dual color, dual fusion translocation probe is a mixture of the BCR probe, usually labeled in green, and the ABL probe, usually labeled in orange. The ABL probe has a genomic target of approximately 650 kb, extending from an area centromeric of the argininosuccinate synthetase gene (ASS) telomeric of the last ABL exon. The BCR probe target spans a genomic distance of about 1.5 Mb within the breakpoint cluster region.

The triple color ASS/ABL/BCR probe is used to detect deletions of the ASS gene on the derivative chromosome 9q34 locus that have been associated with the BCR/ABL translocation. The probe can be labeled with aqua or gold.

Abnormal results may occur in which chromosome 9 containing a 5' ABL deletion is present but is cryptic by chromosome analysis. Patients with a cytogenetic variant Philadelphia translocation also have been shown to have a three-fold greater frequency of an ASS gene deletion. This variant signal configuration is useful for evaluation of patients to monitor disease.

Translocation (9;22) is found in approximately 95% of CML cases. FISH analysis using an unstimulated peripheral blood sample rather than bone marrow could be considered as a less invasive way to monitor this patient's progress/remission. When monitoring treatment, the more sensitive Q-RT PCR BCR/ABL assay may provide information when the FISH result is negative.

Looking especially for polymorphonucleated (PMN) cells is important in analysis of CML cases, as most abnormal cells will appear in this cell type (Figure 29.6).

It should also be noted that in CML, most of the BCR/ABL gene rearrangements occur in the major breakpoint cluster region, between exons 12 and 16, which differs from the minor breakpoint cluster region which is generally observed in acute lymphoblastic leukemia (ALL) cases.

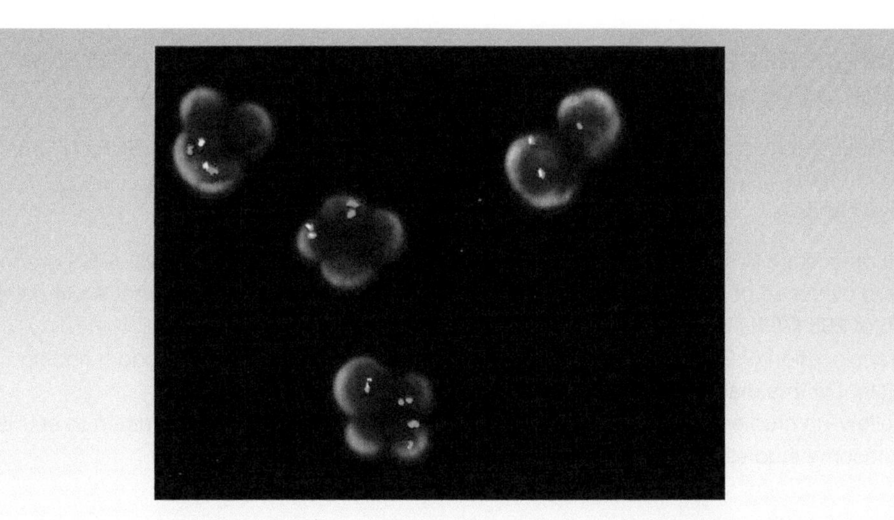

Figure 29.6 CML – polymorphonucleated cells showing two normal signals for BCR and ABL probes. Courtesy of Mehdi Jamehdor MD, Kaiser Permanente Regional Laboratory, Southern California.

Normal signal pattern for chromosomes 9 and 22 using a triple color, dual fusion probe
nuc ish 9q34(ABL1,ASS)x2,22q11.2(BCRx2)[200]

Abnormal results with a (9;22) translocation using a triple color, dual fusion probe
nuc ish 9q34(ABL1,ASS)x3,22q11(BCRx3),(ABL1,ASS con BCRx1)[ ]

Abnormal results with a (9;22) translocation and ASS gene deletion
nuc ish 9q34(ABL1x3),(ASSx1),22q11(BCRx3),(ABL1,ASS con BCRx1), 9q34(ASSx1)[ ]

EXAMPLE REPORTING OF t(9;22) – TRIPLE COLOR, DUAL FUSION PROBE

ISCN Result: nuc ish 9q34(ABL1,ASS)x3,22q11(BCRx3),(ABL1,ASS con BCRx1)[156/200]

Interpretation

Fluorescence *in situ* hybridization (FISH) analysis with the t(9;22) BCR/ABL/ASS probe set showed evidence of a t(9;22)/BCR/ABL gene rearrangement in 156 of 200 (78%) interphase nuclei scored. No evidence of a 9q34 deletion was observed.

Translocation (9;22) is a recurrent, non-random finding consistent with a diagnosis of chronic myeloid leukemia.

Follow-up studies to monitor for this abnormality by FISH analysis in addition to standard chromosome studies are suggested.

EXAMPLE REPORTING OF t(9;22) – TRIPLE COLOR, DUAL FUSION PROBE WITH 9q34 DELETION

ISCN Result: nuc ish 9q34(ABL1x3),(ASSx1),22q11(BCRx3),(ABL1,ASS con BCRx1)[156/200]

Interpretation

Fluorescence *in situ* hybridization (FISH) analysis with the t(9;22) BCR/ABL/ASS probe set showed evidence of a t(9;22)/BCR/ABL gene rearrangement, with a concurrent 9q34 deletion, in 156 of 200 (78%) interphase nuclei scored.

Translocation (9;22) and deletion of the ASS gene is a recurrent, non-random finding consistent with a diagnosis of chronic myeloid leukemia.

Follow-up studies to monitor for this abnormality by FISH analysis in addition to standard chromosome studies are suggested.

29.1.8 Chromosome 11

Rearrangements/deletion of 11q23

SPECIFIC FEATURES

The MLL dual color, breakapart rearrangement probe is designed to detect 11q23 rearrangements associated with various translocations involving the MLL gene. Translocations disrupting the MLL gene are among the most common cytogenetic abnormalities observed in hematopoietic malignancies. Although over 30 variant translocations have been reported involving MLL translocations, the most common abnormalities are t(4;11)(q21;q23), t(9;11)(p22;q23) and t(11;19)(q23;p13). In approximately 25% of 11q23 translocations, a region beginning at the MLL breakpoint and extending distally is deleted. The dual color probe provides a better indication of the presence of the 11q23 translocation than a single color probe configuration.

Rearrangements of the MLL gene are common in both *de novo* and treatment-related acute leukemias as well as MDS and biphenotypic leukemias. When observed in association with AML, rearrangements of MLL are particularly common in acute myelomonocytic/acute monocytic leukemias (FAB subtypes M4/M5). Rearrangements of the MLL gene are generally associated with a poor prognosis.

Normal signal pattern for chromosome 11
nuc ish 11q23(MLLx2)[200]

Abnormal signal pattern for an MLL rearrangement
nuc ish 11q23(MLLx2)(5'MLL sep 3'MLLx1)[ ]

EXAMPLE REPORTING OF 11q23 REARRANGEMENT – DUAL COLOR, BREAKAPART PROBE

ISCN Result: nuc ish 11q23(MLLx2)(5'MLL sep 3'MLLx1)[156/200]

Interpretation

Fluorescence *in situ* hybridization (FISH) analysis with the MLL breakapart probe showed evidence of a 11q23 rearrangement involving the MLL locus in 156 of 200 (78%) interphase nuclei scored. The partner of this rearrangement is not revealed by this FISH analysis.

This rearrangement is a recurrent, non-random abnormality consistent with the diagnosis of acute myeloid leukemia.

Follow-up studies to monitor for this abnormality by FISH analysis in addition to standard chromosome studies are suggested.

EXAMPLE REPORTING OF 11q23 DELETION – DUAL COLOR, BREAKAPART PROBE

ISCN Result: nuc ish 11q23(MLLx1)[156/200]

Interpretation

Fluorescence *in situ* hybridization (FISH) analysis with the MLL breakapart probe showed evidence of an 11q23 deletion involving the MLL locus in 156 of 200 (78%) interphase nuclei scored.

This rearrangement is a recurrent, non-random abnormality finding consistent with the diagnosis of acute myeloid leukemia.

Follow-up studies to monitor for this abnormality by FISH analysis in addition to standard chromosome studies are suggested.

Translocation (11;19)(q23;p13.1)

SPECIFIC FEATURES

The MLL-ELL uses a dual color probe to detect a fusion gene resulting from the translocation (11;19)(q23;p13.1) that is associated with *de novo* and therapy-related AML. This translocation is distinct from another translocation (11;19) with a 19p13.3 breakpoint that results in the fusion of MLL to the ENL gene. ELL was named EAF1 for ELL-associated Factor 1. Translocations involving chromosome band 11q23 are frequent in both acute lymphoblastic and acute myeloid leukemias and are the most common genetic alteration in infants with leukemia. In all age groups and all phenotypes of leukemia, an 11q23 translocation is generally associated with a poor prognosis.

Normal signal pattern for chromosomes 11 and 19
nuc ish 11q23(MLLx2),19p13.1(ELLx2)[200]

Abnormal signal pattern for an (11;19) rearrangement
nuc ish 11q23(MLLx3),19p13.1(ELLx3),(MLL con ELLx2)[ ]

EXAMPLE REPORTING OF t(11;19) (q23;p13.1) – DUAL COLOR, DUAL FUSION PROBE

ISCN Result: nuc ish 11q23(MLLx3),19p13.1(ELLx3),(MLL con ELLx2)[156/200]

Interpretation

Fluorescence *in situ* hybridization (FISH) analysis with the MLL/ELL probes showed evidence of an MLL/ELL gene rearrangement in 156 of 200 (78%) interphase nuclei scored.

Translocation (11;19)(q23;p13.1) results in the fusion of the MLL gene at 11q23 with the ELL gene at 19p13.1, and is a recurrent, non-random abnormality observed in acute myeloid leukemia (AML). AML with t(11;19)(q23;p13.1) is generally associated with a poor prognosis.

Follow-up studies to monitor for this abnormality by FISH analysis in addition to standard chromosome studies are suggested.

Translocation (11;19)(q23;p13.3)

SPECIFIC FEATURES

The MLL/ENL rearrangement uses a dual color probe to detect a fusion gene from the t(11;19)(q23;p13.3), which is one of the common chromosomal translocations in acute leukemias involving MLL rearrangements. Patients with AML and the MLL/ENL translocation carry a poor prognosis, but non-infant children with ALL and MLL/ENL fusion may have a favorable prognosis. The MLL/ENL translocation results in the generation of a fusion protein that retains the MLL N-terminus, including both an A-T hook domain and a region similar to mammalian DNA methyltransferase.

Normal signal pattern for chromosomes 11 and 19
nuc ish 11q23(MLLx2),19p13.1(ENLx2)[200]

Abnormal signal pattern for an (11;19) rearrangement
nuc ish 11q23(MLLx3),19p13.1(ENLx3),(MLL con ENLx2)[ ]

EXAMPLE REPORTING OF t(11;19)(q23;p13.3) – DUAL COLOR, DUAL FUSION PROBE

ISCN Result: nuc ish 11q23(MLLx3),19p13.3(ENLx3),(MLL con ENLx2)[156/200]

Interpretation

Fluorescence *in situ* hybridization (FISH) analysis with the MLL/ENL probes showed evidence of a MLL/ENL gene rearrangement in 156 of 200 (78%) interphase nuclei scored.

Translocation (11;19)(q23;p13.3) results in the fusion of the MLL gene at 11q23 with the ENL gene at 19p13.3 and is a recurrent, non-random abnormality observed in both acute myeloid leukemia (AML) and B-cell lymphoblastic leukemia(B-cell ALL). This rearrangement is generally associated with a poor prognosis.

Follow-up studies to monitor for this abnormality by FISH analysis in addition to standard chromosome studies are suggested.

29.1.9 Chromosome 15

Translocation (15;17)(q24;q21)

SPECIFIC FEATURES

The PML/RARA dual color translocation probe is used to detect the standard (15;17) translocation. Translocation (15;17) results in the fusion of the PML gene on chromosome 15 with the RARA gene on chromosome 17 and is associated with acute promyelocytic leukemia (APL). Patients with this translocation may benefit from all trans-retinoic receptor acid (ATRA) therapy. Variant translocations are also observed, mainly involving chromosomes 5q32, 11q13 and 11q23, which may include three-way translocations. However, all translocations result in a RARA gene rearrangement to form a novel fusion protein product.

Normal signal pattern for chromosomes 15 and 17
nuc ish 15q24(PMLx2),17q21(RARAx2)[200]

Abnormal signal pattern for an (15;17) rearrangement
nuc ish 15q24(PMLx3),17q21(RARAx3),(PML con RARAx2)[ ]

EXAMPLE REPORTING OF t(15;17) – DUAL COLOR, DUAL FUSION PROBE

ISCN Result: nuc ish 15q24(PMLx3),17q21(RARAx3),(PML con RARAx2)[156/200]

Interpretation

Fluorescence *in situ* hybridization (FISH) analysis with the t(15;17) dual color, dual fusion (PML/RARA) probe showed evidence of a PML/RARA gene rearrangement in 156 of 200 (78%) interphase nuclei scored.

Translocation (15;17) resulting in a PML/RARA gene rearrangement is a recurrent, non-random finding consistent with a diagnosis of acute promyelocytic leukemia.

Follow-up studies to monitor for this abnormality by FISH analysis in addition to standard chromosome studies are suggested.

29.1.10 Chromosome 16

Rearrangements of 16q22

SPECIFIC FEATURES

The CBFB (core binding factor beta-subunit) dual color, breakapart rearrangement probe is designed to detect inversions of chromosome 16, including inv(16)(p13q22) and the related translocation (16;16)(p13;q22). The inv(16) fuses the CBFB gene on 16q22 with the MYH11 gene on 16p13, giving rise to a chimeric protein.

The CBFB probe on chromosome 16q22 can detect inversions, translocations and deletions of the CBFB gene that are associated with acute myelomonocytic leukemia with eosinophilia (FAB type $M4_{EO}$). Abnormalities involving the CBFB gene are seen in approximately 5–10% of AML cases. Often, inv(16) is not easily detected by standard cytogenetics, especially if metaphase preparations reveal poor banding and morphology. Therefore, FISH analysis is suggested to confirm a possible abnormality when AML is suspected.

It is also possible to use a CBFB-MYH11 dual color, dual fusion probe to detect inversions and translocations of chromosome 16. The MYH11 probe is located at 16p13. Therefore, it may also be used to detect deletions of either 16p13 or 16q22 regions.

Also, it is known that AML with chromosome 16 rearrangements have a higher predisposition to FLT3 and NPM gene mutations. Therefore, molecular analysis for these genes is suggested in these cases.

Normal results
nuc ish 16q22(CBFBx2)[200]

Abnormal results
nuc ish 16q22(CBFBx2)(5'CBFB sep 3'CBFBx1)[ ]

EXAMPLE REPORTING OF 16q22 REARRANGEMENT – DUAL COLOR, BREAKAPART PROBE

ISCN Result: nuc ish 16q22(CBFBx2)(5'CBFB sep 3'CBFBx1)[156/200]

Interpretation

Fluorescence *in situ* hybridization (FISH) analysis with the CBFB breakapart probe showed evidence of a 16q22 rearrangement corresponding to either an inversion 16 or 16;16 translocation, in 156 of 200 (78%) interphase nuclei scored. The partner of this rearrangement is not revealed by this FISH analysis.

This rearrangement is a recurrent, non-random abnormality consistent with the diagnosis of acute myeloid leukemia (AML). AML with inv(16) or t(16;16) is generally associated with a favorable prognosis.

Follow-up studies to monitor for this abnormality by FISH analysis in addition to standard chromosome studies are suggested.

29.1.11 Chromosome 17

Deletion of 17p

SPECIFIC FEATURES

The TP53 probe is used to identify deletions or amplifications of the short arm of chromosome 17. The TP53 probe maps to the 17p13.1 region that contains the tumor suppressor gene. Monoallelic deletion of TP53 is common in many disorders. Loss of heterozygosity of TP53 has been identified in many tumors. Deletion of the TP53 gene is reported to have a poor prognosis in hematological malignancies and is often a secondary change associated with progression of disease.

Normal results
nuc ish 17p13(TP53x2)[200]

Abnormal results
nuc ish 17p13(TP53x1)[ ]

EXAMPLE REPORTING OF DELETION 17p

ISCN Result: nuc ish 17p13(TP53x1)[156/200]

Interpretation

Fluorescence *in situ* hybridization (FISH) analysis with the TP53 probe showed evidence of a deletion of 17p13.1 in 156 of 200 (78%) interphase nuclei scored.

Deletions of 17p are observed in both *de novo* and secondary myelodysplastic syndromes (MDS) and acute myeloid leukemia (AML). Deletions of chromosome 17p, resulting in loss of the TP53 gene, are generally associated with a poor prognosis.

Follow-up studies to monitor for this abnormality by FISH analysis in addition to standard chromosome studies are suggested.

Rearrangements of 17q21

SPECIFIC FEATURES

The RARA dual color, breakapart rearrangement is used to detect rearrangements involving the RARA gene at 17q21. Due to its breakapart design, this probe may be used to detect rearrangements of the RARA gene on chromosome 17 which may occur as a result of the (15;17) translocation as well as with other variant RARA translocations associated with APL. Patients with some RARA variant translocations may also benefit from ATRA therapy. Cytogenetic analysis is suggested to define the translocation partner in these cases.

Normal signal pattern for chromosome 17
nuc ish 17q21(RARAx2) [200]

Abnormal signal pattern for a RARA rearrangement
nuc ish 17q21(RARAx2)(5'RARA sep 3'RARAx1)[ ]

EXAMPLE REPORTING OF RARA REARRANGEMENT – DUAL COLOR, BREAKAPART PROBE

ISCN Result: nuc ish 17q21(RARAx2)(5'RARA sep 3'RARAx1)[[156/200]

Interpretation

Fluorescence *in situ* hybridization (FISH) analysis with the17q21/RARA probe showed evidence of a RARA gene rearrangement in 156 of 200 (78%) interphase nuclei scored.

RARA rearrangements involving the retinoic acid receptor gene are recurrent, non-random findings consistent with a diagnosis of acute promyelocytic leukemia.

Follow-up studies to monitor for this abnormality by FISH analysis in addition to standard chromosome studies are suggested.

29.1.12 Chromosome 20

Deletion of 20q

SPECIFIC FEATURES

The D20S108 single or dual color probe is used to identify deletions within the long arm of chromosome 20. A tumor suppressor gene is believed to reside within 20q12. Deletions of chromosome 20 are most commonly observed in myeloid malignancies, including polycythemia vera, idiopathic myelofibrosis, MDS and AML.

Normal results
nuc ish 20q12(D20S108x2)[200]

Abnormal results
nuc ish 20q12(D20S108x1)[ ]

EXAMPLE REPORTING OF DELETION 20q – SINGLE/DUAL COLOR, DELETION PROBE

ISCN Result: nuc ish 20q12(D20S108x1)[156/200]

Interpretation

Fluorescence *in situ* hybridization (FISH) analysis with the D20S108/20q12 probe showed evidence of a deletion of 20q12 in 156 of 200 (78%) interphase nuclei scored.

Chromosome 20q deletions are recurrent, non-random abnormalities observed in myeloid malignancies, including myelodysplasia, myeloproliferative disorders and acute myeloid leukemia.

Follow-up studies to monitor for this abnormality by FISH analysis in addition to standard chromosome studies are suggested.

29.1.13 Chromosome 21

Trisomy 21

SPECIFIC FEATURES

The chromosome 21 AML1/RUNX1 dual color probe can be used to detect rearrangements of the RUNX1 gene at 21q22. This probe can also be used to determine copy number changes of chromosome 21 in interphase and metaphase cells. Another probe commonly used for detection of chromosome 21 copy number changes is the centromeric chromosome 21 probe, but this cannot be used for any possible rearrangements.

Normal results
nuc ish 21q22(RUNX1x2)[200]

Abnormal results
nuc ish 21q22(RUNX1x3)[ ]

EXAMPLE REPORTING OF TRISOMY 21 – DUAL COLOR PROBE

ISCN Result: nuc ish 21q22(RUNX1x3)[156/200]

Interpretation

Fluorescence *in situ* hybridization (FISH) analysis with the chromosome 21 RUNX1 probe showed evidence of trisomy 21 in 156 of 200 (78%) interphase nuclei scored.

Trisomy 21 is a recurrent, non-random abnormality observed in myelodysplastic syndromes (MDS) and acute myeloid leukemia (AML). When it occurs as the sole cytogenetic abnormality, trisomy 21 is generally associated with a poor prognosis.

Follow-up studies to monitor for this abnormality by FISH analysis in addition to standard chromosome studies are suggested.

29.2 Biphenotypic and therapy-related abnormalities

29.2.1 Rearrangements of 12p13

EXAMPLE REPORTING OF 12p13 ABNORMALITIES – DUAL COLOR, BREAKAPART PROBE

ISCN Result: nuc ish 12p13(ETV6x2)(5'ETV6 sep 3'ETV6x1)[156/200]

Interpretation

Fluorescence *in situ* hybridization (FISH) analysis with the ETV6 probe showed an abnormal result with evidence of an ETV6/12p13 gene rearrangement in 156 of 200 (78%) interphase nuclei scored.

Rearrangements of 12p13/ETV6 are recurrent, non-random findings in a broad spectrum of hematological malignancies, including acute lymphoblastic (ALL), acute myeloid leukemia (AML), myelodysplastic syndromes (MDS), myeloproliferative neoplasms, and non-Hodgkin lymphomas. Chromosome 12p13 rearrangements are characteristic findings in secondary leukemia after prior mutagenic exposure and are generally associated with a poor prognosis.

Follow-up studies to monitor for this abnormality by FISH analysis with the ETV6 probe in addition to standard chromosome studies are suggested.

29.2.2 Combined chromosome 5 and 7 abnormalities

EXAMPLE REPORTING OF THERAPY-RELATED CHROMOSOME 5, 7 AND 11 ABNORMALITIES

ISCN Result: nuc ish 5q31(EGR1x1)[24/200],7cen(D7Z1x2),7q31(D7S486x1)[34/200], 7cen(D7Z1x1),7q31(D7S486x1)[100/200],11q23(MLLx2)(5'MLL sep 3'MLLx1)[50/200]

Interpretation

Fluorescence *in situ* hybridization (FISH) analysis with the EGR1/5q31, CEP7, 7q31 and MLL/11q23 probes showed an abnormal result with evidence of a deletion of 5q31, a deletion of 7q31, monosomy 7 and an 11q23 rearrangement. Del(5q31), del(7q31), monosomy 7 and 11q23 rearrangements are recurrent, non-random abnormalities observed in therapy-related myeloid malignancies.

Monosomy 7 and del(7q) are generally associated with a poor prognosis. Rearrangements involving MLL are often observed in myelodysplasias following treatment with anti-topoisomerase II (epipodophyllotoxins) and are also associated with a poor prognosis.

Follow-up studies to monitor for these abnormalities by FISH analysis in addition to standard chromosome studies are suggested.

29.3 Panels of probes

The most common FISH abnormalities associated with hematological malignancies may be identified by FISH analysis as a panel of tests. MDS, AML, chronic lymphocytic leukemia (CLL), eosinophilia, myeloproliferative neoplasms (MPNs), multiple myeloma (MM) and non-Hodgkin lymphoma (NHL) are all examples where FISH panels of probes are useful in identifying common cytogenetic anomalies in a single test. Combining the most common abnormalities detectable by FISH analysis as a panel of tests is effective in terms of cost and laboratory time, and improves patient care by the identification of all possible known chromosomal abnormalities in a fast and relatively easy approach. Often these abnormalities are seen in conjunction with one another, rather than as a sole anomaly; therefore, this section provides examples of more than one abnormality per panel.

There are various ways to write normal and abnormal results in a report, and the following is one approach in which normal and abnormal probes are listed in chromosome order (for a different approach in which normal and abnormal probes are separated from each other, see Section 30.2 for panels of lymphoid disorders). Example slide preparations for panels of probes is seen in Figure 29.7. For a list of common FISH probes for each panel for hematological malignancies, see Table 29.1. Panels of lymphoid probes will be described in Chapter 30.

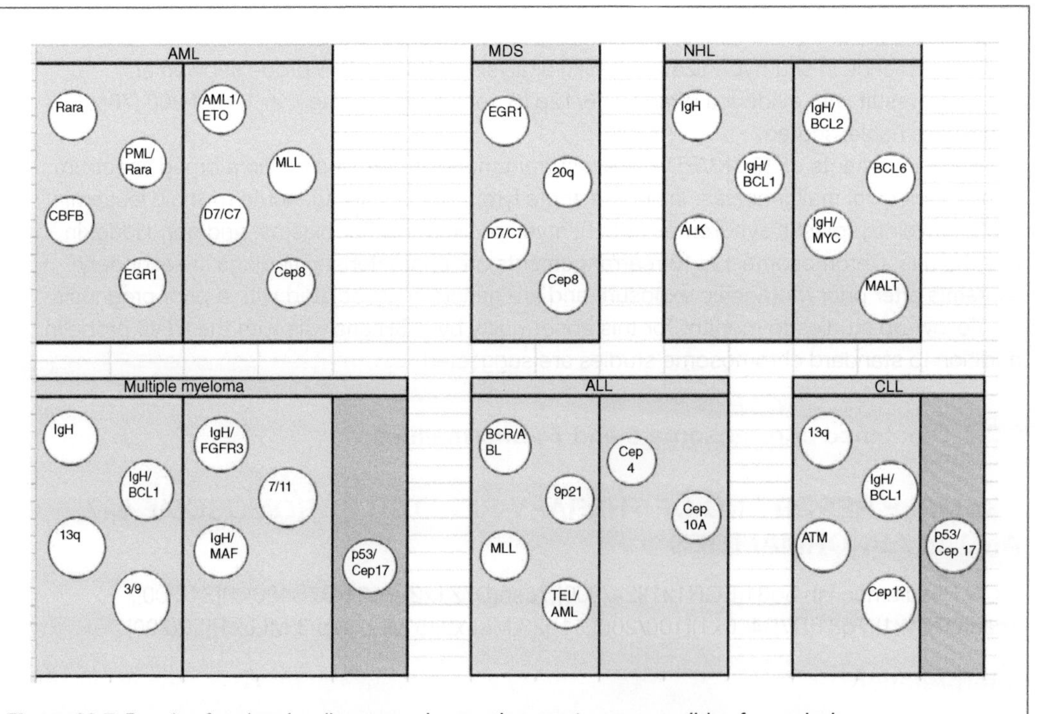

Figure 29.7 Panels of probes by disease and example ways to prepare slides for analysis.

Table 29.1 Common FISH panels of probes for hematological disorders

MDS	AML	MPN	ALL	CLL	NHL – B-cell	MM
5q deletion EGR1/D5S23	5q deletion EGR1	4q12/CHIC2/ FIP1L1/PDGFRa deletion	t(9;22) BCR/ABL translocation	11q22 ATM deletion	t(11;14) IgH/BCL1 translocation	t(4;14) IgH/FGFR3 translocation
7q deletion D7S522/CEP 7	7q deletion D7S522	5q33 PDGFRb rearrangements	11q23 MLL rearrangement	Centromere 12 CEP12	t(8;14) IgH/MYC translocation	t(11;14) IgH/BCL1 translocation
Centromere 8 D8Z2	Centromere 8 D8Z2	8p12/FGFR1 rearrangement	t(12;21) ETV6/RUNX1 (TEL/AML1) translocation	13q DLUE1/DLUE2 (D13S319/LAMP1) deletion	t(14;18) IgH/BCL2 translocation	14q32 IGH rearrangement
20q deletion D20S108	t(8;21) RUNX1/RUNX1T (AML/ ETO) translocation	Centromes 8/9 trisomies	9p21 P16/CEP 9 deletion	17p13.1 TP53 deletion	14q32 IGH rearrangement	13q/ DLUE1/DLUE2 (D13S319/LAMP1) deletion
11q23 MLL rearrangement	11q23 MLL rearrangement	BCR/ABL/t(9;22) rearrangement	Centromeres 4/10/17 trisomies	t(11;14) IgH/BCL1 translocation	3q27 BCL6 rearrangement	17p13.1 TP53 deletion
17p13.1 TP53 deletion	16q22 CBFB rearrangement	13q DLUE1/DLUE2 (D13S319/ LAMP1) deletion	t(8;14) MYC/IGH/CEP8 rearrangement	14q32 IGH rearrangement	18q21 MALT1 rearrangement	Centromeres 3/7/9/11 hyperdiploidy
	t(15;17) PML/RARA translocation	20q deletion D20S108	t(1;19) PBX1/TCF3 (E2A-PBX1) rearrangement	6q deletion MYB rearrangement	2q23 ALK rearrangement	t(14;16) IgH/MAF1 translocation
	17q22 RARA rearrangement	5q deletion EGR1	t(4;11) MLL/AF4 rearrangement		t(8;14) MYC/IGH rearrangement	1q21 CKS1B gain
	D20S108 20q deletion	7q deletion D7S522				
	t(9;22) BCR/ABL/ASS rearrangement					

ALL, acute lymphoblastic leukemia; AML, acute myeloid leukemia; CLL, chronic lymphocytic leukemia; MDS, myelodysplastic disorder; MM, multiple myeloma; MPN, myeloproliferative neoplasm; NHL, non-Hodgkin lymphoma.

29.3.1 MDS panel of probes

The four most common abnormalities associated with MDS may be identified by either cytogenetic or FISH analyses and include chromosome 5 and 7 for monosomies and deletions of the long arm, trisomy 8 and deletions of the long arm of chromosome 20.

EXAMPLE REPORTING OF NORMAL RESULTS

ISCN Result: nuc ish 5q31(EGR1x2),7cen(D7Z1x2),7q31(D7S486x2),8cen(CEP8x2), 20q12(D20S108x2)[200]

Interpretation

Fluorescence *in situ* hybridization (FISH) analysis was performed with the MDS panel of probes: EGR1, D7Z1, D7S486, CEP8 and D20S108. This analysis showed normal results with no evidence of deletion 5q31, monosomy 5 or 7, deletion 7q31, trisomy 8 or deletion 20q12 in 200 cells scored for each probe.

EXAMPLE REPORTING OF ABNORMAL RESULTS

ISCN Result: nuc ish 5q31(EGR1x2)[200],7cen(D7Z1x2),7q31(D7S486x1)[156],8cen(CEP8x3)[34], 20q12(D20S108x2)[200]

Interpretation

Fluorescence *in situ* hybridization (FISH) analysis was performed with the MDS panel of probes: EGR1, D7Z1, D7S486, CEP8 and D20S108. This analysis showed evidence of deletion 7q31 in 78% and trisomy 8 in 17% of interphase nuclei scored, respectively.

The remaining probes showed normal results in 200 interphase nuclei scored for each probe.

Chromosome 7q deletion and trisomy 8 are recurrent, non-random abnormalities observed in myeloid malignancies, including myelodysplasia, myeloproliferative disorders and acute myeloid leukemia.

Follow-up studies to monitor for these abnormalities by FISH analysis in addition to standard chromosome studies are suggested.

29.3.2 AML panel of probes

The most common abnormalities associated with AML detectable by FISH analysis tend to be translocations plus the common abnormalities described in the MDS panel. The translocation abnormalities include t(8;21), MLL/11q23 rearrangements, t(15;17), inv(16) and t(16;16).

EXAMPLE REPORTING OF NORMAL RESULTS

ISCN Result: nuc ish 8q22(RUNX1T1x2),21q22(RUNX1x2),11q23(MLLx2),15q22(PMLx2), 17q21.1(RARAx2),16q22(CBFBx2)[200]

Interpretation

Fluorescence *in situ* hybridization (FISH) analysis with the RUNX/RUNX1T1, MLL, CBFB and PML/RARA probes showed normal results with no evidence of translocation (8;21), translocation (15;17) or rearrangements involving the MLL or CBFB loci in 200 cells scored for each probe.

EXAMPLE REPORTING OF ABNORMAL RESULTS (ANY POSSIBLE ABNORMALITY)

ISCN Result: nuc ish 8q22(RUNX1T1x2),21q22(RUNX1x2)(RUNX1T1 con RUNX1x1)[_], 11q23(MLLx2)(5'MLL sep 3'MLLx1)[_],15q22(PMLx3),17q21.1(RARAx3)(PML con RARAx2), 16q22(CBFBx2)(5'CBFB sep 3'CBFBx1)[_]

Interpretation

Fluorescence *in situ* hybridization (FISH) analysis with the RUNX1/RUNX1T1, MLL, CBFB and PML/RARA probes showed evidence of t(8;21), rearrangements involving the MLL, CBFB locus or t(15;17) in interphase nuclei scored.

FISH analysis with the remaining probes showed no evidence of abnormalities involving these loci in 200 interphase nuclei scored for each probe.

These abnormal findings are consistent with a diagnosis of acute myeloid leukemia.

Follow-up studies to monitor for these abnormalities by FISH analysis in addition to standard chromosome studies are suggested.

29.3.3 MPN panel of probes

Myeloproliferative neoplasm abnormalities that are detectable by FISH analysis include the FIP1L1, CHIC2, PDGFRA gene deletions and the FGFR1 rearrangements. Also included is the t(9;22) that is common in MPNs and CML in particular.

EXAMPLE REPORTING OF NORMAL RESULTS

ISCN Result: nuc ish 4q12(FIP1L1,CHIC2,PDGFRA)x2, 5q33(PDGFRBx2), 8p12(FGFR1x2), 9q34(ABL1x2),22q11.2(BCRx2)

Interpretation

Fluorescence *in situ* hybridization (FISH) analysis with the MPD panel of probes, PDGFRA, PDGFRB, FGFR1 and BCR/ABL1, showed normal results with no evidence of a deletion of the CHIC2 locus, which would result in a fusion of the PDGFRA and FIP1L1 loci at 4q12, a rearrangement of the PDGFRB locus at 5q33, a rearrangement of the FGFR1 locus at 8p12, or a 9;22 translocation in 200 cells scored for each probe.

EXAMPLE REPORTING OF ABNORMAL RESULTS

ISCN Result: nuc ish 4q12(CHIC2x1),(FIP1L1x2),(PDGFRAx2) (FIP1L1 con PDGFRAx1) [ ], 5q33(PDGFRBx2)(5'PDGFRB sep 3'PDGFRBx1) [ ], 8p12(FGFR1x2)(5'FGFR1 sep 3'FGFR1x1) [ ], 9q34(ABL1x3),22q11.2(BCRx3)(ABL1 con BCRx2) [ ]

Interpretation

Fluorescence *in situ* hybridization (FISH) analysis with the MPD panel of probes, PDGFRA, PDGFRB, FGFR1 and BCR/ABL1, showed abnormal results with evidence of a deletion of the CHIC2 locus resulting in a fusion of PDGFRA and FIP1L1 loci at 4q12, a rearrangement of the PDGFRB locus at 5q33, a rearrangement of the FGFR1 locus at 8p12 and a t(9;22) in interphase nuclei scored.

FISH analysis with the remaining MPD panel of probes, PDGFRA, PDGFRB, FGFR1 and BCR/ABL1 showed normal results with no evidence of a rearrangement, deletion of the PDGFR locus at 4q12, a rearrangement of the PDGFRB locus at 5q33, a rearrangement of the FGFR1 locus at 8p12, or a t(9;22) in 200 interphase nuclei scored for each probe.

This finding is consistent with a specific group of myeloproliferative and lymphoid neoplasms, which result from formation of a fusion gene encoding an aberrant tyrosine kinase with rearrangements involving either PDGFRA, PDGFRB or FGFR1.

In PDGFRA-related disease, cases often present as chronic eosinophilic leukemia/ hypereosinophilic syndrome (CEL/HES) but can also present as acute myeloid leukemia or precursor T-cell lymphoblastic lymphoma (T-cell LBL).

Follow-up monitoring for these abnormalities by FISH using the PDGFR alpha probe is suggested as this abnormality is not visible by standard chromosome analysis.

Bibliography

Ageberg M, Drott K, Olofsson T, et al. Identification of a novel and myeloid specific role of the leukemia-associated fusion protein DEK-NUP214 leading to increased protein synthesis. Genes Chromosomes Cancer 2008; 47(4): 276–287.

American College of Medical Genetics. *Guidelines in Clinical Cytogenetics. Section E10 Methods in Fluorescence In situ Hybridization – Interphase/Nuclear Fluorescence In Situ Hybridization*. American College of Medical Genetics, Bethesda, MD.www.acmg.net/StaticContent/SGs/Section_E.html, accessed 28 January 2014.

Atlas of Genetics and Cytogenetics in Oncology and Haematology. http://AtlasGeneticsOncology.org/Anomalies, accessed 28 January 2014.

Chase A, Grand F, Zhang JG, et al. Factors influencing the false positive and negative rates of BCR-ABL fluorescence in situ hybridization. Genes Chromosomes Cancer 1997; 18: 246–253.

Dewald GW, Schad CR, Christensen ER, et al. The application of fluorescent in situ hybridization to detect Mbcr/abl fusion in variant Ph chromosomes in CML and ALL. Cancer Genet Cytogenet 1993; 71: 7–14.

Gersen S, Keagle M (eds). *Principles of Clinical Cytogenetics*. Humana Press, Totowa, New Jersey, 1999.

Grimwade D, Hills RK, Moorman AV, et al. Refinement of cytogenetics classification in acute myeloid leukemia: determination of prognostic significance or rare recurring chromosomal abnormalities among 5879 younger adult patients treated in the United Kingdom Research Council trials. Blood 2010; 116(3): 354–365.

Heim S, Mitelman F. *Cancer Cytogenetics*, 2nd edn. Wiley-Liss, New York, 1995.

Kearney L, Watkins PC, Young BD, et al. DNA sequences of chromosome 21-specific YAC detect the t(8;21) breakpoint of acute myelogenous leukemia. Cancer Genet Cytogenet 1991; 57: 109–119.

Langstrom AP, Tefferi A. Fluorescent in situ hybridization in the diagnosis, prognosis, and treatment monitoring of chronic myeloid leukemia. Leuk Lymphoma 2006; 47: 397–402.

McKenna RW. Acute myeloid leukemia. In: Kjeldsberg C (ed) *Practical Diagnosis of Hematologic Disorders*, 4th edn. ASCP Press, Chicago, 2006, pp. 457–498.

Morel F, Ka C, Le Bris MJ, et al. Deletion of the 5'ABL region in Philadelphia chromosome positive chronic myeloid leukemia: frequency, origin and prognosis. Leuk Lymphoma 2003; 44(8): 1333–1338.

Rooney DE, Czepulkowski BH. *Human Cytogenetics. A Practical Approach*. Oxford University Press, New York, 1992.

Rowley JD, Diaz MO, Espinosa R III, et al. Mapping chromosome band 11q23 in human acute leukemia with biotinylated probes: identification of 11q23 translocation breakpoints with a yeast artificial chromosome. Proc Natl Acad Sci USA 1990; 87: 9358–9362.

Shaffer LG, McGowan-Jordan J, Schmid M (eds). *ISCN 2013: An International System for Human Cytogenetic Nomenclature*. Karger Publishers, Unionville, CT, 2013.

Shurtleff SA, Buijs A, Behm FG, et al. TEL/AML fusion resulting from a cryptic t(12;21) is the most common genetic lesion in pediatric ALL and defines a subgroup of patients with an excellent prognosis. Leukemia 1995; 9: 1985–1989.

Swerdlow SH, Campo E, Harris NL, et al. (eds). *WHO Classification of Tumours of Haematopoietic and Lymphoid Tissues*. IARC Press, Lyon, 2008.

Vance GH. Detection of recurrent cytogenetic abnormalities in acute lymphoblastic and myeloid leukemias using fluorescence in situ hybridization. Methods Mol Biol 2013; 999: 79–91.

Wordsworth S, Papanicolas I, Buchanan J, et al. Molecular testing for somatic cancer mutations: a survey of current and future testing in UK laboratories. J Clin Pathol 2008; 61: 373–376.

Commercial FISH probes are available at these websites

www.AbbottMolecular.com

www.Kreatech.com

www.Rainbowscientific.com

www.empiregenomics.com

www.abnova.com

www.cytocell.com

www.exiqon.com

www.cambridgebluegnome.com

CHAPTER 30

Recurrent FISH abnormalities in lymphoid disorders

30.1 Individual abnormalities in lymphoid disorders by chromosome order

30.1.1 Chromosome 1

Rearrangements of 1p

SPECIFIC FEATURES

Chromosome 1 abnormalities are commonly observed in multiple myeloma (MM), especially amplification of the CKS1B gene at the 1q21 region. Abnormalities involving the CKS1B gene are generally associated with a poor prognosis and disease progression. Reports have shown that 1q21 gain is related to poor survival, while 1q21 amplification is related to disease relapse. Often 1q21 abnormalities are seen in conjunction with IGH translocations, especially t(4;14) and del(13q), which are also associated with a poor prognosis in MM. Chromosome 1p32-36 deletions have also been reported in cases of MM involving the p18/CDKN2C gene at 1p32.3. This gene is often built into the probe mixture with CKS1B as a dual color probe.

Normal results

nuc ish 1p32.3(CDKN2Cx2),1q21(CKS1Bx2)[200]

Abnormal results

nuc ish 1p32.3(CDKN2Cx1),1q21(CKS1Bx3)[ ]

Cytogenetic Abnormalities: Chromosomal, FISH and Microarray-Based Clinical Reporting, First Edition. Susan Mahler Zneimer.

EXAMPLE REPORTING OF 1q21/1p32.3 REARRANGEMENTS – DUAL COLOR PROBE

ISCN Result: nuc ish 1p32.3(CDKN2Cx1),1q21(CKS1Bx3)[156/200]

Interpretation

Fluorescence *in situ* hybridization (FISH) analysis with the CKS1B/CDKN2C probes showed an abnormal result with evidence of loss of the 1p32.3 region including the CDKN2C gene, and the gain of 1q21 region including the CKS1B gene in 156 of 200 (78%) interphase nuclei scored.

Chromosome 1q21 gain and 1p32.3 loss are recurrent, non-random abnormalities detected in multiple myeloma, and are generally associated with a poor prognosis and disease progression.

Follow-up studies to monitor for these abnormalities by FISH analysis in addition to standard chromosome studies are suggested.

30.1.2 Chromosome 2

Translocation (2;5)(p23;q35)

SPECIFIC FEATURES

The ALK (anaplastic lymphoma kinase) dual color, rearrangement probe is designed to detect the known 2p23 rearrangements that occur in translocation (2;5) and their variants. The (2;5) translocation fuses the nucleophosmin (NPM) gene located on chromosome 5q35 with the ALK gene on chromosome 2p23.2. This NPM/ALK gene fusion gives rise to a chimeric protein that is overexpressed. The ALK probe contains two differently labeled fluorochromes on opposite sides of the breakpoint of the ALK gene.

Rearrangements of the ALK gene are most commonly associated with anaplastic large cell lymphoma (ALCL) and are found in approximately 50–60% of patients with this disease. Translocation (2;5) is seen in approximately 70–80% of these patients, with the remainder showing a variant ALK rearrangement. Rearrangements of the ALK gene can also be observed in inflammatory myelofibroblastic tumors.

Normal results
nuc ish 2p23(ALKx2),5q35(NPMx2)[200]

Abnormal results
nuc ish 2p23(ALKx3),5q35(NPMx3),(ALK con NPMx2)[ ]

EXAMPLE REPORTING OF t(2;5) – DUAL COLOR, DUAL FUSION PROBE

ISCN Result: nuc ish 2p23(ALKx3),5q35(NPMx3),(ALK con NPMx2)[156/200]

Interpretation

Fluorescence *in situ* hybridization (FISH) analysis with the ALK/NPM probes showed an abnormal result with evidence of an ALK/NPM gene rearrangement in 156 of 200 (78%) interphase nuclei scored.

Translocation (2;5)(p23;q35) results in the fusion of the ALK gene at 2p23 with the NPM (nucleophosmin) gene at 5q35 and is observed in 50–60% of cases of anaplastic large cell lymphoma (ALCL).

Follow-up studies to monitor for this abnormality by FISH analysis in addition to standard chromosome studies are suggested.

30.1.3 Chromosome 3

Rearrangements of 3q27

SPECIFIC FEATURES

The BCL6 gene rearrangement dual color, breakapart probe is used to detect rearrangements of the BCL6 gene on chromosome 3 at band 3q27. It is also used in detecting chromosome breaks associated with a number of different translocations that involve the BCL6 gene.The BCL6 probe is a mixture of a labeled fluorochrome on the 5'BCL6 region and a differently labeled fluorochrome on the 3'BCL6 region.

Rearrangements of the BCL6 gene on chromosome 3q27 are most commonly observed in B-cell non-Hodgkin lymphomas, including diffuse large cell lymphoma in 20–30%, follicular center lymphoma in approximately 15%, and marginal zone lymphoma in <5% of patients, respectively.

Normal results
nuc ish 3q27(BCL6x2)[200]

Abnormal results
nuc ish 3q27(BCL6x2)(5'BCL6 sep 3'BCL6x1)[ ]

EXAMPLE REPORTING OF 3q27 – DUAL COLOR BREAKAPART PROBE

ISCN Result: nuc ish 3q27(BCL6x2)(5'BCL6 sep 3'BCL6x1) [156/200]

Interpretation

Fluorescence *in situ* hybridization (FISH) analysis with the BCL6 probe showed an abnormal result with evidence of a rearrangement of the BCL6 gene at 3q27 in 156 of 200 (78%) interphase nuclei scored. The chromosomal partner cannot be identified by this assay.

Rearrangements of the BCL6 gene on chromosome 3q27 are most commonly observed in B-cell non-Hodgkin lymphomas, including diffuse large cell lymphoma in 20–30%, follicular center lymphoma in approximately 15%, and marginal zone lymphoma in <5% of patients, respectively.

Follow-up studies to monitor for this abnormality by FISH analysis in addition to standard chromosome studies are suggested.

30.1.4 Chromosome 4

Translocation (4;11)(q21;q23)

SPECIFIC FEATURES

The AFF1-MLL dual color probe corresponding to translocation (4;11) is used to detect rearrangements which are seen in approximately 85% of infant acute lymphoblastic leukemia (ALL) patients. Although there are numerous MLL gene fusion partner chromosomes, the most common is the t(4;11) rearrangement, fusing the MLL gene on chromosome 11q23 with the AFF1 gene on chromosome 4q21. As in all MLL rearrangements, this translocation is generally associated with a poor prognosis.

Normal results
nuc ish 4q21(AFF4x2),11q23(MLLx2)[200]

Abnormal results
nuc ish 4q21(AFF4x3),11q23(MLLx3)(AFF4 con MLLx2)[ ]

EXAMPLE REPORTING OF t(4;11) – DUAL COLOR, DUAL FUSION PROBE

ISCN Result: nuc ish 4q21(AFF4x3),11q23(MLLx3)(AFF4 con MLLx2)[156/200]

Interpretation

Fluorescence *in situ* hybridization (FISH) analysis with the MLL/AFF4 probes showed an abnormal result with evidence of an MLL/AFF4 gene rearrangement in 156 of 200 (78%) interphase nuclei scored.

Translocation (4;11)(q21;q23) results in the fusion of the MLL gene at 11q23 with the AFF4 gene at 4q21, and is a recurrent, non-random finding in B-cell acute lymphoblastic leukemia and lymphoblastic lymphoma (B-cell ALL/LBL). MLL rearrangements are generally associated with a poor prognosis. Infants with MLL translocations, especially those <6 months of age, have a particularly poor prognosis.

Follow-up studies to monitor for this abnormality by FISH analysis in addition to standard chromosome studies are suggested.

Translocation (4;14)(p16.3;q32)

SPECIFIC FEATURES

The IGH/FGFR3 dual color, dual fusion translocation probe is designed to detect translocations of the immunoglobulin heavy chain (IGH) locus and the FGFR3 gene. The IGH probe set generally contains two labeled regions that hybridize to sequences on either side of the IGH J-region breakpoint associated with t(4;14)(p16.3;q32).

Translocation (4;14) is a common IGH rearrangement observed in multiple myeloma. The region of chromosome 14q32 in this translocation generally involves the MMSET gene region (Multiple Myeloma SET domain), which results in a fusion gene product between MMSET and the IGH locus and overexpression of the FGFR3 gene.

When an IGH gene rearrangement is seen that does not include the FGFR3 gene, a different signal configuration is observed. This result shows two signals of FGFR3 and three signals of IGH, since the IGH gene is interrupted by a partner chromosome and so is split in two, giving rise to an extra IGH signal (located on an unknown chromosome). Since it is not the FGFR3 region, there is no fusion probe. With an extra IGH signal, it appears to be trisomy 14; however, the vast majority of cases are actually IGH gene rearrangements and not gain of IGH.

Normal results
nuc ish 4p16.3(FGFR3x2),14q32(IGHx2)[200]

Abnormal results
nuc ish 4p16.3(FGFR3x3),14q32(IGHx3),(FGFR3 con IGHx2) [ ]

Abnormal results of an IGH gene rearrangement – without FGFR3
nuc ish 4p16.3(FGFR3x2),14q32(IGHx3)[ ]

EXAMPLE REPORTING OF t(4;14) – DUAL COLOR, DUAL FUSION PROBE

ISCN Result: nuc ish 4p16.3(FGFR3x3),14q32(IGHx3),(FGFR3 con IGHx2) [156/200]

Interpretation

Fluorescence *in situ* hybridization (FISH) analysis showed an abnormal result with evidence of an IGH/FGFR3 gene rearrangement in 156 of 200 (78%) interphase nuclei scored.

Translocation (4;14)(p16.3;q32) results in the fusion of the IGH gene at 14q32 with the FGFR3 gene at 4p16.3. This rearrangement is a recurrent, non-random rearrangement observed in lymphoproliferative disorders, especially plasma cell dyscrasias, where it is detected in up to 10% of abnormal cases.

Follow-up studies to monitor for this abnormality by FISH analysis in addition to standard chromosome studies are suggested.

30.1.5 Chromosome 6

Deletions of 6q

SPECIFIC FEATURES

The MYB dual color probe is used to detect deletions of the long arm of chromosome 6. Chromosome 6q deletions are recurrent abnormalities in lymphoproliferative disorders, most commonly in chronic lymphocytic leukemia (CLL). Breakpoints of the deletions vary from bands q13-q23. A chromosome 6 centromere probe is often used as a second color control probe. Deletions of chromosome 6q are also seen in myeloid and mixed-lineage cells.

Normal results
nuc ish 6cen(D6Z1x2),6q23.3(MYBx2) 200]

Abnormal results
nuc ish 6cen(D6Z1x2),6q23.3(MYBx1)[ ]

EXAMPLE REPORTING OF 6q – DUAL COLOR, DELETION PROBE

ISCN Result: nuc ish 6cen(D6Z1x2),6q23.3(MYBx1)[156/200]

Interpretation

Fluorescence *in situ* hybridization (FISH) analysis with the MYB probe showed an abnormal result with evidence of a deletion of the MYB gene at 6q23.3 in 156 of 200 (78%) interphase nuclei scored.

Deletions of chromosome 6q are recurrent, non-random findings in pre-B-cell acute lymphoblastic leukemia and lymphoproliferative disorders, including chronic lymphocytic leukemia. They are also seen in myeloid and mixed-lineage cells.

Follow-up studies to monitor for this abnormality by FISH analysis in addition to standard chromosome studies are suggested.

30.1.6 Chromosome 8

Rearrangements of 8q24

SPECIFIC FEATURES

The MYC dual color, breakapart rearrangement probe is a mixture of two probes that hybridize to opposite sides of the MYC gene region on chromosome 8q24, which allows for the detection of the vast majority of breakpoints including t(8;22)(q24;q11), t(2;8)(p12;q24) and t(8;14)(q24;q32). Rearrangements of the MYC gene on chromosome 8q24 are generally seen in non-Hodgkin lymphomas. Rearrangements of the MYC gene and MYC amplification are associated with dysregulation of MYC expression and have been reported in association with disease progression in multiple myeloma. Acute lymphoblastic leukemia with rearrangement of the MYC gene is almost always of L3 morphology.

Normal results
nuc ish 8q24(MYCx2) [200]

Abnormal results
nuc ish 8q24(MYCx2)(5'MYC sep 3'MYCx1) [ ]

EXAMPLE REPORTING OF 8q24 – DUAL COLOR, BREAKAPART PROBE

ISCN Result: nuc ish 8q24(MYCx2)(5'MYC sep 3'MYCx1) [156/200]

Interpretation

Fluorescence *in situ* hybridization (FISH) analysis with the MYC probe showed an abnormal result with evidence of a rearrangement of the MYC gene at 8q24 in 156 of 200 (78%) interphase nuclei scored. The chromosomal partner cannot be identified by this assay.

Translocations involving the MYC gene are recurrent, non-random findings in pre-B-cell acute lymphoblastic leukemia and lymphoproliferative disorders, including non-Hodgkin lymphomas and Burkitt lymphoma in particular.

Follow-up studies to monitor for this abnormality by FISH analysis in addition to standard chromosome studies are suggested.

Translocation (8;14)(q24;q32)

SPECIFIC FEATURES

The IGH/MYC, CEP 8 dual or triple color, dual fusion translocation probe is designed to detect the juxtaposition of IGH locus and MYC gene region sequences. Inclusion of a third fluorochrome for CEP 8 in this probe mixture serves as a control for chromosome 8 for tricolor probes. Control for chromosome 8 is important when MYC amplification and loss of the der(8) chromosome occur.

The translocation t(8;14) (q24;q32) involving IGH at 14q32 and the MYC region at 8q24 is the most frequently observed MYC translocation. The remaining cases have one of the variant MYC translocations, either t(8;22)(q24;q11) or t(2;8)(p12;q24), that involve one of the two light chain immunoglobulin loci (kappa on chromosome 2 or lambda on 22). These three translocations are believed to disrupt the normal regulation of the MYC transcription factor by bringing the MYC gene under control of a regulatory element from one of the immunoglobulin loci.

Translocation (8;14) is associated with a broad spectrum of B-cell lymphomas. It is particularly common in Burkitt lymphoma. Rearrangements of the MYC gene and MYC amplification are also associated with dysregulation of MYC expression and have been reported in association with disease progression in multiple myeloma. Acute lymphoblastic leukemia with rearrangement of the MYC gene is almost always of L3 morphology.

When an IGH gene rearrangement is seen that does not include the MYC gene, a different signal configuration is observed. This result shows two signals of MYC and three signals of IGH, since the IGH gene is interrupted by a partner chromosome and so is split in two, giving rise to an extra IGH signal (located on an unknown chromosome). Since it is not the MYC region, there is no fusion probe. With an extra IGH signal, it appears to be trisomy 14; however, the vast majority of cases are actually an IGH gene rearrangement and not gain of IGH.

Normal results
nuc ish 8q24(MYCx3),14q32(IGHx2)[200]

Abnormal results
nuc ish 8q24(MYCx3),14q32(IGHx3),(MYC con IGHx2)[ ]

Abnormal results of an IGH gene rearrangement – without MYC
nuc ish 8q24(MYCx2),14q32(IGHx3)[ ]

EXAMPLE REPORTING OF t(8;14) – DUAL COLOR, DUAL FUSION PROBE

ISCN Result: nuc ish 8q24(MYCx3),14q32(IGHx3)(MYC con IGHx2)[156/200]

Interpretation

Fluorescence *in situ* hybridization (FISH) analysis with the MYC/IGH probes showed an abnormal result with evidence of an MYC/IGH gene rearrangement in 156 of 200 (78%) interphase nuclei scored.

Translocation (8;14) results in the fusion of the MYC gene at 8q24 with the immunoglobulin heavy chain gene (IGH) at 14q32 and is a recurrent, non-random abnormality observed in Burkitt lymphoma, but may also be seen in diffuse large B-cell lymphoma and in transformed lower grade lymphomas.

Follow-up studies to monitor for this abnormality by FISH analysis in addition to standard chromosome studies are suggested.

30.1.7 Chromosome 9

Del(9)(p21)

SPECIFIC FEATURES

The p16/CEP 9 dual color probe is designed to detect 9p21 deletions. The p16/CDKN2A and p15/CDNK2B tumor suppressor genes have been found to lie within the 9p21 region, which is commonly deleted in lymphoid disorders. Deletions of the short arm of chromosome 9 are observed in approximately 10% of both childhood and adult acute lymphoblastic leukemias. Deletion of 9p has been found to be an adverse prognostic risk factor in pediatric B-cell lineage but not in adult or T-cell ALL.

Normal results
nuc ish 9cen(D9Z1x2),9p21(CDKN2A,CDKN2B)x2[200]

Abnormal results with heterozygous loss of 9p21
nuc ish 9cen(D9Z1x2),9p21(CDKN2A,CDKN2B)x1[ ]

Abnormal results with homozygous loss of 9p21
nuc ish 9cen(D9Z1x2),9p21(CDKN2A,CDKN2B)x0[ ]

EXAMPLE REPORTING OF 9p21 – DUAL COLOR, DELETION PROBE

ISCN Result: nuc ish 9cen(D9Z1x2),9p21(CDKN2A,CDKN2B)x1 [156/200]

Interpretation

Fluorescence *in situ* hybridization (FISH) analysis with the CDKN2A/B probes showed an abnormal result with evidence of a deletion of the CDKN2A/B genes at 9p21 in 156 of 200 (78%) interphase nuclei scored.

Deletions of 9p21 result in the loss of the CDKN2A and CDKN2B genes, and are recurrent, non-random abnormalities observed in lymphoblastic leukemia and lymphoblastic lymphoma (ALL/LBL). Deletions of the CDKN2A gene are associated with a poor prognosis in pediatric B-cell lineage but not in adult or T-cell lineage ALL.

Follow-up studies to monitor for this abnormality by FISH analysis in addition to standard chromosome studies are suggested.

Heterozygous and homozygous loss of 9p21

EXAMPLE REPORTING OF 9p21 – DUAL COLOR, DELETION PROBE

ISCN Result: nuc ish 9cen(D9Z1x2),9p21(CDKN2A,CDKN2B)x1[25/200],9cen(D9Z1x2),9p21(CDKN2A,CDKN2B)x0[50/200]

Interpretation

Fluorescence *in situ* hybridization (FISH) analysis with the CDKN2A/B probes showed an abnormal result with evidence of a heterozygous deletion of the CDKN2A/B genes at 9p21 in 25 of 200 (12.5%) interphase nuclei scored, and a homozygous deletion of the CDKN2A/B genes at 9p21 in 25% interphase nuclei scored.

Deletions of 9p21 result in the loss of the CDKN2A and B genes and are recurrent, non-random abnormalities observed in lymphoblastic leukemia and lymphoblastic lymphoma (ALL/LBL). Deletions of the CDKN2A gene are associated with a poor prognosis in pediatric B-cell lineage, but not in adult or T-cell ALL.

Follow-up studies to monitor for these abnormalities by FISH analysis in addition to standard chromosome studies are suggested.

Translocation (9;22)(q34;q11.2)

SPECIFIC FEATURES

This dual color, dual fusion probe hybridizes to the BCR and ABL1 regions on chromosome 22q11 and 9q34, respectively, and is used to detect the (9;22) translocation associated with chronic myeloid leukemia (CML), ALL and acute myeloid leukemia (AML). The BCR/ABL1 probe mix varies per manufacturer in its signal configuration. The example below uses a dual fusion mix; however, single fusion with or without an extra signal may be used, as well as a triple color with the ASS1 gene on chromosome 9q34.

Translocation (9;22) is found in approximately 15–20% of ALL cases. It is the most common cytogenetic abnormality observed in ALL and is most often associated with L2 morphology and sometimes L1, but generally not L3. Since a deletion of the ASS1 gene on 9q34 may also be present, the tricolor, dual fusion probe may be used to establish whether a BCR/ABL1 gene rearrangement is present with a simultaneous deletion of the ASS1 gene (for an example result with the tricolor probe, see section 29.1.7.1). Cases of a BCR/ABL1 chimeric fusion, especially in the presence of a co-deletion of ASS1, are generally associated with a very poor prognosis in ALL.

It should be noted that if molecular studies are to be performed to identify a BCR/ABL1 fusion gene product, then most of the rearrangements are located within the minor breakpoint cluster region, between exons 1 and 2, in ALL.

Normal results

nuc ish 9q34(ABL1x2),22q11.2(BCRx2)[200]

Abnormal results

nuc ish 9q34(ABL1x3),22q11.2(BCRx3),(ABL1 con BCRx2)[ ]

EXAMPLE REPORTING OF t(9;22)

ISCN Result: 9q34(ABL1x3),22q11.2(BCRx3),(ABL1 con BCRx2)[156/200]

Interpretation

Fluorescence *in situ* hybridization (FISH) analysis with the BCR/ABL1 gene rearrangement probes showed evidence of a (9;22) translocation in 156 of 200 (78%) of interphase nuclei scored.

The presence of a BCR/ABL1 rearrangement is consistent with a diagnosis of acute lymphoblastic leukemia and generally associated with a poor prognosis.

Follow-up studies to monitor for this abnormality by FISH analysis in addition to standard chromosome studies are suggested.

30.1.8 Chromosome 11

Rearrangements of 11q22

SPECIFIC FEATURES

The ATM dual color probe is designed to detect ATM gene deletions with usually the centromere as an internal control or as a breakapart probe. The ataxia telangiectasia mutation gene has been found to lie within the 11q22.3 region, and deletions are commonly associated with CLL. It is the second most common abnormality detected in CLL patients, preceded only by deletions of chromosome 13. Deletions of 11q22-23 are also observed in up to 25% of cases of non-Hodgkin lymphomas (NHL), with the highest incidence occurring in mantle cell lymphoma and diffuse large cell lymphoma. Deletion of the ATM gene is generally associated with a poor prognosis and relatively aggressive disease in CLL.

Normal results
nuc ish 11q22(ATMx2)[200]

Abnormal results
nuc ish 11q22(ATMx2)(5'ATM sep 3'ATMx1)[ ]

EXAMPLE REPORTING OF 11q22 – DUAL COLOR, BREAKAPART PROBE

ISCN Result: nuc ish 11q22(ATMx2)(5'ATM sep 3'ATMx1)[156/200]

Interpretation

Fluorescence *in situ* hybridization (FISH) analysis with the ATM probe showed an abnormal result with evidence of a rearrangement of the ATM gene at 11q22 in 156 of 200 (78%) interphase nuclei scored. The chromosomal partner cannot be identified by this assay.

Deletions of 11q22-23 are observed in up to 25% of cases of non-Hodgkin lymphomas (NHL), with the highest incidence occurring in mantle cell lymphoma and diffuse large cell lymphoma. It is also a common abnormality detected in CLL patients. Loss of the ATM gene at 11q22 is reported to be an independent adverse prognostic factor in B-CLL, associated with relatively aggressive disease.

Follow-up studies to monitor for this abnormality by FISH analysis in addition to standard chromosome studies are suggested.

Rearrangements of 11q23

SPECIFIC FEATURES

The MLL dual color, breakapart rearrangement probe is designed to detect 11q23 rearrangements associated with various translocations involving the MLL gene. Translocations disrupting the MLL gene are among the most common cytogenetic abnormalities observed in hematopoietic malignancies. Although over 30 variant translocations have been seen involving MLL translocations, the most common abnormalities are t(4;11)(q21;q23), t(9;11)(p22;q23) and t(11;19)(q23;p13). In approximately 25% of 11q23 translocations, a region beginning at the MLL breakpoint and extending distally is deleted. This probe can provide a better indication of the presence of the 11q23 translocation than a single color probe design.

Rearrangements of the MLL gene are common in both *de novo* and treatment-related acute leukemias as well as MDS and biphenotypic leukemia. In acute leukemia, MLL rearrangements appear to be equally as common in lymphoblastic as myeloid leukemias. Rearrangements of the MLL gene are generally associated with a poor prognosis.

Normal results
nuc ish 11q23(MLLx2)[200]

Abnormal results
nuc ish 11q23(MLLx2)(5'MLL sep 3'MLLx1)[ ]

EXAMPLE REPORTING OF 11q23 – DUAL COLOR, BREAKAPART PROBE

ISCN Result: nuc ish 11q23(MLLx2)(5'MLL sep 3'MLLx1)[156/200]

Interpretation

Fluorescence *in situ* hybridization (FISH) analysis with the MLL probe showed an abnormal result with evidence of a balanced translocation of the MLL gene at 11q23 in 156 of 200 (78%) interphase nuclei scored. The chromosomal partner cannot be identified by this assay.

Rearrangements involving the MLL gene are recurrent, non-random findings in pre-B-cell acute lymphoblastic leukemia, which can fuse with a large number of partners. Infants with MLL translocations, especially those <6 months of age, have a particularly poor prognosis.

Follow-up studies to monitor for this abnormality by FISH analysis in addition to standard chromosome studies are suggested.

Translocation (11;14)(q13;q32)

SPECIFIC FEATURES

The IGH/CCND1 dual color, dual fusion translocation probe is designed to detect translocations of the IGH locus and the cyclin D1 gene (CCND1/BCL1). The IGH probe set generally contains two labeled fluorochromes that hybridize to sequences on either side of the IGH J-region breakpoint associated with t(11;14). The CCND1 probe has a hybridization target that spans the common breakpoint region (MTC major translocation cluster) that lies centromeric to CCND1.

Translocation (11;14) is most commonly associated with mantle cell lymphoma, and it is also the most common IGH rearrangement observed in multiple myeloma. It is detected in up to 30% of abnormal myeloma cases by FISH analysis. Translocation (11;14) results in overexpression of cyclin D1. Myeloma patients with this translocation reportedly respond better to treatment than patients with other abnormalities.

When an IGH gene rearrangement is seen that does not include the CCND1 gene, a different signal configuration is observed. This result shows two signals of CCND1 and three signals of IGH, since the IGH gene is interrupted by a partner chromosome and so is split in two, giving rise to an extra IGH signal (located on an unknown chromosome). Since it is not the CCND1 region, there is no fusion probe. With an extra IGH signal, it appears to be trisomy 14; however, the vast majority of cases are actually an IGH gene rearrangement and not gain of IGH.

Normal results
nuc ish 11q13(CCND1x2),14q32(IGHx2)[200]

Abnormal results
nuc ish 11q13(CCND1x3),14q32(IGHx3),(CCND1 con IGHx2)[ ]

Abnormal results of an IGH gene rearrangement – without CCND1
nuc ish 11q13(CCND1x2),14q32(IGHx3)[ ]

EXAMPLE REPORTING OF t(11;14) – DUAL COLOR, DUAL FUSION PROBE

ISCN Result: nuc ish 11q13(CCND1x3),14q32(IGHx3),(CCND1 con IGHx2) [156/200]

Interpretation

Fluorescence *in situ* hybridization (FISH) analysis showed an abnormal result with evidence of an IGH/CCND1 gene rearrangement in 156 of 200 (78%) interphase nuclei scored.

Translocation (11;14)(q13;q32) results in the fusion of the IGH at 14q32 with the CCND1 gene at 11q13. This rearrangement is a recurrent, non-random rearrangement observed in large B-cell lymphomas and is most commonly associated with mantle cell lymphoma. It is also observed in plasma cell dyscrasias, where it is detected in up to 30% of abnormal multiple myeloma cases by FISH. Myeloma patients with this translocation reportedly respond better to treatment than patients with other abnormalities.

Follow-up studies to monitor for this abnormality by FISH analysis in addition to standard chromosome studies are suggested.

Translocation (11;18)(q21;q21)

SPECIFIC FEATURES

The API2-MALT1 dual color, dual fusion probe is designed to detect translocation (11;18)(q21;q21), which is associated with mucosa-associated lymphoid tissue (MALT) lymphoma, generating a chimeric fusion product. This translocation is frequently detected in gastric and pulmonary MALT lymphomas.

Studies have suggested that this chromosomal aberration is associated with intractability of *Helicobacter pylori* eradication treatment in patients with gastric MALT lymphomas. However, the role of t(11;18)(q21;q21) in the pathogenesis of gastrointestinal MALT lymphomas is still unclear.

Normal results

nuc ish 11q21(API2x2),18q21(MALTx2)[200]

Abnormal results

nuc ish 11q21(API2x3),18q21(MALTx3),(API2 con MALTx2)[ ]

EXAMPLE REPORTING OF t(11;18) – DUAL COLOR, DUAL FUSION PROBE

ISCN Result: nuc ish 11q21(API2x3),18q21(MALTx3),(API2 con MALTx2)[156/200]

Interpretation

Fluorescence *in situ* hybridization (FISH) analysis with the API2/MALT probes showed evidence of an API2/MALT rearrangement in 156 of 200 (78%) interphase nuclei scored.

Translocation (11;18) results in the fusion of the API2 at 11q21 with the MALT1 gene at 18q21. This rearrangement is a recurrent, non-random abnormality observed in extranodal marginal zone lymphoma of mucosa-associated lymphoid tissue (MALT) type.

Follow-up studies to monitor for this abnormality by FISH analysis in addition to standard chromosome studies are suggested.

Translocation (11;19)(q23;p13.3)

SPECIFIC FEATURES

As discussed in the FISH myeloid section, (11;19)(q23;p13.3) is one of the most common chromosomal translocations in acute leukemias involving the MLL gene. This translocation generates MLL/ENL fusion transcripts. Patients with AML and the MLL/ENL translocation generally have a poor prognosis, but non-infant children with ALL and MLL/ENL fusion may have a favorable prognosis.

Normal results
nuc ish 11q23(MLLx2),19p13.3(ENLx2)[200]

Abnormal results
nuc ish 11q23(MLLx3),19p13.3(ENLx3),(MLL con ENLx2)[ ]

EXAMPLE REPORTING OF t(11;19) – DUAL COLOR, DUAL FUSION PROBE

ISCN Result: nuc ish 11q23(MLLx2),19p13.3(ENLx3),(MLL con ENLx2)[156/200]

Interpretation

Fluorescence *in situ* hybridization (FISH) analysis with the MLL/11q23 and ENL/19p13.3 probes showed an abnormal result with evidence of a (11;19) translocation in 156 of 200 (78%) interphase nuclei scored.

Translocation (11;19)(q23;p13.3) results in the fusion of the MLL gene at 11q23 with the ENL gene at 19p13.3. It is a recurrent, non-random abnormality observed in acute myeloid leukemia (AML), and in both B-cell and T-cell lymphoblastic leukemia and lymphoblastic lymphoma (B-cell ALL/LBL).

Follow-up studies to monitor for this abnormality by FISH analysis in addition to standard chromosome studies are suggested.

30.1.9 Chromosome 12

Translocation (12;21)(p13;q22)

SPECIFIC FEATURES

The dual color, dual fusion probe corresponding to translocation (12;21) results in the fusion of the TEL gene on chromosome 12p13 with the AML (RUNX1) gene on chromosome 21q22. It is the most common rearrangement in childhood ALL and is a submicroscopic or cryptic abnormality that cannot be detected with chromosome analysis; therefore, FISH analysis is necessary for the identification of this abnormality. T(12;21), unlike other translocations associated with ALL, is associated with a good prognosis. Co-deletion of the homologous TEL gene may also seen with the translocation event. Another probe that may be used is the RUNX1 dual color breakapart probe to detect any rearrangment of chromosome 21q22 that might be present.

Normal results
nuc ish 12p13(ETV6x2),21q22(RUNX1x2)[200]

Abnormal results
nuc ish 12p13(ETV6x3),21q22(RUNX1x3),(ETV6 con RUNX1x2)[ ]

EXAMPLE REPORTING OF t(12;21) – DUAL COLOR, DUAL FUSION PROBE

ISCN Result: nuc ish 12p13(ETV6x3),21q22(RUNX1x3),(ETV6 con RUNX1x2)[156/200]

Interpretation

Fluorescence *in situ* hybridization (FISH) analysis with the ETV6/12p13 and RUNX1/21q22 probes showed an abnormal result with evidence of a balanced (12;21) translocation in 156 of 200 (78%) interphase nuclei scored.

This translocation is a recurrent, non-random finding in pre-B-cell acute lymphoblastic leukemia, and generally associated with a good prognosis.

Since this is a cytogenetic cryptic abnormality, standard chromosome analysis will not aid in the detection of this rearrangement. Therefore, follow-up studies to monitor for this abnormality by FISH analysis are suggested.

Trisomy 12

SPECIFIC FEATURES

The chromosome 12 centromere probe is often used as an adjunct to standard karotyping to identify extra copies of chromosome 12 in nuclei of cells obtained from peripheral blood lymphocytes or bone marrow in patients with B-cell CLL. In multi-site clinical trials, the CEP 12 analysis of interphase nuclei was 100% sensitive and 91% specific compared to traditional cytogenetic analysis. Trisomy 12 is present in 30% of reported cases, making it the most common cytogenetic abnormality in B-CLL. It has also been observed in approximately 25% of cases of mantle cell lymphoma and is observed in other B-cell lymphoproliferative disorders. Trisomy 12 has been associated with decreased overall survival and the need for early treatment.

Normal results
nuc ish 12cen(D12Z2x2)[200]

Abnormal results
nuc ish 12cen(D12Z2x3)[ ]

EXAMPLE REPORTING OF TRISOMY 12 – CENTROMERIC PROBE

ISCN Result: nuc ish 12cen(D12Z2x3)[156/200]

Interpretation

Fluorescence *in situ* hybridization (FISH) analysis with the chromosome 12 centromeric probe (D8Z2) showed evidence of trisomy 12 in 156 of 200 (78%) interphase nuclei scored.

Trisomy 12 is one of the most common chromosome abnormalities in chronic lymphocytic leukemia (CLL). It has also been observed in approximately 25% of cases of mantle cell lymphoma and is observed in other B-cell lymphoproliferative disorders. Trisomy 12 is generally associated with an intermediate to poor prognosis.

Follow-up studies to monitor for this abnormality by FISH analysis in addition to standard chromosome studies are suggested.

30.1.10 Chromosome 13

Del(13q)

SPECIFIC FEATURES

The chromosome 13q14.3 region, including the D13S319 and D13S25 dual color probe, is used to identify deletions of this region with a 13q34/LAMP1 gene probe often used as a control probe. There are other probe configurations as well that use different probes within the long arm of chromosome 13 to detect deletions. The locus D13S319 is located between RB1 and the D13S25 loci and is a commonly deleted marker. A candidate tumor suppressor gene resides telomeric of the RB1 gene at 13q14.

Deletions of chromosome 13 as detected by FISH analysis are reported in a number of B-lineage chronic lymphoproliferative disorders with the following frequencies:

- chronic lymphocytic leukemia (40%)
- atypical CLL (<40%)
- splenic marginal zone lymphoma (12–47%)
- mantle cell lymphoma (40–60%)
- multiple myeloma (15–20%).

Deletion of chromosome 13q14 is generally associated with a good prognosis in B-CLL but in MM, it is reported to be associated with a poor prognosis.

Normal results

nuc ish 13q14(D13S319x2),13q34(LAMP1x2)[200]

Abnormal results – del(13q)

nuc ish 13q14(D13S319x1),13q34(LAMP1x2)[ ]

EXAMPLE REPORTING OF DELETION OF 13q – DUAL COLOR, DELETION PROBE

ISCN Result: nuc ish 13q14(D13S319x1),13q34(LAMP1x2)[156/200]

Interpretation

Fluorescence *in situ* hybridization (FISH) analysis with the chromosome 13q14 D13S319 gene probe showed evidence of a deletion of chromosome 13q14 in 156 of 200 (78%) interphase nuclei scored.

Deletion of 13q is a recurrent abnormality in many chronic lymphoproliferative diseases. In chronic lymphocytic leukemia (CLL), deletion of 13q is generally associated with a favorable prognosis; however, it is associated with a poor prognosis in multiple myeloma.

Follow-up studies to monitor for this abnormality by FISH analysis in addition to standard chromosome studies are suggested.

30.1.11 Chromosome 14

Rearrangements of 14q32

SPECIFIC FEATURES

The IGH dual color, breakapart rearrangement probe is designed to detect chromosomal breakage at J or switch regions of the IGH locus that is associated with 14q32 translocations involving a variety of other loci. Breakpoints within the IGH locus may occur at either the J segments (e.g. breakpoints commonly observed with t(14;18)) or within switch sequences located within the constant gene segments.

The IGH breakapart probe is a mixture of two probes that hybridize to opposite sides of the J segments through constant regions of the IGH locus. The hybridization target of the IGH 3' flanking probe is located completely 3' to the IGH locus. As a result of this probe configuration, any translocation with a breakpoint at the J segments or within switch sequences should produce separate orange and green signals.

Rearrangements of the IGH gene are observed in 40–50% of all B-cell NHLs, as well as other lymphoproliferative disorders, with numerous translocation partners for IGH rearrangements identified.

Normal results

nuc ish 14q32(IGHx2)[200]

Abnormal results

nuc ish 14q32(IGHx2)(5'IGH sep 3'IGHx1)[ ]

EXAMPLE REPORTING OF 14q32 – DUAL COLOR, BREAKAPART PROBE

ISCN Result: nuc ish 14q32(IGHx2)(5'IGH sep 3'IGHx1)[156/200]

Interpretation

Fluorescence *in situ* hybridization (FISH) analysis with the IGH probe showed an abnormal result with evidence of rearrangement of the IGH gene at 14q32 in 156 of 200 (78%) interphase nuclei scored. The chromosomal partner cannot be identified by this assay.

Translocations involving the immunoglobulin heavy chain gene (IGH) at14q32 are recurrent, non-random abnormalities observed in 40–50% of B-cell non-Hodgkin lymphomas and are also seen in other lymphoproliferative disorders.

Follow-up studies to monitor for this abnormality by FISH analysis in addition to standard chromosome studies are suggested.

Translocation (14;16)(q32;q22)

SPECIFIC FEATURES

The IGH/MAF dual color, dual fusion translocation probe is designed to detect the translocation of the IGH locus and MAF gene, t(14;16)(q32;q22). Translocation (14;16) is a common chromosome abnormality associated with MM, seen in approximately 25% of cases. The fusion product of this translocation results in overexpression of the MAF gene and is generally associated with a poor prognosis in MM patients.

When an IGH gene rearrangement is seen that does not include the MAF gene, a different signal configuration is observed. This result shows two signals of MAF and three signals of IGH, since the IGH gene is interrupted by a partner chromosome and so is split in two, giving rise to an extra IGH signal (located on an unknown chromosome). Since it is not the MAF region, there is no fusion probe. With an extra IGH signal, it appears to be trisomy 14; however, the vast majority of cases are actually an IGH gene rearrangement and not gain of IGH.

Normal results
nuc ish 14q32(IGHx2),16q22(MAFx2)[200]

Abnormal results
nuc ish 14q32(IGHx3),16q22(MAFx3),(IGH con MAFx2)[ ]

Abnormal results – IGH gene rearrangement, without MAF
nuc ish 14q32(IGHx3),16q22(MAFx2)[ ]

EXAMPLE REPORTING OF t(14;16) – DUAL COLOR, DUAL FUSION PROBE

ISCN Result: nuc ish 14q32(IGHx3),16q22(MAFx3),(IGH con MAFx2)[156/200]

Interpretation

Fluorescence *in situ* hybridization (FISH) analysis showed an abnormal result with evidence of an IGH/MAF gene rearrangement in 156 of 200 (78%) interphase nuclei scored.

Translocation (14;16)(q32;q22) results in the fusion of the IGH at 14q32 with the MAF gene at 16q22. This rearrangement is observed in 25% of cases of multiple myeloma and is generally associated with a poor prognosis.

Follow-up studies to monitor for this abnormality by FISH analysis in addition to standard chromosome studies are suggested.

Translocation (14;18)(q32;q21)

SPECIFIC FEATURES

The IGH/BCL2 dual color, dual fusion translocation probe is designed to detect the translocation of the IGH locus and BCL2 gene, t(14;18)(q32;q21). Translocation of an IGH transcriptional enhancer next to the BCL2 gene as a result of the t(14;18) is thought to cause constitutive overexpression of the anti-apoptotic BCL2 protein.

The breakpoints at 14q32 occur at the IGH-J segments, and about 75% of the breaks at 18q21 occur in either the 2.8 kb major breakpoint region (MBR) 3' of BCL2 exon 3 or in the minor cluster region (MCR) located about 30 kb 3' to the MBR. The remaining BCL2 breakpoints are thought to lie between the MBR and MCR regions, or 5' of the BCL2 gene.

This rearrangement is observed in 80–90% of cases of follicular lymphoma and approximately 30% of cases of large B-cell lymphoma.

When an IGH gene rearrangement is seen that does not include the BCL2 gene, a different signal configuration is observed. This result shows two signals of BCL2 and three signals of IGH, since the IGH gene is interrupted by a partner chromosome and so is split in two, giving rise to an extra IGH signal (located on an unknown chromosome). Since it is not the BCL2 region, there is no fusion probe. With an extra IGH signal, it appears to be trisomy 14; however, the vast majority of cases are actually an IGH gene rearrangement and not gain of IGH.

Normal results
nuc ish 14q32(IGHx2),18q21(BCL2x2)[200]

Abnormal results
nuc ish 14q32(IGHx3),18q21(BCL2x3),(IGH con BCL2x2)[ ]

Abnormal results – IGH rearrangement, without BCL2
nuc ish 14q32(IGHx3),18q21(BCL2x2)[ ]

EXAMPLE REPORTING OF t(14;18) – DUAL COLOR, DUAL FUSION PROBE

ISCN Result: nuc ish 14q32(IGHx3),18q21(BCL2x3),(IGH con BCL2x2) [156/200]

Interpretation

Fluorescence *in situ* hybridization (FISH) analysis showed an abnormal result with evidence of an IGH/BCL2 gene rearrangement in 156 of 200 (78%) interphase nuclei scored.

Translocation (14;18)(q32;q21) results in the fusion of the IGH at 14q32 with the BCL2 gene at 18q21. This rearrangement is observed in 80–90% of cases of follicular lymphoma and in approximately 30% of cases of large B-cell lymphoma.

Follow-up studies to monitor for this abnormality by FISH analysis in addition to standard chromosome studies are suggested.

30.1.12 Chromosome 17

Del(17)(p13.1)

SPECIFIC FEATURES

The TP53 dual color probe located at chromosome 17p13.1 contains the TP53 tumor suppressor gene. Monoallelic deletion of TP53 is common in many disorders, and loss of heterozygosity of TP53 has been identified in many tumors. This probe is used to detect deletions and amplification of the TP53 locus.

Deletion of the TP53 gene on chromosome 17p is one of the four most common abnormalities observed in B-cell chronic lymphocytic leukemia and is detected in approximately 10–15% of B-CLL patients by FISH analysis. Deletion of the TP53 gene is considered one of the most important independent prognostic indicators in B-CLL, and it is associated with poor survival. Deletion of the TP53 gene on chromosome 17p is also observed in approximately 30% of patients with newly diagnosed myeloma and in up to 55% of relapsed cases. Deletion of the TP53 gene is also associated with poor survival in multiple myeloma.

Normal results
nuc ish 17cen(D17Z1x2),17p13.1(TP53x2)[200]

Abnormal results with del(17p)
nuc ish 17cen(D17Z1x2),17p13.1(TP53x1)[ ]

EXAMPLE REPORTING OF DELETION 17p – DUAL COLOR, DELETION PROBE

ISCN Result: nuc ish 17cen(D17Z1x2),17p13.1(TP53x1)[156/200]

Interpretation

Fluorescence *in situ* hybridization (FISH) analysis with the TP53 probe showed evidence of a deletion of 17p13.1 in 156 of 200 (78%) of interphase nuclei scored.

Deletion of the TP53 gene on chromosome 17p is a common abnormality observed in lymphoproliferative disorders, including B-cell chronic lymphocytic leukemia and multiple myeloma, and is generally associated with poor prognosis.

Follow-up studies to monitor for this abnormality by FISH analysis in addition to standard chromosome studies are suggested.

30.1.13 Chromosome 18

Rearrangements of 18q21

SPECIFIC FEATURES

The MALT1 (18q21) breakapart rearrangement probe will identify 18q21 translocations with multiple unidentified chromosome partners. The probe will also identify aneuploidy of chromosome 18. The MALT1 probe consists of a mixture of two DNA probes. The first probe flanks the 5' side of the MALT1 gene, while the second probe flanks the 3' side of the MALT1 gene, with a gap between the two probes. The known breakpoints within the MALT1 gene region are found in introns 2, 3 and 8.

Chromosome rearrangements involving the MALT1 (MALT lymphoma-associated translocation 1) gene on chromosome 18q21 have been observed in several types of lymphoma, especially MALT lymphoma. In t(11;18) (q21;q21), a well-documented translocation involving the MALT1 gene, a gene fusion is produced on the der(11) chromosome between a 5' portion of the API2 (11q21) gene and a 3' portion of the MALT1 gene (18q21). The resulting API2/MALT1 fusion gene encodes an abnormal chimeric protein.

Normal results
nuc ish 18q21(MALT1x2)[200]

Abnormal results
nuc ish 18q21(MALT1x2)(5'MALT1 sep 3'MALT1x1)[ ]

EXAMPLE REPORTING OF DELETION 18q21 REARRANGMENT – DUAL COLOR, BREAKAPART PROBE

ISCN Result: nuc ish 18q21(MALT1x2)(5'MALT1 sep 3'MALT1x1)[156/200]

Interpretation

Fluorescence *in situ* hybridization (FISH) analysis with the MALT1 probe showed evidence of a translocation involving the MALT1 gene at 18q21 in 156 of 200 (78%) interphase nuclei scored.

Translocations involving the MALT1 gene on chromosome 18q21 have been observed in several types of lymphomas, especially mucosa-associated lymphoid tissue (MALT) lymphoma.

Follow-up studies to monitor for this abnormality by FISH analysis in addition to standard chromosome studies are suggested.

30.1.14 Chromosome 19

Rearrangements of 19p13.3

SPECIFIC FEATURES

The t(1;19)(q23;p13.3) is one of the most common chromosomal abnormalities in B-cell precursor acute lymphoblastic leukemia (BCP-ALL) and usually gives rise to the TCF3/E2A-PBX1 fusion gene. The TCF3 gene is also involved in several other chromosome translocations, including t(19;19) and inv(19) with the TFPT gene, t(12;19) with the ZNF384 gene and t(17;19)(q22;p13.3) with the HLF gene. Therefore, using the TCF3 dual color, breakapart probe is useful in identifying any fusion gene partner. This is a subtle rearrangement and may be missed with standard chromosome analysis. Thus, FISH analysis is suggested to confirm a 19p13.3 rearrangement.

Normal results
nuc ish 19p13.3(TCF3x2)[200]

Abnormal results with a balanced translocation
nuc ish 19p13.3(5'TCF3x2,3'TCF3x1),(5'TCF3 con 3'TCF3x1][156/200]

Abnormal results with an unbalanced translocation
nuc ish 19p13.3(TCF3x2)(5'TCF3 sep 3'TCF3x1)[156/200]

EXAMPLE REPORTING OF 19p13.3 – DUAL COLOR, BREAKAPART PROBE – BALANCED TRANSLOCATION

ISCN Result: nuc ish 19p13.3(5'TCF3x2,3'TCF3x1),(5'TCF3 con 3'TCF3x1][156/200]

Interpretation

Fluorescence *in situ* hybridization (FISH) analysis with the TCF3 probe showed an abnormal result with evidence of a balanced translocation of the TCF3 (E2A) gene at 19p13.3 in 156 of 200 (78%) interphase nuclei scored. The chromosomal partner cannot be identified by this assay.

Translocation of TCF3 at 19p13.3 is a recurrent, non-random finding in pre-B-cell acute lymphoblastic leukemia.

Follow-up studies to monitor for this abnormality by FISH analysis are suggested.

EXAMPLE REPORTING OF 19p13.3 – DUAL COLOR, BREAKAPART PROBE – UNBALANCED TRANSLOCATION

ISCN Result: nuc ish 19p13.3(TCF3x2)(5'TCF3 sep 3'TCF3x1)[156/200]

Interpretation

Fluorescence *in situ* hybridization (FISH) analysis with the TCF3 probe showed an abnormal result, which is suggestive of an unbalanced translocation involving TCF3 with the loss of the other derivative chromosome in 156 of 200 (78%) interphase nuclei scored.

This unbalanced rearrangement of TCF3 at 19p13.3 is a recurrent, non-random finding in pre-B-cell acute lymphoblastic leukemia.

Follow-up studies to monitor for this abnormality by FISH analysis are suggested.

30.1.15 Chromosome 21

Trisomy 21

SPECIFIC FEATURES

Trisomy 21 is the most frequent aneuploidy observed in both childhood and adult ALL and lymphoblastic lymphoma. It is observed in approximately 15% of ALL cases, usually in conjunction with other abnormalities. In adults, trisomy 21 commonly occurs with a (9;22) translocation. Trisomy 21 has a good prognosis as a sole abnormality, or in children with a 47–50 chromosome count without other unfavorable structural rearrangements. Trisomy 21 may be detected using a single color centromeric probe or more commonly with the RUNX1/AML1 dual color probe at locus 21q22.

Normal results
nuc ish 21q22(RUNX1x2)[200]

Abnormal results
nuc ish nuc ish 21q22(RUNX1x3)[ ]

EXAMPLE REPORTING OF TRISOMY 21 – DUAL COLOR PROBE

ISCN Result: nuc ish 21q22(RUNX1x3)[156/200]

Interpretation

Fluorescence *in situ* hybridization (FISH) analysis with the chromosome 21q22 RUNX1 probe showed evidence of three RUNX1 signals consistent with trisomy 21 in 156 of 200 (78%) interphase nuclei scored.

Trisomy 21 is the most frequent aneuploidy observed in childhood and adult B-cell lymphoblastic leukemia and lymphoblastic lymphoma (B-cell ALL/LBL). It is observed in approximately 15% of B-cell ALL/LBL cases, usually in conjunction with other abnormalities. When seen as the sole abnormality, trisomy 21 is generally associated with a good prognosis, or in children with a hyperdiploid chromosome count without other unfavorable structural rearrangements.

Follow-up studies to monitor for this abnormality by FISH analysis in addition to standard chromosome studies are suggested.

30.2 Panels of probes

Panels of probes are commmonly used for indications of ALL, CLL, NHL and MM. For a comprehensive list of possible probes to be included in a lymphoid disease panel, see Table 29.1. There are various ways to write normal and abnormal results from a panel of probes in a report, and the following examples are one approach in which normal and abnormal probes are separated to make the report easier to read (for a different approach in which all probes are listed in chromosome order, see section 29.3 for panels of myeloid disorders).

When using a probe panel, more than one abnormality may be present, therefore, there are examples of abnormal generic templates that list all probes in the panel in an abnormal results section, with places that can be filled in with the number of cells per probe, depending on which probes are abnormal. The rest of the probes that are normal can be removed from this abnormal result section. All of the probes are then listed in a normal result section to be used for those probes that are normal. The abnormal probes can be deleted from this normal result section. The interpretation also lists all the probes and can be modified based on which probes are abnormal and which are normal. This template is provided as a tool for ease in writing results with a large set of probes in a single test, that would otherwise be cumbersome when probes are written singularly.

30.2.1 ALL panel of probes

The most common abnormalities associated with ALL may be identified by FISH analyses, including hyperdiploidy, MYC/8q24 rearrangements, MLL/11q23 deletions or rearrangements, TCF3/19p13.3 deletions or rearrangements, t(12;21) and t(9;22). Hyperdiploidy may be detected by analyzing any number of chromosomes in which gains are commonly observed, such as chromosomes 4, 10 and 17.

NORMAL ALL RESULTS

ISCN Result: nuc ish 4cen(CEP4x2),8q24(MYCx2),9q34(ABL1x2),10cen(CEP10x2),11q23 (MLLx2), 12p13(ETV6x2),17cen(D17Zx2),19p13.3(TCF3x2),21q22(RUNX1x2),22q11.2(BCRx2)[200]

Interpretation

Fluorescence *in situ* hybridization (FISH) analysis with chromosome 4, 10 and 17 centromere probes and the locus-specific probes for MYC, BCR/ABL, ETV6/RUNX1, TCF3 and MLL gene regions showed no evidence of abnormalities in 200 cells scored for each probe.

Hyperdiploid positive ALL

EXAMPLE REPORTING OF HYPERDIPLOIDY AS THE SOLE ANOMALY (FIGURE 30.1)

Abnormal results

nuc ish 4cen(CEP4x3)[24/200],10cen(CEP10x3)[48/200]

Normal results

nuc ish 8q24(MYCx2),9q34(ABL1x2),11q23(MLLx2),12p13(ETV6x2),17cen(D17Zx2),19p13.3 (TCF3x2),21q22(RUNX1x2),22q11.2(BCRx2)[200]

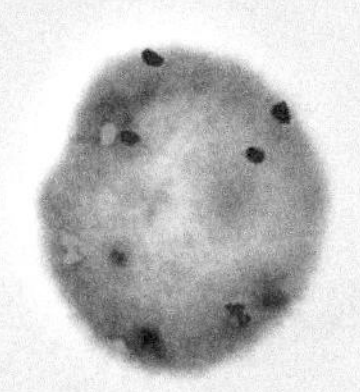

Figure 30.1 Hyperdiploidy, showing three copies of blue and green signals, corresponding to trisomy 4 and 10, respectively. Two copies of red signals correspond to disomy for chromosome 17. Courtesy of Lauren Jenkins PhD, Kaiser Permanente Regional Laboratory, Northern California.

Interpretation

Fluorescence *in situ* hybridization (FISH) analysis with chromosome 4 and 10 centromere probes showed evidence of trisomy 4 in 12% and trisomy 10 in 24% of interphase nuclei scored for each probe, respectively.

FISH analysis with the remaining probes showed no evidence of trisomy 17, 9;22 or 12;21 translocations, 8q24, 19p/TCF3 or 11q23/MLL deletion or rearrangements, in 200 cells scored for each probe.

Hyperdiploidy is a recurrent, non-random finding in childhood B-cell lymphoblastic leukemia (B-cell ALL), observed in approximately 25% of cases. B-cell ALL with hyperdiploidy is generally associated with a good prognosis in children, especially if their risk profile is otherwise favorable.

Follow-up studies to monitor for these abnormalities by FISH analysis in addition to standard chromosome studies are suggested.

Gene rearrangement positive ALL

EXAMPLE REPORTING OF t(9;22) AS THE SOLE ANOMALY IN ALL

Abnormal results

9q34(ABL1x3),22q11.2(BCRx3),(ABL1 con BCRx2)[100/200]

Normal results

nuc ish 4cen(CEP4x2),8q24(MYCx2),10cen(CEP10x2),11q23(MLLx2),17cen(D17Z1x2), 19p13.3(TCF3x2)[200]

Interpretation

Fluorescence *in situ* hybridization (FISH) analysis with the BCR/ABL1 gene rearrangement probes showed evidence of a 9;22 translocation in 50% of interphase nuclei scored.

FISH analysis with the remaining probes showed no evidence of a 12;21 translocation, or an 8q24, 11q23 or 19p13.3 deletion or rearrangement in 200 cells scored for each probe.

The presence of a BCR/ABL1 rearrangement is consistent with a diagnosis of acute lymphoblastic leukemia and generally associated with a poor prognosis.

Follow-up studies to monitor for this abnormality by FISH analysis in addition to standard chromosome studies are suggested.

EXAMPLE REPORTING WITH A GENERAL TEMPLATE FOR ANY REARRANGEMENT

Abnormal results

nuc ish
8q24(MYCx2)(5'MYC sep 3'MYCx1)[_]
9q34(ABL1x3),22q11.2(BCRx3),(ABL1 con BCRx2)[_]
11q23(MLLx2)(5'MLL sep 3'MLLx1)[_]
19p13.3(TCF3x2)(5'TCF3 sep 3'TCF3x1)[_]

Normal results

nuc ish
4cen(CEP4x2),10cen(CEP10x2),17cen(D17Z1x2)
8q24(MYCx2)
9q34(ABL1x2),22q11.2(BCRx2)
11q23(MLLx2)
19p13.3(TCF3x2)

Interpretation

Fluorescence *in situ* hybridization (FISH) analysis with the MYC, BCR/ABL1, MLL and TCF3 probes showed an abnormal result with evidence of 8q24 rearrangement involving the MYC locus/9;22 translocation, an 11q23 deletion/rearrangement involving the MLL locus or a 19p13.3 rearrangement involving the TCF3 locus in (_%) cells scored for this probe.

FISH analysis with the remaining probes showed no evidence of hyperdiploidy, 8q24 rearrangement involving the MYC locus, 9;22 translocation, an 11q23 deletion, rearrangement involving the MLL locus, or a 19p13.3 deletion or rearrangement involving the TCF3 locus in 200 cells scored for these probes.

The presence of this _ rearrangement is consistent with a diagnosis of acute lymphoblastic leukemia.

Follow-up studies to monitor for this abnormality by FISH analysis in addition to standard chromosome studies are suggested.

30.2.2 CLL panel of probes

The most common abnormalities associated with CLL may be identified by FISH analyses, including ATM deletions of 11q22.3, trisomy 12, deletions of the long arm of chromosome 13 and deletion of the short arm of chromosome 17.

EXAMPLE REPORTING OF NORMAL RESULTS

ISCN Result: nuc ish 11q22.3(ATMx2),12cen(D12Z3x2),13q14.3(D13S319x2),13q34(LAMP1x2), 17cen(D17Z1x2),17p13.1(TP53x2)

Interpretation

Fluorescence *in situ* hybridization (FISH) analysis with the CLL panel of probes ATM, D12Z3, D13S319, LAMP1 and TP53 showed no evidence of ATM, 13q14, TP53 deletions or trisomy 12 in 200 cells examined for eavch probe.

EXAMPLE REPORTING OF ABNORMAL RESULTS – GENERIC TEMPLATE

Abnormal results

nuc ish 11q22.3(ATMx1)[_],12cen(D12Z3x3)[_],13q14.3(D13S319x1),13q34(LAMP1x2)[_], 13q14.3(D13S319x0),13q34(LAMP1x2)[_],13q14.3(D13S319x1)[_],13q34(LAMP1x1)[_], 17cen(D17Z1x1)[_],17p13.1(TP53x2)[_]

Normal results

nuc ish 11q22.3(ATMx2),12cen(D12Z3x2),13q14.3(D13S319x2),13q34(LAMP1x2),17cen (D17Z1x2),17p13.1(TP53x2)[200]

Interpretation

Fluorescence *in situ* hybridization (FISH) analysis with the chronic lymphocytic leukemia (CLL) panel of probes showed evidence of:

- a deletion 11q22.3 (ATM) in _% of cells scored
- trisomy 12 in _/_00 (_%) of cells scored
- a deletion 13q (D13S319) in _% of cells scored
- a deletion of 13q (D13S319) on both homologs in _% of cells scored
- monosomy 13 (loss of both D13S319 and LAMP1) in _% of cells scored
- a deletion 17p (TP53) in _% of cells scored.

FISH analysis with the remaining probes showed no evidence of a deletion of the ATM locus, 13q14, TP53 deletion or trisomy 12.

Deletion of 11q22.3/ATM gene is a recurrent finding in CLL and other B-cell lymphoproliferative disorders. Prognosis tends to be poor in CLL.

Trisomy 12 is a recurrent finding seen in CLL and other B-cell lymphoproliferative disorders. Prognosis tends to be intermediate in CLL.

Deletions of the chromosome 13 long arm are a recurrent finding seen in lymphoid and myeloid hematological malignancies. Prognosis tends to be good in CLL.

Deletions of the chromosome 17 short arm are a recurrent finding seen in lymphoid and myeloid hematological malignancies. Prognosis tends to be poor in CLL.

Follow-up studies to monitor for this abnormality by FISH analysis in addition to standard chromosome studies are suggested.

30.2.3 NHL panel of probes

The most common abnormalities associated with NHL may be identified by FISH analyses, including ALK/2p23, BCL6/3q27 and MYC/8q24 rearrangements, t(11;14) and t(14;18).

EXAMPLE REPORTING OF NORMAL RESULTS

ISCN Result: nuc ish 2p23(ALKx2),3q27(BCL6x2),8q24(MYCx2),11q13(CCND1x2),
14q32(IGHx2),18q21(BCL2x2)[200]

Interpretation

Fluorescence *in situ* hybridization (FISH) analysis with the non-Hodgkin lymphoma panel of probes, including ALK, BCL6, MYC, IGH, CCND1 and BCL2 genes, showed no evidence of an IGH gene rearrangement, ALK, BCL2 or MYC rearrangements in 200 cells scored for each probe or probe combination.

EXAMPLE REPORTING OF ABNORMAL RESULTS

Abnormal results

nuc ish 2p23(ALKx2)(5'ALK sep 3'ALKx1)[ ],
3q27(BCL6x2)(5'BCL6 sep 3'BCL6x1)[ ],
8q24(MYCx2)(5'MYC sep 3'MYCx1)[ ],
11q13(CCND1x3),14q32(IGHx3),(CCND1 con IGHx2)[ ],
14q32(IGHx3),18q21(BCL2x3),(IGH con BCL2x2)[ ]

Normal results

nuc ish 2p23(ALKx2),3q27(BCL6x2),8q24(MYCx2),11q13(CCND1x2),14q32(IGHx2),
18q21(BCL2x2)[200]

Interpretation

Fluorescence *in situ* hybridization (FISH) analysis with the non-Hodgkin lymphoma panel of probes BCL6, MYC, IGH/CCND1 and IGH/BCL2 showed evidence of:

- a rearrangement involving the ALK locus at 2p23 in _% of cells scored
- a rearrangement involving the BCL6 locus at 3q27 in _% of cells scored
- a rearrangement involving the MYC locus at 8q24 in _% of cells scored
- a 14;18 translocation involving IGH and BCL2 in _% of cells scored
- a 11;14 translocation involving IGH and CCND1 in _% of cells scored.

These findings are consistent with a clinical diagnosis of non-Hodgkin lymphoma.

FISH analysis with the remaining probes showed normal results with no evidence of a 11;14 or 14;18 translocation or other rearrangements involving ALK, BCL6, MYC and IGH in 200 cells scored for each probe.

"Triple hit" lymphoma (THL) shows rearrangements of BCL2, MYC and BCL6 and is associated with a poor prognosis. This combination of abnormalities is observed in B-cell lymphomas with features intermediate between DLBCL and BL.

"Double hit" lymphoma (DHL) is consistent with a rearrangement of both BCL2 (18q21) and MYC (8q24), and is associated with a poor prognosis. This combination of abnormalities is observed in B-cell lymphomas with features intermediate between diffuse large B cell lymphoma (DLBCL) and Burkitt lymphoma (BL).

Follow-up studies to monitor for these abnormalities by FISH analysis in addition to standard chromosome studies are suggested.

30.2.4 MM panel of probes

Trisomies of chromosomes 3, 7, 9 and 11 are among the most commonly observed numerical abnormalities in multiple myeloma. However, chromosome aneuploidy, and particularly trisomy, is also detected in patients with monoclonal gammopathy of unknown significance (MGUS), smoldering myeloma and amyloidosis. In these conditions, the number of trisomic chromosomes tends to accumulate over time with the frequency of numerical abnormalities, showing some correlation with progression of the disease. Therefore, when observed as the only genetic abnormality, aneuploidy should be interpreted cautiously in the context of other clinical, pathological, serological and radiological findings.

Deletion of chromosomes 13q, 17p, and (4;14), (11;14) and (14;16) rearrangements are also commonly observed in myeloma. MM FISH analysis is best performed with plasma cell enrichment by cell sorting. This allows for the detection of abnormal cells to a larger degree. Targeted FISH analysis usually is performed with a 60–100 cell count for a high level of accuracy of results.

When an IGH gene rearrangement is seen that does not include any of the genes in the panel, a different signal configuration is observed. This result shows two signals of the partner gene and three signals of IGH, since the IGH gene is interrupted by a different partner chromosome and so is split in two, giving rise to an extra IGH signal (located on an unknown chromosome). Since it is not the partner gene region, there is no fusion probe. With an extra IGH signal, it appears to be trisomy 14; however, the vast majority of cases are actually an IGH gene rearrangement and not gain of IGH. In this case, the IGH breakapart probe is useful in confirming an IGH gene rearrangement versus trisomy 14.

EXAMPLE REPORTING OF NORMAL RESULTS

ISCN Result: nuc ish
4p16(FGFR3x2),9q34(ASSx2),11q13(CCND1x2),13q14(D13S319x2),13q34(LAMP1x2),
14q32(IGHx2),15q22(PMLx2),16q23(MAFx2),17cen(D17Z1x2),17p13.1(TP53x2)[100]

Interpretation

Fluorescence *in situ* hybridization (FISH) analysis performed on CD138+ sorted cells with the multiple myeloma panel of probes, with the t(4;14)/FGFR3/IGH, 13q14/D13S319, 14q32/IGH, t(11;14)/CCND1/IGH, t(14;16)/IGH/MAF, 15q22/PML and 17p13.1/TP53 genes showed no evidence of an IGH gene rearrangement, 13q or TP53 deletion, or gain of chromosomes 9, 11 or 15, in 100 cells scored with each probe.

EXAMPLE REPORTING OF ABNORMAL RESULTS – t(11;14)

Abnormal results

nuc ish 11q13(CCND1x3),14q32(IGHx3),(CCND1 con IGHx2)[40/100]

Normal results

nuc ish 4p16(FGFR3x2),9q34(ASSx2),13q14(D13S319x2),13q34(LAMP1x2),15q22(PMLx2),
16q23(MAFx2),17cen(D17Z1x2),17p13.1(TP53x2)[100]

Interpretation

Fluorescence *in situ* hybridization (FISH) analysis was performed on CD138+ sorted cells with the multiple myeloma panel of probes, with the t(4;14)/FGFR3/IGH, 13q14/D13S319, 14q32/IGH, t(11;14)/CCND1/IGH, t(14;16)/IGH/MAF, 15q22/PML and 17p13.1/TP53 genes.

This analysis showed evidence of an 11;14 translocation in 40 of 100 cells scored (40%). This abnormal finding is consistent with a diagnosis of multiple myeloma.

FISH analysis with the remaining probes showed normal results with no evidence of hyperdiploidy or abnormalities of any of these loci in 100 cells scored for each probe or probe combination.

Follow-up studies to monitor for this abnormality by FISH analysis in addition to standard chromosome studies are suggested.

EXAMPLE REPORTING OF ABNORMAL RESULTS – AN IGH REARRANGEMENT WITH t(11;14) OR t(4;14) OR t(14;16)

Abnormal results

nuc ish 14q32(IGHx2)(5'IGH sep 3'IGHx1)[30/100],11q13(CCND1x3),14q32(IGHx3),(CCND1 con IGHx2)[40/100]

or

nuc ish 14q32(IGHx2)(5'IGH sep 3'IGHx1)[30/100],4p16(FGFR3x3),14q32(IGHx3),(FGFR con IGHx2)[40/100]

or

nuc ish 14q32(IGHx2)(5'IGH sep 3'IGHx1) [30/100], 14q32(IGHx3),16q23(MAFx3), (IGH con MAFx2)[40/100]

Normal results

nuc ish 9q34(ASSx2),13q14(D13S319x2),13q34(LAMP1x2),15q22(PMLx2),17cen(D17Z1x2), 17p13.1(TP53x2)[100]

Interpretation

Fluorescence *in situ* hybridization (FISH) analysis was performed on CD138+ sorted cells with the multiple myeloma panel of probes, including 9q34/ASS, 14q32/IGH, t(11;14)/CCND1/IGH, t(4;14)/FGFR3/IGH, t(14;16)/IGH/MAF,15q22/PML and 17p13.1/TP53 genes.

This analysis showed evidence of an 11;14/4;14/14;16 translocation in 40 of 100 (40%) cells scored confirmed by an IGH breakapart rearrangement in 30/100 (30%) cells scored. This abnormal finding is consistent with a diagnosis of multiple myeloma.

FISH analysis with the remaining probes showed no evidence of hyperdiploidy or abnormalities of any of these loci in 100 cells scored for each probe or probe combination.

Follow-up studies to monitor for this abnormality by FISH analysis in addition to standard chromosome studies are suggested.

EXAMPLE REPORTING OF ABNORMAL RESULTS – IGH GENE REARRANGEMENT – PARTNER UNKNOWN

Abnormal results

nuc ish 14q32(IGHx2)(5'IGH sep 3'IGHx1)[_/_]
4p16(FGFR3x2),14q32(IGHx3)[_/_]
11q13(CCND1x2),14q32(IGHx3)[_/_]
14q32(IGHx3),16q32(MAFx2)[_/_]

Normal results

nuc ish 9q34(ASSx2),13q14(D13S319x2),13q34(LAMP1x2),15q22(PMLx2),17cen(D17Z1x2), 17p13.1(TP53x2)[100]

Interpretation

Fluorescence *in situ* hybridization (FISH) analysis was performed on CD138+ sorted cells with the multiple myeloma panel of probes, including t(4;14)/FGFR3/IGH, 9q34/ASS,14q32/IGH, t(11;14)/CCND1/IGH, t(14;16)/IGH/MAF,15q22/PML and 17p13.1/TP53 genes.

This analysis showed evidence of an IGH gene rearrangement in _% cells scored. The t(11;14), t(4;14) and t(14;16) showed signal configurations of an IGH rearrangement, which do not involve the chromosome partners of the probe set. The IGH breakapart probe confirmed an IGH gene rearrangement. These abnormal findings are consistent with an IGH gene rearrangement whose partner remains unknown, and is associated with a diagnosis of multiple myeloma.

FISH analysis with the remaining probes showed normal results with no evidence of hyperdiploidy or abnormalities of any of these loci in 100 cells scored for each probe or probe combination.

Follow-up studies to monitor for this abnormality by FISH analysis in addition to standard chromosome studies are suggested.

EXAMPLE REPORTING OF ABNORMAL RESULTS – DELETIONS OF 13q AND 17p

Abnormal results

nuc ish 13q14.2(D13S319x1)[25/100],17p13.1(TP53x1)[60/100]

Normal results

nuc ish 4p16(FGFR3x2),9q34(ASSx2),11q13(CCND1x2),13q34(LAMP1x2),14q32(IGHx2), 15q22(PMLx2),16q23(MAFx2),17cen(D17Z1x2)[100]

Interpretation

Fluorescence *in situ* hybridization (FISH) analysis was performed on CD138+ sorted cells with the multiple myeloma panel of probes, including 9q34/ASS, 14q32/IGH, t(11;14)/CCND1/IGH, t(4;14)/FGFR3/IGH, t(14;16)/IGH/MAF, 15q22/PML and 17p13.1/TP53.

This analysis showed evidence of deletions of the D13S319 locus at 13q14 in 25%, and TP53 locus at 17p13 in 60% of cells scored, respectively.

These abnormal findings are consistent with a diagnosis of multiple myeloma.

FISH analysis with the remaining probes showed no evidence of hyperdiploidy or abnormalities of any of these loci in 100 cells scored for each probe or probe combination.

Follow-up studies to monitor for these abnormalities by FISH analysis in addition to standard chromosome studies are suggested.

EXAMPLE REPORTING OF ABNORMAL RESULTS – HYPERDIPLOIDY

Abnormal results

nuc ish 9q34(ASSx3),11q13(CCND1x3)[45/100],15q22(PMLx3)[40/100]

Normal results

nuc ish 4p16(FGFR3x2),14q32(IGHx2),16q23(MAFx2),17cen(D17Z1x2),17p13.1(TP53x2)[100]

Interpretation

Fluorescence *in situ* hybridization (FISH) analysis was performed on CD138+ sorted cells with the multiple myeloma panel of probes, including 9q34/ASS, 14q32/IGH, t(11;14)/IGH/CCND1, t(4;14)/IGH/FGFR3, t(14;16)/IGH/MAF, 15q22/PML and 17p13.1/TP53.

This analysis showed evidence of hyperdiploidy with the gains of chromosomes 9 and 15 in 40% and gain of chromosome 11 in 45% of cells scored, respectively.

This abnormal finding is consistent with a diagnosis of multiple myeloma.

FISH analysis with the remaining probes showed no evidence of abnormalities of any of these loci in 100 cells scored for each probe or probe combination.

Follow-up studies to monitor for these abnormalities by FISH analysis in addition to standard chromosome studies are suggested.

Bibliography

Ahmann GJ, Jalal SM, Juneau AL, et al. A novel three-color, clone-specific fluorescence in situ hybridization procedure for monoclonal gammopathies. Cancer Genet Cytogenet 1998; 101(1): 7–11.

American College of Medical Genetics. *Guidelines in Clinical Cytogenetics. Section E10 Methods in Fluorescence In situ Hybridization – Interphase/Nuclear Fluorescence In Situ Hybridization*. American College of Medical Genetics, Bethesda, MD. www.acmg.net/StaticContent/SGs/Section_E.html, accessed 28 January 2014.

Atlas of Genetics and Cytogenetics in Oncology and Haematology. http://AtlasGeneticsOncology.org/Anomalies, accessed 28 January 2014.

Avet-Loiseau H. Fish analysis at diagnosis in acute lymphoblastic leukemia. Leuk Lymphoma 1999; 33(5-6): 441–449.

Chan JKC. Practical lymphoma diagnosis: a simplified approach. Presented at the 111th CTTR Semi-Annual Seminar, December 2, 2001.

Chase A, Grand FH, Cross NC. Activity of TKI258 against primary cells and cell lines with FGFR1 fusion genes associated with the 8p11 myeloproliferative syndrome. Blood 2007; 110(10): 3729–3734.

Dewald G, Brockman S, Paternoster S, et al. Chromosome anomalies detected by interphase fluorescence in situ hybridization: correlation with significant biological features of chronic lymphocytic leukemia. Br J Haematol 2003; 121: 287–295.

Dohner H, Stilgenbauer S, Dohner K, Bentz M, Lichter P. Chromosome aberrations in B-cell chronic lymphocytic leukemia: reassessment based on molecular cytogenetic analysis. J Mol Med 1999; 77(2): 266–281.

Erben P, Gosenca D, Müller MC, et al. Screening for diverse PDGFRA or PDGFRB fusion genes is facilitated by generic quantitative reverse transcriptase polymerase chain reaction analysis. Haematologica 2010; 95(5): 738–744.

Fu JF, Der-Cherng L, Lee-Yung S. Analysis of acute leukemias with MLL/ENL fusion transcripts: identification of two novel breakpoints in ENL. Am J Clin Pathol 2007; 127: 24–30.

Gersen S, Keagle M (eds). *Principles of Clinical Cytogenetics*. Humana Press, Totowa, New Jersey, 1999.

Hanson CA, Steensma D, Hodnefield J, et al. Isolated trisomy 15: a clonal chromosome abnormality in bone marrow with doubtful hematologic significance. Am J Clin Pathol 2008; 129: 478–485.

Heim S, Mitelman F. *Cancer Cytogenetics*, 2nd edn. Wiley-Liss, New York, 1995.

Kjeldsberg C (ed). *Practical Diagnosis of Hematologic Disorders*, 3rd edn. ASCP Press, Chicago, 2000.

Metzgeroth G, Walz C, Score J, et al. Recurrent finding of the FIP1L1-PDGFRA fusion gene in eosinophilia-associated acute myeloid leukemia and lymphoblastic T-cell lymphoma. Leukemia 2007; 21(6): 1183–1188.

Meyer C, Kowarz E, Hofmann J, et al. New insights to the MLL recombinome of acute leukemias. Leukemia 2009; 23: 1490–1499.

Nordgren A. Hidden aberrations diagnosed by interphase fluorescence in situ hybridisation and spectral karyotyping in childhood acute lymphoblastic leukaemia. Leuk Lymphoma 2003; 44(12): 2039–2053.

Rooney DE, Czepulkowski BH. *Human Cytogenetics. A Practical Approach*. Oxford University Press, New York, 1992.

Shaffer LG, McGowan-Jordan J, Schmid M (eds). *ISCN 2013: An International System for Human Cytogenetic Nomenclature*. Karger Publishers, Unionville, CT, 2013.

Shou Y, Martelli M, Gabrea A, et al. Diverse karyotypic abnormalities of the c-myc locus associated with c-myc dysregulation and tumor progression in multiple Myeloma. Proc Natl Acad Sci USA 2000; 97(1): 228–233.

Sorour Y, Dalley CD, Snowden JA, Cross NC, Reilly JT. Acute myeloid leukaemia with associated eosinophilia: justification for FIP1L1-PDGFRA screening in cases lacking the CBFB-MYH11 fusion gene. Br J Haematol 2009; 146(2): 225–227.

Swerdlow SH, Campo E, Harris NL, et al. (eds). *WHO Classification of Tumours of Haematopoietic and Lymphoid Tissues*. IARC Press, Lyon, 2008.

Wiktor AE, van Dyke DL, Stupca PJ, et al. Preclinical validation of fluorescence in situ hybridization assays for clinical practice. Genet Med 2006; 8: 16–23.

Commercial FISH probes are available at these websites

www.AbbottMolecular.com

www.Kreatech.com

www.Rainbowscientific.com

www.empiregenomics.com

www.abnova.com

www.cytocell.com

www.exiqon.com

www.cambridgebluegnome.com

CHAPTER 31

Integrated reports with cytogenetics and FISH in hematological malignancies

Many cytogenetic studies include the combination of chromosome and fluorescence *in situ* hybridization (FISH) analyses. Chromosome analysis gives a full karyotypic picture of all the genetic changes that are present in the patient at any one time, which may include abnormalities not detectable by FISH analysis. However, if FISH probes are available for a particular chromosomal abnormality observed, FISH analysis is suggested to confirm the abnormality. FISH will also be able to monitor the abnormal finding performed on subsequent specimens on a routine basis without subjecting the patient to periodic bone marrow biopsies, as is usually needed for metaphase chromosome analysis. FISH analysis has a rapid result time, is less expensive than chromosome analysis and is therefore useful in monitoring a patient's progress of disease once a diagnosis is made. Specific examples of common cytogenetic abnormalties that utilize both chromosome and FISH analyses are described below.

31.1 Translocation (9;22) with BCR/ABL1 FISH analysis

EXAMPLE REPORTING OF t(9;22) WITH CHROMOSOME AND BCR/ABL FISH ANALYSIS

ISCN Result: 46,XX,t(9;22)(q34;q11.2)[20].
nuc ish 9q34(ABL1x3),22q11.2(BCRx3),(ABL con BCRx2)[156/200]

Interpretation

Cytogenetic analysis revealed an abnormal female chromosome complement in all cells examined with a translocation between the long arms of chromosomes 9 and 22,

Cytogenetic Abnormalities: Chromosomal, FISH and Microarray-Based Clinical Reporting, First Edition. Susan Mahler Zneimer.

resulting in a Philadelphia chromosome as the sole anomaly. Translocation (9;22) results in the fusion of the ABL1 gene at 9q34 and the BCR gene at 22q11.2 and is consistent with the diagnosis of chronic myeloid leukemia (CML). Fluorescence *in situ* hybridization (FISH) analysis confirmed a BCR/ABL gene rearrangement in 156 of 200 (78%) cells scored.

Follow-up FISH analysis is suggested to help monitor the status of the disease in this patient.

31.2 Monosomy 7 with a marker chromosome and chromosome 7 FISH analysis

EXAMPLE REPORTING OF MONOSOMY 7 WITH A MARKER CHROMOSOME KNOWN TO BE CHROMOSOME 7 BY FISH ANALYSIS (FIGURE 31.1)

ISCN Result: 46,XY,der(7)del(7)(p11.2)del(7)(q11.2q36)[20].
nuc ish 7cen(D7Z1x1),7q31(D7S486x1)[16/200],
nuc ish 7cen(D7Z1x2),7q31(D7S486x1)[145/200]

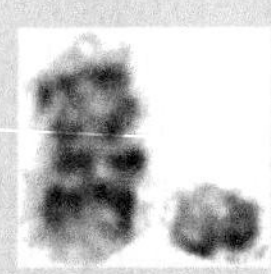

Figure 31.1 Partial karyotype showing monosomy 7 with a marker chromosome known to be chromosome 7 by FISH analysis: 46,XX,der(7)del(7)(p11.2)del(7)(q11.2).nuc ish 7cen(D7Z1x2),7q31(D7S486x1). Courtesy of Sarah South PhD, ARUP Laboratories.

Interpretation

Chromosome analysis revealed an abnormal male chromosome complement with one chromosome 7 containing deletions within both the short and long arms.

Fluorescence *in situ* hybridization (FISH) analysis confirmed this small chromosome is of chromosome 7 origin, showing monosomy 7 in 8% and del(7q) in 72.5% of interphase cells scored, respectively. Del(7q) and monosomy 7 are non-random, recurrent abnormalities commonly observed in the myelodysplastic syndromes and acute myeloid leukemia, occurring in both *de novo* and secondary disease prior to exposure to chemical mutagens or chemotherapy treatments with alkylating agents. Monosomy 7 is generally associated with a poor prognosis.

31.3 Complex abnormalities with the MDS FISH panel

EXAMPLE REPORTING OF COMPLEX CHROMOSOME ABNORMALITIES WITH CORRESPONDING FISH ANALYSIS IN MDS (FIGURE 31.2)

ISCN Result: 43,XY,del(5)(q22q35),add(11)(q23),-16,-18,-21[4]/44,sl,+8[3]/46,XY[14].
nuc ish 5p15.2(D5S721,D5S23x2),5q31(EGR1x1)[167/200],
nuc ish 8cen(D8Z2x3)[48/200]

Interpretation

Cytogenetic analysis revealed two related abnormal male cell lines. The first cell line (stemline), seen in 4 of 21 metaphase cells, contained the following clonal changes:

- hypodiploidy with the loss of chromosomes 16, 18 and 21
- a deletion within the long arm of chromosome 5, confirmed by fluorescence *in situ* hybridization (FISH) analysis in 83.5% of interphase cells
- and additional material of unidentified origin on the long arm of chromosome 11.

The second cell line (sideline), seen in three cells, contained the stemline abnormalities plus an extra copy of chromosome 8, confirmed by FISH analysis in 24.0% of interphase nuclei scored.

The remaining 14 cells showed a normal male chromosome complement.

Del(5q), trisomy 8 and rearrangements involving 11q23 are observed in myelodysplastic syndromes (MDS) and acute myeloid leukemia (AML). Patients with complex cytogenetic abnormalities in addition to deletion of 5q generally have a poor prognosis.

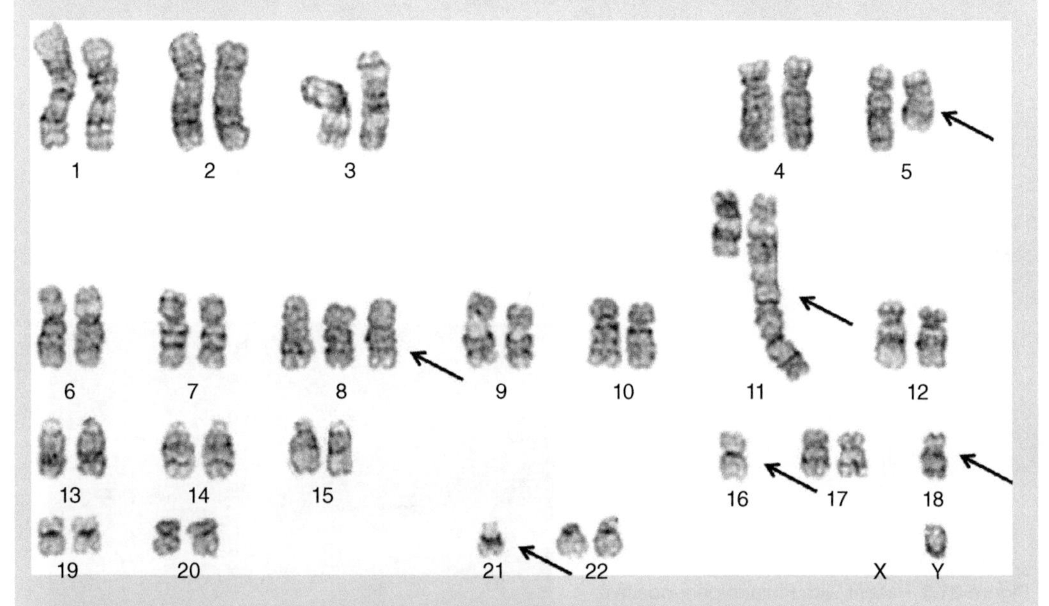

Figure 31.2 Complex abnormalities with corresponding FISH analysis in a patient with anemia to rule out myelodysplasia: 43,XY,del(5)(q22q35),+8,add(11)(q23),-16,-18,-21 (loss of the X chromosome is random loss).

31.4 Complex abnormalities with ALL FISH panel

EXAMPLE REPORTING OF COMPLEX CHROMOSOME ABNORMALITIES WITH CORRESPONDING FISH ANALYSIS IN ACUTE LYMPHOBLASTIC LEUKEMIA (FIGURE 31.3)

ISCN Result: 46,XY,t(9;22)(q34;q11.2)[11]/57,sl,+X,+Y,+4,+6,+14,+15,+17,+18,+21,+der(22)t(9;22),+mar[4]/46,XY[5].
nuc ish 9q34(ABL1x3),22q11.2(BCRx3),(ABL1 con BCRx2)[40/200],
nuc ish 9q34(ABL1x4),22q11.2(BCRx4),(ABL1 con BCRx3)[20/200],
nuc ish 4cen(CEP4x3)[52/200]

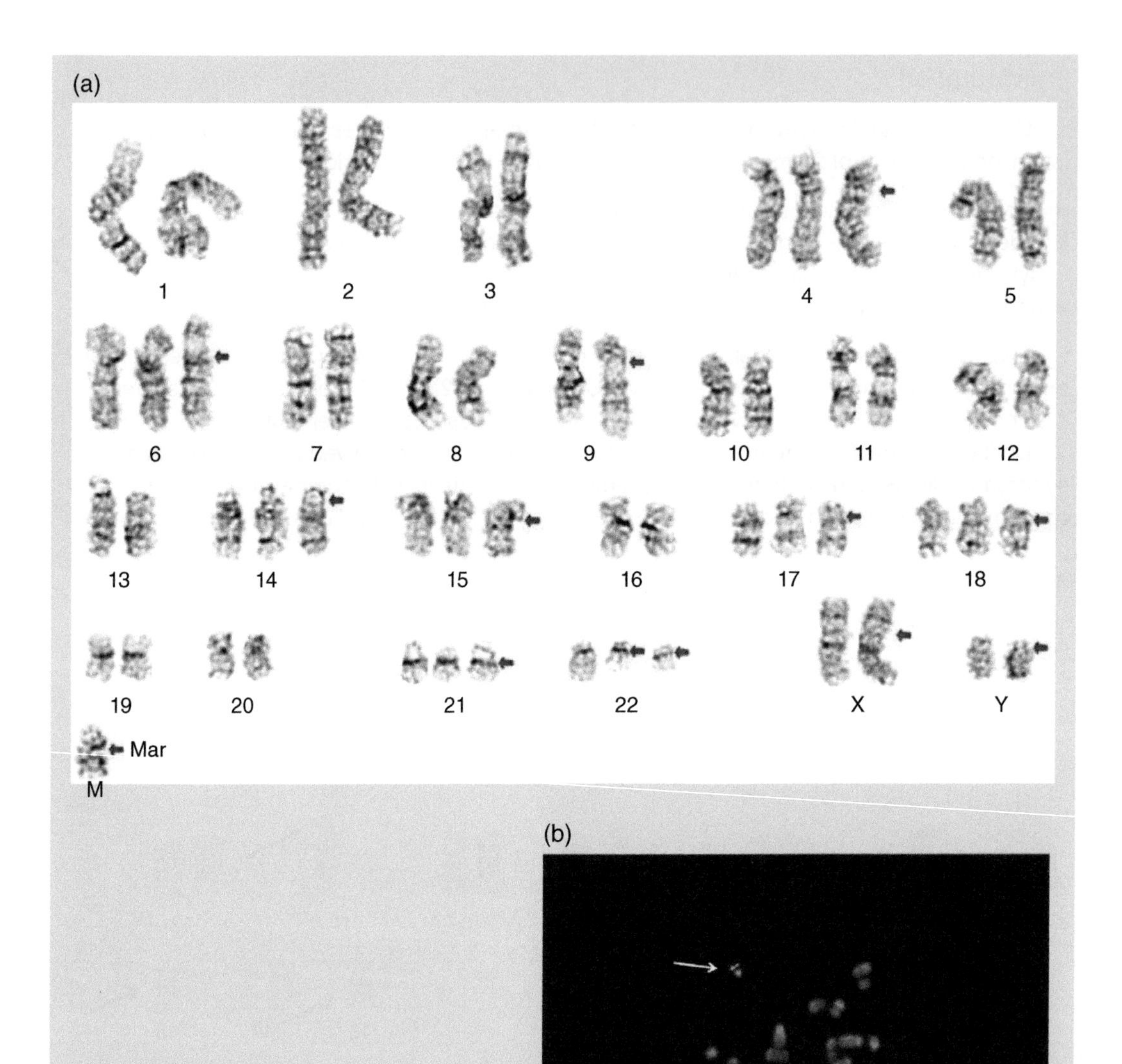

Figure 31.3 Patient with Philadelphia-positive acute lymphoblastic leukemia. (a) Sideline with the following karyotype: 57,XY,+X,+Y,+4,+6,t(9;22)(q34;q11.2),+14,+15,+17,+18,+21,+der(22)t(9;22),+mar. (b) FISH analysis showing two copies of the BCR/ABL1 gene rearrangement (*arrows*).

Interpretation

Cytogenetic analysis revealed an abnormal male chromosome complement with two related abormal male cell lines. The first cell line (stemline), seen in 11 of 20 metaphase cells examined, contained a translocation between the long arms of chromosomes 9 and 22, resulting in a Philadelphia (Ph) chromosome as the sole anomaly. Fluorescence *in situ* hybridization (FISH) analysis confirmed a BCR/ABL gene rearrangement in 20% of cells scored.

The second cell line (sideline), seen in 4 of 20 metaphase cells examined, contained the stemline t(9;22) plus the following clonal abnormalities:

- hyperdiploidy with the gain of chromosomes X, Y, 4, 6, 14, 15, 17, 18 and 21. Trisomy 4 was confirmed by FISH analysis in 26.0% of interphase nuclei scored
- an extra copy of the Philadelphia chromosome, confirmed by FISH analysis in 10% of cells scored
- and gain of an unidentified marker chromosome.

The remaining five cells showed a normal male chromosome complement.

Translocation (9;22) results in the fusion of the ABL1 gene at 9q34 and the BCR gene at 22q11.2, and is consistent with the diagnosis of Ph-positive leukemia, including acute lymphoblastic leukemia (ALL).

Translocation (9;22) is generally associated with a poor prognosis in ALL. Hyperdiploidy, especially with gains of chromosomes 4, 6, 14, 18 and 21, and gain of the der(22), are frequent secondary changes associated with t(9;22)-positive B-ALL.

31.5 Complex abnormalities with MM FISH panel

EXAMPLE REPORTING OF COMPLEX CHROMOSOME ABNORMALITIES WITH CORRESPONDING FISH ANALYSIS IN MULTIPLE MYELOMA (FIGURE 31.4)

ISCN Result: 50,XX,+7,der(10)t(1;10)(q12;p13),t(14;16)(q32;q23),+17,+19,+21[4]/46,XX[16].
nuc ish 7cen(D7Z1x3),7q31(D7S486x3)[92/100],
nuc ish 17p13.1(TP53x3),17cen(D17Z1x3)[86/100],
nuc ish 14q32(IGHx3),16q22(MAFx3),(IGH con MAFx2)[79/100]

Interpretation

Cytogenetic analysis revealed an abnormal female chromosome complement in 4 of 20 cells analyzed with the following clonal changes:

- gain of chromosomes 7, 17, 19 and 21. Trisomy 7 and 17 were confirmed by fluorescence *in situ* hybridization (FISH) analysis in 92.0% and 86.0% of plasma cells scored, respectively
- an unbalanced translocation between the long arm of chromosome 1 and the short arm of chromosome 10, resulting in partial trisomy for chromosome 1q
- and a translocation between the long arms of chromosomes 14 and chromosome 16. An IGH/MAF gene rearrangement was detected by FISH analysis in 79.0% of plasma cells scored.

The remaining 16 cells showed a normal female chromosome complement.

Trisomy 1q, 7, 17 and 19, and translocation (14;16), are recurrent, non-random abnormalities observed in plasma cell disorders. IGH/MAF rearrangements are found in approximately 5–10% of patients with multiple myeloma by FISH analysis and are generally associated with a poor prognosis.

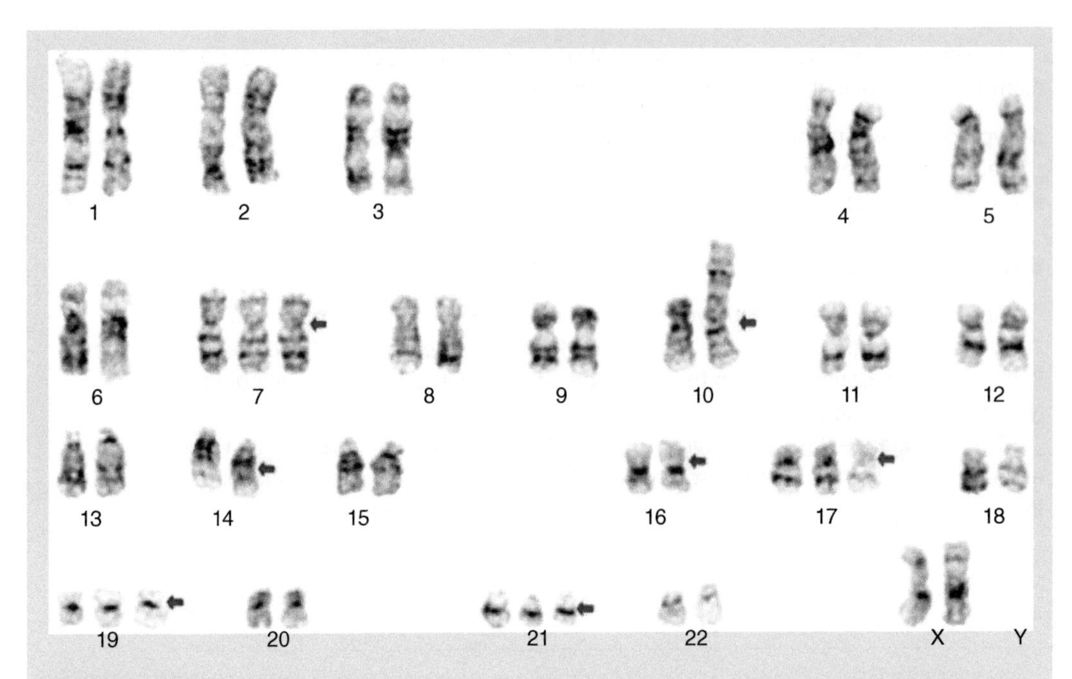

Figure 31.4 Karyotype showing hyperdiploidy, t(14;16) and trisomy 1q, with additional abnormalities, in a patient with multiple myeloma: 50,XX,+7,der(10)t(1;10)(q12;p13),t(14;16)(q32;q23),+17,+19,+21.

31.6 Complex abnormalities with AML FISH panel

EXAMPLE REPORTING OF COMPLEX CHROMOSOME ABNORMALITIES WITH CORRESPONDING FISH ANALYSIS WITH PANCYTOPENIA (FIGURE 31.5)

ISCN Result: 45,XY,-5,del(7)(q22q32),+8,der(14;21)(q10;q10),-16,add(17)(p12),add(21)(p11.2), +22[14]/46,sl,+6[3]/46,XY[3].
nuc ish 5p15.2(D5S23,D5S721x2),5q31(EGR1x1)[160/200],
nuc ish 7cen(D7Z1x2),7q31(D7S486x1)[178/200],
nuc ish 8cen(D8Z2x3)[167/200]

Interpretation

Cytogenetic analysis revealed a male chromosome complement with two related abnormal cell lines. The first cell line (stemline), seen in 14 cells, showed the following clonal changes:

- loss of chromosomes 5 and 16;
- gain of chromosomes 8 and 22
- a deletion within the long arm of chromosome 7;
- a whole arm translocation between the long arms of chromosomes 14 and 21
- and added material on the short arms of chromosomes 17 and 21, resulting in deletion of 17p.

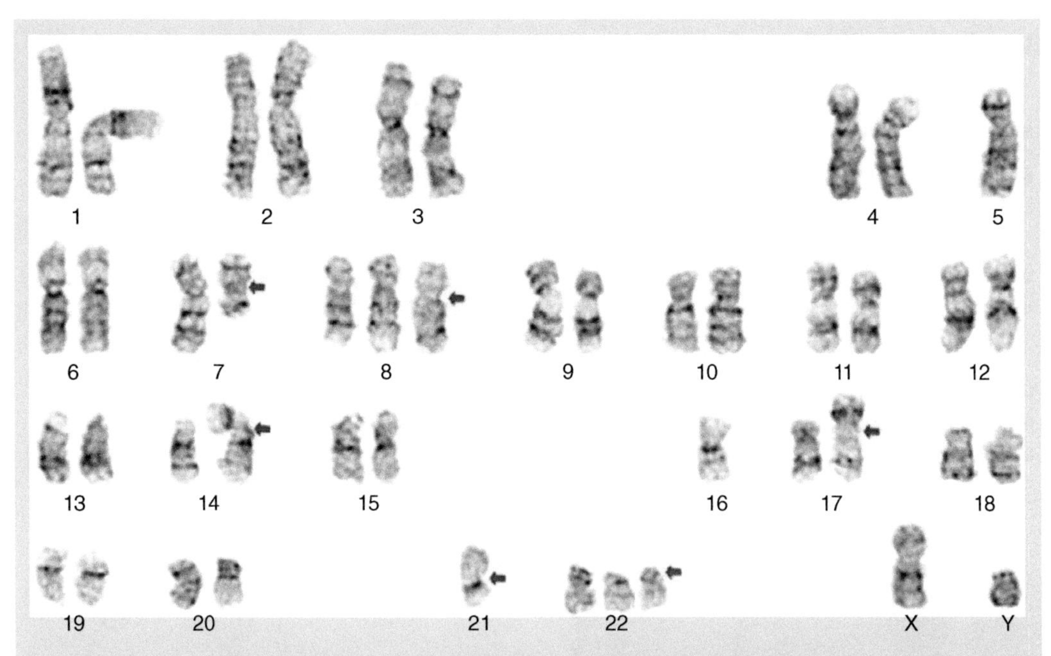

Figure 31.5 Complex abnormalities including monosomy 5, del(7q) and trisomy 8, with FISH confirmation, in a patient with pancytopenia: 45,XY,-5,del(7)(q22q32),+8,der(14;21)(q10;q10),-16,add(17)(p12),add(21)(p11.2),+22.

The second cell line (sideline), seen in three cells, showed the stemline aberrations plus the gain of chromosome 6.

Fluorescence *in situ* hybridization (FISH) analysis confirmed del(5q), del(7q), +8 and -16, and showed the gain of chromosome 21, which may be represented by the additional material of unknown origin.

The remaining three cells showed a normal male chromosome complement.

Deletion of chromosomes 5q and 7q and 17p, and trisomy 8 are recurrent abnormalities in both *de novo* and secondary myelodysplasia (MDS) and acute myeloid leukemia (AML). Patients with complex cytogenetic abnormalities involving chromosomes 5 and 7 generally have a poor prognosis.

31.7 Complex abnormalities with AML FISH panel in therapy-related disease

EXAMPLE REPORTING OF COMPLEX CHROMOSOME ABNORMALITIES WITH CORRESPONDING FISH ANALYSIS WITH THERAPY-RELATED DISEASE (FIGURE 31.6)

ISCN Result: 46,XX,del(5)(q11.2),del(7)(q31q34),+8,del(10)(q24q26),del(12)(p11.2p13),-18[20].
nuc ish 5p15.2(D5S23x2),5q31(EGR1x1)[189/200],
nuc ish 7cen(D7Z1x2),7q31(D7S486x1)[158/200],
nuc ish 8cen(D8Z2x3)[194/200]

Interpretation

Cytogenetic analysis revealed an abnormal female chromosome complement in all cells examined showing the following clonal changes:

- deletions within the long arms of chromosomes 5, 7 and 10, and within the short arm of chromosome 12. Del(5q) and del(7q) were confirmed by fluorescence *in situ* hybridization (FISH) analysis in 94.5% and 79.0% of interphase nuclei scored, respectively
- gain of chromosome 8, confirmed by FISH analysis in 97.0% of interphase nuclei scored
- and loss of chromosome 18.

Deletions of chromosomes 5q, 7q and 12p, and trisomy 8 are recurrent, non-random abnormalities in myelodysplasia (MDS) and acute myeloid leukemia (AML), often in therapy-related disease. Patients with complex cytogenetic abnormalities involving chromosomes 5 and 7 generally have a poor prognosis.

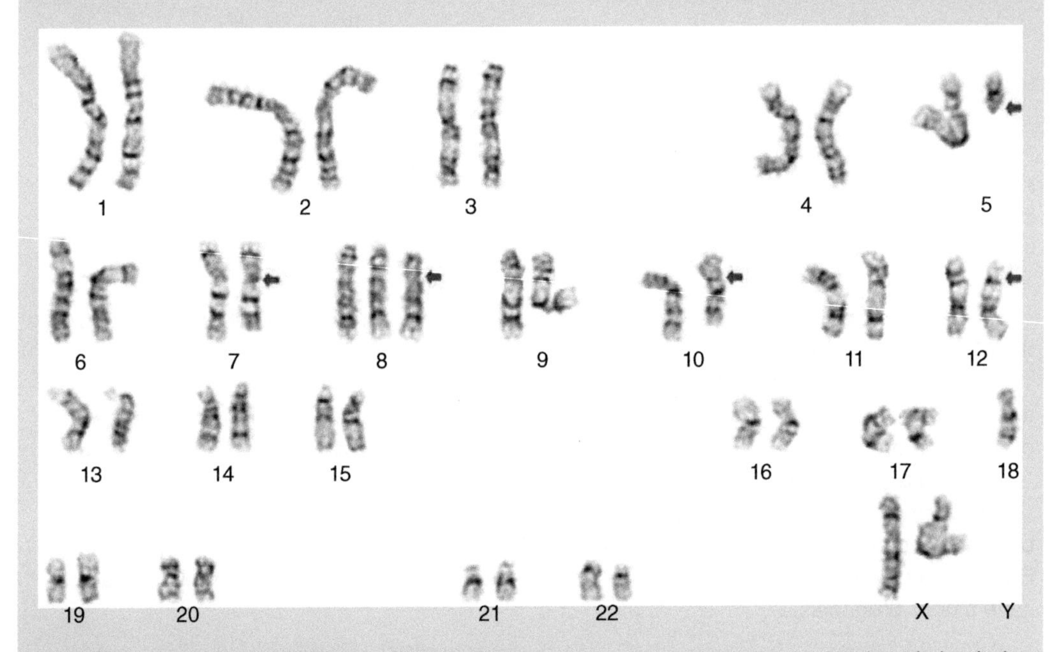

Figure 31.6 Complex abnormalities with FISH confirmation in a patient with therapy-related myelodysplasia: 46,XX,del(5)(q11.2),del(7)(q31q34),+8,del(10)(q24q26),del(12)(p11.2p13),-18.

Bibliography

American College of Medical Genetics. *Guidelines in Clinical Cytogenetics. Section E10 Methods in Fluorescence In situ Hybridization – Interphase/Nuclear Fluorescence In Situ Hybridization*. American College of Medical Genetics, Bethesda, MD. www.acmg.net/StaticContent/SGs/Section_E.html, accessed 28 January 2014.

Atlas of Genetics and Cytogenetics in Oncology and Haematology. http://AtlasGeneticsOncology.org/Anomalies, accessed 28 January 2014.

Gersen S, Keagle M (eds). *Principles of Clinical Cytogenetics*. Humana Press, Totowa, New Jersey, 1999.

Heim S, Mitelman F. *Cancer Cytogenetics*, 2nd edn. Wiley-Liss, New York, 1995.

Rooney DE, Czepulkowski BH. *Human Cytogenetics. A Practical Approach*. Oxford University Press, New York, 1992.

Shaffer LG, McGowan-Jordan J, Schmid M (eds). *ISCN 2013: An International System for Human Cytogenetic Nomenclature*. Karger Publishers, Unionville, CT, 2013.

Swerdlow SH, Campo E, Harris NL, et al. (eds). *WHO Classification of Tumours of Haematopoietic and Lymphoid Tissues*. IARC Press, Lyon, 2008.

Commercial FISH probes are available at these websites

www.AbbottMolecular.com

www.Kreatech.com

www.Rainbowscientific.com

www.empiregenomics.com

www.abnova.com

www.cytocell.com

www.exiqon.com

www.cambridgebluegnome.com

CHAPTER 32

Recurrent FISH abnormalities in solid tumors using paraffin-embedded tissue

Sections from formalin-fixed paraffin-embedded blocks are commonly used for fluorescence *in situ* hybridization (FISH) analysis. The sections are pretreated to deparaffinize the tissue and digest away any cytoplasmic components from the chromatin. The chromatin of these cells is then denatured to a single-stranded configuration and allowed to hybridize with the probes of interest. Following hybridization, the unbound probe is removed by a series of washes and the nuclei are counterstained with DAPI (4,6 diamidino-2-phenylindole), a DNA-specific stain that fluoresces as a blue color. Hybridization of probes is viewed using an epifluorescence microscope equipped with the appropriate emission and excitation filters to reveal the fluorescence signals.

Fluorescence *in situ* hybridization on paraffin-embedded tissue poses technical challenges not commonly associated with fresh peripheral blood and bone marrow samples. The limitations are mainly due to cellular truncation and long exposure to embedding fixatives, which can cause difficulties in scoring and interpretation of fluorescent signal patterns. However, FISH on paraffin-embedded tissue was developed to increase sensitivity and specificity in detecting genetic anomalies in tissue preparations.

Successful *in situ* hybridization on paraffin-embedded tissue is dependent on two important factors: effective disruption of the methylene bridge and penetration of the probe of interest to the tissue. Formalin fixation of tissue results in the formation of the methylene bridges between amino groups in the DNA and/or proteins. Fixatives such as formalin and B5 are immunohistochemical reagents used to preserve nuclear morphology and cellular composition, but have adverse effects when applied to molecular genetic techniques. While the integrity of the tissue is maintained with fixative procedures, formalin inhibits binding of the DNA to the tissue and reduces the binding sites available to the DNA. To reverse the effects of the methylene bridge or cross-linking, tissues are exposed to high temperature and a buffering solution. The high temperature provides enough energy to break the methylene bridge. Chelating agents in the buffering system bind heavy metals, including calcium, magnesium and divalent cations intimately associated with the DNA and/or proteins. Buffering systems stabilize the pH of the solution by inhibiting nuclease activity when heavy metals are precipitated and prevents

Cytogenetic Abnormalities: Chromosomal, FISH and Microarray-Based Clinical Reporting, First Edition. Susan Mahler Zneimer.

denaturing of the DNA. Co-denaturation of both the probe and tissue at increased temperature further obviates the protein cross-linking and permits annealing of the probe to complementary sequences, making FISH on paraffin-embedded tissue a good technique to identify genetic abnormalities.

Nuclear truncation can occur during sectioning of the tissue when sections from invasive areas are not sectioned entirely from the block and all or part of the invasive section is missing from the original tissue block. The section on the slide must contain the entire tissue from the block since the hematoxylin and eosin (H&E)-stained slide is used to elucidate the representative tissue section.

The amount of time the tissue spends in fixative may also have adverse effects on hybridization and consequently signal scoring. Since greater exposure of the tissue to embedding fixative increases cross-linking, minimizing the length of time in fixatives produces better FISH results.

The thickness and depth across the tissue can also be a factor during sectioning. Some areas on the tissue may be further embedded in paraffin during paraffinization, and the invasive area may be truncated from the rest of the tissue.

Sample-to-sample variation can result from a number of different factors, mainly variation in exposure times in embedding fixatives, the type of tissue, the depth and thickness across the section, and also the charge on the slide.

Focal plane distortions may lead to difficulties in scoring and interpretation of signals. Samples with high autofluorescence may also interfere with signal scoring. Exposure of the specimens to acids, strong bases or extreme heat should also be avoided. Such conditions are known to damage DNA and may result in FISH assay failure.

For a list of probes used for common tumors, see Table 32.1. For a comprehensive list of all available probes for hematological disorders and tumors, see Table 32.2.

Table 32.1 Common FISH probes for tumors

Tumor	Probe			
Bladder cancer	P16/9p21 homozygous loss	Centromere 3/7/17 trisomies		
Breast cancer	Her2/ERBB2/ 17q.11.2-q12 amplification -	Centromere 17 control		
Colorectal cancer	EGFR/17p12 amplification	Centromere 7 control		
Ewing sarcoma	EWSR1/22q12 rearrangement			
Myxoid liposarcoma	DDIT3/CHOP/12q13 rearrangement			
Neuroblastoma	N-MYC/2p24.3 amplification			
Non-small cell lung carcinoma	ALK/2p23 rearrangements	EGFR/17p12 amplification	ROS1 rearrangements	RET1 rearrangements
Oligodendroglioma	1p/19q deletions			
Rhabdomyosarcoma	FOX01/13q14 rearrangement			
Synovial sarcoma	SYT/18q11.2 rearrangement			

Table 32.2 Comprehensive list of individual probes for hematological and tumor disorders, by chromosome order

Probe location	Gene region	Disease of significance
1p/19q	t(1;19) unbalanced rearrangements	Glioma, astrocytoma
1p36	TP58	Neuroblastoma, glioblastoma deletion 1p36
1p32	SIL-TAL 1	T-cell ALL
2p23-24	MYCN	Neuroblastoma amplification status
t(2;13) or variant	FKHR	Rhabdomyosarcoma
t(2;5)	ALK/NPM	Anaplastic large cell lymphoma
t(3q26)	EVI1	Myeloid leukemia
3q27	BCL6	DLBCL
4q10	CEP 4	Pediatric precursor B-ALL
4q12 deletion	CHIC2	Hypereosinophilia/mast cell disease
t(4:14)	FGFR3/IGH	Multiple myeloma
5q31	EGR1/D5S23	Myeloid disorders deletion 5q/monosomy 5
5q33	PDGFRB	Myeloproliferative neoplasms
6q24/6q23	SHPRH*/MYB/CEP6	Myeloma, Waldenstrom macroglobulinemia
6q27	MYB	Lymphoma, myeloma, Waldenstrom macroglobulinemia
t(6;9)	DEK/CAN*	AML
7q10	CEP7	Myeloid or lymphoid leukemia, NHL
7p11-p12	EGFR	Non-small cell lung cancer, glioma amplification status
7q31	D7S486/CEP7	Myeloid disorders del 7q/monosomy 7
8q10	CEP8 trisomy	Myeloid leukemia or MDS
8q24	MYC break apart	Lymphoma, leukemias: amplification, rearrangement
t(8;14)	MYC/IGH	Burkitt lymphoma
t(8;21)	RUNX1T1/RUNX1 (ETO/AML1)	AML M2
9q10	CEP9	Lymphoid or myeloid leukemia, MDS
9p21	TP16	T-cell lymphoblastic leukemia, pediatric ALL
t(9;22)	BCR/ABL1	CML, AML and ALL
10q10	CEP 10	Pediatric precursor B-ALL
10q	PTEN	Glioma
11p15	WT1	Wilms tumor
11p15	NUP98	AML
11q22.3	ATM	CLL
t(11q23)	MLL	Mixed lineage, AML, ALL, deletion, rearrangement or amplification
t(11;14)	CCND1/IGH (BCL1/IGH)	Mantle cell lymphoma, multiple myeloma

(*Continued*)

Table 32.2 *(Continued)*

Probe location	Gene region	Disease of significance
t(11;18) or variant	MALT1	MALT lymphoma
t(11;19)	ELL/ENL	ALL or AML. Identify gene on 19p involved in t(11;19)
t(11;22) or variant	EWSR 1	Ewing sarcoma
12q10	CEP 12 trisomy	CLL
t(12;16) or t(12;22)	CHOP	Liposarcoma
t(12;21) (cryptic)	ETV6/RUNX1 (TEL/AML1)	Pediatric precursor B-ALL
13q14	RB1	Retinoblastoma deletion
13q14	D13S25/D13S319	CLL, MM, MPD deletion
14q32	IGH gene rearrangement	Lymphoma, MM, HCL
14q11.2	TCRTM*	T-cell ALL, PLL
t(14;16)	MAF/IGH	Multiple myeloma
t(15;17)	PML/RARA	AML M3
16q22	CBFB	AML M2/M4
17q10	CEP 17	Pediatric precursor B-ALL
17p13	TP53	Solid tumors, lymphoid or myeloid leukemias, MM
20q12	D20S108	Myeloid disorders del 20q
21q22	LSI21	Myeloid or lymphoid leukemia, NHL
Xq10, Yq10	CEP X and Y	post gender mismatched BMT or SCT
t(X;18)	SYT	Synovial sarcoma

ALL, acute lymphoid leukemia; AML, acute myeloid leukemia; BMT, bone marrow transplant; CLL, chronic lymphoid leukemia; CML, chronic myeloid leukemia; DLBCL, diffuse large B-cell lymphoma; HCL, hairy cell leukemia; MALT, mucosa-associated lymphoid tissue; MDS, myelodysplastic syndrome; MM, multiple myeloma; MPD, myeloproliferative disorder; NHL, non-Hodgkin lymphoma; PLL, prolymphocytic leukemia; SCT, stem cell transplant.

32.1 Ewing sarcoma

SPECIFIC FEATURES

Approximately 90% of Ewing tumors show a t(11;22)(q24;q12). This translocation results in the fusion of the EWSR1 gene at 22q12 with the transcription factor gene FLI1 at 11q24, leading to an oncogenic chimeric protein. Variant translocations also exist, including t(21;22)(q12;q12) with the ERG gene and t(7;22)(p22;q12) with the ETV1 gene. Therefore, FISH analysis with the EWSR1 breakapart probe is useful in identifying any chromosome 22q12 rearrangements. Additional anomalies in Ewing tumors mainly consist of chromosomal gains, including trisomy 8 (most frequently), but also trisomy for chromosomes 1q, 2, 5, 7, 9 and 12.

EWSR1 rearrangements are also seen in peripheral primitive neuroectodermal tumors, desmoplastic small round cell sarcomas, malignant melanoma of soft parts and myxoid chondrosarcomas.

EXAMPLE REPORTING OF 22q12 – EWING SARCOMA (FIGURE 32.1)

ISCN Result: nuc ish 22q12(EWSR1x2)(5′EWSR1 sep 3′EWSR1x1)[60/100]

Interpretation

Fluorescence *in situ* hybridization (FISH) analysis on paraffin-embedded tumor with the EWSR1 probe showed evidence of a 22q12 rearrangement in 60% of interphase nuclei scored.

EWSRI rearrangements are recurrent, non-random abnormalities associated with Ewing sarcomas, peripheral primitive neuroectodermal tumors, desmoplastic small round cell sarcomas, malignant melanoma of soft parts and myxoid chondrosarcomas.

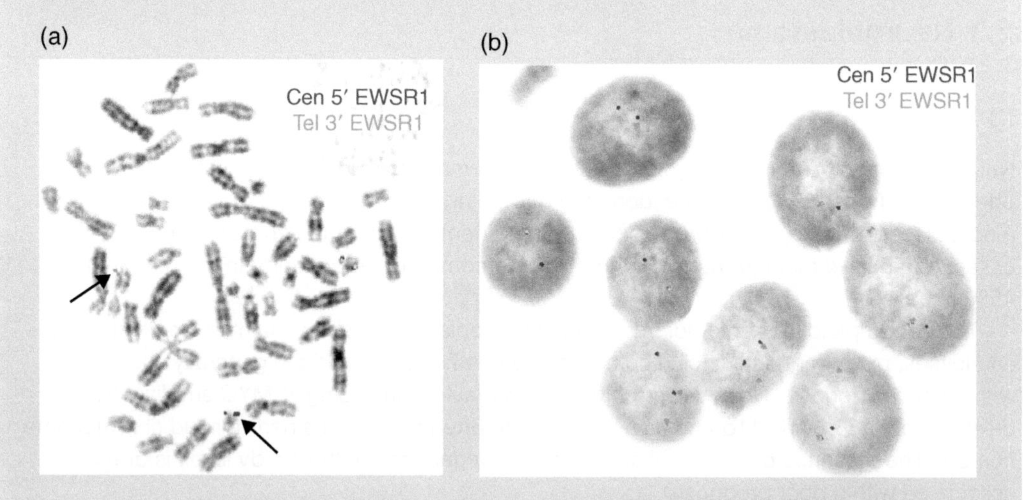

Figure 32.1 Ewing's tumor with a breakapart probe with (a) metaphase cell showing EWSR1 rearrangement (*arrows* with split red/green signals on chromosome 22 and partner chromosome), and (b) interphase cells showing one fusion signal on the normal chromosome 22 and split red/green signals with an EWSR1 rearrangement. Courtesy of Lauren Jenkins PhD, Kaiser Permanente Regional Laboratory, Northern California.

32.2 Liposarcoma

SPECIFIC FEATURES

Myxoid liposarcomas show a distinct translocation, t(12;16)(q13;p11), in approximately 95% of cases. This rearrangement results in the fusion of the CHOP (DDIT3) gene on chromosome 12q13 with the TLS (FUS) gene on chromosome 16p11. A variant translocation, t(12;22)(q13;q12), involving the CHOP gene with the EWS gene on chromosome 22q12 has also been observed in a minority of cases. The CHOP probe is a dual color, breakapart probe that will ascertain any gene rearrangement with chromosome 12q13.

EXAMPLE REPORTING OF 12q13 – LIPOSARCOMA

ISCN Result: nuc ish 12q13(CHOPx2)(5′CHOP sep 3′CHOPx1)[60/100]

Interpretation

Fluorescence *in situ* hybridization (FISH) analysis on a paraffin-embedded tumor with the CHOP probe showed evidence of a 12q13 rearrangement in 60% of interphase nuclei scored.

CHOP rearrangements are recurrent, non-random abnormalities associated with myxoid liposarcomas and are seen in approximately 95% of cases.

32.3 Neuroblastoma

SPECIFIC FEATURES

Neuroblastomas have distinct genomic DNA abnormalities that predict the clinical phenotype. Two main groups are identified according to the tumor cell ploidy: near-triploid (and pentaploid/hexaploid) chromosome complement, occurring in approximately 55% of cases, and near-diploid (tetraploid) chromosome complement occuring in the remaining 45% of cases.

The near-triploid (or pentaploid/hexaploid) neuroblastomas show mainly numerical gains and losses, most frequently including chromosomes 7 and 17, and are generally associated with a good prognosis. However, structural abnormalities, including N-MYC amplification, deletions of 1p36.3 and 11q and gains of 17q, are often seen with a near-triploid chromosome number. The presence of structural abnormalities, regardless of the ploidy level, is often associated with a poor prognosis.

Diploid/tetraploid tumors are generally associated with a very poor prognosis and often include unbalanced structural rearrangements. One rearrangement is N-MYC amplification which is often seen as double minute chromosomes (dmins) and can also be seen with 1p deletions and 17q gain. Cases without N-MYC amplification may show an 11q deletion and 17q gain, often in conjunction with 3p deletions. Identification of N-MYC amplification is an important criterion for making clinical subgroup decisions.

EXAMPLE REPORTING OF 2p24.1 – NEUROBLASTOMA (FIGURE 32.2)

ISCN Result: nuc ish nuc ish 2p24.1(N-MYCx3-10),2cen(D2Z1x2)[60/100]

Interpretation

Fluorescence *in situ* hybridization (FISH) analysis on a paraffin-embedded tumor with the N-MYC probe showed evidence of amplification of the N-MYC gene in 60% of interphase nuclei scored.

N-MYC amplification is a recurrent, non-random abnormality observed in neuroblastomas and generally associated with a poor prognosis.

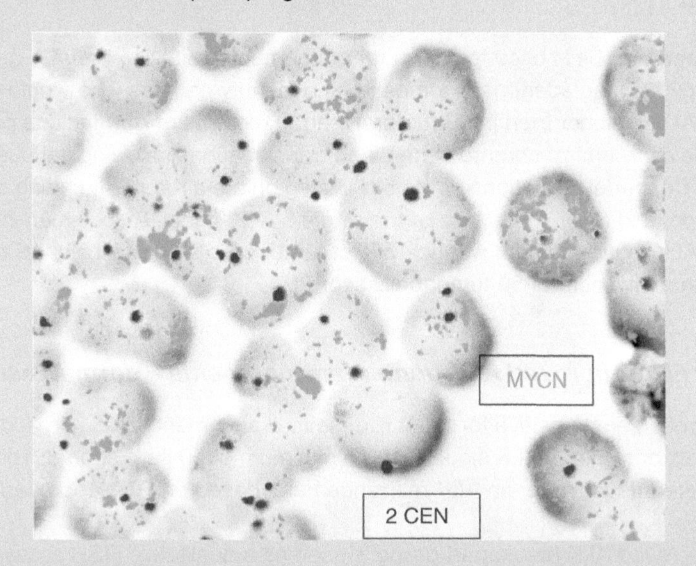

Figure 32.2 N-MYC amplification (seen as green signals) in a patient with neuroblastoma. Courtesy of Lauren Jenkins PhD, Kaiser Permanente Regional Laboratory, Northern California.

32.4 Non-small cell lung cancer

SPECIFIC FEATURES

Non-small cell lung cancer (NSCLC) tumors generally include TP53/17p13.1, CDKN2A/9p21.3, ALK, EGFR/7p11.2 and KRAS/12p12.1 abnormalities. Mutations in the epidermal growth factor receptor (EGFR), KRAS and anaplastic lymphoma kinase (ALK) genes are mutually exclusive in patients with NSCLC, and the presence of one of these mutations can influence response to targeted therapy. Up to 6% of all NSCLC carry an ALK rearrangement in which ALK is fused to the EML4 gene, which is seen as inv(2)(p21;p23). The inv(2)(p21;p23) also shows multiple variants depending on the breakpoint on the EML4 gene.

Other genes involved in NSCLC show gene amplification but in relatively low frequencies, and may be studied by FISH analysis, including CCND1/11q13, TP73L/3q28, KRAS2/12p12, MYC/8q24.21, EGFR/7p11.2, RET/10q11.2 and ROS1/6q22 genes.

The RET1 gene/10q11.2, seen as a pericentric inversion of chromosome 10, is known to increase expression of the RET gene by fusion with KIF5B. Translocations with other fusion partners have also been described. Elevated expression of RET is observed in NSCLC, in which tyrosine kinase-based therapies are used to inhibit the fusion proteins. Translocations involving RET have also been described in thyroid carcinomas.

Approximately 2% of lung tumors have ROS1 abnormalities. Several different ROS1 rearrangements have been described in NSCLC. These include SLC34A2-ROS1, CD74-ROS1, EZR-ROS1, TPM3-ROS1 and SDC4-ROS1; therefore, using a breakapart probe is the FISH method of choice for ROS1 abnormalities.

32.4.1 ALK gene rearrangements

SPECIFIC FEATURES

The ALK breakapart probe is used to detect rearrangements involving the ALK gene at chromosome 2p23 in lung adenocarcinomas to help identify patients who would benefit from treatment with ALK-inhibitor therapy. Structural rearrangements of ALK include paracentric inversions of the short arm of chromosome 2 and translocations involving various partner chromosomes. Concurrent deletions of the 5' ALK region are a recurrent finding in association with the inversion, and result in the loss of the single 5'ALK signal. A specimen is considered positive for an ALK rearrangement when >15% of tumor cells show split signals or single 3' ALK signals with a minimum of 50 tumor cells scored.

Disclaimer for FDA-approved ALK gene rearrangements – normal results

A specimen is considered positive for an ALK gene rearrangement when >15% of tumor cells show split signals or single 3' ALK signals with a minimum of 50 cells scored. Therefore, this result is considered negative for an ALK rearrangement, based on FDA-approved scoring criteria.

The FDA-approved ALK breakapart probe is used as a qualitative FISH assay to detect rearrangements in formalin-fixed paraffin-embedded lung adenocarcinoma tumors to help identify patients who would benefit from treatment with ALK inhibitor therapy.

Comment for copy number changes of ALK

There was no evidence of an ALK gene rearrangement in the cells examined. However, gain of the ALK gene was observed. Copy number changes of the ALK gene are common findings in lung carcinomas, occurring in a significantly greater frequency than ALK rearrangements. The clinical significance of ALK aneuploidy is unknown in lung cancer at this time, with no correlation between ALK copy number and prognosis of disease.

ALK FISH studies using a ALK dual color breakapart FISH probe (Figure 32.3)

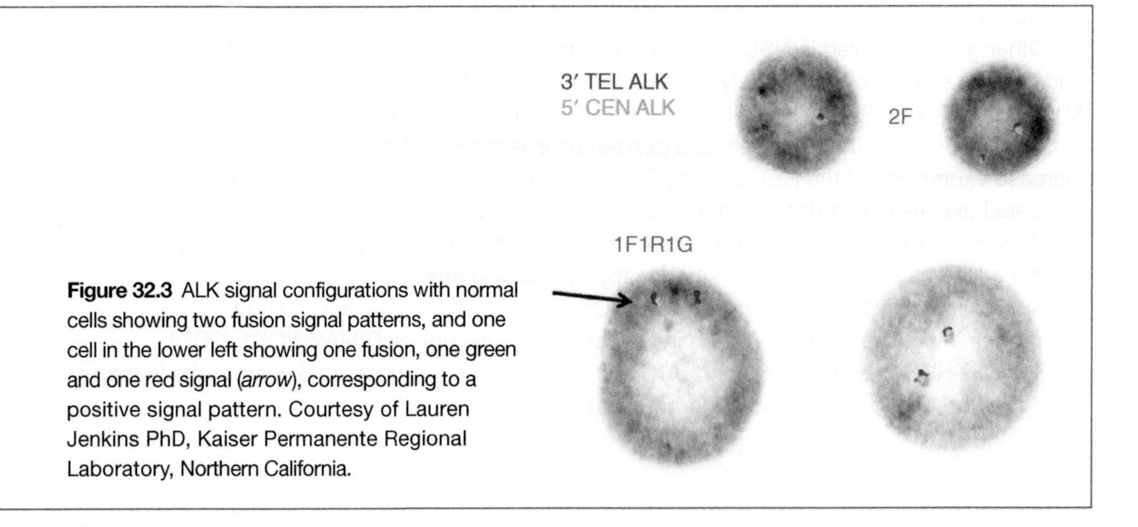

Figure 32.3 ALK signal configurations with normal cells showing two fusion signal patterns, and one cell in the lower left showing one fusion, one green and one red signal (*arrow*), corresponding to a positive signal pattern. Courtesy of Lauren Jenkins PhD, Kaiser Permanente Regional Laboratory, Northern California.

Common negative signal patterns
nuc ish(ALKx2) normal – 2 fusions
nuc ish(5'ALKx2,3'ALKx1)(5'ALK con 3'ALKx1) ALK with 3' deletion – 1 green/1 fusion
nuc ish(ALKx1) loss of intact ALK signal – 1 fusion
nuc ish(ALKx3) gain of intact ALK signals – 3 fusions

Common positive signal patterns
nuc ish(ALKx2)(5'ALK sep 3'ALKx1) ALK rearrangement – 1 green, 1 red, 1 fusion
nuc ish(ALKx2)(5'ALK sep 3'ALKx2) ALK rearrangement x2 – 2 green, 2 red
nuc ish(5'ALKx1,3'ALKx2),(5'ALK con 3'ALKx1) ALK rearrangement with 5' loss – 1 red, 1 fusion
nuc ish(5'ALKx1,3'ALKx2)(5'ALK sep 3'ALKx1) ALK rearrangements x2, one 5' loss – 1 green, 2 red

Normal ALK = fusion
5'ALK = red
3'ALK = green

EXAMPLE REPORTING OF ALK REARRANGEMENTS IN LUNG CANCER – NORMAL

ISCN Result: nuc ish 2p23(ALKx2)[50]

Interpretation

Fluorescence *in situ* hybridization (FISH) analysis on a paraffin-embedded tumor with the ALK probe showed no evidence of an ALK gene rearrangement in 50 interphase nuclei scored.

A specimen is considered positive for an ALK gene rearrangement when >15% of tumor cells show split signals or single 3' ALK signals with a minimum of 50 cells scored. Therefore, this result is considered negative for an ALK rearrangement.

Structural rearrangements of the ALK gene include paracentric inversions of the short arm of chromosome 2 and translocations involving various translocation partners with the ALK gene. Also, deletions of the 5' ALK gene region are a recurrent abnormality in conjunction with inversions and result in the loss of a single 5' ALK signal.

EXAMPLE REPORTING OF ALK REARRANGEMENTS IN LUNG CANCER – ABNORMAL

ISCN Result: nuc ish 2p23(ALKx2)(5'ALK sep 3'ALKx1)[8/50]/
2p23(5'ALKx1,3'ALKx2),(5'ALK con 3'ALKx1)[5/50]/
2p23(ALKx2)(5'ALK sep 3'ALKx2)[3/50]

Interpretation

Fluorescence *in situ* hybridization (FISH) analysis on a paraffin-embedded tumor with the ALK probe showed evidence of an ALK gene rearrangement in 32% interphase nuclei scored.

A specimen is considered positive for an ALK gene rearrangement when >15% of tumor cells show split signals or single 3' ALK signals with a minimum of 50 cells scored. Therefore, this result is considered positive for an ALK rearrangement.

Structural rearrangements of the ALK gene include paracentric inversions of the short arm of chromosome 2, and translocations involving various translocation partners with the ALK gene. Also, deletions of the 5' ALK gene region is a recurrent abnormality in conjunction with inversions and results in the loss of a single 5' ALK signal.

EXAMPLE REPORTING OF ALK REARRANGEMENTS IN LUNG CANCER – WITH COPY NUMBER CHANGES

ISCN Result: nuc ish 2p23(ALKx2)[40/50]/2p23(ALKx3-5)[10/50]

Interpretation

Fluorescence *in situ* hybridization (FISH) analysis on a paraffin-embedded tumor with the ALK probe showed no evidence of an ALK gene rearrangement in 50 interphase nuclei scored. However, gain of the ALK gene (3–5 copies) was observed. Copy number changes of the ALK gene are common findings in lung carcinomas, occurring in a significantly greater frequency than ALK rearrangements. The clinical significance of ALK aneuploidy is unknown in lung cancer at this time, with no correlation between ALK copy number and prognosis of disease.

A specimen is considered positive for an ALK gene rearrangement when >15% of tumor cells show split signals or single 3' ALK signals with a minimum of 50 cells scored. Therefore, this result is considered negative for an ALK rearrangement.

Structural rearrangements of the ALK gene include paracentric inversions of the short arm of chromosome 2 and translocations involving various translocation partners with the ALK gene. Also, deletions of the 5' ALK gene region are a recurrent abnormality in conjunction with inversions and result in the loss of a single 5' ALK signal.

32.4.2 EGFR gene rearrangements

SPECIFIC FEATURES

EGFR mutations are more commonly observed in patients with lung adenocarcinomas with no prior history of smoking as well as those of Asian descent. EGFR mutations in lung carcinomas make the disease more responsive to treatment with tyrosine kinase inhibitors (TKIs). Use of the EGFR-TKIs gefitinib and erlotinib is limited to patients with adenocarcinomas who have known activating EGFR mutations. The EGFR gene is located on chromosome 7p11.2 and is amplified in NSCLC. Some patients with EGFR amplification show improved survival on EGFR tyrosine kinase inhibitors.

Scoring of EGFR abnormalities by FISH includes tumors with four or more copies of the EGFR gene in ≥40% of the cells (high polysomy). Tumors with EGFR gene amplification (gene to chromosome ratio of ≥2 or presence of gene clusters of ≥15 gene copies in ≥10% of the cells) are considered to be EGFR FISH positive. All other tumors are considered to be EGFR FISH negative.

EXAMPLE REPORTING OF EGFR/7p11.2 AMPLIFICATION IN NSCLC

ISCN Result: nuc ish 7p11.2(EGFRx3-10)[60/100]

Interpretation

Fluorescence *in situ* hybridization (FISH) analysis on a paraffin-embedded tumor with the EGFR probe showed evidence of amplification of the EGFR gene in 60% of interphase nuceli scored.

EGFR amplification is a recurrent, non-random abnormality observed in non-small cell lung cancer and may respond well to EGFR kinase inhibitor drugs.

32.4.3 RET gene rearrangements

SPECIFIC FEATURES

The RET gene located at chromosome 10q11.2 uses a breakapart probe and is optimized to detect translocations and inversions involving the RET (ret protooncogene) gene region. Recently, RET gene rearrangements have been identified in 1–2% of lung adenocarcinomas. Pericentric inversions and translocations of chromosome 10 result in increasing the expression of the RET gene by the fusion of RET with KIF5B (10p11) or with genes on other chromosomes. RET gene rearrangements define a distinct molecular subset of NSCLC that is most often mutually exclusive from EGFR mutations, KRAS mutations, ALK and ROS1 rearrangements. RET gene rearrangements associated with NSCLC are observed in patients who are either light smokers or never smoked, and are young (<50 years). Patients with RET gene rearrangements may be responsive to RET TKIs, including sunitinib, sorafenib or vandetanib.

EXAMPLE REPORTING OF RET/10q11.2 REARRANGEMENTS IN NSCLC

ISCN Result: nuc ish 10q11.2(RETx2)(5'RET sep 3'RETx1)[60/100]

Interpretation

Fluorescence *in situ* hybridization (FISH) analysis on a paraffin-embedded tumor with the RET probe showed evidence of a rearrangement of the RET gene in 60% of interphase nuclei scored.

RET rearrangements are recurrent, non-random abnormalities observed in non-small cell lung cancer and may respond well to treatment with the RET kinase inhibitor drugs sunitinib, sorafenib or vandetanib.

32.4.4 ROS1 gene rearrangements

SPECIFIC FEATURES

The ROS1 gene located at chromosome 6q22 uses a dual color probe to detect translocations involving the ROS1 gene. Oncogenic gene fusions involving the ROS1 receptor tyrosine kinase have recently been identified in up to 1–2% of NSCLC. ROS1 gene rearrangements define a distinct molecular subset of NSCLC that are most often mutually exclusive from EGFR, KRAS and ALK rearrangements. ROS1 gene rearrangements are associated with lung adenocarcinomas in patients who are light smokers or never smoked, and are young (<50 years). In recent clinical studies, patients with advanced NSCLC containing ROS1 rearrangements showed a benefit from crizotinib treatment.

EXAMPLE REPORTING OF ROS1/6q22 REARRANGEMENTS IN NSCLC

ISCN Result: nuc ish 6q22(ROS1x2))(5'ROS1 sep 3'ROS1x1)[60/100]

Interpretation

Fluorescence *in situ* hybridization (FISH) analysis on a paraffin-embedded tumor with the ROS1 probe showed evidence of a rearrangement of the ROS1 gene in 60% of interphase nuclei scored.

ROS1 rearrangements are recurrent, non-random abnormalities observed in non-small cell lung cancer and may respond well to treatment with the ROS1 kinase inhibitor drug crizotinib.

32.5 Oligodendroglioma

SPECIFIC FEATURES

Gliomas account for >70% of all primary brain tumors, of which the most common type is glioblastoma. Most gliomas are generally associated with a poor prognosis with the exception of pilocytic astrocytomas. Less than 3% of glioblastoma patients are still alive 5 years after diagnosis, higher age being the most significant predictor of having a poor outcome. Studies have shown two factors for primary brain tumors, including exposure to high doses of ionizing radiation and inherited mutations of highly penetrant genes. Also, polymorphic genes affecting detoxification, DNA repair and cell cycle regulation have been implicated in the development of gliomas.

The most common numerical chromosome abnormality is trisomy 7, which is observed in both astrocytomas and anaplastic astrocytomas (AA) as well as oligodendrogliomas. In AA, other chromosomal gains include trisomy 19 and 20. The most common losses in both AA and oligodendrogliomas include deletion of 9p, monosomy 10, 13, 14, 22 and a sex chromosome, and deletions or translocations involving 22q. Aneuploidies are generally associated with progression of disease. Genetic instability also is common in oligodendrogliomas and results in numerous clonal populations.

Centrosome amplification is also observed, where it has been hypothesized that it drives tumor aneuploidy by increasing the frequency of abnormal mitoses that leads to chromosome missegregation.

In pilocytic astrocytomas, the most common chromosomal aberrations consist of 6q and 7q gain and loss of 9q. In glioblastomas, the most frequently reported chromosomal abnormalities include loss of 17p and 13q, and gains of 1q, 2q, 3q and 17q. Ependymomas most commonly show gain of 1q and 7q and loss of 22q.

Of significance are deletions of 1p and 19q, which are associated with oligodendroglioma tumors. Combined alterations have been observed in up to 70% of oligodendrogliomas. Chromosome 1p and 19q deletions are predictive of radiochemosensitivity in anaplastic oligodendroglial tumors and associated with a better prognosis.

EXAMPLE REPORTING OF 1p AND 19q DELETIONS – OLIGODENDROGLIOMAS

ISCN Result: nuc ish 1p36(TP73x0~1),1q25(ANGPTx1~2)[170/250]/ 19p13(ZNF443x1~2),19q13(GLTSCRx1~2)[180/250]

Interpretation

Fluorescence *in situ* hybridization (FISH) analysis on a paraffin-embedded tumor with chromosomes 1p36/19q13 probes showed evidence of a deletion of chromosome 1p36 in 170 of 250 (68%), and a deletion of chromosome 19q13 in 180 of 250 (72%) interphase nuclei scored, respectively.

The chromosome 1p/1q ratio equals 0.60, and the 19p/19q ratio equals 0.55. A tumor is considered positive for a 1p or 19q deletion when chromosome 1p/1q or 19p/19q ratios are less than 0.80.

These results are consistent with a chromosome 1p deletion with a co-deletion of chromosome 19q. Co-deletions of 1p and 19q are observed in 70% of oligodendrogliomas, but are rare in other types of gliomas, particularly astrocytomas. Generally, co-deletions of chromosomes 1p and 19q are considered good genetic markers for predicting response to treatment.

EXAMPLE REPORTING OF 1p BUT NOT 19q DELETION – OLIGODENDROGLIOMAS (FIGURE 32.4)

ISCN Result: nuc ish 1p36(TP73x0~1),1q25(ANGPTx1~2),19p13(ZNF443x1~2)[160/250]/ 19q13(GLTSCRx1~2)[200/250]

Interpretation

Fluorescence *in situ* hybridization (FISH) analysis on a paraffin-embedded tumor with chromosomes 1p36/19q13 probes showed evidence of a deletion of chromosome 1p36 in 160 of 250 (64%), but was negative for a deletion of chromosome 19q13 in 250 cells scored.

The chromosome 1p/1q ratio equals 0.75 whereas the ratio for 19p/19q equals 0.91. A tumor is considered positive for a 1p or 19q deletion when chromosome 1p/1q or 19p/19q ratios are less than 0.80.

These results are consistent with a chromosome 1p deletion without a co-deletion of chromosome 19q. Co-deletions of 1p and 19q are observed in 70% of oligodendrogliomas but are rare in other types of gliomas, particularly astrocytomas. Generally, co-deletions of chromosomes 1p and 19q are considered good genetic markers or predicting response to treatment. Therefore, due to a single chromosome 1p deletion, clinical-pathological correlation is suggested.

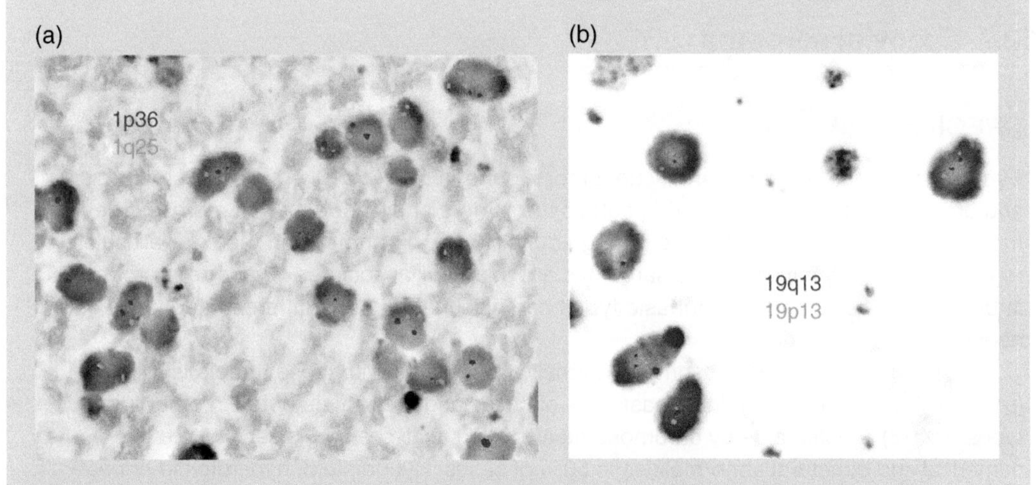

Figure 32.4 Oligodendroglioma showing different signal patterns at all gene loci examined. (a) 1p/19q deletions. (b) 19p/19q deletions. Courtesy of Lauren Jenkins PhD, Kaiser Permanente Regional Laboratory, Northern California.

32.6 Rhabdomyosarcoma

SPECIFIC FEATURES

Rhabdomyosarcomas are the most common soft tissue sarcomas found in children, including alveolar and embryonal subtypes. Alveolar rhabdomyosarcomas (ARMS) are associated with t(2;13)(q35;q14) as well as t(1;13)(p36;q14), which results in fusion of the FKHR (FOXO1) gene on chromosome 13q14 with either the PAX3 on chromosome 2q35 or PAX7 gene on chromosome 1p36.

EXAMPLE REPORTING OF CHROMOSOME 13q14 REARRANGEMENT IN RHABDOMYOSARCOMAS – DUAL COLOR BREAKAPART PROBE

ISCN Result: nuc ish 13q14(FOXO1x2)(5'FOXO1 sep 3'FOXO1x1)[60/100]

Interpretation

Fluorescence *in situ* hybridization (FISH) analysis on a paraffin-embedded tumor with the FOXO1 probe showed evidence of a FOXO1 gene rearrangement in 60% of interphase nuclei scored.

The most common FOXO1 rearrangement is the translocation (2;13)(q35;q14), which is a recurrent, non-random abnormality associated with rhabdomyosarcomas; however, t(1;13)(p36;q14) has also been observed.

32.7 Synovial sarcoma

SPECIFIC FEATURES

Synovial sarcomas are rare soft tissue tumors found in 5–8% of all soft tissue sarcomas, and are the most common pediatric soft tissue tumors after rhabdomyosarcomas. They are most prevalent in adolescents and young adults, occurring primarily in the para-articular regions of the extremities. These sarcomas are generally associated with a poor prognosis in both the biphasic and monophasic types and also in the poorly differentiated small cell neoplasm.

The most common chromosome abnormality associated with these tumors is t(X;18)(p11.2;q11.2), which is seen in at least 80% of cases and is rarely detected in other tumor types. T(X;18) is detectable by chromosome and FISH analyses and is seen with other numerical and structural abnormalities in 50% of cases. Hypodiploid, hyperdiploid or near-tetraploid chromosome complements have been described but without a common chromosomal pattern.

EXAMPLE REPORTING OF CHROMOSOME 18q11.2 REARRANGEMENT IN SYNOVIAL SARCOMAS – DUAL COLOR BREAKAPART PROBE

ISCN Result: nuc ish 18q11.2(SYTx2)(5'SYT sep 3'SYTx1)[60/100]

Interpretation

Fluorescence *in situ* hybridization (FISH) analysis on a paraffin-embedded tumor with the SYT probe showed evidence of an SYT gene rearrangement in 60% of interphase nuclei scored.

The most common SYT rearrangement is the translocation (X;18)(p11.2;q11.2), which is a recurrent, non-random abnormality associated with synovial sarcomas and is seen in up to 90% of tumor specimens.

Bibliography

American College of Medical Genetics. *Guidelines in Clinical Cytogenetics. Section E10 Methods in Fluorescence In situ Hybridization – Interphase/Nuclear Fluorescence In Situ Hybridization.* American College of Medical Genetics, Bethesda, MD. www.acmg.net/StaticContent/SGs/Section_E.html, accessed 28 January 2014.

Arber DA. Molecular diagnostic approach to non-Hodgkin's lymphoma. J Mol Diagn 2000; 2: 178–190.

Aster JC, Longtine JA. Detection of BCL2 rearrangements in follicular lymphoma. Am J Pathol 2002; 160: 759–762.

Atlas of Genetics and Cytogenetics in Oncology and Haematology. http://AtlasGeneticsOncology.org/Anomalies, accessed 28 January 2014.

Barrans SL, Evans PAS, O'Connor SJM, et al. The t (14; 18) is associated with germinal center-derived diffuse large B-cell lymphoma and is strong predictor of outcome. Clin Cancer Res 2003; 9: 2133–2139.

Braziel RM, Arber DA, Slovak ML, et al. The Burkitt-like lymphomas: a Southwest Oncology Group study delineating phenotypic, genotypic, and clinical features. Blood 2001; 97: 3713–3720.

Bull JH, Harnden P. Efficient nuclear FISH on paraffin-embedded tissue sections using microwave pretreatment. Biotechniques 1999; 26: 416–422.

Chin SF, Daigo Y, Huang HE, et al. A simple and reliable pretreatment protocol facilitates fluorescent in situ hybridization on tissue microarrays of paraffin wax embedded tumor samples. J Clin Pathol 2003; 56: 275–279.

Gersen S, Keagle M (eds). *Principles of Clinical Cytogenetics.* Humana Press, Totowa, New Jersey, 1999.

Heim S, Mitelman F. *Cancer Cytogenetics*, 2nd edn. Wiley-Liss, New York, 1995.

Hirsch F, Herbst R, Olsen C, et al. Increased EGFR gene copy number detected by fluorescent in situ hybridization predicts outcome in non-small-cell lung cancer patients treated with cetuximab and chemotherapy. J Clin Oncol 2008; 26: 3351–3357.

Izquierdo DS, Buchonnet G, Siebert R, et al. MALT1 is deregulated by both chromosomal translocation and amplification in B-cell non-Hodgkin lymphoma. Neoplasia 2003; 101: 4539–4545.

Johnson B, Janne P. Selecting patients for epidermal growth factor receptor inhibitor treatment: a FISH Story or a tale of mutations? J Clin Oncol 2005; 23: 6813–6816.

Kwak E, Bang Y, Camidge D, et al. Anaplastic lymphoma kinase inhibition in non-small-cell lung cancer. N Engl J Med 2010; 363: 1693–1703.

Leers MPG, Schutte B, Theunissen PHMH, Ramaekers FCS, Nap M. Heat pretreatment increases resolution in DNA flow cytometry of paraffin-embedded tumor tissue. Cytometry 1999; 35: 260–266.

Li JY, Gaillard F, Moreau A, et al. Detection of translocation t(11; 14)(q13; q32) in mantle cell lymphoma by fluorescence in situ hybridization. Am J Pathol 2002; 154: 1449–1452.

Lipson D, Capelletti M, Yelensky R, et al. Identification of new ALK and RET gene fusions from colorectal and lung cancer biopsies. Nature Med 2012; 18: 382–384.

Mosse YP, Diskin SJ, Wasserman N, et al. Neuroblastomas have distinct genomic DNA profiles that predict clinical phenotype and regional gene expression. Genes Chromosomes Cancer 2007; 46(10): 936–949.

Olaussen K, Dunant A, Fouret P, et al. DNA repair by ERCC1 in non-small-cell lung cancer and cisplatin-based adjuvant chemotherapy. N Engl J Med 2006; 355: 983–991.

Pao W, Miller V, Politi K, et al. Acquired resistance of lung adenocarcinomas to gefitinib or erlotinib is associated with a second mutation in the EGFR kinase domain. PLoS Med 2005; 2(3): e73.

Pao Wm Wang T, Riely G, et al. KRAS mutations and primary resistance of lung adenocarcinomas to gefitinib or erlotinib. PLoS Med 2005; 2(1): e17.

Paternoster SF, Brockman SR, McClure RF, Remstein ED, Kurtin PJ, Dewald GW. A new method to extract nuclei from paraffin-embedded tissue to study lymphomas using interphase fluorescence in situ hybridization. Am J Pathol 2002; 160: 1967–1972.

Remstein ED, Kurtin PJ, James CD, Wang XY, Meyer RG, Dewald GW. Mucosa-associated lymphoid tissue lymphomas with t(11; 18)(q21; q21) and mucosa-associated lymphoid tissue lymphomas with aneuploidy develop a long different pathogenetic pathways. Am J Pathol 2002; 161: 63–7S1.

Rooney DE. *Human Cytogenetics, Malignancy and Acquired Abnormalities*, 3rd edn. Oxford University Press, New York, 2001.

Rooney DE, Czepulkowski BH. *Human Cytogenetics. A Practical Approach.* Oxford University Press, New York, 1992.

Sasaki H, Shimizu S, Tani Y, et al. RET expression and detection of KIF5B/RET gene rearrangements in Japanese lung cancer. Cancer Med 2012; 1: 68–75.

Scagliotti G, Hanna N, Fossella F, et al. The differential efficacy of pemetrexed according to NSCLC histology: a review of two phase III studies. Oncologist 2009; 14: 253–263.

Schurter MJ, LeBrun DP, Harrison KJ. Improved technique for fluorescence in situ hybridization analysis of isolated nuclei from archival, B5 or formalin fixed, paraffin wax embedded tissue. J Clin Pathol Mol Pathol 2002; 55: 121–124.

Shaffer LG, McGowan-Jordan J, Schmid M (eds). *ISCN 2013: An International System for Human Cytogenetic Nomenclature.* Karger Publishers, Unionville, CT, 2013.

Shaw A, Camidge D, Engelman J, et al. Clinical activity of crizotinib in advanced non-small cell lung cancer (NSCLC) harboring ROS1 gene rearrangement. J Clin Oncol 2012; 30(suppl): abstr 7508.

Shaw AT, Yeap B, Mino-Kenudson M, et al. Clinical features and outcomes of patients with non-small-cell lung cancer who harbor EML4-ALK. J Clin Oncol 2009; 27: 4247–4253.

Slovak ML, Tcheurekdjian F, Zhang FF, Murata-Collins JL. Simultaneous detection of multiple genetic aberrations in single cells by spectral fluorescence in situ hybridization. Cancer Res 2001; 61: 831–836.

Swerdlow SH, Campo E, Harris NL, et al. (eds). *WHO Classification of Tumours of Haematopoietic and Lymphoid Tissues*. IARC Press, Lyon, 2008.

National Comprehensive Cancer Network. The NCCN Non-Small Cell Lung Cancer Clinical Practice Guidelines in Oncology (Version 3.2012). National Comprehensive Cancer Network, 2012. www.nccn.org, accessed 3 February 2014.

Viardot A, Moller P, Hogel J, et al. Clinicopathologic correlations of genomic gains and losses in follicular lymphomas. J Clin Pathol 2002; 20: 4523–4530.

Winton T, Livingston R, Johnson D, et al. Vinorelbine plus cisplatin vs. observation in resected non-small-cell lung cancer. N Engl J Med 2005; 352: 2589–2597.

Commercial FISH probes are available at these websites

www.AbbottMolecular.com

www.Kreatech.com

www.Rainbowscientific.com

www.empiregenomics.com

www.abnova.com

www.cytocell.com

www.exiqon.com

www.cambridgebluegnome.com

CHAPTER 33

Breast cancer – HER2 FISH analysis

The HER2 fluorescence *in situ* hybridization (FISH) assay is designed to detect amplification of the HER2 (also called ERBB2) gene using FISH analysis of formalin-fixed, paraffin-embedded (FFPE) human breast cancer tissue specimens. HER2 amplification arises early in breast malignancies, occurring in almost 50% of ductal *in situ* carcinomas and in approximately 20% of all breast tumors. HER2 amplification is generally associated with a poor prognosis. HER2 overexpressed breast tumors have shown up to 25–50 copies of the HER2 gene and up to a 40–100-fold increase in HER2 protein. Therefore, the differential HER2 expression between normal breast tissue and breast tumor aids in defining HER2 as a treatment target. Trastuzumab has been developed as a targeted therapy for overexpression of the HER2 gene, and the results from this FISH assay are used in addition to clinical and pathological information for the assessment of patients for whom trastuzumab treatment is being considered.

The HER2 gene is also overexpressed in several human tumors, playing a significant role in the tumorigenesis and progression of disease, including bladder, colorectal, gastric and prostate cancers. Generally, overexpression of the HER2 gene in these malignancies is also associated with a poor prognosis.

SPECIFIC FEATURES

The probes used for HER2 FISH testing usually consist of a dual color, two probe mixture of DNA sequences on specific regions of chromosome 17, including the centromere of chromosome 17 (CEP17) and the HER2 gene region at chromosome 17q12. The analysis of these probes provides information on chromosome copy number of the HER2 gene with respect to the CEP17 probe, giving a ratio of signals for each region.

For breast carcinomas, the guidelines for establishing an amplified or equivocal result has changed based on the College of American Pathologist (CAP) and the American Society of Clinical Oncology ASCO) guidelines, effective January 2014. The old guidelines stated that HER2/CEP17 ratio <1.8 is considered a normal result, >2.2 is considered an amplified abnormal result, and a ratio of 1.8-2.2 is considered an equivocal result.

Cytogenetic Abnormalities: Chromosomal, FISH and Microarray-Based Clinical Reporting, First Edition. Susan Mahler Zneimer.

The new CAP guidelines for dual color probes states the following:

A normal result occurs when:

- the HER2/17 centromere signal ratio <2.0
- the average HER2 copy number <4.0 signals per cell
- that these observations are within a homogenously contiguous population
- and with >10% of invasive tumors cells present.

An equivocal result occurs when:

- the HER2/17 centromere signal ratio <2.0
- the average HER2 copy number is >4.0 and <6.0 signals per cell
- that these observations are within a homogenously contiguous population
- and with >10% of invasive tumors cells present.

An amplified result occurs when:

- the HER2/17 centromere signal ratio >2.0 with an average HER2 copy number that less than, equal to or greater than 4.0 signals per cell
- or with the HER2/17 centromere signal ratio <2.0 if the average HER2 copy number >6.0 signals per cell
- that these observations are within a homogenously contiguous population
- and with >10% of invasive tumors cells present.

Genetic heterogeneity may be present in HER2 analyses, in which there are either scattered or clustered areas of tumor between 5–50% of cells showing HER2 amplification, while other parts of the tumor show a normal result. Genetic heterogeneity is further discussed in section 33.3.

Report formats vary greatly between laboratories; however, there is information that is commonly reported.
The following includes examples of specific relevant information for a HER2 FISH report.

- Specimen type: paraffin-embedded tissue
- Indication for testing: breast cancer
- Specimen fixative type: formalin
- Duration of fixation: 13.0 hours
- Time of duration: tissue biopsy fixation time was between 6 and 48 hours
- Fixative disclaimer: 10% neutral buffered formalin
- Cells analyzed by FISH: 60 interphase cells
- Number of technologists performing analysis: two

33.1 Common report comments

33.1.1 Disclaimer comments

- The PathVysion® HER2 DNA Probe Kit has been approved by the US Food and Drug Administration (FDA) for detecting amplification of the HER2 gene via fluorescence *in situ* hybridization (FISH) in formalin-fixed, paraffin-embedded human breast cancer tissue specimens. The performance characteristics of this test have been established by [name of laboratory].

- Results from HER2 FISH testing are intended to be used as an adjunct test for the prognosis of stage II, node-positive breast cancer patients. This test is indicated as an aid in the assessment of treatment for patients in whom trastuzumab is being considered.

33.1.2 No signals present

- The HER2 FISH assay performed on this patient's specimen showed suboptimal hybridization after repeated attempts; therefore, no result was obtained.

33.1.3 Insufficient invasive component – quantity not sufficient/no tumor present

- The HER2 FISH assay requires a sufficient number of invasive carcinoma cells for analysis. The pathologist's review of this patient's specimen revealed an insufficient component of invasive carcinoma cells to obtain a result. If possible, please provide a block containing a larger invasive carcinoma component for analysis.
- The specimen submitted for HER2 FISH testing was reviewed by a pathologist and revealed no invasive or metastatic breast cancer tumor component present. Therefore, this test was not performed.

33.1.4 Decalcification

- The HER2 FISH assay performed on this patient's specimen showed poor chromatin integrity, precluding a full analysis. Certain specific fixation methodologies yield a suboptimal or no result, including treatment with an acid decalcification. This treatment is known to provide suboptimal specimens for FISH testing. It is suggested that a specimen be processed in 10% neutral buffered formalin.

33.2 Example HER2 reports

33.2.1 Normal – not amplified

EXAMPLE REPORT – NOT AMPLIFIED HER2 (FIGURE 33.1)

ISCN Result: nuc ish 17cen(D17Z1x2.4),17q12(HER2x3.3)[60]

Interpretation

Not amplified
HER2:CEP17 ratio = 1.4
The average of HER2 copies per cell: 3.3
The average of 17cen copies per cell: 2.4
This result is considered normal for the HER2 gene.

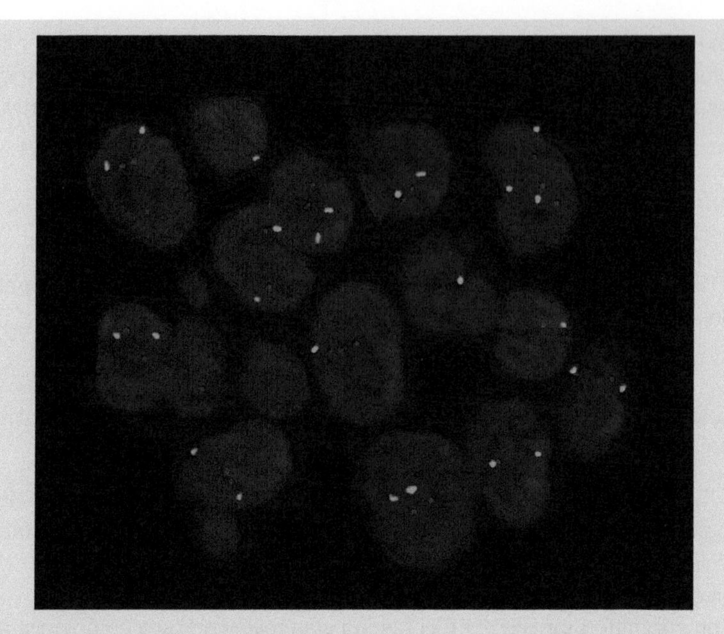

Figure 33.1 HER2 FISH analysis showing most cells with a normal signal pattern of two red, two green signals for HER2 and chromosome 17 centromeres, respectively, resulting in a normal HER2/CEP17 ratio. Courtesy of Mehdi Jamehdor MD, Kaiser Permanente Regional Laboratory, Southern California.

33.2.2 Abnormal – amplified

EXAMPLE REPORT – AMPLIFIED HER2 (FIGURE 33.2)

ISCN Result: nuc ish 17cen(D17Z1x2.3),17q12q12(HER2x10.4)[60]

Figure 33.2 HER2 FISH analysis showing cells with an amplified signal pattern for the HER2 gene resulting in a HER2/CEP17 ratio of 4.5.

Interpretation

Amplified
HER2:CEP17 ratio=4.5
The average of HER2 copies per cell: 10.4
The average of 17cen copies per cell: 2.3
This result is considered amplified for the HER2 gene.

33.2.3 Equivocal results

EXAMPLE REPORT – EQUIVOCAL HER2 RATIO

ISCN Result: nuc ish 17cen(D17Z1x2.2),17q12(HER2x4.2)[90]

Interpretation

Equivocal
HER2:CEP17 ratio=1.9
The average of HER2 copies per cell: 4.2
The average of 17cen copies per cell: 2.2

This specimen is considered equivocal for the HER2 gene as defined by the 2013 ASCO/CAP guidelines. An additional 30 cells were analyzed throughout the designated tumor area to confirm this finding. Comparison with results from another methodology such as immunohistochemistry (IHC) or FISH on another block is suggested.

33.3 Genetic heterogeneity

Intratumor genetic heterogeneity (GH) of the HER2 gene copy number has been well documented and represents clonal diversity within the tumor. The incidence of intratumor HER2 genetic heterogeneity has been reported to range from 5% to 30%. In 2007, the American Society of Clinical Oncology (ASCO) and the College of American Pathologists (CAP) published guidelines for HER2 testing in breast cancer to improve the accuracy and reproducibility of testing. The CAP agreed to convene an expert panel to address the issue of genetic heterogeneity and publish separate recommendations, which were completed in 2009. The panel defined HER2 GH as when there are more than 5% but less than 50% of infiltrating tumor cells with a HER2/CEP17 ratio higher than 2.2. For example, if 20 cells are scored by FISH and at least one cell is identified with a HER2/CEP17 signal ratio greater than 2.2, this specimen contains GH (5.0%). If 40 cells are examined and two or more cells have a ratio higher than 2.2, then GH is present. For FISH using a probe for HER2 alone, without a control probe, GH exists if there are between 5% and 50% of infiltrating tumor cells with greater than six HER2 signals per cell.

When reporting GH, there should be a statement in the interpretation that HER2 GH is present. In cases of GH, based on the panel recommendations, the report must include the percentage of invasive tumor showing HER2 amplification, whether the amplified cells are scattered throughout the tumor or whether cells are in a cluster, and the ratio and number of CEP17 and HER2 signals per cell or area. A comment should also be included on whether the area is histologically distinct. If the examination was performed on a needle core biopsy, the report should indicate that the evaluation may not be representative of the entire tumor, and it might be desirable to perform FISH testing on a resection sample.

33.3.1 GH Normal – not amplified

EXAMPLE REPORT – NOT AMPLIFIED HER2 WITH GH

ISCN Result: nuc ish 17cen(D17Z1x2.4),17q12(HER2x3.3)[60]

Interpretation

Not amplified with genetic heterogeneity
HER2:CEP17 ratio = 1.4
This result is considered normal with genetic heterogeneity for the HER2 gene.

HER2 genetic heterogeneity (GH) exists if there are more than 5% but less than 50% of infiltrating tumor cells with a ratio higher than 2.2. This HER2 analysis showed genetic heterogeneity with amplified cells scattered within the tumor area. Of 60 cells analyzed, six cells (10.0%) showed amplification with a HER2 ratio of 2.9.

Results of all cells analyzed:

The average of HER2 copies per cell: 3.3
The average of CEP17 copies per cell: 2.4

Results of cells with genetic heterogeneity:

Abnormal cells = 6/60 (10.0%)
The average of HER2 copies per cell: 3.7
The average of centromere 17 copies per cell: 1.3
Abnormal cell result ratio of HER2/CEP17 ratio = 2.9

33.3.2 GH Equivocal results

EXAMPLE REPORT – EQUIVOCAL HER2 RATIO WITH GH (FIGURE 33.3)

ISCN Result: nuc ish 17cen(D17Z1x2.2),17q12(HER2x4.2)[90]

Interpretation

Equivocal with genetic heterogeneity
HER2:CEP17 ratio = 1.9
This result is considered equivocal for the HER2 gene.

This specimen is considered equivocal with genetic heterogeneity for the HER2 gene as defined by the 2013 ASCO/CAP guidelines. An additional 30 cells were analyzed throughout the designated tumor area to confirm this finding. Comparison with results from another methodology, such as immunohistochemistry (IHC) or FISH, on another block is suggested.

HER2 genetic heterogeneity (GH) exists if there are more than 5% but less than 50% of infiltrating tumor cells with a ratio higher than 2.2. This HER2 analysis showed genetic heterogeneity with amplified cells scattered within the tumor area. Of 90 cells analyzed, 37 cells (41.1%) showed amplification with a HER2/CEP17 ratio of 2.7.

Results of all cells analyzed:

The average of HER2 copies per cell: 4.0
The average of 17cen copies per cell: 1.9

Results of cells with genetic heterogeneity:

Abnormal cells = 37/90 (41.1%)
The average of HER2 copies per cell: 4.9
The average of centromere 17 copies per cell: 1.8
Abnormal cell result ratio of HER2/CEP17 ratio = 2.7

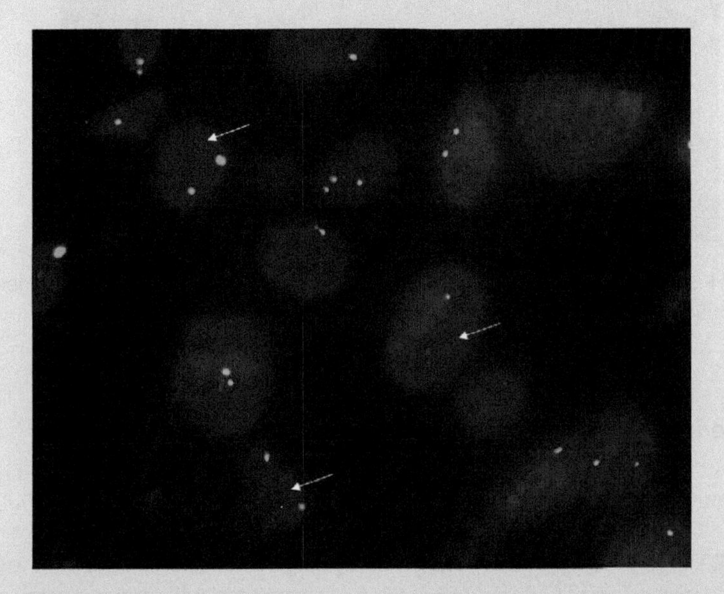

Figure 33.3 HER2 FISH analysis showing cells with an equivocal signal pattern for the HER2 gene resulting in a HER2/CEP17 ratio of 2.1. Genetic heterogeneity is also present, in which some cells exhibit an amplified HER2 ratio (*arrows*), while other cells show a normal signal pattern.

33.3.3 Clustered genetic heterogeneity – report comment

HER2 genetic heterogeneity (GH) exists if there are more than 5% but less than 50% of infiltrating tumor cells with a ratio higher than 2.2. This HER2 evaluation showed genetic heterogeneity observed as clusters of amplified cells. Of 60 cells analyzed, three cells (7.5%) showed amplification of HER2 with a ratio of 2.8. Dr Pathologist reexamined this population of cells and determined the invasive cells to be histologically distinctive. A minimum of 20 cells were scored within the clusters.

33.3.4 Polysomy

Polysomy in HER2 gene assessment refers to gains of both the CEP17 and HER2 signals in analyzed cells, which may result in a normal or equivocal result. These cases generally do not result in overexpression of the HER2 mRNA or protein; however, the clinical significance of increased copy gains of polysomy is unknown, as is the response to trastuzumab therapy.

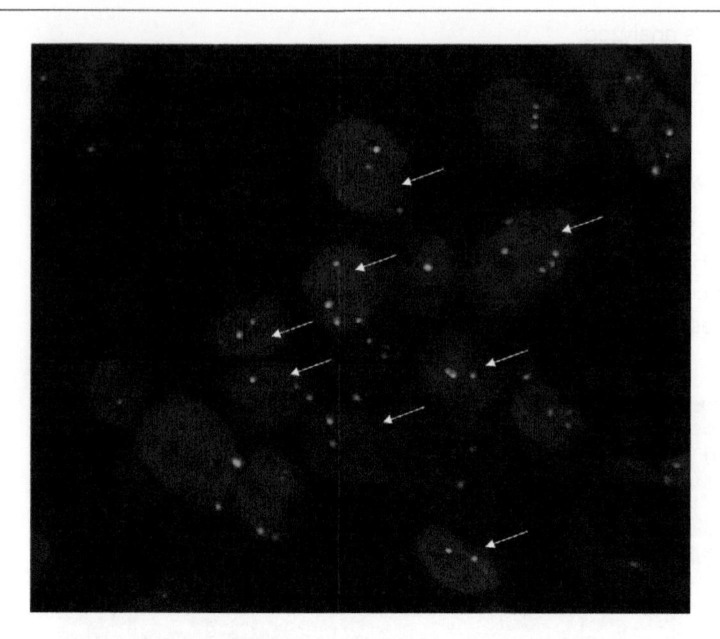

Figure 33.4 HER2 FISH analysis showing cells with a polysomy signal pattern for the HER2 gene resulting in a HER2/CEP17 ratio of 1.1, but with gains of both the HER2 gene and CEP17.

Report comment for polysomy (Figure 33.4)

- Normal or equivocal findings may result from the detection of increased copy numbers of chromosome 17 centromere or other control probes (>3 signals per cell), with concurrent gain of the HER2 gene. Since HER2/CEP17 analysis gives a ratio as a result, these findings are not considered as amplified and generally not associated with overexpression of the HER2 protein. Gain of CEP17 may be associated with partial or full polysomy 17. Therefore, the clinical significance of the presence of polysomy 17 is unknown, and the potential benefit of trastuzumab therapy is not well described.

Bibliography

American College of Medical Genetics. *Guidelines in Clinical Cytogenetics. Section E10 Methods in Fluorescence In situ Hybridization – Interphase/Nuclear Fluorescence In Situ Hybridization.* American College of Medical Genetics, Bethesda, MD. www.acmg.net/StaticContent/SGs/Section_E.html, accessed 28 January 2014.

Atlas of Genetics and Cytogenetics in Oncology and Haematology. http://AtlasGeneticsOncology.org/Anomalies, accessed 28 January 2014.

Bonadonna G, Valagussa T, Zambetti M. Milan adjuvant trials for stage I-II breast cancer. In: Slamon SW (ed) *Adjuvant Chemotherapy of Cancer V.* Grune and Stratton, Orlando, FL, 1987, pp.211–221.

Borg A, Tandon AK, Sigurdsson H, et al. HER-2/neu amplification predicts poor survival in node-positive breast cancer. Cancer Res 1990; 50: 4332–4337.

Fisher B, Bauer M, Wickermann LL, et al. Relationship of number of positive axillary nodes to the prognosis of patients with primary breast cancer: an NSABP update. Cancer 1983; 52: 1551–1557.

Fisher ER, Redmond C, Fisher B, et al. Pathologic findings from the national surgical adjuvant breast and bowel projects (NSABP). Cancer 1990; 65: 2121–2128.

Fujii H, Marsh C, Cairns P, Sidransky D, Gabrielson E. Genetic divergence in the clonal evolution of breast cancer. Cancer Res 1996; 56: 1493–1497.

Glockner S, Buurman H, Kleeberger W, Lehmann U, Kreipe H. Marked intratumoral heterogeneity of c-myc and cyclinD1 but not of c-erbB2 amplification in breast cancer. Lab Invest 2002; 82: 1419–1426.

Guilick WJ, Love SB, Wright C, et al. cerbB-2 protein overexpression in breast cancer is a risk factor in patients with involved and uninvolved lymph nodes. Br J Cancer 1991; 63: 434–438.

Gusterson BA, Gelber RD, Goldhirsch A, et al. Prognostic importance of c-erbB-2 expression in breast cancer. J Clin Oncol 1992; 10: 1049–1056.

Heim S, Mitelman F. *Cancer Cytogenetics*, 2nd edn. Wiley-Liss, New York, 1995.

Lewis JT, Ketterling RP, Halling KC, et al. Analysis of intratumoral heterogeneity and amplification status in breast carcinomas with equivocal (2_) HER- 2 immunostaining. Am J Clin Pathol 2005; 124: 273–281.

Mansour EG, Gray R, Shatila AH, et al. Efficacy of adjuvant chemotherapy in high-risk node-negative breast cancer. N Engl J Med 1989; 320: 485–490.

Mass RD, Sanders C, Kasian C, et al. The concordance between the clinical trial assay (CTA) and fluorescence in situ hybridization (FISH) in the herceptin pivotal trials (abstract). Am Soc Clin Oncol 2000; 19: 75a.

Miller DV, Jenkins RB, Lingle WL, et al. Focal HER2/neu amplified clones partially account for discordance between immunohistochemistry and fluorescence in situ hybridization results: data from NCCTG N9831 intergroup adjuvant trial. 2004 ASCO Annual Meeting Proceedings. J Clin Oncol 2004; 22(14S): 568.

Pauletti G, Singh R, Press MF, et al. HER-2/neu gene amplification detected by fluorescence in situ hybridization: a comparative study with other techniques. Abstract 3247. Proc Am Assoc Cancer Res 1994; 35: 545.

Persons DL, Bui MM, Lowery MC, et al. Fluorescence in situ hybridization (FISH) for detection of HER-2/neu amplification in breast cancer: a multicenter portability study. Ann Clin Lab Sci 2000; 30: 41–48.

Press MF, Bernstein L, Thomas PA, et al. HER-2/neu gene amplification characterized by fluorescence in situ hybridization: poor prognosis in node-negative breast carcinomas. J Clin Oncol 1997; 15: 2894–2904.

Shaffer LG, McGowan-Jordan J, Schmid M (eds). *ISCN 2013: An International System for Human Cytogenetic Nomenclature*. Karger Publishers, Unionville, CT, 2013.

Slamon DJ, Clark GM, Wong SG, et al. Human breast cancer: correlation of relapse and survival with amplification of the HER-2/neu oncogene. Science 1987; 235: 177–182.

Swerdlow SH, Campo E, Harris NL, et al. (eds). *WHO Classification of Tumours of Haematopoietic and Lymphoid Tissues*. IARC Press, Lyon, 2008.

Tandon AK, Clark GM, Chamness AU, et al. HER-2/neu oncogene protein and prognosis in breast cancer. J Clin Oncol 1989; 7: 1120–1128.

Tubbs RR, Hicks DG, Cook J, et al. Fluorescence in situ hybridization (FISH) as primary methodology for the assessment of HER2 status in adenocarcinoma of the breast: a single institution experience. Diagn Mol Pathol 2007; 16: 207– 210.

Vance G, Barry T, Bloom K, et al. Genetic heterogeneity in HER2 testing in breast cancer. Arch Pathol Lab Med 2009; 133: 18–43.

Wingo PA, Tong T, Bolden S. Cancer statistics, 1995. CA Cancer J Clin 1995; 45: 8–31.

Wolf AC, Hammond EH, Schwartz JN, et al. American Society of Clinical Oncology/College of American Pathologists Guideline recommendation for human epidermal growth factor receptor 2 testing in breast cancer. Arch Pathol Lab Med 2007; 131: 18–43.

Zarbo RJ, Hammond MEH. Conference summary, Strategic Science Symposium, Her-2/neu testing of breast cancer patients in clinical practice. Arch Pathol Lab Med 2003; 127: 549–553.

Commercial FISH probes are available at these websites

www.AbbottMolecular.com

www.Kreatech.com

www.Rainbowscientific.com

www.empiregenomics.com

www.abnova.com

www.cytocell.com

www.exiqon.com

www.cambridgebluegnome.com

CHAPTER 34

Bladder cancer FISH analysis

The bladder cancer recurrence fluorescence *in situ* hybridization (FISH) assay is designed to detect aneuploidy of chromosomes 3, 7 and 17 as well as homozygotic loss of the 9p21 locus in cells collected from urine specimens and bladder washings from subjects with transitional cell carcinoma of the bladder (TTC). TTC is the most frequent bladder cancer in Europe and the US, seen in 90–95% of urothelial carcinomas of the bladder, while squamous cell carcinomas represent only 5% in these countries. However, squamous cell carcinomas are seen in approximately 70–80% of cases in the Middle East. Urothelial carcinomas occur mainly in older individuals, from 60–80 years; high-risk factors include cigarette smoking and occupational exposure to aniline, benzidine and naphthylamine, even with a latency period of 20–30 years after exposure.

Chromosome abnormalities in urothelial carcinomas consist of recurrent, non-random structural and numerical anomalies which vary with the stage of disease. Monosomy 9 or nullisomy 9, which is seen in approximately 50% of cases, is an abnormality seen early in disease. With disease progression, gains of chromosomes towards pseudo-tetraploidy and complex structural anomalies emerge. Recurrent anomalies include del(11p) or monosomy 11, del(13q) and del(17p). These abnormalities develop later in disease progression and are generally associated with a poor prognosis. Many other chromosomal rearrangements are frequently found. Loss of heterozygosity is also common late in disease and is also associated with worsening prognosis.

Fluorescence *in situ* hybridization assays for urothelial carcinomas generally include probes for chromosome 9p21 to identify loss of chromosome 9, and centromere probes for chromosomes 3, 7 and 17, or other chromosomes to identify gains of chromosomes. Results from the FISH assay are intended for use as a non-invasive method for monitoring tumor recurrence in conjunction with cytoscopy in patients previously diagnosed with bladder cancer.

SPECIFIC FEATURES

The probes that are typically used in this assay include four color probes, in a mixture of DNA sequences homologous to specific regions of chromosomes 3, 7, 9 and 17. The FISH assay is designed generally for use on voided urine specimens and bladder washings. FISH detection of chromosomes 3, 7 or 17 aneuploidy and/or homozygous deletion of chromosome 9p21 is associated with urothelial carcinoma progression.

Cytogenetic Abnormalities: Chromosomal, FISH and Microarray-Based Clinical Reporting, First Edition. Susan Mahler Zneimer.

Typical information on each report should include the following.

- Specimen type: urine
- Indication for testing: bladder cancer recurrence
- Cells analyzed by FISH: 25 interphase cells

34.1 Common report comments

34.1.1 Disclaimers

- This test will not detect abnormalities other than those designated. Therefore, this result needs to be evaluated in conjunction with other clinical and laboratory information.
- This test is devised for monitoring recurrence of transitional bladder cancer carcinoma in conjunction with cystoscopy and may be used to aid in the initial diagnosis of bladder carcinoma in patients with hematuria.
- Positive results in the absence of other signs or symptoms of bladder cancer may be evidence of other urinary tract-related cancers (e.g. cancers of the ureter, urethra, kidney or prostate (in males)) and further patient follow-up is recommended. If test results are not consistent with other clinical findings, consultation with the treating physician is suggested.

34.1.2 Technical failure – no result

- The bladder cancer FISH assay was performed on a representative sampling of cells from this patient's urine. Due to the lack of hybridization of the probes utilized in this assay, no result was obtained. Please submit a new specimen for analysis.

34.1.3 Comment for an insufficient number of cells for analysis

- An insufficient number of cells were available for the bladder cancer FISH assay.
- A small number of urine samples have an inadequate number of cells for analysis. To ensure that a sufficient number of cells are present for this assay, the urine specimen should not be split with other tests and should be preserved immediately in 50% alcohol in a 1:1 dilution.

34.2 Example reports

34.2.1 Normal results

EXAMPLE REPORT – NORMAL (FIGURE 34.1)

ISCN Result: nuc ish 3cen(D3Z1x2),7cen(D7Z1x2),9p21(p16x2),17cen(D17Z1x2)[25]

Interpretation

Fluorescence *in situ* hybridization (FISH) analysis revealed a normal result, with no evidence of aneuploidy of chromosomes 3, 7 and 17, or homozygous loss of chromosome 9p21.

This result is not indicative of bladder cancer recurrence.

A positive result is defined as four or more cells with gains of multiple chromosomes (3, 7 and/or 17) or 12 or more cells with homozygous loss of 9p21.

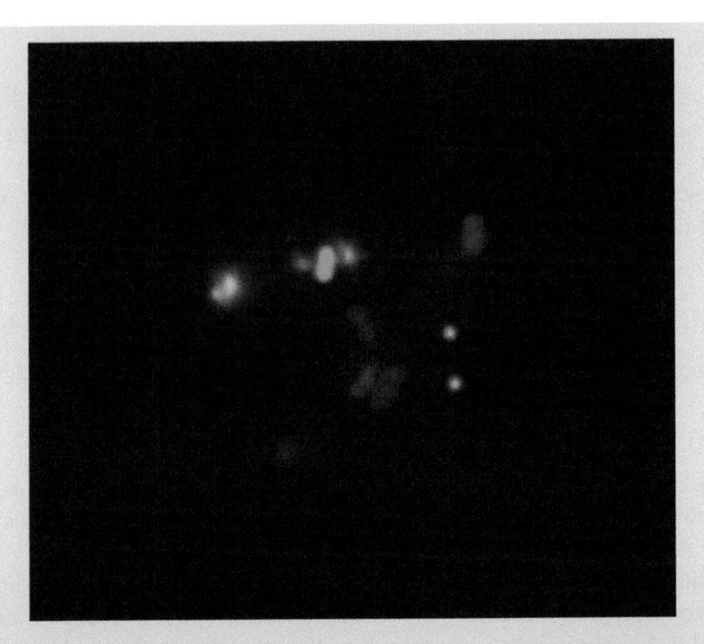

Figure 34.1 Bladder cancer FISH analysis showing two copies of chromosomes 3 (*red*), 7 (*green*), 9 (*gold*) and 17 (*aqua*), corresponding to a normal result.

34.2.2 Abnormal results with hyperdiploidy

EXAMPLE REPORT – ABNORMAL WITH HYPERDIPLOIDY (FIGURE 34.2)

ISCN Result: nuc ish 3cen(D3Z1x2~7),7cen(D7Z1x2~5),9p21(p16x2),17cen(D17Z1x2~6)[24/25]

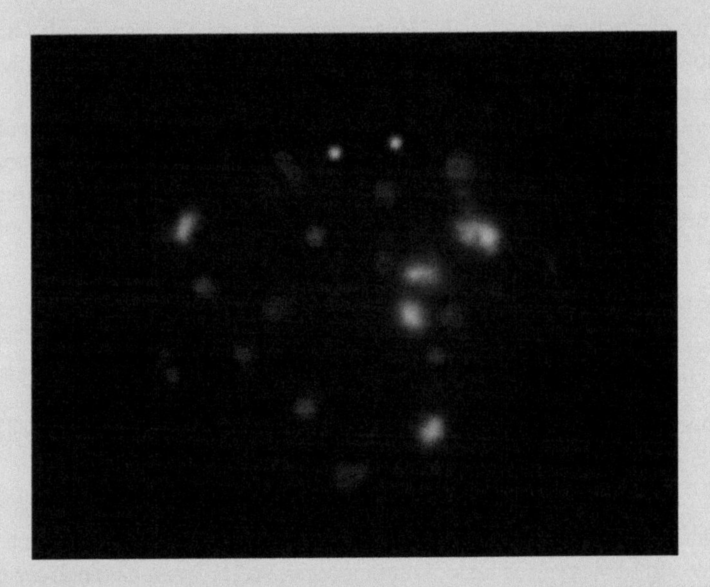

Figure 34.2 Bladder cancer FISH analysis showing gains of chromosomes 3 (*red*), 7 (*green*) and 17 (*aqua*), and two copies of chromosome 9p21 (*gold*), corresponding to an abnormal hyperdiploid result.

Interpretation

Fluorescence *in situ* hybridization (FISH) analysis revealed an abnormal result with 24 of 25 (96%) cells showing gain of chromosomes 3, 7 and 17. These findings are consistent with bladder cancer recurrence.

A positive result is defined as four or more cells with gains of multiple chromosomes (3, 7 and/or 17) or 12 or more cells with homozygous loss of 9p21.

EXAMPLE REPORT – ABNORMAL WITH HYPERDIPLOIDY IN 4 CELLS

ISCN Result: nuc ish 3cen(D3Z1x2~7),7cen(D7Z1x2~5),9p21(p16x2),17cen(D17Z1x2~6)[4/25]

Interpretation

Fluorescence *in situ* hybridization (FISH) analysis revealed an abnormal result with 4 of 25 (16%) cells showing gain of chromosomes 3, 7 and 17. These findings are consistent with bladder cancer recurrence.

Since a positive result is defined as four or more cells with gains of multiple chromosomes (3, 7 and/or 17) or 12 or more cells with homozygous loss of 9p21, this result is at the cut-off for an abnormal result. The results were reenumerated by a second technologist, who verified these findings. However, these results should be interpreted with caution.

EXAMPLE REPORT – HYPERDIPLOIDY IN 3 CELLS

ISCN Result: nuc ish 3cen(D3Z1x2~7),7cen(D7Z1x2~5),9p21(p16x2),17cen(D17Z1x2~6)[3/25]

Interpretation

Fluorescence *in situ* hybridization (FISH) analysis revealed a normal result with 3 of 25 (12%) cells showing gain of chromosomes 3, 7 and 17. This result is not indicative of bladder cancer recurrence.

Since a positive result is defined as four or more cells with gains of multiple chromosomes (3, 7 and/or 17) or 12 or more cells with homozygous loss of 9p21, this result is near the cut-off for an abnormal result. The results were reenumerated by a second technologist, who verified these findings. However, these results should be interpreted with caution.

34.2.3 Abnormal results with homozygous loss of 9p21

EXAMPLE REPORT – ABNORMAL WITH HOMOZYGOUS LOSS OF 9p21

ISCN Result: nuc ish 3cen(D3Z1x2),7cen(D7Z1x2),9p21(p16x0),17cen(D17Z1x2)[20/25]

Interpretation

Fluorescence *in situ* hybridization (FISH) analysis revealed an abnormal result with 20 of 25 (80%) cells showing homozygous loss of chromosome 9p21. These findings are consistent with bladder cancer recurrence.

Chromosome 9p21 loss is one of the most common genetic changes associated with urothelial carcinoma (UC) and generally occurs early in the development of papillary UC and *in situ* UC.

A positive result is defined as four or more cells with gains of multiple chromosomes (3, 7 and/or 17) or 12 or more cells with homozygous loss of 9p21.

EXAMPLE REPORT – ABNORMAL WITH HOMOZYGOUS LOSS OF 9p21 IN 12 CELLS

ISCN Result: nuc ish 3cen(D3Z1x2),7cen(D7Z1x2),9p21(p16x0),17cen(D17Z1x2)[12/25]

Interpretation

Fluorescence *in situ* hybridization (FISH) analysis revealed an abnormal result with 12 of 25 (48%) cells showing homozygous loss of chromosome 9p21. These findings are consistent with bladder cancer recurrence.

Chromosome 9p21 loss is one of the most common genetic changes associated with urothelial carcinoma (UC) and generally occurs early in the development of papillary UC and *in situ* UC.

Since a positive result is defined as four or more cells with gains of multiple chromosomes (3, 7, and/or 17) or 12 or more cells with homozygous loss of 9p21, this result is at the cut-off for an abnormal result. The results were reenumerated by a second technologist who verified these findings. However, these results should be interpreted with caution.

EXAMPLE REPORT – ABNORMAL WITH HOMOZYGOUS LOSS OF 9p21 IN 10 OR 11 CELLS

ISCN Result: nuc ish 3cen(D3Z1x2),7cen(D7Z1x2),9p21(p16x0),17cen(D17Z1x2)[10/25]

Interpretation

Fluorescence *in situ* hybridization (FISH) analysis revealed a normal result with 10 of 25 (40%) cells showing homozygous loss of chromosome 9p21. These findings are not consistent with bladder cancer recurrence.

Chromosome 9p21 loss is one of the most common genetic changes associated with urothelial carcinoma (UC) and generally occurs early in the development of papillary UC and *in situ* UC.

Since a positive result is defined as four or more cells with gains of multiple chromosomes (3, 7, and/or 17) or 12 or more cells with homozygous loss of 9p21, this result is near the cut-off for an abnormal result. The results were reenumerated by a second technologist, who verified these findings. However, these results should be interpreted with caution.

34.2.4 Abnormal results with hyperdiploidy and homozygous loss of 9p21

EXAMPLE REPORT – ABNORMAL WITH HYPERDIPLOIDY AND HOMOZYGOUS LOSS OF 9p21 (FIGURE 34.3)

ISCN Result: nuc ish 3cen(D3Z1x2~7),7cen(D7Z1x2~5),9p21(p16x0),17cen(D17Z1x2~6)[24/25]

Interpretation

Fluorescence *in situ* hybridization (FISH) analysis revealed an abnormal result with 24 of 25 (96%) cells showing gain of chromosomes 3, 7 and 17, and homozygous loss of chromosome 9p21. These findings are consistent with bladder cancer recurrence.

A positive result is defined as four or more cells with gains of multiple chromosomes (3, 7 and/or 17) or 12 or more cells with homozygous loss of 9p21.

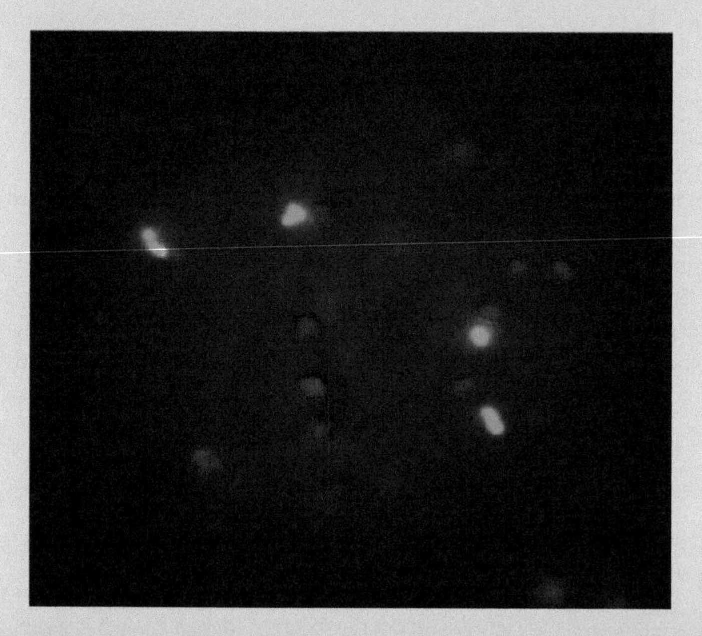

Figure 34.3 Bladder cancer FISH analysis showing gains of chromosomes 3 (*red*), 7 (*green*) and 17 (*aqua*), and no signals for chromosome 9p21 (*gold*), corresponding to an abnormal hyperdiploid and nullisomy 9p21 result.

34.2.5 Normal results with aneuploidy of a single chromosome

EXAMPLE REPORT – NORMAL WITH ANEUPLOIDY OF A SINGLE CHROMOSOME

ISCN Result: nuc ish 3cen(D3Z1x2~7),7cen(D7Z1x2),9p21(p16x2),17cen(D17Z1x2)[24/25]

Interpretation

Fluorescence *in situ* hybridization (FISH) analysis revealed a normal result, with no evidence of aneuploidy of chromosomes 3, 7 and 17, or homozygous loss of chromosome 9p21.

However, of 25 cells examined, 24 (96%) cells showed gain of chromosome 3 as the sole anomaly.

Since a positive result is defined as four or more cells with gains of multiple chromosomes (3, 7 and/or 17) or 12 or more cells with homozygous loss of 9p21, this result is considered a normal result. The results were reenumerated by a second technologist, who verified these findings. However, these results are atypical and should be interpreted with caution.

34.2.6 Tetraploidy

EXAMPLE REPORT – ABNORMAL RESULTS WITH TETRAPLOIDY

ISCN Result: nuc ish 3cen(D3Z1x4),7cen(D7Z1x4),9p21(p16x4),17cen(D17Z1x4)[5/25]

Interpretation

Fluorescence *in situ* hybridization (FISH) analysis revealed an abnormal result with 5 of 25 (20%) cells showing tetraploidy of chromosomes 3, 7, 9 and 17. These findings are consistent with bladder cancer recurrence.

This result is considered positive for aneuploidy. However, the finding of a low level of tetraploid cells may represent normal umbrella cells and not be indicative of bladder cancer. Therefore, these results should be interpreted with caution. Follow-up studies are recommended to help clarify these findings.

A positive result is defined as four or more cells with gains of multiple chromosomes (3, 7 and/or 17) or 12 or more cells with homozygous loss of 9p21.

34.2.7 Monosomy 9p21

EXAMPLE REPORT – NORMAL WITH MONOSOMY 9p21

ISCN Result: nuc ish 3cen(D3Z1x2),7cen(D7Z1x2),9p21(p16x1),17cen(D17Z1x2)[15/25]

Interpretation

Fluorescence *in situ* hybridization (FISH) analysis revealed a normal result, with no evidence of aneuploidy of chromosomes 3, 7 and 17, or homozygous loss of chromosome 9p21. This result is not indicative of bladder cancer recurrence.

However, of 25 cells examined, 15 (60%) cells showed loss of one copy of chromosome 9p21.

Since a positive result is defined as four or more cells with gains of multiple chromosomes (3, 7 and/or 17) or 12 or more cells with homozygous loss of 9p21, this result is considered a normal result. The results were reenumerated by a second technologist, who verified these findings. However, these results are atypical and should be interpreted with caution.

Bibliography

American College of Medical Genetics. *Guidelines in Clinical Cytogenetics. Section E10 Methods in Fluorescence In situ Hybridization – Interphase/Nuclear Fluorescence In Situ Hybridization.* American College of Medical Genetics, Bethesda, MD. www.acmg.net/StaticContent/SGs/Section_E.html, accessed 28 January 2014.

Chung KY, Shia J, Kemeny NE, et al. Cetuximab shows activity in colorectal cancer patients with tumors that do not express the epidermal growth factor receptor by immunohistochemistry. J Clin Oncol 2005; 23(9): 1803–1810.

Dohner H, Stilgenbauer S, Benner A, et al. Genomic aberrations and survival in chronic lymphocytic leukemia. N Engl J Med 2000; 343: 1910–1916.

Halling KC, King W, Sokolova IA, et al. A comparison of cytology and fluorescence in situ hybridization for the detection of urothelial carcinoma. J Urol 2000; 164: 1768–1775.

Hebbar M, Wacrenier A, Desauw C, et al. Lack of usefulness of epidermal growth factor receptor expression determination for cetuximab therapy in patients with colorectal cancer. Anticancer Drugs 2006; 17: 855–857.

Heim S, Mitelman F. *Cancer Cytogenetics*, 2nd edn. Wiley-Liss, New York, 1995.

Hirsch FR, Herbst RS, Olsen C, et al. Increased EGFR gene copy number detected by fluorescent in situ hybridization predicts outcome in non-small cell lung cancer patients treated with cetuximab and chemotherapy. J Clin Oncol 2008; 26(20): 3351–3357.

Hyo R, Yakushijin Y, Masaki Y, et al. Clinicopathological features of lymphoma/leukemia patients carrying both BCL2 and MYC translocations, Haematologica 2009; 94: 935–943.

Jensen KC, Turbin DA, Leung S, et al. New cutpoints to identify increased HER2 copy number: analysis of a large, population-based cohort with long-term follow-up. Breast Cancer Res Treat 2008; 112 453–459.

Johnson BE, Jänne PA. Selecting patients for epidermal growth factor receptor inhibitor treatment: a FISH story or a tale of mutations? J Clin Oncol 2005; 23(28): 6813–6816.

Ladanyi M, Pao W. Lung adenocarcinoma: guiding EGFR-targeted therapy and beyond. Mod Pathol 2008(suppl 2): S16–22.

Moroni M, Sartore-Bianchi A, Veronese S, Siena S. EGFR FISH in colorectal cancer: what is the current reality? Lancet Oncol 2008; 9(5): 402–403.

Sartore-Bianchi A, Moroni M, Veronese S, et al. Epidermal growth factor receptor gene copy number and clinical outcome of metastatic colorectal cancer treated with panitumumab. J Clin Oncol 2007; 25(22): 3238–3245.

Shaffer LG, McGowan-Jordan J, Schmid M (eds). *ISCN 2013: An International System for Human Cytogenetic Nomenclature*. Karger Publishers, Unionville, CT, 2013.

Sokolova IA, Halling KC, Jenkins RB, et al. The development of a multitarget multicolor fluorescence in situ hybridization assay for the detection of urothelial carcinoma in urine. J Mol Diagn 2000; 2(3): 116–123.

Swerdlow SH, Campo E, Harris NL, et al. (eds). *WHO Classification of Tumours of Haematopoietic and Lymphoid Tissues*. IARC Press, Lyon, 2008.

Wolff AC, Hammond ME, Schwartz JN, et al. American Society of Clinical Oncology/College of American Pathologists guideline recommendations for human epidermal growth factor receptor 2 testing in breast cancer. J Clin Oncol 2007; 25(1): 118–145.

Commercial FISH probes are available at these websites

www.AbbottMolecular.com

www.Kreatech.com

www.Rainbowscientific.com

www.empiregenomics.com

www.abnova.com

www.cytocell.com

www.exiqon.com

www.cambridgebluegnome.com

Section 3 Chromosomal Microarray Analysis (CMA)

CHAPTER 35

Chromosomal microarray analysis for hematological disorders

35.1 Introduction

The standard of care for genetic testing for hematological malignancies has always been chromosome analysis with a minimum of 20 metaphase cells. Since chromosome analysis can detect recurrent balanced and unbalanced rearrangements that are associated with different subtypes of disease, this testing has been used for both the diagnosis and prognosis of disease. Additionally, some genetic aberrations detected by chromosome analysis can predict sensitivity to drug treatments, such as imatinib for chronic myeloid leukemia showing a (9;22) translocation or all trans-retinoic acid for acute promyelocytic leukemia with a (15;17) translocation. However, standard chromosome analysis has a very low level of DNA resolution, especially in neoplastic cells, often with a minimum of 5–10 Mb of DNA abnormalities present to be visibly identified. Additionally, standard chromosome analysis requires cell culturing and the analysis of 20 metaphase cells, which are often hard to achieve in spontaneously dividing or quiescent cells for most hematological malignancies. Thus, it is often the case that standard chromosome analysis is unable to detect many abnormalities that are seen by fluorescence *in situ* hybridization (FISH) and array methodologies.

Fluorescence *in situ* hybridization analysis also has its limitations in that it is generally a targeted test using DNA probes specific to regions in the genome known to be associated with particular hematological disorders. This is all well and good if it is known which chromosomal regions need to be assayed; however, no information regarding the rest of the genome will be obtained. Since it is well established that, in many cases, more than one cytogenetic abnormality may be present in a patient, and may often contain abnormalities that are not available for targeted gene regions, FISH analysis alone may be unable to detect many abnormalities that may be present in a patient's specimen.

Chromosomal microarrays (CMAs) have the advantage of identifying abnormalities throughout the genome in one assay and at a very high level of DNA resolution. They also do not require actively dividing cells. However, one limitation is having enough abnormal neoplastic cells to identify this cell population in a background of normal cells. That is, mosaicism for an abnormal cell line may not be observed if it is not present in a high enough proportion to be detected on the array platform.

Cytogenetic Abnormalities: Chromosomal, FISH and Microarray-Based Clinical Reporting, First Edition. Susan Mahler Zneimer.

Since bacterial artificial chromosome (BAC) and oligoarrays can only detect copy number losses and gains, single nucleotide polymorphism (SNP) arrays have become a platform of choice for hematological disorders. Often the combination of oligo- and SNP arrays, specific to regions in the genome known to contain common, non-random abnormalities, is paired in a single platform. This combination allows for the detection of both copy number changes (CNCs), copy number neutral loss of heterozygosity (LOH), and specific non-random aberrations of known regions in the cancer genome.

Chromosomal microarray analysis does have some limitations, however. For example, the American College of Medical Genetics and Genomics (ACMG), in its Standards and Guidelines, has not recommended CMAs for tumor surveillance or detection of minimal or residual disease due to the test's lack of sensitivity for low levels of disease. The ACMG recommends other methods (FISH or qPCR) to monitor patient response to treatment and for residual disease detection.

35.2 Categories of abnormalities

35.2.1 Pathogenic

Acquired CNCs are designated as pathogenic with a documented, clinically significant or disease-associated clonal genetic aberration. Rare constitutional CNCs, when detected, should be noted and investigated for clinical significance. For example, cancer predisposing gene aberrations and deletions or duplications associated with a known constitutional syndrome should be noted. The ACMG Standards and Guidelines have set criteria for the interpretation and reporting of constitutional CNCs.

35.2.2 Variant of uncertain significance (VOUS)

This category may include CNCs that are not known to be associated with disease but meet the reporting criteria established by the laboratory. A CNC in this category is not clearly pathogenic, and there is insufficient evidence for an unequivocal determination of clinical significance. If reported, CNCs may be categorized as one of the following.

- Uncertain clinical significance, acquired, likely pathogenic. The eventual understanding of the clinical significance will depend on accumulation of sufficient information and correlation with clinical features.
- Uncertain clinical significance, likely constitutional. For reporting purposes, this category can revert to the ACMG Standards and Guidelines for interpretation and reporting of constitutional CNCs.
- Uncertain clinical significance, not otherwise specified. This category is used when a CNC is detected that meets the laboratory parameters for reporting but has no features to categorize it further.

35.2.3 Benign

Reporting benign CNCs is at the discretion of the laboratory. This category should include CNCs reported in multiple peer-reviewed publications or curated databases as a benign variant and CNCs without relevant genetic content that meet criteria for reporting. It should be recognized, however, that cancer-associated anomalies that occur in known variant regions might not be benign.

35.3 Complex abnormalities throughout the genome, chromothripsis and homozygosity

Abnormal results in hematological malignancies using CMA may be complicated, and some laboratories prefer to report only significant aberrations associated with disease. Other laboratories may also include alterations in copy number of genes known to be involved in cancer, even when their association with any particular malignancy is unclear. Also, results can be complex enough that there is not a precise ISCN nomenclature for each abnormality; however, in some instances, clear base pair descriptions may be given.

Two reporting formats are described below. Examples of writing complex abnormalities throughout the genome and chromothripsis are given. ISCN nomenclature rules for hematologic malignancies are the same as for constitutional abnormalities for the most part and are described in the previous part of the book. ISCN rules for common abnormalities seen almost exclusively in malignancies are described here.

ISCN RULES FOR ABNORMAL CHROMOSOMAL MICROARRAY RESULTS – COMPLEX ABNORMALITIES THROUGHOUT THE GENOME

- First write "arr" to describe any array analysis.
- In a female, then, in parentheses, write "1-22" followed by a comma, followed by "X".
- In a male, in parentheses, write "1-22" followed by a comma, followed by "XY".
- Then write "cx".

For example, a female with multiple aberrations through the genome is written as: arr(1-22,X)cx. A male with multiple aberrations through the genome is written as: arr(1-22,XY)cx.

In a female with two normal X chromosomes or a male with normal X and Y chromosomes, while the autosomes contain complex abnormalities, the ISCN is then written as: arr(1-22)cx. Note that only the abnormal chromosomes are described.

Chromothripsis refers to alternating gains and losses within a single chromosome, and is decribed below.

ISCN RULES FOR CHROMOTHRIPSIS

- First write "arr", followed by a space.
- Next write the chromosome number involved, followed (without a space) by the chromosome band(s) involved.
- Next, in parentheses, write the start nucleotide of the region, followed by a dash, followed by the end nucleotide of the region.
- Then write "cth" to designate multiple changes in copy number of this chromosomal region.

For example, chromosome abnormalities including loss and gain within the region 11q14.2 to q25 would be written as: arr 11q14.2q25(87,102,415-134,373,630)cth.

ISCN RULES FOR HOMOZYGOSITY DETECTED BY A SNP ARRAY

- First write "arr", followed by a space.
- Next write the chromosome number involved, followed (without a space) by the chromosome band(s) involved.
- Next, in parentheses, write the start nucleotide of the region, followed by a dash, followed by the end nucleotide of the region.

- Next write “x2” to designate copies of this chromosomal region, followed by a space.
- Then write “hmz” to designate that the two copies are homozygous.

For example, homozygosity within the region 7q31.33 to q36.3 would be written as: arr 7q31.33q36.3(125,865,971-159,119,220)x2 hmz.

Note: If regions on chromosome 7 show heterozygosity (htz) and homozygosity, nomenclature may be used to distinguish the heterozygosity from the homozygosity.

For example, with most of the region of chromosome 7 as heterozygous and a portion of the chromosome with homozygosity, this may be written as:

arr 7p22.3q31.33(1-125,865,970)x2 htz,7q31.33q36.3(125,865,971-159,119,220)x2 hmz.

35.4 Normal results and disclaimers

Normal results with microarray analysis can be reported in various ways. With a targeted array, some laboratories will not describe any abnormalities that are unclearly identified as non-random abnormalities in a hematological malignancy. Therefore, ISCN nomenclature may not be appropriate, as a whole genome analysis was not performed. Consequently, a normal result, or no significant abnormalities detected, may suffice, followed by an interpretation explaining the result.

EXAMPLE REPORTING OF CMA – NORMAL RESULTS – CHRONIC LYMPHOCYTIC LEUKEMIA (CLL)

ISCN Result: No significant chromosomal copy number changes were detected

Interpretation

Chromosomal microarray analysis revealed a normal result based on a targeted array platform from a DNA sample extracted from this patient. The tumor DNA was referenced to a normal male DNA sample as an internal control.

The absence of detectable chromosomal abnormalities in chronic lymphocytic leukemia is associated with a more favorable prognosis and response to treatment.

Note: Balanced translocations, intragenic mutations and copy number neutral genomic changes are not detected by this test.

Disclaimers

- This chromosomal microarray assay is used to identify chromosomal copy number changes associated with gains and losses of DNA segments within the malignant cancer genome. This test will only detect whole chromosome aneuploidy, deletions and duplications (segmental aneusomy) within the genome.
- This test does not detect point mutations, small intragenic deletions or duplications, balanced chromosomal aberrations, including reciprocal translocations, inversions or balanced insertions. This assay will also not detect imbalances in genomic regions that are not represented on the microarray or copy number neutral loss of heterozygosity (LOH). Mosaicism cannot be detected when present in less than 30% of the sample.

- Normal results do not exclude the presence of a malignancy or syndrome resulting from a genetic mutation that is undetectable by this test. Clinical implications of some of the reported findings may not be known at this time.
- Other commonly observed genomic alterations in hematological malignancies targeted in this array were within normal limits.
- Other chromosomal regions not known to be pathogenic in hematological malignancies are not reported.

35.5 Example abnormal results in hematological malignancies

35.5.1 Lymphoproliferative disorders

For lymphoproliferative disorders, it is often difficult to obtain dividing cells in culture with which to complete a cytogenetic study. Even with dividing cells, analysis often results in normal chromosomes due to lack of the abnormal cell population. Therefore, CMA is advantageous in detecting abnormalities that would otherwise be missed. However, it is critical to have a large enough abnormal cell population in order to visualize mosaicism. Uusally 30% of abnormal cells are required to detect these cells on a CMA.

Chronic lymphocytic leukemia

EXAMPLE REPORTING OF ABNORMAL RESULTS – TRISOMY 12 WITH AN INDICATION OF CLL (FIGURE 35.1)

ISCN Result: arr(12)x3

Interpretation

Chromosomal microarray analysis (CMA) was performed with a whole genome assay to detect acquired genomic abnormalities for chronic lymphocytic leukemia (CLL) and revealed an abnormal result with the gain of one copy of chromosome 12.

Trisomy 12 is a recurrent, non-random abnormality observed in lymphoproliferative disorders including CLL, non-Hodgkin lymphomas and plasma cell disorders. In CLL, trisomy 12 is generally associated with an intermediate prognosis.

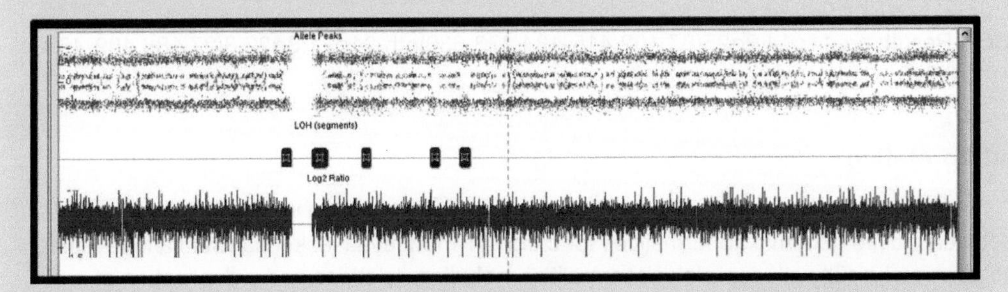

Figure 35.1 SNP array showing trisomy 12. The upper graph shows the alleles consistent with gain of the whole chromosome, while the lower graph shows the gain of one copy from the baseline of zero at the center. Courtesy of Peter Papenhausen PhD, LabCorp, Inc.

EXAMPLE REPORTING OF ABNORMAL RESULTS – del(13q) WITH AN INDICATION OF CLL (FIGURE 35.2)

ISCN Result: arr 13q14.11q14.3(40,919,702-50,227,234)x1

Interpretation

Chromosomal microarray analysis (CMA) was performed with a whole genome assay to detect acquired genomic abnormalities for chronic lymphocytic leukemia and revealed an abnormal result with a deletion of 9.31 Mb of chromosome 13q14.11-q14.3 region, including the SETDB2/PHF11/RCBTB1 gene cluster, and the minimal deleted region (MDR) genes and microRNAs: DLEU1, DLEU2, DLEU7, hsa-mir-15a, hsa-mir-16-1.

Deletions of the chromosome 13q region are recurrent, non-random abnormalities observed in lymphoproliferative disorders, including chronic lymphocytic leukemia (CLL), non-Hodgkin lymphomas and plasma cell disorders. In CLL, del(13q), including RB1 and the SETDB2/PHF11/RCBTB1 gene cluster, is associated with disease progression and a poor prognosis.

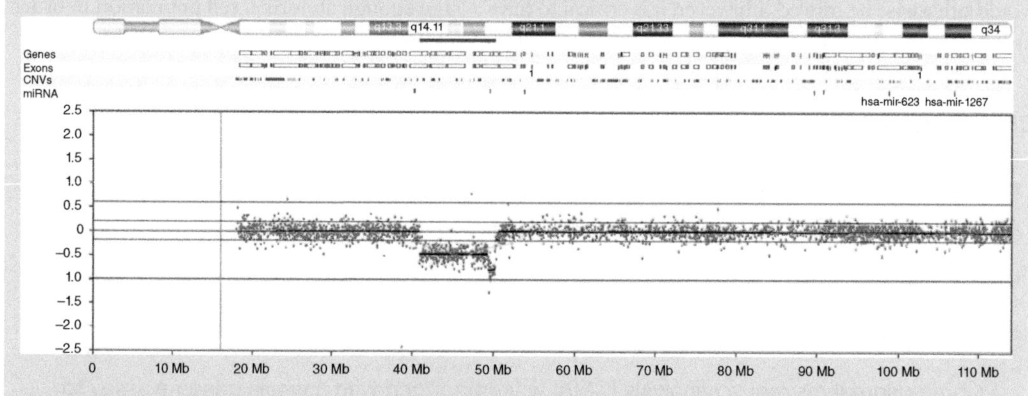

Figure 35.2 Oligoarray showing a deletion within the long arm of chromosome 13 from bands 13q14.11-q14.3. Courtesy of Karine Hovanes PhD, CombiMatrix, Inc.

EXAMPLE REPORTING OF ABNORMAL RESULTS – del(11q) IN CLL (FIGURE 35.3)

ISCN Result: arr 11q14.2q25(87,102,415-134,373,630)cth,21q21.3q22.3(27,186,409-46,914,780)x3

Interpretation

Chromosomal microarray analysis (CMA) was performed with a whole genome assay to detect acquired genomic abnormalities for chronic lymphocytic leukemia (CLL) and revealed an abnormal result with complex abnormalities including deletions and a duplication of chromosome 11q including the following copy number changes:

- a discontinuous deletion of 47.27 Mb of chromosome 11q22.3 region including the ATM gene, and a small duplication within this region
- a duplication of 19.7 Mb of chromosome 21q21.3-q22.3 region.

Deletions of chromosome 11q22.3 region including the ATM gene are recurrent, non-random abnormalities observed in lymphoproliferative disorders, including CLL and non-Hodgkin lymphomas. In CLL, ATM deletions are associated with aggressive disease and are generally responsive to fludarabine, chlorambucil, rituximab (FCR) chemotherapy.

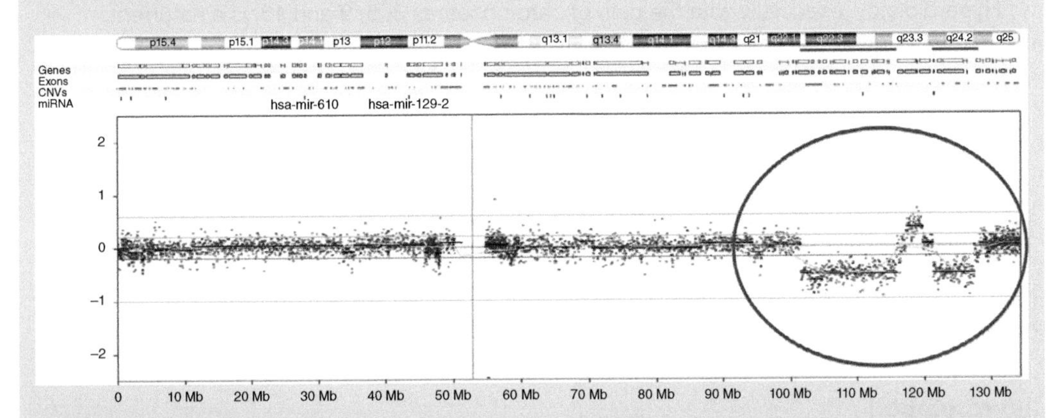

Figure 35.3 Oligoarray showing both deletions and a duplication of 11q. Courtesy of Karine Hovanes PhD, CombiMatrix, Inc.

Multiple myeloma

EXAMPLE REPORTING OF ABNORMAL RESULTS – HYPERDIPLOIDY IN MULTIPLE MYELOMA (FIGURE 35.4)

ISCN Result: Gain of chromosomes 3, 5, 7, 9, 11 and 15

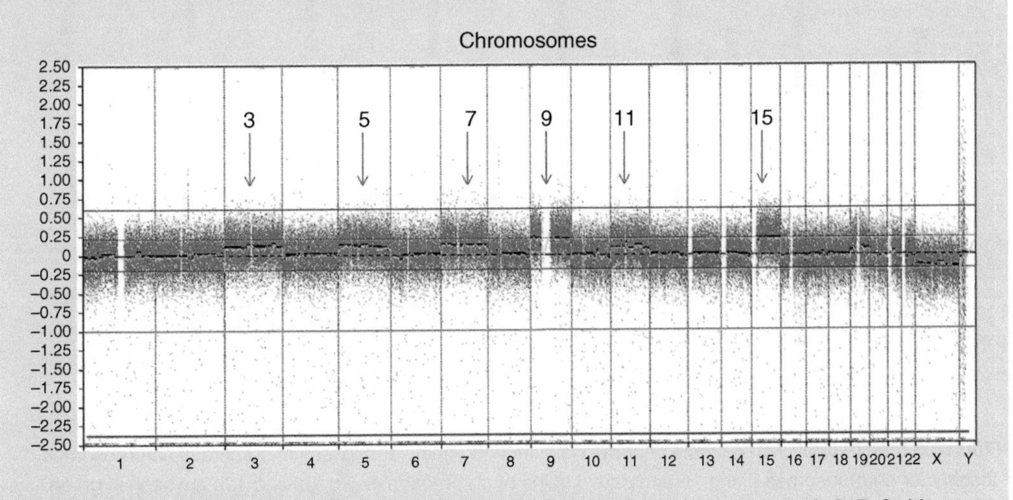

Figure 35.4 Oligoarray with a whole genome view showing trisomy for chromosomes 3, 5, 7, 9, 11 and 15. Courtesy of Karine Hovanes PhD, CombiMatrix, Inc.

Interpretation

Chromosomal microarray analysis (CMA) was performed with a whole genome assay to detect acquired genomic abnormalities for multiple myeloma and revealed an abnormal result with the gain of numerous chromosomes, indicating hyperdiploidy.

Hyperdiploidy, especially with the gain of chromosomes 3, 5, 9 and 15, is a recurrent, non-random abnormality associated with multiple myeloma.

EXAMPLE REPORTING OF ABNORMAL RESULTS – HYPERDIPLOIDY AND MONOSOMY 13 IN MULTIPLE MYELOMA (FIGURE 35.5)

ISCN Result: Gain of chromosomes Xq (partial), 8q (partial), 9, 15 and 19p; loss of chromosomes X (partial), 1p, 8 (partial), 10, 12, 13, 14, 16q, 17, and 20

Interpretation

Chromosomal microarray analysis (CMA) was performed with a whole genome assay to detect acquired genomic abnormalities for multiple myeloma and revealed an abnormal result with aneuploidy of numerous chromosomes.

Gains, especially with trisomy of chromosomes 9 and 15, and monosomy 13, are recurrent, non-random abnormalities associated with multiple myeloma. Monosomy 13 is generally associated with a poor prognosis in multiple myeloma.

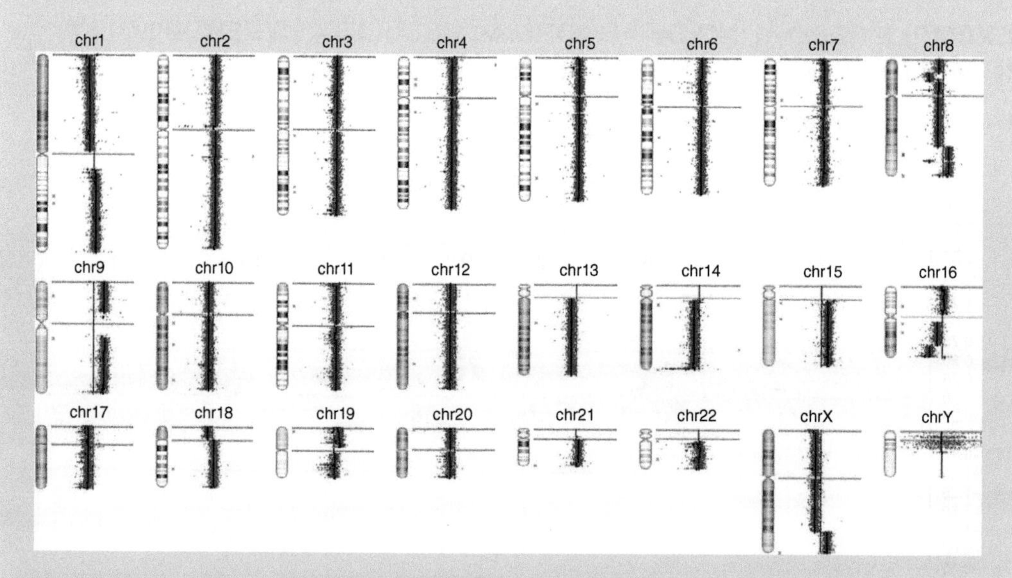

Figure 35.5 SNP array with a whole genome view showing gains and losses of numerous chromosomes. Trisomy for chromosomes 9 and 15 and partial gains of chromosomes Xq, 8q and 19p are seen in green, and losses of chromosomes 10, 12, 13, 14, 17 and 20 and partial losses of chromosomes X, 1p, 8 and 16q are seen in red. Courtesy of Karine Hovanes PhD, CombiMatrix, Inc.

EXAMPLE REPORTING OF ABNORMAL RESULTS – TRISOMY 1q IN MULTIPLE MYELOMA (FIGURE 35.6)

ISCN Result: Gain of chromosome1q (trisomy 1q)

Interpretation

Chromosomal microarray analysis (CMA) was performed with a whole genome assay to detect acquired genomic abnormalities for multiple myeloma and revealed an abnormal result with the gain of the long arm of chromosome 1.

Trisomy 1q is a recurrent, non-random abnormality associated with multiple myeloma and is generally associated with a poor prognosis.

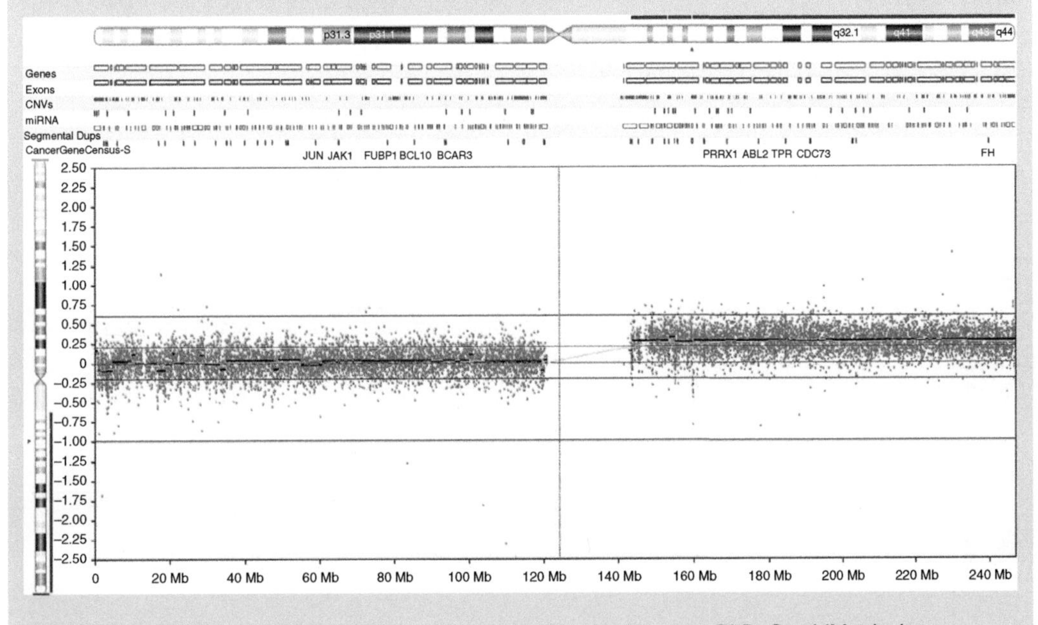

Figure 35.6 Oligoarray showing trisomy 1q. Courtesy of Karine Hovanes PhD, CombiMatrix, Inc.

35.5.2 Myeloid disorders

Acute myeloid leukemia

Since acute myeloid leukemia (AML) shows recurrent aberrations critical to the diagnosis and prognosis of disease, many of which contain copy number variations, CMA has become a very useful methodology to detect acquired DNA changes. Often, standard chromosome analysis yields only a normal karyotype, which is generally associated with an intermediate risk factor in AML. Therefore, further testing to detect submicroscopic DNA alterations helps in assessing risk factors as well as treatment protocols for these patients. Up to 50% of AML patients are classified into an intermediate risk group, but many of these patients have undetectable genetic mutations. For example, normal chromosome analysis that harbors a FLT3 tandem duplication confers a poor prognosis, while NPM and CEBPA mutations with a normal chromosome analysis have a better prognosis. Another example is the presence of translocation (8;21) or inversion of chromosome 16, which as the sole abnormality is generally associated with a good prognosis; however, with the presence of a KIT mutation, it will revert to a worsening prognosis. Other copy number changes that may be present with or without visible chromosome changes will clearly redefine the implications of diagnosis, prognosis and treatment strategies for these patients.

EXAMPLE REPORTING OF ABNORMAL RESULTS – DELETIONS IN ACUTE MYELOID LEUKEMIA (FIGURE 35.7)

ISCN Result: arr 2p11.2(86,971,798-89,908,726)x1,14q32.33(104,459,891-106,252,332)x1,21q22.12(34,996,788-35,390,390)x1

Interpretation

Chromosomal microarray analysis (CMA) was performed with a whole genome assay to detect acquired genomic abnormalities for acute myeloid leukemia and revealed an abnormal result with deletions of chromosomes 2p, 14q and the RUNX1 gene on chromosome 21.

Rearrangements involving the RUNX1 gene on chromosome 21 are often seen in acute myeloid leukemia; however, the clinical significance of a deletion is unclear.

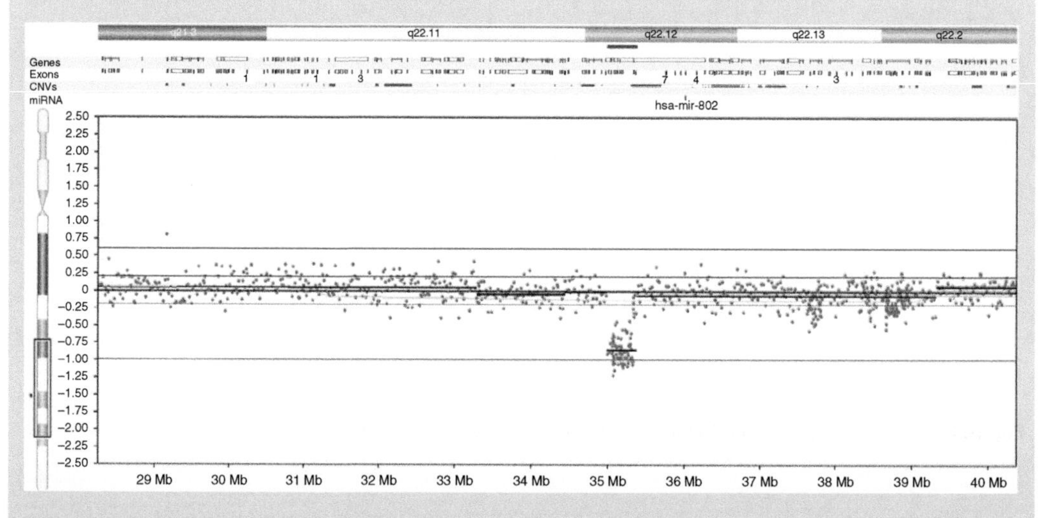

Figure 35.7 Oligoarray showing a RUNX1 deletion on chromosome 21. Courtesy of Karine Hovanes PhD, CombiMatrix, Inc.

EXAMPLE REPORTING OF ABNORMAL RESULTS – HOMOZYGOSITY IN ACUTE MYELOID LEUKEMIA (FIGURE 35.8)

ISCN Result: arr 7q31.33q36.3(125,865,971-159,119,220)x2 hmz

Interpretation

Chromosomal microarray analysis (CMA) was performed with a whole genome assay to detect acquired genomic abnormalities for acute myeloid leukemia and revealed an abnormal result with a region of homozygosity or loss of heterozygosity (LOH) of 33.25 Kb within the long arm of chromosome 7 from bands 7q31.33 to 7q36.3.

Loss of heterozygosity of chromosome 7q is a recurrent, non-random abnormality in myeloid disorders, including acute myeloid leukemia and myelodysplasia, and is generally associated with a poor prognosis.

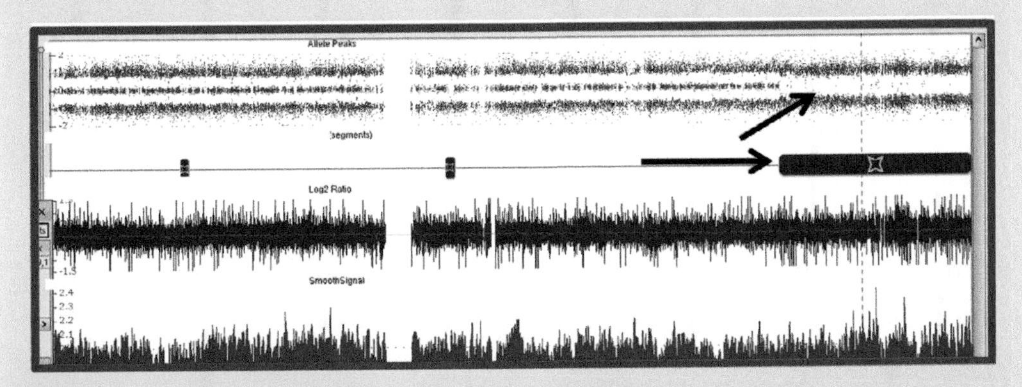

Figure 35.8 SNP array showing loss of heterozygosity for chromosome 7q (*arrows*). Courtesy of Peter Papenhausen PhD, LabCorp, Inc.

Myelodysplasia

Myelodysplastic syndromes (MDS) often result in clearly visible cytogenetic changes, the most common of which include loss of chromosomes 5 and 7, trisomy 8 and deletions of the long arms of chromosomes 5, 7 and 20. However, many other genetic aberrations, which are recurrent, non-random changes, have also been identified. Approximately 50% of patients show clonal aberrations, which leaves an equal 50% of patients with normal karyotypes. These patients most likely contain genomic alterations which are too subtle to detect microscopically and require a more precise analysis, such as CMA, for detection. For example, studies have shown recurrent chromosomally cryptic cytogenetic abnormalities including 4q24 deletions involving the TET2 gene, deletions within chromosome 21q22 involving the RUNX1 gene, and small interstitial long arm deletions of chromosomes 5 and 7. These abnormalities are significant in changing the prognosis and treatment protocols for these patients.

EXAMPLE REPORTING OF ABNORMAL RESULTS IN MYELODYSPLASIA (FIGURE 35.9)

ISCN Result: arr 5q13.2q34(70,345,552-162,985,867)x1,21q22.12(34,996,788-35,390,390)x3

Interpretation

Chromosomal microarray analysis (CMA) was performed with a whole genome assay to detect acquired genomic abnormalities for myelodysplasia and revealed an abnormal result with numerous copy number changes, most significantly including a deletion of chromosome 5q13.2-q34, and duplication of chromosome 21q including the RUNX1 gene.

Deletions of chromosome 5q and duplications involving the RUNX1 gene on chromosome 21 are recurrent, non-random abnormalities in myelodysplastic syndromes.

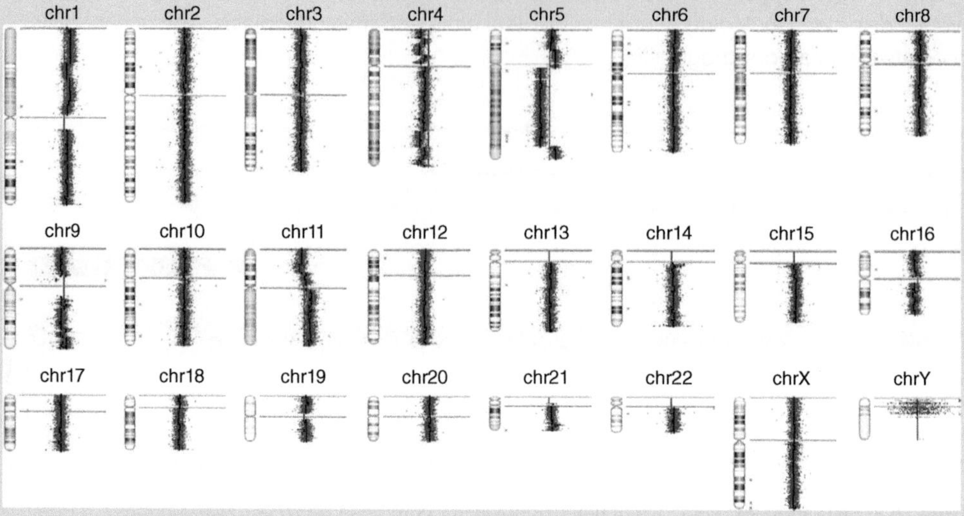

chr1
chr2
chr3
chr4
chr5
chr6
chr7
chr8
chr9
chr10
chr11
chr12
chr13
chr14
chr15
chr16
chr17
chr18
chr19
chr20
chr21
chr22
chrX
chrY

Figure 35.9 Deletion of 5q and other complex copy number changes (gains are seen in green and losses are seen in red). Courtesy of Karine Hovanes PhD, CombiMatrix, Inc.

EXAMPLE REPORTING OF ABNORMAL RESULTS – TET2 GENE DELETION IN MYELODYSPLASIA (FIGURE 35.10)

ISCN Result: arr 4q24(105,771,492-106,433,607)x1

TET2
PPA2
GSTCD
NPNT
TBCK
TACR3
PPA2
GSTCD
NPNT
TBCK
DKK2

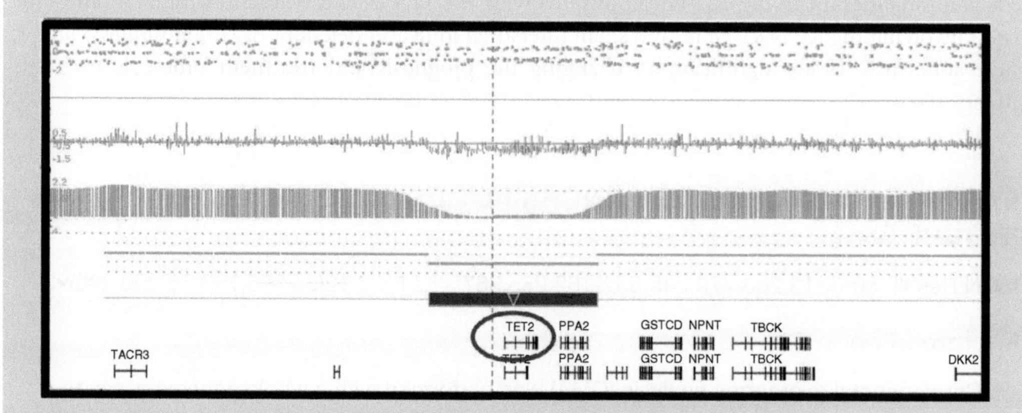

Figure 35.10 SNP array showing a chromosome 4q24-TET2 deletion. Courtesy of Peter Papenhausen PhD, LabCorp, Inc.

Interpretation

Chromosomal microarray analysis (CMA) was performed with a whole genome assay to detect acquired genomic abnormalities for myelodysplasia and revealed an abnormal result with a deletion of chromosome 4q24 including the TET2 gene.

Deletion of chromosome 4q involving the TET2 gene is a recurrent, non-random abnormality in myeloid disorders, including acute myeloid leukemia, myeloproliferative neoplasms and myelodysplasia, often with normal chromosomes, and is generally associated with a poor prognosis.

Bibliography

Avet-Loiseau H, Li C, Magrangeas F, et al. Prognostic significance of copy-number alterations in multiple myeloma. J Clin Oncol 2009; 27: 4585–4590.

Cooley LD, Lebo M, Li MM, Slovak ML, Wolff DJ, for the Working Group of the American College of Medical Genetics and Genomics (ACMG) Laboratory Quality Assurance Committee. American College of Medical Genetics and Genomics technical standards and guidelines: microarray analysis for chromosome abnormalities in neoplastic disorders. Genet Med 2013; 15: 484–494.

Davies JJ, Wilson IM, Lam WL. Array CGH technologies and their applications to cancer genomes. Chromosome Res 2005; 13: 237–248.

Giertlova M, Hajikova M, Vaskova J, et al. Cytogenetic abnormalities predict treatment-free interval and response to therapy in previously untreated chronic lymphocytic leukemia patients. Neoplasma 2011; 58(1): 82–85.

Graubert TA, Payton MA, Shao J, et al. Integrated genomic analysis implicates haploinsufficiency of multiple chromosome 5q31.2 genes in de novo myelodysplastic syndromes pathogenesis. PLoS One 2009; 4(2): e4583.

Gutierez N, Garcia J, Hernandez J, et al. Prognostic and biologic significance of chromosomal imbalances assessed by comparative genomic hybridization in multiple myeloma. Blood 2004; 104: 2661–2666.

Heim S, Mitelman F. *Cancer Cytogenetics*, 2nd edn. Wiley-Liss, New York, 1995.

Heinrichs S, Look AT. Identification of structural aberrations in cancer by SNP array analysis. Genome Biol 2007; 8(7): 219.

Heinrichs S, Li C, Look AT. SNP array analysis in hematologic malignancies: avoiding false discoveries. Blood 2010; 115(21): 4157–4161.

Jacoby MA, Walter MJ. Detection of copy number alterations in acute myeloid leukemia and myelodysplastic syndromes. Expert Rev Mol Diagn 2012; 12(3): 253–264.

Kearney HM, Thorland EC, Brown KK, Quintero-Rivera F, South ST, for the Working Group of the American College of Medical Genetics Laboratory Quality Assurance Committee. American College of Medical Genetics standards and guidelines for interpretation and reporting of postnatal constitutional copy number variants. Genet Med 2011; 13: 680–685.

Ley TJ, Ding L, Walter MJ, et al. DNMT3A mutations in acute myeloid leukemia. N Engl J Med 2010; 363(25): 2424–2433.

Lockwood WW, Chari R, Chi B, Lam WL. Recent advances in array comparative genomic hybridization technologies and their applications in human genetics. Eur J Hum Genet 2006; 14: 139–148.

Mestre-Escorihuela C, Rubio-Moscardo F, Richter JA, et al. Homozygous deletions localize novel tumor suppressor genes in B-cell lymphomas. Blood 2007; 109: 271–280.

O'Keefe CL, Tiu R, Gondek LP, et al. High-resolution genomic arrays facilitate detection of novel cryptic chromosomal lesions in myelodysplastic syndromes. Exp Hematol 2007; 35(2): 240–251.

Parker H, Rose-Zerilli M, Parker A, et al. Cytogenetics and molecular genetics: 13q deletion anatomy and disease progression in patients with chronic lymphocytic leukemia. Leukemia 2011; 25: 489–497.

Paulsson K, Heidenblad M, Strombeck B, et al. High-resolution genome-wide array-based comparative genome hybridization reveals cryptic chromosome changes in AML and MDS cases with trisomy 8 as the sole cytogenetic aberration. Leukemia 2006; 20(5): 840–846.

Pinkel D, Albertson DG. Array comparative genomic hybridization and its applications in cancer. Nat Genet 2005; 37: 511–517.

Pinkel D, Segraves R, Sudar D, et al. High resolution analysis of DNA copy number variation using comparative genomic hybridization to microarrays. Nat Genet 1998; 20: 207–211.

Pollack JR, Perou CM, Alizadeh AA, et al. Genome-wide analysis of DNA copy-number changes using cDNA microarrays. Nat Genet 1999; 23: 41–46.

Rucker FG, Bullinger L, Schwaenen C, et al. Disclosure of candidate genes in acute myeloid leukemia with complex karyotypes using microarray-based molecular characterization. J Clin Oncol 2008; 24(24): 3887–3894.

Shaffer LG, McGowan-Jordan J, Schmid M (eds). *ISCN 2013: An International System for Human Cytogenetic Nomenclature.* Karger Publishers, Unionville, CT, 2013.

Shinawi M, Erez A, Shardy DL, et al. Syndromic thrombocytopenia and predisposition to acute myelogenous leukemia caused by constitutional microdeletions on chromosome 21q. Blood 2008; 112: 1042–1047.

Solinas-Toldo S, Lampel S, Stilgenbauer S, et al. Matrix-based comparative genomic hybridization: biochips to screen for genomic imbalances. Gene Chromosome Cancer 1997; 20: 399–407.

Starczynowski DT, Vercauteren S, Telenius A, et al. High-resolution whole genome tiling path array CGH analysis of CD34+ cells from patients with low-risk myelodysplastic syndromes reveals cryptic copy number alterations and predicts overall and leukemia-free survival. Blood 2008; 112(8): 3412–3424.

Suela J, Alvarez S, Cigudosa JC. DNA profiling by arrayCGH in acute myeloid leukemia and myelodysplastic syndromes. Cytogenet Genome Res 2007; 118: 304–309.

Swerdlow SH, Campo E, Harris NL, et al. (eds). *WHO Classification of Tumours of Haematopoietic and Lymphoid Tissues.* IARC Press, Lyon, 2008.

Tefferi A, Sirhan S, Sun Y, et al. Oligonucleotide array CGH studies in myeloproliferative neoplasms: comparison with JAK2V61-7F mutational status and conventional cytogenetic analysis. Leuk Res 2009; 33: 662–664.

Thiel A, Beier M, Ingenhag D, et al. Comprehensive array CGH of normal karyotype myelodysplastic syndromes reveals hidden recurrent and individual genomic copy number alterations with prognostic relevance. Leukemia 2011; 25(3): 387–399.

Walker B, Leone P, Chiecchio L, et al. A compendium of myeloma-associated chromosomal copy number abnormalities and their prognostic value. Blood 2010; 116: e56–e65.

Microarray database resources

Database of Genomic Variants (http://projects.tcag.ca/variation/)

Online Mendelian Inheritance in Man (www.ncbi.nlm.nih.gov/omim/)

DECIPHER (www.sanger.ac.uk/research/areas/)

dbVar – database of Structural Variation (www.ncbi.nlm.nih.gov/dbvar)

dbGaP – database of Genotypes and Phenotypes (www.ncbi.nlm.nih.gov/gap)

Memorial Sloan-Kettering Cancer Center (http://cbio.mskcc.org/CancerGenes)

Cancer Genome Anatomy Project (www.ncbi.nlm.nih.gov/ncicgap/)

UCSC Genome Bioinformatics (http://genome.ucsc.edu/cgi-bin/hgGateway)

Cancer Genome Atlas (http://cancergenome.nih.gov/)

Ensembl (http://uswest.ensembl.org/Homo_sapiens/Gene/Summary)

International Standards for Cytogenomics Arrays Consortium (www.iscaconsortium.org/)

Wellcome Trust Sanger Institute (www.sanger.ac.uk/genetics/CGP/cosmic/)

www.genpathdiagnostics.com/oncology/hematopathology

CHAPTER 36

Chromosomal microarrays for tumors

36.1 Introduction and disclaimers

Cytogenetic analysis provides valuable information regarding the diagnosis and prognosis of various tumor neoplasms. However, standard chromosome analysis has a low level of DNA resolution and is hard to culture *in vitro*, while fluorescence *in situ* hybridization (FISH) analysis provides only a limited, targeted approach to identifying specific DNA segments in the genome. With the advancement of chromosomal microarray analysis (CMA), it is now possible to detect copy number changes (CNCs) throughout the genome on uncultured cells with a much greater level of precision of genomic breakpoints as well as regions of loss of heterozygosity (LOH), a fairly common feature in tumor malignancies. Most arrays offer BAC, oligonucleotide and/or single nucleotide polymorphisms (SNP) for the detection of genomic changes. SNP arrays have an advantage over oligo- and bacterial artificial chromosome (BAC) arrays in their ability to detect abnormalities at a much higher level of DNA resolution and to detect acquired copy neutral LOH (also called somatic uniparental disomy), but may be difficult to interpret complex abnormal results. There is usually a higher density of DNA probes within cancer-relevant regions, which allows for a more precise definition of small genomic variations that have clinical relevance. Limitations of this type of analysis, however, include the inability to detect balanced rearrangements and inversions that may be common in specific tumor disorders.

Since cancer is genetic, and all cancers have an underlying genetic defect that results in disease, tumors will have some form of clonal changes in their genome. However, most of these changes have yet to be precisely determined for each tumor type. CMA has a great advantage in identifying these abnormalities, even if many of them have little or no known clinical significance as yet. This section discusses only a few of the possible tumors that have been studied by CMA but these are the most common tumors studied to date, including breast, lung, colon and prostate cancers. Examples of abnormal results are given for these tumor types, including a generalized version of results versus precise, descriptive ISCN nomenclature. Since tumors may show complex genetic changes, it may not be feasible to provide a detailed interpretation of every CNC or LOH region detected. In such cases,

Cytogenetic Abnormalities: Chromosomal, FISH and Microarray-Based Clinical Reporting, First Edition. Susan Mahler Zneimer.

it may be best to provide a basic result and a more detailed interpretation to describe variants and their clinical significance rather than a specific ISCN nomenclature.

Since this type of analysis is in the relatively early stages of tumor diagnosis and prognosis, only a few examples are given rather than trying to describe the myriad possible aberrations that may be present in any tumor type.

36.1.1 Disclaimers

- This chromosomal microarray (CMA) was developed and its performance determined by this laboratory for the identification of gains and losses of DNA segments within the tumor genome. The microarray detects chromosomal aneuploidy in addition to deletions and duplications (segmental aneusomy) within the entire tumor genome.
- This test does not detect point mutations and small intragenic deletions and duplications, balanced chromosomal aberrations, including reciprocal translocations, inversions and balanced insertions, and less than 30% mosaicism.
- Normal findings do not preclude a diagnosis of any disorder due to genetic abnormalities that may be undetectable by this assay.
- Clinical implications of some copy number alterations may be unknown at the time of analysis.
- Consultation with a genetic professional is recommended for test interpretation.

36.2 Breast cancer

Studies on breast cancer have recently emerged using the array technology, rather than solely by immunohistochemistry (IHC) and FISH analysis. Chromosomal microarrays have shown many abnormalities previously undetectable by IHC or FISH. With FISH, for example, since the ratio method of copy number of the centromere to the HER2 gene is used, false results may be obtained if there is a gain of copy number of HER2 at 17q12 with an equal gain of copy number of the centromere, or even loss of copy number of these regions.

EXAMPLE CMA REPORT FOR BREAST CANCER – NORMAL HER2 RESULTS

Result: Negative for HER2 amplification
HER2 value = 1.1

Interpretation

Chromosomal microarray analysis (CMA) was performed with a whole genome assay to detect acquired genomic abnormalities for breast cancer and revealed a normal HER2 gene copy number.

Reference values based on the mean of fluorescence intensity ratios for HER2 probes:

- deletion <0.8
- normal = 0.8–1.25 HER2 negative
- amplified = 1.26–3.5 HER2 positive.

EXAMPLE CMA REPORT FOR BREAST CANCER – NORMAL RESULTS FOR HER2 WITH OTHER ABNORMAL RESULTS (FIGURE 36.1)

Result: Negative for HER2 amplification
HER2 value = 1.1

Interpretation

Chromosomal microarray analysis (CMA) was performed with a whole genome assay to detect acquired genomic abnormalities for breast cancer and revealed a normal HER2 gene copy number.

However, amplification of the following gene regions was observed:

- chromosome 1q24.1-q31.2, including the CKS1B gene
- chromosome 8, including the MYC gene at 8q24
- and chromosome 9, including the CDKN2A gene at 9p21.

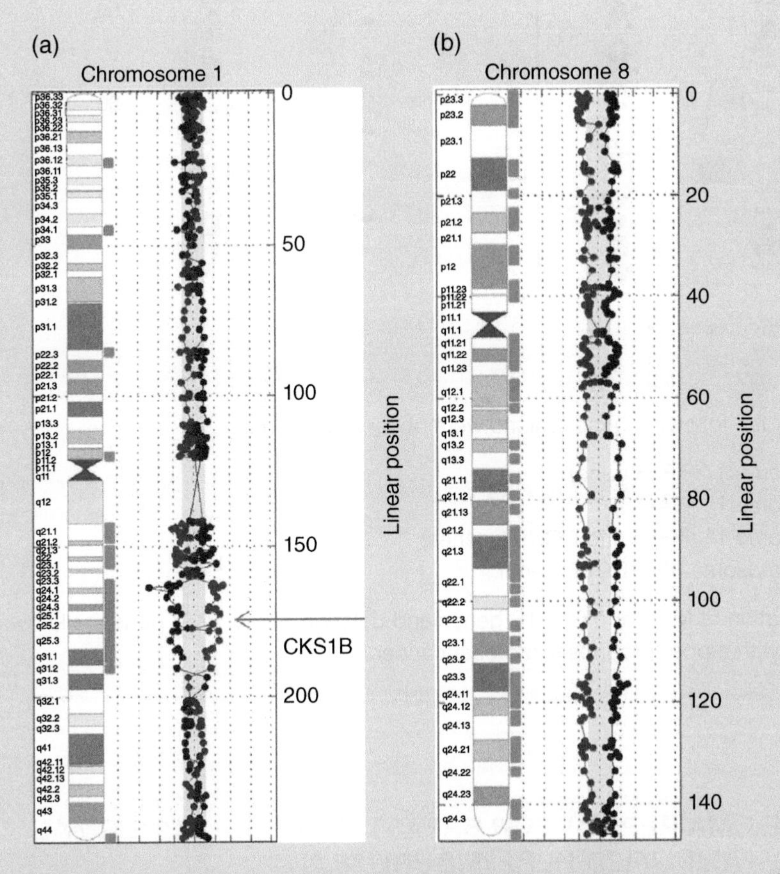

Figure 36.1 BAC array testing for breast cancer showing amplification of (a) chromosome 1 (CKS1B gene), (b) chromosome 8 and (c) chromosome 9, and deletion of (d) chromosome 17 (TP53 gene). Note that the HER2 gene region is normal. Courtesy of Karine Hovanes PhD, CombiMatrix, Inc.

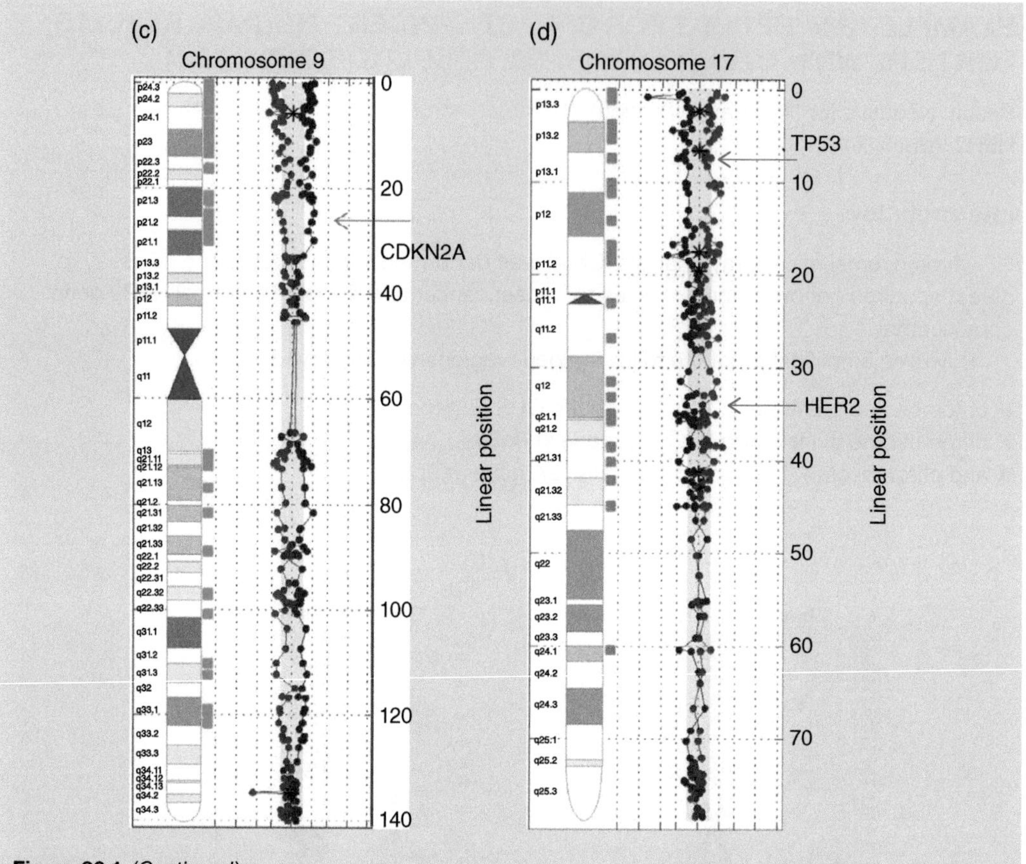

Figure 36.1 (*Continued*)

Deletions of the following gene regions were observed:

- chromosome 17 (TP53 gene)
- chromosome 10 (GATA3)
- chromosome 11 (FGF4 and CCND1 genes)
- and chromosome 22 (PDGFB gene).

Amplification of MYC and CKS1B genes and deletion of the TP53 gene are known to be associated with a poor prognosis in breast cancer.

EXAMPLE CMA REPORT FOR BREAST CANCER – ABNORMAL RESULTS – AMPLIFIED HER2 (FIGURE 36.2)

Result: Positive for HER2 amplification
HER2 value = 1.3

Interpretation

Chromosomal microarray analysis (CMA) was performed with a whole genome assay to detect acquired genomic abnormalities for breast cancer and revealed amplification of the HER2 gene region on chromosome 17q12.

Amplification of the HER2 gene is known to have a beneficial association with treatment using trastuzumab/lapatinib.

Reference values based on the mean of fluorescence intensity ratios for HER2 probes:

- deletion <0.8
- normal = 0.8–1.25 HER2 negative
- amplified = 1.26–3.5 HER2 positive.

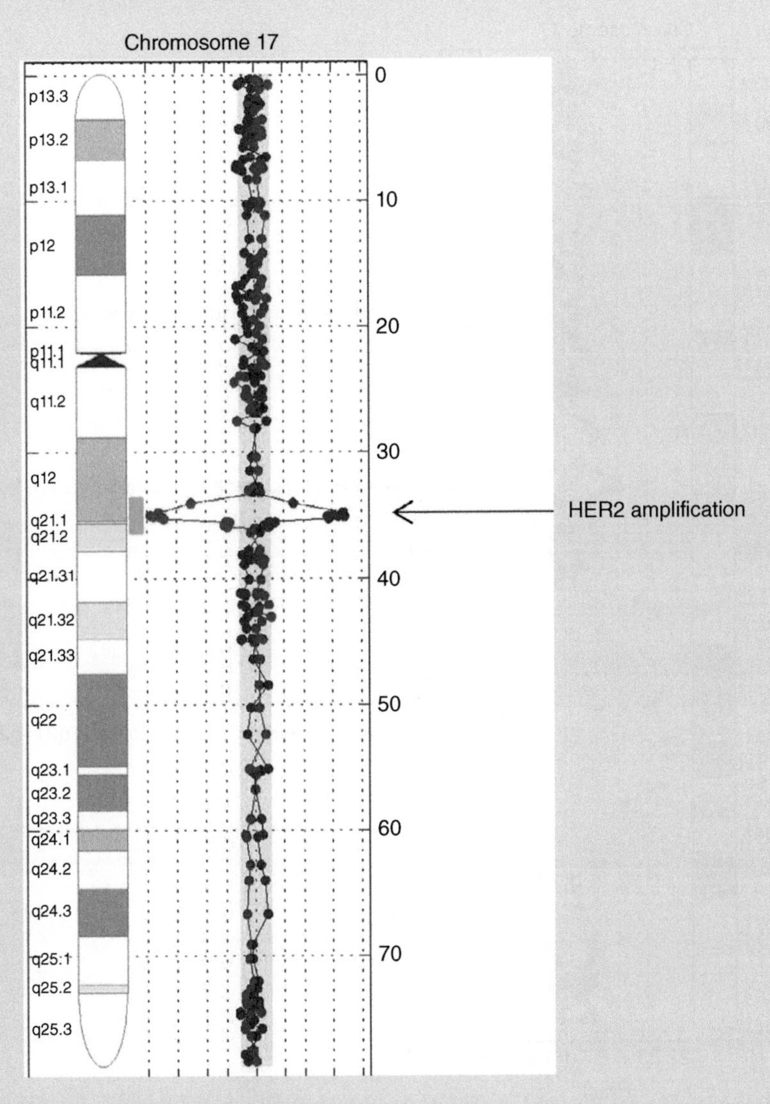

Figure 36.2 BAC array showing HER2 amplification. Courtesy of Karine Hovanes PhD, CombiMatrix, Inc.

EXAMPLE CMA REPORT FOR BREAST CANCER – ABNORMAL RESULTS – AMPLIFIED HER2 AND TOP2A GENES (FIGURE 36.3)

Result: Positive for HER2 and TOP2A amplification
HER2 value = 1.3

Interpretation

Chromosomal microarray analysis (CMA) was performed with a whole genome assay to detect acquired genomic abnormalities for breast cancer and revealed amplification of the HER2 and TOP2A gene regions on chromosome 17q12 and 17q21, respectively.

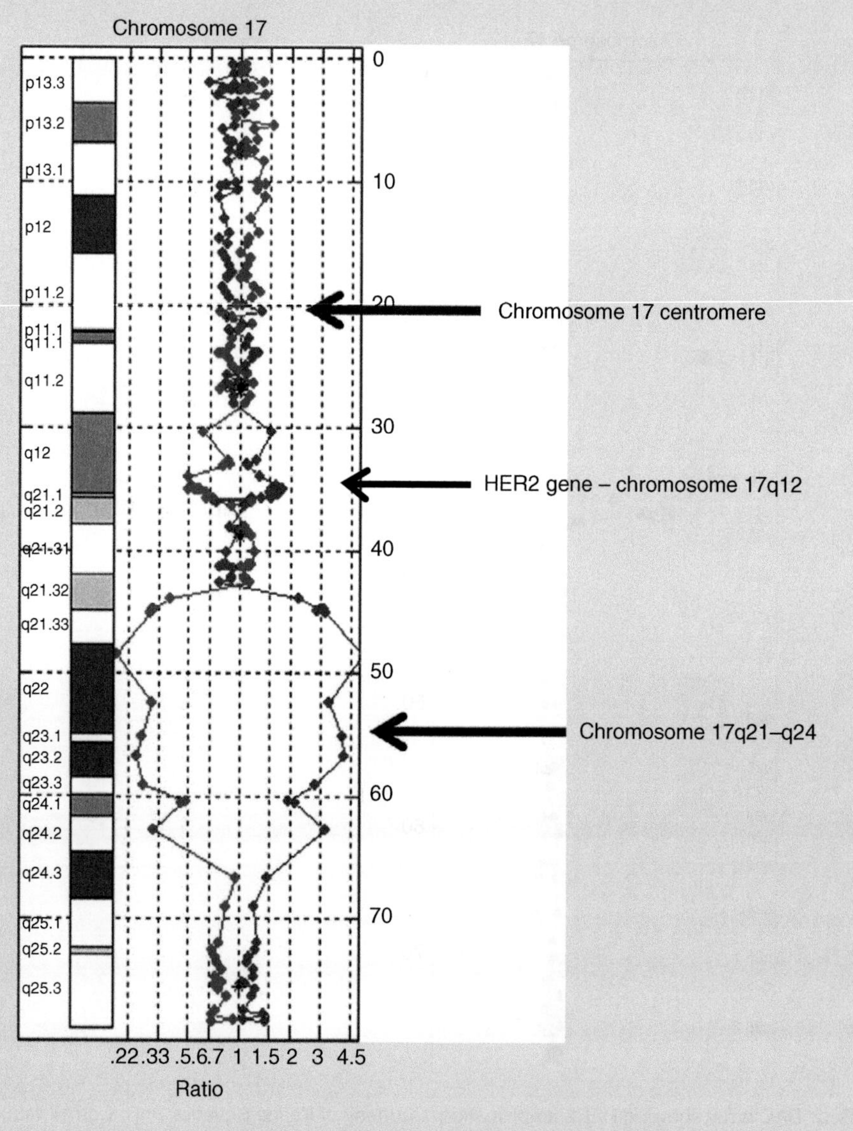

Figure 36.3 Breast cancer showing gain for HER2 and TOP2A genes. Courtesy of Karine Hovanes PhD, CombiMatrix, Inc.

Amplification of the HER2 gene is known to have a beneficial association with treatment using trastuzumab/lapatinib. Amplification of the TOP2A gene is known to have a beneficial association with treatment with anthracycline.

Reference values based on the mean of fluorescence intensity ratios for HER2 probes:

- deletion <0.8
- normal = 0.8–1.25 HER2 negative
- amplified = 1.26–3.5 HER2 positive

EXAMPLE CMA REPORT FOR BREAST CANCER – AMPLIFIED HER2 WITH CO-AMPLIFICATION OF THE CENTROMERE (FIGURE 36.4)

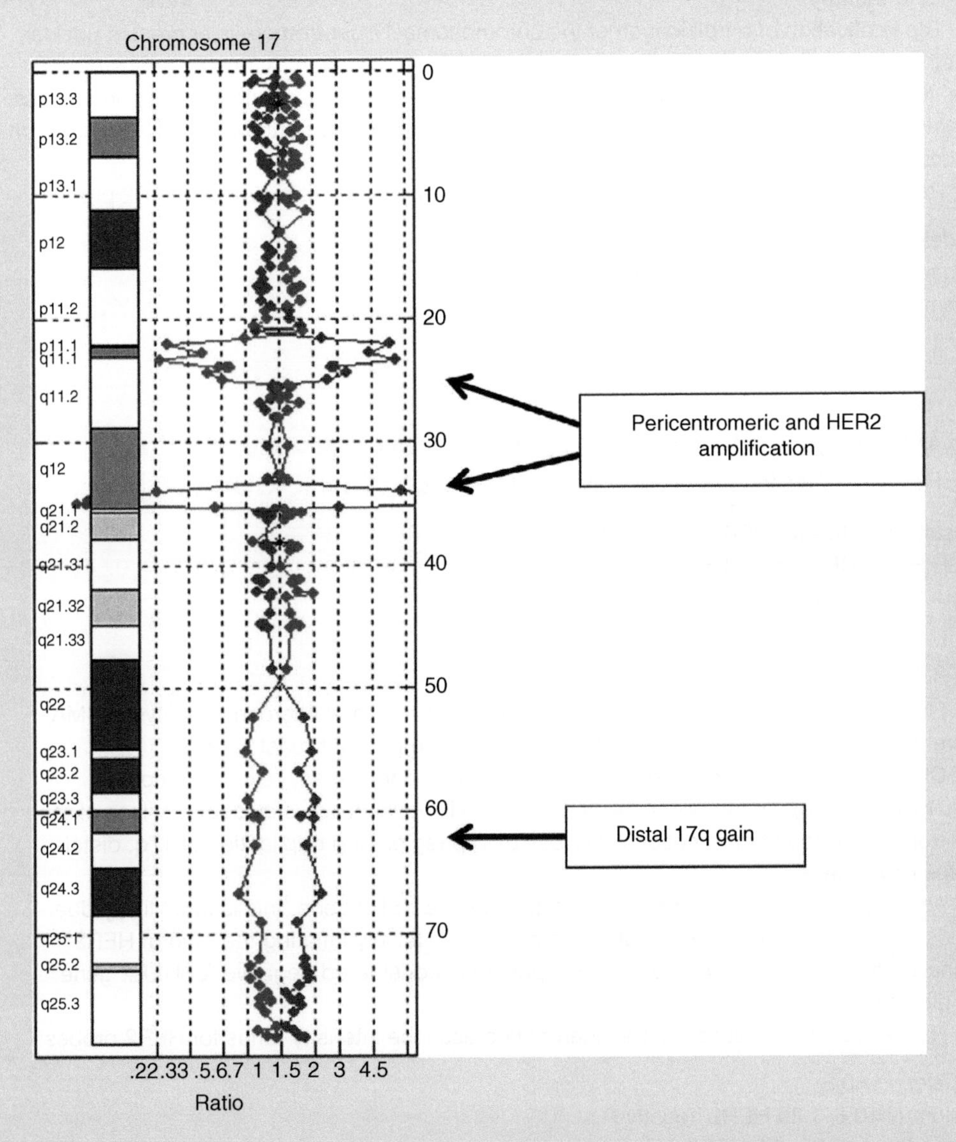

Figure 36.4 BAC array showing HER2 positive with co-amplification of the centromeric region and distal 17q region. Courtesy of Karine Hovanes PhD, CombiMatrix, Inc.

Result: Positive for HER2 amplification
HER2 value = 1.3

Interpretation

Chromosomal microarray analysis (CMA) was performed with a whole genome assay to detect acquired genomic abnormalities for breast cancer and revealed amplification of the HER2 gene.

Concurrent amplification of the pericentromeric region of chromosome 17 and distal 17q was also observed. Based on the amplified ratio for the HER2 gene, this patient is considered positive for HER2 gene amplification and should be eligible for treatment with anti-HER2 targeted therapy.

The implication of amplification of the chromosome 17 centromere is at present unclear, but it may represent a subset of patients with HER2 amplification. A recent study suggests that the presence of centromere 17 aneuploidy may be associated with non-responsiveness to therapy, independent of HER2 amplification status, and may contribute to the progression of breast cancer through accumulation of genomic aberrations.

Reference values based on the mean of fluorescence intensity ratios for HER2 probes:

- deletion <0.8
- normal = 0.8–1.25 HER2 negative
- amplified = 1.26–3.5 HER2 positive.

EXAMPLE CMA REPORT FOR BREAST CANCER – AMPLIFIED HER2 BY FISH WITH NORMAL ARRAY RESULTS (FIGURE 36.5)

Result: Negative for HER2 amplification
FISH HER2/CEP ratio = 3.3
CMA HER2 value = 1.1

Interpretation

Fluorescence *in situ* hybridization (FISH) and chromosomal microarray analysis (CMA) were performed for the assessment of HER2 copy number for breast cancer.

Chromosomal microarray analysis with a whole genome assay revealed a normal result with two copy numbers of the HER2 gene. However, there is evidence of loss of the centromeric region of chromosome 17, loss of 17p region and duplication of 17q, distal to the HER2 gene.

Although FISH analysis appears amplified for the HER2 gene, this is most likely due to loss of the centromere region of chromosome 17, falsely inflating the ratio of HER2 signals. Therefore, by CMA analysis, this patient is considered negative for HER2 gene amplification.

Reference values based on the mean of fluorescence intensity ratios for HER2 probes:

- deletion <0.8
- normal = 0.8–1.25 HER2 negative
- amplified = 1.26–3.5 HER2 positive.

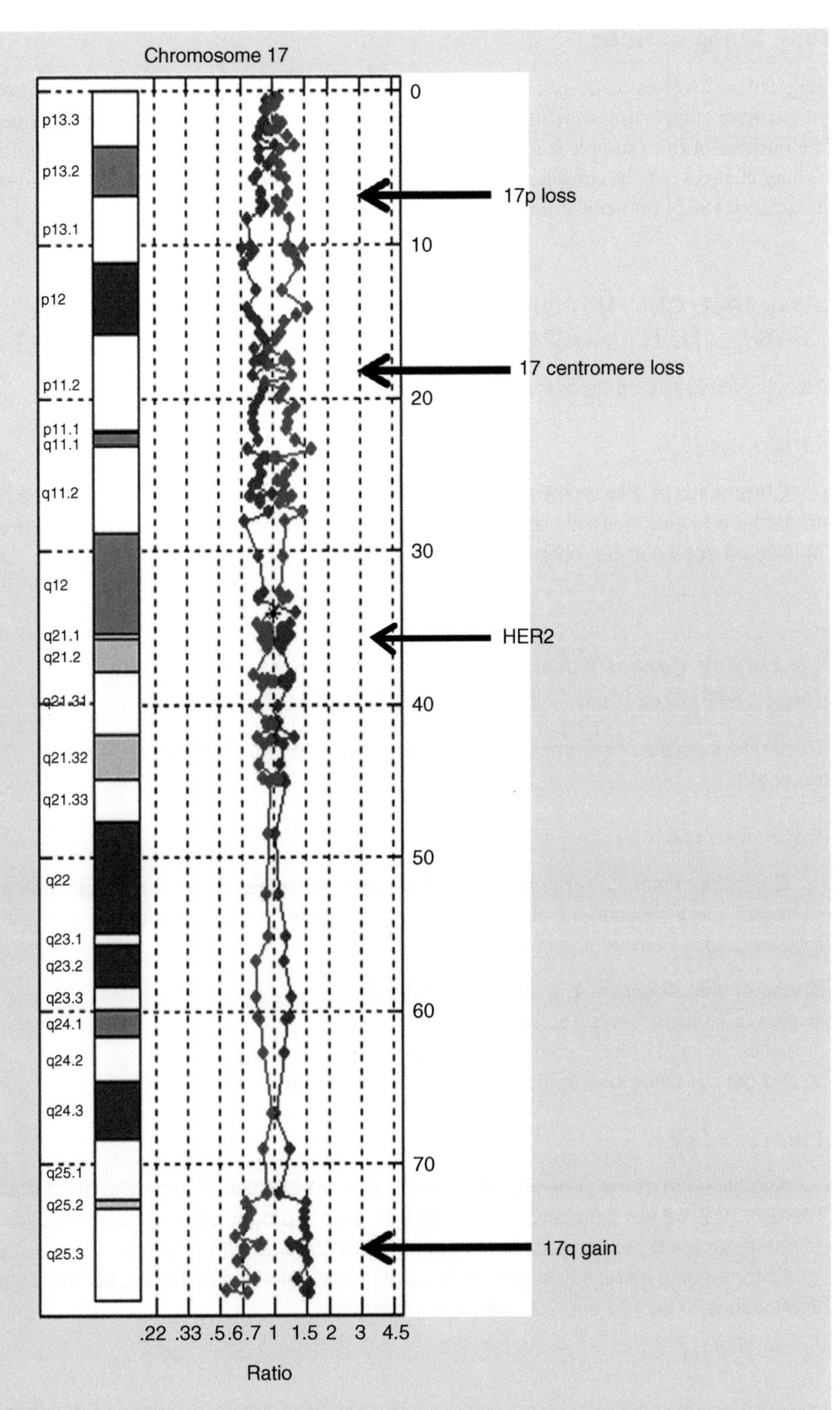

Figure 36.5 HER2 false positive by FISH based on BAC array testing. Courtesy of Karine Hovanes PhD, CombiMatrix, Inc.

36.3 Lung cancer

Lung cancer involves many gene alterations, both for initiation and progression of disease. Below is one example of reporting copy number changes in lung cancer, though many examples could be given. The purpose of this example is to give a prototype for reporting any possible abnormality. Since copy number changes may be complex, a simpler version of writing the abnormal result is given rather than the detailed ISCN nomenclature.

EXAMPLE CMA REPORT FOR TUMOR ANALYSIS – ADENOCARCINOMA OF THE LUNG – NORMAL RESULTS

Result: No significant chromosomal copy number changes were detected.

Interpretation

Chromosomal microarray analysis (CMA) was performed with a whole genome assay to detect acquired genomic abnormalities in adenocarcinoma of the lung and revealed no significant copy number changes.

EXAMPLE CMA REPORT FOR TUMOR ANALYSIS – LUNG CANCER – ABNORMAL RESULTS (FIGURE 36.6)

Result: High levels of chromosomal instability and copy number changes were detected.

Interpretation

Chromosomal microarray analysis (CMA) was performed with a whole genome assay to detect acquired genomic abnormalities in adenocarcinoma of the lung and revealed the following copy number changes:

- loss of chromosomes 1, 2, 4, 6, 13, 18 and 21
- gain of chromosomes 15, 17 and 19
- biallelic loss of chromosome 9p, including the CDKN2A gene
- and gain of chromosome 8q including the MYC gene.

Interpretation

Amplification of the genes CCNE1/19q12 and MYC/8q24.21 and deletion of the gene SMAD4/18q21.2 are generally associated with a poor prognosis in tumors. A biallelic loss of chromosome 9p encompassing the gene CDKN2A was also observed.

Chromosome instability evidenced by additional deletions of 1p, 2q, 3p and 3q, and duplications of 8q,19q and 20p was also observed.

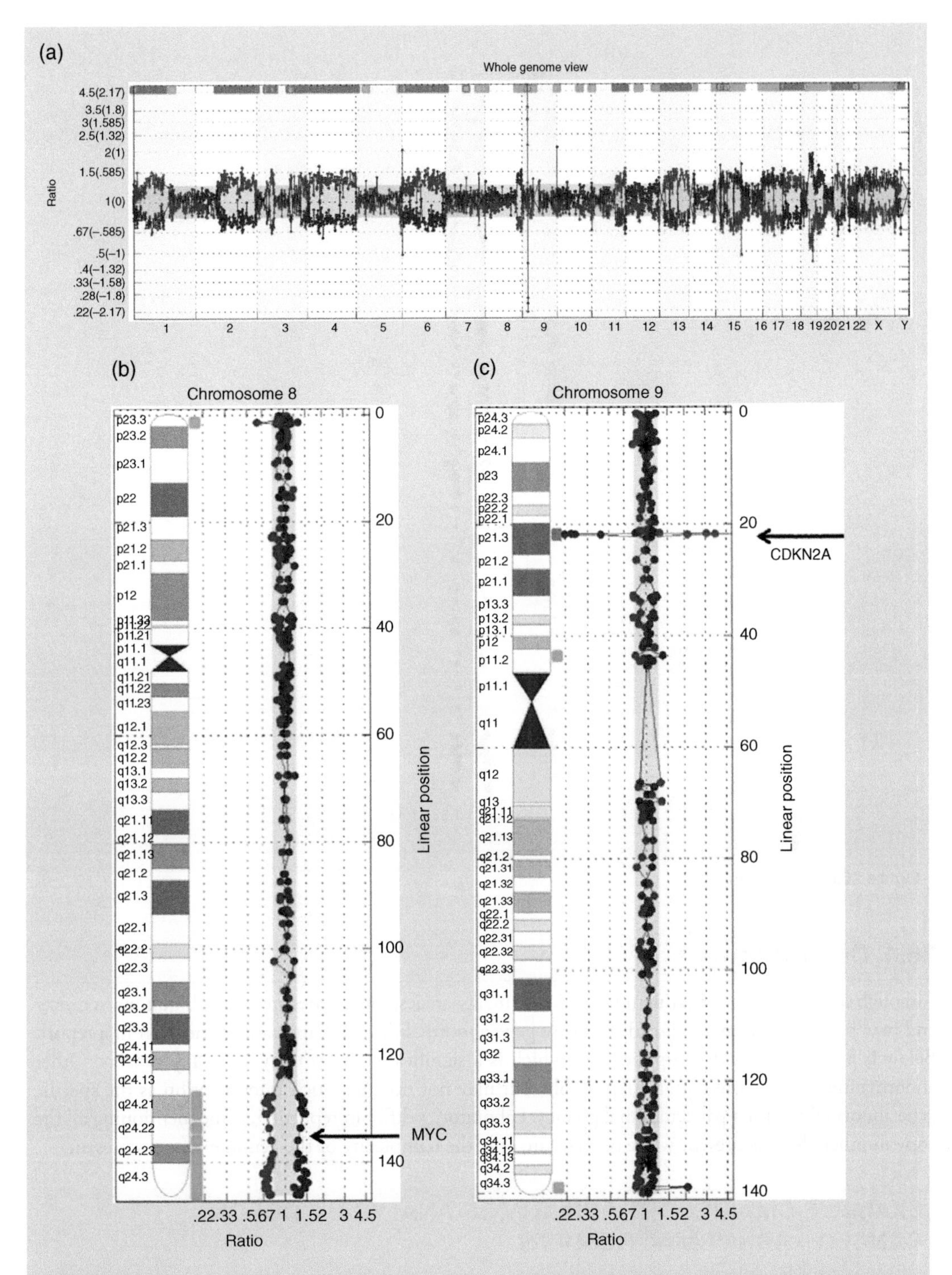

Figure 36.6 BAC array of a lung carcinoma showing (a) a whole genome view, (b) MYC amplification on chromosome 8, (c) CDKN2A deletion on chromosome 9, and (d) chromosome 17 gain. Courtesy of Karine Hovanes PhD, CombiMatrix, Inc.

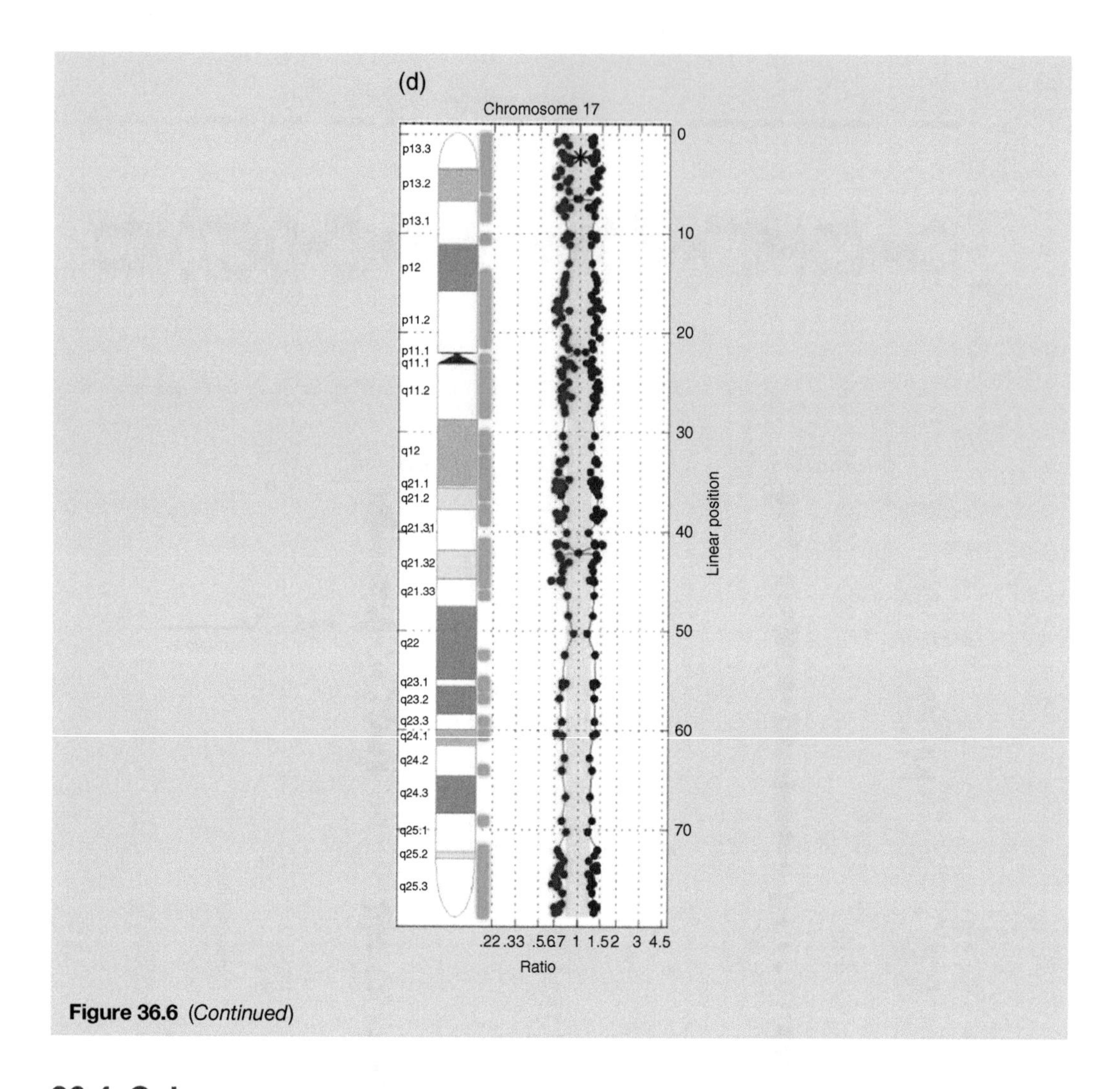

Figure 36.6 *(Continued)*

36.4 Colon cancer

Although specific genes are pertinent to particular malignancies, some genes are more ubiquitous in cancer and may be present in many disorders. These gene abnormalities are important to mention in a report. Below is a case with MYC amplification, which is a significant abnormality for many cancers. Other abnormalities may be present and reportable, but may not need as much interpretation as the specific gene abnormalities for a particular malignancy. Laboratories differ in reporting, some mentioning all the copy number changes present while others state only the complexity of the genome changes present.

EXAMPLE CMA REPORT FOR TUMOR ANALYSIS – COLON CANCER – ABNORMAL RESULTS

Result: Chromosomal instability and copy number changes were detected.

Interpretation

Chromosomal microarray analysis (CMA) was performed with a whole genome assay to detect acquired genomic abnormalities in colon cancer and revealed duplications of chromosomes 7p, 8q (MYC gene) and 20, and a small deletion of 19q13.33 involving the BAX gene.

Amplification of the MYC gene is known to be associated with a poor prognosis in colon cancer.

Other chromosomal regions not known to be pathogenic in malignancies are not reported.

36.5 Prostate cancer

Numerous cytogenetic abnormalities have been reported in prostate cancer, mostly by SNP microarrays. The most common abnormality is the gain of chromosome 8q24 involving the MYC and FAM84B genes which is generally associated with aggressive tumors and poor prognoses. Other common chromosome copy number changes include hyperdiploidy and aneusomy of chromosomes 7 and 17; however, complex changes of many other chromosomes have been reported.

EXAMPLE CMA REPORT FOR TUMOR ANALYSIS – PROSTATE CANCER – ABNORMAL RESULTS

Result: Chromosomal instability and copy number changes were detected.

Interpretation

Chromosomal microarray analysis (CMA) was performed with a whole genome assay to detect acquired genomic abnormalities in prostate cancer and revealed duplications of chromosomes 7, 8q (MYC gene) and 17.

Gain of chromosomes 7, 8q24 and 17 is known to be associated with a poor prognosis in prostate cancer.

Other chromosomal regions not known to be pathogenic in malignancies are not reported.

36.6 Unspecified tumor present

If a tumor type is not indicated when a specimen is received, it is often difficult to intepret the results when they are abnormal, especially if there are complex changes. It may be necessary to add information to a report that describes the diagnostic or prognostic implications of the copy number detected when known and give a general report, rather than specific ISCN nomenclature for each abnormality observed.

Although many genes are involved in the initiation and progression of disease, specific genes are pertinent to mention in a report. Below is a case with a PTEN deletion, which is a significant abnormality for different cancers. Other abnormalities may be present, and reportable, but may not need as much interpretation as the specific gene abnormalites for a particular malignancy.

EXAMPLE CMA REPORT FOR TUMOR ANALYSIS – MALIGNANT NEOPLASM – ABNORMAL RESULTS

Result: High levels of chromosomal instability and copy number changes were detected.

Interpretation

Chromosomal microarray analysis (CMA) was performed with a whole genome assay to detect acquired genomic abnormalities for an acquired malignancy and revealed a high level of chromosomal instability, amplification of the CCND1 gene at 11q13, and deletion of the PTEN tumor suppressor gene at 10q24.

The PTEN gene deletion in cancer is generally associated with a high genetic risk of disease.

Genomic instability reference ranges:

- low: <5 significant genomic changes without amplifications
- moderate: 5–10 significant genomic changes and/or one amplified region
- high: >10 significant genomic changes and/or >1 amplified region.

Other chromosomal regions not known to be pathogenic in malignancies are not reported.

EXAMPLE REPORT OF CMA FOR TUMOR ANALYSIS – MALIGNANT NEOPLASM – ABNORMAL RESULTS

Result: High levels of chromosomal instability and copy number changes were detected.

Interpretation

Chromosomal microarray analysis (CMA) was performed with a whole genome assay to detect acquired genomic abnormalities for an acquired malignancy and revealed a high level of chromosomal instability and the following copy number changes:

- gain of chromosomes 1q, 5 and 7
- a deletion of chromosome 6p involving the CCND3 gene.

Additional amplification of the gene regions of clinical significance was also observed.

Oncogene	Cytoband	Tumor association	Prognosis
SHH	7q36.3	Gastric	Poor
EGFR	7p11.2	Various	Unknown
PDGFRB	5q33	Various	Unknown
AKT3	1q44	Various	Unknown
RAB25	1q22	Ovarian	Poor
CKS1B	1q21.3	Breast	Poor

Bibliography

Abouantoun TJ, Castellino RC, MacDonald TJ. Sunitinib induces PTEN expression and inhibits PDGFR signaling and migration of medulloblastoma cells. J Neurooncol 2011;101:215–226.

Alazzouzi H, Alhopuro P, Salovaara R, et al. SMAD4 as a prognostic marker in colorectal cancer. Clin Cancer Res 2005; 11: 2606–2611.

Badoux X, Keating M, Wierda W. What is the best frontline therapy for patients with CLL and 17p deletion? Curr Hematol Malig Rep 2011; 6: 36–46.

Birrer MJ, Johnson ME, Hao K, et al. Whole genome oligonucleotide-based array comparative genomic hybridization analysis identified fibroblast growth factor 1 as a prognostic marker for advanced-stage serous ovarian adenocarcinomas. J Clin Oncol 2007; 25: 2281–2287.

Cappuzzo F, Hirsch F, Rossi E, et al. Epidermal growth factor receptor gene and protein and gefitinib sensitivity in non-small-cell lung cancer. J Natl Cancer Inst 2005; 97: 643–655.

Cheng KW, Lahad JP, Kuo WL, et al. The RAB25 small GTPase determines aggressiveness of ovarian and breast cancers. Nat Med 2004; 10(11): 1251–1256.

Davies JJ, Wilson IM, Lam WL. Array CGH technologies and their applications to cancer genomes. Chromosome Res 2005; 13: 237–248.

Guiu S, Gauthier M, Coudert B, et al. Pathological complete response and survival according to the level of HER-2 amplification after trastuzumab-based neoadjuvant therapy for breast cancer. Br J Cancer 2010; 103(9): 1335–1342.

Gunn S, Yeh I, Lytvak I, et al. Clinical array-based karyotyping of breast cancer with equivocal HER2 status resolves gene copy number and reveals chromosome 17 complexity. BMC Cancer 2010; 10: 396.

Iwakawa R, Kohno T, Kato M, et al. MYC amplification as a prognostic marker of early stage lung adenocarcinoma identified by whole genome copy number analysis. Clin Cancer Res 2011; 17: 1481–1489.

Jong K, Marchiori E, van der Vaart A, et al. Cross-platform array comparative genomic hybridization meta-analysis separates hematopoietic and mesenchymal from epithelial tumors. Oncogene 2007; 26: 1499–1506.

Kallioniemi A. CGH microarrays and cancer. Curr Opin Biotechnol 2008; 19: 36–40.

Keyomarsi K, Tucker SL, Buchholz TA, et al. Cyclin E and survival in patients with breast cancer. N Engl J Med 2002; 347(20): 1566–1575.

Kim SW, Kim JW, Kim YT, et al. Analysis of chromosomal changes in serous ovarian carcinoma using high-resolution array comparative genomic hybridization: potential predictive markers of chemoresistant disease. Genes Chromosomes Cancer 2007; 46: 1–9.

Kransikas A, Bartlett D, Cieply K, Dacic S. CDKN2A and MTAP deletions in peritoneal mesotheliomas are correlated with loss of p16 protein expression and poor survival. Mod Pathol 2010; 23: 531–538.

Kunivasu H, Yasui W, Kitadai Y, Yokozaki H, Ito H, Tahara E. Frequent amplification of the c-met gene in scirrhous type stomach cancer. Biochem Biophys Res Commun 1992; 189(1): 227–232.

Lockwood WW, Chari R, Chi B, Lam WL. Recent advances in array comparative genomic hybridization technologies and their applications in human genetics. Eur J Hum Genet 2006; 14: 139–148.

Mestre-Escorihuela C, Rubio-Moscardo F, Richter JA, et al. Homozygous deletions localize novel tumor suppressor genes in B-cell lymphomas. Blood 2007; 109: 271–280.

Miller CT, Moy JR, Lin L, et al. Gene amplification in esophageal adenocarcincomas and Barrett's with high-grade dysplasia. Clin Cancer Res 2003; 9: 4819–4825.

Mure H, Matsuzaki K, Kitazato K, et al. Akt2 and Akt3 play a pivotal role in malignant gliomas. Neurol Oncol 2010; 12(3): 221–232.

Nahta R, Yuan LX, Du Y, Esteva FJ. Lapatinib induces apoptosis in trastuzumab-resistant breast cancer cells: effects on insulin-like growth factor I signaling. Mol Cancer Ther 2007; 6(2): 667–674.

Nakajima M, Sawada H, Yamada Y, et al. The prognostic significance of amplification and overexpression of c-met and c-erb B-2 in human gastric carcinomas. Cancer 1999; 85(9): 1894–1902.

Naylor TL, Greshock J, Wang Y, et al. High resolution genomic analysis of sporadic breast cancer using array-based comparative genomic hybridization. Breast Cancer Res 2005; 7: R1186–R1198.

Petroni S, Addati T, Mattioli E, et al. Centromere 17 copy number alteration: negative prognostic factor in invasive breast cancer? Arch Pathol Lab Med 2012; 136: 993–1000.

Pinkel D, Albertson DG. Array comparative genomic hybridization and its applications in cancer. Nat Genet 2005; 37: 511–517.

Pinkel D, Segraves R, Sudar D, et al. High resolution analysis of DNA copy number variation using comparative genomic hybridization to microarrays. Nat Genet 1998; 20: 207–211.

Pollack JR, Perou CM, Alizadeh AA, et al. Genome-wide analysis of DNA copy-number changes using cDNA microarrays. Nat Genet 1999; 23: 41–46.

Shaffer LG, McGowan-Jordan J, Schmid M (eds). *ISCN 2013: An International System for Human Cytogenetic Nomenclature.* Karger Publishers, Unionville, CT, 2013.

Villman K, Sjöström J, Heikkilä R, et al. TOP2A and HER2 gene amplification as predictors of response to anthracycline treatment in breast cancer. Acta Oncol 2006; 45(5): 590–596.

Solinas-Toldo S, Lampel S, Stilgenbauer S, et al. Matrix-based comparative genomic hybridization: biochips to screen for genomic imbalances. Genes Chromosomes Cancer 1997; 20: 399–407.

Soung YH, Lee J, Nam S, et al. Mutational analysis of AKT1, AKT2 and AKT3 genes in common human carcinomas. Oncology 2006; 70(4): 285–289.

Souzaki R, Tajiri T, Souzaki M, et al. Hedgehog signaling pathway in neuroblastoma differentiation. J Pediatr Surg 2010; 45(12): 2299–2304.

Swerdlow SH, Campo E, Harris NL, et al. (eds). *WHO Classification of Tumours of Haematopoietic and Lymphoid Tissues.* IARC Press, Lyon, 2008.

Wang XC, Tian LL, Tian J, Wu HL, Meng AM. Overexpression of Cks1 is associated with poor survival by inhibiting apoptosis in breast cancer. J Cancer Res Clin Oncol 2009; 135(10): 1393–1401.

Wiedemeyer W, Dunn I, Quayle S, et al. Pattern of retinoblastoma pathway inactivation dictates response to CD4/6 inhibition in GBM. Proc Natl Acad Sci USA 2010; 107: 11501–11506.

Microarray database resources

Database of Genomic Variants (http://projects.tcag.ca/variation/)

Online Mendelian Inheritance in Man (www.ncbi.nlm.nih.gov/omim/)

DECIPHER (www.sanger.ac.uk/research/areas/)

dbVar – database of Structural Variation (www.ncbi.nlm.nih.gov/dbvar)

dbGaP – database of Genotypes and Phenotypes (www.ncbi.nlm.nih.gov/gap)

Memorial Sloan-Kettering Cancer Center (http://cbio.mskcc.org/CancerGenes)

Cancer Genome Anatomy Project (www.ncbi.nlm.nih.gov/ncicgap/)

UCSC Genome Bioinformatics (http://genome.ucsc.edu/cgi-bin/hgGateway)

Cancer Genome Atlas (http://cancergenome.nih.gov/)

Ensembl (http://uswest.ensembl.org/Homo_sapiens/Gene/Summary)

International Standards for Cytogenomics Arrays Consortium (www.iscaconsortium.org/)

Wellcome Trust Sanger Institute (www.sanger.ac.uk/genetics/CGP/cosmic/)

www.genpathdiagnostics.com/oncology/hematopathology

CHAPTER 37

Integrated reports with chromosomes, FISH and microarrays

Below are a few examples of cases reported with all three technologies, chromosome, fluorescence *in situ* hybridization (FISH) and microarray analyses, with an indication of a hematological malignancy. It is rare to have all three technologies performed on tumor analyses, as most tumors are now studied on formalin-fixed paraffin-embedded (FFPE) samples, and chromosome analyses are only performed on fresh tissue. Additionally, laboratories can now decide if microarray will replace FISH analysis, rather than doing both analyses, unless it is to confirm or clarify specific findings and the following are such examples. Only pertinent images are displayed for these cases, for the purposes of visualizing the results. Futhermore, when cases show complex microarray results, ISCN nomenclature is often replaced by a general result in the report, in order to make the report easier to understand and interpret.

The first example is a homozygous loss of chromosome 9p21 with other copy number changes in acute lymphoblastic leukemia. This example is given to show the homozygous loss that is seen only by FISH and array analyses, and not by chromosome analysis.

37.1 Homozygous deletion of 9p21 identified by FISH and CMA

EXAMPLE REPORTING OF ABNORMAL RESULTS – CHROMOSOME 9p LOSS (FIGURE 37.1)

Result: Positive for a homozygous loss of chromosome 9p21

Interpretation

Chromosome analysis revealed a normal female chromosome complement in all 20 cells examined. However, chromosomal microarray analysis (CMA) was performed with a whole genome assay to detect acquired genomic abnormalities in acute lymphoblastic leukemia and revealed an abnormal result with the following copy number changes:

- gain of chromosome 4
- homozygous loss of chromosome 9p21 including the CDNK2A gene.

Cytogenetic Abnormalities: Chromosomal, FISH and Microarray-Based Clinical Reporting, First Edition. Susan Mahler Zneimer.

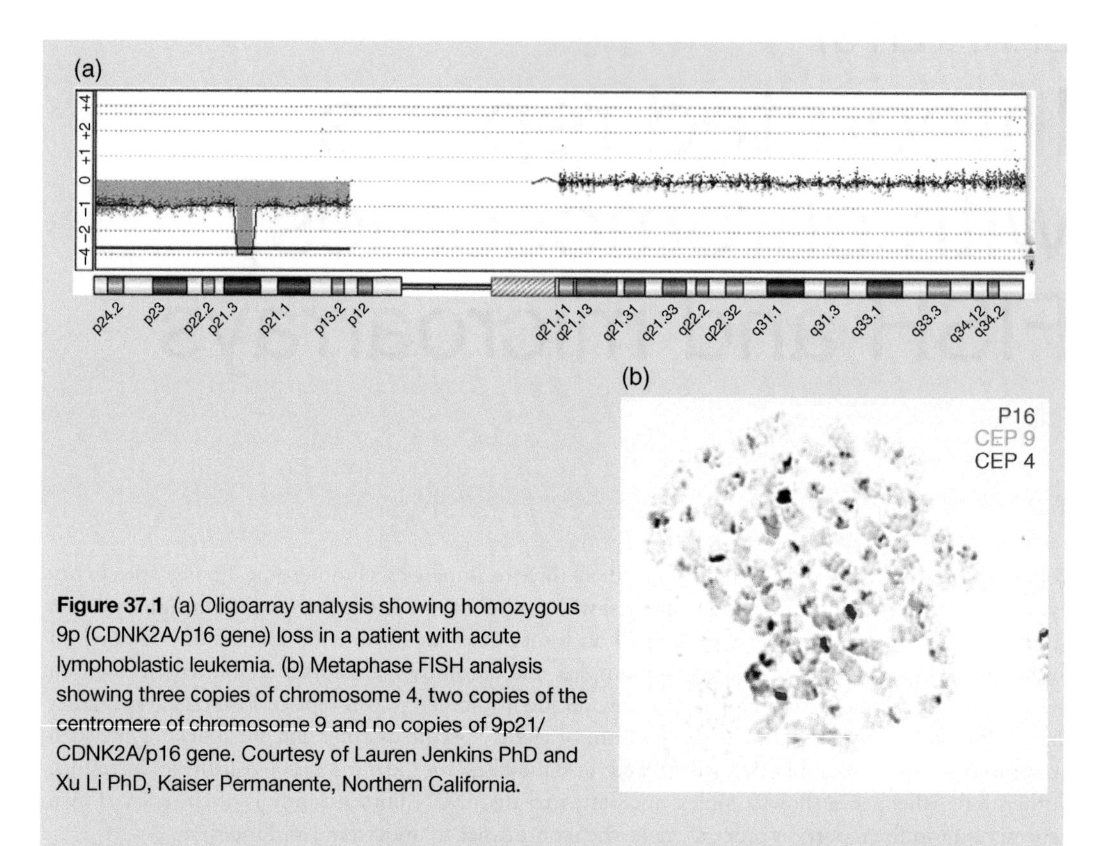

Figure 37.1 (a) Oligoarray analysis showing homozygous 9p (CDNK2A/p16 gene) loss in a patient with acute lymphoblastic leukemia. (b) Metaphase FISH analysis showing three copies of chromosome 4, two copies of the centromere of chromosome 9 and no copies of 9p21/ CDNK2A/p16 gene. Courtesy of Lauren Jenkins PhD and Xu Li PhD, Kaiser Permanente, Northern California.

Fluorescence *in situ* hybridization (FISH) analysis confirmed trisomy 4 and del(9)(p21) in interphase nuclei scored.

Trisomy 4 and homozygous deletions of chromosome 9p21 region, including the CDNK2A gene, are recurrent, non-random, abnormalities observed in acute lymphoblastic leukemia.

Other commonly observed genomic alterations in hematological malignancies targeted in this array were within normal limits. Other chromosomal regions not known to be pathogenic in hematological malignancies are not reported.

37.2 Identifying marker chromosomes by chromosome analysis, FISH and CMA

Unidentifiable marker chromosomes are common in hematological malignancies. When chromosome analysis is performed solely, there is no possibility of clarifying the origin of these chromosomes which may have diagnostic and prognostic implications. The example below shows the gain of 2–4 copies of the same marker chromosome by chromosome analysis. It appeared that the marker could be of chromosome 7 or 21 origin. Single nucleotide polymorphism (SNP) arrays were done to identify the origin of the marker chromosomes, the results of which had significant clinical implications for the patient.

EXAMPLE REPORTING OF MARKER CHROMOSOMES WITH CHROMOSOME, FISH AND ARRAY ANALYSES (FIGURE 37.2)

ISCN Result: 48~50,XY,del(7)(q11.2),-21,+2~4mar[cp20]. nuc ish 7cen(D7Z1x2),7q31(D7S486x1)[176/200]. arr(21)x5-6

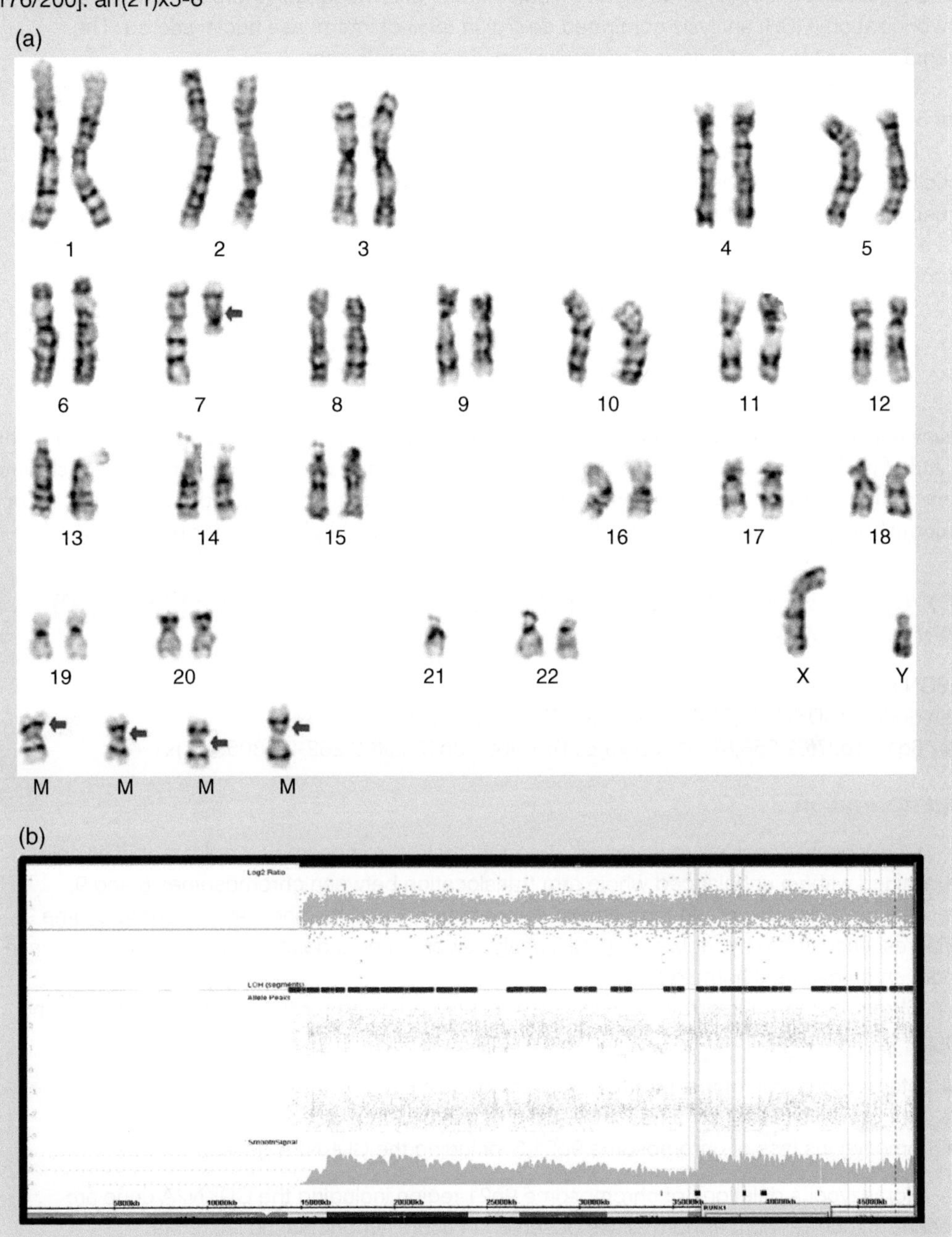

Figure 37.2 Identification of a marker chromosome by SNP microarray analysis. (a) Karyotype showing del(7q) and four identical marker chromosomes. (b) SNP array showing gain of chromosome 21 and the RUNX1 gene corresponding to the marker chromosomes.

Interpretation

Cytogenetic analysis revealed an abnormal male chromosome complement in 19 of 20 metaphase cells with a deletion of the long arm of chromosome 7, loss of chromosome 21 and a gain of 2–4 copies of an unidentifiable marker chromosome. Fluorescence *in situ* hybridization (FISH) analysis confirmed del(7q) in 88% of interphase nuclei scored. The remaining cell showed a normal male chromosome complement.

Chromosomal microarray analysis (CMA) confirmed an abnormal result with the presence of 5–6 copies of chromosome 21 including the RUNX1 gene.

Deletions of 7q are associated with both myeloid and lymphoid malignancies, including myelodysplasias, acute myeloid leukemia and B-cell, low-grade lymphoid malignancies, in particular splenic marginal zone lymphomas. RUNX1 gene amplification is a recurrent anormality in myeloid malignancies and is generally associated with a poor prognosis.

37.3 Unbalanced translocation identification by chromosomes, FISH and CMA

Below is an example of a chromosome result that showed an unbalanced whole arm translocation, but in and of itself did not clearly define the abnormality at the band resolution of bone marrow chromosome analysis. Further FISH and microarray analyses were better able to identify the submicroscopic abnormalities present, enhancing the diagnostic and prognostic implications for this patient.

EXAMPLE REPORTING OF ABNORMAL RESULTS – der(6;9)(p10;q10) (FIGURE 37.3)

ISCN Result: 45,XY,der(6;9)(p10;q10)[20].
ish der(6;9)(CDKN2A-,D9Z1+),del(9)(CDKN2A-,D9Z1+).
arr 6q11.1q27(62,358,874-170,748,691)x1,9p24.3p13.2(406,262-38,805,471)x1

Interpretation

Chromosome analysis revealed an abnormal male chromosome complement in all cells examined with an unbalanced whole arm translocation between chromosomes 6 and 9, resulting in a deletion of the long arm of chromosome 6 and the short arm of chromosome 9. Fluorescence *in situ* hybridization (FISH) analysis confirmed a deletion of the CDKN2A gene in both chromosome 9 homologs.

Chromosomal microarray analysis (CMA) confirmed an abnormal result with the following copy number changes:

- heterozygous loss of the long arm of chromosome 6
- heterozygous loss of chromosome 9 from bands p13.2 to p24.3
- homozygous loss of chromosome 9p21.3 including the CDKN2A gene.

Homozygous deletions of chromosome 9p21 region including the CDKN2A gene are recurrent, non-random abnormalities observed in acute lymphoblastic leukemia.

Other commonly observed genomic alterations in hematological malignancies targeted in this array were within normal limits. Other chromosomal regions not known to be pathogenic in hematological malignancies are not reported.

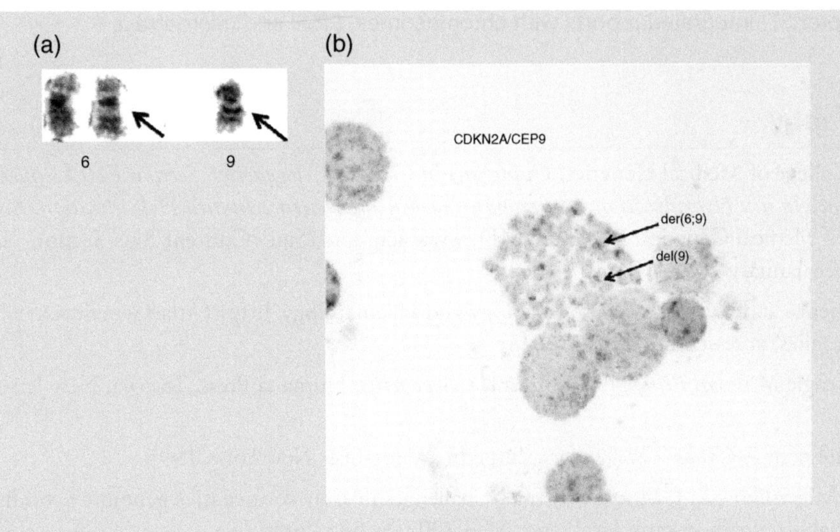

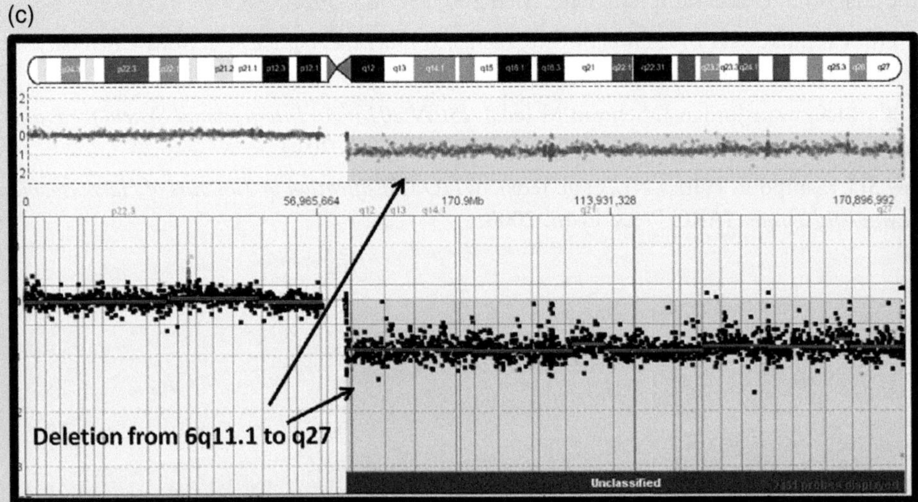

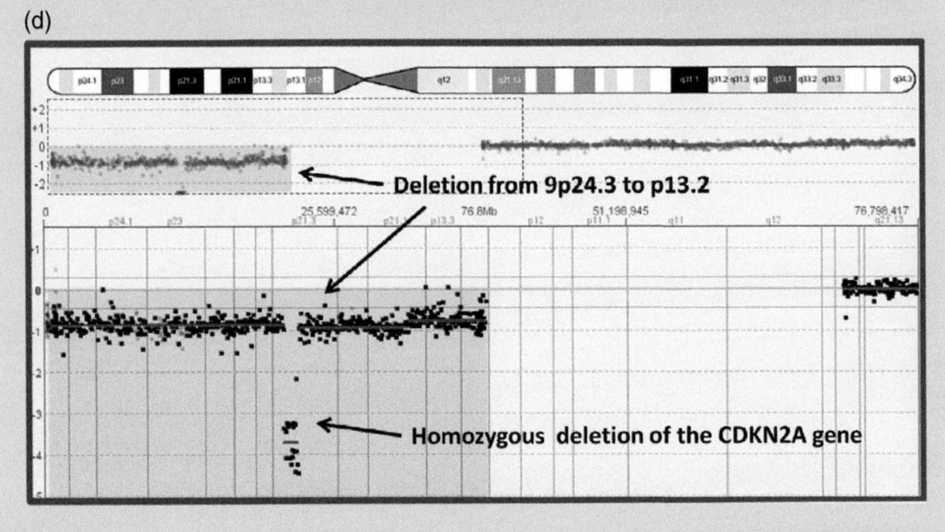

Figure 37.3 T-cell acute lymphoblastic leukemia in a 9-year-old male with (a) partial karyotype showing der(6;9)(p10;q10), (b) G-band to FISH: CDKN2A (*red*)/D9Z1 (*green*), showing the deletion of CDKN2A in both chromosomes 9 (loss of red signals), (c) oligoarray showing the deletion of 108.4 Mb of chromosome 6 from 6q11.1 to q27, and (d) oligoarray showing the deletion of 38.4 Mb of chromosome 9 from 9p24.3 to p13.2, including a homozygous deletion of the CDKN2A gene. Courtesy of Lauren Jenkins PhD and Xu Li PhD, Kaiser Permanente, Northern California.

Bibliography

American College of Medical Genetics. *Guidelines in Clinical Cytogenetics. Section E10 Methods in Fluorescence In situ Hybridization – Interphase/Nuclear Fluorescence In Situ Hybridization*. American College of Medical Genetics, Bethesda, MD. www.acmg.net/StaticContent/SGs/Section_E.html, accessed 28 January 2014.

Atlas of Genetics and Cytogenetics in Oncology and Haematology. http://AtlasGeneticsOncology.org/Anomalies, accessed 28 January 2014.

Gersen S, Keagle M (eds). *Principles of Clinical Cytogenetics*. Humana Press, Totowa, New Jersey, 1999.

Heim S, Mitelman F. *Cancer Cytogenetics*, 2nd edn. Wiley-Liss, New York, 1995.

Nowak NJ, Miecznikowski J, Moore SR, et al. Challenges in array comparative genomic hybridization for the analysis of cancer samples. Genet Med 2007; 9: 585–595.

Rooney DE, Czepulkowski BH. *Human Cytogenetics. A Practical Approach*. Oxford University Press, New York, 1992.

Shaffer LG, McGowan-Jordan J, Schmid M (eds). *ISCN 2013: An International System for Human Cytogenetic Nomenclature*. Karger Publishers, Unionville, CT, 2013.

Swerdlow SH, Campo E, Harris NL, et al. (eds). *WHO Classification of Tumours of Haematopoietic and Lymphoid Tissues*. IARC Press, Lyon, 2008.

APPENDIX 1
Example assay-specific reagent (ASR) FISH validation plan for constitutional disorders and hematological malignancies on fresh tissue

Summary

This validation is to be performed to verify the performance of assay-specific fluorescence *in situ* hybridization (ASR FISH) tests. The validation testing process will strictly adhere to all established standard operating procedures.

Test methodology

Assay-specific FISH tests use FISH technology to detect copy numbers of DNA probes to specific loci in the genome. FISH involves the annealing of a single-stranded, fluorochrome-labeled DNA probe to the DNA strand complementary to the target DNA sequences. Hybridization of the DNA probe in the cell is visible by direct detection using fluorescent microscopy.

List of reagents used in the validation

Add specific laboratory information here.

List of equipment used in the validation

Add specific laboratory information here.

Other materials used in the validation

Add specific laboratory information here.

Sample types used in the validation

- Peripheral blood specimens are used for constitutional disorders.
- Peripheral blood, bone marrow and lymph node tissues may be used for hematological malignancies.

Cytogenetic Abnormalities: Chromosomal, FISH and Microarray-Based Clinical Reporting, First Edition. Susan Mahler Zneimer.

Number of samples required to validate the range of testing from normal to abnormal

Normal samples are samples which show no criteria for disease. For hematological diseases, normal is based on normal cytogenetics and flow cytometry results.

Five normal male peripheral blood samples (and/or bone marrow samples for hematological malignancies) will be used for probe localization and analytical sensitivity/specificity studies. Three normal male peripheral blood (and/or bone marrow for hematological malignancies) samples will be used for stability and reproducibility studies.

A minimum of two (one normal, 1–5 abnormal) peripheral blood (and/or bone marrow for hematological malignancies) samples will be used for familiarization.

A total of 20 normal male peripheral blood samples (5–10 normal male peripheral blood samples, 5–10 normal male bone marrow samples, and 5–10 normal male lymph node samples for hematological malignancies) will be used to establish a normal database. This database will be used to calculate the normal cut-off values.

Accuracy

Accuracy is defined as the extent to which the measured amount of a substance agrees with the actual amount of the substance. Another definition is the degree to which a measurement represents a true value of the action that is being measured, or the degree of conformity of a measured or calculated quantity to its actual or true value.

For FISH testing, accuracy is conducted by performing probe localization on metaphase chromosomes. This process verifies that the probe hybridizes to the appropriate chromosome locus to be tested and to no other chromosome region.

Procedure to measure accuracy

Select five normal male samples (normal by both cytogenetics and flow cytometry for hematological malignancies) and using G-banding to FISH techniques, analyze one metaphase from each case to verify that the probe hybridizes to the specified chromosome target.

Acceptability criteria equal ≥95% of the expected signal pattern at the correct chromosome location.

Stability

Analyte stability in a sample is a function of storage conditions, chemical properties of the analyte and the containment system. It is appropriate to establish required conditions of specimen storage and transport to ensure that the measured concentrations reflect *in vivo* concentrations at the time of collection.

Procedure to measure stability

DNA probes will be placed at room temperature for an hour before use for three consecutive days on three separate patient samples. Results of each sample will be scored and compared to assess the quality of hybridization for each day.

Table A1 Stability assay

Case no.	Day 1 result	Day 2 result	Day 3 result	Concordance (Y/N)
Sample 1				
Sample 2				
Sample 3				

Acceptability criteria equal ≥95% concordance between these runs and no evidence of reduced quality.

Precision/reproducibility

The precision or reproducibility of an analytical method is defined as the closeness of the individual measure of an analyte when the procedure is applied repeatedly to multiple aliquots of a single homogenous volume of sample. Another definition is the degree to which measurements will show the same or similar results, or the degree of mutual agreement among a series of individual measurements or results. These results may be expressed by the standard deviation.

Intraassay precision

Intraassay precision is defined as the extent to which a set of measurements for a sample within a single run agrees with the mean concentration. It describes an assay's variability within a single run.

Procedure to measure intraassay precision

Intraassay precision and reproducibility will be assessed by selecting three known normal samples which show no criteria for disease (normal by cytogenetics and flow cytometry for hematological malignancies) and running sample triplicates of each case in a single run. Samples will be deidentified and results scored.

Table A2 Intraassay reproducibility

Case no.	Triplicate 1 result	Triplicate 2 result	Triplicate 3 result	Concordance (Y/N)
Sample 1				
Sample 2				
Sample 3				

Acceptability criteria equal ≥95% concordance between each triplicate.

Interassay precision

Interassay precision is defined as the extent to which a set of measurements for a sample from different runs agrees with the mean concentration. It describes an assay's variability over multiple runs.

Procedure to measure interassay precision

Interassay precision and reproducibility will be assessed by running the same three known normal samples which show no criteria for disease (normal by cytogenetics and flow cytometry for hematological malignancies) on three different days. Samples will be deidentified and results scored.

Table A3 Interassay reproducibility

Case no.	Day 1 result	Day 2 result	Day 3 result	Concordance (Y/N)
Sample 1				
Sample 2				
Sample 3				

Acceptability criteria equal ≥95% concordance between each run.

Sensitivity

Sensitivity is defined as the smallest amount of analyte that can be distinguished from the absence of that analyte. Assay sensitivity or limit of quantitation (LOQ) defines the lower end of the reportable range. For FISH probes analytical sensitivity is defined as the percentage of metaphases with the expected signal pattern at the correct chromosomal location, that can be visually identified by fluorochrome probes. Sensitivity defines the false negative rate.

$$\text{Sensitivity} = \frac{\text{Number of true positives}}{\text{Number of true positives} + \text{number of false negatives}}$$

Procedure to measure sensitivity

A total of 100 metaphase cells from five cases will be scored and the signal pattern of each metaphase cell is documented. Acceptability criteria equal ≥95% concordance.

Specificity

Specificity is defined as the ability of the FISH probe to differentiate and quantitate the analyte in the presence of other constituents in the sample. For FISH, specificity is defined by the percentage of signals that hybridize to the correct locus and no other location. Specificity defines the false positive rate.

$$\text{Specificity} = \frac{\text{Number of true negatives}}{\text{Number of true negatives} + \text{number of false positives}}$$

Procedure to measure specificity

A total of 100 metaphase cells from five cases will be scored and the signal pattern of each metaphase cell is documented. Acceptability criteria equal greater than or equal to 95% concordance.

Familiarization

Before a new ASR probe is introduced in the lab, staff performing FISH testing should be familiar with the probe's parameters, including signal intensity, signal patterns of normal and abnormal samples, and any cross-hybridization that is likely to confound test results. Familiarization will also serve as the initial experience with the method and help create scoring criteria, assess equipment and evaluate potential interfering factors.

Procedure to assess familiarization

One known negative sample and up to five known positive samples will be used for familiarization. For each sample, two scorers will analyze 100 interphase cells each (for a total of 200 cells) from each case and document the signal configurations observed. One representative image from each sample will be captured for documentation purposes.

Reference range

Reference intervals developed by the manufacturer or taken from the literature may be dependent on variables such as subject population, which may vary from one laboratory to another. It is therefore appropriate that the reference intervals be verified for normal subjects encountered by a laboratory as part of the method validation process.

Procedure to establish the reference range

Establish a normal FISH database by scoring 20 normal male peripheral blood samples (5–10 normal male peripheral blood samples, 5–10 normal male bone marrow samples, and 5–10 normal male lymph node samples for hematological malignancies). For each sample, score 200 nuclei and record all signal patterns observed. The normal cut-off for an analysis of 200 nuclei will be calculated using the Microsoft Excel β inverse function, = BETINV (95% confidence level, false-positive cells plus one, 200 cells analyzed).

Reportable range

Reportable ranges for normal and abnormal results are based on at least a 95% confidence interval achieved from 20 normal individuals in the database. A normal result is defined as the percentage of nuclei that are normal within the 95% confidence interval based on the database of normal individuals. An abnormal result is defined as the percentage of nuclei that are abnormal outside the 95% confidence interval based on the database of normal individuals.

Glossary

Abnormal in genetic terms, something considered atypical, unusual or uncommon. When used in reference to chromosomes, an abnormal chromosome complement differs from a normal female (46,XX) or normal male (46,XY) karyotype.

Acrocentric chromosomes those chromosomes, specifically chromosomes 13, 14, 15, 21 and 22, that are made up of long arm material with only satellite genetic material on the short arm.

Acute lymphoblastic leukemia (ALL) a hematological malignancy affecting white blood cells (lymphocytes) characterized by the excess of undifferentiated lymphocytes (lymphoblasts).

Acute myeloid leukemia (AML) a hematological malignancy affecting the myeloid lineage of blood cells, characterized by rapid growth of white blood cells in bone marrow, which interferes with the production of normal blood cells in bone marrow.

Advanced maternal age women who are age 35 years or older at delivery are at an increased risk for chromosome abnormalities.

Alleles different forms of each gene. Different alleles produce variations in inherited characteristics such as eye color or blood type. Each individual inherits two alleles of each gene, one maternal and one paternal in origin.

Alphafetoprotein (AFP) a protein produced by the fetus, which is excreted into the amniotic fluid and into the mother's bloodstream through the placenta throughout pregnancy. AFP levels, both in the maternal blood and in the amniotic fluid, vary at particular periods during the pregnancy and may be associated with the presence of neural tube defects or chromosomal abnormalities in the fetus.

Amniocentesis a prenatal diagnostic procedure performed from 12 weeks in gestation through the second trimester, which involves the removal of a small amount of amniotic fluid, which can be examined to look for chromosomal, biochemical or gene alterations in the fetus.

Amniocytes cells found in the amniotic fluid which are obtained by an amniocentesis procedure.

Amniotic fluid the fluid that surrounds the fetus, held inside the amniotic sac.

Amniotic sac the membrane-bound compartment in the uterus that holds the fetus and amniotic fluid intact.

Amplification the gain of copy number of genes or a chromosomal region in a cell. With chromosome analysis, this may include double minute chromosomes or homogeneously staining regions.

Anaphase lag an error during cell division when one chromosome fails to be inserted into a daughter cell. This process is one mechanism for trisomic rescue.

Anemia a decreased level of red blood cells in circulation, or less than the normal quantity of hemoglobin in blood.

Aneuploidy the gain or loss of one or more chromosomes from the human chromosome complement of 46 chromosomes.

Cytogenetic Abnormalities: Chromosomal, FISH and Microarray-Based Clinical Reporting, First Edition. Susan Mahler Zneimer.

Angelman syndrome a disorder characterized by a chromosome 15q11.2 abnormality resulting in the clinical manifestations of severe mental impairment, developmental and growth delay and frequent spastic and uncontrollable movements and speech.

Apoptosis programmed cell death in which a sequence of events leads to the degradation of cells without releasing harmful substances into the surrounding area. Aberrant levels or processing of apoptosis play a role in developing diseases, including cancer.

Array comparative genomic hybridization (CGH) a technique that utilizes comparative genomic hybridization of chromosomes on a microarray. DNA fragments of the whole genome use bacterial artificial chromosomes, shorter single-stranded oligonucleotides or single nucleotide polymorphism on an array platform.

Autosomes chromosomes that are not sex chromosomes. In humans, the autosomes are the chromosomes numbered from 1 to 22, where chromosome 1 is the largest and chromosome 22 is the smallest.

Bands chromosomal regions that stain either darkly or lightly when chemically treated with certain enzymes and dyes. Each chromosome has its own characteristic banding pattern that elucidates the specific identification of homologous chromosomes. The laboratory procedure used to achieve this result is called banding. Examples of types of banding include Geimsa-banding, centromeric, reverse, quinacrine, nuclear organizer region, bromodeoxyuridine, DAPI, and sister chromatid exchange.

Basophils basophil granulocytes are a small subset of white blood cells, accounting for approximately 0.01–0.03% of total white blood cells in circulation. The name is derived from the fact that they are susceptible to staining by a base dye.

Benign tumor a non-malignant tumor that remains localized and does not spread to other sites.

Birth defect an abnormality of structure, function or metabolism, either genetic or environmentally caused, which was formed *in utero*, often resulting in a physical or mental handicap.

Blast an immature precursor cell derived from either a lymphocyte (lymphoblast), or granulocyte (myeloblast), which does not normally appear in circulating peripheral blood. Blast cells can be recognized by their large size and primitive nuclei. When present in the blood in high enough numbers, they are associated with acute leukemia.

Blastocyst a very early stage of embryonic development, beginning at day 5 after fertilization in humans, consisting of approximately 70–100 cells. These cells include an inner cell mass, which subsequently forms the embryo, and an outer layer of cells, or trophoblast, surrounding the inner cell mass and a fluid-filled cavity known as the blastocele.

Blastomeres cells produced by the cleavage or cell division of a fertilized egg that forms a zygote and are an essential part of blastula formation.

Blood part of the circulatory organ system in humans composed of seven different types of cells: red blood cells (erythrocytes), platelets (thrombocytes) and five types of white blood cells (lymphocytes, monocytes, eosinophils, basophils and neutrophils).

Bone core a biopsy in which a piece of bone and its associated marrow are removed from the body for pathological and laboratory analyses.

Bone marrow the soft spongy tissue that lies within the hollow interior of long bones. In adults, marrow in large bones contains immature stem cells that produce new blood cells.

Bone marrow aspirate a sample of bone marrow removed with a syringe and needle for pathological and laboratory analyses.

Breakpoint a point on a chromosome that is the site of a DNA rearrangement with either another chromosome or a different part of the same chromosome.

Carcinoma cancer that is derived from putative epithelial cells whose genome has become altered or damaged in which the cells become transformed and exhibit abnormal malignant properties.

Carrier a person who has one normal gene and one mutated gene for a recessively inherited disease, or a person with a balanced chromosomal rearrangement. Carriers do not usually develop manifestations of a disease, but can pass on the mutated copy of a gene or an unbalanced form of the chromosome rearrangement to their offspring.

Cell the basic subunit of living beings.

Cell culture used to achieve cell growth and division, leading to a good mitotic index for harvesting cells during the prometaphase or metaphase stages of mitosis. Cell culture methods vary with the tissue of origin, e.g. bone marrow, peripheral blood, solid tumors, and cell lines of various tissue origins.

Cell cycle synchronization the manipulation of cells so they are all brought to the same phase of the cell cycle at the same time.

Cell division the process of mitosis in somatic cells, whereby cells grow and divide, resulting in the growth of tissues or organs. Cell division which produces the reproductive (gametic) cells is called meiosis.

Centromere the part of a chromosome which joins the short and long arms. The short arm is called the 'p' arm; the long arm is called the 'q' arm.

Chorion the outermost protective membrane around a fetus *in utero*. The cells of the chorion are sampled during chorionic villus sampling.

Chorionic villi small, flower-like projections that emerge from the outer sac which surrounds a fetus *in utero*. Chorionic villi are of fetal origin and eventually form the placenta. The cells of the chorionic villi are sampled during chorionic villus sampling.

Chorionic villus sampling (CVS) a prenatal diagnosis procedure performed at 10–12 weeks' gestation which involves obtaining a sample of chorionic villi. These cells generally represent the cells of the fetus, and can be examined to look for chromosomal, biochemical or gene alterations in the fetus.

Chromatid one of a pair of strands of DNA on a chromosome as part of two sister chromatids, each of which contains an identical copy of its original chromosome. The sister chromatids eventually separate to create two daughter cells identical to the first in mitosis.

Chromatin the complex of DNA and proteins that make up chromosomes which is located in the nucleus of the cell.

Chromosomal mosaicism the presence of two or more cell lines in an individual that has developed from a single fertilized egg.

Chromosomal satellite region (also called satellite II) a small mass of chromatin attached to the stalk of the short arm of each chromatid of human acrocentric chromosomes. They do not always stain darkly by G-banding and may be difficult to see. Satellites on different chromosomes are often attracted to one another, causing the acrocentric chromosomes to be in satellite association. Not to be confused with satellite DNA (satellite I).

Chromosome derived from the Greek for "colored body". Thread-like structures composed of DNA, RNA and proteins containing the genetic information from both parents within the nucleus of each cell of the body. Different kinds of organisms have different modal chromosome numbers, in each cell.

Chromosome arm chromosomes are divided into two parts by the centromere. The shorter part is called the short arm (p - petite) and the longer part is referred to as the long arm (q).

Chromosome number varies among species. In humans, the chromosome number is 23 pairs for a total of 46. Half of the chromosomes, 23 (22 autosomes and one X), are inherited from the mother and half (22 autosomes and one X or Y), from the father.

Chronic lymphocytic leukemia (CLL) a hematological malignancy of white blood cells, in which there is a chronic proliferation of mature lymphocytes that function abnormally and cause a malignancy.

Chronic myeloid leukemia (CML) a hematological malignancy affecting the myeloid lineage of blood cells, characterized by rapid growth of white blood cells in bone marrow, which interferes with the production of normal blood cells in bone marrow leading to chronic leukemia.

Clonal as it pertains to clinical cytogenetics, the presence of two or more cells with the same gains or structurally rearranged chromosomes, or three or more cells with the same loss of chromosomes.

Clone a population of cells derived from a single progenitor cell, including at least two cells in which the abnormality contains the same chromosomal gain or structural change, or three cells with the same chromosomal loss. Subclones evolve as cytogenetically related changes occur from a primary clone as the tumor proliferates. A stemline (sl) is the most basic or primary clone of a tumor cell population, and other subclones derived from the stemline are called sidelines (sdl).

Colcemid the brand name of synthetic colchicine used in cytogenetic cell culture for the synchronization of cell division at metaphase.

Colchicine an alkaloid derived from members of the plant genus *Colchicum* that irreversibly binds tubulin and prevents the chromosomes of mitotic cells from entering anaphase.

Complete medium a basal culture medium supplemented with nutrients to enhance the growth of cells.

Composite karyotype the presence of chromosomal variation from cell to cell in a tumor cell population that cannot clearly be defined into subclonal cell lines. Composites contain in the ISCN nomenclature all the clonally occurring abnormalities in the cells examined, even when certain cells do not contain each abnormality present.

Conception the fusion of egg and sperm to create an embryo.

Confined embryonic mosaicism a condition where chromosomally abnormal cells are found in the fetus but not in the placenta.

Confined mosaicism a condition where chromosomally abnormal cells are confined to a certain tissue.

Confined placental mosaicism a condition where chromosomally abnormal cells are found in the placenta but not in the fetus.

Congenital present at birth, but not necessarily inherited.

Congenital heart defect a condition present at birth is considered congenital. Heart defects occur early in pregnancy while the heart is developing. An example of a congenital heart defect is atrial septal defect, which means that the wall between the left and right upper chambers of the heart (the atria) does not close completely.

Constitutional present at birth.

Copy number analysis (CNA) the process of analyzing genomic data for the presence of DNA copy number changes in a patient's sample. CNA detects chromosomal copy number variation that may cause or increase the risk of various genetic disorders.

Copy number variation (CNV) copy number changes that are found in a segment of DNA in an individual by comparing two or more genomes.

Counterstain a second stain or dye that is of contrasting color to a first specific stain or dye used to identify a particular target. Counterstains can be used to stain cells, nuclei or DNA, depending on the assay. As it pertains to FISH analysis, counterstains consisting of fluorochromes are used to stain DNA, with a contrasting color to distinguish the background DNA from the specifically labeled probe hybridized to the target DNA. Commonly used counterstains for FISH include DAPI and propidium iodide.

Crossing over an event that occurs during meiosis, in which homologous pairs of chromosomes exchange genetic material, resulting in a part of the maternal chromosome exchanging places with the corresponding part of the paternal chromosome. This is also referred to as a recombination event.

Cystic fibrosis an autosomal recessive genetic disorder, caused by a genetic mutation of the CTFR gene on chromosome 7, which causes the body to produce excessively thick, sticky mucus that clogs the lungs and pancreas, impairing breathing and digestion.

Cytogenetics the study of chromosomes and how changes in chromosome structure and number affect the growth, development and health of individuals.

Cytopenia a deficiency of any type of blood cell.

DAPI 4,6 diamino-2-phenylindole, a DNA dye which fluoresces a blue color when exposed to ultraviolet light (UV) and is used to stain DNA in FISH assays. DAPI is a good counterstain for FISH when using red, green and gold labeled chromosome painting probes.

Daughter cell identical cells resulting from a parent cell during cell division.

Deletion loss of a whole chromosome or part of a chromosome. A terminal deletion refers to the breakage and loss of the end of a chromosome, and an interstitial deletion refers to the loss of genomic material within a chromosome.

Denaturation the separation of complementary bases along the double helix DNA strands into single DNA strands. The break-up of the chemical bonds involved in complementary base pairing is produced by the combination of either heat and/or chemicals. Denaturation of genomic DNA is necessary for hybridization of FISH probes.

de novo a chromosome abnormality that occurs in an individual that was not inherited from either parent.

Diandry the presence in a triploid chromosome complement of two paternal and one maternal haploid chromosome sets.

Dicentric a chromosome that possesses two centromeres, either from the same chromosome or from non-homologous chromosomes.

Differentiation the development of a progenitor blast cell to become a specialized or mature cell that has a function. It also pertains to stem cell differentiation in which a cell that has the potential to become any cell in the human body permanently becomes a specialized cell (for example, a white blood cell).

Digynic the presence in a triploid chromosome complement of two maternal and one paternal haploid chromosome sets.

Diploid(y) a complete genome consisting of two copies of each chromosome pair. In humans, the diploid number is 46 chromosomes with a homologous pair of each chromosome, consisting of one of each of the chromosomes from each parent.

Disomy the presence of two copies of each chromosome in the nucleus of a cell. This is the normal chromosome complement in humans.

Dispermy fertilization of a single oocyte by two separate sperm. It is the most common cause of triploidy.

DNA (deoxyribonucleic acid) a large molecule in the nucleus of a cell consisting of the basic material of heredity. It carries the genetic information packaged as chromosomes for cells to replicate and produce proteins. DNA is made up of nucleotide bases, linked together in a spiral-shaped chain called a double helix. DNA contains the genetic instructions used in the development and functioning of organisms.

Double minutes (dmin) variable numbers of small unidentifiable acentric chromatin bodies corresponding to the amplification of a small chromosomal segment of the genome.

Down syndrome trisomy 21, with the presence of three copies of chromosome 21.

Duplication the gain of a whole chromosome or part of a chromosome. A terminal duplication refers to the breakage and gain of the end of a chromosome, and an interstitial duplication refers to the gain of genomic material within a chromosome.

Egg the female reproductive cell, also called the ovum. The egg carries 23 chromosomes, 22 autosomes and an X chromosome. The egg fuses with the sperm at conception to produce an embryo.

Embryo the term used to describe a developing human from conception to 8 weeks of development.

Embryonic progenitor cells the original 1–5 cells of a blastocyst that are destined to develop into a fetus.

Eosinophils granulocytic cells that develop during hematopoiesis in the bone marrow before migrating into the blood. They constitute the immune system component of white blood cells responsible for combating infections. Together with mast cells, they control the mechanism associated with allergy and asthma.

Euchromatin chromosomal DNA which is genetically active and rich in genes. During interphase it is uncoiled and often transcriptionally active whereas heterochromatin is condensed and inactive. Euchromatin comprises over 90% of the human genome.

Extraembryonic tissue the tissues that develop from a fertilized egg that do not involve the actual fetus, but include the placenta and the membranes.

Familial disease the prevalence of a disease that is present in more members of a family than is expected by chance.

Fertilization the fusion of the egg and sperm at conception to create an embryo.

Fertilized egg — an egg, or oocyte, becomes fertilized once its genetic complement has fused with that of the sperm which has entered it. A fertilized egg contains 23 chromosomes from the mother, and 23 chromosomes from the father, that arrived in the sperm. Once fertilized, an egg is known as a zygote.

Fetal blood sampling — a prenatal diagnostic procedure by obtaining a blood sample from the fetus *in utero* and analyzing it for chromosomal, biochemical or gene alterations in the fetus.

Fetus — term used to describe a developing baby from the ninth week of development to birth.

Fibroblasts — the type of cell that make up skin.

FISH (fluorescence *in situ* hybridization) — a molecular cytogenetic technique that uses DNA segments or specific genes or loci of interest, when combined with a fluorescent dye to become a DNA probe, to be visualized on a metaphase chromosome or interphase nucleus. FISH analysis can detect either numeric changes in genetic material, including the gain, amplification or deletion of parts of or whole chromosomes, as well as balanced and unbalanced chromosomal rearrangements.

Fixative — a solution used to preserve cells for microscopic analysis. The most common fixative for cytogenetic analysis is Carnoy's fix, composed of a 3:1 methanol:acetic acid mixture. It is used to stop the exposure of hypotonic solution during harvesting of chromosome spreads and promotes the degradation of the cytoplasmic membrane and cytoplasmic parts of the cell.

Flow cytometry — a technique used to count and separate cells by suspending them in a stream of fluid and passing them through electronic detection. This technique is routinely used to diagnose hematological malignancies.

Fluorescent microscopy — utilizes an array of fluorochromes to identify segments of DNA within the nucleus of cells with a high degree of specificity and sensitivity. Through the use of multiple fluorescent labeling, different probes can simultaneously identify several target molecules.

G1 Phase — The phase of the cell cycle between the end of mitosis (M phase) and the beginning of DNA synthesis (S phase).

G2 Phase — The phase of the cell cycle between the end of the DNA synthesis (S phase) and the initiation of mitosis (M phase).

Gametes — germline cells which each contain 23 chromosomes, consisting of sperm in males and eggs in females.

Gender mismatched bone marrow transplantation — in hematopoietic malignancies, monitoring of residual disease which is commonly done by FISH testing of donor cells in host tissues that can be discerned by the XX/XY - female/male DNA probes.

Gene — a unit of DNA that codes for the synthesis of a specific protein located on chromosomes. Each gene has one or more specific effects on an individual's phenotype, and can contain a mutation that results in various allelic forms.

Gene rearrangement — a change in the gene or chromosomal structure or molecular arrangement within a chromosome. For example, ABL/BCR rearrangement occurs as a result of a translocation between two chromosomes 9 and 22. This translocation is annotated t(9;22). Also, for example, hematopoietic cells destined to become T-cells or B-cells rearrange their T-cell receptor genes and immunoglobulin genes in a fashion that allows for the generation of receptor diversity.

Genetic counseling education sessions that help individuals, couples or families understand genetic information, including recurrence risks and reproductive choices that may be available to them.

Genetic disorder disease resulting from an abnormality of part of the genome of an individual. The genetic abnormality can be caused by a small mutation in a single gene or within a chromosome, or by the addition or deletion of an entire chromosome or set of chromosomes.

Genome refers to the total genetic material of the chromosomes from a cell of an organism.

Genotype the total genomic information carried by an individual.

Germ cell a reproductive cell, consisting of sperm in males and eggs in females.

Gestation the carrying of an embryo or fetus during a pregnancy. Most pregnancies are single gestations but multiple gestations can occur (twins or more fetuses). A 40-week gestational period is considered normal and a gestational period of less than 37 weeks is considered to be premature.

Giemsa a complex dye mixture used for staining metaphase chromosomes, or as a reagent in chromosomal banding protocols.

Growth factors cell culture additives that enhance cell growth and proliferation.

GTG-band the common banding pattern produced with trypsin and Giemsa stain and used to routinely identify chromosomes.

Haploid a single set of a chromosome complement, which in humans is half of the full set of genetic material in a cell or organism. In humans, the haploid number of chromosomes is 23.

Hematological malignancy a cancer of the blood or bone marrow, such as leukemia or lymphoma.

Hematopoiesis the production and maturation of blood cells, occurring in the bone marrow, spleen, lymph nodes and other organs. Cells and tissues that produce blood cells are known as hemopoietic.

Heterochromatin chromosomal material which is highly condensed and transcriptionally inactive during interphase. It consists of repetitive DNA sequences that are relatively rich in AT base pairs and is late replicating in the cell cycle. In metaphase chromosomes, it is dark staining with G- and C-banding.

Heterodisomy one of two types of uniparental disomy (the other is isodisomy), when both copies of a particular chromosome are inherited from the same parent, but the two copies are composed of different homologs, and each locus has two different alleles.

Heterozygote having two different alleles for a specific gene.

Histones proteins within chromosomes whose function is to ensure the condensation of a chromosome.

Histopathology the examination of a tissue specimen under a microscope in order to detect the manifestation of a disease.

Hodgkin lymphoma a cancer of the lymphatic system, originating from white blood cells (lymphocytes), whose result is a specific form of lymphoma.

Homogeneously staining region (HSR) chromosomal region that stains uniformly and represents amplified copies of a DNA segment.

Homologous chromosomes a pair of chromosomes of the same length, centromere position, staining pattern and shape. Each chromosome in a pair is inherited from a different parent.

Homozygote having two identical alleles for a specific gene.

Hybridization base pairing of two single strands of DNA or RNA by hydrogen bonding between complementary nucleotides.

Hypotonic solution a solution having a lower ionic concentration relative to another solution. A cell placed in a hypotonic solution will swell and possibly burst with time.

Ideogram a schematic drawing of chromosomes when stained with a particular stain, such as G-banding. Chromosomes contain discretely distinct regions called light and dark bands, giving each chromosome a unique appearance, and are given specific band nomenclature by the ISCN.

Immunophenotyping the analysis of a heterogeneous population of cells using antibody markers to detect the proteins expressed in each cell.

Implantation approximately 6–8 days after fertilization, the trophoblast cells of the blastocyst attach themselves to the endometrial lining of the uterus, and the placenta begins to form.

Imprinting when a gene or part of a chromosome is "turned off" depending on which parent it was inherited from. During development of the sperm or egg, some genes or parts of chromosomes are "paternally stamped" when inherited from the father and some are "maternally stamped" when inherited from the mother.

Inner cell mass after fertilization, the conceptus begins to divide mitotically (each daughter cell is identical to the parent cell) and the cells begin to differentiate into trophoblast cells with the inner cell mass creating a blastocyst. The inner cell mass will eventually become the embryo and subsequent fetus.

Insertion the addition of a piece of genetic material into a chromosome where it is not normally found.

***In situ* hybridization (ISH)** the base pairing of a nucleotide sequence to metaphase chromosomes and/or interphase nuclei.

Interphase a stage of the cell cycle between two mitoses, comprising G1, S and G2. The nuclear DNA is not highly condensed and individual chromosomes cannot be distinguished.

Inversion when a chromosome breaks in two places, and the segment turns 180 degrees and rejoins, creating an inverted segment.

***In vitro* fertilization (IVF)** the process in which an egg is fertilized with a sperm outside the body and then transplanted into a woman's uterus.

ISCN International System of Cytogenetic Nomenclature, providing the terminology used to describe chromosome abnormalities.

Isochromosome a chromosome with a centromere and two identical arms, which is a mirror-image chromosome. These chromosomes result in one duplicated arm, and the other arm is absent.

Isodisomy one of two types of uniparental disomy (the other is heterodisomy), when both copies of a particular chromosome are inherited from the same parent, and the two copies are composed of the same homolog, and each locus has two identical alleles.

Karyotype	the arrangement of an individual's chromosomes organized in a standard format. The 23 pairs are organized according to size, location of the centromere and the patterns of the dark and light bands. In humans, a normal karyotype includes 22 pairs of autosomes and one pair of sex chromosomes (XX or XY).
Kinetochore	a protein structure at the centromere to which the spindle fibers attach during cell division.
Klinefelter syndrome	a cytogenetic disorder characterized by two X chromosomes and one Y chromosome (47,XXY), resulting in a male development and phenotype, exhibiting problems with fertility and some mild learning difficulties.
Leukemia	malignancies of several acute or chronic forms involving a neoplastic process of the bone marrow with the occurrence of unrestrained proliferation of white blood cells. Symptoms often include anemia, impaired blood clotting or enlargement of lymph nodes, spleen or liver.
Leukocyte (see also lymphocyte)	white blood cells of any type.
Leukocytosis	an increase in the number of leukocytes above the normal range.
Leukopenia	a decrease in the number of leukocytes below the normal range.
Low copy repeat (LCR)	highly homologous, chromosome-specific, repetitive DNA sequences.
Lymphocyte (see also leukocyte)	a type of white blood cell. Lymphocytes consist of large granular natural killer cells. Small lymphocytes consist of T (thymus) and B (bone) cells.
Lymphocytopenia (see also leukopenia)	an abnormally decreased number of lymphocytes in the blood.
Lymphocytosis (see also leukocytosis)	an increased proportion of lymphocytes in blood shown by complete blood cell count.
Lymphoma	cancer affecting the lymphatic system, usually lymphocytic, including the spleen and lymph nodes. Lymphomas are divided into Hodgkin lymphoma and non-Hodgkin lymphoma.
Malignant tumor	a tumor with properties of invasion and metastasis which can cause a malignant disease.
Marker chromosome	a part of a chromosome of unidentifiable chromosomal origin.
Maternal	originating from the mother.
Maternal serum screen	a screening test which provides a calculated risk of fetal abnormalities such as neural tube defects and Down syndrome for a woman during pregnancy. The calculation is based on the levels of various analytes in the mother's blood during pregnancy, including alphafetoprotein, estriol and human chorionic gonadotropin.
Maternal uniparental disomy (UPD)	when both members of a chromosome pair are inherited from the mother, rather than one chromosome of the pair from each parent.
Megakaryocyte	a bone marrow cell responsible for the production of blood thrombocytes (platelets), essential for blood clotting.
Meiosis	cell division of reproductive cells resulting in gametes (eggs and sperm cells), each containing the haploid (23) number of chromosomes as in.

Meiotic non-disjunction when a chromosome pair fails to separate correctly during meiosis, resulting in reproductive cells which have missing or extra chromosomes.

Meiotic origin originating during the development of the sperm or egg.

Metacentric a subdivision of chromosomes which have centromeres in the center of the short and long arms, resulting in equal length of the p and q arms.

Metaphase stage of cell division in mitosis or meiosis when the chromosomes have reached their maximal condensation and individual chromosomes can be individually identified. Metaphase spreads refer to the view of a cell's chromosomes at the metaphase stage which are spread on a microscope slide.

Metaphase harvest a process of isolating cells in the metaphase stage of mitosis involving three major steps: blocking cell division, hypotonic treatment of cells, and fixation.

Miscarriage loss of a pregnancy prior to 20 weeks. Most miscarriages occur in the first trimester, prior to 12 weeks of pregnancy, and are often caused by a genetic abnormality.

Mitogens chemical compounds used to initiate cell division in a tissue, when spontaneous cell division is low or absent. Specific mitogens are used to divide T- and/or B-cells.

Mitosis cell division of somatic cells resulting in two daughter cells that have the same genetic composition as the parental cell. Human somatic cells contain a diploid chromosome complement of 46.

Mitotic Index the number of cells in the mitosis stage in a total cell population of cells.

Mitotic non-disjunction when a chromosome pair fails to separate correctly during somatic cell division or mitosis, resulting in daughter cells which have missing or extra chromosomes.

Modal number the chromosome count that is characteristic of cells from a particular species or cell line. For example, the normal human modal number is 46 chromosomes.

Molecular cytogenetics a field of genetics involving a combination of standard cytogenetics and molecular genetic techniques. Generally, it involves applying FISH techniques to metaphase chromosomes or interphase cells, and microarray technology.

Monocyte a type of white blood cell that forms part of the human body's immune system. Its roles include replenishing macrophages and dendritic cells to respond to inflammation signals that results in an immune response.

Monolayer culture single layer of cells attached to substrate.

Monosomy a cytogenetic abnormality where cells have lost one or more chromosomes from the normal chromosome complement.

Morula one of the earliest stages of a fertilized egg, when there are 12–15 blastomeres.

Mosaic (mosaicism) the presence of two or more cell lines within an individual or tissue, with one or more chromosome abnormalities, while the rest of the cells have either the normal chromosomal constitution or a second abnormal cell line.

Mutation process by which a gene or DNA segment experiences a change in its normal DNA composition.

Myelodysplastic syndrome (MDS) improperly formed, non-functioning myeloid cells in the hematopoietic system.

Myeloid cells white blood cells, specifically granulocytes and monocytes.

NCCN Clinical Practice Guidelines in Oncology guidelines which are widely recognized and used as the standard for clinical policy in oncology by clinicans and payors. (www.nccn.org/professionals/physician_gls/f_guidelines.asp)

Neutropenia	an abnormally low number of neutrophils (a type of white blood cell) in the blood.
Neutrophil	neutrophil granulocytes are the most abundant white blood cells in mammals and form an essential part of the immune system. They form part of the polymorphonuclear cell family together with basophils and eosinophils.
Non-disjunction	a chromosome pair that fails to separate correctly during cell division, resulting in daughter cells which have missing or extra chromosomes.
Non-Hodgkin lymphoma (NHL)	a cancer of lymphocytes that has a tendency to spread to other parts of the body. There are many subtypes and disease entities of NHL, including mantle cell lymphoma, follicular cell lymphoma, Burkitt lymphoma and diffuse large B-cell lymphoma.
Nucleosome	the primary structural unit of chromatin consisting of DNA wrapped around a core of histone proteins.
Nullisomic	a cell in which there are no copies of a specific chromosome or chromosomal region.
Oncogene	a dominant-acting gene that is able to transform normal cells by causing them to grow in an uncontrolled manner, and is responsible for tumor development. Mutation, overexpression or amplification of oncogenes in somatic cells can lead to cellular transformation and cause disease.
Ovum	the female reproductive cell, also called the egg. The ovum carries 23 chromosomes, 22 autosomes and an X chromosome. It fuses with the sperm at conception to produce an embryo.
"p" arm	the short arm of a chromosome as opposed to the long arm (see "q" arm). The p arm is divided from the q arm by the joining of the centromere.
Pancytopenia	a condition in which there is a decrease of white blood cells and platelets.
Paternal	originating from the father.
Paternal uniparental disomy (UPD)	when both members of a chromosome pair are inherited from the father, rather than one chromosome from each parent.
Phenotype	the observable characteristics of an individual which are determined by their gene composition along with other modifying genes and the environment.
Phytohemagglutinin (PHA)	a specific mitogen for T-cell lymphocyte growth. Other mitogens include TPA, PWM, LPS, Protein A, IL-6, EBV, DPS-30, IL-4 and IL-2, which stimulate both B- and T-cells to divide.
Placenta	the organ that provides the fetus with nutrients during development *in utero*. It is attached to the wall of the uterus and connects the mother and fetus.
Plasma cells	white blood cells which produce large volumes of antibodies. They are transported by the blood plasma and lymphatic system. Originally in bone marrow, they migrate as B-cells before terminal differentiation into plasma cells in lymph nodes.
Platelets	platelets or thrombocytes are small, irregularly shaped anuclear fragments, derived from megakaryocytes, and function to form blood clots.
Pluripotent cell	a cell that is capable of becoming any type of differentiated cell.
Polymorphism	a variation in the DNA or chromosome that occurs in more than 1% of the population and is thus considered to be a normal variation. These normal variants are inherited within families without causing problems in development or producing phenotypic consequences.

Polyploidy entire extra sets of chromosomes, where n equals a haploid set of chromosomes. Examples include trisomy (3n), tetrasomy (4n), pentasomy (5n), etc.

Postzygotic after fertilization.

Prader–Willi syndrome a genetic disorder characterized by obesity and insatiable appetite, mental impairment, small genitalia and short stature. It may be caused by a deletion on chromosome 15 or maternal UPD15.

Precursor cell an original cell.

Pregnancy termination intervention to ensure that a pregnancy does not continue.

Prenatal diagnosis the procedure of examining fetal cells taken from the amniotic fluid (amniocentesis), placenta (CVS) or umbilical cord to detect fetal abnormalities during pregnancy.

Probe a small segment of DNA of known genetic composition, designed to recognize complementary DNA on a target chromosome. The probe is labeled with a tag, such as a fluorescent dye, which can then be used to confirm the presence or absence of the target DNA region.

Protease an enzyme that breaks down proteins.

Protooncogene a normal gene that can become an oncogene due to mutations or increased expression. Protooncogenes code for proteins that help regulate cell growth and differentiation. Upon activation, a protooncogene becomes a tumor-inducing agent, an oncogene. Examples of protooncogenes include ABL, RAS, MYC and TRK.

"q" arm the long arm of a chromosome as opposed to the short arm (see "p" arm). The q arm is divided from the p arm by the joining of the centromere.

Recessive a form of a gene or allele,that when present in two copies, will result in a genetic condition, or when present in one copy, is a carrier of a genetic condition.

Reciprocal translocation the breakage and exchange of genetic material between two chromosomes without any gain or loss of genetic material (balanced). When the translocated chromosomes are passed on to offspring in a way which results in gain or loss of material, the translocation is described in the offspring as "unbalanced".

Recombination during meiosis when homologous pairs of chromosomes exchange genetic material, in which part of the maternal chromosome crosses over and exchanges places with the corresponding part of the paternal chromosome.

Recurrence risk the chances of a genetic abnormality recurring in family members.

Relapsed disease an incomplete response to treatment of cancer, resulting in the recurrence of disease.

Remission absence of disease. At times, total remission occurs but relapse can occur even after years of therapy and remission.

Repetitive DNA DNA sequences that are present in multiple copies in the genome.

Replication during the cell cycle when an identical copy of the DNA is made to become a double-stranded molecule.

Reproductive cells also called germ cells, consisting of the egg and sperm cells. Each reproductive cell carries a single set of 23 chromosomes.

Residual disease/ minimal residual disease when some cancer cells remain during or after therapy causing a malignancy.

Ring chromosome a portion of a chromosome which has broken off and fuses at the ends to form a circle or ring. There is generally loss of genetic material at the ends of the chromosome prior to the fusion.

Robertsonian translocation the fusion of the whole long arms of two acrocentric chromosomes, creating a translocation that results in the loss of the p arms of the two chromosomes. These translocations may be balanced, resulting in 45 chromosomes, or unbalanced, resulting in the gain of a chromosome.

S phase the phase of the cell cycle during which DNA synthesis takes place.

Satellite DNA DNA containing many tandem repeats of a short basic repeating unit. There are different types of tandem repeats throughout the genome. Not to be confused with the term "chromosomal satellite" which is a small mass of chromatin at the end of the short arm of each chromatid of an acrocentric chromosome.

Secondary constriction a negatively staining heterochromatic region of a chromosome.

Sex cells a germ cell or reproductive cell, consisting of sperm in males and eggs in females.

Sex chromosome the chromosomes responsible for sex determination. Females have two X chromosomes, while males have one X and one Y chromosome each.

Sister chromatids two identical copies of a chromosome attached together by the centromere.

Solid tumor any malignancy that is not derived from the hematopoietic system.

Somatic cells all the cells in the body excluding the reproductive cells.

Somatic mutation a change occurring in any cell in the body that is not destined to become a reproductive cell.

Sperm the male reproductive cell, consisting of 23 chromosomes, 22 autosomes and an X or a Y chromosome. The sperm fuses with the ovum at conception to produce an embryo.

Stem cell undifferentiated cell that has the potential to differentiate into many cell types.

Stroma connective tissue cells in an organ or tumor. These may be fibroblasts, blood vessels or normal cells that provide a supportive matrix to the organ or tumor.

Submetacentric chromosomes with centromeres which are not centrally placed and whose short and long arms are not of equal length.

Syndrome a recognizable pattern or group of characteristics or symptoms resulting in a disorder.

Telomere DNA sequences located at the ends of chromosomes.

Tetraploidy four copies of every chromosome present in a cell, resulting in 92 chromosomes, rather than the typical diploid number of 46 chromosomes in humans.

Tetrasomy four copies of a single chromosome present in a cell, resulting in 48 chromosomes, rather than the typical number of 46 chromosomes in humans.

Thrombocytopenia a decrease in the normal level of blood cells and platelets.

Thrombocytosis the presence of highly abnormal level of platelets in the blood.

Tissue a group of cells that have a similar structure and function. Many different tissue types can make up an organ.

Totipotent a cell that has the potential to become any specialized cell in the body. Totipotent cells are found in the zygote, and begin to specialize a few days after fertilization and after several cycles of cell division.

Translocation a chromosome rearrangement where segments of two chromosomes break off and each join to the other chromosome. An individual with a balanced translocation has the normal amount of chromosomal material. A person with an unbalanced translocation will have a loss or gain of chromosomal material.

Triploidy three copies of every chromosome in a cell, resulting in 69 chromosomes, instead of the typical diploid number of 46 chromosomes in humans.

Trisomic rescue when a fertilized egg initially contains a meiotic non-disjunction resulting in 47 chromosomes (instead of the normal 46), then loses the extra chromosome in subsequent cell divisions. The trisomic cell has been "rescued" and is now a disomic cell.

Trisomy three copies of one particular chromosome present in each cell, resulting in 47 chromosomes, rather than the typical number of 46 chromosomes.

Trisomy 13 (Patau syndrome) three copies of chromosome 13 in each cell.

Trisomy 18 (Edward syndrome) three copies of chromosome 18 in each cell.

Trisomy 21 (Down syndrome) three copies of chromosome 21 in each cell.

Trophoblast cells after fertilization, the zygote begins to divide mitotically and differentiate into the inner cell mass and trophoblast cells, creating a blastocyst. Trophoblast cells are located around the outside of the blastocyst and are the cells responsible for implantation into the uterine wall. They will eventually differentiate further into all the extraembryonic tissues, including the placenta, amnion and chorion.

Tumor suppressor gene a normal gene involved in the regulation of cell growth that when mutated can lead to a neoplastic process and tumor development.

Turner syndrome only one copy of the X chromosome and no Y chromosome, resulting in 45 chromosomes instead of the typical number of 46 chromosomes, with only one sex chromosome.

Ultrasound the use of sound waves to obtain an image of body tissues and structures. Prenatal ultrasound provides an image of the fetus in order to assess growth and development.

Uniparental disomy the inheritance of both copies of a homologous chromosome pair of chromosomes or all chromosomes from a single parent rather than one chromosome of each pair being inherited from each parent.

Uterus the female organ where the fetus develops into a baby, also known as the womb.

Villus stroma one of two cell types in the inner layer of the chorionic villi of the placenta (the layer closer to the amniotic sac and developing fetus). Trophoblast cells make up the outer later of the chorionic villi.

White blood cells also termed leukocytes, these are cells of the immune system which protect against foreign bodies and infectious disease. There are five different kinds of leukocytes: lymphocytes, monocytes, eosinophils, basophils and neutrophils.

X chromosome a sex chromosome; normal females carry two X chromosomes and normal males carry one X and one Y chromosome.

X-inactivation a process early in the development of a normal female, in which one of the X chromosomes in each cell is randomly inactivated. All daughter cells of that cell will have the same X chromosome inactivated. Consequently, all females contain a mixture of some cells which express the paternally derived X chromosome and other cells with express the maternally derived X chromosome.

XYY syndrome a genetic condition in males with an extra Y chromosome. These individuals are tall in stature (generally over 6 feet) and may experience problems with fertility, developmental delay and/or learning problems.

Y chromosome a sex chromosome; normal males carry one Y and one X chromosome and females carry no Y chromosome.

Zygote a fertilized egg, representing a single cell with 46 chromosomes in humans resulting from the fertilization of an ovum (23 chromosomes) by a sperm (23 chromosomes). Through cell division, the zygote develops into a multicellular embryo and then into a fetus.

Sources for glossary

http://cancergenetics.com/cgi-research/glossary-of-terms/
www.kumc.edu/gec/glossary.html
Glossary of Genetic Terms, Genetics Education Center, University of Kansas Medical Center
Glossary of Genetics (illustrations/oral descriptions), National Human Genome Research Institute
Understanding Gene Testing Glossary, US Department of Health and Human Services, Public Health Service, National Institutes of Health, National Cancer Institute
Illustrated Glossary, GeneReviews
Dictionary of Genetic Terms, Genomics and Its Impact on Medicine and Society,
DNA From the Beginning, experiments and people behind concepts, Cold Spring Harbor Lab
Dictionary of Cell and Molecular Biology
GeneCards, Weizmann Institute of Science
Genome Glossary, Department of Energy (DOE)
Genetic Glossary, cross-referenced, Biology Teaching Organisation (BTO), University of Edinburgh
Genomics Glossaries & Taxonomies, Cambridge Healthtech Institute (CHI)
Genome Glossary, Cracking the Code of Life
Genomics Lexicon, PhRMA (Pharmaceutical Research and Manufacturers of America)
Genomes Online Terms, GOT It!™, Foundation for Genetic Medicine
Glossary, Emery's Elements of Medical Genetics
Glossary, GenomePrairie, Canada
Glossary, Birth Defects Information Directory
Glossary, Genetics Modules, Pennsylvania Health Department
Glossary and Definitions of words describing genetic disorders and birth defects, GAPS
List of terms, University of Medicine and Dentistry, New Jersey
Glossary, Access Excellence Program, Genentech, Inc., San Francisco, CA
Glossary Diving Into The Gene Pool, The Exploratorium, San Francisco, CA
Glossary, DNA Science
Glossary of Genetic Terms, Glossary of Malformations and Glossary of Abbreviations, Clinical Genetics: Self Study Guide, University of South Dakota Medical School
Glossary of Genetic Terms, Greenwood Genetic Center
Glossary of Genetic Terms, Virtual Hospital
Glossary of Terms, Vysis
Inherited Disorders & Birth Defects Glossary, NetWellness Consumer Health Information
Medical Terminology (Medicine - Mythology - Arts - Sciences)
Mother and Child Glossary, Health on the Net (HON), genetic screening, also birth defects, genetic basics, childhood diseases, A-Z
Pandora's Wordbox (birth defects, congenital malformations, dysmorphic syndromes). International Birth Defects Information Systems (I.B.I.S.)

Terms and Definitions, Office of Rare Diseases, National Institutes of Health (NIH)
Terminology Page, American Society of Gene Therapy
What Do the Words Mean, Genetic Glossary: Definitions for Kids Only, geneCRC
Genetics Glossary from Growth, Genetics and Hormones Journal
Glossary: Mitochondrial and Metabolic Diseases, Mitochondrial and Metabolic Disease Center, University of California San Diego
Glossary of Parentage Testing Terms, Family Law Advisor, Fairfax Identity Laboratories
Glossary of Terms, MD Anderson Cancer Center
Glossary of Terms, Cancer, EXACTA Sciences
Infertility Glossary: definitions on infertility, multiple miscarriage and pregnancy loss, terms and acronyms, International Council on Infertility information dissemination
Common Medical Terms, the Doctors' Doctor, terms used in pathology reports
Dictionary of Cell and Molecular Biology
Glossary, Cancerpage.com
Life Science Dictionary, BioTech, Indiana Institute for Molecular and Cellular Biology
Multilingual Glossary of Technical and Popular Medical Terms, nine European Languages, European Commission (DG III), Heymans Institute of Pharmacology and Mercator School, Department of Applied Linguistics
Old Disease Names and their Modern Definition, Cape Cod Genealogy
Online Medical Dictionary, CancerWeb
Online Medical Dictionaries, National Library of Medicine

Index

Printed and bound by CPI Group (UK) Ltd, Croydon, CR0 4YY
16/04/2025
14658540-0001